FV

REINFORCED CONCRETE FUNDAMENTALS

REINFORCED CONCRETE FUNDAMENTALS

FIFTH EDITION

PHIL M. FERGUSON
THE LATE T. U. TAYLOR PROFESSOR EMERITUS OF CIVIL ENGINEERING
THE UNIVERSITY OF TEXAS AT AUSTIN

JOHN E. BREEN
THE NASSER I. AL-RASHID CHAIR IN CIVIL ENGINEERING
THE UNIVERSITY OF TEXAS AT AUSTIN

JAMES O. JIRSA
THE PHIL M. FERGUSON PROFESSOR OF CIVIL ENGINEERING
THE UNIVERSITY OF TEXAS AT AUSTIN

WILEY

JOHN WILEY & SONS
NEW YORK   CHICHESTER   BRISBANE   TORONTO   SINGAPORE

Library of Congress Cataloging in Publication Data:

Ferguson, Phil Moss, 1899–1986
    Reinforced concrete fundamentals / Phil M. Ferguson, John E.
Breen, James O. Jirsa.—5th ed.

        p.   cm.
    Includes index.
    ISBN 0-471-80378-2
    1. Reinforced concrete construction.   I. Breen, J. E. (John
Edward), 1932–   . II. Jirsa, J. O. (James Otis), 1938–   III. Title.

TA683.2.F4 1988
624.1'8341—dc19   87–20727
                                                                        CIP

Printed in the United States of America

10 9 8 7 6 5 4 3 2 1

To Phil Ferguson—teacher, researcher, engineer, friend.

PREFACE

The fifth edition of *Reinforced Concrete Fundamentals* maintains the basic Ferguson approach in which design procedures stem from and provide the basis for a clear understanding of the behavior of reinforced concrete. Throughout the book the behavior of reinforced concrete members and assemblages at all load stages is illustrated by figures and photos. The physical behavior is related to calculation models to provide both student and practitioner with a solid foundation for assessing various design situations.

Numerous examples are provided to fully develop both the visualization of reinforced concrete behavior and the calculation processes necessary in design. Many examples that were very complex in previous editions are simplified or clarified in the present revision. Much user confusion with the fourth edition came from the widespread use of SI units in the early examples. Most of those examples are restored to customary or English units in this edition. The overriding goal is to improve clarity and understanding, and to keep the general scope about the same as in the previous edition.

Practical design of reinforced concrete in office practice utilizes computers and many versatile software packages. The authors feel strongly that use of such programs should be encouraged only *after* an understanding of the fundamentals of reinforced concrete design is developed. Throughout the text the authors indicate how calculation procedures can be easily extended for computer use.

The following major changes are made in this fifth edition.

1. Modernization to conform to the technical changes in the 1983 and 1986 revisions to the ACI Building Code.
2. De-emphasis of SI units in examples. In contrast to the fourth edition which had most of the introductory course-related examples in SI notation, only a few examples are in SI units. The conversion to SI practice in the United States, which seemed likely when the fourth edition was written, now appears remote. Enough SI examples are given to make readers aware of the ACI 318M Code in SI units and the general approach to SI calculations.

3. Reorganization of material to better match probable class usage. The fourth edition Chapter 8 on serviceability is eliminated as a free-standing unit. Service load analysis of stresses, computation of deflections and distribution of reinforcement to control crack widths is incorporated into Chapters 3 and 4, which treat analysis and design of flexural members. The design of stocky and of slender columns is always covered in a first course on reinforced concrete. In previous editions these topics were covered following an extensive treatment of slabs. Many first courses do not treat two-way slab design. Column design is moved to Chapters 6 and 7 to immediately follow flexure, shear, and torsion. Emphasis on joint importance is provided by a new Chapter 10 on detailing of joints that considerably amplifies the material on detailing introduced in Chapters 8 and 9. This emphasis reflects the growing incidence of structural collapse because of inadequacy of joint details. Arrangement of the text material now permits a first semester course to basically cover Chapters 1 to 11 with an option for further design applications from other chapters, such as Chapter 18 on retaining walls or Chapter 19 on footings. Students should have the ACI Building Code and Commentary available. The authors always allow students to use Code and Commentary in quizzes and examinations.

4. Revision and expansion of seismic design related topics. A new Chapter 22 on shear walls is added and Chapter 23 on seismic design and detailing is expanded.

5. Emphasis on conceptual models for design. Shear and torsion design emphasize the visualization of the truss analogy as the fundamental basis for design rules and proportioning of web reinforcement. The truss model concept is further extended in the chapter on joint detailing to develop a method for visualization of the general pattern of force transfer between members.

The behavior of reinforced concrete still has a prominent place in this textbook, with many pictures of members at failure. Because reinforced concrete is largely semiempirical, design engineers who understand how reinforced concrete behaves in approaching ultimate resistance have an advantage in assessing the many situations that they face day to day.

Some limits in the scope of this book have been necessary. Only a simple introduction to prestressed concrete is included, composite member coverage is kept brief, shearheads and brackets are outlined rather than covered in depth, and Vierendeel trusses are omitted. Service load analysis, as in the fourth edition, starts with transformed areas and deflections in the flexure analysis chapter and is summarized with the treatment of flexure in Appendix A.

Although this book is now better arranged for the beginning student in reinforced concrete, it covers material adequate for a second semester of work. Many detailed examples appear in the text. Because it is primarily a book on basic philosophy, behavior, and theory, design is included chiefly as a teaching tool.

The authors are greatly indebted to their many friends, students, and colleagues who have so willingly shared their time and their ideas in countless discussions of reinforced concrete behavior and design. The skill and dedication of Carol Booth in typing this manuscript is gratefully acknowledged.

Readers in the past have been most cooperative in reporting any errors, and these comments permit errors to be corrected at each reprinting. This is most helpful and appreciated.

This text was originally developed by the senior author, Professor Ferguson, as a means of introducing the strength design approach to reinforced concrete design when the ACI Code first recognized the ultimate limit state in 1956. Phil Ferguson was a true pioneer in explaining and popularizing reinforced concrete design approaches that were given a solid conceptual foundation based on the behavior of reinforced concrete structures at failure. His visionary work in the areas of reinforcement development, combined shear and torsion, and slender column action in frames resulted in many of the present-day design code provisions. His leadership in teaching, research, and professional service set a standard of excellence that inspired countless students and colleagues to focus on better ways of understanding reinforced concrete behavior and to translate this understanding into design approaches. Professor Ferguson died on August 18, 1986 as this text revision was nearing completion. We are dedicating this edition in his memory in the hope that through our efforts we will be able to bring his love for teaching and his understanding of reinforced concrete behavior to new generations of engineers.

John E. Breen
James O. Jirsa

Austin, Texas
February 1987

ABOUT THE AUTHORS

Phil M. Ferguson

The late Phil M. Ferguson was T. U. Taylor Professor Emeritus at the University of Texas at Austin and the sole author of the first four editions of this text. His association with the University of Texas as a student, teacher, and departmental chairman spanned more than 60 years. In 1979, the University of Texas named its structural engineering facility the Phil M. Ferguson Structural Engineering Laboratory. He was a longtime member of the ACI Building Code Committee 318. He was named an Honorary Member of both the American Concrete Institute and the American Society of Civil Engineers in recognition of his long and distinguished service to those societies. Both the University of Texas and the University of Wisconsin recognized him as a distinguished graduate. He was recognized for his contributions to research by ASCE (Huber Research Prize), ACI (Lindau Award and several Wason Medals), CRSI (Honor Award) and RCRC (Boase Award). He was elected to the National Academy of Engineering in 1973. He died in 1986.

John E. Breen

John E. Breen holds the Nasser I. Al-Rashid Chair in Civil Engineering at the University of Texas at Austin. He is past-Director of the Phil M. Ferguson Structural Engineering Laboratory. Professor Breen received his doctorate from the University of Texas. He has served as a U.S. Navy Civil Engineer Corps officer and has worked as a structural designer and consultant. He has received numerous teaching excellence awards at Texas as well as professional achievement awards from Marquette, Missouri, and Texas. Dr. Breen's reinforced and prestressed concrete research and development work has earned numerous awards from the American Concrete Institute, the American Society of Civil Engineers, the Prestressed Concrete Institute, and the Reinforced Concrete Research Council. He was elected to the National Academy of Engineering in 1976. A registered professional engineer in both Missouri and Texas, Prof. Breen is the Chairman of the ACI Building Code Committee 318. He is also Vice Chairman of the Working Commission on Concrete Structures of the International Association for Bridge and Structural Engineering.

James O. Jirsa

James O. Jirsa is the Phil M. Ferguson Professor of Civil Engineering and Director of the Phil M. Ferguson Structural Engineering Laboratory at the University of Texas at Austin. Prof. Jirsa received his Ph.D. from the University of Illinois, Champaign-Urbana in 1962. He was a Fulbright Scholar in France in 1963–1964. After teaching at the University of Nebraska and Rice University, he joined the faculty of the University of Texas at Austin in 1972. He has conducted research projects that have led to changes in design codes and recommendations in the area of floor slab design, anchorage and development of reinforcement, joints in reinforced concrete frames, and structures subjected to seismic loads. He has received awards from the American Concrete Institute and the American Society of Civil Engineers. A registered professional engineer in Texas, Dr. Jirsa is currently Chairman of the American Concrete Institute Technical Activities Committee and a member of the Board of Directors. He is a member of ACI Building Code Committee 318. He is a member of several Comite Euro-International du Beton Commissions and is a member of the International Association for Bridge and Structural Engineering and the Earthquake Engineering Research Institute.

CONTENTS

1

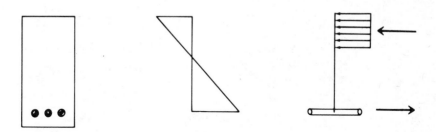

INTRODUCTION

1.1 Reinforced Concrete as a Building Material

Reinforced concrete is one of the principal building materials used in engineered structures. The economy, efficiency, durability, moldability, and rigidity of reinforced concrete make it an attractive material for a wide range of structural applications. It is well suited for bridges and buildings, such as those shown in Figs. 1.1, 1.2, and 1.3, and in essential but less glamorous utility structures, such as that shown in Fig. 1.4. As a building material it has unique characteristics. The very low tensile strength of plain concrete is offset by the high tensile efficiency of the encased steel bars, wires, and strands. The relatively high unit cost of steel is offset by reducing the quantity of steel required through the use of concrete for those portions of the structure that are in compression or are relatively unstressed. Such lightly stressed zones include many deck and wall elements where the structure also serves as a space divider and encloser. The steel filaments in reinforced concrete limit the crack widths and give ductility to the structural elements. The concrete in turn provides both corrosion protection and fireproofing to the more susceptible steel bars. Prestressing allows very high strength steels to be used efficiently and safely.

Proper utilization of reinforced concrete requires knowledge, effort, and cooperation from everyone involved in a project. The design engineer, the materials suppliers, the constructor's superintendent, and the construction workers all have an important role. Although this book is primarily intended for both students and practicing engineers involved in reinforced concrete design, the reinforced concrete construction process must be understood and carefully considered throughout the design and detailing steps.

If a reinforced concrete project is to be a success, it must meet the full expectations of both the owner and the users. The structure must be strong and safe. The proper application of the fundamental engineering principles of analysis, the laws of equilibrium, and consideration of the structural properties of the

Figure 1.1 The Salginatobel Bridge in Switzerland. A beautiful 295-ft clear span reinforced concrete arch designed by Robert Maillart and constructed in 1929–1930.

Figure 1.2 Dallas Municipal Center. (Courtesy Jack E. Rosenlund & Co., Consulting Engineers.)

Figure 1.3 The first precast segmental post-tensioned concrete bridge in the United States under construction in Corpus Christi, Texas.

Figure 1.4 Reinforced concrete water treatment tank.

component materials should result in a sufficient margin of safety to guard against collapse under possible overloads. The structure must be stiff and appear unblemished. Care must be taken to control deflections under usual loads as well as to limit to acceptable widths the possible cracking of the concrete that is often necessary to permit the steel reinforcement to work efficiently. The designer can ensure the serviceability of the structure by attention and care in the design and detailing process. The structure must be durable. Materials must be chosen and specified carefully so that the structure can resist those forces of nature or use that tend to erode, progressively shatter, or decompose the structure. The structure must be economical. Materials should be used efficiently. Attention should be given to the entire construction process. One observer of concrete construction remarked that the process often requires building one structure (the falsework and formwork shown in Fig. 1.5) to support the concrete structure, building the concrete structure, and then demolishing the first structure. He points out that the first and third of these steps, although necessary, are not directly productive in terms of the final structure. The experienced designer realizes that true economy in reinforced concrete construction does not come from precision tailoring of each member to the calculated applied forces. True economy comes from consideration of the over-

Figure 1.5 Formwork and falsework for the University of Texas Performing Arts Center. (Courtesy Jack E. Rosenlund & Co., Consulting Engineers.)

all structure and selection of member sizes to permit maximum reuse of formwork with minimum modification. Concrete members can be "fine tuned" for strength by relatively simple variations in reinforcement quantities. Economy also results from clear and careful detailing. The detailing should reflect not only the arrangement of reinforcement required to provide the tensile and compressive force paths to carry the loads on their proper way, but an awareness of how a constructor must assemble the reinforcement, as shown in Fig. 1.6. It is not enough for a design to provide strength, serviceability, and durability. If a structure is to give value and economy, it also must provide for constructability. The final design should reflect an awareness of the construction process. Form reuse should be enhanced. Reinforcement fabrication should be simplified by provision of standardized, repetitive bar bending and assembly patterns wherever possible. Adequate cover, spacing between bars and bar layers, and spacing of secondary reinforcement such as ties and stirrups should be used. The sensitive designer always should be aware of how and where the builder can place the concrete in the forms, vibrate or otherwise compact it, and then cure it. Constructors are noted for their ingenuity and can in fact build "unbuildable designs." They cannot do it economically. A highly successful

Figure 1.6 Assembling reinforcement for Dallas Municipal Center. (Courtesy Jack E. Rosenlund & Co., Consulting Engineers.)

project requires sensitivity in design to the limitations of the constructor and appreciation in construction of the requirements of the designer. Throughout this book, an attempt is made to encourage sensitivity to and awareness of constructability and good detailing. An excellent companion reference is the *ACI Detailing Manual* (ACI SP-66).[1] It provides a clear presentation of necessary details and has a wealth of information on good detailing procedures. Although it is a guide to layout and drafting of engineering and reinforcement placement drawings, it calls attention to many important considerations in designing reinforced concrete.

1.2   Safety Philosophy

The principal role of the structural designer in the reinforced concrete construction process is to determine the form, dimensions, and details of the structure, and to specify the required material properties for all structural components. To adequately safeguard against failure, the designer must visualize and quantify all of the critical loading conditions that might produce failure.[2] This requires both the ability to foresee the worst loading cases on the structure and an understanding of the basic methods and magnitudes of resistance to load of each section of the structure. Many new engineers, overly impressed by the power, speed, and apparent accuracy of modern structural analysis computational procedures think less and less about equilibrium and details. It is essential that the designer envision the overall structural action and sketch the general load paths for each critical loading condition. The structure must be viewed as an entity, as shown in Fig. 1.7. Proper ties and connections must be provided so that the structure will possess the continuity, ductility, and redundancy required to ensure general structural integrity.[3] Such integrity ensures that if some unforeseen loading or accident does occur, the resulting damage is limited in proportion to the cause. Without such integrity there is a likelihood that a local accident will propagate in a chain reaction, as shown in Fig. 1.8.

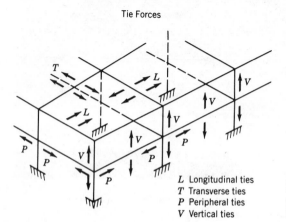

Tie Forces

L Longitudinal ties
T Transverse ties
P Peripheral ties
V Vertical ties

Figure 1.7 Tie forces to develop overall structural integrity.

Figure 1.8 Reinforced concrete apartment house. During concrete placement, a large portion collapsed when top floor forms gave way.

The premature removal of form supports for a top story allowed fresh and partially cured concrete to impact on the floor below. The lower floor could not carry the debris load and failed. A chain reaction occurred that resulted in the death of a large number of construction workers. Detailing procedures that ensure development of catenary action to suspend the initial debris several feet above the next floor slab could stop such chain reactions.[4] If the designer envisions such an occurrence, it is possible to prevent or minimize it. A basic problem in reinforced concrete design is that excessive attention is paid to member strength and too little thought is given to the ties and connections between elements.

Structural safety, in a somewhat oversimplified view, as shown in Fig. 1.9, may be determined by considering as normally distributed random variables both the loads that come on to a structure and the resistance of the various elements of the structure to those loads. The chance of a certain load U coming on to the structure can be reasonably defined if there is a good knowledge of the likelihood of various loadings to which the structure might be subjected,[3,5,6] as well as a good means of analysis of the structure. Similarly, the probable resistance of the structural elements R can be predicted, based on the knowledge of their probable dimensions, proportions, material qualities, and the construction accuracy. The chance or probability that the actual load might exceed the actual resistance and cause a failure can be shown as the shaded areas where these curves overlap.[2] The smaller this overlap is, the smaller the chance of failure. Safety factors can be applied through use of load factors and resistance factors, or of other methods to space these curves apart to minimize the probability of failure to some acceptable level. This conceptual representative of safety can lead to a clearer understanding of how safety factors are set in reinforced concrete design. Minimum levels of safety are established in the controlling codes or design criteria. The individual designer may wish to increase the level of safety where particularly important structures or uncertain loading conditions exist.

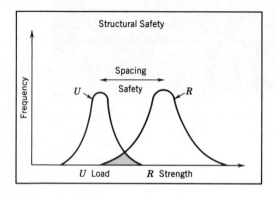

Figure 1.9 Simplified representation of structural safety.

1.3 Design Codes

A specification for reinforced concrete design may take the form of a code or a recommended practice. A code is written in the form of a law for enactment by public bodies such as city councils. It represents, usually, the minimum requirements necessary to protect the public from danger. It makes no attempt to specify the best practice, although it usually attempts to eliminate the most common mistakes, especially those involving safety. A recommended practice, on the other hand, attempts to define the best practice or at least to state satisfactory design assumptions and procedures. It may state reasons as well as methods.

The most significant reinforced concrete code in the United States is the "Building Code Requirement for Reinforced Concrete" (ACI 318-83) (revised 1986). The Code is available from the American Concrete Institute,* Box 19150, Redford Station, Detroit, Mich., 48219. This code will be quoted frequently in this book, usually as the Code, or the ACI Code. Frequently, a section reference number to a specific portion of the ACI Code will be provided.

A Commentary on the 1986 Code is also available from the same address. The Commentary is not a legal part of the Code but it attempts to indicate some of the reasoning behind the various sections of the Code. The ACI Building Code and Commentary undergo major revisions on approximately a six-year cycle as well as less extensive revisions midway between major revisions. This text is updated to include the 1986 revisions to the 1983 Code and Commentary. Strictly speaking, the correct reference to the present Code is ACI 318-83 (Revised 1986)[7], but for convenience the updated Code and Commentary will be referred to as ACI 318-86. In 1983 ACI issued complete SI versions of the ACI Code and Commentary[8,9], which were also updated in 1986. The updated SI Code and Commentary are referred to in this text as ACI 318M-86. (The M is for metric.)

The ACI Code has an extensive notation that has been incorporated in this text. All Code symbols are defined in each Code chapter and summarized in Code Appendix C.

Several model general codes have considerable usage: the "Uniform Building Code," the "Basic Building Code," the "Southern Standard Building Code," and the "National Building Code." These model codes generally use the ACI Code for most of their reinforced concrete provisions, either by reference or inclusion. In addition, many cities write their own codes, making some modifications in these more standard codes. Designers must at the beginning of any project determine the code under which they are required to operate.

Highway bridges are normally designed under specifications prepared by the American Association of State Highway and Transportation Officials (AASHTO).

---

* ACI members receive a nominal discount on the price.

In Europe, in addition to the various national codes, a set of "Principles and Recommendations" valuable to code writing commissions has been developed. The CEB–FIP "Model Code for Concrete Structures" using SI units was issued in April 1978 by Comite Euro-International du Beton as part of CEB Bulletin No. 124/125. Explanatory and supplementary "Complements" to the Code are available in CEB Bulletin No. 139.

1.4   Stress Versus Strength Concepts

In basic mechanics courses emphasis tends to be placed on analysis of members and cross sections in which the material properties are linear. Stress is assumed to be proportional to strain. The curved stress-strain diagram for concrete, with all strains increasing with time because of creep, complicates theoretical analyses. At low stress levels elastic analysis is reasonable, that is, stress varies almost linearly with the distance from the neutral axis. However, the reader should not confuse this with a feeling that such calculated stresses are therefore either elastic or exact. Shrinkage (usually nonuniform), creep, and cracking all complicate the stresses. Shrinkage tends to reduce compressive stress or increase tensile stress in typical cases. Creep increases the strains, whether tensile or compressive, in effect reducing the modulus of the concrete. Stresses computed elastically can only be index numbers, not real stresses.

Early investigators (before 1910) sensed this behavior and favored calculations based on ultimate strength, usually using the parabolic stress distribution of Fig. 1.10a. However, the apparent simplicity of the triangle of stress shown in Fig. 1.10b was strong enough to lead to the adoption of the so-called *working-stress design* (WSD). In contrast to predicting behavior at failure load levels, it was idealized as a prediction of stresses at ordinary or normal load levels.

In the 1930s field studies of strains in columns showed that the elastic concept of adjacent stresses in steel and concrete in proportion to the moduli values was not tenable; creep shifted much more of the initial load from con-

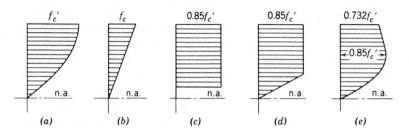

Figure 1.10   Stress distributions assumed on compression side of beam. (a) Simple parabola (ultimate). (b) Straight line (working stress). (c) Rectangular stress block (ultimate). (d) Trapezoidal stress block (ultimate). (e) Parabolic, with straight line (ultimate).

crete to steel. Tests on beams showed extreme face strains in compression, much in excess of the strain at peak stress, which can generally be taken as approximately 0.002. It became clear that the descending part of the stress-strain curve for $\varepsilon > 0.002$ was also important. In 1942 C. S. Whitney* presented a paper emphasizing this fact and showing how a probable stress-strain curve like that of Fig. 1.10e could, with reasonable accuracy,† be replaced with an artificial rectangular stress block simplification, as in Fig. 1.10c. With the rectangular stress block simplification, the 1956 ACI Code added an Appendix permitting *ultimate strength design* (USD) as an alternate to WSD. The 1963 Code gave both methods equal standing.

Because of the nonlinear response of columns and prestressed concrete, engineers have been pushed toward ultimate strength design. Since 1971 the Code has been almost totally a *strength* code with strength meaning ultimate. However, an alternate method which for beams and slabs (not columns) was effectively a working stress method remained in the 1971 Code and constitutes Appendix B of the 1986 Code.

For strength design the Code permits any reasonable stress distribution to be assumed, if it agrees with tests, such as those in Figs. 1.10c to 1.10e or others. Practice has moved to the rectangular stress block because of its simplicity. Research often uses the shaded stress distribution of Fig. 1.10e. The use of the computer has also encouraged the development of a continuous function rather than this half parabola plus a straight line.

This book emphasizes the strength method and its advantages. Nevertheless, it also includes working stress analysis briefly in Chapters 3, 4, 17, and Appendix A for two reasons. First, its assumption of linear stress distribution is appropriate for calculations of deflections under service load conditions and crack investigations at service loads. Second, a minimum understanding of working stress design methods will improve communications between those trained in today's best methods and those versed in older procedures.

1.5   Safety Provisions

(a) Factor of Safety   A *factor of safety* is a concept of long standing, some reserve strength to care for unusual loads beyond those in the design. It has often been thought of as the ratio of yield stress to service load stress, although this concept is invalid where nonlinear responses exist. Correctly defined, the factor of safety is the ratio of the load that would cause collapse to the service or working load. Factor of safety is now such a misused term it almost requires a definition each time it is used.

---

* Whitney was not the first to suggest the rectangular stress block, but he was the first in the United States to achieve some general acceptance.
† Accuracy with beams is excellent. With columns the results are more sensitive to the exact stress-strain curve assumed.

(b) Factors Determining Safety   Strength design as presented in the Code recognizes that a factor of safety is necessary for a number of reasons that are individually important but have no real interrelation. (1) Some are matters beyond the control of the engineer, such as the possibility of unforeseen future loads. (2) Others are in part under the engineer's control, such as the quality of materials used, which the engineer can at least partially control by specifications and inspection; the usual dimensional tolerances in the sizing of beams and areas of bars and the placement of reinforcing, which can be closely inspected but have practical limitations inherent in manufacture and field construction. (3) Others relate to the importance of the member in maintaining the integrity of the structure; a local slab failure is not as serious as the collapse of a column. (4) Others recognize that the level of safety should vary with the level of warning at impending collapse. Ductile-type failures with large cracking and deflections can permit evacuation and temporary shoring of distressed structures. Such failure types might be designed for smaller safety factors than brittle, sudden failure types. (5) Still other factors vary with the degree of safety factor that is justified, such as the hazards to life and limb from the collapse of a school building compared to that of a simple shed for storage of equipment or material; these relative hazards are considered as already reasonably recognized in the size of the live load required by all general building codes.

(c) Partial Factors of Safety   Safety has always been clearer when thought of in terms of partial safety factors. In design these might well be in terms of loads, construction practices, the quality of the materials used, and the importance of the member or structure.

The loads specified in codes are usually not the average or mean loads, but more nearly the maximum loads. The European concept of characteristic loads is excellent. A characteristic load is one that has only a 5% probability of ever being exceeded. If one used such a load the strength above this load could be smaller than if one started from the mean or average load concept. The difficulty lies in identifying and specifying this characteristic load.

Construction practices introduce an element of uncertainty. If columns can be built to a $\frac{1}{2}$-in. tolerance (Code 7.5.2.1), it is well to note that a $11\frac{1}{2}$-in. square column is some 8% short of the area of a 12-in. square column.

Materials are also variables. The designer will not always find actually present at a critical spot the particular concrete strength that was assumed in design, in spite of good controls.

All the various overloads or deficiencies are less likely to occur together. Probability theory can predict the reduced probability of all bad aspects occurring together if the probability of each occurring separately is known; however, the input data cannot be clearly established in many cases.

(d) Code Safety Philosophy   The ACI Code separates safety provisions into two parts in the hope that ultimately (not in the present Code) different factors may be established for different qualities of specifications and inspection, and so on. The two factors now prescribed are load factors and strength reduction or ϕ factors. The same general approach was recently introduced for structural steel design through the LRFD format (Load and Resistance Factor Design). It is hoped that eventually common load factors will be used in structural design with all structural materials.[5,6]

Load factors (greater than unity) attempt to assess the possibility that prescribed service loads may be exceeded. Obviously, a specified live load is more apt to be exceeded than a dead load that is largely fixed by the weight of the construction. The ultimate strength of the members must care for the total of all service loads, each multiplied by its respective load factor; and the load factors are different in magnitude for dead load, live load, and wind or earthquake loading.

The ϕ factors (less than unity) are provided to allow for variations in materials, construction dimensions, and calculation approximations, that is, matters at least partially under the engineer's control. At present* ϕ varies only with the type of stress or member considered, recognizing that a column failure, for instance, would be more serious than a beam failure in most cases.

The general building code prescribes live loads that are assumed to cover adequately the relative hazards resulting from failure. For example, the prescribed live load for a school auditorium is properly oversafe compared to that for an office building. The general code also usually considers the probability of all areas being fully loaded at the same time.

The basic requirement in the Code for strength design may be expressed as

Required Strength $\gtrless$ Design Strength

$$\left.\begin{array}{c} P_u \\ V_u \\ M_u \\ T_u \end{array}\right] = U \gtrless \phi \text{ [Nominal Strength]} = \left[\begin{array}{c} \phi P_n \\ \phi V_n \\ \phi M_n \\ \phi T_n \end{array}\right.$$

The "required strength" U is determined from the structural analysis of the load effects of the factored loads. Factored loads increase the ordinary or service loads to reflect the possibility of overloads or analysis limitations. The best estimate of the idealized strength of a member, cross section, or connection calculated in accordance with the requirements and assumptions of the Code is termed the "nominal strength." The "design strength" represents a reduction in the nominal strength to reflect possible understrength or undesirable failure mode. As the design strength must be equal or greater than the required strength determined from the factored loads and structural analysis and as ϕ factor values are always less than unity, the nominal strength must always equal or exceed the required strength.

---

* At some future time the ϕ factor might have a range of values for different type specifications or inspection provisions.

For design purposes it is often convenient to consider the nominal strength needed by rearranging the previous equation.

$$\begin{bmatrix} P_n \\ V_n \\ M_n \\ T_n \end{bmatrix} = [\text{Nominal Strength}] \geqslant \frac{U}{\phi} = \begin{bmatrix} P_u/\phi \\ V_u/\phi \\ M_u\phi \\ T_u/\phi \end{bmatrix}$$

Thus, the designer will often proportion his members to provide a nominal strength or resistance to match factored load effects that also have been divided by the applicable ϕ factor to simplify computation. This text will generally use this method of including the ϕ factor in the design process.

(e) Load Factors For dead and live loads the Code specifies that factored loads, factored shears, and factored moments be obtained from service loads by using the relation:

$$U = 1.4\,D + 1.7\,L \qquad \text{(Code Eq. 9.1)}$$

where U represents any of these required strengths for factored loads, D represents that for service dead load, and L that for service live load.

When wind W is included, the total loading may be worse, but the chance of maximum wind occurring when an overload in both dead and live load exists is less than the chance of the overloads existing alone. Hence, the Code uses an 0.75 coefficient on the sum of all three:

$$U = 0.75(1.4\,D + 1.7\,L + 1.7\,W) \qquad \text{(Code Eq. 9.2)}$$

where L must be considered *both* as zero and as its maximum value. If L does act to reduce the total, this also suggests that D and W might act in an opposite sense. If so, an overlarge D is on the unsafe side and the third Code condition becomes necessary:

$$U = 0.9\,D + 1.3\,W \qquad \text{(Code Eq. 9.3)}$$

This equation underestimates D when it opposes W. (Note that $1.3\,W$ is approximately $0.75 \times 1.7\,W$, as in Eq. 9.2, with only two significant digits.) The Code (9.2) also gives load factors for earthquake, earth pressure, fluid pressure, impact, and special effects such as settlement, creep, shrinkage, or temperature change, but these special cases are not essential to the discussion here.

(f) Code ϕ Factors The idealized or nominal strength is multiplied by the ϕ factor to obtain a reasonably dependable strength. The nominal strength equation might be said to give the "ideal" strength assuming materials are as strong as specified, sizes are as shown on the drawings, bars are of full weight, that calculations of load, shear, and moment are exact, and that the strength equation itself is scientifically correct. The practical dependable strength of a particular member will often be something less, probably sometimes very much less, as all these factors vary over some range. How low the strength may go can only be determined by probability theory; and the variables are not yet clearly enough defined to make this approach clearly better. Furthermore, the adequacy of theory varies, being lower for shear than for flexure. The Code provides for these variables by using these simple ϕ factors for the several cases listed (Code 9.3.2).

Flexure, without axial load	0.90
Axial tension and axial tension with flexure	0.90
Axial compression and axial compression with flexure:	
Members with spiral reinforcement conforming to Sec. 10.9.3	0.75*
Other reinforced members	0.70*
Shear and torsion	0.85
Bearing on concrete (See also Code 18.13 for prestress tendon anchorage)	0.70

The ϕ is largest for flexure because the variability of steel is less than that of concrete; all flexural members are specified to be designed for failure in tension, that is, in the steel; thus, the ductility of such members should be large. The ϕ values for columns are lowest (favoring the toughness of spiral columns a little over tied columns) because columns fail in compression where concrete strength is critical and ductility is often absent; there is some small danger that the analysis may miss the worst combination of axial load and moment; the column is critical in the building. For shear and torsion ϕ is intermediate as they depend on concrete strength, but on $\sqrt{f'_c}$ rather than f'_c itself; the validity of shear and torsion theory is also still questionable to some degree.

(g) Improved Code Notation The 1977 Code clarified the notation associated with member strengths. This was carried over to the 1986 Code and is used throughout this book. The subscript *u* is reserved for *required* strengths, that is, the M_u, V_u, P_u, and so on, required for the factored loads. The theoretical strengths provided are designated by the subscript *n* for nominal (or ideal or theoretical) strengths and the design strengths or usable portions thereof as ϕM_n, ϕV_n, ϕP_n, to be compared to the required values:

$$M_u \gtrless \phi M_n, \qquad V_u \gtrless \phi V_n, \qquad P_u \gtrless \phi P_n, \qquad T_u \gtrless \phi T_n$$

(h) Authors' Evaluation of Code Safety Provisions The approximate factor of safety, which is given by the load factor divided by ϕ, is essentially sound. Philosophically, the authors feel that if the same ratio had been held with each term a little lower, there would be more encouragement to improve design practice, materials, and inspection. Any of these would justify an increase in the ϕ value which is now somewhat constrained by the very high 0.90 value of ϕ for flexure. By the time allowance is made for necessary tolerances in construction dimensions and placement of reinforcing bars, underweight (but legal) bars, and the like, the 10% margin between 0.90 and 1.00 looks none too large. In fact, in the 1971 Code, to recognize improvement in design through new

---

* Provisions are included for the gradual increase to 0.90 as the axial compression drops below ϕP_n of $0.10 f'_c A_g$ (Code Sec. 9.3.2c).

Code provisions, better quality control of materials, and additional research and experience, the ϕ was left unchanged and the load factors lowered. This gave the intended reduction in the necessary factor of safety; only the philosophy that ϕ (not load factor) was to reflect such matters had to suffer.

Considerable representation has been made that the level of fabrication tolerances, material controls, and inspection is generally higher in many precast and prestressed concrete plants than in field cast-in-situ construction. The AASHTO Bridge Specifications have recognized this by increasing the ϕ value slightly for flexure in precast prestressed members where the plant is part of an ongoing certification program.

A complete reevaluation of ϕ factors and load factors is underway within ACI Committee 318 in response to proposals in ANSI Standard A58.1 for common load factors for all structural engineering applications. Should the ANSI A58[3] suggested load factors be adopted, most of the present ACI Code ϕ factors would need reduction to maintain the desired level of safety.

1.6 Ductility in Reinforced Concrete Construction

Provision for ductility in reinforced concrete could, with considerable logic, be considered part of the safety provisions, but probably deserve the emphasis of separate consideration. Ductility in a structure or a member means the maintenance of strength while sizeable deformation or deflection occurs. The engineer seeks to build ductility into structures for several reasons. In typical indeterminate structures, ductility permits a heavily stressed portion to continue to carry its capacity load while deforming enough to bring lesser stressed neighboring portions more definitely into the resistance pattern. In slabs and beams ductility means a warning of overloads will be present in the form of excessive cracking and deflection. Where energy must be absorbed, as in blast and earthquake situations, ductility is particularly important.

Although concrete in compression is far from being elastic at higher stresses, it is basically a brittle material that crushes in compression. In beams and slabs ductility is obtained by (1) using reinforcing steel, which is not brittle, to carry tension and then (2) limiting the amount of tensile steel to that which will yield before the concrete crushes in compression. This method of control is discussed in Chapter 3.

Reinforced concrete columns are basically brittle under vertical load, although spiral columns (Chapter 6) do introduce limited ductility. Under lateral loads (sideway) considerable ductility can be built into a column by proper detailing and the use of spirals. The detailing of reinforcement at the joints between beams and columns is critical in the development of ductility and the capacity to survive shock loadings.

Where ductility cannot be achieved in a detail or a design, added strength is necessary to insure that any potential failure be initiated at a more ductile section or in a more ductile manner. In the Code this concern shows, for example, in more conservative ϕ values for columns and in longer bar laps for tension splices. In this text Chapter 10 emphasizes the development of continuity and ductility through improved detailing. Chapter 23 based on Code Appendix A gives special provisions for seismic design details and is largely directed toward building ductility into the structural frame.

1.7 Limit Design

Present strength design is based on moments and shears calculated from elastic analysis of members and frames. Strength evaluations of the cross sections, nevertheless, involve inelastic action at the critical design sections. The elastic analysis gives moments and shears generally on the safe side, but their use involves at least a philosophical inconsistency. Although it is not usually computed, it also should be noted that the influence of the cracking of some of the members and the influence of the joints where beam and column have a common section can modify the usual* computed moments more than most designers realize.

In view of these problems, limit design concepts look very attractive. A limited arbitrary reassignment of moments is now permitted, but limit design as such is still in the development stage in the United States. A number of countries in Europe have moved further in this direction. Limit design is discussed further in Chapter 11.

1.8 Calculation Accuracy

Reinforced concrete is still not a precise theory and, hence, judgment is always an important factor in the application of this theory.

A distinguished engineer and the chairman of the ACI committee responsible for the 1963 ACI Building Code, Raymond C. Reese, introduced his *CRSI Design Handbook* with these comments on the accuracy of calculations (slightly abbreviated here):

> If . . . involved mathematical methods . . . lead one to believe that the design of reinforced concrete structures requires a high degree of precision, the reverse is the case. Concrete is a job-made material, and control cylinders that do not vary more than 10% are remarkably good. Reinforcing bars are shop-made; yet variations in strength characteristics run 3% to 5%; rolled weights can vary $3\frac{1}{2}$%. Formwork is field-built; frequently a 2×8 or 2×10 (measuring, respectively, $7\frac{5}{8}$ in. and $9\frac{1}{2}$ in.) is used to form the soffit of an 8-in. or 10-in. beam. Bars that are held in place to an accuracy of between $\frac{1}{8}$ in. and $\frac{1}{4}$ in. are extremely well placed. Two-figure accuracy is sufficient for almost all problems in reinforced concrete design.
> . . . the time and effort of the designer is best spent in recognizing and providing for . . . tensions wherever they may exist, not in striving for a high degree of precision by carrying figures to an unmeaning number of significant places.

---

* Based on members with a constant EI, often based on gross concrete sections.

On the other hand, . . . when numbers are subtracted, significant figures are often lost. It is, therefore, recommended, more for control of the computations, for ready checking, and to keep the computer alert, rather than for any effect on the completed structure, that figures be carried to three significant places. . . . The following table is suggested.

Record Values to the Following Precision

Loads to nearest 1 psf; 10 plf; 100 lb concentration
Span lengths to about 0.01 ft ($\frac{1}{8}$ in. = 0.01 ft)
Total loads and reactions to 0.1 kip
Moments to nearest 0.1 kip-in., if readable
Individual bar areas to 0.01 in.^{2}
Concrete sizes to $\frac{1}{2}$ in.
Bar spacings to $\frac{1}{2}$ in. (supports are crimped at 1-in. intervals)
Effective beam depth to 0.1 in.

As a practical limit the designer should note the construction tolerances on effective depth d and clear cover specified under Code 7.5.2.1:

d of 8 in. or less	$\pm\frac{3}{8}$ in.	Min. cover $-\frac{3}{8}$ in.
$d > 8$ in.	$\pm\frac{1}{2}$ in.	Min. cover $-\frac{1}{2}$ in.

"except that tolerance for the clear distance to formed soffits shall be minus $\frac{1}{4}$ in. and tolerance for cover shall not exceed minus one-third of the concrete cover required in the contract drawings or in the specifications."

1.9   Handbooks

Because this book is written primarily as a textbook, it does not concern itself greatly with office practice and the use of design aids. Nevertheless, the student should be aware that curves and tables can speed up design considerably.

The *Design Handbook in Accordance with the Strength Design Method of ACI 318-77, Vol. 1—Beams, Slabs, Brackets, Footings and Pile Caps,*[10] published by ACI in 1982 and *Vol. 2—Columns,*[10] published by ACI in 1980, have many useful tables and charts. They are currently under revision and should be available in revised form to agree with the 1986 Code.

The 1984 *CRSI Handbook*[11], published by the Concrete Reinforcing Steel Institute has many tables of allowable design loads on various types of members, some with details of the reinforcing steel.

In the detailing of structures the *ACI Detailing Manual*[1] is most helpful as a guide to the clear presentation of the necessary details. It is a drafting rather than a design manual, although it calls attention to some important considerations in designing reinforced concrete.

1.10   The Use of SI Units in Some Examples

Unlike the previous edition, SI units are used in only a few elementary examples. The limited use of SI units is to make the student aware that concrete design is truly international but that the empirical nature of many provisions makes the calculation procedures very unit dependent.

It is now assumed that nearly all engineering students at the junior-year level have used SI units earlier and will find their limited use here creates no problem. Nevertheless, Sec. 1.13 summarizes SI units and prefixes and outlines a few general good practices for using them in calculations and on drawings. The aim is primarily to refresh the reader's memory in case he or she has not been using SI calculations recently or has not been using them in the area of forces and stresses, mass, weights, and energy.

1.11   Limits on the Use of SI Units

With few exceptions other than the United States, design practices are based on SI units. Material properties and dimensions are based on SI units. The United States is moving very slowly towards the use of the SI system. Both because of its international membership and because American designers are heavily involved in projects outside of the United States, the American Concrete Institute has adopted a complete SI version of the ACI Building Code (ACI 318M-86).

Reinforced concrete is not the best area in which to introduce SI units because of its complexity. In many respects it is an empirical field, because there is no mathematical basis that can be extended to many areas. It is nonlinear in behavior. Concrete is reinforced with steel because it is weak in tension and in most structural usage is uneconomical as plain concrete. To make reinforcement work economically, the concrete must crack in tension at service loads (fortunately, if well designed, with cracks too small to see without searching). But such physical responses mean there are many empirical constants from research studies that enter into designs.

In the United States and Canada most of these constants were established using customary units and the allowable values in many cases involve judgment decisions. For example, one of the unit shear values in psi is limited to $2\sqrt{f_c'}$, where f_c' is the concrete strength in compression in psi. If you transform this mathematically into megapascals (MPa), equal to newtons per square millimeter (N/mm²), you obtain something like 0.166 as the constant. But you know that the input was 2 and not 2.0 or 2.00. Hence, 0.166 seems unreasonable, somewhat like estimating a distance as 600 ft and having someone say you estimated it as 0.1136 miles (600/5280 = 0.113 636). The 0.166 is actually no more accurate than the starting 2; hence a $\frac{1}{6}\sqrt{f_c'}$ in SI usage (as in ACI 318M-86) is at least as accurate as the original input, and conveys an idea of the actual precision involved. Examples are even more vivid in the dimensional field using essentially judgment decisions. For example, in Code Sec. 7.6.5 a maximum reinforcing bar spacing of 18 in. is given. In the SI conversion with ACI 318-77 the "theoretically" correct value of 457 mm is given. However, in the conversion that resulted in ACI 318M-86, the value was set at the more practical 500 mm. What was in reality about one-half of a yard in the customary measurement system is thus set at one-half of a meter in the SI system. Throughout this book, all SI values will be based on ACI 318M-86.

Specified SI concrete strengths will almost certainly be C20, C25, C30, and so on, where the number is the cylinder strength in megapascals. One pascal is one newton per square meter or N/m^2.

Likewise, new bar sizes have been written into a new ASTM standard. The smallest deformed bar would have a cross section of 100 square millimeters and be called a #10 bar (very roughly 10 mm diameter), the largest #55 with an area of 2500 mm^2. Grades of reinforcing bars will be Grade 300 and Grade 400 with the number the yield strength in megapascals (MPa). These new standards provide a reasonably sound basis for flexure calculations with SI units.

Totally outside the questions of concrete strength and reinforcement strengths and sizes, there are many other interlocking construction materials. Lumber sizes now nominally in full inches, such as 2 × 12 in. nominal size, $1\frac{5}{8}$ by $11\frac{1}{2}$ in. actual size, will probably differ in SI units; lumber sizes often determine beam widths for economy. Joist forms will probably change size as they move to SI units. Loadings must change from pounds per square foot to newtons per square meter, very different quantities. For distributed loads kPa = kN/m^2 has been recommended as very convenient. Slabs now are typically designed for one foot strips because that width fits loads per square foot. The SI designer will use one meter widths.

1.12   The Status of SI Units

The United States is committed to change to SI units, but no calendar or schedule has been adopted. Hence, one should say "changing very slowly." Canada, on the other hand, has made the changeover and has a new code to match. England has been using an SI code for years.

A number of countries do not want to give up kilograms for force even though it also means mass; the SI system is kilogram for mass *only*, newton for force, and pascal for unit stress. The SI usage discourages the use of centimeters, because either millimeters or meters can be written with less confusion from decimal places. Drawings for structures may be entirely in millimeters to avoid decimals, a fraction of a millimeter being negligible in most construction cases.

Australia, New Zealand, and South Africa have moved to SI units. In the United States ACI has issued an SI version of the 1983 Code (updated in 1986) and has deleted the 1977 Code Appendix D (SI equivalents) and Appendix E (MKS–metric equivalents) from ACI 318-83.

1.13 SI Units Frequently Applicable to Reinforced Concrete Design

The SI standard has one (and only one) unit for each physical quantity. Each of these units are subdivided into three classes of terms, which include the following related to structures:[12,13]

		Symbol	Formula
Basic Units (out of a total of seven)			
length	meter	m	
mass	kilogram	kg	
time	second	s	
Supplementary Units (total of two)			
plane angle	radian	rad	
solid angle	steradian	sr	
Derived units (very large group)			
acceleration	meter per second squared		m/s^2
area	square meter		m^2
energy	joule	J	$N \cdot m$*
force	newton	N	$kg \cdot m/s^2$
moment of force	newton meter		$N \cdot m$
moment of inertia of an element of area	meter to the fourth power		m^4
moment of inertia	kilogram meter squared		$kg \cdot m^2$
pressure, stress	pascal	Pa	N/m^2
elastic modulus	pascal	Pa	N/m^2
torque	newton meter		$N \cdot m$
velocity	meter per second		m/s
volume	cubic meter		m^3
work	joule	J	$N \cdot m$

* Note that a period at midheight is a multiplication sign.

Rather than a large number of digits, prefixes attached to units have been standardized. Multiples for 100, 10, 0.1, and 0.001 are available, but discouraged. (Centimeter usage is thus discouraged.) The prefixes most often useful in structural calculations are:

1 000 000† = 10^6	mega	M
1 000 = 10^3	kilo	k
0.001 = 10^{-3}	milli	m
0.000 001 = 10^{-6}	micro	μ

† Note that commas are *not* used in the open gaps. The gaps can be either full spaces or half spaces.

Preferred practice is to use the prefixes to keep isolated numbers in the range between 0.1 and 1000; but *not* in tabulations of data, where a given unit should be maintained even if the data scatter widely.

The units spelled out are kept in lowercase, even though as abbreviations they may be capitalized, such as newton with the symbol N, pascal with the symbol Pa, and so on. The form newton/meter is discouraged for all cases where a unit is spelled out; use either newton per meter or N/m.

For more than three digits use no space for four places, but for more places skip a place to keep numbers in groups of three: 1213, but 12 130 and 1 121 300 and similarly in decimals 0.1213, but 0.012 13 and 0.012 130 4. Note that commas are not used, as we are accustomed to do in customary units, because the comma in some areas means a decimal point.

Selected References

1. *ACI Detailing Manual,* SP-66, Amer. Concrete Inst., Detroit, 1980.

2. J. G. MacGregor, "Safety and Limit States Design for Reinforced Concrete," *Canadian Jour. of Civil Engineering, 3,* No. 4, 1976, pp. 484–513.

3. ANSI Standard A58.1-1982, *Minimum Design Loads for Buildings and Other Structures,* Amer. National Standards Inst., New York, 1982, 100 pp.

4. PCI Committee on Precast Bearing Walls, "Considerations for the Design of Precast Bearing-Wall Buildings to Withstand Abnormal Loads," *Jour. Prestressed Concrete Inst., 21,* No. 2, March–April 1976, pp. 46–69.

5. T. V. Galambos, B. Ellingwood, J. G. MacGregor, and C. A. Cornell, "Probability-Based Load Criteria Assessment of Current Design Practice," *ASCE Jour. Struc. Div., ST-5, 108,* May 1982, pp. 959–977.

6. B. Ellingwood, J. G. MacGregor, T. V. Galambos, and C. A. Cornell, "Probability-Based Load Criteria—Load Factors and Load Combinations," *ASCE Jour. Struc. Div., ST-5, 108,* May 1982, pp. 978–997.

7. "Revision—Building Code Requirements for Reinforced Concrete" (ACI 318-83), *Jour. of the Amer. Concrete Inst.,* No. 3, *Proc. 83,* May–June 1986, pp. 524–531.

8. *Building Code Requirements for Reinforced Concrete* (ACI 318M-83), Amer. Concrete Inst., Detroit, 1983.

9. *Commentary on Building Code Requirements for Reinforced Concrete* (ACI 318RM-83), Amer. Concrete Inst., Detroit, 1983.

10. *Design Handbook in Accordance with the Strength Design Method of ACI 318-77, Vol. 1—Beams, Slabs, Brackets, Footings and Pile Caps* (1982), and *Vol. 2—Columns* (1980), Amer. Concrete Inst., Detroit.

11. *CRSI Handbook 4th ed.,* Concrete Reinforcing Steel Inst., Chicago, 1984.

12. American National Standards Institute, *ASTM/IEEE Standard Metric Practice,* ASTM E 380-76, IEEE Std 268-1976.

13. *Recommended Practice for the Use of Metric (SI) Units in Building Design and Construction,* National Bureau of Standards, NBS Technical Note 938, 1977.

2

MATERIALS

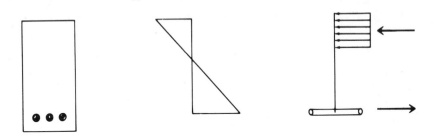

2.1 Concrete Materials and Production

Concrete for reinforced concrete consists of inert aggregate particles bound together by a paste made from portland cement and water. The paste fills the voids between the aggregate particles, and after the fresh concrete is placed, it hardens as a result of exothermic chemical reactions between cement and water to form a solid and durable structural material. A typical cross section of hardened concrete is shown in Fig. 2.1a.

Although there are a number of standard portland cements, most concrete for buildings is made from Type I ordinary or standard cement (for concrete where the critical strength is needed in something like 28 days) or from Type III high-early strength cement (for concrete where strength is required in a few days). The heat generated by the different types of cements during the setting and hardening process varies widely, as indicated in Fig. 2.2. Where shrinkage and temperature stresses are important in the design, the volumetric change associated with these heat differences becomes significant. Air-entraining cement or admixtures for entraining air in the concrete are frequently used for greater workability or durability. ACI Code Sec. 3.6.4 requires that air-entraining admixtures meet ASTM C260.

In recent years there has been a substantial increase in the use of other chemical additives for cement dispersion, acceleration or retardation of time of initial set, and improvement of workability. The so-called "superplasticizers" are being used in many applications where high strength concrete with substantial slump is desired. ACI Code Sec. 3.6.5 requires that most water reducing and set regulating admixtures meet ASTM C494. Expansive cement to limit shrinkage is also on the market.

Aggregate consists of both fine and coarse aggregate, usually sand for the fine and gravel or crushed stone for the coarse aggregate. Lightweight aggregate made from expanded shale, slate, or clay has become increasingly impor-

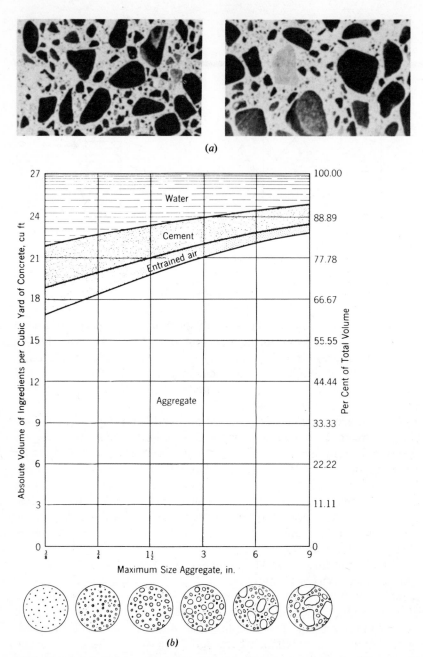

(a)

Maximum Size Aggregate, in.

(b)

Figure 2.1 Components of a concrete mix. (*a*) Cross sections of concrete showing coarse and fine aggregate separated by cement paste. (Courtesy Bureau of Reclamation.) (*b*) Quantities of each material in 1 cu yd of concrete. (From Reference 1, Bureau of Reclamation.)

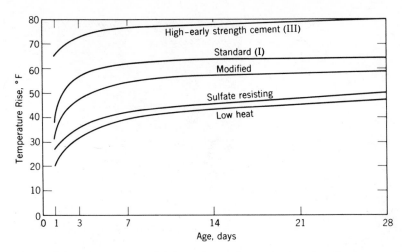

Figure 2.2 Temperature rise in concrete for various types of cement, when no heat is lost. One barrel of cement per cubic yard. (From References 1 and 2, ACI and Bureau of Reclamation.)

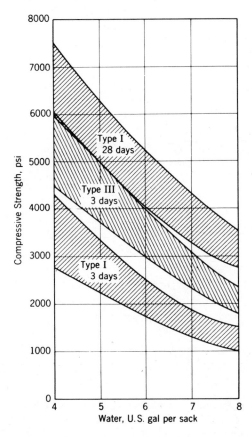

Figure 2.3 Effect of water-cement ratio on strength of non-air-entrained concrete at different ages. All specimens moist cured at 70°F. (Modified from Reference 5, Portland Cement Assn., earlier edition.)

(a)　　　　　　　　　　　　　　　　(b)

Figure 2.4 Testing for workability of concrete. (*a*) Slump test. (Courtesy Portland Cement Assn.) (*b*) Kelly ball test. The "ball" penetration is read on the graduated shaft by the late Professor Kelly of the University of California. This mix is quite stiff.

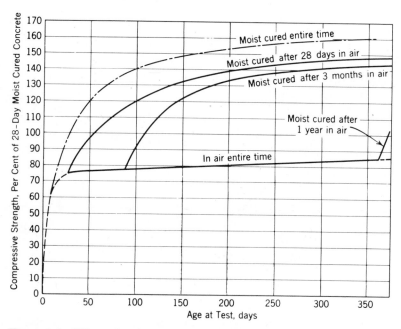

Figure 2.5 Effect of curing conditions on strength of concrete. (From earlier edition of Reference 5, Portland Cement Assn.)

tant. Other aggregates, such as expanded slag, are also used. The size and the grading of aggregate have an important influence on the amount of cement and water required to make a given unit of concrete of a given consistency (Fig. 2.1b). It also exerts a major influence on bleeding, ease of finishing, shrinkage, and permeability.

The quantity of water relative to that of the cement is the most important item in determining concrete strength. The effect of the water-cement ratio on strength is indicated in Fig. 2.3. The water is sometimes controlled indirectly and approximately by specifying the cement content in terms of sacks per cubic yard of concrete.

It is important that concrete have a workability adequate to assure its consolidation in the forms without excessive voids. This property is usually indirectly measured in the field by the slump test (Fig. 2.4a) or the Kelly ball test (Fig. 2.4b). The necessary slump may be small when vibrators are used to consolidate the concrete. For methods of designing concrete mixes the student is referred to the American Concrete Institute's "Standard Practice for Selecting Proportions for Normal, Heavyweight and Mass Concrete" (ACI 211.1-81),[3] "Standard Practice for Selecting Proportions for Structural Lightweight

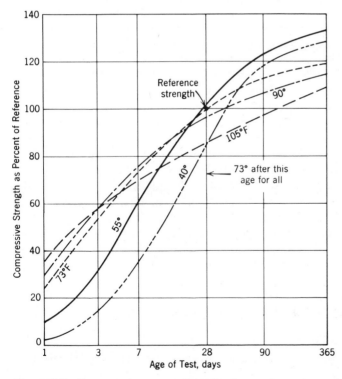

Figure 2.6 Compression strength attained at various ages and temperatures as percent of 28-day strength under curing at 73°F. Type I cement. (Modified from Reference 5, Portland Cement Assn.)

Concrete" (ACI 211.2-81),[4] or the Portland Cement Association booklet, "Design and Control of Concrete Mixtures."[5]

Proper curing of concrete requires that the water in the mix not be allowed to evaporate from the concrete until the concrete has gained the desired strength. Figure 2.5 is representative of the variations in strength that can result from differences in curing. Note that air-dried concrete is able to gain strength if it is moistened at a later time. Temperature is also an important element in the rate at which concrete gains strength: low temperatures slow up the process but raise the potential strength if normal temperature is restored, as indicated in Fig. 2.6. Early high temperatures lead to rapid setting and some permanent loss of strength potential.

2.2 Compressive Strength

Depending on the mix (especially the water-cement ratio) and the time and quality of the curing, compressive strengths of concrete can be obtained up to 14,000 psi or more. Commercial production of concrete with ordinary aggregates is usually in the 3000 to 12,000 psi range with the most common ranges for cast-in-place buildings from 3000 to 6000 psi. On the other hand, precast and prestressed applications often expect strengths of 4000 to 8000 psi. Because of the difference in aggregates and to a lesser degree in cements, the same mix proportions result in substantially lower strengths in some sections of the country than in others. In these sections a lower water-cement ratio must be used.

Compressive strength f'_c is based on standard 6-in. $\times$ 12-in. cylinders cured under standard laboratory conditions and tested at a specified rate of loading at 28 days of age.* In some applications it may be considerably later before substantial portions of the dead and live loads are imposed on a member. ACI Code Sec. 4.1.3 does allow test ages other than 28 days, but requires the age to be indicated in design drawings or specifications. The designer should note that building concrete cured in place on the job will rarely develop as much strength as standard cured cylinders. Separate cylinders should be made to check on the quality of curing if this is desired. These same cylinders, however, are not suitable for checking the quality of the mix.

The ACI Code specifies the average of two cylinders from the same sample tested at the same age (usually 28 days) for a strength test. It specifies the frequency of such tests and sets up this criterion:

4.7.2.3—Strength level of an individual class of concrete shall be considered satisfactory if both of the following requirements are met:

a. Average of all sets of three consecutive strength tests equal or exceed f'_c.
b. No individual strength test (average of two cylinders) falls below f'_c by more than 500 psi.

---

* Test procedure is covered in detail in ASTM C-39.

Such a simple criterion cannot always be perfect. The Code Commentary states that, even with the strength level and uniformity satisfactory, an occasional low strength may be indicated (at least once in 100 tests) and makes suggestions about allowance for these.

The average concrete strength for which a concrete mix must be designed must exceed f'_c by an amount which depends on the uniformity of plant production, that is, on how well the variations in operation are minimized. Where production records are available, the Code in Sec. 4.3 specifies the mix must be designed for an average strength f'_{cr} that is greater than the specified compressive strength, f'_c, using the larger value of:

$$f'_{cr} = f'_c + 1.34\ s \qquad \text{(Code Equation 4.1)}$$

or

$$f'_{cr} = f'_c + 2.33\ s - 500 \qquad \text{(Code Equation 4.2)}$$

Code Sec. 4.3.1 indicates how the value of the production facility standard deviation s is to be established. If there is a lack of sufficient data and the standard deviation cannot be established, then Code Sec. 4.3.2.2 requires average target compressive strengths from 1000 to 1400 psi greater than the specified compressive strength, f'_c.

It must be emphasized that f'_c for design is *not* to be considered as the average strength of job cylinder tests. The design f'_c is nearer a minimum than an average, but it is not an absolute minimum. An individual test (average of two cylinders) may be nearly 500 psi low and still be acceptable if all averages of three consecutive tests are satisfactory.

With lightweight aggregates a mix design should definitely be made on the basis of trial batches. Many lightweight aggregates produce 3000 psi concrete and some easily give 6000 psi concrete under proper control.

A special treatment of concrete after it has set, which can lead to extremely high potential strengths, is currently in an interesting stage of development. Polymer impregnated concrete is made by impregnating ordinary concrete with a liquid organic monomer* that is subsequently polymerized by radiation or thermal catalytic methods. The result is greatly increased tension and compression strengths (up several hundred percent) accompanied by much improved impermeability, hardness, and durability.

2.3   Tensile Strength

The tensile strength of concrete is relatively low, about 10 to 15% of the compression strength, occasionally 20%. This strength is more difficult to measure and the results vary more specimen to specimen than those from compression cylinders. The modulus of rupture as measured from standard 6-in. square

---

* Such as methyl methacrylate.

beams somewhat exceeds the real tensile strength. The value of $7.5\sqrt{f'_c}$ is often used for the modulus of rupture, in psi when f'_c is in psi. The cylinder splitting test, often called the Brazilian test, is the most respected test for tensile strength. A cylinder is placed horizontally in a testing machine. A uniform strip compression force is applied full length along the upper surface and the cylinder is supported full length on a strip support along the connecting diameter. This splits the cylinder open along the connecting diameter and the approximately uniform tensile stress is easily calculated (See ASTM C496).

Lightweight concrete in many, but not all cases, has a lower tensile strength than ordinary weight concrete. The ACI Code accepts either of the following approaches in computing shear and development of reinforcement. If the splitting tensile strength is specified and the concrete is proportioned under Code 4.2, the designer substitutes $f_{ct}/6.7 \leqslant 1$ for $\sqrt{f'_c}$ in computing the allowable. If f_{ct} is not specified, the assumed shear resistance may simply be reduced by a factor of 0.75 for ''all-lightweight'' concrete and 0.85 for ''sand-lightweight'' concrete. The latter refers to lightweight concrete containing natural sand for the fine aggregate. Linear interpolation may be used for mixtures of natural sand and lightweight fine aggregate. Development length (Chapter 8) is modified in a similar way, with the ACI Code factors $6.7\sqrt{f'_c}/f_{ct}$, 1.33, and 1.18, as splitting strength enters here as a reciprocal.

2.4   Significance of Low Tensile Strength

In a homogeneous elastic beam subjected to bending moment, one can calculate the bending stresses from $f = Mc/I$. The extreme fiber on one face carries compression and tension on the opposite face. If the beam is rectangular (or of any shape symmetrical about the centroidal axis), the maximum tensile stress equals the maximum compressive stress. In concrete construction, except in massive structures such as gravity dams and sometimes heavy footings, it is not economical to accept the low tensile strength of plain concrete as a limit on beam strength. It is generally more economical to design a beam such that compressive bending stresses are carried by concrete and tensile bending stresses are carried entirely by steel reinforcing bars. It is not possible for concrete to cooperate with steel in carrying these tensile stresses except at very low and uneconomical values of steel stress. Except in prestressed concrete, when the steel stress reaches about 6000 psi, the tensile concrete starts to crack and the steel soon thereafter must pick up essentially all the tension necessary to provide for the applied moment. Hence, in ordinary reinforced concrete beams the tensile concrete is not assumed to assist in resisting the moment. Still, tension in the concrete does reduce the beam deflection considerably.

2.5   Shear Strength

The shear strength of concrete is large, variously reported as from 35 to 80% of the compression strength. It is difficult to separate shear from other stresses in testing and this accounts for some of the variation reported. The lower values represent attempts to separate friction effects from true shears. The shear value is significant only in rare cases, as shear must ordinarily be limited to much lower values in order to protect the concrete against diagonal tension stresses (Chapter 5).

Diagonal tension stresses are often referred to as shear stresses, but this is actually a misnomer. If the student will keep in mind that true shear strength is rarely in question, it will not matter that the term "shear" is often loosely used for diagonal tension.

2.6   Stress-Strain Curve

Typical stress-strain curves for concrete cylinders on initial loadings are shown in Fig. 2.7. The first part of each curve is nearly a straight line, but there is some

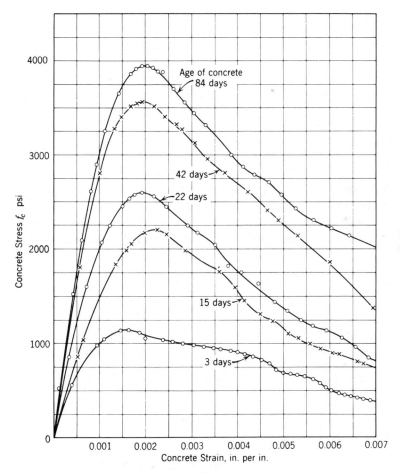

Figure 2.7   Concrete stress-strain curves from compression cylinders. (From Reference 6, Bureau of Reclamation.)

curvature at f_c equal to half the maximum value.* The maximum stress is designated as f'_c (but note the statistical nature of the f'_c value to be used in design, as given in Sec. 2.2). The curve for low strength concrete has a long and relatively flat top. For high strength concrete the peak is sharper. Special techniques are necessary to establish these curves on the steep downward sections beyond the peak stress. Otherwise the testing machine characteristics result in sudden instead of gradual failure. Figure 2.8b illustrates this difficulty with high strength concretes.

At strains beyond the peak value of stress, considerable strength still exists. It will be noted that the cylinder strain occurring near maximum stress is nearly the same for all strengths of concrete, being roughly 0.002 in./in.† In a cylinder a maximum strain of something like 0.0025 measures the useful limit for all concretes except those of low strength or those made with lightweight aggregate. In beams of ordinary reinforced concrete, the shape of the cross section is a factor and the steeper the strain gradient the greater is the usable edge strain for a given f'_c. Observations show that unit strains of 0.0030 to 0.0045 normally occur before a beam fails; with f'_c over 6000 psi, the maximum observed strains are from 0.0025 to 0.0040. Tests[7] have proved conclusively that the stress-strain curve for the compression face of a beam is essentially identical with that for a standard test cylinder when stress is applied at the same rate, as shown in Fig. 2.8. However, confinement of the concrete by a spiral or by heavy ties can greatly increase the extreme strain value in compression, as can a steep strain gradient.[8-10]

Strictly speaking, concrete on initial loading has no fixed ratio of f_c/ε that truly justifies the term "modulus of elasticity." The initial slope of the stress-strain curve defines the initial or tangent modulus used with the parabolic stress method and occasionally elsewhere. The slope of the chord (up to about $0.5 f_c$) determines the secant modulus of elasticity that is generally used in straight-line stress calculations (Fig. 2.9). When E or the term "modulus of elasticity" is used without further designation, it is usually the secant modulus which is intended.

The secant modulus in psi is taken in Code Sec. 8.5.1 as

$$E_c = w^{1.5} 33 \sqrt{f'_c}$$

where w is the weight of concrete in pounds per cubic foot for values between 90 and 155 pcf. Both f'_c and $\sqrt{f'_c}$ have units of psi. For normal weight concrete, w may be taken as 145 pcf, which leads to $E_c = 57{,}000 \sqrt{f'_c}$.

The corresponding empirical relationship for E_c in MPa (or MN/m²) is

$$E_c = w^{1.5} 0.043 \sqrt{f'_c}$$

---

* Nonlinearity results from the formation of microscopic internal cracks that lower the stiffness.
† Lightweight concrete with its lower modulus will give higher values.

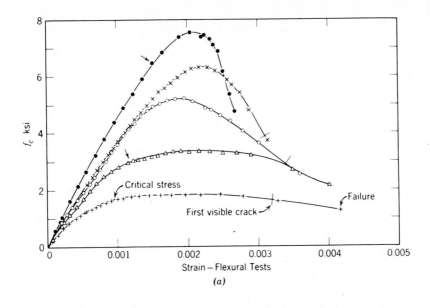

(a)

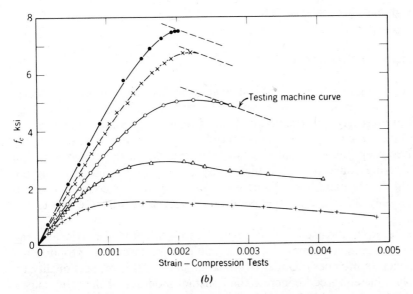

(b)

Figure 2.8 Compression stress-strain curves at age of 28 days. (From Reference 7, ACI.) (a) From flexural tests on 5 × 8 × 16-in. prisms. (b) From direct compression tests on 6- × 12-in. cylinders for the concretes shown in (a).

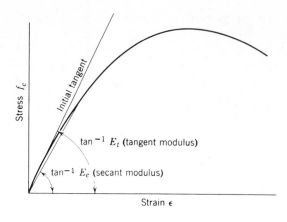

Figure 2.9 Tangent and secant modulus of elasticity determinations.

where w is the weight of concrete in the range between 1500 kg/m³ and 2500 kg/m³, and both f_c' and $\sqrt{f_c'}$ are in MPa (or MN/m²). For normal weight concrete, E_c may be considered $4700\sqrt{f_c'}$.

The use of a reduced modulus is one way to account for the time effects discussed in Sec. 2.7 and Sec. 2.8.

2.7   Creep of Concrete

The initial strain in concrete on first loading at low unit stresses is nearly elastic, but this strain increases with time even under constant load (Fig. 2.10a). This increased deformation with time is called creep and under ordinary conditions it may amount to more than the elastic deformation. Factors tending to increase creep include loading at an early age (while the concrete is still "green"), using concrete with a high water-cement ratio, and exposing the concrete to drying conditions. Concrete completely wet or completely dried out creeps only a little, and in general creep decreases with the age of the concrete.

At stresses up to the usual service level, creep is directly proportional to the unit stress; hence, elastic and creep deformations are essentially proportional in plain concrete members. Under overload conditions this proportionality no longer holds; in reinforced concrete the constant modulus of the steel causes strain readjustments with time. Although these creep strains are minor for tension steel, compression steel stresses may be more than doubled by creep.

The creep rate is more rapid when the load is first applied and decreases somewhat exponentially with time, as shown in Fig. 2.10b. Although unloading allows an immediate (but partial) elastic recovery (Fig. 2.10a), a further creep recovery occurs with time; but this always leaves a substantial residual deformation.

Creep is also much reduced by delaying the loading until the concrete is more mature, as indicated by Fig. 2.10c. This figure shows that both the initial elastic strain and the creep are reduced when the loading is delayed. However, temperatures from 120°F up substantially increase creep.

Creep is one common cause of deflections that increase with time. In reinforced concrete, without compressive steel, the final deflection will usually be from 2.5 to 3.0 times the initial deflection. It has been suggested that long-time deflections be calculated on the basis of a reduced modulus having a value $\frac{1}{3}$ that

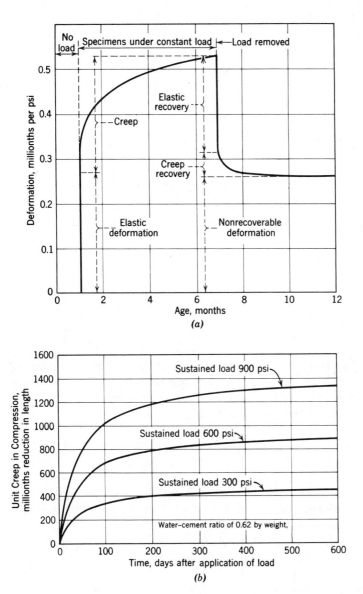

Figure 2.10 Creep of concrete. (From Reference 1, Bureau of Reclamation.) (*a*) Creep and elastic deformation. (*b*) Effect of unit stress on unit creep for identical concretes. (*c*) Effect of age at loading. (From Reference 11, ACI.)

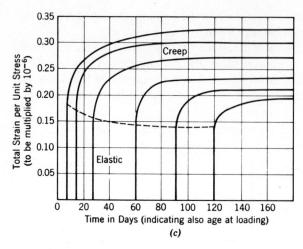

Figure 2.10 (cont.)

of the instantaneous modulus, but see Chapter 3 for better procedures. The Code requires (9.5.2.5)* that the additional long-time deflection for the sustained load be obtained by multiplying the immediate deflection caused by the sustained load by the factor

$$\lambda = \frac{\xi}{1 + 50\rho'}$$

where ρ' is the ratio of nonprestressed compression reinforcement, A'_s/bd. ξ is a time-dependent factor for sustained loads that ranges from a value of 1.0 for a loaded period of 3 months to a value of 2.0 for a loaded period of 5 years or more.

Creep deformation is not always harmful. Creep relaxes the stress effect of early deformation loadings. For example, concrete stresses set up by differential settlement may be nearly eliminated where settlement occurs gradually but early in the life of the concrete, while it is still "green."

2.8   Shrinkage of Concrete

As concrete loses moisture by evaporation, it shrinks. Because moisture is never uniformly withdrawn throughout the concrete, the differential moisture changes cause differential shrinkage tendencies and internal stresses. Stresses owing to differential shrinkage can be quite large and this is one of the reasons for insisting on moist curing conditions. The larger the ratio of surface area to member cross section, the larger will be the resulting shrinkage. Therefore, large specimens shrink much less than small ones.

---

* The section number in the 1986 ACI Code is often listed in this book in this manner.

In plain concrete completely unrestrained against contraction, a uniform shrinkage would cause no stress; but complete lack of restraint and uniform shrinkage are both theoretical terms, not ordinary conditions. With reinforced concrete, because of the internal restraint caused by the reinforcement that does not shrink, even uniform shrinkage causes stresses. These stresses are compression in the steel and tension in the concrete.

Expansive cement is sometimes used to minimize these shrinkage stresses. Because of expansive material in the cement, this concrete first expands a little. If it is partially restrained by embedded reinforcement, tension builds up in this steel and compression in the concrete. As the concrete shrinks and cools, it comes to equilibrium with less change from its initial length.

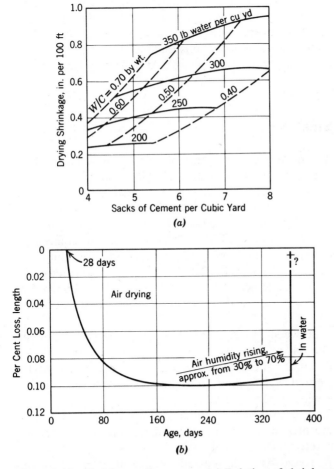

Figure 2.11 Shrinkage of concrete. (*a*) Relation of shrinkage to water content. (From Reference 2, ACI.) (*b*) Typical shrinkage-time curve starting with water-cured specimen (3 × 3 × 40-in.) 28 days old. Note recovery when again placed in water; constant volume reached in about 24 hours of soaking. (Replotted and modified from Reference 12, The Engineering Foundation.)

In ordinary concrete the amount of shrinkage will depend on the exposure and the concrete. Exposure to wind greatly increases the shrinkage rate. A humid atmosphere will reduce shrinkage; a low humidity will increase shrinkage. Shrinkage is usually expressed in terms of the shrinkage coefficient s, which is the shortening per unit length. This coefficient varies greatly, with values commonly 0.0002 to 0.0006 and sometimes as much as 0.0010. An indication of how shrinkage varies with the water and cement content is given in Fig. 2.11a with shrinkage there expressed in terms of inches per 100 ft. This figure can only show trends because the amount of shrinkage differs with materials and drying conditions. In lightweight concrete early shrinkage is noticeably reduced by the water contained in the pores of the lightweight aggregate.

Shrinkage is, to a considerable extent, a reversible phenomenon. If concrete is soaked after it has shrunk, it will expand to nearly its original size, as indicated in Fig. 2.11b. The recovery is now known not to be *total*.[13]

Shrinkage is one common cause of deflections that increase with time. Only symmetrical reinforcement can prevent curvature and deflection from shrinkage (Chapter 3).

2.9   Reinforcing Steel

Reinforcing bars are made in many grades. These principal steels have the yield strengths and ultimate strengths shown in Table 2.1. The 1986 ACI Code provides that for any nonprestressed reinforcement with a specified yield strength f_y exceeding 60,000 psi, f_y will be the stress corresponding to a strain of 0.35%. Although most reinforcing bars tend to be Grade 60, the Code permits bars up to Grade 80 provided that they meet both the 0.35% yield strain requirement and the bend test requirements of ASTM A615. Proposed grades for reinforcing bars under the SI system are shown in Table 2.1. The modulus of elasticity E_s is usually taken as 29×10^6 psi (Code Sec. 8.5.2), or 200 000 MPa in SI units. For design purposes it is usually assumed that the stress-strain curve of steel is elastic-plastic.

To increase the bond between concrete and steel, projections called deformations are rolled on the bar surface, as shown in Fig. 2.12, with a pattern varying with the manufacturer. The deformations shown must satisfy ASTM Spec. A615-85 to be accepted as *deformed* bars. The deformed wire has indentations pressed *into* the wire or bar to serve as deformations.

Except for spirals, only deformed bars, deformed wire, and fabric made from cold drawn or deformed wire may be counted for strength in nonprestressed reinforced concrete under the ACI Code.

All standard bars are round bars, designated by size as #3 to #18; this number corresponding roughly to the bar diameter in one-eighths of an inch. The bar weights and nominal areas, diameters, and circumferences or perimeters are tabulated in Table 2.2, including the bars that will be used with SI units. Because the area is calculated from the weight, including that of the deforma-

TABLE 2.1 Reinforcement Grades and Strengths

Referenced ASTM Standard	Types		Min Yield Point or Yield Strength f_y	Ultimate f_u	Size Restrictions
A615-85	Billet steel:	Grade 40	40 ksi	70 ksi	#3–#6 only
		Grade 60	60	90	
A616-85*	Rail steel:	Grade 50	50	80	#3–#11 only
		Grade 60	60	90	#3–#11 only
A617-85	Axle steel:	Grade 40	40	70	#3–#11 only
		Grade 60	60	90	#3–#11 only
A706-84a	Low alloy: (weldable)	Grade 60	60 (78 max)	80	
A496-85	Deformed wire:	Reinf.	75	85	
A497-79		Fabric	70	80	
A82-85	Cold-drawn wire:	Reinf.	70	80	
A185-85		Fabric	65,56	75,70	
A615M-84a	Tentative SI bar grades				
	Grade 300	300 MPa = 300 $\times$ 10^{6}Pa			
	Grade 400	400 MPa = 400 $\times$ 10^{6}Pa			

* Including Supplementary Requirement S1.

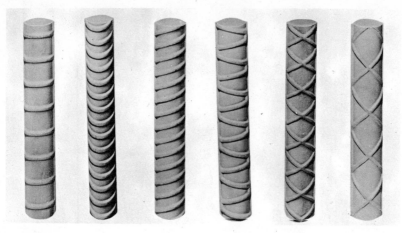

Figure 2.12 Various types of deformed bars. (Courtesy Concrete Reinforcing Steel Inst.)

tions, it is a nominal area as far as minimum cross section is concerned. The perimeter is the circumference of this nominal circular area. The SI bar sizes are organized around convenient bar areas, but the bar numbering is a measure of the bar diameter (rather roughly).

In recent years, corrosion protection of reinforcing bars has assumed growing importance. Reinforcing bars can be ordered with galvanized coating under ASTM Specification A679-79 or with epoxy coating under ASTM Specification A775-84.

Welded wire fabric is increasingly important for slab and pavement reinforcing. Fabric is made of cold-drawn wires running in two directions and welded together at intersections. A more recent development is deformed wire fabric and the use of larger rod sizes made up into fabric mats. Plain wire fabric depends largely on the welded crosswires for development. Deformed wire fabric usually has fewer crosswires with lighter welds and depends more, but not wholly, on the wire deformations for development. Deformed wire up to about 0.6-in. in diameter is available.

2.10   Prestressing Steel

Since its introduction in the United States around the middle of the twentieth century, the use of prestressed concrete has grown rapidly. In many respects the mechanics of prestressed concrete in the failure states are very similar to those of nonprestressed concrete. Although textbooks have generally treated the behavior and analysis of prestressed concrete as a wholly different subject, in this text the sharp demarcation is avoided where possible.

In prestressed concrete, a high initial tension strain and corresponding tensile stress is developed in the prestressing steel tension element or tendon by

TABLE 2.2 Weight, Area, and Perimeter of Individual Bars

Bar Designation No.	Weight per Foot, lb	1985 Standard[14] Nominal Dimensions—Round Sections		
		Diameter, d_b, in.	Cross-Sectional Area, A_b, in.^{2}	Perimeter in.
3	0.376	0.375	0.11	1.178
4	0.668	0.500	0.20	1.571
5	1.043	0.625	0.31	1.963
6	1.502	0.750	0.44	2.356
7	2.044	0.875	0.60	2.749
8	2.670	1.000	0.79	3.142
9	3.400	1.128	1.00	3.544
10	4.303	1.270	1.27	3.990
11	5.313	1.410	1.56	4.430
14	7.65	1.693	2.25	5.32
18	13.60	2.257	4.00	7.09

Proposed Metric No.	Weight kg/m	Area, A_b mm^2	1984 SI Standard[15] Diameter		Perimeter mm
			mm	mm*	
10	0.79	100	11.3	11	35.5
15	1.57	200	16.0	16	50.3
20	2.36	300	19.5	20	61.3
25	3.93	500	25.2	25	79.2
30	5.50	700	29.9	30	93.9
35	7.85	1000	35.7	36	112
45	11.8	1500	43.7	44	137
55	19.6	2500	56.4	56	177

* Rounded off for use in this book.

jacking. This jacking may take place either before the concrete is placed (pretensioning) or after the concrete has cured but with the tendons free to elongate in a duct (post-tensioning). The compressive reaction for these high tensile forces is the member cross section that is thus placed in precompression before external loads are placed on the structure. Because substantial stresses are lost both in manufacturing operations (elastic shortening, tendon seating) and owing to long-time effects (creep, shrinkage, and relaxation), the use of prestressed concrete requires the availability of very high strength steels. With such steels, the losses are a small fraction of the usable stresses and permanent prestressing becomes a reality.

A variety of prestressing tendon materials are available. In the United States the principal tendon materials are seven-wire strand, smooth wire, smooth bars, and deformed bars. For pretensioned applications the seven-wire strand is almost universally used. In post-tensioned applications all four types have

TABLE 2.3 Prestressing Reinforcement—Grades and Strengths

Referenced ASTM Standard	Type	Min Yield Strength* f_{py} (ksi)	Breaking Strength* f_{pu} (ksi)
ASTM A421-80 (1985)	Stress-relieved wire	204	240
ASTM A421-80 (S1-LRW)	Stress-relieved—low relaxation wire	216	240
ASTM A416-85	Stress-relieved seven-wire strand:		
	Grade 250	212	250
	Grade 270	229	270
ASTM A416-85 (S1-LRW)	Stress-relieved—low relaxation seven-wire strand:		
	Grade 250	225	250
	Grade 270	243	270
ASTM A779-80	Stress-relieved—compacted seven-wire strand:		
	Grade 245	215	247
	Grade 260	229	263
	Grade 270	235	270
ASTM A722-75 (1981)	Prestress bar:		
	Type I (plain)	128	150
	Type II (deformed)	120	150

* Based on values for $\frac{1}{4}$-in. diameter wire or $\frac{1}{2}$-in. diameter strand.

seen use with a wide variety of anchorage devices and systems available.[16] The yield strengths and tensile strengths are shown in Table 2.3.

The number of sizes of individual prestressing wires, bars, or strands, as shown in Table 2.4, is considerably more limited than the number of reinforcing bar sizes. However, this is not really significant because individual tendons in a post-tensioned member may be made up of a number of wires or strands within a single duct.

TABLE 2.4 Size, Area, and Weight of ASTM Standard
Prestressing Tendons (From Reference 17)

Type*	Nominal diameter, in.	Nominal area, sq in.	Nominal weight lb per ft
Seven-wire strand	1/4 (0.250)	0.036	0.12
(Grade 250)	5/16 (0.313)	0.058	0.20
	3/8 (0.375)	0.080	0.27
	7/16 (0.438)	0.108	0.37
	1/2 (0.500)	0.144	0.49
	(0.600)	0.216	0.74
Seven-wire strand	3/8 (0.375)	0.085	0.29
(Grade 270)	7/16 (0.438)	0.115	0.40
	1/2 (0.500)	0.153	0.53
	(0.600)	0.215	0.74
Prestressing wire	0.192	0.029	0.098
	0.196	0.030	0.10
	0.250	0.049	0.17
	0.276	0.060	0.20
Prestressing bars	3/4	0.44	1.50
(smooth)	7/8	0.60	2.04
	1	0.78	2.67
	1-1/8	0.99	3.38
	1-1/4	1.23	4.17
	1-3/8	1.48	5.05
Prestressing bars	5/8	0.28	0.98
(deformed)	3/4	0.42	1.49
	1	0.85	3.01
	1-1/4	1.25	4.39
	1-3/8	1.56	5.56

* Availability of some tendon sizes should be investigated in advance.

Selected References

1. *Concrete Manual,* U.S. Bureau of Reclamation, Denver, Colo, 7th ed., 1963.
2. *ACI Manual of Concrete Inspection,* ACI SP-2, Detroit, 1974.
3. ACI Committee 211, "Standard Practice for Selecting Proportions for Normal, Heavyweight, and Mass Concrete," (ACI 211.1-81), Amer. Concrete Inst., Detroit, 1981, 32 pp.
4. ACI Committee 211, "Standard Practice for Selecting Proportions for Structural Lightweight Concrete," (ACI 211.1-81), Amer. Concrete Inst., Detroit, 1981, 18 pp.
5. "Design and Control of Concrete Mixtures," Portland Cement Assoc., Chicago, 11th ed., 1968.

6. D. Ramaley and D. McHenry, "Stress-Strain Curves for Concrete Strained Beyond Ultimate Load," *Lab. Rep. No. Sp-12,* U.S. Bureau of Reclamation, Denver, Colo., 1947.

7. E. Hognestad, N. W. Hanson, and D. McHenry, "Concrete Stress Distribution in Ultimate Strength Design." *ACI Jour, 27,* Dec. 1955 *Proc., 52,* p. 455.

8. W. G. Corley, "Rotational Capacity of Reinforced Concrete Beams," *ASCE Proc., ST-5,* Vol. 92, Oct. 1966, pp. 121–146.

9. P. H. Karr and W. G. Corley, "Properties of Confined Concrete for Design of Earthquake Resistant Structures," *Proc.,* Sixth World Conference on Earthquake Engineering, New Delhi, Jan. 1977, pp. 9–17.

10. J. S. Ford, D. C. Chang, and J. E. Breen, "Behavior of Concrete Columns Under Controlled Lateral Deformation," *ACI Jour, 78,* No. 1, Jan.–Feb. 1981, pp. 3–20.

11. L. Zetlin, C. H. Thornton, and I. P. Lew, "Internal Straining of Concrete," *Designing for Effect of Creep, Shrinkage, and Temperature in Concrete Structures,* SP-27, Amer. Concrete Inst., Detroit, 1970, p. 323.

12. J. L. Savage, I. E. Houk, H. J. Gilkey, and F. Vogt, *Arch Dam Investigation,* Vol. II, *Tests of Models of Arch Dams and Auxiliary Concrete Tests,* Engineering Found., New York, 1934.

13. R. A. Helmuth and D. H. Turk, "The Reversible and Irreversible Drying Shrinage of Hardened Portland Cement and Tricalcium Silicate Pastes," *J. PCA Research and Development Laboratories, 9,* No. 2, May 1967.

14. "Standard Specification for Deformed and Plain Billet-Steel Bars for Concrete Reinforcement," *ASTM Spec. A615-85,* ASTM, Philadelphia, 1985.

15. "Standard Specification for Deformed and Plain Billet-Steel Bars for Concrete Reinforcement [Metric]," ASTM *Spec.* A615M-84a, ASTM, Philadelphia, 1984.

16. *Post Tensioning Manual,* 4th ed., Post Tensioning Inst., Phoenix, 1985, pp. 406.

17. ACI Committee 318, "Building Code Requirements for Reinforced Concrete," (ACI 318-83), Amer. Concrete Inst., Detroit, 1983, 111 pp.

3

FLEXURAL ANALYSIS AND SERVICEABILITY OF BEAMS

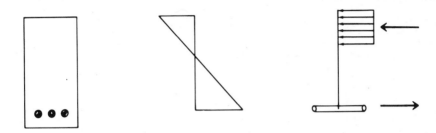

3.1   Introduction

If a lightly reinforced simply supported concrete beam (Fig. 3.1a) has steadily increasing concentrated loads applied at the third points, the cracking and deflection patterns are as shown in Figs. 3.1b through f. To show actual deflections, a reference line is stretched between supports and a horizontal line is marked on the beam near midheight. Midspan deflections are plotted as a function of load in Fig. 3.1f. Cracks are highlighted with a marking pen because they are barely visible to the eye. At very low load levels ($P < P_{cr} = 5$ k), deflections will be very small and tensile strains and stresses will be within the tensile capacity of the concrete. The section will remain uncracked. Stresses across the cross section can be determined from linear elastic expressions such as $f = Mc/I$. When the maximum tensile strain and stress on the outer tensile fiber exceeds the concrete tensile capacity, the outer fiber will rupture. This in effect overstresses the next inner fiber, which will also immediately rupture, and a rapid chain reaction takes place, resulting in formation of one or more cracks as shown in Fig. 3.1b for $P = P_{cr} = 5$ k. Principle tension resistance now will be provided by the longitudinal reinforcement bars that cross the cracks.

With modern deformed reinforcement, the concrete and steel are effectively bonded together between cracks, and the same average strain takes place in the surrounding concrete and the steel. As load increases above P_{cr}, the deflection per unit load can be seen to increase appreciably (Figs. 3.1c and f) reflecting the substantial reduction in member stiffness owing to the loss of much of the tensile concrete section. Cracks will continue to extend, forcing the neutral axis (or plane of zero strain) upward. The increasing load and resulting increased curvature and deflection imposes increased demand on the reinforcement tensile force contribution. At $P = P_y = 35$ k (Fig. 3.1d), the stress in the reinforcement reaches the yield strength of the bars. Above this load, the stiffness of the section is greatly reduced because the reinforcement cannot

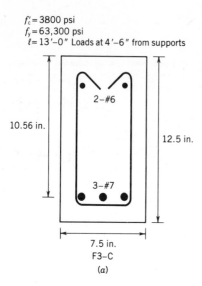

$f'_c = 3800$ psi
$f_y = 63,300$ psi
$\ell = 13'-0''$ Loads at $4'-6''$ from supports

10.56 in.

12.5 in.

2–#6

3–#7

7.5 in.

F3–C

(a)

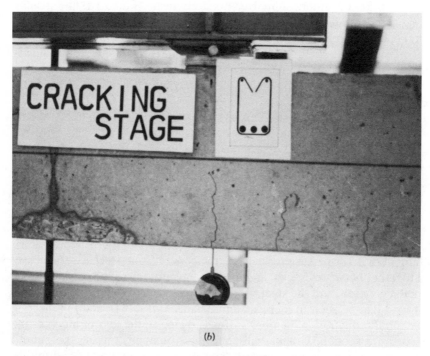

(b)

Figure 3.1 Cracking and deflection behavior of a simply supported under-reinforced concrete beam. (a) Cross section. (b) First cracking at $P = 5$ k (c) Cracking and deflection at $P = 14$ k. (d) Cracking and deflection at yielding of reinforcement with $P = 35$ k (e) Compression zone crushing at ultimate with $P = 36$ k. (f) Load versus centerline deflection. (Courtesy N. H. Burns.)

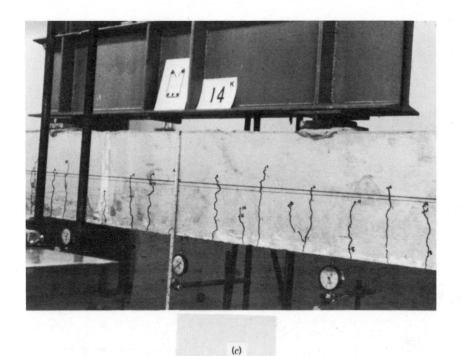

(c)

(d)

Figure 3.1 (cont.)

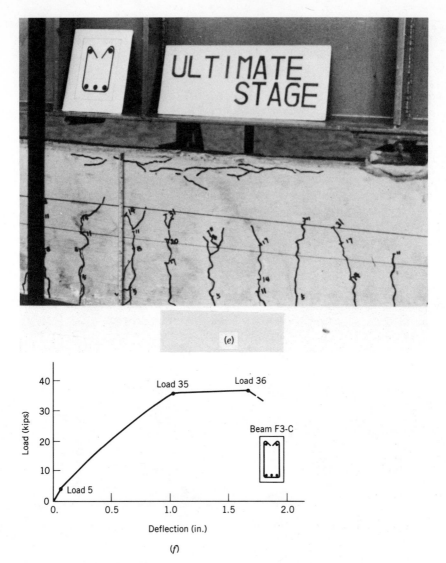

(e)

(f)

Figure 3.1 (cont.)

develop any higher tensile force. Further load increase can only be resisted by minor increases in the lever arm between the tensile force acting at the centroid of the tension reinforcement and the equilibrating compression force acting at the effective centroid of the compressive concrete zone. The cracks continue to extend upward and widen appreciably. The member deflects greatly. The visual appearance of the wide cracks and substantial deformation give clear warning to the public that this member is in distress.

The area under the load-deflection curve is an index of the toughness or energy absorption capacity of the section. The substantial ductility or deformation capacity beyond first yielding shown in Figs. 3.1e and f is typical of *under-reinforced beams*, that is, beams that have a reinforcement tension yield capacity less than the flexural concrete compression capacity. In these beams the reinforcement yields before the concrete compression zone is destroyed. Such beams give substantial warning before failure. This type of behavior is required by the ACI Code provisions in the unlikely event that an actual member fail in service. At deformation substantially above the deflection at first yielding and under an increased load, $P_{max} = P_n = 36$ k, the concrete compression zone is destroyed by crushing, as shown in Fig. 3.1e. The beam can no longer resist load and collapses. The corresponding moment is the value designated as M_n, the nominal moment capacity of the beam with known cross-sectional dimensions and material properties. The actual failure takes place when the cracks have extended quite high, effectively reducing the depth of the compression zone to a minimum and thus requiring the compressive stress to be a maximum. Careful measurements of the actual strains at various levels on the critical cross section indicate that the strain profile from top to bottom of the section remains linear (when sufficiently large gage lengths are used to average out several crack widths). These strain measurements indicate that the strain measured on the topmost compression fiber will dependably equal or exceed 0.003 in./in. at failure or time-of-compression concrete crushing.

The general behavior shown in this beam is the basis of the ACI Code provisions for flexure. Code 10.2.1 requires consideration of equilibrium and compatibility of strains. Code 10.2.2 states that except for the special case of very deep beams, a linear strain profile is assumed across the section. Code 10.2.3 establishes the maximum concrete strain on the extreme compression fiber as 0.003. Code 10.2.4 adopts a simple bilinear stress strain relation for reinforcement ($f_s = \varepsilon_s E_s$ up to f_y and a flat-top branch $f_s = f_y$ for any strain greater than $\varepsilon_y = f_y/E_s$). The insignificant contribution of the tensile strength of the uncrushed concrete regions above and between the cracks is neglected under Code 10.2.5.

In the succeeding sections, further details of the calculation of the dependable resisting moment at ultimate load and of the deflections at various load stages will be developed. However, the reader should frequently return to Figs. 3.1a–f for an overall perspective and increased understanding of the flexural behavior of a ductile reinforced concrete member.

3.2 The Resisting Couple in a Beam

Statics shows that an external bending moment on any beam must be resisted by internal stresses that can be indicated as a resultant tension N_t and a resultant compression N_c. Unless there is axial load, summation of horizontal forces (Fig. 3.2) indicates that N_t must equal N_c and that together they form a couple. In a reinforced concrete beam the reinforcing steel is assumed to carry all the tension N_t and thus N_t is located at the level of the steel. The compressive force N_c is the resultant of compression stresses over some depth of beam and thus

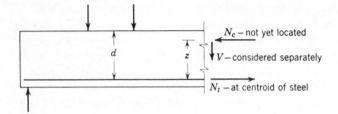

Figure 3.2 Internal resisting couple in a beam.

its location is not established from statics alone. Let the arm or distance between these N_t and N_c forces be designated as z, as shown in Fig. 3.2. Evaluation of z is first presented as a portion of Sec. 3.6. The further depth below the steel is chiefly to fireproof the steel, to protect it from moisture, and to assist in bonding the reinforcing steel to the concrete; it does not influence present flexural strength calculations. However, the significance of cover (and bar spacings) in the development of reinforcement tensile stress will be considered in Chapter 8.

The resisting couple idea can be applied to homogeneous beams. It is simpler for rectangular beams with straight-line distribution of stresses than is the use of $f = Mc/I$. For steel I-beams it is more awkward, but is useful for rough estimates based on neglect of the web area. For reinforced concrete beams it has the definite advantage of using the basic resistance pattern.

3.3   Distribution of Compressive Stress

In reinforced concrete beams the compressive stress varies from zero at the neutral axis to a maximum at or near the extreme fiber. How it actually varies and where the neutral axis actually lies depend both on the amount of the load and on the history of past loadings. This variation is the result of several factors: (1) The spacing and depth of tension cracks depend on whether the beam has been loaded before, and how heavily. (2) Shrinkage stresses and creep of concrete are important factors in relation to stress distribution, factors very difficult to include in an analysis. (3) Most important, the stress-strain curve for concrete as indicated in Figs. 3.3 and 2.9 is not a straight line.

Experiments confirm that, even for reinforced concrete cracked on the tension side of the beam, unit strains vary as the distance from the neutral axis,* the maximum compressive strain increasing with the moment from ε_1 to ε_2, to ε_3, ... on to failure. Although the ACI Code completely discourages the use of *over-reinforced beams,* beams that have a reinforcement tension yield capacity greater than the flexural concrete compression capacity, study of the progressive development of the concrete compression zone in such an over-

---

* Exactly at a crack this is not quite true, but over an average gage length it is statistically true.

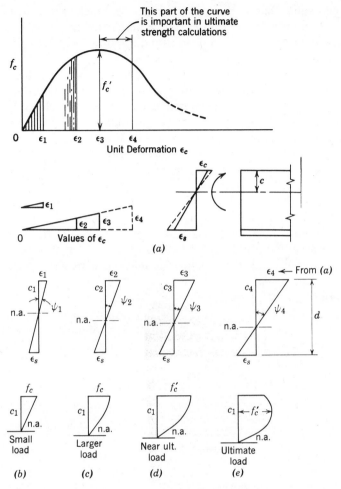

Figure 3.3 (*a*) Stress-strain curve for concrete of medium strength and possible strain-stress combinations. (*b–e*) Shifting strain and stress patterns with increasing moment up to compression failure ($f_s < f_y$) The triangles show the unit rotations ψ at each stress level.

reinforced beam is quite useful in understanding beam behavior. The curve of Fig. 3.3*a* thus indicates that the compressive stress distribution in such an over-reinforced beam on the first application of loading (without any shrinkage) would go through several stages. The neutral axis would shift location with the stress pattern to keep $N_t = N_c$. For example, the small shaded triangle $O\varepsilon_1$ would represent the early stresses, as shown more conveniently in Fig. 3.3*b*. (Actually, at the very start of the first loading before the beam has cracked in tension the neutral axis would be lower, usually below middepth.) Similarly, ε_2 (Fig. 3.3*a*) would limit a large nontriangular stress distribution as in (*c*). The maximum unit stress would be f'_c in (*d*), with strain at ε_3 in (*a*). The maximum total compression in (*e*) would occur at some strain value ε_4 greater than ε_3. A detailed analysis would show that as the higher strains are reached, the stiffness of the concrete decreases. When the tension is all carried in the elastic range by the steel, the neutral axis has to drop as indicated in Fig. 3.2b–e to keep N_c increasing as rapidly as N_t.

Design for this compression type of failure is not permitted, because the ψ unit rotations here are restricted to those relatively small values of the upper triangles of Fig. 3.2b–e. Member stiffness and deflection are directly related to the angle changes per unit length, and these control ductility (Sec. 3.4).

Ductility, measured simply as the rotation ψ per unit length of member, is important in reinforced concrete design and is discussed further in the next section (3.4). In the usual continuous type member, ductility under overload conditions is evidenced by widening tension cracks and increased deflections that warn of approaching failure. At the same time the member is absorbing energy from the loading and its local reduced stiffness results in shifting the resisting moment to lesser strained sections. This ability to adjust to local overstress is a valuable asset for continuous reinforced concrete construction. The next section shows how much this ductility is influenced by the design Code.

Under service loads it probably is not feasible, and it may not even be possible, to do more than estimate rather crudely the approximate magnitude of real stresses. Certainly these are complicated by shrinkage and creep and depend on how much cracking has been induced by previous loadings. Specifications formerly emphasized stresses at working loads, but laboratory tests of reinforced concrete beams show that actual deformations and stresses at working loads only faintly resemble values conventionally calculated on the basis of straight-line stresses (Fig. 3.3b). Design procedures have gradually shifted to ultimate strength methods, with checks at service load for deflection and cracking.

Fortunately the *ultimate* strength of reinforced concrete beams can be predicted or calculated with quite satisfactory accuracy.[1] Design on the basis of (ultimate) strength is now standard in the ACI Code.

Reference 2 by Park and Paulay includes a more thorough coverage of the research background in this area.

3.4 Tension and Compression Failures Owing to Bending Moment; Member Ductility

Beams may initially fail from moment because of weakness in the tension steel or the compression concrete. Final flexural destruction of all concrete beams occurs with rupture and crushing of the compression concrete. Failure from stresses primarily related to shear will be considered later and separately, as in the case of steel construction, as it is only in very deep and relatively short beams that these influence bending strength.

Most beams are weaker in their reinforcing steel than in their compression concrete. Both the Code and economy require such design. Such beams fail

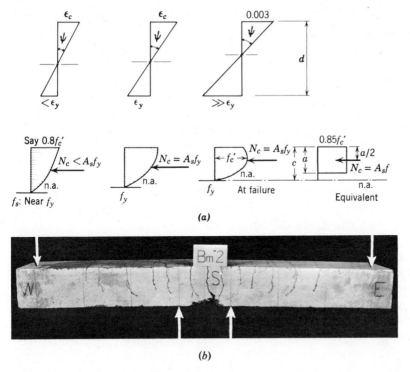

Figure 3.4 Under-reinforced beam with initial flexural failure in tension. (*a*) Variation in strains and compressive stresses as beam approaches failure in tension. (*b*) Beam after failure under negative moment (tension on top).

under a load slightly larger than that which makes $N_t = f_y A_s$, where f_y is the yield point strength of the steel and A_s is the area of the steel. Because $f_y A_s$ represents the nominal usable strength of steel, the further increase in moment resistance needs explanation. When the steel first reaches its yield point, the compressive stress distribution may be like that previously illustrated in Fig. 3.3*c* or *d*. A slight additional load at this stage causes the steel to stretch a considerable amount. The increasing steel deformation in turn causes the neutral axis to rise (when the tension is on the bottom) and the center of compression N_c therefore moves upward, Fig. 3.4*a*. This increase in the arm z between N_c and N_t gives an increased resisting moment $N_t z$ even though N_t is essentially unchanged.* The rising of the neutral axis also reduces the area under compression and thereby increases the unit compressive stress required to develop the nearly constant value of N_c. This process continues until the reduced area fails in compression, as a secondary effect, at about the same ε_4 as indicated in Fig. 3.3. These changes (neglecting any tension in the concrete) are

---

* When the steel has no sharp yield point, N_t increases above the yield strength enough to make calculations based on a yielding steel conservative. The same problem arises when a short yield plateau introduces a strain hardening region.

summarized in Fig. 3.4*a*. This type of failure is shown in Figs. 3.1*e* and 3.4*b* (inverted because it is in a negative moment region).

As the ultimate *c* to neutral axis is smaller in Fig. 3.4 than in Fig. 3.3 and the maximum ε_c is about the same, the unit rotation ψ at ultimate is greater, usually by 35% to over 200%. Such an *under-reinforced* beam (Fig. 3.5*a*) shows greatly increased deflection after the steel reaches the yield point, giving adequate warning of approaching beam failure; the steel, being ductile, will not actually pull apart even at failure of the beam. The increase in *z* will be rarely more than a few percent, say 3 to 8%. If the concrete stress is very high before the steel stress f_s reaches the yield point value of f_y, the increase in *z* may be very small.

If the concrete reaches its full compressive strength just as the steel reaches its yield point stress, the beam is said to be a *balanced* beam (at failure) (Code 10.3.2). Such a beam requires very heavy steel, is rarely economical, and is not allowed by the Code (Code 10.3.3).

Beams in which the full compressive strength is reached before the steel yields are termed *over-reinforced beams*. Such beams (Fig. 3.5*b*) show no ductility and are not allowed by the Code (Code 10.3.3), because the reinforcement required is substantially above the ¾ balanced amount that is the Code maximum. Before such a beam fails initially from weakness in compression, the top elements of the beam shorten considerably under the final increments of load, causing the neutral axis to move lower down in the beam. Such move-

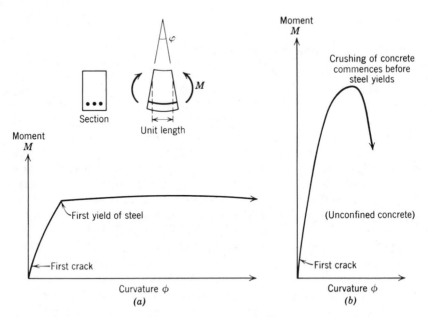

Figure 3.5 Moment-curvature relationships for singly reinforced beam sections. (*a*) Under-reinforced section failing in tension, $\rho < \rho_b$. (*b*) Over-reinforced section failing in compression, $\rho > \rho_b$. From Reference 2. (Reproduced by permission of the publisher.)

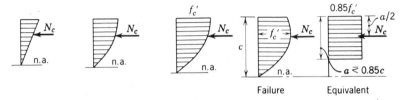

Figure 3.6 Variation in compressive stresses as over-reinforced beam approaches failure in compression. See text for values of a. (Corresponding strains are similar to those in Fig. 3.3b–e.)

ment of the neutral axis increases the area of concrete carrying compression and importantly increases the total N_c that can be carried (Fig. 3.6). The greatly increased N_c is offset slightly by a reduced arm z. The concrete finally fails suddenly and often explosively in compression, and the steel stress remains below the yield point unless the beam is exactly balanced. These stress changes, again ignoring any tension in the concrete, are summarized in Fig. 3.6.

In a plot (Fig. 3.5) of resisting moment M against curvature ϕ for a given cross section, energy absorbed per unit length is proportional to the area under the curve. Such a curve for a balanced beam would be much like the curve on the right of the figure with the peak at a point where f'_c and f_y are reached simultaneously. The limited area under the curve, up to the peak value, measures the energy absorption capacity available before damage becomes severe. The curve shown at the left would be for a lower A_s, possibly 40% of balanced reinforcement.* The lower the A_s, the lower the M capacity and the longer the slightly rising flat portion of the curve created by the rising neutral axis (Fig. 3.4a), while the reinforcement continues to yield without damage, except for wider tension cracks. This "flat" extension stops at the ϕ value corresponding to the ε_c value causing the concrete to crush. The larger area under this curve indicates this member's greater ductility. The ratio of curvature at ultimate to that at yield is called the ductility factor. As A_s is increased, the higher the M at first yielding becomes and the shorter the near-horizontal length, with a lower ductility resulting. Code 10.3.3 assures some ductility by limiting the maximum A_s to 75% of the balanced A_s. Limitation of deflections or economy of design often demand larger members that need less A_s; 40 to 50% of balanced A_s is a very common range.

For design purposes real final stress distribution may be replaced adequately by an equivalent rectangle of stress (pioneered in the United States by Whitney) of intensity $0.85\,\mathbf{f'_c}$ and depth a as shown in the final sketches in Figs. 3.4a and 3.6. For rectangular beams, the shaded area of the rectangular stress block should equal that of the real stress block and their centroids should be at the

---

* Also see Figs. 11.1 and 11.2.

same level. The value of a given in Code 10.2.7 is intended to give this result and is recommended:

For $f'_c \lessgtr 4000$ psi $\quad a = 0.85c$
For $f'_c > 4000$ psi, reduce the 0.85 factor* linearly at the rate of 0.05 per 1000 psi excess over 4000 psi, but not to any value less than 0.65.

The ACI 318M Code in SI units uses the following:

For $f'_c \lessgtr 30$ MPa (C30 concrete) $\quad a = 0.85c$
For $f'_c > 30$ MPa, reduce the 0.85 factor linearly at the rate of 0.008 per 1 MPa, but not to any value less than 0.65.

3.5   Analysis or Review Versus Design

In analysis or review of a given design, whether for actual stresses or permissible moments, the engineer deals with given beams, known for both dimensions and steel. The engineer has no control over the location of the neutral axis, which lies at a definite but initially undefined depth.

In design, loads and ultimate stresses are known and some or all the dimensions remain to be fixed. In this case designers have some control over the location of the neutral axis. They can shift it where they want it, to the extent that the change in dimensions changes the depth of rectangular stress block or magnitude of N_t.

The student should understand clearly this fundamental difference between analysis (or review) and design problems. Analysis for various types of beams is discussed first and then computation of deflections, which is also an analysis procedure. Design is deferred to Chapter 4.

3.6   The Balanced Rectangular Beam

The balanced beam in ultimate strength design is not a practical beam, but the concept is fundamental to the philosophy of the Code.
The Code 10.3.2 defines:

Balanced strain conditions exist at a cross section when tension reinforcement reaches the strain corresponding to its specified yield strength just as the concrete in compression reaches its assumed ultimate strain of 0.003.

The Code 10.3.3 specifies:

For flexural members . . . the ratio of reinforcement ρ provided shall not exceed 0.75 of the ratio ρ_b that would produce balanced conditions for the section under flexure without axial load. For members with compression reinforcement, the portion of ρ_b balanced by compression reinforcement need not be reduced by the 0.75 factor.

---

* This factor is designated β_1.

(The first use and discussion of the second sentence in Code 10.3.3 quoted is in Sec. 3.15 on double-reinforced beams.) The ratio $\rho = A_s/bd$, where A_s is the area of steel reinforcement and b and d are shown in Fig. 3.7a. The Code (10.2.3) requires that the concrete strain not exceed 0.003, as also shown in Fig. 3.7a for the balanced condition.

These regulations represent an attempt to assure the ductile failure produced by yielding of steel compared to the brittle type of failure occurring when the failure is in compression. (See also Sec. 3.7.)

The analysis of the balanced beam starts from the strain triangles at failure as shown in Fig. 3.7a. The steel strain will be $\varepsilon_s = \varepsilon_y = f_y/E_s$ and the maximum concrete strain 0.003, which is conservatively in agreement with test observations. The neutral axis can be located from the strain triangles, most simply by considering the compression strain triangle and the large dotted triangle.

$$\frac{0.003}{c_b} = \frac{0.003 + \varepsilon_y}{d}$$

$$c_b = \frac{0.003}{0.003 + f_y/E_s} d$$

For $E_s = 29 \times 10^6$ psi:

$$c_b = \frac{87,000}{87,000 + f_y} d \tag{3.1}$$

For $E_s = 200\,000$ MPa and f_y in MPa:

$$c_b = \frac{600}{600 + f_y} d \tag{3.1a}$$

Experimental work has established the necessary depth a of stress block to go with a given neutral axis, or conversely the neutral axis distance c to go with a given stress block depth a. The Code requirement for this $a = \beta_1 c$ has already been stated in the last paragraph of Sec. 3.4 above. With a_b established, the balanced reinforcement ratio $\rho_b = A_{sb}/bd$ can now be found from summation of

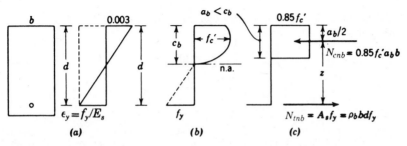

Figure 3.7 Balanced beam. (a) Strains. (b) Stresses. (c) Equivalent stresses and forces.

horizontal forces $N_{tnb} = N_{cnb}$*, as indicated in Fig. 3.7c. The n subscript on each term indicates that each is now a nominal (or ideal) value; the strength reduction factor ϕ must still be applied to N_n values or to M_n values to establish the *design* strength provided. Then $N_{tb} = \phi N_{tnb}$, $N_{cb} = \phi N_{cnb}$, and $M_{un} = \phi M_{nb}$. Usually only the nominal values of N_c or N_b are needed, because the change to design strength can conveniently be applied to the moment term.

The following numerical approach, presented first in English units and then in SI units is recommended to the student as it presents a clearer picture.

For $f'_c = 3000$ psi, $f_y = 40,000$ psi, Sec. 3.4 shows $a_b = 0.85 c_b$. Then for a rectangular beam of depth d and width b:

From similar triangles

$$\frac{c_b}{0.003} = \frac{d}{0.003 + \varepsilon_y}$$

$$c_b = \frac{0.003\,d}{0.003 + \varepsilon_y} = \frac{0.003}{0.003 + \dfrac{40,000}{29,000,000}} = 0.685\,d$$

or alternatively using Eq. 3.1

$$c_b = \frac{87,000}{87,000 + 40,000}\,d = 0.685\,d$$

The authors recommend the use of sketches such as Fig. 3.7, development of all equations from first principles such as strain diagrams and compatibility, and avoidance of simplifying equations like Eq. 3.1 until the reader has completely mastered the fundamentals.

$$a_b = 0.85 c_b = 0.85 \times 0.685\,d = 0.582\,d$$

$$N_{cnb} = 0.85 f'_c a_b b = 0.85 f'_c (0.582d) b = 0.495 f'_c bd$$

Since $N_{ncb} = N_{tnb}$,

$$0.495 f'_c bd = A_{sb} f_y = (\rho_b bd) f_y$$

$$0.495 \times 3000 = \rho_b \times 40,000$$

$$\rho_b = 0.037$$

Algebraically, this is also given in ACI Commentary Fig. 10.3.2(a) with β_1 representing a/c:

$$\rho_b = 0.85 \beta_1 \frac{f'_c}{f_y} \frac{87,000}{87,000 + f_y}.$$

---

* Normally, triple subscripts can be avoided. For example, the b subscript is needed only to emphasize that a balanced beam is being considered. The subscript n emphasizes that these are nominal (or ideal) values that do *not* include the strength reduction value ϕ (equally applicable to both terms).

Although there is little practical use for the balanced moment in beam design, it is simple to establish from the resisting couple of Fig. 3.7c:

$$z = d - a_b/2 = d - 0.29d = 0.71d$$

$$M_{nb} = N_{tnb}z = (\rho_b bd)f_y z = (0.037bd)f_y(0.71d)$$

$$= 0.026f_y bd^2 = 1040bd^2$$

or

$$M_{nb} = N_{cnb}z = 0.85f_c'(0.58d)b(0.71d)$$

$$= 0.35f_c'bd^2 = 1050bd^2$$

$$\doteq 1040bd^2, \text{ to the accuracy carried.}$$

Design strength $= \phi M_{nb} = 0.9 \times 1050bd^2 = 940bd^2$

where ϕM_{nb} is in in.-lb when b and d are in inches and f_c' and f_y in psi.

For $f_c' = 25$ MPa, $f_y = 300$ MPa, the closing comments in Sec. 3.4 show $a_b = 0.85c_b$. Then for a rectangular beam of width b:

$$c_b = \frac{600}{600 + 300}d = 0.667d$$

$$a_b = 0.85c_b = 0.85 \times 0.667d = 0.567d$$

$$N_{cnb} = 0.85f_c'a_b b = 0.85f_c' \times 0.567db = 0.482f_c'bd$$

If b and d are in m and f_c' is in MPa, the force N_{cnb} is in MN (meganewtons).

Since $N_{cnb} = N_{tnb} = A_{sb}f_y$

$$0.482f_c'bd = (\rho_b bd)f_y$$

$$0.482 \times 25 = \rho_b \times 300, \qquad \rho_b = 0.040$$

This could be written algebraically, with β_1 representing a/c

$$\rho_b = 0.85\beta_1 \frac{f_c'}{f_y} \frac{600}{600 + f_y}$$

with all stresses in MPa. Although there is little practical use for the balanced moment in beam design, it is simple to establish from the nominal resisting couple.

$$z = d - a_b/2 = d - 0.567d/2 = 0.717d$$

$$\text{(If } d \text{ is in meters, } z \text{ is in meters)}$$

$$M_{nb} = N_{tnb}z = (\rho_b bd)f_y z = (0.040bd)300 \times 0.717d = 8.60bd^2$$

or

$$M_{nb} = N_{cnb}z = (0.482f'_c bd)0.717d = 0.346 \times 25 bd^2$$
$$= 8.64 bd^2 \cong 8.60 bd^2, \text{ to the accuracy carried}$$

Design strength $= \phi M_{nb} = 0.90 \times 8.60 bd^2 = 7.74 bd^2$

where ϕM_{nb} is in M N $\cdot$ m if b and d are in m and f'_c and f_y in MPa.

This example has been followed through in detail because the methods also apply to nonbalanced beams when either c, N_{nt}, or a is known.

Other shapes of beams such as T-beams, double-reinforced beams, and irregular shapes, can likewise be evaluated as balanced beams. The neutral axis is at the same relative depth in every case, being established from strains rather than stresses. The procedures with varying widths and with compression reinforcement are discussed starting in Sec. 3.12.

3.7 Tensile Steel Limitation—For Ductility*

Because a balanced beam will ultimately fail suddenly in compression, the Code, as quoted in Sec. 3.6, limits the tensile reinforcement to a maximum of $0.75\rho_b$ both for beams and for some lightly loaded columns (Code 10.3.3). In beams this is equivalent to requiring $N_t = 0.75N_{cb}$. Thus, when the reinforcement reaches yield, the concrete will still have roughly an added $\frac{1}{3}$ reserve strength in compression. The usual ductile reinforcement assures member ductility, with wide cracking and more deflection as a warning of approaching capacity.

It is rarely economical to make beams so small that this rule becomes an operative limit. Both deflections and costs normally point toward larger beams with even smaller steel ratios.

3.8 Code Maximum and Minimum Moments For Rectangular Beams

The limitation to $0.75\rho_b$ as a Code maximum is equally well described by $N_t = 0.75N_{tb}$ or $N_c = 0.75N_{cb}$ and for the single case of a rectangular beam also by $c = 0.75c_b$ or $a = 0.75a_b$. For the rectangular beam the last is the most direct form. Then the Code maximum moment and reinforcement ratio for any combination of materials can be found.

---

* Reference 2 has a very extensive treatment of ductility.

For example, for $f'_c = 3000$ psi and $f_y = 40{,}000$ psi:
From similar triangles

$$c_b = 0.685\,d,$$
$$a_b = 0.582\,d, \text{ (from Sec. 3.6)}$$
$$\text{Max } a = 0.75 \times 0.582\,d = 0.437\,d, \text{ say } 0.44\,d$$
$$\text{Max } N_c = 0.85 f'_c(0.44\,d)b = 0.37 f'_c bd$$
$$z = d - a/2 = d - 0.22\,d = 0.78\,d$$
$$\text{Nominal } M_n = N_c z = 0.37 f'_c bd(0.78\,d) = 0.29 f'_c bd^2$$
$$\text{Dependable moment} = \phi M_n = 0.9(0.29 f'_c bd^2)$$
$$= 0.26 f'_c bd^2 = 780\,bd^2 = k_m bd^2$$

Since $N_t = N_c$,

$$\rho bdf_y = N_c = 0.37 f_c bd$$
$$\text{Max } \rho = 0.37 f'_c/f_y = 0.37 \times 3/40 = 0.0278$$
$$\text{Nominal } M_n = N_t z = (0.0278\,bdf_y)(0.78\,d) = 0.022 f_y bd^2$$
$$\text{Design moment} = \phi M_n = 0.9 \times 0.022 f_y bd^2 = 0.020 f_y bd^2 = 800\,bd^2$$

This dependable moment ϕM_n would, with more decimal places, be the same whether based on N_c or N_t because M_n represents a couple. For convenience, designate the constant 780* as k_m and say design moment $\phi M_n = k_m bd^2$. Similarly, values of k_n are usable with $M_n = k_n bd^2$.

Each combination of f'_c and f_y has a unique constant for each value of ρ. When $\rho = \rho_{\max} = \frac{3}{4}\rho_b$, these constants are called the maximum values.

Table 3.1 lists a number of values of *maximum* k_m, k_n, ρ and a/d for easy reference. (This table is repeated as Table B.5 in Appendix B.) The reader should note that the k_m just calculated as 780 is listed as 783; it was tabulated here to match many printed tables. Nevertheless, the accuracy of the basic input data limits the accuracy of the output†; the extra digits on the computer results do not reflect improved accuracy. The last digit for any of the three or four digit quantities in this table adds nothing in the direction of true accuracy; a zero would serve just as well, except possibly where the last digit represents several percent of the total.

The significance of entries in Table 3.1 for $\rho = 0.18 f'_c/f_y$ is discussed near the end of Sec. 3.9.

Except for the slightly different grades for concrete and steel strengths to fit SI units, the algebraic values for ϕM_n would be unchanged up to the point

---

* This constant carries a unit of psi for b and d in inches. The symbol k_m is in accord with the new ACI notation, but is not yet standardized; the notation K or K_w has also been used.
† f'_c is often given in ksi units. In this case the value would be 3 ksi. Using 3000 in psi units gives a misleading sense of accuracy. The reader is referred back to Sec. 1.8 for a further discussion of calculation accuracy.

TABLE 3.1 Useful Design Constants for Rectangular Beams

English

Concrete	f_y = 40,000 psi				f_y = 50,000 psi				f_y = 60,000 psi			
	(1) k_m	(2) k_n	(3)	(4)	(1) k_m	(2) k_n	(3)	(4)	(1) k_m	(2) k_n	(3)	(4)
f'_c psi	psi	psi	$100\rho_{max}$	a/d	psi	psi	$100\rho_{max}$	a/d	psi	psi	$100\rho_{max}$	a/d
f'_c = 3000	783	870	2.78	0.44	740	822	2.06	0.40	705	783	1.61	0.38
f'_c = 4000	1047	1164	3.72	0.44	987	1097	2.75	0.40	937	1041	2.14	0.38
f'_c = 5000	1244	1383	4.36	0.41	1181	1312	3.24	0.38	1115	1238	2.53	0.36
f'_c = 6000	1424	1582	4.90	0.39	1346	1496	3.64	0.36	1281	1423	2.83	0.34

$$\rho = 0.18 f'_c/f_y \qquad 0.145 f'_c \qquad 0.161 f'_c \qquad 0.212.$$
$$(5) \qquad\qquad (6) \qquad\quad (7) \qquad\quad (8)$$

Same for both systems of units and all f_y grades

SI units

Concrete	f_y = 300 MPa				f_y = 400 MPa			
	(1) k_m	(2) k_n	(3)	(4)	(1) k_m	(2) k_n	(3)	(4)
Gr. f'_c MPa	MPa	MPa	100ρ	a/d	MPa	MPa	100ρ	a/d
C20 20	5.11	5.68	2.40	0.43	4.73	5.25	1.62	0.38
C25 25	6.40	7.11	3.01	0.43	5.92	6.58	2.03	0.38
C30 30	7.68	8.53	3.61	0.43	7.10	7.89	2.44	0.38
C35 35	8.65	9.61	4.01	0.41	7.98	8.87	2.71	0.36
C40 40	9.51	10.56	4.36	0.39	8.77	9.74	2.94	0.35

(1) $M_u = \phi M_n = k_m bd^2$
with $\rho = \frac{3}{4}\rho_b$

(2) $M_n = k_n bd^2$
with $\rho = \frac{3}{4}\rho_b$

(3) $\rho_{max} = \frac{3}{4}\rho_b$

(4) $a/d = \frac{3}{4}a_b/d$

(5) ρ about midrange between ρ_{min} and ρ_{max}

(6) $M_u = \phi M_u = k_m bd^2$
with $\rho = 0.18 f'_c/f_y$

(7) $M_n = k_n bd^2$
with $\rho = 0.18 f'_c/f_y$

(8) a/d with
$\rho = 0.18 f'_c/f_y$

where the value of f'_c or f_y is inserted. However, changes in f'_c or f_y can change the initial value of c_b and all that follows.

Thus for $f'_c = 25$ MPa and $f_y = 300$ MPa, the values of c_b and a_b from Sec. 3.6 lead to the following:

$$c_b = 0.667\,d, \qquad a_b = 0.85\,c_b = 0.567\,d,$$
$$\text{Max. } a = 0.75 \times 0.567\,d = 0.425\,d$$
$$\text{Max. } N_c = 0.85\,f'_c(0.425\,d)b = 0.361\,f'_c\,bd$$
$$z = d - a/2 = d - 0.212\,d = 0.788\,d$$
$$\text{Max. } M_n = N_c z = 0.361\,f'_c\,bd(0.788\,d) = 0.284\,f'_c\,bd^2$$
$$\text{Design moment strength} = \phi M_n = 0.9(0.284\,f'_c\,bd^2) = 0.25\,f'_c\,bd^2$$
$$= 0.256 \times 25\,bd^2 = 6.40\,bd^2 = k_m\,bd^2$$
$$\text{Alternatively, since } N_t = N_c, \; \rho b d f_y = N_c = 0.361\,f'_c\,bd$$
$$\text{Max. } \rho = 0.361\,f'_c/f_y = 0.361 \times 25/300 = 0.0301$$
$$\text{Max. } M_n = N_t z = (0.301\,bd f_y)0.788\,d = 0.0237\,f_y\,bd^2$$
$$\text{Design moment strength} = \phi M_n = 0.9 \times 0.0237\,f_y\,bd^2$$
$$= 0.0213 \times 300\,bd^2$$
$$= 6.39\,bd^2$$
$$\doteq 6.40\,bd^2 \text{ to the accuracy carried.}$$

For convenience designate the constant 6.40 as k_m and say $\phi M_n = k_m bd^2$, where k_m is in MPa if b and d are in m and f'_c and f_y in MPa. Moments are in MN · m. Table 3.1 also tabulates *maximum* k_m, k_n, ρ, and a/d for SI units, using the unit for k_m as MPa. The earlier comments on accuracy with customary units apply equally here; the last digit of the three digit terms is *not* a significant digit, except possibly where the third digit is more than 2 or 3% of the whole.

The legal Code limit for ρ is definitely $0.75\,\rho_b$. However, a designer can be as logical in permitting a greater ρ (with moment capacity of the member limited to that for $0.75\,\rho_b$) as in permitting the use of an actual f_y of 46 ksi where Grade 40 steel is specified. Sometimes in the analysis of an existing building, the reinforcement actually in place may exceed the current Code maximum $\frac{3}{4}\rho_b$. The designer must then *discount* the actual reinforcement and count only $\frac{3}{4}\rho_b$ as effective in the load rating of the structure. Accordingly, the authors limit the k_m value quite strictly and are less concerned (in nonseismic work) about extra steel that can be ignored for strength. In most designs, economy calls for something like one-half of balanced reinforcement. A very useful midrange percentage of reinforcement is $\rho \cong 0.18\,f'_c/f_y$. The design constants for this value, which is about midway between Code maximum and minimum percentages, are also shown as a separate line in Table 3.1. They will be derived in the next section.

In a very lightly reinforced section, there is a danger that the tensile force carried by the uncracked concrete could exceed the tensile capacity of the reinforcement at yield strength. This would mean that the section could not ductilely resist the full moment that caused first cracking. The beam would fail brittlely when the cracking moment is reached. To safeguard against this, Code 10.5 requires a minimum percentage of tension reinforcement at every section except where the area of reinforcement provided is at least one third greater than that required by analysis. This somewhat arbitrary percentage of reinforcement was derived by making assumptions as to the probable tensile cracking strength and determining an equivalent amount of reinforcement to replace the tensile force carried by the concrete before cracking. The Code committee is working on new expressions that are more sensitive to member shape and a range of material strengths. The 1986 Code 10.5.1 requirement is $\rho_{min} = 200/f_y$ unless *all* the flexural reinforcement for the member is at least $\frac{4}{3}$ that required by analysis. This means ρ_{min} of 0.005 for Grade 40 and 0.0033 for Grade 60 bars. For SI units, Code 318M 10.5.1 specifies $\rho_{min} = 1.4/f_y$ with f_y in MPa, leading to 0.005 for Grade 300 and 0.0035 for Grade 400. For joists or T-beams with webs in tension, ρ_{min} is based on web width rather than flange width. For slabs of uniform thickness the minimum drops back to temperature and shrinkage requirements (Code 7.12.2) of 0.0020 of *gross* slab area for Grade 40 bars and 0.0018 for Grade 60 bars, with maximum bar spacing of five times the slab thickness but not more than 18 in. The SI code uses the same 0.0020 of *gross* slab area for Grade 300 and 0.0018 for Grade 400. The maximum spacing limit is five times the slab thickness, but not more than 500 mm.

The effect of the Code-imposed reinforcement limits is readily seen in Fig. 3.8, which shows the general relation between moment capacity and reinforcement percentage for a rectangular beam with $f'_c = 4$ ksi and Grade 60 reinforcement. The moment capacity increases substantially with increasing reinforcement percentage until the balanced beam condition is attained. At rein-

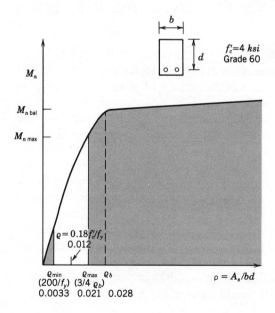

Figure 3.8 Moment capacity as a function of reinforcement percentage.

forcement percentages below ρ_b, the compression force can readily match any increase in tension force. The increase in moment is not linear because the lever arm z is shortened by the increasing a necessary to develop the compression force. Above the balanced reinforcement percentage, any increase in moment is small because the growth in the compression force is counteracted a great deal by the decreasing lever arm. Because the steel does not reach f_y, the strength is only slightly increased by adding reinforcement. The maximum and minimum reinforcement limits are shown. The shaded areas under the curve cannot be used because of these limits. The useful area for design lies between $\rho_{\min}$ and $\rho_{\max}$. A wide variety of reinforcement percentages and corresponding moment capacities exist in this allowable range. The approximate midrange value of $\rho = 0.18 f'_c/f_y$ is also shown, although it is certainly not a distinctly advantageous value.

In checking any beam, the actual percentage of reinforcement, $\rho = A_s/bd$ should be checked against the $\rho_{\max}$ and $\rho_{\min}$ limits before any other computations are performed. This check is handled automatically in design by selecting an approximate percentage to begin the design that falls within the Code limits.

3.9 Analysis of Rectangular Beams

Only under-reinforced beams, as discussed in Sec. 3.7, are permitted by the Code and their analysis is simple. First establish whether the given ρ, which is simply A_s/bd, is less than (or equal to) $0.75\rho_b$, the maximum ρ, tabulated in Table 3.1. (Of course, the equivalent check can be made later, at some risk to earlier calculations, in terms of k_m or of maximum a/d.) Then check to make sure the given ρ is greater than (or equal to) $\rho_{\min} = 200/f_y$, the minimum ρ. If $\rho < 0.75\rho_b$, the beam obviously fails primarily in tension. The resulting stretching of the steel will raise the neutral axis until the final secondary compression failure occurs at the compression strain ε_4 in Fig. 3.3, taken in the Code as 0.003. The usual analysis, however, avoids the necessity of using this strain directly for simple cases.

Since $N_t = N_c$, the ultimate moment is reached when the compressive area can just support a N_c equal to $A_s f_y$. By using the Code simplification as shown in Fig. 3.9, the compression N_c can be evaluated as an equivalent block of uniform stress, of intensity $0.85 f'_c$ and height a (Fig. 3.9b) acting on the rectangular beam cross-sectional area ab. Then $N_c = N_t$ becomes:

$$0.85 f'_c ab = A_s f_y$$

$$a = \frac{A_s f_y}{0.85 f'_c b} = \frac{\rho f_y d}{0.85 f'_c}$$

where $\rho = A_s/bd$. The arm between N_c and N_t is $d - a/2$, because the centroid of the stress block on a beam of constant width is at $a/2$.

$$M_n = N_t z = A_s f_y \left(d - \frac{a}{2} \right)$$

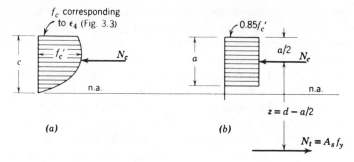

Figure 3.9 The Code (or Whitney) compressive stress block. (*a*) Probable stress distribution. (*b*) Simplified model for calculations.

M_n is the nominal moment strength under ideal conditions and the value of a is found from the previous relation. The design moment $= \phi M_n = 0.9 M_n$.

An alternate form for the formula for ϕM_n in terms of $\omega = \rho f_y / f'_c$ is:

$$\phi M_n = \phi f'_c b d^2 \omega (1 - 0.59\omega)$$

This is sometimes convenient and involves only an algebraic transformation of the previous process. In terms of the steel, ϕM_n also could be written:

$$\phi M_n = \phi A_s f_y d \left(1 - \frac{0.59 \rho f_y}{f'_c}\right)$$

The authors recommend that students avoid the formulas and concentrate on the basic equilibrium relations until they are quite familiar with the subject.

The steel yield strength in calculations is limited to 80 ksi (Code 9.4). The Code specifies (3.5.3.2) that if the yield strength exceeds 60,000 psi the f_y used shall be that corresponding to a strain of 0.0035 determined by tests on full size bars.

The ratio of $\beta_1 = a/c$ used is decreased 0.05 for each 1000 psi of f'_c above 4000 psi (Code 10.2.7.3), as already noted at the close of Sec. 3.4, to reflect the more triangular or less curved nature of the stress-strain curve for these higher strength concretes (Fig. 2.8*a*).

Under service loads, deflection of members must be considered. Deflection calculations are discussed later in this chapter. When the length–depth ratios of Table 3.2 (a copy of ACI Code Table 9.5*a*) are not satisfied or when the member supports partitions that might be cracked, deflections must be calculated. This criterion does not necessarily separate members without deflection troubles from those with excessive deflections. It simply reflects, in a somewhat over-simplified fashion, a boundary where the designer must certainly become aware of deflection.

The 1963 Code required that deflections be checked where ρ exceeded $0.18 f'_c / f_y$ (the approximately midrange value shown on Fig. 3.8). There then was a greater design experience with members designed with approximately this steel ratio under working stress methods; often it is still an economical ratio to use. Larger reinforcement ratios imply smaller beam sizes and hence

TABLE 3.2 (Code Table 9.5a). Minimum Thickness of Beams or One-way Slabs Unless Deflections Are Computed*

	Minimum Thickness, h			
	Simply Supported	One End Continuous	Both Ends Continuous	Cantilever
Member	Members not supporting or attached to partitions or other construction likely to be damaged by large deflections.			
Solid one-way slabs	$l/20$	$l/24$	$l/28$	$l/10$
Beams or ribbed one-way slabs	$l/16$	$l/18.5$	$l/21$	$l/8$

* The span length l is in inches.
  Values given shall be used directly for members with normal weight concrete ($w = 145$ pcf) and Grade 60 reinforcement. For other conditions, the values shall be modified as follows:
  (a) For structural lightweight concrete having unit weights in the range 90–120 lb per cu ft, the values in the table shall be multiplied by $1.65 - 0.005 w_c$ but not less than 1.09, where w_c is the unit weight in lb per cu ft.
  (b) For f_y other than 60,000 psi, the values shall be multiplied by $0.4 + f_y/100,000$.

larger deflections. Short-span beams can use larger ratios without deflection problems; long-span beams can present deflection problems even when a smaller ratio is used. Long-span beams, cantilever construction, and shallow members always require special attention to deflections. As a guidepost from past experience, the criterion still has value.

As a sample of the determination of beam constants for a given ρ, consider Fig. 3.10 and the following for $\rho = 0.18 f_c'/f_y$:

$$N_t = A_s f_y = \rho b d f_y = 0.18 f_c' b d$$
$$a = N_t/(0.85 f_c' b) = (0.18 f_c' b d) \div (0.85 f_c' b) = 0.212 d$$
$$z = d - 0.5 \times 0.212 d = 0.894 d$$
$$M_n = N_t z = 0.18 f_c' b d \times 0.894 d = 0.161 f_c' b d^2$$
$$\text{Design moment} = \phi M_n = 0.9 \times 0.161 f_c' b d^2 = 0.145 f_c' b d^2$$
$$k_m = 0.145 f_c'$$

In this form the constants are the same for both customary and SI units.

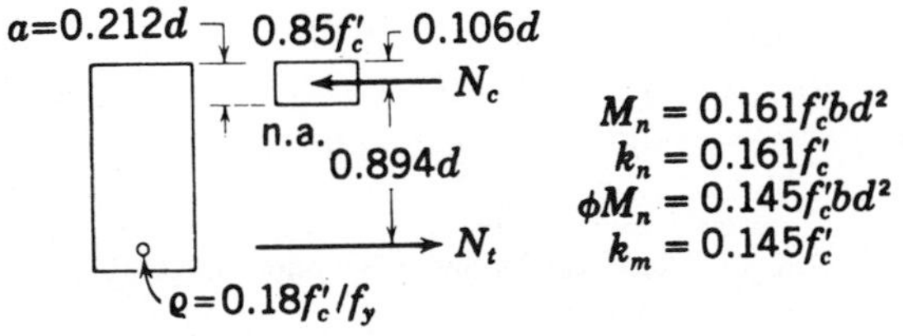

$a=0.212d$ $0.85f_c'$ $0.106d$

N_c

n.a.

$0.894d$

N_t

$\rho = 0.18 f_c'/f_y$

$M_n = 0.161 f_c' b d^2$
$k_n = 0.161 f_c'$
$\phi M_n = 0.145 f_c' b d^2$
$k_m = 0.145 f_c'$

Figure 3.10 Design constants for $\rho = 0.18 f_c'/f_y$, usable with either customary or SI units.

3.10   Rectangular Beam Examples

(a) A rectangular beam has $b = 12$ in., $d = 20$ in., $A_s = 3 - \#8$ bars, $(A_b = 0.79$ si for a #8 bar), $f'_c = 4$ ksi, grade 60 reinforcement $(f_y = 60$ ksi). Calculate design moment $= \phi M_n$

Solution

$$\rho_{\min} = 200/f_y = 200/60,000 \text{ (in psi units)} = 0.0033$$

$\rho_{\max}$ may be found from Table 3.1 for $f'_c = 4000$ psi, $f_y = 60,000$ psi as 0.0214.

Until the reader is familiar with the subject, $\rho_{\max}$ is better found from first principles as $\frac{3}{4}\rho_b$. ρ_b is determined from strain triangles of Fig. 3.7a and compatibility as

$$\frac{c_b}{0.003} = \frac{d}{0.003 + \varepsilon_y} \qquad \varepsilon_y = \frac{f_y}{E_s} = \frac{60}{29 \times 10^3} = 0.00207$$

$$c_b = [0.003/(0.003 + 0.00207)]20 = 11.83 \text{ in.}$$

$$\text{For } f'_c = 4000 \text{ psi}, \ \beta_1 = 0.85$$

$$a_b = \beta_1 c_b = (0.85)(11.83) = 10.06 \text{ in.}$$

$$N_{cb} = 0.85 f'_c a_b b = (0.85)(4)(10.06)(12) = 410.4 \text{ k}$$

$$N_{tb} = N_{cb} \therefore A_{sb} f_y = \rho_b b d f_y = \rho_b (12)(20)(60) = 410.4 \text{ k}$$

$$\rho_b = 0.0285 \qquad \rho_{\max} = \tfrac{3}{4}\rho_b = 0.0214$$

Checking the actual reinforcement

$$\rho = A_s/bd = (3)(0.79)/(12)(20) = 0.0099$$

$$\therefore \rho_{\min} \lessgtr \rho \lessgtr \rho_{\max}(0.0033 < 0.0099 < 0.0214)$$

Beam is **O.K.***

This check indicates that tension controls and $f_s = f_y$ at ultimate. Alternate forms of this check are possible. The actual a could have been determined from equilibrium and checked against $\frac{3}{4}$ of the a_b value. A_s could have been checked against $\frac{3}{4}A_{sb}$ without using percentages. The designer will choose the system that is most convenient for practice.

Once this important check has been made, the actual properties of the member are used for evaluation of the moment capacity. As shown in Fig. 3.11

$$\text{Ultimate } N_{tn} = A_s f_y = 3 \times 0.79 \times 60 = 142.2 \text{ k}$$

$$N_{cn} = 0.85 f'_c ab = 0.85 \times 4 \times a \times 12 = N_{tn} = 142.2 \text{ k}$$

$$a = 3.48 \text{ in.} \qquad (c = a/\beta_1 = 3.48/0.85 = 4.10 \text{ in.})$$

$$z = d - a/2 = 20 - (3.48/2) = 18.26 \text{ in.}$$

$$M_n = N_{tn} z = (142.2)(18.26) = 2597 \text{ k-in. Say 2600 k-in.}$$

---

* Designers in the United States often use O.K. and N.G. (or OK and NG without the periods) to mean satisfactory for use or "no good" and to be discarded.

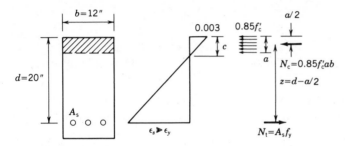

Figure 3.11 Beam for Example 3.10a.

Design strength $M_u = \phi M_n = 0.9 M_n = 0.9 \times 2600 = 2340$ k-in. $= 195$ k-ft. Because $a < 0.212 d = 4.24$ in. (Fig. 3.10), deflections are probably no problem. The length-depth ratio is not available for checking deflection.

(b) If in **(a)** the steel is changed to $4 - \#11 = 6.24$ si, calculate the design moment capacity and corresponding f_s.

Solution

With the steel area increased to 6.24 si both ρ and a from Example (a) are greatly increased and may exceed the permitted limits of Table 3.1. The simpler check is on ρ.

$$\rho = A_s/bd = 6.24/(12 \times 20) = 0.026 > 0.0214 \text{ of Table 3.1} \qquad \textbf{N.G.}$$

The beam does not satisfy the Code Sec. 10.3.3 requirement for ductility. If ductility is not a critical concern in this case, the authors would accept this beam for a design moment based on limiting values defined by $\rho_{max} = \frac{3}{4}\rho_b$ as given in Table 3.1, as discussed at the close of Sec. 3.8. This would leave the steel much understressed at the design load. The easiest evaluation on this basis is from the k_m value of Table 3.1, 0.937 ksi.

$$\text{Design moment} = \phi M_n = k_m bd^2 = 0.937 \times 12 \times 20^2 = 4500 \text{ k-in.}$$

The nominal f_s at this value of ϕM_n is easily found from the limiting a used here, which Table 3.1 shows to be $0.38 d = 7.6$ in. Then $z = d - a/2 = 20 - 7.6/2 = 16.2$ in.

$$f_s = \phi M_n/(\phi A_s z) = 4500/(0.9 \times 6.24 \times 16.2) = 49.5 \text{ ksi}$$

Because the steel operates at only 80 percent of f_y, it appears that $3 - \#11$ plus a small bar would be as good as the $4 - \#11$ assumed to be used.

Deflections may be large at this moment because $\rho \gg 0.18 f'_c/f_y = 0.012$, but the lower f_s would help some.

(c) How much steel can be effectively used in the beam of **(a)**? What is the design moment, ϕM_n, for this area of steel?

Solution

The first-principles solution for $\rho_{max} = \frac{3}{4}\rho_b$ in (**a**) or the use of Table 3.1 shows the maximum percentage of reinforcement as 2.14% and the corresponding k_m as 0.937 ksi. Hence,

$$\text{Max. } A_s = 0.0214 \times 12 \times 20 = 5.14 \text{ si}$$
$$\text{Design moment} = \phi M_n = k_m b d^2 = 0.937 \times 12 \times 20^2 = 4500 \text{ k-in.}$$

As a preferred alternate, M_n could be evaluated from basic equilibrium principles just as in (**a**) except that $A_{s\,max}$ is used. a, N_{tn}, and design moment $= \phi M_n = \phi N_{tn}(d - a/2)$ are then calculated successively.

Deflection would also need investigation.

3.11   Slabs

A one-way slab is simply a wide, shallow, rectangular beam insofar as analysis is concerned. The reinforcing steel is usually spaced uniformly over its width. For convenience, a 1-ft width is generally taken for analysis or design, since loads are frequently specified in terms of load per square foot; on a 1-ft strip, this unit load becomes the load per linear foot. Since the slab can average the effect of steel over some width, the effective A_s may well correspond to a fractional number of bars in the 1-ft width. For example, #5 bars at 8-in. spacing, as in Fig. 3.12a, give $A_s/ft = 0.31 \times 12/8 = 0.465$ in.^{2}/ft for the 1-ft design strip, Fig. 3.12b. For SI units the typical design width chosen is 1 m because loads are very often given in terms of load per square meter.

3.12   Analysis of T-Beams

Because slabs and beams are ordinarily cast together as shown in Fig. 3.13, the beams are automatically provided with an extra width at the top that is called a flange. Such beams are known as T-beams. The portion below the flange is called the web.

Flange bending stress is not uniform from beam to beam. It is largest over the web and tends to drop off with distance from the web. In part this results from the necessity of transferring all the flange flexural stress by longitudinal

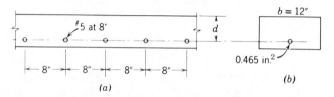

Figure 3.12   Design strip and effective steel area in slab analysis. (a) Slab cross section. (b) Design strip.

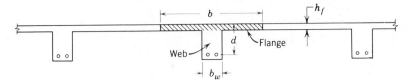

Figure 3.13 T-beam as part of a floor system.

shearing stresses on vertical sections parallel to the web. Codes use a reduced effective flange width at an assumed uniform stress distribution laterally in lieu of a wider actual slab with nonuniform stresses. The Code (8.10.2) for symmetrical monolithically cast T-beams limits the usable flange width for calculations to a maximum projection of eight times the slab thickness on each side of the web, or half way to the next beam, or a total width of one-fourth of the span, whichever is smallest.

T-beams are analyzed in much the same way as rectangular beams. The Code intends the same limitation (used in Sec. 3.7) to be placed on tension steel; that is, the tension steel should be limited to 0.75 that of the balanced section for the T-beam. This is usually no problem because the large flange area normally keeps compression on the concrete quite low and economy prevents the engineer from using excessive steel. Hence it is usually safe to assume (and check later) that $N_{tn} = A_s f_y$ and calculate the depth of stress block a which will provide an equal N_{cn}. If a, the resulting depth of stress block, is less than the slab (flange) thickness h_f, as shown in Fig. 3.14a, the entire analysis is identical to that of a very wide rectangular beam of width b, except that the limiting a for ductility will usually not be $0.75 a_b$.

If the area within the flange depth does not provide enough compression, the form of the calculation must be modified to take account of the narrower web width b_w below the flange (Fig. 3.14b). The total tension can be equated to the total compression to establish the depth of the stress block that acts on the summation of the shaded areas in Fig. 3.14b.

$$A_s f_y = 0.85 f'_c [ab_w + h_f(b - b_w)]$$
$$a = \frac{A_s f_y - 0.85 f'_c h_f(b - b_w)}{0.85 f'_c b_w}$$

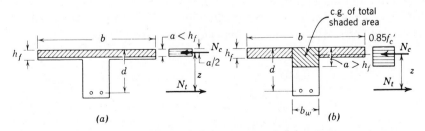

Figure 3.14 Design of T-beams. (*a*) Acts as rectangular beam. (*b*) Direct analysis as T-beam.

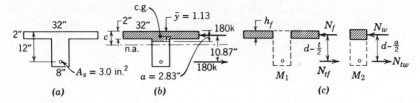

Figure 3.15 T-beam of Sec. 3.13.

The resultant compression N_c acts at the centroid of the total shaded area in Fig. 3.14b.

Similarly, in checking A_s against the balanced reinforcement, the only change from the rectangular beam procedure is in noting that the area which establishes the balanced N_{cb} is not a single rectangle. The neutral axis, being determined from strains, is located at the same depth c_b as for the rectangular beam. The depth of stress block a_b is $0.85 c_b$ or less,* since tests indicate that the ratio a/c used for rectangular beams is accurate enough. The total compression N_{cb} is still $0.85 f'_c$ times the compression area and this N_{cb} can be equated to $N_{tb} = A_s f_y$ to establish the balanced A_s; or, more simply, the usable $N_t = 0.75 N_{cb} = 0.75 N_{tb}$. The rectangular beam limitation of a to $0.75 a_b$ is not usable here because of the varying width.

3.13 T-Beam Examples

For the T-beam shown in Fig. 3.15a, find the design moment if $A_s = 3.00$ in.^{2}, steel is Grade 60, and $f'_c = 3000$ psi.

Solution

As neither span, nor beam spacing is given, the only available check on flange width is in terms of flange thickness. First check ρ_{min} (Code 10.5.1).

$$\rho_w = A_s/b_w d = 3.00/(8 \times 12) > \rho_{min} = 200/f_y = 0.0033 \quad \textbf{O.K.}$$
$$\text{Max. flange overhang} = 8h_f = 16 \text{ in.} > \text{actual} \quad \textbf{O.K.}$$
$$\text{Ultimate } N_t = 3.00 \times 60,000 = 180,000 \text{ lb}$$

Check the ductility requirement $\rho \gtrless 0.75 \rho_b$, equivalent to limiting N_c to $0.75 N_{cb}$. From similar strain triangles for the balanced strain conditions of Fig. 3.7a, which apply to beams of any shape

$$c_b = \frac{0.003}{0.003 + \varepsilon_y}$$

$$d = \frac{87,000}{87,000 + 60,000} \times 12 = 7.10 \text{ in.}$$

$$a_b = 0.85 \times 7.10 = 6.04 \text{ in.}$$

---

* Less than 0.85 for $f'_c > 4000$ psi, Code 10.2.7.3. For SI units see closing portion of Sec. 3.4.

The stress block extends substantially below the top flange for the balanced condition. Applying the rectangular stress block to the concrete in the cross section down to the depth a_b:

$$N_{cb} = 0.85 \times 3000(32 \times 2 + 8 \times 4.04)$$
$$= 246{,}000 \text{ lb}$$
$$\text{Maximum permissible } N_c = 0.75 \times 246{,}000 = 184{,}000 \text{ lb}$$
$$> 180{,}000 \quad \textbf{O.K.}$$

Note that a check on "0.75 balanced compression" is the same as "0.75 balanced reinforcement." Having determined that the actual reinforcement falls between the minimum and maximum amounts, the remainder of the problem deals with the actual N_t and N_c. The compressive capacity of the flange or potential N_c within depth $h_f = 0.85 \times 3000 \times 32 \times 2 = 163{,}000$ lb.

Because the required tension $N_t = 180{,}000 > 163{,}000$, the stress block extends below the flange far enough to pick up the remaining compression of $180{,}000 - 163{,}000 = 17{,}000$ lb.

$$17{,}000 = 0.85 \times 3000 \times 8(a - 2)$$
$$a - 2 = 0.83 \text{ in.,} \qquad a = 2.83 \text{ in.}$$

The resultant N_c acts at the centroid of the compression area, which establishes z and leads to $M_u = N_t z$ or $N_c z$, as indicated in Figs. 3.14b and 3.15b.

$$\bar{y} = \frac{32 \times 2 \times 1 + 8 \times 0.83 \times 2.41}{64 + 6.64} = \frac{64 + 16}{70.6} = 1.13 \text{ in.}$$
$$z = 12 - 1.13 = 10.87 \text{ in.}$$
$$M_n = 180{,}000 \times 10.87/1000 = 1957 \text{ k-in.} = 163 \text{ k-ft}$$
$$\text{Design } M = \phi M_n = 0.9 \times 163 = 147 \text{ k-ft.}$$

As a convenient alternate computational procedure the resisting moment is calculated as the sum of two couples. The compression $N_f{}^*$ on the outstanding flanges, as shown by the shaded areas in Fig. 3.15c, pairs with a corresponding tension to form couple M_1. The remainder of the tension pairs with the compression N_w in the 8-in. web width to form the second couple M_2.

$$N_f = (32 - 8)2 \times 0.85 \times 3000/1000 = 122 \text{ k}$$
$$M_1 = 122(12 - 2/2)/12 = 112 \text{ k-ft}$$
$$N_w = 8 \times 2.83 \times 0.85 \times 3000/1000 = 58 \text{ k}$$
$$M_2 = 58(12 - 2.83/2)/12 = 51 \text{ k-ft}$$
$$M_n = M_1 + M_2 = 112 + 51 = 163 \text{ k-ft}$$
$$\text{Design moment} = \phi M_n = 0.9 \times 163 = 147 \text{ k-ft}$$

---

* The complete symbol N_{cf} will be shortened here to N_f because there is no likely confusion. Because in flexure $N_c = N_t$, it is probable that subscripts c and t can often be dropped where a particular emphasis is not demanded.

The minimum thickness (overall depth) should be checked against Table 3.2 to determine whether deflection is a probable concern. The old standard of $\rho = A_s/bd \gtrless 0.18 f'_c/f_y$ (now discarded from the Code) also gives a rough idea. The corresponding ρ in terms of the M_2 couple alone (Fig. 3.15c) is often convenient for design tables.

3.14   Analysis of Beams With Compression Steel

Beams are occasionally restricted in size to such an extent that steel is needed to help carry the compression. Compression reinforcement A'_s is also excellent for reducing the long-time deflection. For this type of member, the *design of* A_s and A'_s for a required M_u (Sec. 4.5) is simpler than the *analysis* problems that follow. For analysis it is convenient to consider artificially dividing the beam into two couples M_1 and M_2 similar to the approach used for the T-beam in Fig. 3.15c. The M_1 couple has part of the tension steel (A_{s1}) equilibrated by the concrete in compression. The M_2 couple has the remainder of the tension steel (A_{s2}) equilibrated by the compression steel force ($A'_s f'_s$).

If the member is relatively deep the M_2* steel couple based on A'_s and A_{s2} should be evaluated first. Strictly, the extra compression available because of A'_s must also account for the loss of compression concrete displaced by A'_s. Then

$$M_2 = A'_s(f_y - 0.85f'_c)(d - d') \quad \text{and since} \quad N_{t2} = N_{c2}$$
$$A_{s2} = A'_s(f_y - 0.85f'_c)/f_y$$

If A'_s is small and f'_c is low, it is often close enough to neglect the displaced concrete, as follows (in Fig. 3.16):

$$M_2 = A'_s f_y(d - d') \quad \text{and} \quad A_{s2} = A'_s$$

The remaining tension steel is the basis for the M_1 couple.

$$A_{s1} = A_s - A_{s2} \qquad M_1 = A_{s1}f_y(d - a/2)$$
$$\text{Design } M_u = \phi M_n = \phi(M_2 + M_1) = 0.9(M_2 + M_1)$$

The depth of the stress block a is calculated from $N_{c1} = N_{t1}$ or $0.85f'_c ba = A_{s1}f_y$.

If the member is shallow, especially a slab, there is a reasonable chance that the strain at the A'_s level will not develop f_y in compression and this must be verified early in the analysis:

1.   Assume $A_{s2} = A'_s$. Then $A_{s1} = A_s - A'_s$.
2.   Calculate $a = A_{s1}f_y/(0.85f'_c b)$.

---

* The subscript 2 is used because the *usual* problem in *design* is where the M_2 couple is the second operation. It is better not to vary the numbering between the two cases. Both M_1 and M_2 are nominal moments but the omission of the n subscript should cause no confusion here.

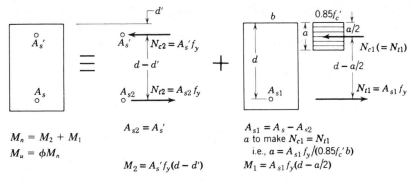

$$M_n = M_2 + M_1$$
$$M_u = \phi M_n$$

$$A_{s2} = A_s'$$

$$M_2 = A_s' f_y (d - d')$$

$$A_{s1} = A_s - A_{s2}$$
$$a \text{ to make } N_{c1} = N_{t1}$$
$$\text{i.e., } a = A_{s1} f_y / (0.85 f_c' b)$$
$$M_1 = A_{s1} f_y (d - a/2)$$

Figure 3.16 Analysis of double-reinforced beam if all steel yields.

3. Calculate c, using the ratios between c and a at the close of Sec. 3.4.
4. Use the strain triangle of Fig. 3.17 to calculate

$$\varepsilon_s' = 0.003(c - d')/c$$

5. If $E_s \varepsilon_s' > f_y$, proceed to evaluate M_1 and M_2, using f_y with A_s'.

If $f_s' < f_y$ it becomes more important to evaluate N_{c2} and A_{s2} considering the displaced concrete, which in turn modifies the original assumed A_{s1}—an iteration problem that quickly converges. For the more mathematically inclined, the second example sets up a possible quadratic equation that can be used. The iteration procedure is recommended for the physical feel it gives.

When checking reinforcement limits in a doubly reinforced beam, care must be taken. The total A_s is effective in resisting tension. Hence, the minimum reinforcement requirement is checked using the total A_s. However, in checking the maximum reinforcement amount, Code 10.3.3 specifically points out that only that portion of the total tension steel (A_{s1}) equilibrated by compression in the concrete need to be limited to $0.75 A_{sb}$.

Deflections are determined by top and bottom strains on the beam and can be measured by the M_1 couple developed by A_{s1} and the concrete, ignoring the M_2 couple developed by A_s' and A_{s2}.

Ties (or stirrups) are required around A_s' in beams, as in columns, Code 7.11.1.

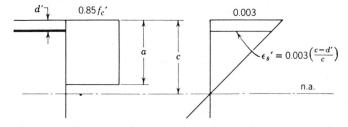

Figure 3.17 Check on compression steel stress.

3.15 Double-Reinforced Beam Examples

(a) A rectangular beam has $b = 11$ in., $d = 20$ in., $A_s = 5\text{-}\#11 = 7.80$ in.^{2}, $f_c' = 3000$ psi, Grade 40 steel, $A_s' = 2\text{-}\#11 = 3.12$ in.^{2}, with d' (cover to center line of A_s') of 2 in. Calculate the design moment capacity.

Solution

$$A_{s\,min} = (200/f_y)bd = (200/40,000)(11)(20) = 1.10 \text{ in.}^2 < A_s$$
$$= 7.80 \text{ in.}^2 \quad \textbf{O.K.}$$

Neglecting concrete displaced by A_s', consider $A_{s2} = A_s' = 3.12$ in.^{2} and assume $f_s' = f_y$. This assumption must be checked later and revised if necessary.

$$M_2 = A_s' f_y (d - d')$$
$$= 3.12 \times 40,000(20 - 2) \div 12,000 = 187 \text{ k-ft}$$

The remainder of the tension steel, $A_{s1} = A_s - A_{s2} = 7.80 - 3.12 = 4.68$ in.^{2}, works with the compression concrete, as in a rectangular beam without A_s' to develop M_1. Since $N_t = N_c$,

$$4.68 \times 40,000 = 0.85 \times 3000 \times 11a$$
$$a = 6.67 \text{ in.} < 0.44d = 8.80 \text{ in. (See Table 3.1)} \therefore A_{s\,max} \quad \textbf{O.K.}$$

This check verifies that the beam has A_{s1} less than 0.75 of the balanced ratio for a simple rectangular beam. Extra A_{s2} and A_s' do not themselves lead to a brittle failure because this steel forms an auxiliary "beam" exhibiting great toughness, provided A_s' develops its f_y. This is the intent of Code 10.3.3 which considers A_{s1} the only reinforcement subject to the $0.75\rho_b$ limitation.

The stress in A_s' will now be checked to verify the assumption $f_s' = f_y$.

$$\text{For } f_c' = 3000, \beta_1 = a/c = 0.85$$
$$c = 6.67/0.85 = 7.85 \text{ in.}$$
$$\varepsilon_s' = 0.003(c - d')/c = 0.003 \times 5.85/7.85 = 0.0022$$
$$\varepsilon_s' = 0.0022 > \varepsilon_y = f_y/E_s^* = 40,000/29,000,000 = 0.00138$$

Therefore $f_s' = f_y$ and the calculations so far are correct.

$$z_1 = d - a/2 = 20 - 6.67/2 = 16.66 \text{ in.}$$
$$M_1 = A_{s1} f_y z_1 = 4.68 \times 40,000 \times 16.66/12,000 = 260 \text{ k-ft}$$
$$M_n = M_1 + M_2 = 260 + 187 = 447 \text{ k-ft}$$
$$M_u = \phi M_n = 0.9 \times 447 = 402 \text{ k-ft}$$

---

* E_s is given in Code 8.5.2. For nonprestressed reinforcement it may be taken as 29,000,000 psi. The SI value is 200 000 MPa.

(b) A rectangular beam has $b = 11$ in., $d = 20$ in., $A_s = 3$-#11 $= 4.68$ in.^{2}, $f'_c = 5000$ psi, Grade 60 steel, $A_s = 1$-#11 $= 1.56$ in.^{2}, with $d' = 2$ in. Calculate the design moment capacity.

Solution

$$A_{s\,min} = (200/f_y)bd = (200/60,000)(11)(20)$$
$$= 0.73 \text{ in.}^2 < 4.68 \text{ in.}^2 \quad \textbf{O.K.}$$

Again neglecting concrete displaced by A'_s and assuming $f'_s = f_y$:

$$A_{s2} = A'_s = 1.56 \text{ in.}^2$$
$$A_{s1} = A_s - A_{s2} = 4.68 - 1.56 = 3.12 \text{ in.}^2$$
$$a = A_{s1}f_y/(0.85f'_cb) = 3.12 \times 60/(0.85 \times 5 \times 11) = 4.00 \text{ in.}$$

From Sec. 3.4,

$$\text{For } f'_c = 5000, \beta_1 = a/c = 0.85 - 0.05 = 0.80$$
$$c = a/\beta_1 = 4.00/0.80 = 5.00 \text{ in.}$$

From Fig. 3.17,

$$\varepsilon'_s = 0.003 \times (5.00 - 2.00)/5.00 = 0.00180$$
$$\varepsilon_y = 60/29,000 = 0.00207 > 0.00180, f'_s < f_y$$

The assumption that $f'_s = f_y$ must now be revised. An algebraic solution will be set up later but a trial-and-error technique seems more efficient. Because $f'_s < f_y$, it is easy also to recognize the concrete displaced by A'_s. The compression concrete is valued at $0.85 \times 5 = 4.25$ ksi. The effect of the displaced concrete is often neglected.

Try $f'_s = 52$ ksi (about f_y times the ratio $\varepsilon'_s/\varepsilon_y$)
Effective $f''_s = 52 - 4.25 = 47.8$ ksi
$A'_s f''_s = A_{s2}f_y$, $1.56 \times 47.8 = A_{s2} \times 60$
$A_{s2} = 1.24$ in^2 $A_{s1} - A_{s2} = 4.68 - 1.24 = 3.44$ in.^{2}
Table 3.1 $\rho_{max} = 0.0252$ $A_{s1} < 0.0252 \times 11 \times 20 = 5.54$ in.^{2} **O.K.**

$$a = A_{s1}f_y/(0.85f'_cb) = 3.44 \times 60/(0.85 \times 5 \times 11)$$
$$= 4.42 \text{ in.}$$
$$c = 4.42/0.80 = 5.52 \text{ in.}$$
$$\varepsilon'_s = 0.003 \times (5.52 - 2.00)/5.52 = 0.0019$$
$$f'_s = 0.0019 \times 29\,000 = 55.1 \text{ ksi vs. 52 ksi assumed.}$$

This is probably close enough. Take $f'_s = 55$ ksi as accurate.

Since the trial started with f'_s assumed as 52 ksi, it is probably close enough to use f'_s as 55 ksi. This leads to an effective $f'_s = 55 - 4.25$ or 51 ksi.

$$N_{c2} = f'_s A'_s = 51 \times 1.56 = 79.56 \text{ k}$$
$$M_2 = N_{c2}(d - d') = (79.56)(20 - 2) = 1430 \text{ k-in.}$$

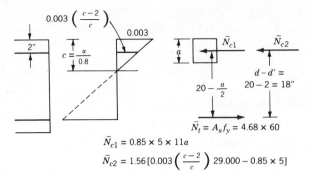

$$\bar{N}_{c1} = 0.85 \times 5 \times 11a$$

$$\bar{N}_{c2} = 1.56\left[0.003\left(\frac{c-2}{c}\right) 29.000 - 0.85 \times 5\right]$$

Figure 3.18 Basic relations in algebraic solution for Sec. 3.15b.

Since $A'_s f'_s = A_{s2} f_y$, $A_{s2} = A'_s f'_s / f_y = 1.56 \times 51/60 = 1.33$ in.^{2}

$A_{s1} = A_s - A_{s2} = 4.68 - 1.33 = 3.35$ in.^{2}

$a = (3.35)(60)/(0.85 \times 5 \times 11) = 4.30$ in., $a/2 = 2.15$ in.

$M_1 = 3.35 \times 60(20 - 2.15) = 3590$ k-in.

$M_u = \phi(M_1 + M_2) = 0.9(3590 + 1430) = 4520$ k-in. $= 376$ k-ft

The solution can also be set up algebraically after it is discovered that $\varepsilon_y > \varepsilon'_s$, as shown in Fig. 3.18.*

$$N_{c1} + N_{c2} = N_t$$

$$0.85 f'_c b(\beta_1 c) + A'_s[E_s \times 0.003(c - d')/c - 0.85 f'_c] = N_t$$

$$0.85 \times 5 \times 11(0.80c) + 1.56[0.003 \times 29{,}000(c - 2)/c - 4.25] = 4.68 \times 60$$

This results in a quadratic in c and is rather cumbersome.

$c = 5.41$ in., $a = 0.8 \times 5.41 = 4.32$ in.

$M_1 = 0.85 \times 5 \times 11 \times 4.32(20 - 0.5 \times 4.32)12 = 300$ k-ft

$M_2 = 1.56[0.003 \times 29\,000(5.41 - 2)/5.41 - 0.85 \times 5](20 - 2)/12$

$\quad = 118$ k-ft

$M_u = \phi(M_1 + M_2) = 0.9(300 + 118) = 376$ k-ft (Same as trial and error)

3.16 Analysis of Special Beam Shapes

The general analysis presented in Secs. 3.6 and 3.8 and adapted to the T-beam in Secs. 3.12 and 3.13 is adaptable to any shape of cross section. The strain triangles can always be used to establish the balanced section neutral axis.†

---

* The algebraic simultaneous equation cannot be used if $\varepsilon'_s > \varepsilon_y$ because the relation between stress and strain is no longer linear.

† Some uncertainty can arise when there are several layers or depths of tension reinforcing. Rather arbitrarily, the authors use the yield strain on the steel most distant from the compression face to establish the balanced section.

The stress block depth may be fixed from the $a/c = \beta_1$ ratios of Sec. 3.4, and N_{cb} evaluated even for very irregular member outlines. The usable N_c or N_t should be restricted to $0.75N_b$ as an upper limit. For this maximum N_t or such smaller A_sf_y as may exist, the required depth of stress block a can be established, the resultant compression located, and the moment couple calculated. Because a also establishes c, values of $f_s < f_y$ can be taken into account when specific bar locations will not develop strains as large as ε_y. A sample analysis for one shape follows the next paragraph.

Tests show that when a nonrectangular compressive stress area is used, the ε_c Code limit of 0.003 could easily be larger for a peaked area, as in the example of Fig. 3.19, or could be slightly more limited for an inverted triangular area. The collapse of the extremely narrow compression top is apparently delayed by the bracing effect of the adjacent wider horizontal strips at lower strain levels. However, these variations are not reflected in the Code.

Example

(a) Given the beam of Fig. 3.19a with $f_c' = 4000$ psi, Grade 60 bars. What is the maximum A_s the Code would allow to be counted as effective if no A_s' is used?

Solution

The maximum A_s would be limited to $0.75A_{sb}$ for the balanced section. The balanced strains are shown in Fig. 3.19a, with $\varepsilon_y = f_y/E_s = 60/29\,000 = 0.00207$

$$c_b = 26 \times 0.003/(0.003 + 0.00207) = 15.38 \text{ in.}$$
$$\text{Depth of stress block} = a_b = 0.85c_b = 13.07 \text{ in.}$$
$$\text{(since } \beta_1 = 0.85 \text{ for } f_c' = 4000 \text{ psi)}$$
$$\text{Area under stress block} = A_c = 13.07 \times 12 - 2 \times 0.5 \times 6 \times 6$$
$$= 156.8 - 36$$
$$= 120.8 \text{ in.}^2$$

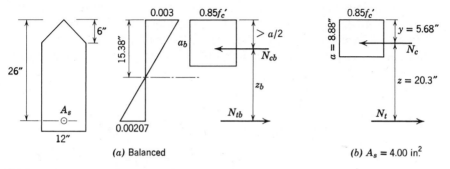

(a) Balanced   (b) $A_s = 4.00$ in.²

Figure 3.19 Example of special beam shape.

$$N_{tb} = N_{cb}, \qquad A_{sb} \times 60\ 000 = 0.85 f'_c A_c = 0.85 \times 4000 \times 120.8$$
$$A_{sb} = 6.85 \text{ in.}^2$$

With irregular shapes it appears simpler to work with A_{sb} rather than with ρ_b.

$$\text{Max. allowable } A_s = 0.75 A_{sb} = 5.14 \text{ in.}^2$$

(b) If the beam in Fig. 3.19 has A_s of $4 - \#9$ bars, what is the permissible design moment?

Solution

$$N_{cn} = 0.85 f'_c A_c = N_{tn} = A_s f_y, \qquad 0.85 \times 4000 A_c = 4 \times 1.00 \times 60\ 000$$

$A_c = 70.6$ in., which shows that the stress block extends below the upper triangle.

$$12a - 2 \times 0.5 \times 6 \times 6 = 12a - 36 = 70.6 \qquad a = 8.88 \text{ in.}$$

Since N_{cn} acts more than $a/2$ from the top, locate it at the centroid of A_c. If y = distance to centroid from top,

$$70.6y = (2 \times 0.5 \times 6 \times 6)4 + 2.88 \times 12(6 + 1.44) = 401 \qquad y = 5.68 \text{ in.}$$
$$z = 26 - 5.68 = 20.3 \text{ in.}$$
$$M_n = N_{tn}z = 4.0 \times 60\ 000 \times 20.3/12\ 000 = 406 \text{ k-ft}$$
$$\text{Allowable design moment} = \phi M_n = 0.9 \times 406 = 365 \text{ k-ft}$$

3.17   Tables and Charts

Sample helpful tables and charts in customary units are included in Appendix B for rectangular beams and slabs, without compression steel. Tables B.1 and B.2 augment Table 3.1, with Table B.1 giving k_m, c/d, a/d, and z/d for several common materials combinations. These are in terms of $\omega (= \rho f_y / f'_c)$ and also in terms of ρ. The use of ω permits the grouping of data into a very compact form, with k_m tabulations here in psi units, over the entire usable range. Table B.2 shows values of coefficients that can be multiplied by f'_c to give k_n values, or by $0.9 f'_c$ to give k_m values, each quite accurate in terms of ω.

For example, consider the following, similar to the example in Sec. 3.10a.

Calculate M_u for a rectangular beam having $b = 12$ in., $d = 20$ in., $A_s = 3 - \#8 = 2.37$ in.^{2}, $f'_c = 4000$ psi, $f_y = 60\ 000$ psi.

$\rho = 2.37/(12 \times 20) = 0.0099$, $\omega = \rho f_y / f'_c = 0.0099 \times 60,000/4000 = 0.148$

Enter Table B.1 with ω and read for given f'_c and f_y, $k_m = 462$ (for $\omega = 0.140$) plus 0.8×30 (for the last 0.008), a total $k_m = 486$. (Note that this table stops at the maximum allowable k_m for $\rho = 0.75\rho_b$, as already tabulated in Table 3.1 in Sec. 3.8.)

$$M_u = k_u bd^2 = 486 \times 12 \times 20^2 = 2,333,000 \text{ lb-in.} = 194.4 \text{ k-ft}$$

Sometimes the values of c, a, and z tabulated at the right of the table are useful.

Alternatively, Tables B.2 may be used, but note that it does not indicate the $0.75\rho_b$ limit. For $\omega = 0.148$ as before, read directly a coefficient of 0.1351. Multiply by $0.9f'_c$ to obtain $k_m = 486$, as before. Graphical values of k_m reading less accurately are also given in Figs. B.1 and B.2.

The ACI Handbook in the references at the end of Chapter 4, although primarily directed toward design, has many charts and tables that are also useful in analysis.

3.18 Increasing Importance of Serviceability

Although emphasis in this chapter has been largely on the dependable resistance at ultimate load, only a miniscule fraction of beams or slabs will ever experience such loading. Almost all such flexural members will be loaded to near the unfactored or service load levels. The deflection and cracking behavior of the members at service load levels is generally referred to as serviceability. As design procedures become more accurate and more efficient, practice tends to move into the realm of smaller members; greater deflections and more need for attention to serviceability result.

Strength design methods were significantly introduced into practice in the United States in the middle 1950s. These improved strength evaluations made smaller columns possible. The increased column slenderness led to emphasis on column deflections and increased secondary moments in the 1971 Code.

The changes in slabs and beams were less pronounced because (1) steel costs increased substantially as member thickness was reduced, and (2) long-term deflections had already been realized as a problem, at least, in slabs. But the possibility of reduced depths without loss of strength, coupled with the possibility of offsetting increased steel costs with smaller areas of high strength steel, made desirable a more definitive statement of what constitutes poor behavior. For serviceability is almost as much a problem of proper standards as of proper calculations. What constitutes poor serviceability?

Aside from corrosion and poor weathering or wearing properties, which are normally avoided by proper controls in mixing and placing, poor serviceability usually relates to excess deflections, and occasionally to extensive cracking or excessive crack width.

3.19 The Deflection Problem

Excessive deflections can lead to crushing of partitions, bulging and buckling of glass enclosure walls or metal partitions, sagging of floors, and unsightly drooping of overhanging canopies. ACI Committee 435 on deflections lists[3] four categories of deflection problems for which it discusses proper limits.

1. Sensory problems that include vibrations (a function of member stiffness) and appearance of droopy members.
2. Serviceability problems such as roofs that do not drain and floors that are not plane enough for their intended use, exemplified by the requirements for bowling alleys or sensitive equipment installations.
3. Effects on nonstructural elements such as masonry and plaster, or movable partitions. This category must include beam deflections caused by lateral building deflection and vertical movements of columns from temperature differentials.
4. Effects on structural behavior, instability, etc., that are not a serviceability but a strength problem.

The report suggests proper limits for each category.

Several factors unique to reinforced concrete make exact deflection prediction very difficult. (a) Unequal top and bottom reinforcing automatically leads to shrinkage deflections as a normal phenomenon. (b) Creep of concrete under stress leads to a gradual increase in deflection of members left under loads. (c) Even the somewhat uncertain stage at which reinforced concrete begins to crack significantly adds uncertainty as to the actual deflections to be expected at service loads.

Many slabs or beams are designed for use where deflections may be less critical. The ACI Code (Sec. 9.5.2.1) provides a table listing minimum thickness for slabs or beams "not supporting or attached to partitions or other construction likely to be damaged by large deflections." This table is reproduced here as Table 3.2 in Sec. 3.9, along with a few comments on its use. Its simple limits often take the place of the lengthy deflection calculations presented in the following sections.

3.20   The Code Solution for Deflection

The Code takes an overall approach in terms of the immediate deflection plus the expected overall percentage increase with shrinkage and time effects.

The immediate deflection that is the starting point is quite sensitive to whether the member is uncracked or cracked, and if cracked how severely cracked. This severity of cracking necessarily varies along the span as the moment changes. The load-deflection curve shown in Fig. 3.20a is typical for a moderately reinforced slab that is simply supported and loaded for the first time. Diagrammatically it could be simplified about as shown in Fig. 3.20b, where the importance of the percentage of tensile steel is shown. The deflections of interest are usually around a steel stress of 50 to 60% of the yield stress or less.

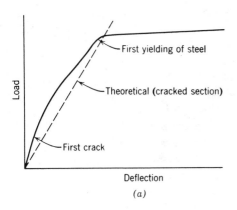

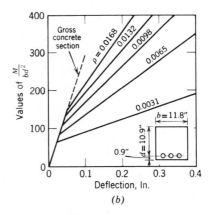

Figure 3.20 (*a*) Simple span slab deflection under increasing load. (*b*) Schematic of influence of ρ on deflection. (From Reference 4, Portland Cement Assn.)

The Code uses an *average* as the effective moment of inertia and then applies the usual methods of elastic analysis to compute the immediate deflection. The Code then requires that the deflection from the sustained portion of the load be multiplied by a factor that reflects the ratio of A'_s/A_s to determine the added long-term effects from creep and shrinkage.

The average moment of inertia used in Sec. 3.22 involves the transformed area concept which must first be explained in general and then applied for the beam. The application to deflections is picked up again in Sec. 3.22.

3.21   Transformed Area Concept—Elastic Analysis

(a) General   Where concrete and steel are deformed to the same unit strain ε, their initial stresses vary as their modulus of elasticity.

$$f_c = \varepsilon E_c \quad \text{and} \quad f_s = \varepsilon E_s = f_c(E_s/E_c) = nf_c$$

where the modular ratio n is defined as the ratio E_s/E_c. The total immediate stress on a bar of area A_s then can be written as

$$f_s A_s = nf_c A_s = f_c(nA_s)$$

The last form associates n with the steel area instead of with the steel stress. It shows that the steel area A_s acts initially* as would a concrete area nA_s. This concrete area is called the *transformed area* of the steel. When the steel is thus replaced in a member by its transformed area, the result is a total area of homogeneous material, in this case concrete, that is relatively easy to analyze for stress or deformation. To be exact, the effective added area is $(n - 1)A_s$ because A_s displaces an equal volume of concrete, but the difference between $n - 1$ and n is beyond the accuracy otherwise attainable in deflection calculations.

---

* The time effects of concrete creep and shrinkage modify this relationship.

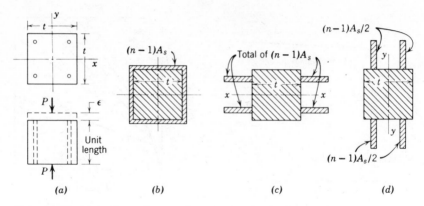

Figure 3.21 Transformed area of a column, elastic theory.

In the symmetrical column of Fig. 3.21a, the transformed area of the column steel may be sketched as shown in Fig. 3.21b, c, or d, that is, at any point around the cross section provided that symmetry is maintained and only an axial load is considered. This is true only because the unit deformation ε is everywhere the same. Normally, eccentricity or moment must be considered.

With bending added, f_s again equals nf_c provided both stresses correspond to the same strain ε. The nA_s area must then be spread *parallel* to the neutral axis to hold to the same level of strain. In Fig. 3.21a for moment or eccentricity about axis $x - x$ the transformed area would have to be sketched parallel to the axis of bending as shown in Fig. 3.21c and as shown in Fig. 3.21d for moment or eccentricity about axis $y - y$. Bars in different planes would then carry different stresses in accordance with their locations.

(b) Applied to Beams   With bending, either in beams or columns, any tension concrete is normally assumed to be cracked and omitted from the calculations, but the tension nA_s area is assumed active in tension to any level. The transformed area of a double-reinforced beam would be as shown in Fig. 3.22a with A_s' replaced by an effective transformed area of $(n - 1)A_s'$ and A_s replaced by nA_s, there being no useful concrete displaced on the tension side. Prior to cracking with the concrete still carrying tension, the area is that of Fig. 3.22b.

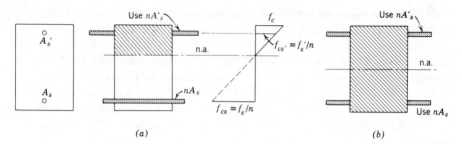

Figure 3.22   Transformed area concept for effect of short-time loads on a beam. (a) Usual cracked section. (b) Before cracking.

As shown in Fig. 3.22, the exactness of considering the displaced concrete or the $(n - 1)$ term is often neglected as shown by the statement "Use nA_s".

(c) Neutral Axis of Beam—Mechanics of Elastic Analysis or Review   The transformed area concept makes it possible to replace any reinforced concrete member (under moderate loads) with an equivalent member of homogeneous elastic material to which basic strength of materials relationships apply. For flexure without axial loading, the neutral axis must lie at the centroid of the effective cross section to make the total tension equal to the total compression and thus provide a resultant couple. The moment and stress are related by the equation

$$f = Mc/I$$

where M = bending moment, usually in.-lb or in.-k

 c = distance to extreme fiber, in.

 f = bending stress, psi or ksi, at extreme fiber (at distance c from neutral axis). This term is used as s or σ in engineering mechanics notation, but f is standard notation in reinforced concrete in the United States.

 I = area moment of inertia, $\int y^2 dA$, about the neutral axis, in.[4]

(d) Review Example with Transformed Area

Find f_c and f_s for the rectangular beam shown in Fig. 3.23a for $M = 130$ kN $\cdot$ m $= 130 \times 10^{-3}$ MN $\cdot$ m, C30 concrete, Grade 400 reinforcement, $E_s = 200\ 000$ MPa.

Solution

From Sec. 2.6, $E_c = 4700\sqrt{f'_c} = 4700\sqrt{30} = 25\ 740$ MPa.

$$n = E_s/E_c = 200\ 000/25\ 740 = 7.77, \text{ say } 8$$

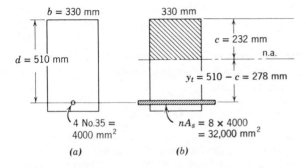

(a)   (b)

Figure 3.23   Analysis of a rectangular beam.

This transformed area is first sketched (Fig. 3.23b) and the neutral axis indicated at the unknown depth c. The neutral axis lies at the centroid of the transformed area. The moment of the areas about the neutral axis* gives a quadratic equation, for which a solution by completing the squares is recommended.

$$330c \cdot c/2 = 32\ 000(510 - c)$$
$$165c^2 + 32\ 000c = 16.32 \times 10^6$$
$$c^2 + 194c + (194/2)^2 = 98\ 900 + (194/2)^2 = 108\ 300$$
$$c + 97 = \pm 329, \qquad c = 232 \text{ mm (or negative)} = 0.232 \text{ m}$$

From Fig. 3.23b the moment of inertia is found from $\Sigma(I_0 + A\bar{d}^2)$ as

$$I_0 \text{ concrete: } 330 \times 232^3/12 = 343 \times 10^6$$
$$A\bar{d}^2 \text{ concrete: } 330 \times 232 \times 116^2 = 1030 \times 10^6$$
$$A\bar{d}^2 \text{ steel: } 32\ 000(510 - 232)^2 = \underline{2473 \times 10^6}$$
$$\text{Total} \qquad\qquad 3850 \times 10^6 \text{ mm}^4 = 3850 \times 10^{-6} \text{ m}^4 \text{ or}$$
$$3.850 \times 10^{-3} \text{ m}^4$$

The transformed area of the steel is considered negligibly thin, so that its moment of inertia about its centroidal axis can be neglected.

$$f_c = Mc/I = 130 \times 10^{-3} \times 0.232/(3.85 \times 10^{-3}) = 7.83 \text{ MPa}$$
$$f_s/n = My_t/I = 130 \times 10^{-3} \times 0.278/(3.85 \times 10^{-3}) = 9.39 \text{ MPa}$$
$$f_s = 8 \times 9.39 = 75.1 \text{ MPa}$$

3.22   Deflection Computations Following the Code

(a) Algebraic Recommendations   The actual moment curvature relation for an under-reinforced beam is highly nonlinear as shown in Fig. 3.5a. Elementary mechanics that assume linear elastic relationships assume linear moment curvature relationships, $\phi = M/EI$. This is very convenient for calculation of deflections that must doubly integrate the curvatures by some computational means such as the familiar moment-area theorems. To facilitate such computations, the Code approach is to establish an effective moment of inertia I_e that varies with level of applied moment. This allows the approximation of the nonlinear moment-curvature relationship in conventional, pseudo-elastic calculation procedures.

The effective moment of inertia is specified as an *average* value to be used all across a simple span and is a weighted value dependent on the extent of probable cracking under moment (Code Eq. 9.7):

$$I_e = (M_{cr}/M_a)^3 I_g + [1 - (M_{cr}/M_a)^3]I_{cr} \gtrless I_g$$

---

* Although moments about an axis not yet located may seem awkward, it should be noted that this gives a simpler equation. In general, $\bar{y}\int dA = \int y\,dA$. If y is measured from the centroid, $\bar{y} = 0$ and the equation becomes simply $\int y\,dA = 0$.

in which $M_{cr} = f_r I_g / y_t = 7.5\sqrt{f_c'} I_g / y_t$* except that f_r must be modified (Code 9.5.2.3a or b) when lightweight concrete is used.

M_a = max. M in member at loading stage considered in lb-in. (or MN · m)

I_{cr} = I based on transformed area of cracked section in in.^{4} (or m^4)

I_g = I based on gross area of concrete section (without steel), in in.^{4} (or m^4)

y_t = distance from centroid to extreme fiber, in in. (or m)

The ACI Committee 435 report noted that I_g in this relation would be more accurate if it included the transformed area of the reinforcement, especially where this was heavy. The Code committee, in the interest of simplicity and recognizing the inherent scatter in deflection computations, specifically defined I_g as "neglecting reinforcement." This does not preclude making a more detailed calculation including the reinforcement if desired. The example in (c) neglects this transformed area.

The above equation can be more simply written as

$$I_e = I_{cr} + (M_{cr}/M_a)^3 (I_g - I_{cr})$$

In this form notice that M_{cr}/M_a is most important when M_a is not too far different from M_{cr}.

The use of I_e gives the immediate deflection when used in the usual formulas for elastic deflections. Both shrinkage and creep will result in substantial increases unless the load is primarily a transient one. The Code requires the following multiplier to establish the *increase* in the deflection (added to the basic deflection) caused by the usual load carried:

$$\lambda = \xi/(1 + 50\rho') \quad \text{(Code Eq. 9.10)}$$

ρ' is the compression reinforcement percentage at midspan except for cantilever beams where it is determined at the support. ξ is a time-dependent factor for sustained loads that varies from 1.0 at 3 months to 2.0 at 5 years or more.

In computing deflections for continuous beams, I_e may be computed at midspan and at supports and the average used in calculating the immediate deflection (Code 9.5.2.4). Branson's Code discussion[5] indicates that a weighting of the two I_e values in terms of the proportion of negative to positive moments is also possible.

A later study by Committee 435[6] reports that "for controlled laboratory conditions, there is a 90% chance that the deflections of a particular beam will be within the range of 20% less to 30% more than the calculated value" as given by the 1971 Code and retained in the 1986 Code.

(b) Practical Complications   The Code procedure covers the simplest case, an immediate unfactored sustained load and its time effects, plus a later live load regarded as transient. The designer will recognize practical complications.

---

* For MPa SI units for both f_r and $\sqrt{f_c'}$, this becomes $f_r = 0.7\sqrt{f_c'}$; for customary units, both f_r and $\sqrt{f_c'}$ are in psi.

How much will the immediate deflection and time effects be increased by other cracking induced by normal construction loading? Should not much of the time effect be based on cracking that goes with the normal full live load, even if such loading is itself transient? When live load is heavy, does this not imply manufacturing or storage usage where much of this live load is probably applied *after* the early initial period, which means this portion of the creep starts on an older concrete.

For the heavy live load case, the graphs of Fig. 3.24 show the complications to be expected. In the long run the dead load deflection becomes that based on the maximum cracking condition plus any accumulated time effects. Hence, early smaller calculated deflections are useful only for (1) evaluating the maximum increase in deflections that partitions must accept and (2) for evaluating time effects. As nonexperts in this area the authors wonder whether the sustained load deflection base with I_e based on M_a for that sustained load is actually the best for determining time effects. Early construction loads or transient live loads will often induce cracking that lowers I_e, even where it is not a

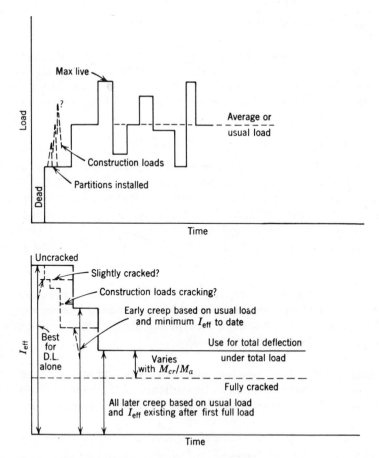

Figure 3.24 Relation between loading and I_{eff} in diagrammatic form.

part of the sustained load. It appears that the Code procedure might give an overly precise value of the early I_e, one that may well be too high in view of the actual physical complications.

(c) Example Based on Cantilever Beam

Given the beam and cross section shown in Fig. 3.25a and b with $f'_c = 3$ ksi and Grade 40 reinforcement.

1. Determine M_{cr}, I_g, and I_{cr}.
2. Assuming 20% of the live load is a long-term sustained load, calculate the long-term sustained load deflection of the free end tip.
3. Assuming that the full live load has been previously repeated numerous times, compute the instantaneous tip deflection as the load is increased from $1DL + 0.2LL$ to $1DL + 1LL$.
4. Assuming that $1DL + 0.2LL$ is in place before attachment of fragile partitions that might be damaged by large deflections, are such partitions likely to be damaged when the member is fully loaded with design live load and creep effects?

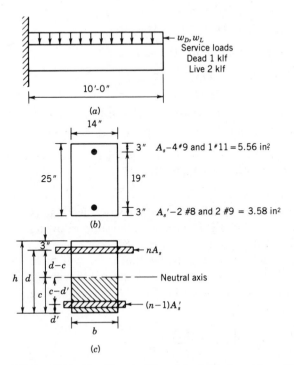

w_D, w_L
Service loads
Dead 1 klf
Live 2 klf

10'-0"

(a)

14"

3" A_s–4 #9 and 1 #11 = 5.56 in?

25"   19"

3" A_s'–2 #8 and 2 #9 = 3.58 in²

(b)

nA_s

$d-c$

h d Neutral axis

c $c-d'$

$(n-1)A_s'$

d'

b

(c)

Figure 3.25 Details considered in deflection of cantilever beam. (a) Support and loading. (b) Cross section. (c) Transformed cracked section.

Solution

(1) $M_{cr} = f_r I_g / y_t$ (Code Eq. 9.8)

$f_r = 7.5\sqrt{f_c'} = 7.5\sqrt{3000} = 411$ psi

$I_g = bh^3/12$ (neglecting the reinforcement as permitted in the Code Sec. 9.0 definition of I_g and y_t)

$= 14 \times 25^3/12 = 18,230$ in.^{4} Say, $I_g = 18,200$ in.^{4}

$y_t = h/2 = 25/2 = 12.5$

$M_{cr} = 411 \times 18\ 230/12.5 = 599,400$ lb-in. $= 599.4$ k-in.

$= 49.9$ k-ft

I_{cr} is the moment of inertia of the cracked, transformed section shown in Fig. 3.25c. The modular ratio n is used to transform the reinforcement into equivalent concrete.

$n = E_s/E_c = 29 \times 10^6/(57,000\sqrt{f_c'}) = 29 \times 10^6/(3.122 \times 10^6) = 9.3$

The transformed tension reinforcement $nA_s = 9.3 \times 5.56 = 51.71$ in.^{2} The transformed compression reinforcement corrected for the displaced concrete $(n - 1)A_s' = 8.3 \times 3.58 = 29.71$ in.^{2} Summing moments of the shaded areas about the neutral axis in Fig. 3.25c allows the depth to the neutral axis c to be determined.

$$nA_s(d - c) = (n - 1)A_s'(c - d') = (bc \cdot c/2)$$
$$(51.71)(22 - c) = (29.71)(c - 3) + (14/2)c^2$$

Solving the quadratic equation by completing the square

$$c^2 + 11.63(+ 5.82^2) = 175.25(+ 5.82^2)$$
$$c = 8.64 \text{ in.}$$
$$I_{cr} = bc^3/12 + bc(c/2)^2 + (n - 1)(A_s')(c - d')^2 + nA_s(d - c)^2$$
$$= 14 \times 8.64^3/12 + 14 \times 8.64 \times 4.32^2 + 29.71 \times (8.64 - 3)^2$$
$$+ 51.71 \times (22 - 8.64)^2$$
$$= 752 + 2257 + 945 + 9230 = 13,184 \text{ in.}^4$$

Say $I_{cr} = 13,200$ in.^{4}

(2) If 20% of the live load is sustained load, the effective sustained load is $w_s = w_D + 0.2w_L = 1 + 0.2(2) = 1.4$ klf. The moment at the support owing to this load is $M_s = w_s \ell^2/2 = 1.4 \times 10^2/2 = 70$ k-ft $= 840$ k-in.

Thus $M_s > M_{cr} = 599.4$ k-in. This means that the beam will have cracked under the sustained loading and the short-time cantilever tip deflection $\Delta = w\ell^4/(8EI)$ must be determined based on I_e corresponding to $M_a = M_s$. From Code Eq. 9.7

$$I_e = \left(\frac{M_{cr}}{M_a}\right)^3 I_g + \left[1 - \left(\frac{M_{cr}}{M_a}\right)^3\right] I_{cr}$$

$$I_e = \left(\frac{599.4}{840}\right)^3 18,200 + \left[1 - \left(\frac{599.4}{840}\right)^3\right] 13,200 = 15,000 \text{ in.}^4$$

Short time $\Delta_{inst} = w_s \ell^4/(8EI)$. Choosing all units in kips and inches, $\Delta_{inst} = (1.4/12)(10 \times 12)^4/(8 \times 3.122 \times 10^3 \times 15,000) = 0.065$ in. This short-term deflection is supplemented by the long-term deflection resulting from creep and shrinkage. This additional deflection is taken as the multiplier $\lambda = \xi/(1 + 50\rho')$ times the deflection under the sustained load. For this case, which has a very large amount of compression reinforcement, $\rho' = A_s'/bd = 3.58/14 \times 22 = 0.0116$. The values of ξ are tabulated in Code Sec. 9.5.2.5 and shown in Fig. 3.26. For very long-term loading ξ may be taken as 2.0 so that $\lambda = 2.0/(1 + 50 \times 0.0116) = 1.27$. Thus; time adds $\Delta_t = 1.27$ $\Delta_{inst} = 1.27 \times 0.065 = 0.082$ in. This *adds* to the initial or instantaneous deflection computed previously to give $\Delta = 0.065 + 0.082 = 0.147$ in. as the tip deflection under sustained load. In this very stiff beam the total sustained load deflection is only $\ell/816$ so that there is no need to further check the much larger deflection limits of Code Table 9.5b for this loading case.

(3) The repeated application of design live load $w_L = 2$ klf means that the beam must be well cracked. M_a for maximum load of $w_D + w_L$ is $(1.0 + 2.0)(10)^2/2 = 150$ k-ft $= 1800$ k-in. To find the instantaneous tip deflection as the load is increased from the sustained load level, $1DL + 0.2LL$ (A) to full load $1DL + 1LL$ (B), the differential $\Delta = \Delta_B - \Delta_A$ must be determined. Δ_A is the instantaneous deflection found in part (2), that is, $\Delta_A = 0.065$ in. Δ_B will depend on I_e for $M_a = 1800$ k-in. Using Code Eq. 9.7

$$I_e = \left(\frac{599.4}{1800}\right)^3 18,200 + \left[1 - \left(\frac{599.4}{1800}\right)^3\right] 13,200 = 13,400 \text{ in.}^4$$

At this service load level, for all practical purposes, $I_e = I_{cr}$ in this example. Code Eq. 9.7 has been frequently criticized as being very complex for members with substantial load or shrinkage cracks when I_{cr} is more realistic and considerably simpler.

$$\Delta_B = [(1.0 + 2.0)/12](10 \times 12)^4/(8 \times 3.122 \times 10^3 \times 13,400)$$
$$= 0.155 \text{ in.}$$

The instantaneous deflection when the live load is increased from $0.2LL$ to $1.0LL$ is found as

$$\Delta = \Delta_B - \Delta_A = 0.155 - 0.065 = 0.09 \text{ in.}$$

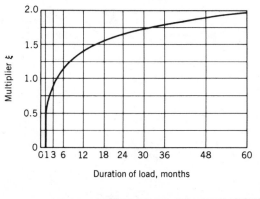

Figure 3.26 Factor for long-term deflection. (From Code Commentary.)

Such a deflection in this stiff member represents a value of $\ell/1330$ and is unlikely to cause any difficulty.

(4) Code Table 9.5b indicates that the deflection limitation for members supporting or attached to nonstructural elements likely to be damaged by large deflections is $\ell/480 = 10 \times 12/480 = 0.25$ in. In this case, the additional sustained load deflection after a long time owing to $1.0DL + 0.2LL$ was found in part (2) to be $\Delta_t = 0.082$ in. The instantaneous deflection when the load is increased from $1.0DL + 0.2LL$ to full service load of $1.0DL + 1.0LL$ was found in part (3) to be $\Delta = 0.09$ in. The loading case specified in Code Table 9.5b for this case corresponds to $\Delta_{crit} = \Delta + \Delta_t = 0.09 + 0.082 = 0.172$ in. This is less than the permissible $\ell/480 = 0.25$ in., therefore, there is no problem.

If this final Δ were more then $\ell_n/480$, a reduction would be proper to recognize the initial dead load deflection including creep occurring before partitions were built in. Also, a more exact calculation might replace the Code multiplier, as creep from loads applied several months after the beam becomes self supporting is considerably reduced, as indicated in Fig. 3.27.

It has already been noted at the close of Sec. 3.22a that deflections cannot be as accurately calculated as the previous numbers might suggest. The designer must make many estimates in computing a beam deflection, and the end result will only by chance be as good as indicated in Ref. 6. Some questions have already been noted concerning the loading sequence and the age at loading, neither of which is under the designer's control. It is also the *average* concrete strength that determines the E rather than the design f'_c. To the extent that curing differences do not completely obscure the picture, the average cylinder strength is the most usable value for deflection calculations. Much more data and study are needed in this area.

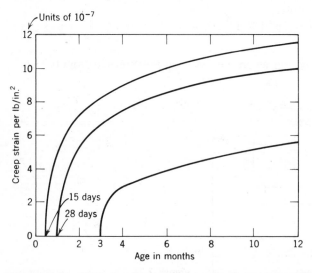

Figure 3.27 Variation in creep with age of concrete at loading.

TABLE 3.3 Multipliers for Additional Deflection
Owing to Shrinkage-plus-Creep

Duration of Loading	$A_s' = 0$	$A_s' = 0.5A_s$	$A_s' = A_s$
1 month	0.53	0.42	0.27
3 months	0.95	0.77	0.55
6 months	1.17	0.95	0.69
1 year	1.42	1.08	0.78
3 years	1.78	1.18	0.81
5 years	1.95	1.21	0.82

3.23 Further Data on Combined Creep and Shrinkage

In 1960 Yu and Winter[7] presented Table 3.3 to show their evaluation of the proper multiplier to be used in terms of the duration of loading and the amount of compressive reinforcement. As in the Code, these multipliers lead to the *additional* (not the total) deflection when applied to an initial deflection calculated from the cracked section I_{cr}. The 1983 ACI Code used essentially the values indicated for $A_s' = 0$ for its ξ values in Code 9.5.2.5.

ACI Committee 435 on Deflections is reported[8] as suggesting the Table 3.4 multipliers to give the additional deflection for either normal or lightweight concrete with original deflections based on I_{eff}, but adding: "For unusual cases (particularly for very shallow members such as canopies) or when very early application of load is necessary, it is suggested that shrinkage and creep deflections be considered separately and that the choice of appropriate shrinkage and creep coefficients be made by the designer (preferably from local test data)."*

3.24 Separate Calculation of Creep Deflection

The previous discussions have combined creep and shrinkage deflections because the numerical values of the variables are not easily determined in advance and exact methods are even more involved. The age at which loading starts is very important as Table 3.4 suggests. Extreme care must be taken when stress is imposed at very early ages as frequently happens with prestressed concrete. The practical simplification is to say that the creep deflection from two cumulative loadings is the sum of the separately computed deflections. But one must beware of any case where f_c becomes much in excess of $0.5f_c'$, because above $0.5f_c'$ to $0.6f_c'$ the influence of microcracking begins to build up and creep increases more rapidly than the stress increases. Fortunately, most sustained loads fall below this level.

---

* See p. 732 of Reference 8.

TABLE 3.4 Multiplier for Additional Long-Time Deflections Owing to Shrinkage and Creep

Concrete Strength f'_c at 28 days	Average Relative Humidity, Age When Loaded								
	100%			70%			50%		
	≤7d	14d	≥28d	≤7d	14d	≥28d	≤7d	14d	≥28d
2500 to 4000 psi	2.0	1.5	1.0	3.0	2.0	1.5	4.0	3.0	2.0
>4000 psi	1.5	1.0	0.7	2.5	1.8	1.2	3.5	2.5	1.5

For the period:	1 month or less	3 months	1 year	5 years or more
Use:	25%	50%	75%	100% of table values

The 50% humidity values may normally be used for lower relative humidities, as in a heated building, for example.

The designer in the past has frequently used a reduced modulus, sometimes called the sustained or effective modulus as a creep approximation. In effect, this is equivalent to assuming E_c effectively reduced to $E_c/2$ or $E_c/2.5$, which is a cruder approximation than those already discussed.

An excellent treatment of all of the factors affecting time-dependent deformations has been published by ACI Committee 209.[9]

3.25 Shrinkage Deflection Philosophy

Shrinkage deflection is discussed at some length here to emphasize the importance of nonload effects in structural concrete. For cantilever slabs, where deflection can be particularly serious, the empirical method of Sec. 3.26 is recommended. The semielastic method of Sec. 3.27, although not as accurate in predicting the results from available test data[10] on 14 beams, is still reasonable, considering the accuracy of the shrinkage coefficient that must be assumed in most cases. The error in predictions may be of interest, compared in terms of the distribution of each group:

Percent Error	0–9%	10–13%	14–20%	21–25%	Over 25%	Max	Mean
Empirical	11	0	1	1	1	29% low	0.989
Semielastic	2	4	4	2	1	44% high	1.064

The unequal intervals are selected to emphasize the different groupings of the results in the two cases.

The semielastic method is flexible in use and able to handle any shape, although sections cracked to an unknown depth become a trial-and-error solution after a first estimate of the extent of the cracking. The method can handle any assumed distribution of shrinkage, although the basic estimate of shrinkage will normally be a crude one at best. The basic process is equally usable for uniform or nonuniform temperature changes.

3.26 Branson's Empirical Method for Shrinkage of Uncracked Sections

This method is quite suitable for slabs with or without compression steel. The cover over the bars is not one of the variables included, and the equations probably fit unusual cover conditions less accurately. However, the design value of the assumed shrinkage (Sec. 2.8) is the largest probable source of error for any method.

For ρ expressed as a percent of steel rather than a ratio, the curvature is expressed for a singly reinforced member in terms of the shrinkage strain ε_{cs} as

$$\theta_{cs} = 0.7 \frac{\varepsilon_{cs}}{h} \sqrt[3]{\rho} = \text{(coefficient)} \; \varepsilon_{cs}/h$$

or

ρ	Coefficient
0.5%	0.56
1	0.70
1.5	0.80
2	0.88

With θ_{cs} known and uniform over the length, this leads to the very simple deflection calculation of $\theta_{cs}\ell^2/8$ for a simple span and $\theta_{cs}\ell^2/2$ for a cantilever.

With compression reinforcement the curvature is

$$\theta_{cs} = 0.7 \frac{\varepsilon_{cs}}{h} \sqrt[3]{\rho - \rho'} \; \sqrt{(\rho - \rho')/\rho} \quad \text{for} \quad \rho - \rho' \gtrsim 3\%$$

$$\theta_{cs} = \varepsilon_{cs}/h \quad \text{for} \quad \rho - \rho' > 3\%$$

The method gives no information as to stresses that, because of creep, are rather uncertain.

3.27 Semielastic Method for Shrinkage or Temperature

(a) General Theory   The shrinkage deflection can be approximated for a given uniform shrinkage by using a reduced modulus of elasticity in lieu of creep

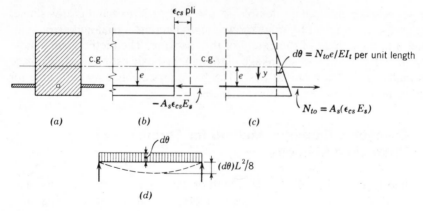

Figure 3.28 Shrinkage deflection calculation. (*a*) Transformed area. (*b*) Free shrinkage of concrete, forced shortening of steel. (*c*) Elimination of forcing force. (*d*) Deflection from uniform shrinkage.

effects, as suggested in Sec. 2.7. The usual transformed area under elastic stresses can be used in several ways under this artificial assumption of elastic behavior of the beam. A method of superposition is used here.

If a member is symmetrically reinforced, it shortens as the concrete shrinks but does not become curved or deflect laterally. Consider a member with unsymmetrical reinforcing as shown in Fig. 3.28*a*. Assume the shrinkage ε_{cs} (inch per inch) shortens the concrete uniformly, without restraint from the steel,* as in Fig. 3.28*b*; this can be accomplished by applying an external compressive force $A_s(\varepsilon_{cs}E_s)$ on the steel to shorten it an equal amount. This gives $f_{c1} = 0$ and $f_{s1} = -\varepsilon_{cs}E_s$, with the minus sign representing compression. Then release the restraint of this unwanted external force by applying an equal and opposite (tensile) force N_{t0} to the entire transformed area (A_t, I_t) as shown in Fig. 3.28*c*. The stresses for this loading at any distance y from the centroid are given, as for any eccentrically loaded member, by

$$f_{cy} = \frac{N_{t0}}{A_t} + \frac{(N_{t0}e)y}{I_t}$$

for tension positive and y positive toward the steel. The total concrete stresses are the algebraic sum of the corresponding stresses from the two cases and the steel stress can be obtained similarly, of course, with $f_{s2} = nf_c$ at the level of the steel.

For curvature, the moment $N_{t0}e$ is the only term of significance, leading to $d\theta = M \, ds/EI_t = N_{t0}e \, ds/EI_t$ or $N_{t0}e/EI_t$ per unit length. For constant reinforcing the member shape is a circular curve which, for a simply supported member

---

* The basic idea is to retain or develop a plane section, and an unrotated plane simplifies the later steps. In the given case the elongation of the concrete to match the unstressed steel also would be an easy starting step.

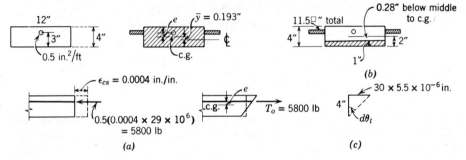

Figure 3.29 Shrinkage deflection calculation. (*a*) With uncracked section. (*b*) Transformed area of cracked section. (*c*) Temperature deformation pli.

as in Fig. 3.28*d* leads to maximum deflection $y = (d\theta)\ell^2/8$ or for a cantilever $y = (d\theta)\ell^2/2$. No significant modification of the method is needed for steel at more than one level. The uncertainties of the method come from (1) the assumption of a uniform ε_{cs} and its magnitude, (2) the assumption of the value of n to include the effect of creep, and (3) the necessity of assuming the member either cracked or uncracked in establishing A_t and I_t for the calculation of $d\theta$.

For beams or slabs with varying A_s over the length, deflections can be calculated by area-moment methods by noting that a $d\theta$ curve simply takes the place of an M/EI curve. The sign of $d\theta$ is positive where A_s is in the bottom of the beam, negative with top tension steel.

It should be noted that compression steel reduces shrinkage deflection, eliminating it completely for uniform shrinkage in slabs and rectangular bams when $A'_s = A_s$. Good practice calls for A'_s in *all* cantilever slabs to avoid shrinkage deflection and to reduce creep deflection from dead load.

(b) Example of Shrinkage Deflection, Uncracked Section

A canopy 4 in. thick (Fig. 3.29) extends 10 ft as a cantilever. If $E = 3.1 \times 10^6$ psi, $n = 9$, find the deflection at the outer end from uniform shrinkage $\varepsilon_{cs} = 0.0004$, using $2.5n$, say, 23 to approximate the creep effects and assuming an uncracked slab section and a horizontal tangent at the support.

Solution

Transform the area as in Fig. 3.29*a* and locate the centroid:

$$y = \frac{11.5 \times 1.0}{11.5 + 12 \times 4} = 0.193 \text{ in. above middle}$$

$$N_{t0} = 0.5(0.0004 \times 29 \times 10^6) = 5800 \text{ lb at } e = 3 - 2.19 = 0.81 \text{ in.}$$

$$M = N_{t0}e = -5800 \times 0.81 = -4680 \text{ in.-lb}$$

$$I_t = 1/12 \times 12 \times 4^3 + (12 \times 4)0.193^2 + 11.5 \times 0.81^2$$

$$= 64 + 1.8 + 7.5 = 73 \text{ in.}^4$$

Effective $E = 3.1 \times 10^6/2.5 = 1.24 \times 10^6$ psi
$$d\theta = M/EI_t = -4680/(1.24 \times 10^6 \times 73)$$
$$= -52 \times 10^{-6} \text{ uniform over } \ell$$
$$y = (d\theta)\ell^2/2 = -52 \times 10^{-6} \times 120^2/2 = -0.38 \text{ in.}$$

More significant digits are not justified.

(c) Stresses from Shrinkage   Shrinkage stresses in structural members are rarely large unless member deformations are restrained, but shrinkage stresses add to flexural tension to induce member cracking earlier than would otherwise be expected. In investigating shrinkage stresses, it is the tensile stresses that are the more important.

Before cracking it would be proper to consider all the transformed area shown in the upper sketch of Fig. 3.29a. In this uncracked slab the shrinkage stresses can be found as the sum of those indicated at the bottom of Fig. 3.29a. In the first sketch, $f_{c1} = 0$, $f_{s1} = -5800/0.5 = -11,600$ psi (minus representing compression). In the second sketch, for axial load:

$$f_{c2} = N/A = +5800/(11.5 + 12 \times 4) = +97.5 \text{ psi}$$
$$f_{s2} = nf_{c2} = +23 \times 97.5 = +2240 \text{ psi}$$

and for flexure, with y positive toward the compression face:

$$f_{c3 \text{ top}} = My/I = -4680(-1.81)/73 = +116 \text{ psi}$$
$$f_{c3 \text{ bott}} = -4680 \times 2.19/73 = -140 \text{ psi}$$
$$f_{s3} = 23(-4680)(-0.81)/73 = +1190 \text{ psi}$$
$$\text{Total } f_{c,\text{top}} = 0 + 97.5 + 116 = +214 \text{ psi (tens.)} < \text{cracking}$$
$$f_{c,\text{bott}} = 0 + 97.5 - 140 = -43 \text{ psi (comp.)}$$
$$f_s = -11,600 + 2240 + 1190 = -8260 \text{ psi (comp.)}$$

However, if other loading has cracked the concrete, the shaded transformed area in Fig. 3.29a is not valid; all resulting tension areas already cracked must be omitted.

(d) Example of Shrinkage Deflection, Cracked Section

The above canopy may be considered as cracked to a depth of 3 in. from earlier live loading. Calculate shrinkage deflection.

Solution

For the transformed area* of Fig. 3.29b, with all the concrete below the crack as effective. $A = 12 \times 1 + 11.5 = 23.5$ in.^{2}

---

* Again the steel is transformed as $2.5 n A_s$ to recognize the earlier precompression and creep although once cracked the value $n A_s$ is correct at the crack locations.

$\bar{y}$ from middepth $= (12 \times 1.5 - 11.5 \times 1)/23.5 = 0.28$ in. (below)

$I = 12 \times 1^3/12 + 12(1.72 - 0.50)^2 + 1.28^2 = 37.7$, say, 38 in.^{4}

$M = N_{t0}e = -5800 \times 1.28 = -7440$ in.-lb.

Stress at bottom of original crack $= N/A + My/I = +5800/23.5 - 7440 \times 0.72/38 = 247 - 141 = 106$ psi (tens.)

This stress is less than enough to crack the concrete (usually taken as $7.5\sqrt{f'_c}$), and no tension can exist at the crack itself. Assumed section is O.K. The deflection calculation could be next, but the other stresses will be calculated as a matter of possible interest.

$$f_{c,bott} = 247 - 7440 \times 1.72/38 = 247 - 337 = -90 \text{ psi(comp.)}$$
$$f_s = -5800/0.5 + 23 \times 247 + 23 \times 7440 \times 1.28/38$$
$$= -11,600 + 5670 + 5750 = -180 \text{ psi(comp.)}$$

Notice that with a cracked section the shrinkage stresses tend to dissipate when no A'_s is present, but this very fact means an increase in curvature.

$$\theta_{cs} = M/EI = -7440/(1.24 \times 10^6 \times 38) = -157 \times 10^{-6}$$
$$y = \theta_{cs}\ell^2/2 = -157 \times 10^{-6} \times 120^2/2 = -1.13 \text{ in.}$$

The probable deflection lies between -1.13 and the original -0.38 in., because the entire length of the beam will probably not be cracked from loading (probably nearer the larger value because the most influential part of the beam is the one which load will crack).

Deflection will be further increased by a differential temperature from the sun on the top surface. For example, assume that the direct sun raises the temperature of the top to 30°F above the underside with a linear temperature gradient between top and bottom. The coefficient of expansion of steel is 6.5×10^{-6} per degree (F) and of concrete around 5.5×10^6, making this difference of little importance in most cases. As shown in Fig. 3.29c,

$$d\theta_t = 5.5 \times 10^{-6} \times 30/4 = 4.1 \times 10^{-5}$$
$$\text{Temperature } y = -d\theta_t\ell^2/2 = 4.1 \times 10^{-5} \times 120^2/2 = -295 \times 10^{-3}$$
$$= -0.29 \text{ in.}$$

The total addition to deflection from dead or live load and creep is thus $-1.13 - 0.29 = -1.42$ in.

(e) Temperature Curvature and Stresses   As a sample for discussion, consider a roof frame where the upper part of the beam is exposed above the roof for architectural reasons and the lower part is within the enclosure. With seasonal temperature changes outside and air conditioning inside, the concrete temperature shifts around. Assume an analysis is needed when the internal concrete temperature is that of Fig. 3.30. Any distribution, straight line or curving, uniform or nonuniform, is possible under the method used previously.

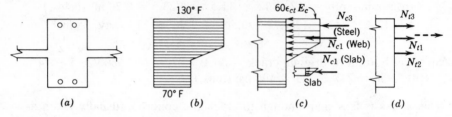

Figure 3.30 Temperature stress example. (*a*) Beam. (*b*) Temperature in concrete. (*c*) Step 1 forces. (*d*) Step 2 forces.

As part of a frame, both changes in length and curvature are important; both should be relative to the basic analysis conditions. For simplicity here, but not essential to the method, consider that creep has adjusted the frame to an average 70° temperature. Step 1 would add on the transformed area the forces required to change the beam and slab section to a plane one, as if all were at 70°, that is, $f_{c1} = \varepsilon_{ct} (130 - 70)E_c$ at the top. E_c is the most troublesome term, probably the elastic value for daily variations, a reduced value for seasonal variations. (In an extreme case parts of the temperature could be assigned to each.) Step 2 is to apply reversed forces (or their resultant) to the transformed area, cracked or uncracked as is realistic. For a statically determinate system, elongation, angle change, and stresses of interest are determined from these reversed forces. For a continuous frame one must introduce these elongations and angle changes into the system as part of the applied load. Always, for deformations, deflections, or stresses it is the sum of step 1 and step 2 values that must be considered, plus any statical redundants induced by them.

3.28 Extra Deflection of Cantilevers and Overhanging Ends

In contrast to the reduction in deflection from continuity, cantilevers and overchanging ends often have their normal deflections y_w increased by rotations of their supported ends, as indicated in Fig. 3.31.

3.29 Effect of Inelastic Action on Deflection

The calculation of deflections is more complex when inelastic action occurs. If the relation between moment and angle change is known, as in Fig. 11.1, it is

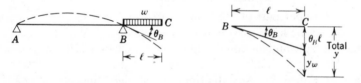

Figure 3.31 Extra cantilever deflection caused by rotation of supports.

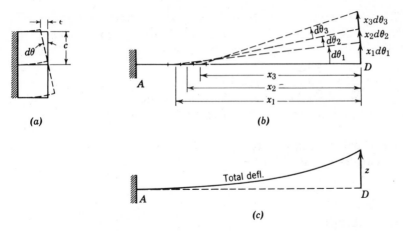

Figure 3.32 Deflection from deformation of beam elements, positive moment.

possible to calculate deflections by a summation process in which the real $d\theta$ (or ϕ) for each given moment is obtained from the curve. Each of these $d\theta$ values contributes to the member deflection just as the $M \, ds/EI$ increments do in ordinary area-moment procedures. The area-moment calculation simply changes from $\Sigma x(M \, ds/EI)$ to $\Sigma x \, d\theta$, as indicated in Fig. 3.32. Of course, in indeterminate structures the process is more involved because the moment itself is influenced (shifted) by inelastic deformations.

When the reinforcing steel reaches the yield point, ε_s and $d\theta$ increase rapidly with small changes in load. The increasing steel deformation crowds the neutral axis toward the compression face at this section and brings about increasing concrete deformations and finally a secondary compression failure. For an ordinary simple-span slab the difference between the load at steel yield point and the failure load will be only a few percent, say, 5% or possibly 10%. The deflection before failure will be many times that at the first yielding of the steel. The total relative deflection will depend on the percentage of steel, the span, and the type of loading, but ten times the yield point deflection would be quite usual. This adding deflection will be almost entirely due to the large angle change θ_2 at the yielding section, the slab tending to fold as in Fig. 3.33 about this point as though it constituted a stiff hinge. This is the general basis for limit design (Chapter 11) and the yield-line method (Chapter 13).

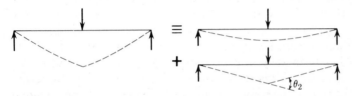

Figure 3.33 Beam deflection after a plastic hinge forms equals elastic deflection plus hinge deflection.

Except for the general uncertainties as to partially elastic action prior to yielding, the total deflection is easy to calculate for any given θ_2 value. If the moment-ϕ diagram became truly horizontal, θ_2 would increase indefinitely to failure. If, however, a definite known relation exists between M and the angle change $d\theta$ or ϕ per unit length, as in Fig. 11.1 the simple beam moment diagram permits $d\theta$ to be established for each section and particularly for those critical sections near the point of maximum moment, that is, at the so-called plastic hinge section.

The area-moment theorems are restated in Reference 11 in a manner consistent with inelastic and plastic hinge action.

3.30 Crack Width

(a) Code Requirement Related to Crack Width   Both for appearance and for resistance to corrosion, the Code discourages large crack widths without dealing directly in terms of this width.

In 1968 Gergely and Lutz developed an equation[12] for expected maximum crack width at the tension face of a flexural member, as follows:

$$w = 0.076\beta f_s \sqrt[3]{d_c A}$$

where w = expected maximum crack width in 0.001 in. units

  β = ratio of distances to the neutral axis from the extreme tension fiber and from the centroid of A_s

  f_s = steel stress in ksi

  d_c = cover of outermost bar of A_s measured to the center of bar

  A = tension area per bar measured as in Fig. 3.34, that is, equal to $\frac{1}{5}$ of the shaded area for the 5 bars used, centered on c.g. of bar areas.

The Code (10.6.4) uses this equation with β taken as 1.2 as the basis for computing limiting values of

$$z = f_s \sqrt[3]{d_c A}$$

which are set at 175 kips per inch (30 MN/m) for interior exposure and 145 kips per inch (20 MN/m) for exterior exposure. These values correspond to w values of 16 and 13, respectively, that is, to 0.016 and 0.013 in. Because crack widths scatter over $\pm 50\%$, the object of specifying z is to push toward more and smaller bars rather than toward fewer large ones. In increasing from Grade 40 to Grade 60 bars it has been easy to keep bars as large as before and to use fewer of them. Better crack widths result if smaller bars are used[18], and this is the remedy where the requirement for z is not met. Both positive and negative moment bars must be checked. For mixed bar sizes use the number of bars as A_s/(area of largest bar).

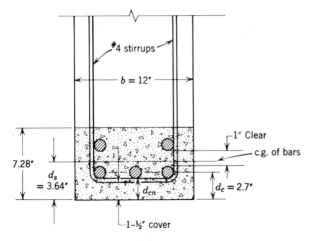

Figure 3.34 For computing the area A, the shaded area is divided by the number of bars which is five. (Modified from ACI Commentary.)

This limitation on z was never intended to control crack width between the A_s level and the neutral axis. For d greater than 30 in., see the face steel requirements discussed in Sec. 4.9.

The Code permits the use of $0.6f_y$ in lieu of a calculated f_s if the designer wishes, and it does not require this check when Grade 40 bars are used. Appendix Fig. B.8 uses the $0.6f_y$ and gives curves that fix the maximum value of the third variable when any two are known.[13]

The value of β in the Gergely–Lutz equation may logically be used in establishing the limiting z in lieu of the 1.2 value used in fixing the Code limits. The Commentary suggests for one-way slabs that 1.35 would be better than 1.2, which would mean that the permissible z given would be multiplied by the ratio 1.2/1.35, to hold the same nominal crack width. Something similarly done for a beam, using a more "exact" β, might be regarded as simply using a more refined method than that in the Code.

Research by Nawy[14,15] has indicated that cracking in two-way slabs follows a pattern somewhat different, a pattern much influenced by the spacing of the reinforcement grid in the slab. This agrees with the authors' observation from research testing that, with transverse steel somewhat widely spaced, flexural cracks nearly always develop first over the stirrups or ties rather than between them. Transverse steel exerts a considerable influence on crack spacing.

(b) Crack Research at University of Texas   The internal shape of cracks were investigated[16] by holding a beam under known loads while flexural cracks in a constant moment region were filled with a colored epoxy. The beams were then sawed to cut across the cracks directly over the negative moment bars and permit measurements with a microscope. Cracks varied greatly in thickness and contours, and the mean values were used for plotting. Such mean widths for six beams are superimposed in two groups (30 ksi and 20 ksi stresses) in Fig. 3.35. The surface crack width varies almost linearly with the stress and the

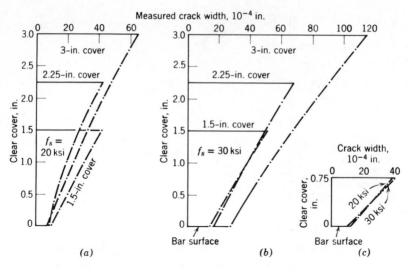

Figure 3.35 Measured widths of cracks from bar surface to tension face of member. The shape of crack varied more than these comparisons suggest. (*a*) Beams at 20 ksi. (*b*) Beams at 30 ksi. (*c*) Thick slabs.

cover over the bar. The interior crack width *at the surface of the bar* is very narrow in comparison. Twenty cycles of loading made little difference in any of these findings.

Corrosion studies related to daily salt spraying of stressed beams (outdoors), a very serious exposure, are reported in Reference 17. A water-cement ratio of 6.25 gallons per sack (94 lb) and 2 in. clear cover showed some corrosion in two years, only 10 or 15% more in stressed sections than in unloaded and uncracked sections. It appeared that a *w/c* ratio not exceeding 5.5 gallons per sack is needed for protection and that the ratio of clear cover to bar diameter is also critical. The indication, based on observing #6, #8, and #11 bars, was that cover of 2.5 to 3.0 bar diameters might be necessary, although this was from extrapolated data, especially with respect to time, with tests running only two years. The results are necessarily limited to chloride (saltwater) corrosion and probably do not apply to carbonation problems.

Selected References

1. A. H. Mattock and L. B. Kriz, "Ultimate Strength of Nonrectangular Structural Concrete Members," *ACI Jour., Proc., 57,* Jan. 1961; p. 737.

2. R. Park and T. Paulay, *"Reinforced Concrete Structures,"* John Wiley & Sons, New York, 1975.

3. ACI Committee 435, "Allowable Deflections," *ACI Jour., 65,* No. 6, June 1968, p. 433.

4. *Deflection of Reinforced Concrete Members,* Bulletin ST-70, Portland Cement Assn., 1947.

5. D. E. Branson, Discussion of "Proposed Revision of the ACI 318-63: Building Code Requirements for Reinforced Concrete," by ACI Committee 318, *ACI Jour., 67,* No. 9, Sept. 1970, p. 692.

6. ACI Committee 435, "Variability of Deflections of Simply Supported Reinforced Concrete Beams," *ACI Jour., 69,* No. 1, Jan. 1972, p. 29.

7. W. W. Yu and G. Winter, "Instantaneous and Long-Time Deflections of Reinforced Concrete Beams Under Working Loads," *ACI Jour., 57,* No. 1, July 1960, p. 29.

8. D. E. Branson, "Design Procedures for Computing Deflections," *ACI Jour., 65,* No. 9, Sept. 1968, p. 730.

9. ACI Committee 209, "Prediction of Creep, Shrinkage, and Temperature Effects in Concrete Structures," *Designing for Creep and Shrinkage in Concrete Structures,* ACI Publication SP-76, Amer. Concrete Inst., Detroit, 1982, pp. 193–300.

10. ACI Committee 435, "Deflections of Reinforced Concrete Flexural Members," *ACI Jour., 63,* No. 6, Jan. 1966, p. 637.

11. G. C. Ernst, "Ultimate Slopes and Deflections—A Brief for Limit Design," *ASCE Trans., 121,* 1956, p. 605.

12. P. Gergely and L. A. Lutz, "Maximum Crack Width in Reinforced Concrete Flexural Members," *Causes, Mechanism, and Control of Cracking in Concrete,* SP-20, Amer. Concrete Inst., Detroit, 1968, p. 87.

13. P. Gergely, "Distribution of Reinforcement for Crack Control," *ACI Jour., 69,* No. 5, May 1972, p. 275.

14. E. G. Nawy and G. S. Orenstein, "Crack Width Control in Reinforced Concrete Two-Way Slabs," *ASCE Proc., 96,* ST-3, Mar. 1970, p. 701.

15. E. G. Nawy and K. W. Blair, "Further Studies in Flexural Crack Control in Structural Slab Systems," *Cracking, Deflection, and Ultimate Load of Concrete Slab Systems,* SP-30, Amer. Concrete Inst., Detroit, 1971, p. 1.

16. S. I. Husain and P. M. Ferguson, "Flexural Crack Width at Bars in Reinforced Concrete Beams." Research *Report 104-1F,* Center for Highway Research, The Univ. of Texas at Austin, June 1968, 35 pp.

17. E. Atimtay and P. M. Ferguson, "Early Chloride Corrosion of Reinforced Concrete—A Test Report," *ACI Jour., 70,* No. 9, Sept. 1973, p. 606.

18. G. D. Base, J. B. Read, A. W. Beeby, and H. J. P. Taylor, "An Investigation of the Crack Control Characteristics of Various Types of Bars in Reinforced Concrete Beams," Research Report No. 18, Cement and Concrete Assn., London, 1966.

Problems

GENERAL NOTE *Tables 2.1 and 2.2 in Sec. 2.9 list bar sizes, areas, and strengths. Necessary load factors are given in Sec. 1.5e (usually Code Eq. 9.1 unless otherwise specified) and ϕ factors in Sec. 1.5f. Beam constants for many cases are given in Table 3.1 and Tables B.1 and B.2 in Appendix B.*

PROB. 3.1. If $f'_c = 3000$ psi and steel is Grade 60, find ϕM_n, the moment capacity of the beam of Fig. 3.36 if the A_s is:

(a) 2-#8. (c) 4-#9.
(b) 4-#8. (d) 4-#10.

Figure 3.36 Beam for Prob. 3.1.

PROB. 3.2. Find moment capacity ϕM_n of the beams of Fig. 3.37, assuming sketch (a) for a and b, sketch (b) for c, d.

(a) Grade 40 steel. (c) Grade 300 steel, $f_y = 300$ MPa.
(b) Grade 60 steel. (d) Grade 400 steel, $f_y = 400$ MPa.

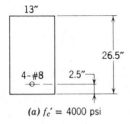

(a) $f'_c = 4000$ psi (b) $f'_c = 25$ MPa **Figure 3.37** Beams of Prob. 3.2.

PROB. 3.3. If $f'_c = 4000$ psi and steel is Grade 60, calculate the moment capacity ϕM_n of the beam of Fig. 3.38, assuming the steel as follows:

(a) 3-#8. (c) 4-#11.
(b) 6-#8. (d) 6-#11.

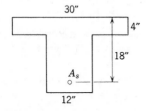

Figure 3.38 Beam of Prob. 3.3.

PROB. 3.4. Calculate ϕM_n on the beams of Fig. 3.39, assuming sketch (a) for a, b, sketch (b) for c, d; A_s as follows:

(a) 4-#35, Grade 300. (c) 4-#11, Grade 40.
(b) 3-#25, Grade 300. (d) 3-#8, Grade 40.

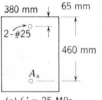

(a) $f'_c = 25$ MPa (b) $f'_c = 4000$ psi **Figure 3.39** Beam of Prob. 3.4.

PROB. 3.5. Check the adequacy of the beam of Fig. 3.40 in flexure assuming $f'_c = 5000$ psi, Grade 40 bars, service live load of 1200 plf and dead load (including beam weight) of 500 plf. (*Suggestion:* Compare calculated ϕM_n with required ϕM_n)

Figure 3.40 Beam for Prob. 3.5.

PROB. 3.6. Find the moment capacity ϕM_n of the joist of Fig. 3.41 if $f'_c = 4000$ psi and A_s is 2-#6 bars of Grade 60.

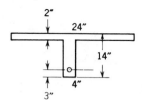

Figure 3.41 Joist of Prob. 3.6.

PROB. 3.7. Find the allowable moment ϕM_n on a 1-ft strip of the slab of Fig. 3.12 if the bars are made #7 at 7 in. on centers and $d = 5$ in., $f'_c = 3000$ psi, Grade 60 steel.
PROB. 3.8. Find allowable moment ϕM_n for the beam of Fig. 3.42a if $f'_c = 5000$ psi and Grade 60 steel is used. Ignore the lack of symmetry.
(*a*) Omitting the 2-#8 bars.
(*b*) Including the 2-#8 bars.
Find the above but use Fig. 3.42b with SI data (slightly different) for $f'_c = 35$ MPa and Grade 400 steel.
(*c*) Omitting top bars.
(*d*) Including top bars.

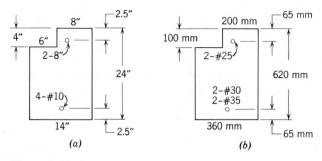

Figure 3.42 Irregular beams for Prob. 3.8.

PROB. 3.9. Prob. 3.8 except consider the beam of Fig. 3.43.
(*a*) Omitting the 2-#9.
(*b*) Including the 2-#9.

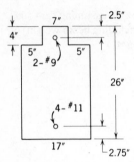

Figure 3.43 Irregular beam for Prob. 3.9.

PROB. 3.10. If $f'_c = 5000$ psi and steel is Grade 60, what is the moment capacity ϕM_n of the beam of Fig. 3.44 when each leg is reinforced with:
(a) 1-#7.　　　(b) 1-#10.

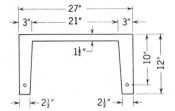

Figure 3.44 Beam for Prob. 3.10.

PROB. 3.11. If the precast section of Fig. 3.45 has $f'_c = 5000$ psi, Grade 60 steel, 1-#8 bar in each leg, calculate the moment capacity ϕM_n. Flange $h_f = 1.5$ in.

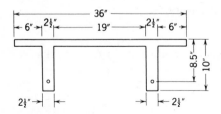

Figure 3.45 Precast section for Prob. 3.11.

PROB. 3.12.
(a) Calculate the moment capacity ϕM_n (Fig. 3.46) of a strip of slab 12 in. wide if $f'_c = 4000$ psi and the steel is Grade 60.
(b) What would be the service or working live load per square foot if the slab were used on a 10-ft simple span? Use the Code load factors. Concrete weight may be taken at 150 pcf which includes the weight of reinforcement.

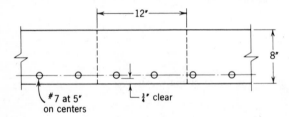

Figure 3.46 Slab for Prob. 3.12.

PROB. 3.13. A rectangular beam has $b = 15$ in., $d = 18$ in., $f'_c = 3000$ psi, Grade 40 steel, $A_s = 6$-#11.

(a) Calculate the moment capacity ϕM_n.

(b) Calculate the moment capacity ϕM_n if one adds $A'_s = 3$-#11 with cover d' to center of steel $= 2.5$ in.

PROB. 3.14.

(a) Determine the long-term deflection of the beam of Prob. 3.5 under service dead load plus a sustained live load of one third the service live load.

(b) Determine the instantaneous deflection that will take place when the remaining portion of the service live load is added to the loading in (a).

(c) Assuming this member is a free-standing exterior cantilever floor that is not supporting or attached to nonstructural elements, is it within the deflection limits of code Table 9.5b?

4

DESIGN FOR FLEXURE

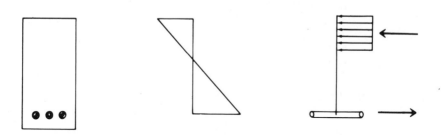

4.1　General Procedures

Analysis and design are related problems that nevertheless involve fundamental differences and require different techniques. It is essential that the student clearly understands these differences.

In analysis or review for allowable loads or moments the engineer deals with given beams with both dimensions and steel known. Each case has a neutral axis unique to those particular dimensions and independent of the engineer's desires.

In design, loads and material strengths are known and some or all the dimensions remain to be fixed. Designers have some control over the location of the neutral axis. They can shift it where they want it to the extent that changes in dimensions can alter the balance between tension and compression areas.

After loads are known and the layout of the structure or structural element has been established,* the maximum bending moment can be determined. The member design for moment involves three separate steps:

1. Choice of beam cross section.
2. Choice of reinforcing steel at point or points of maximum moment.
3. Determination of points where bars are no longer needed for moment, that is, points for bending or stopping bars.

Often factors other than moment determine the choice of the beam cross section, such as shear (Chapter 5). Such interrelationships are discussed later.

---

* The choice of proper design loads and the layout of a structure to support these loads efficiently is a major part of the design. The choice of loads and the general problem of framing layout are not covered in this book. Chapter 9 has a brief discussion of beam-and-girder framing and the proper arrangement of loading for maximum moment on continuous one-way slabs and beams. Various types of slab construction are discussed in Chapters 12, 15, and 16.

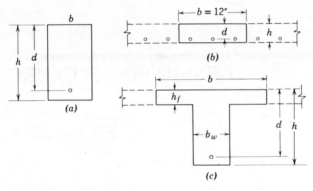

Figure 4.1 Typical reinforced concrete beams. (*a*) Rectangular. (*b*) Slab.
(*c*) T-beam.

In rectangular beams (Fig. 4.1*a*), both width *b* and depth *d* to steel are to be chosen. In slabs (Fig. 4.1*b*), the strip width *b* is fixed and only the depth *d* or the overall thickness *h* can be selected. In T-beams (Fig. 4.1*c*), the flange thickness h_f and flange width *b* are determined, usually by a slab already designed, thus leaving the stem width b_w and the depth *d* to the steel to be selected. The choice of b_w and *d* is not usually determined by the moment alone.

In double-reinforced beams the compression steel is required because *b* and *d* have been chosen smaller than the normal requirements or because it is desirable to reduce creep deflections. However, when compression steel in definite amounts is desired for protection against deflection or otherwise, the size of double-reinforced beams can be based on moment requirements.

For any beam shape the design of steel, once the moment and beam size are known, is dependent on a proper choice of the internal lever arm *z* between resisting tension and compression forces. In rectangular beams (and slabs) and in double-reinforced beams, the analytical solution for *z* is feasible, but the resulting quadratic equation is usually not conducive to an accurate and rapid solution. Hence the authors suggest that the cut-and-try procedure necessary for T-beams and irregular shapes be used for all shapes. The value of *z* is first estimated, the corresponding approximate A_s is calculated, and from this A_s a better value of *z* is established; in turn, this better *z* leads to a better and usually satisfactory value of A_s. Convergence is rapid.

The bending or stopping of bars is determined by the maximum moments to be resisted at various points along the beam. Bending of bars is discussed separately from choice of beam and steel (Sec. 4.11).

4.2 Rectangular Beams—Moment Design

The economical dimensions for a rectangular beam are not sharply defined. A shallow beam is expensive because of the weight of steel required; on long spans it may run into deflection problems. A deep beam is more economical of steel but costs more for side forms and uses up headroom, which means greater story heights. The ratio of overall depths to span listed in Table 3.2 in Sec. 3.8 (copied from Code Table 9.5a) can serve as a guide.

This chapter deals with beam sizes established in various manners. It is almost wholly concerned with strength, slightly concerned with deflection, and not concerned with overall economy. However, it will emphasize methods that conserve steel for the particular beam size chosen. The general beam size will depend on the percentage of steel reinforcement selected. Figure 3.8 shows that a given cross section can accommodate a wide range of moments depending on the ρ used. Under the code provisions ρ can vary widely, from ρ_{min} to ρ_{max}. With $f'_c = 4000$ psi and Grade 60 reinforcement this allows selection of ρ from 0.0033 to 0.0214. Conversely, for a given ρ there are a wide range of interrelated b and d values that will carry a given moment. An experienced designer usually begins the design problem with a general value of ρ in mind to obtain both economy and constructability (by minimizing the reinforcement congestion and concrete placement difficulties that usually accompany high ρ values).

The maximum ρ ratios developed in Sec. 3.8 (Table 3.1) are absolute limits on the steel that may be considered effective because of danger from brittle-type failures.

The minimum reinforcement of $\rho = 200/f_y$ (Code 10.5.1) is necessary to insure that the member does not lose some strength when it first cracks. This requirement does not apply if the A_s used is 4/3 that required by analysis, because this implies a lowered probability that the cracking load will ever be reached, in other words, adequate safety. Slabs of uniform thickness are also exempt, but the authors question this exemption in those locations where shears are near the limiting value. A small ρ and a maximum shear make a bad combination.

Although it is not possible to formalize the design process, the following steps are usually required:*

1. Select the material grades (f'_c, f_y) and the approximate reinforcement percentage.

2. Make a preliminary estimate of beam size so that the beam self-weight can be included in the dead load. As computations proceed and the beam size becomes better established, upgrade this estimate and iterate through the overall process.

3. Using the best estimate of loads, apply the load factors to obtain factored loads. Carry out a conventional structural analysis to determine design

---

* In these introductory problems, the authors will often provide a substantial amount of information in the problem statement to simplify the task for the beginning student. All of these steps may not be required.

moments, M_u. Since $M_u \leqq \phi M_n$, determine the required nominal moment capacity $M_n = M_u/\phi$.

4. Based on the decisions in step 1, determine an approximate design constant k_n where $M_n = k_n bd^2$. k_n can be determined by the basic procedures of Sec. 3.9 for any ρ value selected. Values of k_n corresponding to ρ_{max} and to $\rho = 0.18 f'_c/f_y$ are given in Table 3.1. Other combinations are given in Fig. B.1 and Tables B.1 and B.2.

5. For the required M_n and selected k_n, determine b and d combinations from $M_n = k_n bd^2$. There will be an infinite number of combinations possible so that the designer must arbitrarily select a b, d, or shape (b/d ratio) from economic, framing, deflection, or other considerations. Values of b should be selected for convenience in form dimensions (usually even whole inches are preferable). The need for cover, web reinforcement, and number of layers of reinforcement must be considered for increasing d to a practical overall thickness h.

6. A firm value of b and h are established. The self-weight estimate of step 2 is revised if necessary, and steps 4 and 5 recycled through until a firm estimate of b and h is determined. Then a practical d value is calculated by subtracting the required minimum cover (Code 7.7) from h and making allowance for web reinforcement and main reinforcement dimensions. (See Fig. 4.2.) This d value does *not* have to agree with the theoretical d value established in step 5. It can be larger or smaller than that value except at the reinforcement limits. If ρ is near ρ_{min}, then the actual d must be equal or less than the theoretical value from step 5. If ρ is near ρ_{max}, then the actual d must be equal or greater. This ensures that the ρ_{min} and ρ_{max} limits will be met.

7. For the actual M_n required, and b and d chosen, the two equations of equilibrium ($N_c = N_t$ and $M_n = N_t z = A_s f_y(z^*)$) are used to find the required A_s. Note that it is not correct to determine $A_s = \rho bd$ for the ρ selected in step 4 if *any* change in b or d is made from the theoretical. The authors recommend the simple trial and error method for solution of these two equations. It quickly converges on the correct A_s.

8. Upon determination of the required A_s, actual reinforcement is selected considering the minimum spacing rules of Code 7.6, the cover requirements of Code 7.7, and the flexural reinforcement distribution requirements (z factor check) of Code 10.6, which were explained in Sec. 3.30.

9. Each design of a cross section should end with development of an adequate sketch showing key dimensions and bar placement along with any special notes to the detailer. Good office practices on such engineering sketches are outlined in the ACI Detailing Manual.[1]

---

* $z = d - a/2$ for a rectangular beam.

4.3 Rectangular Beam Examples

(a) Design a minimum size (code maximum reinforcement percentage) beam for service moments of 50 k-ft for D.L. (including beam self-weight) and 100 k-ft for L.L. using f'_c = 3000 psi and Grade 60 steel for interior exposure use.

Solution

The load factors of Code Eq. 9.1 (see Sec. 1.5e) are assumed to govern:

$$M_u = 1.4M_D + 1.7M_L = 1.4 \times 50 + 1.7 \times 100 = 240 \text{ k-ft}$$
$$M_n \geqslant M_u/\phi = 240/0.9 = 267 \text{ k-ft}$$

Since $\rho = \rho_{max}$, it is appropriate to use Table 3.1. For these materials, maximum k_n = 783. The size of the beam can be found by equating $k_n bd^2$ to the external M_n (or $k_m bd^2$ to M_u or ϕM_n).

$$M_n = k_n bd^2 = 783 \, bd^2 = 267 \times 1000 \times 12$$
$$\text{Reqd. } bd^2 = 4100 \text{ in.}^3$$

This requirement can be met by a wide range of sizes, for example:

If b = 7.5 in., $d = \sqrt{547} = 23.4$ in. b = 10, $d = \sqrt{410} = 20.2$
b = 8, $d = \sqrt{512} = 22.6$ b = 11, $d = \sqrt{373} = 19.3$
b = 9, $d = \sqrt{456} = 21.3$ b = 11.5, $d = \sqrt{357} = 18.9$
b = 9.5, $d = \sqrt{432} = 20.8$ b = 12, $d = \sqrt{342} = 18.5$
  b = 13, $d = \sqrt{315} = 17.8$

The 7.5-in., 9.5-in., and 11.5-in. widths are listed because these are convenient actual widths for form lumber;* forms are one of the most costly items in reinforced concrete construction. The narrowest beam shown may be too narrow to accommodate the reinforcing steel, it uses up the most headroom, and requires the greatest area of forms; but it uses the smallest volume of concrete and steel. A d/b ratio from 1.0 or 1.5 to 2.5 or 3.0 is usually considered in the desirable range, with the larger ratios for the bigger beams. Because deep beams often require increased story heights, more wall and partition height, and thus other costs than the cost of the beam, there is no simple textbook answer to the *best* depth. There are a number of satisfactory choices possible. For the purpose of this problem, b = 9.5 in. will be chosen, although a wider beam might be needed for a lower strength steel, with more bars then necessary.

---

* Actual dimensions of a 2 in. $\times$ 10 in. are 1-½ in. $\times$ 9-½ in. If plywood or metal forms are utilized, even whole inches would be chosen.

If not exposed to the weather or the ground such a beam requires 1.5 in. of clear cover (Code 7.7.1c) over the steel. The steel typically includes some stirrups as reinforcement, as shown in Fig. 4.2, and #3 or #4 bars would be a reasonable size for this beam, say, #3 bar stirrups. Then the cross section of the beam would appear as shown in Fig. 4.2a. The minimum overall depth would be $20.8 + 0.5$ (assuming bar diam. $d_b = 1$ in.) + 0.38 (stirrup) + 1.5 (cover) = 23.2 in. This is an impractical dimension to give a carpenter for forms and the only change permitted is to greater dimensions, because maximum k_n was used originally.

USE 9.5 in. $\times$ 23.5 in. (overall)

$$d = 23.5 - 1.5 - 0.38 - 0.5 \text{ (assumed } d_b/2) = 21.1 \text{ in.}$$

With $d = 21.1$ in., the beam is a little deeper than the theoretical minimum and thus has a greater z than the minimum, which means N_t and the balancing compression block depth a will be slightly reduced. Note that Table 3.1 shows maximum $a/d = 0.38$, or $a = 0.38 \times 20.8 = 7.90$ in.

With the beam dimensions fixed, the two equilibrium equations of step 7 can be written and solved to find the actual A_s required. It is most convenient to arrange them as:

(1)

$$N_c = N_t$$
$$0.85 f_c' ab = A_s f_y$$
$$a = A_s f_y / 0.85 f_c' b$$

(2)

$$M_n = A_s f_y z = A_s f_y (d - a/2)$$

$$A_s = M_n / f_y (d - a/2)$$

By estimating a beginning value of $z = d - a/2$, Eq. (2) can be used to determine an approximate A_s. That value can be substituted in Eq. (1) to get a better estimate of a and hence a new z can be determined for substitution in Eq. (2). The process quickly converges to A_s required. Of course, the reader can solve the equations simultaneously which results in a quadratic equation for A_s. To illustrate the trial-and-error solution:

Assume $a = 7.75$ in., $z = 21.1 - 0.5 \times 7.75 = 17.2$ in.

$$M_n = A_s f_y z, \qquad A_s = 267 \times 12/(60 \times 17.2) = 3.10 \text{ in.}^2$$
$$a = A_s f_y / (0.85 f_c' b) = 3.10 \times 60/(0.85 \times 3 \times 9.5) = 7.68 \text{ in.}$$

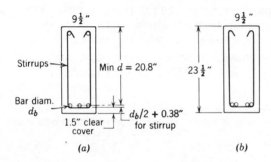

$9\frac{1}{2}''$

Stirrups

Min $d = 20.8''$

$23\frac{1}{2}''$

$9\frac{1}{2}''$

Bar diam. d_b

1.5″ clear cover

$d_b/2 + 0.38''$ for stirrup

(a)

(b)

Figure 4.2 Dimensions for beam of Sec. 4.3. (a) Minimums. (b) Final.

This indicates the original guess is adequate, but this can be proved by completing the cycle:

$$z = 21.1 - 0.5 \times 7.68 = 17.3 \text{ in.}$$
$$A_s = 267 \times 12/(60 \times 17.3) = 3.09 \text{ in.}^2$$

The choice of bars cannot be final without checks on bar development (Chapter 8), but a selection can be made tentatively:

2-#11 = 3.12 in.^{2}	Development apt to be troublesome with large bars. d_b is 1.41 in. and this larger d_b would reduce available d by 0.2 in. or $1^+\%$.
3-#10 = 3.81 in.^{2}	Wasteful
4-#9 = 4.00 in.^{2}	Wasteful, but 3 bars would be 3% short
4-#8 = 3.16 in.^{2}	O.K.
6-#7 = 3.60 in.^{2}	Wasteful
7-#6 = 3.08 in.^{2}	O.K. except probably requires two layers of bars, which requires increase in over-all beam depth

The clear spacing of bars in a layer must not be less than the nominal bar diameter, 1 in., or 4/3 of the nominal maximum aggregate size (Code 7.6.1 and 3.3.3c).

Try 4-#8 for spacing. Clear width inside stirrups is $9.5 - 2 \times 1.5$ cover $- 2 \times 0.38$ stirrups $= 5.75$ in. The four bars cannot be spaced individually, because this would require 4×1 (for bars) $+ 3 \times 1$ (for spaces) $= 7$ in., assuming $\frac{3}{4}$-in. aggregate is used. However, if development is not too troublesome, two bars might be bundled together in each corner of the stirrups as shown in Fig. 4.2b.

Bars up to four in number can be bundled together under special limitations stated in Code 7.6.6, which include limitation to #11 maximum size bars, provision of ties or stirrups enclosing the bundled bars, and so on. For spacing purposes the equivalent diameter of two bundled bars can be found from the diameter of an imaginary bar having the area of the two bundled bars, that is, $2 \times 0.79 = 1.58 \text{ in.}^2$

$$d_b = \sqrt{1.58/0.786} = 1.42 \text{ in.}$$

The clear spacing between the two pairs of bars is $5.75 - 4 \times 1 = 1.75$ in., which exceeds the 1.42 in. minimum. Subject to development, the arrangement is acceptable and economical. The reinforcement distribution (crack control) provisions of Code 10.6.4 must be checked because f_y exceeds 40 ksi. $z = f_s \sqrt[3]{d_c A}$ must be not more than 175 k/in. for this interior exposure. $f_s = 0.6f_y = (0.6)(60) = 36$ ksi. From Fig. 4.2a, $d_c = 1.5 + 0.38 + 1.0/2 = 2.38$ in., the distance from the extreme tension fiber to the centroid of the lowest layer of reinforcement. The area of effective tension concrete per bar $A = 2d_c b/n = (2)(2.38)(9.5)/4 = 11.30$ for bars in a single layer. $z = 36 \sqrt[3]{(2.38)(11.30)} = 108 < 175$ k/in. **O.K.**

Use 4-#8 bars

Deflections must be investigated (Chapter 3).

The student should note that the entire design process could be very compactly presented. This presentation has been greatly expanded by explanations.

(b) Using customary units design a beam similar to that in (a) except limit the reinforcement ratio to $\rho = 0.18 f'_c/f_y$ for which constants in the middle of Table 3.1 or Fig. 3.10 are usable. Use service moments of 50 k-ft (not including self-weight) for D.L. and 100 k-ft for L.L. with $f'_c = 3000$ psi and Grade 60 steel ($f_y = 60$ ksi). The beam is simply supported on a 24-ft span with interior exposure.

Solution

A major difference from (a) is that the dead load of the beam must be estimated to begin the problem. This estimate can be based on a crude guess, on experience, or using some available "rule of thumb" guidelines. Table 3.2 (Code Table 9.5a) gives useful guidelines for beam overall depth-to-span ratios. Deflections usually are acceptable if the beam has a depth at least as large as the values in this table, which offers a good starting point for estimating size. For a simple span beam, Table 3.2 indicates $h_{min} = \ell/16 = 24/16 = 1.5$ ft. Using a preliminary slightly larger estimate, say $h = 20$ in. As previously indicated d/b ratios usually run from 1.0 to 3.0. Trying around midrange, guess $h/b = 2.0$. Say, $b = 10$ in. From this, the first estimate of beam dead load owing to self-weight (assuming normal weight concrete at 150 pcf) is:

$$\text{Self-weight: } w_D = \frac{10 \times 20}{144} \times 0.150 = 0.21 \text{ k-ft}$$

$$M_{sw} = w_D \ell^2/8 = (0.21)(24)^2/8 = 15.1 \text{ k-ft}$$
$$M_D = 50 + 15.1 = 65.1 \text{ k-ft}$$
$$M_u = 1.4 M_D + 1.7 M_L = 1.4 \times 65.1 + 1.7 \times 100 = 261.1 \text{ k-ft}$$
$$= 3134 \text{ k-in.}$$

$$\text{Required } M_n = M_u/\phi = 3134/0.9 = 3482 \text{ k-in.}$$

From basic principles or Fig. 3.10

$$k_n = 0.161 f'_c = 0.483 \text{ (ksi)}$$
$$bd^2 = M_n/k_n = 3482/0.483 = 7210 \text{ in.}^3$$

Assume metal or plywood forms, so that b and h are desired in whole even inches.

$$\text{If } b = 10 \text{ in., } d = \sqrt{721} = 26.85 \text{ in.}$$
$$\text{If } b = 12 \text{ in., } d = \sqrt{601} = 24.51 \text{ in.}$$

Since $h \approx d + 2\text{-}\frac{1}{2}$ to 3 in., beam size may run 10×30 or 12×28. Obviously, not enough self-weight dead load was allowed. In normal calculations the designer would simply strike through previous estimate numbers and write the new estimates above them. In this example, the equations will be rewritten to minimize confusion.

New estimate on size: Try $b = 12$ in., $h = 28$ in.

$$w_D = (12 \times 28)/144 \times 0.150 = 0.35 \text{ k-ft}$$
$$M_{sw} = (0.35)(24)^2/8 = 25.2 \text{ k-ft} \qquad M_D = 50 + 25.2 = 75.2 \text{ k-ft}$$
$$M_u = 1.4 \times 75.2 + 1.7 \times 100 = 275.3 \text{ k-ft} = 3304 \text{ k-in.}$$
$$M_n = 3304/0.9 = 3670 \text{ k-in.}$$

Using $k_n = 0.483$, $bd^2 = 3670/0.483 = 7600 \text{ in.}^3$

If $b = 12$, $d = \sqrt{633} = 25.16$. Allowing 1.5 in. cover, 0.5 in. for a stirrup, and 0.7 in. for $0.5 d_b$, $h = 27.86$ in. Use $b = 12$ in., $h = 28$ in., $d = 28 - 1.5 - 0.5 - 0.7 = 25.3$ in. M_n required is now correctly estimated as 3670 k-in. Because this ρ is about midrange, the designer was free to use an actual d smaller or larger than the value indicated by $M_n = k_n bd^2$. The actual strength required is provided by the equilibrium equations with b and d chosen to determine the A_s required to provide the actual M_n needed.

The equilibrium equations are:

(1)

$$N_c = N_t$$

$$0.85 f'_c ab = A_s f_y$$

$$a = \frac{A_s f_y}{0.85 f'_c b} = \frac{A_s \times 60}{0.85 \times 3 \times 12} = 1.96 A_s$$

(2)

$$M_n = A_s f_y (d - a/2)$$

$$A_s = \frac{M_n}{f_y(d - a/2)}$$

$$A_s = \frac{3670}{60(25.3 - a/2)} = \frac{61.2}{(25.3 - a/2)}$$

Since $\rho \approx 0.18 f'_c/f_y$, a first estimate for $z = d - a/2$ can be obtained from Fig. 3.10 as $0.894 d = 22.6$ in. Begin the trial-and-error process:

$$A_s = 6.12/(22.6) = 2.71 \text{ in.}^2$$
$$a = 1.96(2.71) = 5.31 \qquad A_s = 6.12/(25.3 - 5.31/2) = 2.70 \text{ in.}^2$$
$$a = 1.96(2.70) = 5.30 \qquad A_s = 61.2/(25.3 - 5.30/2) = 2.70 \text{ in.}^2$$

Required $A_s = 2.70 \text{ in.}^2$

This area could be satisfied by several combinations. Try 2 − #10 and 1 − #4 ($A_s = 2.74 \text{ in.}^2$) estimating that #4 stirrups would be used in a beam this size. Checking bar spacing, effective d, and z factor indicates (sketch would be similar to Fig. 4.2, but with revised dimensions and three bars equally spaced):

$$2 \text{ spaces} = \text{width} - 2 \text{ cover} - 2 \text{ stirrups} - 3 \text{ bar diameters}$$
$$= 12 - 2(1.5) - 2(0.5) - (1.27 + 0.5 + 1.27)$$
$$= 4.96 \text{ in.}$$

Each space between bars is 2.48 in. which greatly exceeds the 1 in. required. The centroid of the actual bars would be determined assuming each bar rests on the horizontal stirrup leg which is the way reinforcement is actually fabricated. From the bottom, $\bar{y}$ = 2.60 in. This is very little different from the distance $1.5 + 0.5 + 1.27/2 = 2.63$ in. that would be the distance to the center of the large bars. A designer would ordinarily use the latter number because actual field tolerances make hundredths of inch units meaningless. The effective $d = 28 - 2.60 = 25.4$ in. This small change from the 25.3 used in calculation for A_s is similarly negligible as it only produces 0.25% more d than assumed and might result in about that much less steel if another cycle was made.

In the case of mixed bars, when checking z, d_c is defined as the distance to the center of the bar nearest the outer tension fiber. This could be interpreted as the #4 bar so $d_c = 1.5 + 0.5 + (0.5/2) = 2.25$. A is the effective tension area per bar. For mixed bar size arrangements, the number of bars is defined as the total area divided by the area of the largest size bar. $n = 2.74/1.27 = 2.16$. Thus

$$A = \frac{12 \times 2 \times 2.60}{2.16} \text{ (or 2.63 if the actual centroid is neglected)}$$

$$= 28.92 \text{ in.}^2$$

$$z = f_s \sqrt[3]{d_c A} = 36 \sqrt[3]{2.25 \times 28.92} = 145 \text{ k/in.} < 175 \qquad \textbf{O.K.}$$

As $h = 28 \gg \ell/16 = 18$ in. of Table 9.5a, no deflection check is needed.

(c) Using SI units design a beam similar to that in (a) except that for architectural reasons the beam is to be made 300-mm wide by 600-mm deep overall. Use service moments of 68 kN · m for D.L. (including self-weight) and 136 kN · m for L.L. with C20 concrete ($f_c' = 20$ MPa) and Grade 400 steel ($f_y = 400$ MPa). Assume interior exposure.

Solution

$$M_u = 1.4M_D + 1.7M_L = 1.4 \times 68 + 1.7 \times 136 = 326 \text{ kN} \cdot \text{m}$$
$$\text{Required } M_n = M_u/\phi = 326/0.9 = 362 \text{ kN} \cdot \text{m}$$

The logical first question is whether a beam with $b = 300$ mm (0.3 m) and $h = 600$ mm (0.6 m) can provide the required moment resistance without exceeding ρ_{max}. This can be quickly checked by evaluating the maximum potential moment $M_n = k_n bd^2$ using $k_n = 5.25$ MPa for ρ_{max} from Table 3.1. Estimate d based on 40 mm cover, minimum size stirrups which are #10 with 11-mm diameter, and about #30 bars for main reinforcement ($d_b = 30$ mm). Available $d = 600 - 40 - 11 - 15 = 534$ mm = 0.534 m.

$$\text{Max } M_n = k_n bd^2 = 5.25 \times 0.3 \times 0.534^2 = 0.45 \text{ MN} \cdot \text{m} = 450 \text{ kN} \cdot \text{m}$$
$$> 362 \text{ kN} \cdot \text{m required}$$

Therefore the beam will be under-reinforced and the actual A_s required found simply from the two equilibrium equations:

(1)	(2)

(1)   (2)

$$N_t = N_c \qquad\qquad M_n = A_s f_y (d - a/2)$$

$$A_s f_y = 0.85 f'_c ab \qquad\qquad A_s = M_n/f_y(d - a/2)$$

$$a = A_s f_y/0.85 f'_c b = A_s 400/0.85 \times 20 \times 0.3 \qquad A_s = 0.362/400(0.534 - a/2)$$

$$a = 78.4 A_s \qquad\qquad A_s = 0.905 \times 10^{-3}/(0.534 - a/2)$$

Guessing an initial $z = 0.8d$, $A_s = 0.905 \times 10^{-3}/0.8 \times 0.534 = 2.12 \times 10^{-3}$ m^2

$a = 78.4 \times 2.12 \times 10^{-3} = 0.166$ m $A_s = 0.905 \times 10^{-3}/(0.534 - 0.166/2) = 2.01 \times 10^{-3}$ m^2

$a = 78.4 \times 2.01 \times 10^{-3} = 0.157$ m $A_s = 0.905 \times 10^{-3}/(0.534 - 0.157/2) = 1.99 \times 10^{-3}$ m^2

$a = 78.4 \times 1.99 \times 10^{-3} = 0.156$ m $A_s = 0.905 \times 10^{-3}/(0.534 - 0.156/2) = 1.985 \times 10^{-3}$ m^2

$$= 1985 \text{ mm}^2$$

The solution has converged to 1985 mm^2 for A_s.

USE 4-#25 = 2000 mm^2

Clear inside stirrups = $300 - 2 \times 40$ cover $- 2 \times 11$ stir. = 198 mm

Clear width between bars = $198 - 4 \times 25$ = 98 mm

= 3 spcs. at 33 mm each   **O.K.**

The beam would be subject to deflection and crack width (z) checks as in the other examples.

4.4   Beams with Compression Steel—Moment Design

Compression steel is used wherever a given beam requires more tension steel than the engineer thinks desirable in a rectangular beam, either from consideration of approaching brittle failure conditions or from consideration of excessive deflections. In either case the designer knows how much moment he or she must carry or desires to carry without compression steel and this moment is designated as the M_1* couple with steel A_{s1}. Any necessary additional moment M_2 can then be resisted by added tension steel A_{s2} and compression steel A'_s as sketched in Fig. 4.3. Whether A'_s works at yield stress or a lower stress depends on the ε'_s strain, established by the neutral axis required with the M_1 couple.

When compression capacity requires the compression steel, $\rho_1 = \rho_{max}$ and the M_1 couple is based on the k_n value of Table 3.1. A_{s1} can be calculated from the corresponding ρ value (or from the a value, which establishes z for use in $A_{s1} = M_1/f_y z$). It should be noted that the permissible kbd^2 is normally so large that compression steel is rarely required for strength. When compression steel is used to reduce long-time deflection, M_1 and k_1 can be based on the $\rho = 0.18 f'_c/f_c$ values tabulated in Tables 3.1 and Fig. 3.10, or engineers can set up their own criterion following the method of Sec. 4.5b.

---

* The more correct notations M_{n1} and M_{n2} will be simplified to M_1 and M_2.

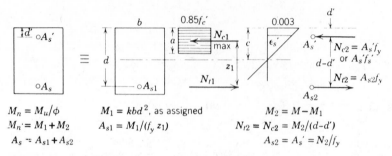

$M_n = M_u/\phi$ $M_1 = kbd^2$, as assigned $M_2 = M - M_1$
$M_n = M_1 + M_2$ $A_{s1} = M_1/(f_y\, z_1)$ $N_{t2} = N_{c2} = M_2/(d-d')$
$A_s = A_{s1} + A_{s2}$ $A_{s2} = A_s' = N_2/f_y$

Figure 4.3   Design of double-reinforced beam.

Whereas *analysis* of beams containing A_s' sometimes requires a cut and try procedure (Sec. 3.15b), the *design* procedure is always direct and relatively simple.

Code 7.11.1 requires ties (or stirrups) around A_s' in beams, as in columns (Fig. 6.19).

4.5   Double-Reinforced Beam Examples

(a) A double-reinforced concrete beam with $b = 11$ in., $d = 20$ in., $f_c' = 3000$ psi, $f_y = 40$ ksi, must carry a service dead load moment of 136 k-ft and live load moment of 150 k-ft. Calculate the required steel using the Code load factors.

Solution

Required $M_u = 136 \times 1.4 + 150 \times 1.7 = 190 + 255 = 445$ k-ft
Required $M_n \geqslant M_u/0.9 = 445/0.9 = 495$ k-ft $= 5940$ k-in.

As a rectangular beam without A_s', the maximum steel for A_{s1} is limited to $0.75\rho_b$, which is tabulated in Table 3.1 as 2.78%, and which develops

$$M_1 = k_n bd^2 = (0.870)11 \times 20^2 = 3830 \text{ k-in.}$$
$$A_{s1} = \rho_1 bd = 0.0278 \times 11 \times 20 = 6.12 \text{ in.}^2$$

Or, using $a = 0.44d$ (from Table 3.1) $= 0.44 \times 20 = 8.8$ in.

$$z = d - a/2 = 20 - 8.8/2 = 15.6 \text{ in.}$$
$$A_{s1} = M_1/f_y z = 3830/(40 \times 15.6) = 6.14 \text{ in.}^2$$

The a value is not quite as accurately stated as the ρ value in the table, but either is as accurate as the loads. The value of a is also needed for f_s' calculations below.

$M_2 = M_n - M_1 = 5940 - 3830 = 2110$ k-in. $N_{t2} = N_{c2} = M_2/(d - d')$

Assume the cover d' over A_s', to center of bars, is 2.5 in., making $d - d' = 20 - 2.5 = 17.5$ in. Then $N_2 = 2110/17.5 = 120.6$ k

$$A_{s2} = N_2/f_y = 120.6/40 = 3.02 \text{ in.}^2$$
$$A_s = A_{s1} + A_{s2} = 6.14 + 3.02 = 9.16 \text{ in.}^2$$

Check to see if A_s' develops the yield strain $\varepsilon_y = 40/29{,}000 = 0.00138$. Use strain triangles based on the previous value of $a = 8.8$ in., $c = a/\beta_1 = 8.8/0.85 = 10.35$ in., and $\varepsilon_s' = 0.003(10.35 - 2.5)/10.35 = 0.0023 > \varepsilon_y$.

Neglecting compression concrete, which is displaced by A_s'

$$A_s' = N_2/f_y = A_{s2} = 3.02 \text{ in.}^2$$

If the displaced concrete is considered, the effective f_s''—that is, the increased value added by A_s' over the loss of the same area of compression concrete—becomes

$$f_y - 0.85f_c' = 40 - 0.85 \times 3 = 37.4 \text{ ksi}$$
$$A_s' = N_2/f_s'' = 120.6/37.4 = 3.22 \text{ in.}^2$$

Choice of bars involves the same procedure used in Sec. 4.3(a).

USE $A_s = 9.16$ in.^{2}, $A_s' = 3.22$ in.^{2}, $d' = 2.5$ in.

The large A_{s1} results from a minimum depth beam for strength and may give a beam with excess deflection. Add cover to the d of 20 in. and check that h against Table 3.2 (Code Table 9.5a), or calculate deflection (Chapter 3). The z check for reinforcement distribution is likely to be difficult in this congested section.

(b) Recalculate the reinforcing steel for the previous case if it is desired to reduce deflection somewhat by limiting A_{s1} to $\rho_1 = 0.18f_c'/f_y,$* $d' = 2.5$ in.

Solution

The total required M_n is 5940 k-in., as calculated before. Table 3.1 and Fig. 3.10 show that this steel limitation gives $M_1 = 0.161\ f_c'bd^2$, $a = 0.212d$, and $z_1 = 0.894d$. Thus, $k_1 = 0.161 \times 3 = 0.483$ ksi.

$$M_1 = k_1bd^2 = 0.483 \times 11 \times 20^2 = 2125 \text{ k-in.}$$
$$A_{s1} = (0.18f_c'/f_y)bd = (0.18 \times 3/40)11 \times 20 = 2.97 \text{ in.}^2$$
$$M_2 = M_n - M_1 = 5940 - 2125 = 3815 \text{ k-in.}$$
$$N_{t2} = N_{c2} = M_2/z = 3815/(20 - 2.5) = 218 \text{ k}$$
$$A_{s2} = N_2/f_y = 218/(40) = 5.45 \text{ in.}^2$$
$$A_s = A_{s1} + A_{s2} = 2.97 + 5.45 = 8.42 \text{ in.}^2$$

---

* Note that the method of Sec. 3.9 and Fig. 3.10 is available for any desired ρ_1, to give k_1. Likewise any desired ρ' can lead to k_2. The sum $k_1 + k_2 = $ total k resulting.

For compression strain triangles, $a = 0.212d = 0.212 \times 20 = 4.24$ in. and $c = a/\beta_1 = 4.24/0.85 = 4.98$ in.

$$\varepsilon_s' = 0.003(4.98 - 2.5)/4.98 = 0.00150 > \varepsilon_y = 40/29{,}000 = 0.00138$$

Considering $f_s' = f_y$ and not allowing for displaced concrete, $A_s' = A_{s2} = 5.45$ in.^{2}

USE $A_s = 8.42$ in.^{2}, $A_s' = 5.45$ in.^{2}

Comparison with the design in (a), which was for maximum M_1 and minimum A_s', shows that A_s has dropped 8 percent because of the increased z given by a smaller A_{s1} and hence a reduced stress block depth for compression. On the other hand, A_s' is increased 69 percent.

With shallow beams, ε_s' may be less than ε_y. In that case, $f_s' = E_s \varepsilon_s'$ is simply inserted for f_y in calculations of A_s'. There is no cut and try needed in *design* of double-reinforced beams.

4.6   T-Beams—Moment Design

The T-beam flange usually has been established by the thickness of the slab design, and the web size (b_w and d) is often determined by shear or requirements other than moment. To some extent b_w, the stem width, is influenced by the number of reinforcing bars used for A_s, but the student probably lacks the experience at this stage to estimate this need in advance. In this section it is assumed that design for moment includes only the choice of the area of steel A_s, a check against brittle failure, and some thought of deflection.

A flange is ordinarily assumed to provide enough compression area to eliminate the possibility of brittle failure in compression. This is frequently confirmed by a neutral axis that falls high in the flange and in effect turns a T-beam design into the design of a wide rectangular beam. Actually, a T-beam must be quite deep relative to the slab and be quite heavily reinforced to make it really act as a T-beam in carrying compression. Even when the balanced condition would lead to a neutral axis deep in the beam (and a rather large value of $a = \beta_1 c$), the brittleness concept leads to restricting the compression stress block area to 0.75 of that at the balanced condition. Because it is rare that more than 0.25 of the total stress block area lies below the bottom of the flange, the usable stress block tends to be restricted to less than the total depth of the flange. Thus most T-beams carrying positive moment are only rectangular beams in flexural behavior.

The cut and try determination of z, as for the rectangular beam, is recommended, as illustrated in Sec. 4.7. The Code provision of transverse steel in the flange (8.10.5) must not be overlooked:

Where primary flexural reinforcement in a slab that is considered as a T-beam flange (excluding joist construction) is parallel to the beam, reinforcement perpendicular to the beam shall be provided in the top of the slab in accordance with the following:

(a) Transverse reinforcement shall be designed to carry the factored load on the overhanging slab width assumed to act as a cantilever. For isolated beams, the full width of overhanging flange shall be considered.

(b) Transverse reinforcement shall not be spaced farther apart than five times the slab thickness, nor 18 in.

4.7   T-Beam Examples

(a) The floor of Fig. 4.4 consists of a 4-in. slab supported by beams of 22-ft span cast monolithically with the slab at a spacing of 8 ft center to center. These beams have webs 12-in. wide and a depth of 19 in. to the center of reinforcement plus the necessary cover over the bars. Service dead load moment is 62 k-ft and live load moment is 120 k-ft. Using $f'_c = 3000$ psi, Grade 60 steel, $f_y = 60$ ksi, and the Code load factors of Code Eq. 9.1, calculate the required area of reinforcement.

Solution

The effective flange width is defined by Code 8.10.2 as the smallest of these three limits (or see Sec. 3.12):

$$\text{span}/4 = 22 \times 12/4 = 66 \text{ in.} \qquad \leftarrow\text{Governs } b$$
$$8h_f \text{ overhang gives } b = 2 \times 8 \times 4 + 12 = 76 \text{ in.}$$
$$\text{Beam spacing} = 8 \text{ ft} = 96 \text{ in.}$$
$$\text{Required } M_u = 1.4D + 1.7L = (1.4 \times 62 + 1.7 \times 120) = 291 \text{ k-ft}$$
$$M_n \geqslant M_u/\phi = 291/0.9 = 323 \text{ k-ft} = 3880 \text{ k-in.}$$

The lever arm z usually will be large and a good trial value usually will be at least $0.9d$ or $d - h_f/2$, whichever is the larger.

$$0.9d = 0.9 \times 19 = 17.1 \text{ in.} \qquad d - h_f/2 = 19 - 4/2 = 17 \text{ in.}$$
$$\text{Trial } A_s = M_n/(f_y z) = 3880/(60 \times 17.1) = 3.78 \text{ in.}^2$$

It must be determined whether the stress block is as deep as the flange. If $a = h_f$, available $N_c = 0.85 \times 3 \times 66 \times 4 = 673$ k, compared to the

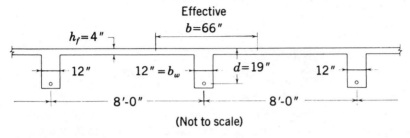

Figure 4.4   Floor of Sec. 4.7.

necessary $N_c = A_s f_y = 3.78 \times 60 = 227$ k. Thus, the neutral axis is obviously high in the flange and the design becomes simply that for a rectangular beam 66-in. wide.

$$a = N_c/(0.85f'_c b) = 227/(0.85 \times 3 \times 66) = 1.35 \text{ in.}$$
$$z = 19 - 0.5 \times 1.35 = 18.32 \text{ in.}$$
$$A_s = M_n/(f_y z) = 3880/(60 \times 18.32) = 3.53 \text{ in.}^2$$

Better $a = 3.53 \times 60/(0.85 \times 3 \times 66) = 1.25$ in.

$$z = 19 - 0.5 \times 1.25 = 18.38 \qquad \text{Close enough}$$

USE $A_s = 3.53$ in.^{2}

There can be no possibility of brittle compression failure with this small an a. Deflection should be checked, bars selected, and z checked.

(b) Calculate A_s if the loads in Example (a) are increased to give $M_u = 1000$ k-ft.

Solution

$$M_n \geqslant M_u/\phi = 1000/0.9 = 1111 \text{ k-ft} = 13,330 \text{ k-in.}$$

If $a = h_f = 4$ in., available $N_c = 673$ k, as in (a). Compare approximate required N_c for $z = 0.9d = 0.9 \times 19 = 17.1$ in.

$$N_c = M_n/z = 13\ 330/17.1 = 780 \text{ k} > \text{available } 673 \text{ k}$$

This means the compression zone required exceeds the slab area.

Since a is greater than h_f, this acts as a true T-beam. As above, try $z = 0.9d$.

Approx. $A_s = 13,330/(60 \times 0.9 \times 19) = 12.99$ in.^{2}
Compression block area $= N_t/(0.85f'_c) = 12.99 \times 60/(0.85 \times 3) = 306$ in.^{2}
Area of flange $= 66 \times 4 = 264$ in.^{2}
Area below flange $= 306 - 264 = 42$ in.^{2}
Depth below flange $= 42/12 = 3.5$ in., total $a = 7.5$ in.

This area is shown in Fig. 4.5 and gives the centroid distance $\bar{y}$ from top as:

$$\bar{y} = (66 \times 4 \times 2 + 12 \times 3.5 \times 5.75)/306 = 2.51 \text{ in.}$$
$$z = d - \bar{y} = 19 - 2.51 = 16.49 \text{ in.},$$
$$A_s = M_n/(f_y z) = 13,330/(60 \times 16.49) = 13.47 \text{ in.}^2$$

Compression block area increases to $(13.47/12.99)306 = 317$ in.^{2}. Area below flange $= 317 - 264 = 53$ in.^{2}, with necessary distance below flange increasing from 3.5 to 4.4 in. The change in y and z (about 0.15 in. more $\bar{y}$, 0.15 in. less z) is reasonably small and could be neglected here. The design

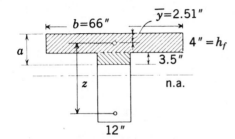

Figure 4.5 Trial compression block.

would be satisfied with $A_s = 13.47$ in.^{2} (more exactly about 1% more, say 13.60 in.^{2}) except for the problem covered in the next paragraph.

The neutral axis is quite low and should be checked with regard to $\rho \gtrless 0.75 \, \rho_b$, where ρ_b is the balanced condition.*

Balanced $c_b = 0.003 \, d/(0.003 + \varepsilon_y) = 0.003 \, d/(0.003 + 0.00207)$

$= 0.592 d = 0.592 \times 19 = 11.24$ in.

$a_b = 0.85 c_b = 0.85 \times 11.24 = 9.55$ in.

Balanced compression stress block area $= 66 \times 4 + 12 \times 5.55 = 331$ in.^{2}

Usable compression area to match $0.75 \rho_b = 0.75 \times 331 = 248$ in.^{2}

This is substantially less than the 317 in.^{2} required previously. Under the Code it is definitely not desirable to use this T-beam for the indicated moment. However, if it is necessary, it can be accomplished by using A_{s1} as required to balance the 248 in.^{2} stress block, in which case this leads to M_1 as for a rectangular beam, because 248 in.^{2} is less than bh_f. The additional moment $M_2 = M_n - M_1$ should then be provided by A_s' and A_{s2} as for any double-reinforced beam.

This T-beam case is far from typical because most are not so heavily loaded with moment. The given moment might well call for a heavier slab design that would automatically increase the moment capacity; otherwise, a deeper beam seems appropriate.

(c) Calculate the required A_s for the beam of (b) if beam depth is increased to 23 in. and the span is reduced to 20 ft-8 in., which limits b to 62 in.†

Solution

Required $M_n \gtreqless M_u/\phi = 1000/0.9 = 1111$ k-ft $= 13,330$ k-in.

---

* Code 10.3.3 limits compression area to 0.75 of the *balanced* compression area as used here. A more consistent value for uniform ductility limit would be to limit c to $0.75 c_b$.
† The 4 in. slab still indicated is probably inappropriate for the larger M_u. It is retained to show how to handle the T-shape of compression block on the very unusual cases where this is necessary—or how to handle other irregular shapes.

Find the required compression stress block.

$$\text{Approximate required } N_c = 13,330/(23 - 4/2) = 635 \text{ k}$$
$$\text{Required area of stress block} = 635/(0.85 \times 3) = 249 \text{ in.}^2$$
$$\text{Area in flange level} = 62 \times 4 = 248 \text{ in.}^2$$
$$\text{Area required below flange} = 249 - 248 = 1 \text{ in.}^2$$
$$\text{Total stress block depth } a = 4 + 1/12 = 4.08 \text{ in.}$$

The centroid is essentially at middepth of the flange. Check this against the ductility requirement that $\rho \gtrsim 0.75\rho_b$, or, if A_c is area under compression, $A_c \gtrsim 0.75 A_{cb}$. Calculate c_b from strain triangles, noting $\varepsilon_y = 60/29,000 = 0.00207$.

$$c_b = 0.003\,d/(0.003 + 0.00207) = 0.003 \times 23/0.00507 = 13.61 \text{ in.}$$
$$a_b = 0.85 \times 13.61 = 11.57 \text{ in.}$$
$$\text{Balanced stress block area} = 62 \times 4 + 12 \times 7.57 = 339 \text{ in.}^2$$
$$\text{Code limit on } A_c = 0.75 \times 339 = 254 \text{ in.}^2 > 249 \text{ in.}^2 \quad \textbf{O.K.}$$

Proceed to locate the centroid of the compression stress block and evaluate z and A_s, using the previous stress block areas, even though it can be seen that the centroid is near middepth of the flange.

$$\text{From top } \bar{y} = [248 \times 4/2 + (0.08 \times 12 \times 4.04)]/(248 + 1) = 2.01 \text{ in.}$$
$$z = 23 - 2.01 = 20.99 \text{ versus 21 assumed (trivial)}$$
$$\text{Required } A_s = 13,330/(60 \times 20.99) = 10.58 \text{ in.}^2$$

It is not mandatory for the designer to find the centroid of the compression block, as done above. For any nonrectangular compression area there is the option of considering it the sum of its parts, but with each part having a common neutral axis. For the previous example the designer could use either the N_1 and N_2 couples of Fig. 4.6 or the N_3 and N_4 couples, much as used earlier in Sec. 3.13 and Fig. 3.15.

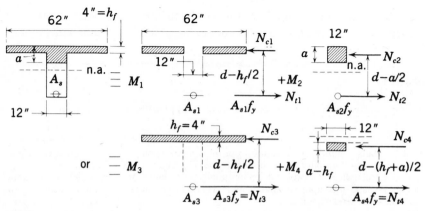

Figure 4.6 Breaking down a total moment into couples convenient for calculation.

As an example, verify the above A_s of 10.58 in.^{2} by using N_{c1} and N_{c2}.

$$N_{c1} = (62 - 12)4 \times 0.85 \times 3 = 510 \text{ k} = N_{t1}$$
$$A_{s1} = N_{t1}/f_y = 510/(60) = 8.50 \text{ in.}^2$$

The approximate a of 4.08 in. from above will be used in calculating A_{s2}.

$$N_{c2} = (12 \times 4.08) \times 0.85 \times 3 = 125 \text{ k} = N_{t2}$$
$$A_{s2} = N_t/f_y = 125/(60) = 2.08 \text{ in.}^2$$

Required $A_s = 8.50 + 2.08 = 10.58$ in.^{2}, the same as earlier.

4.8 Beams of Special Shapes

As presented in the preceding sections, the criterion against brittle failure is perfectly general. The nominal compression must be limited to $0.75 N_{cnb}$. The only Code exception is that the contribution of A_s' to ductility is recognized by the last sentence in Code 10.3.3 and thus in Section 4.5 the 0.75 factor was used only with the M_1 couple, not with the M_2 couple representing A_s' and A_{s2}.

Special shaped beams present small problems for the designer in that the k_m or k_n values, maximum a/d, and the like will not be available in tables unless it happens to be a shape commonly used in precast work or the like. The necessary A_s for a beam such as that of Fig. 3.19 can be found by a cut-and-try process that travels over much of the same ground used in Section 3.16 for finding the allowable design moment to accompany a given A_s. The procedure would be as follows:

1. Assume z, noting that with the narrow top width in Fig. 3.19 N_c acts farther from the top than in a rectangular beam of constant width.
2. Calculate $A_s = M_n/(f_y z)$.
3. Evaluate depth of stress block from $N_{cn} = N_{tn}$, possibly first finding A_c in compression from $A_c = A_s f_y/0.85 f_c'$.
4. Find centroid of A_c to get location of N_{cn} and a better z.
5. Recycle (if the original estimate of z was poor), to the accuracy desired.

An alternate procedure for step 4 would be to break down A_c into simple pieces, as done for the T-beam in Fig. 4.6, each with its individual A_s and couple. This does not, however, so clearly indicate how much error was in the assumed z of step 1 and what would be the better guess for the next cycle.

4.9 Face Steel on Large Beams

The primary Code emphasis on deep beams is for relatively short spans, ℓ_n not more than $5d$. (See coverage in Sec. 5.20 as primarily a shear problem.) However, Code 10.6.7* relates primarily to flexural cracking in any web more than 3-ft deep, regardless of span length, even in constant moment zones where shear is essentially zero. It calls for longitudinal reinforcement "at least 10 percent of the area of the flexural reinforcement—placed near the side faces of the web and distributed in the zone of flexural tension."

A depth of more than 36 in., although an uncommon beam size for buildings, is used in many cases, particularly in architectural applications. Such a depth is common for bridge structures; Code 10.6.7. is written to control flexural cracking on the deep side faces. Unfortunately, it is *not* adequate for this purpose. Actual construction and tests in France[4] and at the University of Texas at Austin[5,6] shows this inadequacy. The basic logic of these requirements is faulty. For example, if a designer must provide a beam with a certain moment capacity, as the beam depth increases the needed area of main reinforcement reduces. The present code requirements then would reduce the required amount of side face reinforcement. Intuition and tests suggest however that as the depth increases, the required area of side face or "skin" reinforcement should also increase, not decrease.

The basic problem is simple, once it is noted. Ordinary beams are designed on the assumption that the strain gradient from flexure is linear, the shear strains being negligible in shallow members. In a deep beam, the flexural strains, top and bottom, are about the same as in a shallow beam, which means the flexural strain gradient varies essentially inversely with depth, thus is very small in very deep members. The shear strains are no longer negligible in comparison. Instead they can almost totally cancel the flexural strains and relax the longitudinal tension in the concrete just above A_s and widen the flexural crack once it gets well above A_s.

A second pattern, less predictable except from experiments, is that all the cracks forming at the level of the bars do *not* rise to equal heights in the web. With deep beams without horizontal web bars there may be 3 or 4 cracks at the A_s level for each crack continuing farther up in the web. In this way also the longitudinal strains can lead to web crack widths that may be several times as wide as those at the A_s level.

The function of longitudinal face bars in the web is to stretch the concrete (help it carry some tension) and to cause more web cracks of narrower average width, rather than fewer wide cracks. The senior author was almost startled by the flexural crack widths in the first 36 in. beam he tested, cracks substantially wider at middepth than at A_s level.

Tests in France[4] reported in 1972 on beams 1-m deep and 0.3-m thick seem to show that bars totaling 0.25 percent of the web area and placed between A_s and the neutral axis limited web cracks to the crack width at the level of A_s. Such bars can also be counted as part of A_s to the extent consistent with their lower strains and shorter internal lever arm for flexure.

The ACI Building Code committee has been balloting a proposal based on

---

* The highway bridge code (AASHTO) discussed in Chapter 17 is essentially the same in this respect.

Reference 6 that would require an amount of side-face skin reinforcement well distributed on each side face for a distance $d/2$ nearest the flexural tension reinforcement. The area of skin reinforcement in sq. in. per foot of height on each side face should be at least $0.012(d - 30)$, where d has the dimensions of inches. The total area of longitudinal skin reinforcement need not exceed one-half of the required flexural tensile reinforcement. This latter provision reflects the need for special treatment of panel-type sections that are very deep for architectural reasons rather than strength requirements. A maximum spacing of 8 in. is suggested. This skin reinforcement can be counted on for tensile strength in moment computations as long as a strain compatibility check is made to determine stresses in each reinforcement level. Studies[6] indicate that with this type of skin reinforcement Grade 40 bars and Grade 60 bars (except when ρ is near ρ_{max}) will always yield at ultimate flexural capacity. This means that there is little economic penalty for distributing the reinforcement to control this side face cracking in large members.

4.10 Beams with Reinforcing at Several Levels

When reinforcing is placed in two layers, it is usually satisfactory to consider the f_y stress at the centroid of the steel. However, if the bars are in many layers or are distributed over all faces, as in a column section, some bars will be near the neutral axis and have a stress less than f_y. In such cases strains as well as stresses must be considered. This is not overly complex when strain triangles are used after the fashion already used for compression steel, as illustrated by Fig. 4.3 and the examples of Sec. 4.5 and Fig. 3.18.

4.11 Stopping or Bending of Bars—For Moment

(a) General Considerations The maximum required A_s for a beam is needed only where the moment is maximum. Insofar as moment is concerned this steel may be reduced at points along the beam where only smaller moments exist. This ability to simply and accurately tailor the moment resistance to fit the moment requirements along the span leads to unique economic advantages for reinforced concrete construction.

Two other stress considerations are also important in fixing the length of bars. First the individual bars require a specific length in which to develop their yield strength. Development length, the subject of Chapter 8, is as important as moment length; the designer must put the two aspects together. Second, when a bar is cut off where the neighboring bars are still carrying tension, a discontinuity in the tensile resistance occurs, flexural cracks open at earlier loads and become wider and more inclined than normal; shear strength is lowered, often with some loss of the beam ductility. Shear strength loss is discussed in Sec. 5.15, which points out the extra shear reinforcement (stirrups) required by Code 12.10.5 near the end of bars that are cut off. This requirement offsets some of the apparent savings from reduced bar length. A bent bar anchored in the compression zone causes no loss in shear strength.

In this chapter only the moment aspect of stopping or bending bars is discussed, and only for statically determinate members. A more detailed discussion for a continuous beam is given in the example beginning in Sec. 9.9.

(b) Maximum Moment Curves   A study of where bars can be stopped or bent away from the tension zone must start with a study of the maximum moments that are possible at all points along the beam.

In some simple cases, the maximum moment diagram is simply the moment diagram for full load. In other cases, as for wheel loads (Fig. 4.7a), the determination of the maximum moment diagram requires the calculation of maximum moments at many points; the maximum moment diagram is then the envelope of the maximum moment values from the several loads, as in Fig. 4.7b for a single wheel load W. Each dashed triangle corresponds to a particular position of the load, the maximum moment at any point being given with the wheel at that point. The envelope in this case happens to be a parabola. With several wheels the curve approximates two half-parabolas separated by a section of constant moment as in Fig. 4.7c.

Only fixed loadings are illustrated in the remainder of this chapter.

(c) Theoretical Bend Point or Cutoff Point   When a variable depth member is considered, the required A_s curve must be established by calculating the required A_s at representative points from the basic relation

$$A_s = \max M_n \div (f_y z)$$

where z varies somewhat like d. This procedure is illustrated in sec. 18.7.

With constant depth beams, the denominator becomes nearly a constant,

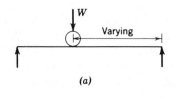

(a)

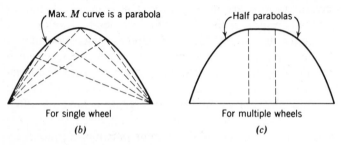

For single wheel

(b)

For multiple wheels

(c)

Figure 4.7   Maximum moment curves.

with the usual small variations in the ratio z/d ignored in this connection. Hence required A_s varies directly with the maximum moment, and the shape of the required A_s curve is identical to that of the maximum moment curve. The maximum moment curve may then be used as the required A_s curve, simply by changing the scale. The maximum ordinate corresponds to the maximum required A_s and this establishes the necessary correlation.

Bars can theoretically be stopped or bent wherever they are no longer needed for moment. The solution can be either graphical (for complex maximum moment diagrams), semigraphical for typical cases, or analytical where desired. The semigraphical process is illustrated in Secs. 4.12 and 4.13.

(d) Shear Complications   Where a bar is cut off, the rather sudden transfer of tension to the continuing bars causes early flexural cracking at the end of the bar. This flexural crack is wider than usual and turns more diagonally to become what is commonly called a shear crack (Fig. 5.4). Shear strength is lowered in a zone around the last portion of the terminated bar (maybe for $0.75d$ from the cutoff). See the bar chart of 34 test results in Fig. 5.18.

Because of this complication, Code 12.10.5 forbids cutting off bars unless *one* of three possible conditions is satisfied:

1.   Shear at cutoff point is not more than 2/3 the allowable.
2.   Extra shear reinforcement is added.
3.   Continuing bar stress is not more than $f_y/2$ *and* shear is not over 3/4 the allowable.

Research showed that bending the bars from one beam face to the other avoided this shear problem.

(e) Arbitrary Requirements for Extending Bars   The ACI Code (12.10.3) requires that each bar be extended a distance d or 12 bar diameters beyond the point where it is "no longer required to resist flexure." This prohibits the cutting off of a bar at the theoretical minimum point, but can be interpreted as permitting bars to be *bent* at the minimum point. Three sound reasons can be advanced for this arbitrary extension, which the authors designate as a, in addition to the shear problem in Sec. 4.11d.

First, when a bar is simply cut off (not bent away from the main steel), there is a large transfer of stress from this bar to those remaining. This stress concentration will cause a moment crack in the concrete at the end of the bar if the beam is carrying its working load. Bending the bar spreads out this concentration; extending the bar, without bending, removes the concentration from a point of maximum steel stress to a point of lower steel stress. (If a bar is cut off at the theoretical minimum point, the remaining bars necessarily work at maximum stress.)

The authors give greater weight to a second reason, well stated in the long obsolete 1940 Joint Committee Specification:

> To provide for contingencies arising from unanticipated loads, yielding of supports, shifting of points of inflection, or other lack of agreement with assumed conditions governing the design of elastic structures, it is recommended that the reinforcement be extended at supports and at other points between supports as indicated.

Thus interpreted, the arbitrary extension of the bar is the result of an envisioned possible extension of the maximum moment diagram. It would follow that bars must not even be bent until most of this extra length has been provided. Tests show, however, that bending bars is less dangerous than cutting them off in tension zones. The authors would be satisfied to see *bends*, not cutoffs, at $d/2$ beyond the theoretical cutoff point. Even this would be a more severe requirement than many engineers observe.

The third and most compelling reason is developed in Sec. 5.5 where it is shown that the truss model for the design of web reinforcement for shear and diagonal tension implicitly increases the longitudinal reinforcement requirements in shear zones to values greater than that required by the moment at the section. This additional longitudinal reinforcement requirement imposed by the inclined shear cracking is automatically satisfied by extending longitudinal reinforcement the distance a before cutoffs and $a/2$ before bends.

The Code (12.12) requires that $\frac{1}{3}$ of the negative moment reinforcement be extended past the point of inflection by the larger of: beam depth d, 12 bar diameters, or $\ell_n/16$ where ℓ_n is the clear span. (In script and on the typewriter it is desirable to use a script ℓ to distinguish it clearly from the number one.)

The Code (12.11) also requires that at least $\frac{1}{3}$ of the positive moment reinforcement extend into the support in the case of simple spans and $\frac{1}{4}$ in the case of continuous beams. Custom usually increases these minimums. However, it must not be overlooked that the Code 12.11.2, for beams that are used as part of the primary lateral load system, requires the $\frac{1}{4}$ into the support be anchored for full f_y to give some ductility to the system. Here it is not acceptable to run twice as many bars into the support anchored for one-half of f_y; this will not give the desired ductility.

(f) Nomenclature   It is common to refer to the first bar cut off or bent, second bar, and so forth. First bar in this context means the one nearest the point of maximum moment, typically closest to the support for top bars and closest to midspan for bottom bars. It is thus possible for the first bar bent up to become the second bar bent down, and the second bent up the first bent down, when only two bars are offset.

Bars are often bent or cut off by pairs instead of singly. The first *pair* can be bent or cut off where the second *bar* is no longer needed, that is, beyond the extensions of Sec. 4.11e, which apply to the second bar.

Bars are sometimes bent at points that make them available as web reinforcement to resist shear (diagonal tension). Because present practice uses fewer bent bars, this value is often ignored in the United States, unless the member is a precast one on a production line.

(g) Authors' Philosophy on Bar Extensions   Beams using the offset bars be-have in a more desirable manner than those with bars cut off, but the labor involved in fabrication and erection often makes them uneconomical. Where members are designed against some earthquake hazard (which should mean over 90% of the United States), the sway motions imposed on a frame may reverse the beam moments at column faces. In such cases, as well as for improved general structural integrity in regions not subject to earthquakes, continuous top and bottom bars with staggered lap splices are strongly recom-mended to make a superior member. If continuous bars become common, the arbitrary extensions of Sec. 4.11e would have less impact because fewer bars would be totally cut off.

The Code phrase quoted in the third paragraph of Sec. 4.11e is of long standing with only minor editorial changes. The authors believe it has now become desirable to start thinking of the arbitrary extensions more as an *actual* potential shift in the moment diagram. As moment diagrams from frame and computer analyses replace the approximate moment coefficients, these more clearly show the omission of the factors the Joint Committee listed; these factors do cause real moments. The *idea* of shifted moment diagram becomes essential in some form if we move toward redistribution of moments (Code 8.4) and limit design.

As far as the lengths of bars *for moment* are concerned, the practical effect of the substitution of a shifted moment diagram for an arbitrary extension is nil; but it makes a large difference in the magnitude of the stress considered to exist next to the cutoff point in the bars that do continue farther. It thus substantially influences the development length demands of the continuing bars as discussed in Chapter 8.

4.12   Examples Involving No Excess Steel or Neglect of the Excess

(a) A uniformly loaded beam of 20-ft simple span requires 5-#9 bars with $d = 15$ in. Considering moment alone, where can the first bar be cut off? The second bar?

Solution

The maximum moment diagram is simply the full load moment diagram and this parabola can be used as the required A_s diagram as in Fig. 4.8. The maximum ordinate becomes five bars. Steel areas could be used as ordinates, but the number of bars provides a more convenient unit when all bars are of the same size.

The first bar can theoretically be cut off at x_1 where only four bars are required. Because offsets to the center tangent vary as the square of the distances from the center,

$$x_1^2/10^2 = 1/5 \qquad x_1 = 10\sqrt{1/5} = 4.47 \text{ ft}$$

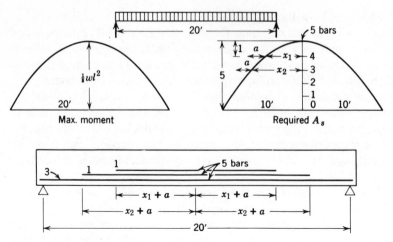

Figure 4.8 Cutting off bars in uniformly loaded beam according to the moment diagram.

Because d of 1.25 ft exceeds $12\,d_b = 12 \times 1.13/12 = 1.13$ ft, use $a = 1.25$ ft for cutoff (or 0.62 for bend). The minimum distance from the center line of the span to the first bar cutoff is $4.47 + a = 4.47 + 1.25 = 5.72$ ft (or 5.09 ft to a bend point).

For the second bar stopped,

$$x_2^2/10^2 = 2/5 \qquad x_2 = 10\sqrt{2/5} = 6.33 \text{ ft}$$

The minimum distance from the center to the second bar cutoff is $6.33 + 1.25$ ft $= 7.58$ ft, or $6.33 + 0.62$ to a bend point.

Cutting off bars in a tension zone, as above, requires special shear calculations (Sec. 5.15). Also development lengths must be checked as in Sec. 7.12c and 9.15.

(b) A uniformly loaded cantilever beam 8-ft long requires 7-#7 bars when d is 18 in. Where can the first pair of bars be bent down for moment?

Solution

The maximum moment diagram is a half-parabola, the maximum ordinate corresponding to seven bars of required A_s (Fig. 4.9). The first pair of bars can be bent where only five bars are required.

$$x^2/8^2 = 5/7 \qquad x = 8\sqrt{5/7} = 6.76 \text{ ft}$$
$$d = 1.50 \text{ ft} \qquad 12\,d_b = 0.875 \text{ ft} \qquad a = 1.50 \text{ ft for cutoff, or } 0.75 \text{ ft for bend}$$

The minimum distance from the support to the bend point for the first pair is $8 - 6.76 + a = 1.24 + 0.75 = 1.99$ ft. With the bar bent instead of cut off there is no loss of shear strength.

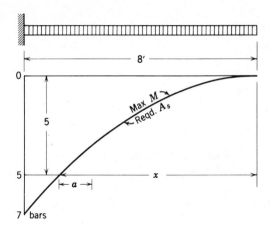

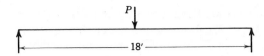

Figure 4.9 Bending bars for cantilever beam.

4.13 Examples Permitting Reduced Lengths Because of Excess Steel

(a) An 18-ft simple span beam with a fixed concentrated load at midspan (negligible uniform load) and $d = 22$ in. requires 4.50 in.^{2} of steel but uses 5-#9 bars $= 5.00$ in.^{2} Where can the first pair of bars be stopped?

Solution

Figure 4.10 shows the maximum moment diagram used as a required A_s curve with the maximum ordinate marked at 4.50 in.^{2} = 4.50 bars required. The first pair of bars can be stopped where only three bars are required.

$$x/9 = 1.5/4.5 \qquad x = 3.0 \text{ ft}$$

$$d = 1.83 \text{ ft} \qquad 12d_b = 1.13 \text{ ft} \qquad a = 1.83 \text{ ft for cutoff, or } 0.92 \text{ ft for bend}$$

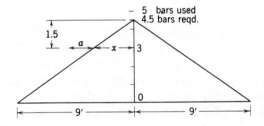

Figure 4.10 Stopping bars, excess steel.

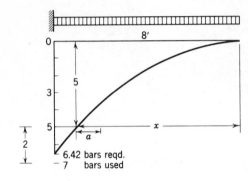

Figure 4.11 Bending bars, excess steel, cantilever beam.

The minimum distance from midspan to the stop point for the first pair is $3.0 + a = 3.00 + 1.83 = 4.83$ ft.

(b) A uniformly loaded cantilever beam 8-ft long with $d = 18$ in. requires 3.85 in.^{2} and uses 7-#7 bars. Where can the first pair of bars be bent down?

Solution

The maximum ordinate of the maximum moment diagram is designated as 3.85 in.$^2 = 3.85/0.60 = 6.42$ bars required (Fig. 4.11). The desired point is where five bars are needed.

$$x^2/8^2 = 5/6.42 \qquad x = 8\sqrt{5/6.42} = 7.06 \text{ ft}$$
$$d = 1.50 \text{ ft} \qquad 12d_b = 0.88 \text{ ft} \qquad a = 1.50 \text{ to cutoff, } 0.75 \text{ ft to bend}$$

The minimum distance from the support to the bend point for the first pair is $8 - 7.06 + a = 0.94 + 0.75 = 1.69$ ft.

4.14   Design Aids

Available design aids for rectangular beams have already been discussed in Sec. 3.17. Figures B.3 and B.4 (see Appendix B) are additional curves, particularly interesting for design. The latter figures are only one pair out of nine in Ref. 2 and cover only the combination of $f_y = 60$ ksi and $f'_c = 3000$ psi, one for slab elements 12-in. wide (the standard width for slab design) and one for an element 10-in. wide. The latter facilitates the use of proportional values for other desired beam widths.

The beam design of Sec. 4.3a is used as a sample. The design M_u, including assumed dead load, is 240 k-ft. (The available k_n or k_m is available several places, for example, in Fig. B.2 for $f_y = 60$ ksi and $f'_c = 3000$ psi k_n shows as 783.) Figure B.4 shows for this moment and the maximum reinforcement ratio a needed effective depth d of about 20 in. with an A_s of 3.25 in.^{2} if a width of 10 in. is used. If one prefers 9.5 in., the width used in Sec. 4.3a, this would mean an equivalent moment on the 10 in. width of $(10/9.5)240 = 253$ k-ft and the use of $9.5/10 = 0.95$ times the A_s shown for the 10 in. width. The 253 k-ft value of M_u on this figure shows minimum d of just under 21 in., compared to 20.8 in. in the example. For $d = 21$ in., the required A_s is 3.25 (from the chart) times $0.95 = 3.08$ in.^{2}. This compares to a final $d = 21.1$ in. and a final required A_s of 3.09 in.^{2} in the example.

The use of Fig. B.3 for slabs is similar except that $b_w = 12$ in. for a slab would not be modified. Revised design tables are available in Reference 3. The CRSI Handbook[7] has extensive tables for many types of construction and curves that appear very helpful. In addition, the rapid increase in the use of personal computers has accelerated the use of computer programs as design aids. In time such programs will probably replace handbooks in many applications. This text attempts to give the reader an understanding of fundamentals necessary before such programs are used.

Selected References

1. *ACI Detailing Manual,* SP-66, Amer. Concrete Inst., Detroit, 1980.
2. *Design Handbook, Vol. 1,* Special Publication No. 17(73), Amer. Concrete Inst., Detroit, 1973.
3. *Design Handbook in Accordance with the Strength Design Method of ACI 318-77, Vol. 1*–Beams, Slabs, Brackets, Footings, and Pile Caps, (1982) and *Vol. 2*—Columns, (1980), Amer. Concrete Inst., Detroit.
4. J. Colonna-Ceccaldi and S. Soretz, "Large Reinforced Concrete Beams with a Main Reinforcement Consisting of Two Thick Bars," Vol. 46 from the series *Betonstahl In Entwicklung,* Tor-Isteg Steel Corp., Luxembourg, 1972.
5. G. C. Frantz and J. E. Breen, "Cracking on the Side Faces of Large Reinforced Concrete Beams," *ACI Jour., Proc., 77,* No. 5, Sept.–Oct. 1980, pp. 307–313.
6. G. C. Frantz and J. E. Breen, "Design Proposal for Side Face Crack Control Reinforcement for Large Reinforced Concrete Beams," *Concrete International,* Oct. 1980, pp. 29–34.
7. *CRSI Handbook,* 4th ed., Concrete Reinforcing Steel Institute, Chicago, 1984.

Problems

NOTE *Problems in SI units start with Prob. 4.21.*

PROB. 4.1. In Prob. 3.5 calculate the exact steel that is needed for the loads given.
PROB. 4.2. In Prob. 3.5 calculate the exact steel that is needed if Grade 60 steel is used.
PROB. 4.3. Calculate steel for the channel section of Prob. 3.10 and Fig. 3.44 if $M_u = 105$ k-ft and the steel is changed to Grade 40.
PROB. 4.4
(a) A simple-span rectangular beam 20-ft long is to carry a service load of 3400 plf made up of 1600 plf dead load (including beam weight) and 1800 plf live load. Use $f'_c = $

4000 psi, Grade 60 steel, and design a beam, subject to later check on shear and bond, for ρ approximately $0.18 f'_c/f_y$. Make $b = 13$ in. and keep d in full inches. Choose bars.

(b) Rework part (a) except assume beam weight is not included in the dead load given.

(c) Rework part (b) except assume that h must be kept in whole, even inches and d selected from cover and stirrup dimensions, assuming #4 stirrups.

PROB. 4.5. Assuming deflections are not considered serious for the usage contemplated, redesign the beam of Prob. 4.4 with $b = 13$ in. and the minimum effective depth (full inches) permitted under the Code.

PROB. 4.6. A continuous rectangular beam with $b = 14$ in., $d = 24$ in., $f'_c = 5000$ psi, Grade 60 steel, must care for $M_u = -1100$ k-ft. Design the necessary steel, using $d' = 3$ in. if A'_s is necessary.

PROB. 4.7. Redesign the depth of Prob. 4.6 to avoid use of A'_s. Design the steel for moment assumed unchanged.

PROB. 4.8. A rectangular beam 13 in. wide × 20 in. deep to center of steel must carry a total factored moment M_u of 250 k-ft with $f'_c = 3000$ psi and Grade 40 steel. Find the required steel.

PROB. 4.9. A slab having $d = 4$ in. must carry a factored moment M_u of 3.9 k-ft per foot width for D.L. and 5.0 k-ft per foot for L.L. Find the required A_s per foot if $f'_c = 3000$ psi and steel is Grade 60. Specify bar size and spacing.

PROB. 4.10. A slab 7.5 in. thick with 1.5-in. cover to center of steel must carry its own weight and a service live load of 100 psf over a 15-ft simple span. Find the required A_s per foot if $f'_c = 3000$ psi and steel is Grade 40. Choose bars.

PROB. 4.11. From basic principles establish the value of the Code maximum steel ratio ρ for a rectangular beam with $f'_c = 3500$ psi and $f_y = 50{,}000$ psi, no A'_s.

PROB. 4.12. A rectangular beam 14 in. wide × 25 in. deep overall with 3-in. cover to center of steel must carry a factored moment of 500 k-ft. Find the necessary steel if $f'_c = 4000$ psi and steel is Grade 60.

PROB. 4.13. Design a double-reinforced beam 13 in. wide for a total $M_u = 500$ k-ft using $f'_c = 3000$ psi, Grade 40 steel, cover 4 in. to center of A_s, and basing the beam size on an approximate k_n of 800. Keep ρ-ρ' not more than $0.18 f'_c/f_y$, as in Sec. 4.5b.

PROB. 4.14. You wish to design a double-reinforced rectangular beam for a known moment (as in Sec. 4.5b) with $f'_c = 4000$ psi and Grade 60 steel. First you want to choose such a beam size that $\rho_1 = 0.18 f'_c/f_y$ will be consistent with the use of $\rho' = 0.01$ for A'_s. Assuming $d'/d = 0.12$ and that A'_s bars yield, find the numerical k_n to use in establishing the beam size. (*Suggestion:* Base k_n on an equation for $M_{n1} + M_{n2}$.)

PROB. 4.15. If the joist of Fig. 3.41 must care for a factored moment of 25 k-ft, calculate the required A_s. Take $f'_c = 5000$ psi and $f_y = 60$ ksi.

PROB. 4.16. A simple span beam with $d = 20$ in. and carrying uniform load over a 22-ft span requires 3.10 in.^{2} of steel and uses 6-#7 bars.

(a) Where can the first pair be bent up? The second pair?

(b) If the excess steel were neglected, where would these bends be permitted?

PROB. 4.17. A 21-ft simple span beam carries a large fixed concentrated load at 9 ft from the left end and requires $A_s = 4.40$ in.^{2} The steel used is 5-#9, $d = 20$ in. If the beam weight is disregarded, where can the first bar be bent up (each side of the load)? The third bar?

PROB. 4.18. A 12-ft cantilever beam carries a large concentrated load at its end, uniform load negligible. Required $A_s = 3.60$ in.^{2}, 4-#9 used, $d = 19$ in. Where can one pair of bars be stopped or bent down?

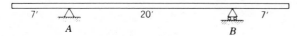

Figure 4.12 Simple beam with overhanging ends for Prob. 4.20.

PROB. 4.19. A 9-ft cantilever beam with $d = 16$ in. carries only uniform load and requires $A_s = 3.50$ in.^{2} If 4-#9 are used, where can one-half of the bars be bent down?

PROB. 4.20. The beam of Fig. 4.12 with $d = 18$ in. is subject to a fixed dead load of 1000 plf and a movable live load of 2000 plf. The required positive moment steel is 4.50 in.^{2} and the required negative moment steel is 2.50 in.^{2} If the positive moment steel used is 5-#9 where can the first pair of positive steel bars be stopped? If the negative moment steel is 4-#8 bars, where can this be reduced to two bars? (Note that distances on *each* side of the support or reaction are needed. See Sec. 1.5e for load factors.)

Problems in SI Units

PROB. 4.21. From basic principles, establish the maximum steel ratio ρ that the Code permits for a rectangular beam (without A_s') when used with $f_c' = 30$ MPa and $f_y = 400$ MPa.

PROB. 4.22. A rectangular beam 350 mm wide × 650 deep overall with 75-mm cover to center of steel must carry a ϕM_n of 700 kN · m. Find the necessary steel if $f_c' = 30$ MPa and $f_y = 400$ MPa.

PROB. 4.23. A rectangular beam 330 mm wide by 500 mm to steel centroid must carry a total ϕM_n of 350 kN · m with $f_c = 35$ MPa and $f_y = 300$ MPa. Find the required steel.

PROB. 4.24. A wide slab with $d = 100$ mm must carry a ϕM_n for dead load of 1800 N · m per 100 mm width of strip and 2300 N · m for live load. Find the required A_s per strip for $f_c' = 35$ MPa and $f_y = 400$ MPa and then specify bar size and spacing.

PROB. 4.25. Considering moment alone, design a rectangular beam 360-mm wide and of minimum depth (using 10 mm as the minimum step in d) to carry a service load M of 580 kN · m with an average load factor of 1.55, using C25 concrete and Grade 400 bars.

5

SHEAR AND TORSION

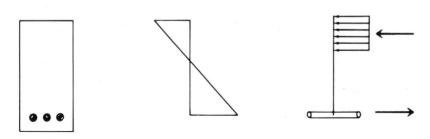

5.1   Introduction

Design procedures for shear and torsion in both reinforced and prestressed concrete members have become quite complex in recent Code revisions. The complexity stems from the highly empirical approach taken by the Codes and from a lack of a unified treatment of shear and torsion, which are actually closely interrelated. In the classical sense, both shear and torsion produce shear flows in the webs of members. For torsion, shear flows also are produced in the flanges. These shear flows can be thought of as developing shear stresses.

Concrete is relatively much weaker in tension than in compression, with real shear strength intermediate between the two. Most failures that would be termed shear failures are diagonal tension failures, occasionally diagonal compression failures. Although long usage has established shear stress as standard nomenclature, the phenomena involved will be more clearly understood if diagonal tension is kept in mind.

Shear stresses as normally computed are, except sometimes in prestressed concrete, stress coefficients only nominally related to the actual critical stresses. Almost the only place a real shear stress is computed is in the shear-friction theory (Sec. 5.19).

5.2   Diagonal Tension Before Cracks Form

In homogeneous beams diagonal stress can be analyzed by well-established relationships. Reinforced concrete beams, prior to the formation of cracks, probably have stresses quite similar to those of a homogeneous beam.* The diagonal tension stresses are emphasized here.

---

* The notation of this section on basic strength of materials does not attempt to follow the new standard ACI notation. C and T are total forces; c, t, v, f_c, and f_t are unit stresses.

In the beam of Fig. 5.1a a small element at the neutral axis at A would be subject to a shearing stress v, but no bending stress. Figure 5.1b shows that such an element will develop unit diagonal tensile and compressive stresses of magnitude v. An element at B in Fig. 5.1a will have a compressive stress f_c in addition to shear, adding a diagonal compression on both diagonals, Fig. 5.1c; combined with the shear of Fig. 5.1b, the diagonal tension on section a-a is reduced and the diagonal compression on section b-b is increased. Vertical loading on top of the beam causes a vertical compression that combines similarly, although rotated 90° to that sketched in Fig. 5.1c.

Similarly, an element at C in Fig. 5.1a will carry a tension stress as well as a

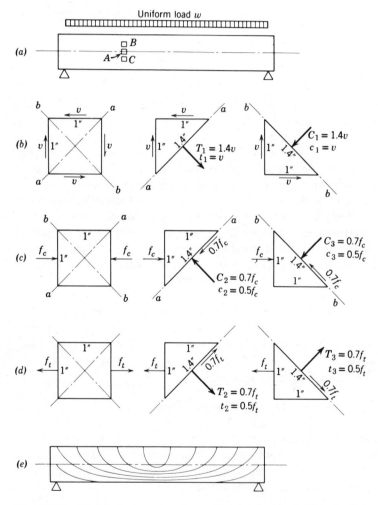

Figure 5.1 Diagonal stress in a homogeneous beam. (a) Typical beam under uniform load. (b) Analysis of stresses at A. (c) Analysis of added stresses at B. (d) Analysis of added stresses at C. (e) Tension stress trajectories.

shear, which adds a diagonal tension on both diagonals, as shown in Fig. 5.1*d*. These tension stresses increase diagonal tension on section *a-a* and reduce diagonal compression on section *b-b*. The combined stresses are maximum on sections other than *a-a* and *b-b*, with tension being maximum on a steeper plane than *a-a* when the (horizontal) direct stress is tension, or on a flatter plane when the (horizontal) direct stress is compression. A load hung from the bottom of the beam adds to the diagonal tension and makes it steeper.

The relation developed in mechanics for maximum diagonal tension, adjusted to the notation used here is:

$$t = f/2 + \sqrt{(f/2)^2 + v^2}$$

In this relation *t* is the unit diagonal tension, *f* is the unit direct stress, taken as positive when it is tension, and *v* is the unit shear. The direction of the maximum diagonal tension is given by the relation:

$$\tan 2\theta = 2v/f$$

where θ is the angle *t* makes with the stress *f*, in this case with the horizontal.

Figure 5.1*e* illustrates the approximate trajectory of maximum tensile stresses in a homogeneous rectangular beam under uniform loading. When the tensile strength of the concrete is exceeded it will crack. Tension cracks would be roughly perpendicular to the trajectories shown in Fig. 5.1*e* for a uniformly loaded beam. In an unreinforced concrete beam cracking would produce failure. In a reinforced concrete beam the maximum tensile stress pattern will be very similar until cracks open. The first cracks to open are usually vertical cracks owing to moment in the lower half of the beam, as shown at *A* in Fig. 5.2*a*. Diagonal tension cracks usually open at approximately 45° with the axis of the beam, starting typically from the top of a moment crack, *B*, but in short shear spans or deep beams starting independently near the neutral axis, *C*.

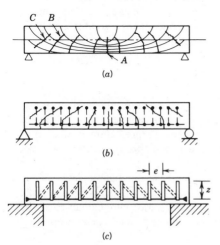

Figure 5.2 Reinforced concrete beam under uniform load. (*a*) Crack formation indicated by tension stress trajectories. (*b*) Typical crack formation in a beam with stirrups. (*c*) Ritter's truss model 1899 (Reference 1).

5.3　Shear Strength—Truss Model and Code Format

Although actual understanding of shear and torsion strength of reinforced and prestressed concrete members has increased dramatically in recent decades, the present Code treatment is largely empirical. It is not always easy to see why one case is algebraically so different from another. There are, however, research data to back up these rules, but some simpler correlation should eventually be found. Some special cases are discussed separately in Secs. 5.17 and 5.21. Strangely enough, the most promising and useful method of correlating many of the complex behavior mechanisms is probably one of the oldest approaches to shear and torsion. In an actual reinforced concrete beam (such as Fig. 5.2b) provided with both longitudinal reinforcement to counter the flexural tension stresses and vertical web reinforcement (or stirrups) to counter the diagonal tension stresses, the crack pattern observed on loading is very similar to the hypothetical pattern of Fig. 5.2a. In 1899, Wilhelm Ritter[1] correctly interpreted the general nature of diagonal tension and presented the concept of the "truss model" for the design of web reinforcement. Ritter introduced his model to dispel the idea that the main function of the stirrups was to resist horizontal shearing stresses by a dowel-type action for which vertical wooden pegs were used in timber beams. Ritter referred to the sketch shown in Fig. 5.2c and stated:

> In this connection, one ordinarily imagines that the stirrups together with the (longitudinal) bars and the concrete form a type of truss in which the stirrups act as the hinged tie bars and the concrete working in the direction of the dashed lines, acts as the diagonal struts. These lines will routinely be assumed up to 45° corresponding to the compressive pressure curve.
>
> On the basis of this view, the stirrups will then be calculated from statics and in fact according to the formula $Q = 2\sigma bd$, where Q indicates the shear force, and b and d are the width and thickness (respectively) of the flat steel bars used as stirrups. The factor 2 is included therein because each stirrup has two legs.
>
> The formula presupposes that the stirrup spacing e is equal to the distance z between the compression and tension centroids. If one makes e greater than z the stresses increase proportionately. It is thus general as
>
> $$\sigma = Qe/2bdz$$
>
> . . . The mode of action of the stirrups however, exists in my opinion as expressed earlier herein, that they resist the tensile stresses acting in the direction of the (diagonal) tension curves and that they prevent the premature formation of cracks. To this end they have to of course be provided approximately at 45°; yet this arrangement would complicate construction.
>
> That the stirrups in a vertical position also increase the load capacity of the beam, one can hardly deny; however, in what way the formula above can make claims to reliability and corresponds to the relative relationship, cannot easily be determined on a theoretical basis. Here, comparative tests might be appropriate.[1]

In almost a century since Ritter's formulation of the basic truss model, thousands of tests have been run to prove or disprove the theory and to find a "better model."

By 1909, based on extensive tests at the University of Wisconsin and the University of Illinois, Withey[2,3] and Talbot[4] examined Ritter's equation and found that it gave stirrup tensile stresses that were too high when compared to measurements made in tests. Talbot[4] concluded:

> Stirrup stresses computed by Ritter's equation appear too high. It is therefore recommended that stirrups be dimensioned for two-thirds of the external shear, the remaining one-third being carried by the concrete in the compression zone.

Decades of study have indicated that although the basic truss model correctly depicts the role of the stirrups and the diagonal compression struts that form between the inclined cracks (see Fig. 5.3*a*), there are several additional

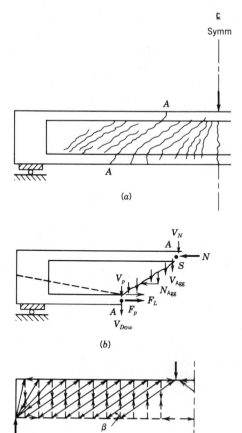

Figure 5.3 The truss model for beams with web reinforcement at failure. (*a*) Crack pattern at failure. (*b*) Free-body diagram. (*c*) Truss model.

load-carrying mechanisms that can supplement the basic truss model. If one cuts a free body along a major inclined crack such as *A-A*, the free-body diagram of Fig. 5.3*b* results. The forces in the truss model, compression (N) and tension (F_L) chords, and the vertical forces in the stirrups (S) can be supplemented by other forces such as V_p, the vertical component of inclined prestressing tendons; V_{Dow}, the dowel action of the longitudinal bars; V_N, the vertical component of the compression chord force; V_{Agg} and N_{Agg}, the aggregate interlock transfer across a jagged crack. These components of force transfer must either be neglected or treated indirectly in the modern truss models and compression field theories that provide a general approach to shear and torsion.[5–7] In the authors' opinion, the generalized truss model (Fig. 5.3*c*) with compression struts at a variable angle of inclination β offers almost unlimited possibilities to the designer and can greatly improve the understanding of design for shear and torsion. In lightly loaded members carrying relatively low levels of shear and torsion, the other force carrying components are significant and some supplementary design mechanism continues to be necessary for economy.

European practice is moving toward fewer stirrups as a result of low measured stirrup stresses in test beams.[15] They reason that (1) the compression members (concrete) act at flatter angles than 45° and thus deliver a smaller component to the stirrup verticals, and (2) the upper chord of the "truss" forms as an inclined thrust line that also leads to lower stress in the verticals. The second reason is similar to ACI's reasoning but is more specific as to *how* the concrete carries part of the shear.

Present ACI Code philosophy for shear design was suggested by ACI Committee 326[8] in 1962 as a logical development in the Ritter–Talbot tradition. It basically assumes:

1. For a beam with no web reinforcement, the shearing force that causes the first diagonal cracking can be taken as the shear capacity of the beam. For a beam that does contain web reinforcement, the concrete (supplementing the truss action) is assumed to carry a constant amount of shear (V_c), and web reinforcement (truss verticals) need only be designed for the shear force (V_s) in excess of that carried by the concrete (supplementary action).
2. The amount of (supplementary) shear (V_c) that can be carried by the concrete at ultimate is at least equal to the amount of shear that would cause diagonal cracking.
3. The amount of shear (V_s) carried by the reinforcement (stirrups) is calculated using the truss analogy with a 45° inclination of the diagonal members.

This philosophy is embodied in the present ACI Code Chapter 11. It produces generally safe designs except that in the authors' opinion at very high levels of shear (near and above present Code V_s maximums) the extensive, wide inclined cracks certainly reduce the V_c component and reliance should be placed primarily on the truss mechanism (V_s). This conclusion is supported by test results. Some foreign codes use a transition expression that reduces V_c for higher values of total shear.

The 1986 Code expresses all allowable shears in terms of the allowable total shear. (The 1971 Code expressed all allowable shears in terms of unit stresses.) There is almost no change in the allowables except the form of presentation. In the author's opinion calculations in one form are just about the same in length as in the other. Some designers have always drawn shear reinforcement curves in terms of unit stresses, although some others, in terms of total shear. In this presentation examples are presented in terms of forces to match the Code format.

Nearly all beams must have web reinforcement for shear, usually stirrups (Fig. 5.8), and occasionally bent bars from flexural reinforcement.

The basic Code statement on shear strength is that the factored shear force V_u shall be equal to or less than the design shear ϕV_n, that is,

$$V_u \geqslant \phi V_n \quad \text{where} \quad V_n = V_c + V_s$$

V_c = nominal shear strength provided by concrete (supplementing the truss)

V_s = nominal shear strength provided by shear reinforcement (usually stirrups proportioned by the truss model)

$\phi = 0.85$

The bulk of Code Chapter 11 is built around permissible values of V_c and V_s, plus definitions of critical sections.

To understand how stirrups improve strength, it is desirable first to understand how beams fail when they contain no web reinforcement such as stirrups.

5.4   Beam Behavior Under Shear Loading—Without Stirrups*

(a) Failure Types by Appearance   A shear failure appears to be least complicated when it occurs away from loads and reaction. It is convenient to classify shear failures in terms of the distance between test load and reaction, a distance designated as shear span or a, Fig. 5.5a.

The simple *diagonal tension failure* of Fig. 5.4 occurs when the shear span is more than $3d$ or $4d$, under which conditions there is ample room for the full crack to develop to failure.

As the shear span is reduced, the load itself begins to influence the diagonal crack more and more in a manner that increases shear capacity. The failure sketched in Fig. 5.5a is called a *shear-compression failure* and occurs when the shear span is from d to $2.5d$, with a rapidly reducing influence in the range of $2.5d$ to $4d$, as indicated in Fig. 5.6a.

---

* Stirrups are vertical U-shaped reinforcement that anchor in the compression face of the beam and enclose the flexural steel bars. See Secs. 5.5 and 5.7.

When the shear span is less than d, the pattern changes to a *splitting failure* or compression failure at the reaction at a still higher load, and a tendency toward sudden failure, as sketched in Fig. 5.5b.

(b) Diagonal Tension Failure Always in the range of a/d above 2, and sometimes at lower a/d values, the diagonal crack starts from the last flexural crack and turns gradually into a crack more and more inclined under the shear loading, as noted in Fig. 5.4a. Such a crack does not proceed immediately to failure, although in some of the longer shear spans this either seems almost to be the case or an entirely new and flatter diagonal crack suddenly causes failure. More typically, the diagonal crack encounters resistance as it moves up

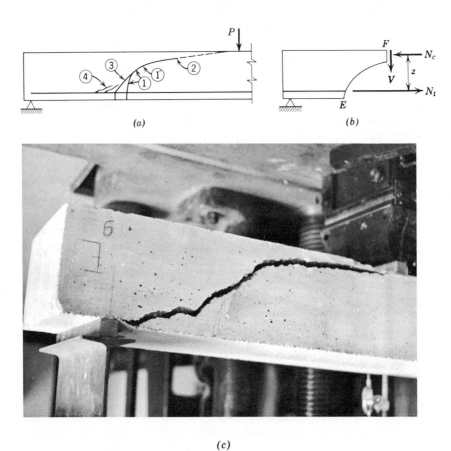

(a)   *(b)*

(c)

Figure 5.4 Development of a diagonal tension crack when loads and reactions are far apart. (a) Diagram showing sequence in crack formation. (b) Equilibrium sketch for portion of beam. (c) Failure of beam. The failure crack developed from the flexural crack faintly seen about one beam from the end. This crack turned gradually into the diagonal crack, as at 1 in the sketch. The final wide crack is comparable to ②–①–③–④ in (a). (The failure picture has been inverted to make this comparison easier.)

into the zone of compression, becomes flatter, and stops at some point such as that marked ① in Fig. 5.4a. With further load, the tension crack extends gradually at a very flat slope until finally sudden failure occurs, possibly from point ②. Shortly before reaching the critical failure point at ② the more inclined lower crack ③ will open back, at least to the steel level, and usually cracks marked ④ will develop. Figure 5.4c illustrates a failure with the start of the crack nearer the end than the usual a/2, with this location resulting because two tension bars were cut off at the crack point. This crack is called a flexure-shear crack.

(c) Shear-Compression Failure   Very often the development of the diagonal crack described previously is stopped by the presence of a nearby load, as indicated in Fig. 5.5a. Then the vertical compressive stresses under the load reduce the possibility of further tension cracking, and the vertical compressive stresses over the reaction likewise limit the bond splitting and diagonal cracking along the steel. Alternatively, a large shear in short shear spans (especially in I-beams) may initiate approximately a 45° crack (called a web-shear crack) across the neutral axis before a flexural crack appears. Such a crack crowds the shear resistance into a smaller depth and, by thus increasing the stresses, tends to be self-propagating until stopped by the load or reaction. With either start,* a compression failure finally occurs adjacent to the load.

This type of failure has been designated as a shear-compression failure because the shaded area also carries most of the shear and the failure is caused by the combination. Such a failure can be expected to occur when the shear span a, as indicated in Fig. 5.5a, is less than four times the beam depth, or possibly a little less for lightweight concrete or very high strength concrete. When the shear span is small, the increased shear strength may be significant, with the ultimate shear over twice as much for $a = 1.5d$ as for $a = 3.0d$ (Fig. 5.6). The width of the critical crack, if there is no crack control steel, becomes large as the load increases, sometimes over $\frac{1}{8}$-in.

Occasionally with inadequate anchorage of the flexural steel beyond the crack, the small diagonal cracks (raveling cracks) will split out the bars before the compression failure occurs—this is called a shear-tension failure. The beam here acts somewhat as a tied arch with a thrust line from the load toward the reaction area, which demands full anchorage of the bars beyond the crack. Stirrups, now required on most beams, almost totally eliminate the raveling.

An overemphasis shows here concerning how flexural compression in the beam chokes off the diagonal tension failure and adds strength in shear (Fig. 5.6, especially at a/d of 1.5 or 2). What happens near a point of inflection (P.I.) where moment is too small to crack the beam? We ran some tests trying to make beams fail in shear near the P.I., but the only shear failure at the P.I. was when we had added shear reinforcement (stirrups as in Sec. 5.5) everywhere

---

* The second case is not always classified as shear-compression; the term web-shear crack is often used.

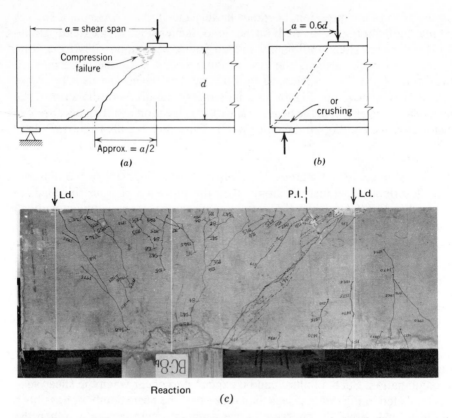

Figure 5.5 (*a*) Shear-compression failure when shear span is small. Shear strength is increased. (*b*) Shear span less than *d*. (*c*) Failure in short shear span of semicontinuous beam from load to reaction.

except around the P.I. This failure was at substantially *higher* unit shear stress. It is safe to say that in nonprestressed beams shear weakness develops from the weakness already introduced by a moment crack and this combination makes it necessary to hold shear stresses low or to add stirrups.

(d) Splitting or True Shear Failure   Where the shear span is less than the effective depth *d*, the shear is carried as an inclined thrust between load and reaction that almost eliminates ordinary diagonal tension concepts. Shear strength is much higher in such cases.[11] The final failure, as in Fig. 5.5*b*, becomes a splitting failure, almost like the vertical splitting of a compression test cylinder (which occurs when end friction is reduced) or it may fail in compression at the reaction. The analysis of such an end section is closely related to the analysis of a deep beam having a span of 2*a*. Reference 11 has shown that it is the clear shear span between bearing and loading plates that is critical.

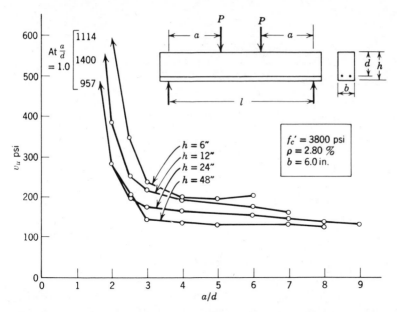

Figure 5.6 Influence of a/d and beam depth on shear resistance, without stirrups. (Adapted from References 9 and 10, ACI.)

(e) Code Shear Strength V_c Because of the objectionable amount of cracking for a member like that in Fig. 5.4, the Code defines the nominal shear strength V_c on the basis of the shear when the diagonal crack first occurs. In many cases just when a crack becomes diagonal is a little vague, as in Fig. 5.4; however, it is not at all vague in Fig. 5.5.

(f) Design Considerations Relating to Shear Span As a result of the shear span effect, the worst position (on the basis of diagonal tension failure) for a concentrated load on the usual simple span test beam is not adjacent to the reaction but at some distance out on the span. The Code uses the distance d from the face of support (Code 11.1.2.1).

In test beams most of the extra resistance beyond initial diagonal cracking is created by the vertical compressive stress under the load and over the reaction. Loads or reaction applied as side face shears, as in Fig. 5.7, create very little vertical compression and add very little shear resistance. If loads are applied below middepth of a girder, say, from precast beams bearing on a lower flange or bracket attached near the bottom of the girder, diagonal cracking occurs at lower loads and ultimate strength is reduced. Extra stirrups (Fig. 5.8) as hangers to pick up such loads are essential in such cases.

Tests have also shown that if the loads at a small a/d are applied as shear loads on the side of the beam, as in Fig. 5.7a, or if the reaction is picked up in shear (as when a beam frames into the side of a girder, as in Fig. 5.7b), not

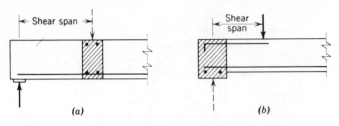

Figure 5.7 Cases where small shear span does not significantly increase shear strength. (*a*) Girder loaded by shear from beam. (*b*) Beam supported by girder.

much increase in shear resistance is obtained as a result of the small shear spans.[13]

5.5   Vertical Stirrups—Influence on Behavior

To understand how stirrups improve beam behavior and strength, consider the stirrup simply as vertical reinforcement spaced along the length of the beam at not more than $0.5d$ on centers, well anchored in the compression zone of the beam and usually bent around the longitudinal tension bars, as shown in Fig. 5.8. In Fig. 5.4 several stirrups, if any had been provided, would have crossed the failure crack. As shown in Fig. 5.3*b*, each stirrup would develop a vertical force S across the crack.

Prior to concrete cracking, the vertical stirrup carries essentially no stress, possibly even a little compression arising from vertical shrinkage of the concrete. However, the stirrup must go into tension as the diagonal crack crosses it; and this tension controls and limits the progress of the crack, delaying the failure of the beam until higher loads are imposed. The next stirrup (toward the reaction) would also hold the flexural steel in place and not allow the bars to be pushed downward by so-called dowel action creating the "raveling out" mentioned as cracks 3 and 4 in Fig. 5.4*a*. Until the stirrup crossing the cracks yields (under higher load), the beam cannot fail in shear.

The Code defines V_s as the nominal shear strength provided by the shear reinforcement, thus the *increase* in total shear capacity above the V_c at first diagonal cracking. The Code equations are given in Sec. 5.6 and the V_s term is based on a truss model with compression diagonals assumed at 45°. The appli-

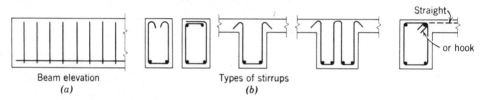

Figure 5.8 Vertical stirrups.

cation of the truss model and the assumption of the development of full plasticity before failure allow the simple equilibrium determination between V_s and the stirrup force $S = A_v f_y$. Because the truss analogy does not recognize the concrete contribution V_c, then $V_n = V_s$. In the following derivations, it will be assumed $V = V_s$. A_v is the cross-sectional area of the vertical shear reinforcement. For instance, in a U-shaped stirrup with two legs, $A_v = 2A_b$, where A_b is the bar cross-sectional area. Figure 5.9 shows the general relation between shear, moment, and the forces in the stirrups S for an assumed inclined crack (direction of compression strut) at an inclination β. When $\beta = 45°$ (the ACI Code basis), cot $\beta = 1$. The number of stirrups required n becomes simply $n = V_s/S$ and the stirrup spacing s is given as $s = z/n = zS/V_s \approx dA_v f_y/V_s$. This is the essential basis of Code Eq. 11.17.

Although the ACI Code format is written in a *cross section* format, implying that V and V_s are shears on a vertical cross section, in reality the actual mechanism is a *member* format with the section cut along GH in Fig. 9.5b. This is important when one considers the force in the tension chord F_L. Note that the force in this chord is not M_A/z as has been assumed in the chapters on flexural design. Summing moments about H shows that $F_L = M_A/z + (V/2)$ cot β. For $\beta = 45°$, $F_L = M_A/z + V/2$. This indicates that in a beam with inclined cracks,

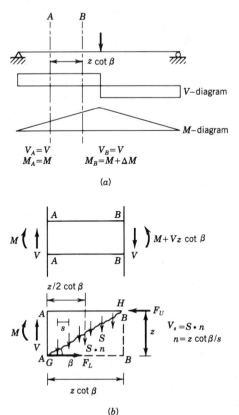

(a)

(b)

Figure 5.9 Forces in the truss model. (a) Actions. (b) Equilibrium of free body.

the tension reinforcement at the base of the crack is more than that calculated by taking a vertical cross section at that point. In reality, it is more like the amount that would be required by the greater moment towards the tip of the inclined crack. If no web reinforcement were present, Fig. 5.9*b* would indicate that $F_L = M_A/z + V$. Then the moment governing the design of the longitudinal steel quantity at *A* would be the moment at *B*. With adequate web reinforcement, the moment governing the longitudinal steel quantity at *A* is the moment midway between *A* and *B* ($\Delta M = Vz/2$). To simplify design, the ACI Code bases determination of flexural reinforcement on the simple vertical cross section and says the longitudinal steel quantity at *A* depends on the moment at *A*. However, in the generalized detailing provisions of Code 12.10.3, all reinforcement is required to extend beyond the point where it is no longer required to resist flexure for a distance at least equal to the effective depth of the member. This provision indirectly ensures that the quantity of reinforcement at *A* will be at least the amount required by the moment at *B*. In Fig. 5.9 the amount of A_s to develop M_B would have to be extended towards *A* for a distance *d* according to Code 12.10.3. This would ensure that the actual A_s at *A* would in fact be A_{sB} after complying with the detailing provisions. This indirect procedure is much easier than calculating longitudinal reinforcement as required for flexure and then adding extra longitudinal reinforcement required for shear. However, the indirect procedure is only valid for combined flexure and shear. Where torsion is present, extra longitudinal reinforcement will have to be explicitly added. Section 5.7 discusses minimum stirrups and maximum stirrup spacings, Sec. 5.8 some Code limits on size of bars used for stirrups, Sec. 5.9 gives some simple examples, and Secs. 5.10 and 5.11 give more detailed background.

One vital function of stirrups for which vertical stirrups are ideal is that of picking up any shear carried by dowel action of the longitudinal steel across a diagonal tension crack. They prevent the raveling out just above the steel at the end of a diagonal crack by transferring the dowel action load into a stirrup stress instead of into a vertical tension stress on the concrete.

Vertical stirrups are adaptable to shear forces acting either upward or downward since, in terms of the truss analogy, only the diagonal crack and the compression diagonal must take a different direction.

5.6 Code Evaluation of V_c and V_s for Beam Containing Vertical Stirrups

(a) **Normal Weight Concrete** With or without stirrups, the nominal shear strength provided by the concrete is V_c.

$$V_c = 2\sqrt{f_c'}\,b_w d \qquad \text{(Code Eq. 11.3)}$$

or

$$V_c = (1.9\sqrt{f_c'} + 2500\rho_w V_u d/M_u)bd \gtrless 3.5\sqrt{f_c'}\,b_w d \qquad \text{(Code Eq. (11.6)}$$

with $V_u d/M_u$ taken not greater than 1.0;

where

b_w = web width

d = effective depth, as for M_n

$\rho_w = A_s/b_w d$ = ratio of longitudinal tension reinforcement

V_u = factored shear force at section

M_u = factored moment at section

The designer may use *either* equation and will soon note that only a few situations give large differences between them. The much simpler Code Eq. 11.3 is generally used except where shear is a substantial governing concern in well-reinforced short, stubby members or well-reinforced members with heavy concentrated loads near the supports. In these cases Code Eq. 11.6 may have some economic value. When one considers the relatively low cost of web reinforcement and the generally small difference between 2 and $3.5\sqrt{f'_c}\,b_w d$ (which is the total range of the V_c term), it is not a big factor even here considering the complexity of application.

Attention is called first to the coefficient 2 in Code Eq. 11.3. It is neither 2.0 nor 2.00, although it is usually used as though it were. It represents a judgment decision of code engineers having much detailed knowledge of research in this area. It would probably be lower if the reinforcement ratio A_s/bd were normally as low as 0.005 or if the Code did not require nominal stirrups wherever V_n was greater than half V_c (Sec. 5.7). Secondly, note that Code Eq. 11.4 uses $1.9\sqrt{f'_c}$ plus another term in parenthesis. The 1.9 is another judgment decision, no more accurate probably than the 2 used in Code Eq. 11.3; it might reasonably be taken as meaning that with many values of ρ and $V_u d/M_u$ this equation should give close to what the other gives, but with large ρ values and $V_u d/M_u$ values it can give higher values.

With SI units the constants become something different. In 1983 the Code was translated into SI units in ACI 318-83M.[14] This is a "soft" or less exact translation that reflects engineering judgment.

When the SI version of the Code is used, $\sqrt{f'_c}$ is in MPa and the values of b_w and d are in mm. The equations at the start of this section then become:

$$V_c = (\sqrt{f'_c}/6)b_w d$$
$$V_c = [(\sqrt{f'_c} + 120\rho_w V_d/M_u)/7]b_w d \gtrless 0.3\sqrt{f'_c}\,b_w d$$

with $V_u d/M_u$ not greater than 1.0.

The 1986 Code is written in a shear *force* format $V_u \gtrless V_n = \phi(V_c + V_s)$. As with flexure design, the authors prefer to remove the ϕ factor early by using $V_u/\phi \gtrless V_n = V_c + V_s$. The needed $V_s = V_n - V_c$ is then easily determined by subtracting the V_c determined from Code Eq. 11.3 or 11.6 from the required V_n.

The nominal shear strength added from vertical stirrups is given in Code 11.5.6:

$$V_s = A_v f_y d/s \quad \text{(Code Eq. 11-17)}$$

where A_v is the shear reinforcement (two bar areas for U-stirrups) and s is the spacing between stirrups in a direction parallel to the axis of the member. This is the same in customary and SI units.

(b) Lightweight Aggregate Concretes   Some lightweight aggregate concretes have a lower shear resistance than ordinary aggregate concrete of the same compressive strength. The Code (11.2) provides alternate procedures: (1) stress $\sqrt{f'_c}$ to be multiplied by an arbitrary 0.75 factor for "all-lightweight" and 0.83 for "sand-lightweight" concrete,* or (2) when f_{ct} is suitably specified the use of $f_{ct}/6.7 \gtrless \sqrt{f'_c}$ instead of $\sqrt{f'_c}$ in stating the allowable. The term f_{ct} is the average splitting tensile strength. In the SI version, $1.8 f_{ct} \gtrless \sqrt{f'_c}$.

5.7   Code Requirements for Minimum Stirrups and Maximum Spacings

Code 11.5.5 requires at least minimum stirrups where the total factored shear force V_u exceeds 1/2 the shear strength provided by the concrete ϕV_c except in:

(a)   Slabs and footings.

(b)   Concrete joist construction as defined in Code 8.11.

(c)   Beams with total depth not more than 10 in., 2.5 times the flange thickness, or one-half the web width.

Where torsion is small (Code 11.5.5.3) the minimum area of shear reinforcement A_v at spacing s is

$$A_v = 50 b_w s / f_y \qquad \text{(Code Eq. 11-14)}$$

for b_w and s in inches. Maximum spacing (Code 11.5.4) of vertical stirrups is $d/2$ in nonprestressed members, but is decreased (Code 11.5.4.3) to $d/4$ where $V_s > 4\sqrt{f'_c} b_w d$. The maximum V_s is limited to $8\sqrt{f'_c} b_w d$ (Code 11.5.6.8).

The 318M Code for A_v in mm² shows minimum $A_v = (1/3) b_w s / f_y$; also $4\sqrt{f'_c}$ and $8\sqrt{f'_c}$ become in MPa $-\sqrt{f'_c}/3$ and $2\sqrt{f'_c}/3$. The 1983 Code vertical stirrup spacing formula gives $s = A_v f_y d / V_s$ or $V_s = A_v f_y d / s$ (Code Eq. 11.17).

Stirrup spacing for stress varies inversely with V_s, with the stirrup located in the middle of the length it serves. The first stirrup should thus be placed at $s/2$ from the support as shown in Fig. 5.10.

The maximum spacings insure that every 45° line representing a potential crack will be crossed in the tension side of the beam by at least one stirrup, as shown in Fig. 5.10a, and by two stirrups where the shear is high. In many beams #3 U-stirrups spaced at $d/2$ will automatically provide more than V_s requires. Where deformed wire is available, its use as web reinforcement may prove economical.

Because the various limits for minimum web reinforcement, maximum stir-

---

* "Sand-lightweight" is concrete using natural sand for fine aggregate.

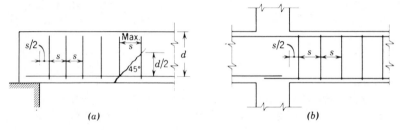

Figure 5.10 Arrangement of vertical stirrups.

rup spacing, and maximum V_s are arranged inconsistently in the ACI Code, the authors urge their students to proportion stirrups based on the type of tabulated limits shown in Table 5.1. The boundaries (in terms of V_n) for four design regions are quickly determined. Each region has different requirements for proportioning stirrups as shown. The table is based on the use of $V_c = 2\sqrt{f'_c}b_w d$. It is invalid if the more complex version of V_c (Code Eq. 11.6) is used to take advantage of increased V_c contribution in stubby or short shear span members.

5.8 Code Limitations on Sizes of Bars Used for Stirrups

Although the general treatment of the development lengths necessary to bond the bar and concrete together is deferred to Chapter 8, a limited discussion

TABLE 5.1 Summary of Limits for Design of Vertical Stirrups Under the ACI Code

Region	V_n Boundaries	A_v	Maximum s	ACI Code Sec.
I	$0 \gtrless V_n \gtrless 1\sqrt{f'_c}b_w d$	$V_n \gtrless V_c/2$ None Required	N.A.	11.5.5.1
II	$1\sqrt{f'_c}b_w d < V_n$	Minimum	$d/2$	11.5.5.1 11.5.5.3
	$V_n \gtrless (2\sqrt{f'_c} + 50)b_w d$	$Av = 50 b_w s/f_y$		11.5.4.1 11.5.4.3
III	$(2\sqrt{f'_c} + 50)b_w d < V_n$	$Av = \dfrac{V_s s}{f_y d}$	$d/2$	11.5.5.3 11.5.6.2
	$V_n \gtrless 6\sqrt{f'_c}b_w d$ $[V_c + V_s = (2 + 4)\sqrt{f'_c}b_w d]$			11.5.4.1 11.5.4.3
IV	$6\sqrt{f'_c}b_w d < V_n$	$Av = \dfrac{V_s s}{f_y d}$	$d/4$	11.5.6.2 11.5.4.3
	$V_n \gtrless 10\sqrt{f'_c}b_w d$ $[V_c + V_s = (2 + 8)\sqrt{f'_c}b_w d]$			11.5.6.8

before other aspects of stirrup capacity are applied to design will be helpful here. Chapter 8 shows that the Code requires a development length between point of zero bar stress and a point of f_y stress which, for #3 to #5 bars, increases directly as the product $d_b f_y$* increases. Thus, it requires a development length ℓ_d 5/3 times as long as to develop a #5 bar to a given f_y stress as it does for a #3 bar.

Code 12.13 deals with development of web reinforcement (stirrups) and only a few subsections are needed here. The stirrup's maximum stress is considered to exist at $d/2$ from the beam compression face. The Code reads:

12.13.1—Web reinforcement shall be carried as close to compression and tension surfaces of member as cover requirements and proximity of other reinforcement will permit.

12.13.2—Ends of single leg, simple U-, or multiple U-stirrups shall be anchored by one of the following means:

12.13.2.1—A standard hook plus an embedment of $0.5\ell_d$. The $0.5\ell_d$ embedment of a stirrup leg shall be taken as the distance between middepth of member $d/2$ and start of hook (point of tangency). (Note: A minimum ℓ_d of 12 in. is not required here, Code 12.2.5.)

12.13.2.2—Embedment $d/2$ above or below middepth on compression side of the member for a full development length ℓ_d but not less than $24d_b$; or for deformed bars or deformed wire, 12 in.

12.13.2.3—For #5 bar and D31 wire, and smaller, bending around longitudinal reinforcement through at least 135 deg. plus, for stirrups with design stress exceeding 40 000 psi, an embedment of $0.33\ell_d$. The $0.33\ell_d$ embedment of a stirrup leg shall be taken as the distance between middepth of member $d/2$ and start of hook (point of tangency).

12.13.2.4 and 12.13.2.5—(omitted here—for welded wire fabric)

12.13.3—Between anchored ends, each bend in the continuous portion of a simple U-stirrup or multiple U-stirrup shall enclose a longitudinal bar.

12.13.4 and 12.13.5—(not reprinted here) cover bent bars acting as web reinforcement and lapping of pairs of U-stirrups to form a closed unit, respectively.

The following ℓ_d values are minimum stirrup values without a hook, in any usual grade of concrete.

Bar	Grade 40	Grade 60	Lightweight
#3	6 in.	9 in.	Multiply by
#4	8 in.	12 in.	factor > 1.0
#5	10 in.	15 in.	See Code 12.2.3.3

---

* Code 12.2.2 gives two basic ℓ_d requirements for #3 to #11 bars: (1) $\ell_d = 0.04 A_b f_y / \sqrt{f_c'}$, (2) $\ell_d = 0.0004 d_b f_y$. For #3 to #5 bars the second is the larger and governs. There are a number of possible multipliers, for lightweight concrete and so on.

If bent at least 135° around a longitudinal bar (close to compression face), the additional development between start of hook and middepth $d/2$ is:

Bar	Grade 40	Grade 60
#3	0	3 in.
#4	0	4 in.
#5	0	5 in.

5.9 Shear Examples in Statically Determined Members

(a) The beam of Fig. 5.11 will be checked for shear assuming $f'_c = 3000$ psi and Grade 40 steel.

Solution

$$\text{Beam wt.} = \left(\frac{10 \times 13}{144}\right) 150 = 136 \text{ plf}$$

$d = 13 - 1.5 \text{ cover} - 0.375 \text{ stir.} - 0.375 \text{ for } 0.5 d_b = 10.75 \text{ in.} = 0.90 \text{ ft}$

$$w_u = 1.4 w_d + 1.7 w_l = 1.4 \times 136 + 1.7 \times 1000 = 1890 \text{ plf}$$

At d from support. $V_u = 1890(6 - 0.90) = 9640 \text{ lb}$ \quad (Code 11.1.2.1)

Largest required $V_n = V_u/\phi = 9640/0.85 = 11,340 \text{ lb}$ \quad (Code Eq. 11.1)

The shear region boundaries given in Table 5.1 are evaluated to determine which of the stirrup detailing rules govern the various sections of the

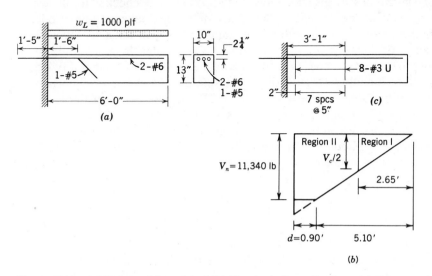

Figure 5.11 (a) Beam of Sec. 5.9a. (b) Nominal shear diagram. (c) Stirrups chosen.

beam. Beginning with region I, boundaries only need to be evaluated until the largest required V_n (here 11,340 lb) is bounded.

Region I $\qquad 0 \gtrless V_n \gtrless 1\sqrt{f'_c}b_w d = (V_c/2)$

(No web reinforcement) $0 \gtrless V_n \gtrless 1\sqrt{3000}(10)(10.75) = 5890$ lb

From similar triangles in Fig. 5.10b, the region I boundary is $(5890/11,340)(5.10) = 2.65$ ft from the free end. No stirrups are required in this region. Stirrups are needed for $(6 - 2.65)(12) = 40$ in. from the support.

Region II $\qquad 5890$ lb $< V_n \gtrless (2\sqrt{f'_c} + 50)b_w d$

(minimum stirrups) 5890 lb $< V_n \gtrless (2\sqrt{3000} + 50)(10)(10.75) = 17,150$ lb

The largest V_n required is 11,340 lb so the remainder of the beam lies in region II in which minimum stirrups govern with a maximum spacing of $d/2 = 5.37$ in. No other region boundaries need be checked.

The smallest stirrup bar is #3 ($A_v = 2 \times 0.11 = 0.22$ in.^{2}) and maximum spacing is $d/2 = 5.37$ in., say, 5 in.

$\qquad$ Min. $A_v = 50b_w s/f_y = 50 \times 10 \times 5/40,000 = 0.062$ in.$^2 \ll 0.22$ in.^{2}

As is often the case, the smallest available deformed bar used as a U stirrup has considerable excess area. It is not practical from a construction standpoint to use single leg stirrups, although use of welded wire fabric could reduce the area. The design will continue for #3 U stirrups.

Each stirrup should be at the center of the length s that it serves. Start the first stirrup at 2 in. from the support (roughly $s/2$) and use a total of 8 stirrups as shown in Fig. 5.11c. In a design this would usually be stated:

USE $8 - $ #3 U (spaced) at 2 in., 7 at 5 in. (caring for $39\frac{1}{2}$ in. $\doteq$ 40 reqd.)

The length of stirrups above middepth ($d/2$) must be checked for development as discussed briefly in Sec. 5.8. For #5 or smaller stirrups Code 12.13.2.3 says bending around a longitudinal bar may be taken as developing an f_y of 40,000 psi (or for a greater f_y an added length $\ell_d/3$ from middepth to start of bend). The #3 stirrup meets this test. Even if it failed such a test, it could be argued that the design showed only a small portion of the #3 bar area was needed (0.062/0.22) and this would justify a decrease in the ℓ_d required in about the same ratio.

Because this is a somewhat stubby beam ($L/h = 5.5$), the more exact equation for V_c (Code Eq. 11.6) possibly could have been used to a small advantage. For instance, at the largest V_n section:

$$V_c = (1.9\sqrt{f'_c} + 2500\rho_w V_u d/M_u)b_w d$$
$$A_s = (2 \times 0.44 + 0.31) = 1.19 \text{ in.}^2$$
$$\rho_w = 1.19/(10 \times 10.75) = 0.011$$
$$M_u = 1890 \times 5.10^2/2 = 24,600 \text{ lb-ft}$$
$$\text{Allowable } V_c = (1.9\sqrt{3000} + 2500 \times 0.011 \times 9640$$
$$\times 0.90/24,600)(10)(10.75) = 12,230 \text{ lb}$$

This compares to $V_c = 2\sqrt{f_c'}b_w d = 11,780$ lb using the simpler Code Eq. 11.3. The approximately 4% increase in V_c does not justify the effort here, especially because V_c would have to be evaluated at several sections since M_u and V_u change throughout the beam. This calculation is shown to illustrate the procedure that would be used if desired.

(b) Determine the stirrups for the beam of Fig. 5.12a, using Grade 40 steel and $f_c' = 4000$ psi.

Solution

The critical section is at $d = 19$ in. from support, 4.42 ft from midspan.

$$\text{Beam wt.} = (9 \times 22/144)150 = 206 \text{ plf}$$
$$\text{D.L. } V_u = 1.4(6500 + 206 \times 4.42) = 10,400 \text{ lb}$$
$$V_u = 10,400 + 1.7 \times 15,000 = 35,900 \text{ lb, largest required}$$
$$V_n = 35,900/0.85 = 42,200 \text{ lb}$$

Just to the left of the load, $V_{u3} = 1.4(6500 + 206 \times 3) + 1.7 \times 15,000 = 35,460$ lb

$$V_n = 35,460/0.85 = 41,700 \text{ lb}$$

Before establishing the rest of the required shear capacity diagram (V_n), the shear region boundaries of Table 5.1 are computed:

Region I (no web reinforcement)
$$0 \geqslant V_n \geqslant 1\sqrt{4000}\,(9)(19) = 10,810 \text{ lb}$$
Region II (minimum stirrups)
$$10,810 \text{ lb} < V_n \geqslant (2\sqrt{4000} + 50)(9)(19) = 30,180 \text{ lb}$$
Region III (V_s stirrups, $d/2$ maximum spacing)
$$30,180 \text{ lb} < V_n \geqslant 6\sqrt{4000}\,(9)(19) = 64,890 \text{ lb}$$

The largest V_n is 42,200 lb so there is no need to evaluate region IV. In region III stirrups need to provide for $V_s = V_n - V_c$. V_c will be $2\sqrt{f_c'}b_w d$ which is $\frac{1}{3}$ of the region III upper boundary or 21,630 lb.

The V_n diagram and the shear regions are shown in Fig. 5.12b. The maximum shear condition between the supports and the concentrated loads occurs with all loads on the span. The V_n in this zone varies from 42,200 lb to 41,700 lb, which is such a small variation that *all* stirrups might as well be designed for $V_n = 42,200$ lb. This falls within region III so $V_s = V_n - V_c = 42,200 - 21,630 = 20,570$ lb. Maximum s is $d/2 = 9.5$ in. Minimum stirrup size is #3.

Try #3 U stirrups, $A_v = 2 \times 0.11 = 0.22$ in.^{2}

$$s = \frac{A_v f_y d}{V_s} = \frac{(0.22)(40,000)(19)}{(20,570)} = 8.13 \text{ in.}$$

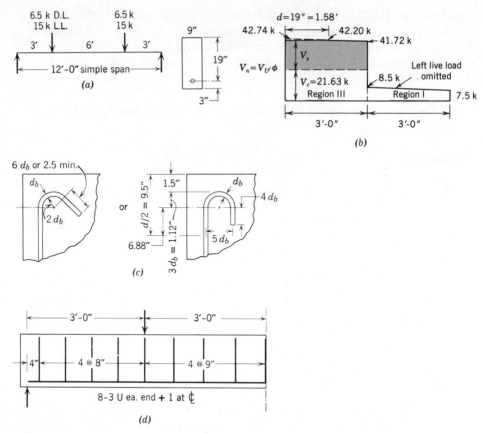

Figure 5.12 Beam of Sec. 5.9b.

It is obvious that #4 U-stirrups would give a calculated spacing greater than the 9.5 in. permissible maximum. (The student might note the value of s varies directly with the bar area and hence #4 would give nearly double the #3 spacing.)

USE #3 U-stirrups.

Sketch (Fig. 5.12c) shows $\ell_d/2$ (from table in Sec. 5.8) = 6/2 = 3 in. required below hook in upper $d/2$. Stirrup bend dimensions are in Code 7.13 and 7.2.2. Stirrup development O.K. Next check beyond the load, the live load to the left being omitted to get maximum shear. This condition will be more demanding than the symmetrically loaded beam. Evaluation of the reactions for the beam loaded unsymmetrically with no live load in the left span shows the left reaction to be 17.20 k. The shear just to the right of the concentrated dead load in the left side of the span is $V_u =$ 17,200 − 1.4(3 × 206 + 6500) = 7230 lb. This is below $V_c/2 =$ 10,810 and, hence, is in region I. As no stirrups are required in region I, the beam could be detailed with only stirrups in the outer shear spans. However, the authors prefer to provide some stirrups in every zone except in very lightly loaded joists. Hence, they arbitrarily would extend the stirrups across the middle length at a value close to the $d/2$ maximum.

A formal stirrup spacing curve could be prepared as in Fig. 5.17, but it is totally unnecessary for this short length and this simple pattern of V_s values. The s value calculated at the critical section must be used back to the support and the small change in V_s to $x = 3$ ft makes further calculations of s totally unnecessary.

USE 8- #3 U at 4 in., 4 at 8 in., 4 at 9 in. at each end with the last one at midspan, total of 17 stirrups. This arrangement is sketched in Fig. 5.12d.

The student should be aware that in proportioning stirrups the arbitrary sign convention between positive and negative shear does not enter into $V_s = V_n - V_c$. Errors have been made by confusing the sign of the V_n term. The equation is more mathematically proper if written as $V_s = |V_n| - V_c$. Thus, vertical stirrups are adaptable to shear forces acting either upward or downward as, in terms of the truss analogy, only the diagonal crack and the compression diagonal must take a different direction.

5.10   Spacing Relations and Comments on Use of Inclined Stirrups and Bars

(a) General Formulation   In practical construction in the United States, vertical stirrups are the prevalent form of web reinforcement. Inclined stirrups and bent bars can be used, however, and are somewhat more efficient in structural action because they are aligned more closely with the principal tension stresses in the web of the beam.

The structural contribution of inclined web reinforcement can be seen most clearly in terms of the truss analogy. When vertical stirrups are present and spaced at a distance s as measured parallel to the longitudinal reinforcement, the stirrup is effective along the shaded zone $z \cot \beta$ of Fig. 5.13a. With the ACI Code truss model that assumes $\beta = 45°$, this effective zone becomes z. In the usual case, multiple stirrups must be in this zone (since $s_{max} = d/2$ or $d/4$); $V_s = nS = (z/s)A_v f_y \approx (d/s)A_v f_y$ (Code Eq. 11.17).

When inclined web reinforcement is present, the general form of the analogous truss of Fig. 5.13b changes appreciably. The tension diagonals (solid) are inclined at an angle α and the compression struts (dashed) are at an angle β. The inclined bar or stirrup is now effective along the shaded zone $z (\cot \alpha + \cot \beta)$. Figure 5.13c shows that each inclined bar or stirrup has a vertical component $A_v f_y \sin \alpha$. Defining the spacing s as measured parallel to the longitudinal reinforcement, $V_s = nS \sin \alpha$. In this case $n = z(\cot \alpha + \cot \beta)/s$ and since $z \approx d$, $V_s = d(\cot \alpha + \cot \beta)A_v f_y \sin \alpha/s$.

For the ACI Code assumed truss analogy with $\beta = 45°$, $V_s = \dfrac{A_v f_y d}{s}$ $(1 + \cot \alpha) \sin \alpha$. Since $1 + \cot \alpha = (\sin \alpha + \cos \alpha)/\sin \alpha$, the previous

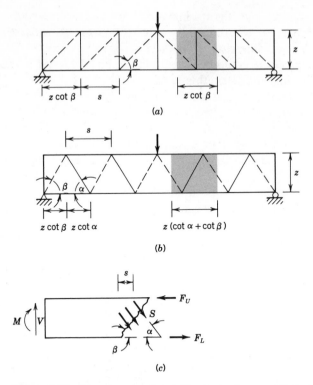

Figure 5.13 Truss analogy. (*a*) Vertical stirrups. (*b*) Inclined stirrups or bent bars. (*c*) Free body along an inclined crack.

equation becomes $V_s = \dfrac{A_v f_y d}{s} (\sin\alpha + \cos\alpha)$ (Code Eq. 11.18). Note that with vertical stirrups, $\alpha = 90°$ and Code Eq. 11.18 reverts to Code Eq. 11.17.

(b) Bent Bars or Inclined Stirrups at 45°   For the 45° bent bars or inclined stirrups most frequently used, the spacing can be easily determined from Code Eq. 11.18 as $V_s = 1.414 A_v f_y d / s$. The greater load carrying efficiency of the individual stirrup is somewhat offset economically by the longer length and increased difficulty in fabricating the reinforcement cages.

Diagonal or inclined stirrups (Fig. 5.14*a*) are aligned more nearly with the principal tension stresses in the beam. They share in carrying this tension at all stages of loading and slightly delay the formation of diagonal tension cracks. Except where the shear force may reverse direction, inclined stirrups would be the preferred type where preassembled cages of steel are used, as in some present construction; but actually they are little used in this country. Where shears reverse, as in earthquake construction, the inclined stirrup would be parallel to the diagonal crack that forms for one case and, hence, ineffective.

Longitudinal bars bent up where they are no longer needed for moment also act as inclined stirrups. Many designs use them, but few designers calculate

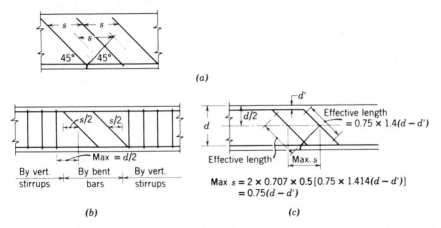

Figure 5.14 Inclined stirrups or bent bars at 45°. (*a*) Basis for calculation. (*b*) Arrangement in combination with vertical stirrups. (*c*) Maximum spacing to intercept cracks.

their value as stirrups. One reason is that usually only a few bars are bent and these may not be convenient for use as web reinforcement. Another reason is that when only a single bar is bent up or when all bent bars are bent at the same point, the value for diagonal tension is discounted to Code Eq. 11.19:

$$V_s = A_v f_y \sin\alpha \gtrless 3\sqrt{f_c'} b_w d$$

This is equivalent to saying that the vertical component of the bar stress carries the shear V_s. Any longitudinal bent bar is considered effective over the center $\frac{3}{4}$ of its inclined length (Code 11.5.6.6) (Fig. 5.14*c*), making maximum $s = 0.75(d - d')$ or one-half this value where V_s exceeds $4\sqrt{f_c'} b_w d$ (Code 11.5.4.2 and 11.5.4.3)

Bent-up bars are normally bent at a 45° angle, but in some members, usually light joists, a flatter angle is used to secure some effect as diagonal tension reinforcement over a greater length.

5.11 Design of Vertical Stirrups

The practical problem of designing stirrups requires:

1. Determination of maximum shears and the length over which stirrups are needed.
2. Choice of desirable size of stirrup bar.
3. Selection of a series of practical spacings.

The maximum shear diagram may involve partial as well as full span loads. Stirrup bar sizes are almost never mixed in a given beam. Hence, with uniform loads, close spacings are required at points of maximum shear, although at points of lesser shear the spacing may be limited by the maximum spacing, that is, the interception of all potential cracks. The stirrup size must be large enough to give a minimum spacing adequate to pass the aggregate readily and, for practical reasons, rarely as small as 2 in., usually 3 in. or more. Larger spacings are more economical if they do not involve too many spaces fixed by crack interception instead of by stress capacity. Also, larger spacings may require stirrup bar sizes that cannot meet the development requirements of Code 12.13, as quoted in Sec. 5.8.

For a member with a constant shear the stirrup spacing would be constant and preferably near the maximum permissible. When the shear varies, as is more usual, a sound general method is to calculate the required stirrup spacing at enough points to establish a stirrup spacing curve, including the specification limits on maximum spacing. From this curve the practical spacings can be worked out. A detailed design of stirrups for a continuous T-beam span is given in Secs. 5.12 and 9.17.

The theoretical number of stirrups required for shear is often useful. For any V_s diagram (not total V_u diagram), such as Fig. 5.15a, the area under the diagram is a direct measure of the number of stirrups theoretically required. Since $V_s = A_v f_y d/s$

$$A_v f_y = V_s s/d = \text{(area of } V_s \text{ diagram for length } s)/d$$

Each stirrup thus cares for an equal area under the V_s diagram and the total number of stirrups needed is

$$n = \frac{\text{(total area of } V_s \text{ diagram)}}{A_v f_y d}$$

This is a theoretical number. Many more stirrups will be required if maximum spacing governs a large portion of the shear zone. Spacings used will be stated in practical units, usually full inches (except for small dimensions, say, under 5 in.), and will thus average less than the theoretical spacings. There also will usually be some spaces kept smaller than the theoretical by maximum spacing rules to intercept potential cracks. Extra stirrups will always be required beyond the theoretical in the length where minimum stirrups govern (region II of Table 5.1). The practical number of stirrups in a carefully designed section will usually be from three to six more than the theoretical number.

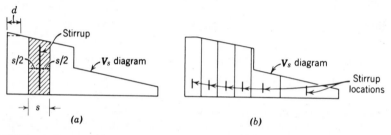

Figure 5.15 Stirrup spacing related to area of V_s diagram.

5.12   Stirrup Design Example—Continuous T-Beam

A continuous T-beam with $b_w = 10$ in., $d = 17.6$ in., $f'_c = 3000$ psi, Grade 60 steel must provide for the factored load shears shown in Fig. 5.16a. (Similar shears are calculated as in Sec. 9.17). The diagram shown represents maximum shear (V_u) conditions in the span and, as shown in Sec. 9.17, combines the shear diagram for a fully loaded span and the shear diagram for partially loaded span into a reasonable approximation of a maximum shear envelope. Design and space the necessary stirrups.

Solution

The maximum shear envelope for factored load shears V_u of Fig. 5.16a is transformed into the required nominal shear capacity V_n diagram of Fig. 5.16b using $V_n = V_u/\phi = V_u/0.85$. The shear region boundaries of Table 5.1 indicate which detailing rules govern the various sections of the beam. The boundaries are:

Region I (no web reinforcement required)
$$0 \gtrless V_n \gtrless V_c/2 = 1\sqrt{f'_c}b_wd = \sqrt{3000}(10)(17.6) = 9640 \text{ lb}$$

No portion of this beam has a V_n this low so web reinforcement will be required throughout.

Region II (minimum web reinforcement required)
$$9640 < V_n \gtrless (2\sqrt{f'_c} + 50)b_wd = (2\sqrt{3000} + 50)(10)(17.6) = 28,080 \text{ lb}$$

The slope of the V_n diagram is $(78,120\text{-}13,180)/120 = 541$ pli. The left boundary of region II may be found from $78,120 - 541x = 28,080$. $x = 92.5$ in. from the support as shown on Fig. 5.16b. In this case the right boundary is at the centerline. Throughout this region web reinforcement will be required for $V_s = 50b_wd = (50)(10)(17.6) = 8800$ lb and the maximum spacing will be $d/2 = 8.8$ in.

Region III (normal web reinforcement required)
$$28,080 < V_n \gtrless 6\sqrt{f'_c}b_wd = 6\sqrt{3000}(10)(17.6) = 28,080 \text{ lb}$$

Using the same procedure, the left boundary at region III is found at $x = 37.5$ in. and is shown on Fig. 5.16b. Throughout this region web reinforcement will be required for $V_s = V_n - V_c$ with a maximum spacing of $d/2 = 8.8$ in. $V_c = 2\sqrt{f'_c}b_wd = 19,280$ lb.

Region IV (heavy web reinforcement required)
$$57,840 < V_n \gtrless 10\sqrt{f'_c}b_wd = 10\sqrt{3000}(10)(17.6) = 94,400 \text{ lb.}$$

The largest V_n that must be provided for is that at the critical section d from the support or 68,590 lb. The initial 37.5 in. falls within region IV where the maximum spacing is limited to $d/4 = 4.4$ in.
   It is helpful to draw the $V_s = V_n - V_c$ diagram of Fig. 5.16c for the beam before selection of the stirrups. In region II $V_s = 50b_wd = 8800$ lb.

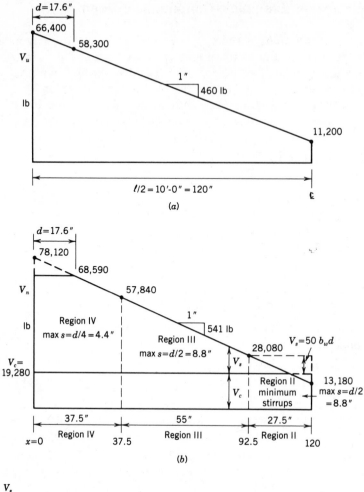

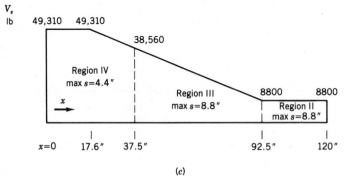

Figure 5.16 V_u and V_n diagrams for beam similar to Sec. 9.17. (a) V_u. (b) V_n. (c) V_s.

At this stage the approximate selection of stirrup size is made with a subsequent detailed stirrup spacing curve to establish final spacings. Usually several stirrup sizes should be checked at this stage. Main concerns are the satisfaction of minimum practical stirrup spacings (say, about 3 in. to allow concrete placement) and the minimization of the zones where stirrup spacings are controlled by the maximum spacing limits to control waste. For vertical U-type stirrups, $A_v = 2A_b$. Code Eq. 11.17 indicates $A_v = V_s s / f_y d$. Checking in the support region, Fig. 5.16c indicates $V_s = 49,310$ lb and the maximum $s = 4.4$ in. A trial $A_v = (49,310)(4.4)/(60,000)(17.6) = 0.206$ si. A #3 U looks good here because $A_v = 2(0.11) = 0.22$ si. At the region III left boundary a similar check for max $s = 8.8$ in. indicates $A_v = (38,560)(8.8)/(60,000)(17.6) = 0.321$ si. A #3 U ($A_v = 0.22$) at closer spacings could be used, but a #4 U ($A_v = 0.40$) would be somewhat wasteful. A final check in the region II minimum web reinforcement zone indicates $A_v = (8800)(8.8)/(60,000)(17.6) = 0.073$ si. Even a #3 U would be wasteful in this region but because it is the smallest deformed bar and U stirrups are very practical for construction, the decision is to use #3 U stirrups throughout the beam. #4 U stirrups would be extremely wasteful in most zones.

The #3 stirrups must be checked for available development length in the upper-half depth of the beam $17.6/2 = 8.8$ in. The required $0.5\ell_d$ between middepth and start of hook (Code 12.13.2.1) is 4.5 in. for a Grade 60 #3 bar as determined by taking half of the development length tabulated in Sec. 5.8. If hooked around longitudinal reinforcement this would become 3 in. A sketch similar to Fig. 5.12c shows available distance is $d/2 - 1.5$ in. cover $- 3d_b$ for hook $= 8.80 - 1.50 - 3 \times 0.375 = 6.18$ in. $>$ 4.5 in. This #3 stirrup is **O.K.**

The detailed stirrup spacing curve will be developed for establishing the final spacings graphically. Theoretical spacings will be calculated at several points by noting that $s = A_v f_y d / V_s = (0.22)(60,000)(17.6)/V_s = 232,320/V_s$. The V_s diagram is redrawn in Fig. 5.17a and the sloping lines arbitrarily divided into several segments.

The theoretical spacing $s = 232,320/V_s$ is calculated for each point and plotted on Fig. 5.17b. A dashed theoretical spacing curve is sketched throughout these points. The maximum spacings drawn in at 4.4 and 8.8 in. show that the theoretical curve governs only over the short range from $x = 37.5$ to 60 in. The actual stirrup spacing diagram is constructed in a semigraphical manner in Fig. 5.17b. Each stirrup is assumed to be effective over a zone s that extends $s/2$ on each side of the stirrup. Starting from the origin, vertical bars with a height and a width equal to s are plotted. The midpoint of each bar is the physical location of the stirrup as shown by the vertical line near the bottom of each bar. The middle of each bar must be no higher than either the dashed theoretical s line or the maximum s line to satisfy Code requirements. The more that the bars can be extended towards these limits, the fewer stirrups required. However, practicality of construction should be considered and stirrup zone widths

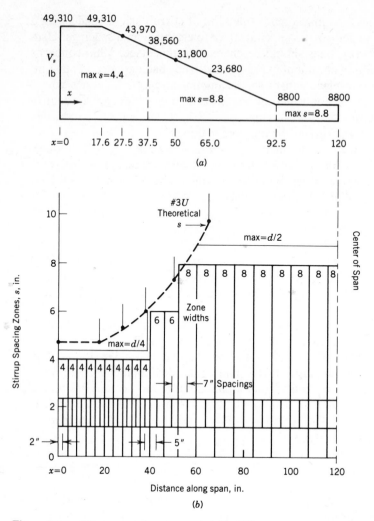

Figure 5.17 Stirrup spacing curve. (*a*) V_s. (*b*) s.

in fractional inches avoided. Because each stirrup is located at the middle of its zone, the actual spacing of the stirrups must be read off carefully. The first stirrup will be located at $s/2$ from the support. (Good detailing would include an extra stirrup the same distance outside the support.) Wherever zone widths change, the stirrup spacing will be the *average* of the two adjacent zones. (Note where 6 and 8 in. zones adjoin, $6/2 + 8/2 = 7$ in.) The final specification of the stirrups between supports would be "Use 20 #3 U at each end plus 1 #3 U at midspan and 1 #3 U outside each support. From the supports use 1 @ 2 in., 9 @ 4 in., 1 @ 5 in., 1 @ 6 in., 1 @ 7 in., 7 @ 8 in. each end. Total 43 #3 U." In this particular case, the last stirrup location at $s = 8$ in. fell conveniently on the centerline. In many cases an extra stirrup may be required if the spacing is not this convenient. Stirrups are relatively inexpensive and the designer cannot usually afford as refined an analysis as presented, but it is given to provide students with the reasoning they must use in whatever procedures they finally adopt.

An alternate procedure for spacing stirrups—less accurate but still satisfactory in the hands of one skilled at it—consists of (1) calculating the theoretical total number of stirrups required, (2) calculating the spacing at a few points, and (3) simply writing down a series of spacings that satisfy (2) and at the same time provide enough extra stirrups to account for the lengths that are governed by maximum spacing instead of stress. Extra stirrups are always necessary as illustrated by the design just completed with 20 stirrups at each end compared to a theoretical

$$n = \frac{(\text{area } V_s \text{ diag.})}{A_v f_y d}$$

$$= \frac{[49{,}310 \times 17.6 + (49{,}310 + 8800) \times 0.5 \times 74.9 + 8800 \times 27.5]}{2 \times 0.11 \times 60{,}000 \times 17.6}$$

$$= 14.2 \text{ stirrups}$$

5.13 Shear in One-Way Slabs

Because a one-way slab is really a wide shallow beam, no shear behavior different from that in beams should occur, but three comments are appropriate.

1. Shear stresses are usually low in one-way slabs, except where heavily loaded as in one-way footings under heavy walls.
2. Stirrups are rarely needed or feasible to use in thin members like most slabs.
3. Concentrated loads require transverse reinforcement and develop shear demands similar to those discussed later in this chapter for two-way slabs.

Two-way slabs supported on beams are even less apt to have shear problems, but two-way slabs supported on columns (without beams) are totally different and have major shear problems.

5.14 Shear in Joists

Joists are defined in Code 8.11.1 as consisting "of a monolithic combination of regularly spaced ribs and a top slab arranged to span in one direction or two orthogonal directions." Clear spacing is limited to 30 in. and any wider spacing disqualifies for the Code title of joists. Two design concessions are made for joists.

Because of their close spacing, joists are better able to distribute local overloads to adjacent joists than are typical beams. Accordingly, joists may be designed for 10% higher shear strength than are permitted for V_c in beams (Code 8.11.8). Joists are also exempted from the requirement that stirrups be used wherever the shear exceeds $0.5\phi V_c$ (Code 11.5.5.1b).

5.15   Shear Loss from Cutting Off Bars in a Tension Zone

As already pointed out in Sec. 4.11d, when flexural tension bars are cut off in the tension zone of a beam a sharp discontinuity in the steel is created; the sharpness is dependent on the percent of bars cut off. The terminated bar opens an early flexural crack at the cutoff point and then appears to act as an eccentric pull on the surrounding concrete that in some way changes the flexural crack into a diagonal crack, again prematurely. The mechanism is not quite clear, but the end result, a reduced shear strength, is well documented, as in Fig. 5.18. The beams represented were designed to reach f_y and the Code shear strength simultaneously; hence the steel stress ratio (ordinate) is also the shear ratio.

This study of 64 beams shows that the usual extension of bars by $12d_b$ or $15d_b$ beyond the moment cutoff point kept the ill effects from being more severe. Without remedial steps only 2 out of 33 beams developed the design ultimate strength, with losses generally in the order of 15% to 25%, a few higher. In contrast, bars bent up caused no losses. Extra stirrups could erase losses, but seemingly at only 50 percent efficiency.

The extra stirrups mentioned in Sec. 4.11d are particularly needed over the tail of the bars cut off because Code 12.10.5 specifies excess stirrups having A_v of at least $60b_w s/f_y$ over the last $0.75d$ of bars cut off, with s limited to not more than $d/8\beta_d$, where β_d is the ratio of bar area cut off to total A_s near the cutoff point. For A_v, b_w, and s in mm, Code 318M uses $A_v \geqslant 0.4b_w s/f_y$.

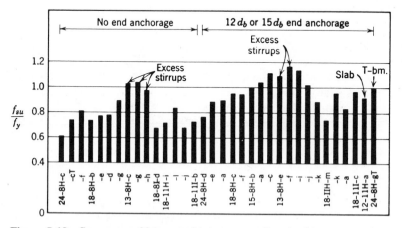

Figure 5.18 Summary of beams with bars cut off and without remedial measures. (The five beams marked with arrows had excess $\rho_n f_y$ of 75 psi to satisfy minimum spacing rule.)

5.16   Members of Varying Depth

The relations for shear, bond, and even moment resistance must be modified for members in which the depth is varying, that is, members in which the bottom and top surfaces are not parallel. The moment effect is not large unless the angle between the faces is at least 10° or 15°, but shear for diagonal tension and bond may be modified as much as 30% by 10° slopes. The 1940 Joint Committee Specification gave the following formula for the effective total shear V_1 to be used for V in the usual relations for v and u:

$$V_1 = V \pm \frac{M}{d} (\tan c + \tan t)$$

where V and M are the external shear and moment to be resisted, d is the depth to tension steel, and t and c are the slope angles of the top and bottom of the beam as shown in Fig. 5.19. The plus sign before the parenthesis is used when the beam depth decreases as the moment increases, as in Fig. 5.19a, and the minus sign is used for the more usual case of increasing depth with increasing moment, as in Fig. 5.19b. When the sum of the angles becomes as much as 30°, such a formula becomes very inexact. The use of the truss analogy for such members is illustrated by Schlaich and Schafer[7].

5.17   Allowable Shear with Axial Stress Superimposed

Axial compression increases the diagonal tension resistance and axial tension lowers it. Overlooked axial tension from shrinkage or temperature variation often plays a prominent part in beam failures that occur in structures, in spite of the relief creep may give.

A quick picture of the influence of axial compression or tension is given in Fig. 5.20, which is adapted from the ACI Commentary. The zero axial load is shown at the bottom with V_c plotted on the vertical dividing line at 2, corresponding to $V_c = 2\sqrt{f'_c} b_w d$.

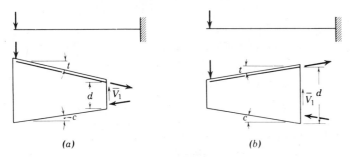

(a)   (b)

Figure 5.19   Shear in beams of varying depth.

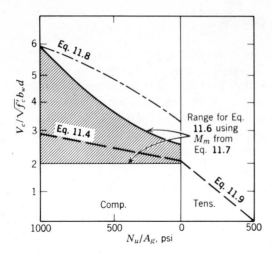

Figure 5.20 Design unit shears with axial load.

1. *Axial tension.* Where axial tension N_u is large, the concrete may be cracked through in tension and unable to help the stirrup-concrete strut truss mechanism carry the tension aspects of shear. The stirrup-concrete strut truss mechanism must then carry the full shear. The permissible V_c can be evaluated from:

$$V_c = 2(1 + 0.002N_u/A_g)\sqrt{f_c'}b_w d \qquad \text{(Code Eq. 11.9)}$$

where N_u is *negative* for tension, with units of psi for N_u/A_g. For 500 psi axial tension this V_c drops to zero. This shows as the sloping dashed line on the right side of Fig. 5.20. The 318M Code for V_c in SI units shows

$$V_c = 2(1 + 0.3N_{uv}/A_g)(\sqrt{f_c'}/6)(b_w d)$$

2. *Axial compression.* The axial compression forces available in a compression strut or column or compression chord of a Vierendeel truss greatly improve its capacity to resist shear. The better analysis is given by the more complex equation for allowable V_c modified by using the M_m of Code Eq. 11.7 for M_u and at the same time not limiting $V_u d/M_u$ to 1.0:

$$V_c = (1.9\sqrt{f_c'} + 2500\rho_w V_u d/M_u)b_w d \qquad \text{(Code Eq. 11.6)}$$
$$M_m = M_u - N_u(4h - d)/8 \qquad \text{(Code Eq. 11.7)}$$

N_u is positive for axial compression. The second term in the M_m equation is approximately the moment of N_u about the center of the resisting compression. Where M_m is negative it is a signal to discard it and go directly to:

$$\text{Max. } V_c = 3.5\sqrt{f_c'}b_w d\sqrt{1 + N_u/500\,A_g} \qquad \text{(Code Eq. 11.8)}$$

In Fig. 5.20 this maximum V_c shows as the uppermost dot-dash curve and the shaded area shows approximately the range of Code Eq. 11.6 using M_m of Code Eq. 11.7 in place of M_u. The 318M SI Code shows

$$\text{Max. } V_c = 0.3\sqrt{f_c'}\sqrt{1 + 0.3N_{uv}/A_g}b_w d$$

Because Code Eq. 11.6 is difficult to apply, the simpler Code Eq. 11.4 (the dashed straight line sloping upward to the left in Fig. 5.20) is permitted and usually gives lower values:

$$V_c = 2(1 + N_u/2000A_g)\sqrt{f_c'}b_w d \qquad \text{(Code Eq. 11.4)}$$

For 500 psi compression this gives $2.5\sqrt{f_c'}$ compared to $2\sqrt{f_c'}$ without compression. This equation is not required if the other is used; it is an alternate. The 318M SI Code shows

$$V_c = (1 + N_{uv}/14A_g)(\sqrt{f_c'}/6)b_w d$$

Where the compression is from prestress, Code 11.4 controls. Comment on the equations there is left to Chapter 20 on prestressed concrete members.

5.18   Brackets and Short Cantilevers

The Code Sec. 11.9 covers brackets and corbels (limited to a/d of unity or less) and the authors include other short cantilevers as an extension of the same general behavior. Their normal resistance near ultimate, shown in Fig. 5.21a, consists of a simple truss model composed of a tension tie across the top with an inclined compression strut forming a triangle, with normal bending making only slight variations. The inclination of this strut determines the tension in the tie by simple truss analysis; the flexural calculation at the face of the column gives essentially the same tension, but the triangular truss idea emphasizes the anchorage problem at the node; and this anchorage problem is the most critical one unless the bar extensions of Fig. 5.21b are possible.

The PCA bracket tests[17] took account of the high shrinkage and expansion stresses frequently occurring from beams supported on brackets. The Code, based on these tests, requires a minimum horizontal force N_{uc} at least equal to $0.2V_u$ (Code 11.9.3.4).

Design of the reinforcement is relatively simple. Code 11.9.3 requires that the section at the support be designed to resist simultaneously the shear V_u, the moment $V_u a + N_{uc}(h - d)$, and the horizontal tensile force N_{uc}. The shear is transferred across the assumed crack of Fig. 5.21d using the shear-friction theory outlined in the next section. This results in determination of an area of shear-friction reinforcement A_{vf} distributed normally to the assumed crack and over 2/3 of the effective depth adjacent to the tension chord. For normal weight concrete this shear-friction strength V_n is limited to $0.2f_c'b_w d$ with a maximum of $800 b_w d$ in pounds. The area of reinforcement in the tension chord A_s is determined by adding the reinforcement area A_f found by normal flexural calculations to the reinforcement area A_n required to resist the direct tension force. Although this quantity $A_s = A_f + A_n$ usually governs, tests[18] show that A_s must also at least equal $A_n + 2/3A_{vf}$ to ensure correct shear transfer. Code 11.9.3.5 and 11.9.4 provide specific guidance and suggest closed stirrups or ties with an area A_h of at least $0.5(A_s - A_n)$ be used for the distributed A_{vf}. This will satisfy both A_s requirements in Code 11.9.3.5.

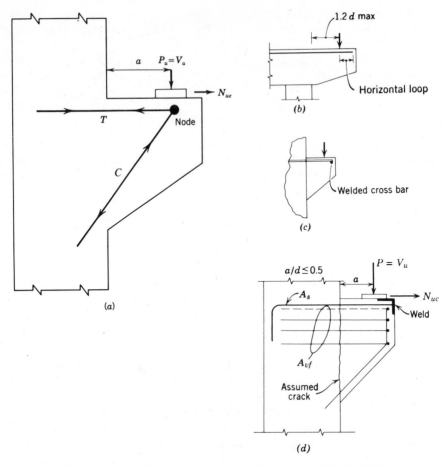

Figure 5.21 Brackets and short cantilevers. (*a*) Truss model. (*b*) Minimum end anchorage. (*c*) Preferred end anchorage in short bracket. (*d*) Shear-friction applied to bracket design.

Area A_h must be distributed over the upper 2/3 of the effective depth. In addition, A_s must not be less than $(0.04 f'_c/f_y)bd$ to preclude sudden failure upon cracking. The Code indirectly limits the effective depth d used at the column face to twice the depth at the outside edge of the bearing (Code 11.9.2) and the authors suggest that this bearing not start closer than 2 in. from the bracket face. In addition to the normal anchorage or development length into the support, the horizontal reinforcement must be developed at the outer node of Fig. 5.21*a*. The problem area in a long bracket is the anchorage of bars because the variable depth (inclined strut idea) makes the bar tension nearly as critical at the load as at the column face. The extension of the tension bars as close to the face as possible and an anchor provided by a welded cross bar of the same diameter is also recommended (Fig. 5.21*c*). In addition to the welded cross bar, Code 11.9.6 allows anchorage by horizontally looped tension bars (Fig. 5.21*b*) or by other means of positive anchorage such as welded or threaded plates and mechanical anchors.

For the longer brackets falling outside the $a/d < 1$ restriction of the Code, the authors would normally calculate the required A_s at the face of column, keep the bars of a diameter that can be anchored into the column (or try for a deeper bracket if necessary to make them smaller), and if a horizontal tension pull were expected add enough bars to resist it. Finally the shear should be checked using shear-friction theory and should rarely control. For a short bracket the shear ma ybe the chief control on the bracket depth. For a heavy bracket, say more than 18 in. deep, additional horizontal U-bars or ties in the upper half will avoid wide cracks, add to the bracket strength, and even help with the moment. Vertical stirrups are essentially useless. The Code requirement is horizontal steel A_h equal to half the top A_s.

5.19   Shear Friction

Shear-friction theory (Code 11.7) is particularly useful for precast assemblies and composite construction.[19-21] The Code suggests this approach may be used for shear transfer in brackets cast-in-place provided $a/d \gtrless 1.0$. This method is especially appropriate where the resistance needed is against pure shear or sliding tendency, rather than primarily against a diagonal tension.

The shear-friction approach is to assume a shear plane already cracked as in Fig. 5.21d, with a coefficient of friction μ given in Code 11.7.4.3 (1.4 if monolithic, 1.0 if placed against intentionally roughened hardened concrete (Code 11.7.9), 0.6 if placed against hardened concrete not intentionally roughened, or 0.7 against as-rolled structural steel. A discount factor λ, reduces these values for lightweight concretes.) The depth must then be such that the shear stress on the area bd will not exceed $0.2f_c'$ or 800 psi. The necessary A_{vf} must be large enough to develop the normal force necessary to support the shear by friction:

$$A_{vf} f_y \mu = V_n$$

Closed stirrups or ties are appropriate for A_{vf}. They must be effectively anchored on both sides of the assumed crack and should be distributed uniformly within $\frac{2}{3}$ of d adjacent to A_s. Any tension across the shear plane owing to flexure or direct tension force must be resisted by additional reinforcement.

5.20   Deep Beams

Deep beams are now covered in Code 11.8. The 1986 Code Revision indicates that some of the provisions of Code 11.8 do not apply to continuous deep beams. (See 1986 Code 11.8.3.) Although the Code deals only with shear, a few general ideas might be helpful. A deep beam is simply a member short enough to make shear deformations important in comparison to pure flexure; the Code considers only simply supported beams with a clear span ℓ_n of up to $5d$.

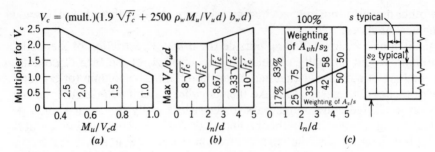

Figure 5.22 Deep beam shear. (*a*) Multiplier for usual V_c. (*b*) Maximum total V_n. (*c*) Weighting of vertical and horizontal shear reinforcement; spacings.

Plane sections in these beams do not remain essentially plane under loading; but this is also true at the end of every simple span beam. If one has a clear picture of how any beam fails in shear at $a/d = 1$, one can get a rough idea of how a deep beam of $\ell_n = 2d$ would fail in shear by imagining two such end sections of beam joined together with the load at the junction. At $a/d = 1$ in a long beam, flexure will not be of much interest; in the deep beam of $\ell_n = 2d$, flexure at the load is the critical flexure. If the load is applied to the top of the deep beam the lever arm z of the internal couple will be smaller than usual, say $0.8d$; if the load is applied to the bottom of the beam in tension, z may drop as low as $0.4d$ or $0.5d$. The Code provisions are limited to loading on top of the beam. The beam behaves more like a truss or tied arch and anchorage of tension steel becomes critical. Leonhardt[15,22] suggests *horizontal* hooks on tension bars (vertical compression helping to avoid splitting) or U-shaped bars lap spliced at midspan.

If such a deep beam is provided with stirrups that are able to deliver the bottom load to the upper part of the beam, the beam will behave nearly like a top loaded beam. It is always desirable in any kind of beam to provide hangers (stirrups) to pick up bottom loads in this fashion. Experimental evidence indicates that stirrups that must act as hangers *and* as web reinforcement need *not* be designed for the sum of the two requirements, but simply for the larger of the two.

Higher shear strengths are available when a/d is small, as already pointed out in Secs. 5.4c and 5.4d. For the deep beams a multiplier is given for application to the usual allowable V_c:

$$V_c = [3.5 - 2.5 M_u/(V_u d)] \times [\text{usual formula values for } V_c]$$

where $[3.5 - 2.5 M_u/(V_u d)]$ has here been designated as the magnifier, to be limited to a maximum of 2.5 and the magnified V_c also to be limited to $6\sqrt{f'_c} b_w d$. The values of this multiplier are plotted in Fig. 5.22*a* and should be applied to the usual allowable of $2\sqrt{f'_c} b_w d$ or to $(1.9\sqrt{f'_c} + 2500 \rho V_u d/M_u) b_w d$.

The critical beam section for shear is to be taken at $0.5a$ for concentrated loads and $0.15\ell_n$ for uniform load. For simple spans these appear quite reasonable values, closely akin to what is done on ordinary beams. The M_u and V_u in the formula are to be taken at this critical section. The shear reinforcement required at the critical section is to be used throughout the span (Code 11.8.10).

The Code also limits V_n to $8\sqrt{f'_c}b_w d$ when $\ell_n/d \gtrless 2$, and for larger ℓ_n/d to

$$V_n = (2/3)(10 + \ell_n/d)\sqrt{f'_c}b_w d$$

which is plotted in Fig. 5.22b. Since V_n on ordinary beams may be used up to $V_n = V_s + V_c = 8\sqrt{f'_c}b_w d + V_c$ at an a/d of 1, the authors regard this deep beam limit as probably too strict.

Vertical stirrups are less effective in these beams because the shear cracks are steeper;[23] *horizontal* bars become more effective as a/d or ℓ_n becomes smaller. The Code (11.8.7) accordingly expresses the shear resistance as the weighted sum of the two types of reinforcement in a modification of the basic equation:

$$V_s = \left[\frac{A_v}{s}\left(\frac{1 + \ell_n/d}{12}\right) + \frac{A_{vh}}{s_2}\left(\frac{11 - \ell_n/d}{12}\right)\right]f_y d$$

The horizontal reinforcement A_{vh} is weighted much the heavier, as shown in Fig. 5.22c. The Commentary suggests the precise weighting is not critical but that the weighted sum should be maintained. In addition the spacing may not be more than 18 in. nor $d/5$ for spacing parallel to the longitudinal reinforcement s or $d/3$ for spacing perpendicular to the longitudinal reinforcement s_2, and the minimum A_v must be at least $0.0015\,bs$, the minimum A_{vh} at least $0.0025\,b_{s_2}$. The designer should also note the requirement of Code 10.6.7 calling for extra side face reinforcement in a longitudinal direction, for *any* web over 3-ft deep, *not* limited to ℓ_n of $5\,d$ or less. The A_{vh} reinforcement should be considered effective as a part of this side face cracking control reinforcement.

5.21   Shear Walls

Shear walls are considered as cantilevers of deep beam type but carrying axial load, with horizontal wall length ℓ_w as beam depth and the wall height h_w as the beam length. Shear is most important in walls having small ratios of h_w/ℓ_w because flexure usually controls for high ratios. Shear walls are covered in Chapter 22.

5.22   Two-Way Footings and Slabs—Shear Strength

When a two-way slab is heavily loaded with a concentrated load or where a column rests on a two-way footing, diagonal tension cracks form that encircle the load or column.[24] These cracks are not visible on the surface, except as flexural cracks. Such cracks extend into the compression area of the slab and encounter resistance near the load similar to the shear-compression condition shown for the beam in Fig. 5.5a. The slab or footing continues to take load and finally fails around and against the load or column, punching out a pyramid of

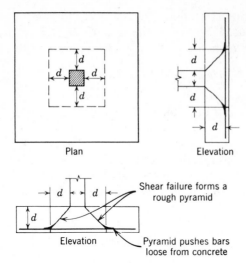

Plan   Elevation

Elevation

Shear failure forms a
rough pyramid

Pyramid pushes bars
loose from concrete

Figure 5.23 A square column tends to shear out a pyramid from a footing.

concrete as indicated in Fig. 5.23. Diagonal cracks do not form further out from the load or column because of the rapid increase in the failure perimeter, which under a square load always totals eight times the distance from the center of load. The initial diagonal cracks thus proceed to failure in shear compression (or punching shear) directly around the load.

In compromising between the initial cracking location and the final punching shear condition at failure for different ratios between column (or load) diameter and footing (or slab) thickness, the Joint Committee[24] recommended a single-strength calculation at a pseudocritical distance $d/2$ from the column face or edge of the load as shown in Fig. 5.24 (Code 11.11.1.2). Because there is no shear failure danger aside from the shear compression type of failure (with the diagonal crack not visible from outside the concrete), these slab shear strengths are based on ultimate strengths rather than cracking loads. Code Eq. 11.36 provides:

$$V_c = (2 + 4/\beta_c)\sqrt{f'_c}\,b_o d$$
$$\text{but Max. } V_c = 4\sqrt{f'_c}\,b_o d, \text{ for } \beta_c \text{ of 2 to 1 or smaller*}$$

where

β_c = ratio of long side to short side of the loaded or reaction area
b_o = perimeter of pseudocritical area, as in Fig. 5.24.

This increased shear strength is available only where two-way bending occurs at the load considered. The 1977 Code introduced the β_c term to warn that a 15 in. × 60 in. wall-like column has a zone on the long side that forces the slab

---

* SI version in ACI 318M is $(1 + 2/B_c)(\sqrt{f'_c}/6)b_o d$ or $(\sqrt{f'_c}/3)b_o d$, whichever is smaller. SI units are: V in newtons; f'_c in MPa, b, d in mm.

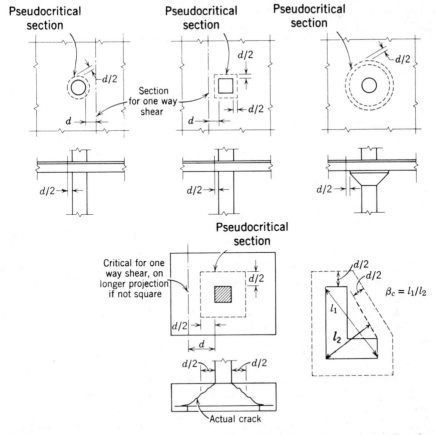

Figure 5.24 Pseudocritical diagonal tension sections in slabs supported directly on columns and in footings.

to act as a one-way slab locally. It would at first appear that a similar behavior would result around a 60 in. square column; it probably would with a thin slab. Slab thickness is obviously an important variable here and more data are needed to cover all cases. Reference 25 is an excellent discussion.

For an L-shaped interior column the present Code would be interpreted rather arbitrarily as shown in Fig. 5.24, using the maximum diagonal of the effective loaded area and the perpendicular width to fix β_c.

Wall and corner columns involve not only large moment transfers (treated for interior columns in Sec. 5.25) but also torsional problems in the slab. The latter must be covered separately.

In special cases footings or slabs may fail from (one-way) beam action at a distance d from the face of the load or support, on a straight section all across the member. Such failure is most apt to happen in long narrow footings or long-span slabs of narrow width. The limiting shear here, as given in Sec. 5.6, is usually $V_c = 2\sqrt{f_c'}\,bd$.

Examples involving shears covered in this section are included in Chapter 15 on flat plates and flat slabs and in Chapter 19 on footings.

5.23   Shear Reinforcement in Slabs

Slabs can be reinforced at interior columns for shear in at least two different fashions. The more common is the use of shearheads (Code 11.11.4). Shearheads are small structural steel channels or I-shapes that can be totally embedded in the slab. They are welded into a crossing pattern (at the same level) with cantileverlike projections into the slab as though they were the start of beams to adjacent columns. These pick up both moment and shear near the column and greatly increase the periphery on which shear on the concrete will be critical, moving it out to a periphery at 0.75 of the shearhead projection. The maximum shear there is limited to $4\sqrt{f_c'}b_o d$, but it may be as much as $7\sqrt{f_c'}b_o$ for V_n at the section near the column.

The other reinforcement pattern (Code 11.11.3) might be called the formation of shallow concrete "beams" of slab depth that have both top and bottom bars around which closed stirrups can be anchored. Here V_n is limited to $6\sqrt{f_c'}b_o d$ and V_c to $2\sqrt{f_c'}b_o d$. Adequate anchorage of the shear reinforcement is a most important element.

The Code Commentary has good discussions of both shearhead and the shallow beam type and these will not be repeated here.

Substantial research has greatly increased the knowledge about shearheads at exterior columns and to some extent corner columns. These are substantially different problems from those at interior columns under vertical loads. The 1986 Code has not been extended to cover exterior columns and its present shearhead requirements should not be extrapolated to such cases.

5.24   Openings in Slabs That Influence Shear Strength

The Code suggests a simple way of handling openings near columns (Code 11.11.5). Radial lines are drawn from the edges of the openings to the center of the column as shown in Fig. 5.25. These radial lines intersect the pseudocritical sections around the column, and all the resisting perimeter caught between such radial lines is considered as ineffective in resisting shears. If too much perimeter is lost, the designer must be sure adequate two-way bending is really present; otherwise the lower shears permitted in beams become the limiting values for the slab.

Openings more than 10 times the slab thickness from the concentrated load or reaction (unless within the column strips of flat slabs) need not be deducted. For slabs with shearheads only half the above deductions are required.

The 1962 Joint Committee Report[26] and Sec. 5.4.3 of the 1974 report[24] both show more detailed background.

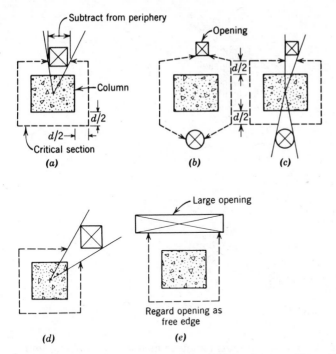

Subtract from periphery

Column

d/2

d/2

Critical section

(a)

Opening

d/2

d/2

(b)

(c)

Large opening

Regard opening as
free edge

(d)

(e)

Figure 5.25 Effect of openings on shear perimeter. (From Reference 24.)

5.25 Transfer of Moment Between Column and Slab

Another condition that adds locally to the shear around the column in flat plate construction arises where a moment is carried to or from a column. The Code requires an analysis of the resulting shears. It recognizes that much of the moment is transferred directly by flexure in the slab, but allows for the extra punching type shear by specifying the following portion γ_v of the moment M_u (transferred) be used in the shear calculation:

$$\gamma_v = \left(1 - \frac{1}{1 + \frac{2}{3}\sqrt{\frac{b_1}{b_2}}}\right)(M_u{}^*)$$

where b_1 is the width of the critical section parallel to span of the slab strip and b_2 is in the perpendicular direction, as in Fig. 5.26. For a square column this moment becomes $0.4M_u$ applied as a torsion on the slab.

---

* Note that the 1986 Code revision changed the notation in this equation to consider more correctly the edge and corner column cases. The previous versions were written in terms of the column width c and slab depth d which were correct only for interior columns.

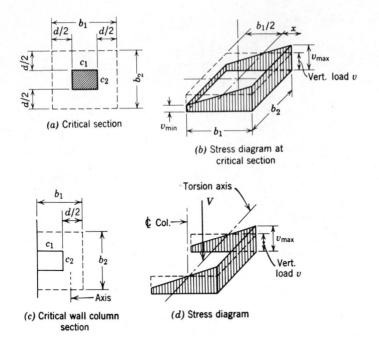

(a) Critical section

(b) Stress diagram at critical section

(c) Critical wall column section

(d) Stress diagram

Figure 5.26 Shears around a column that must introduce moment (and torsion) into a slab or accept moment from the slab.

The ACI-ASCE Committee 326 in 1962 recommended[26] essentially that the slab shear be analyzed as follows.*

1. Consider the usual critical shear perimeter at $d/2$ from the column to resist this shear as a torsion.
2. The resisting area becomes the four vertical sides of the slab cut by the perimeter.

$$A_c = 2d(c_1 + c_2 + 2d)$$

The axis of torsion is a horizontal line, through the column center for the usual interior column.

3. Calculate the polar moment of inertia of the resisting slab areas. From Fig. 5.26a:

$$J = 2(b_1 d^3/12 + db_1^3/12) + 2b_2 d(b_1/2)^2$$

4. Obtain the unit shear from Tx/J where T is the portion of M_u computed above and x is the distance from the torsion axis, with the maximum at $x = b_1/2$.
5. This torsional shear is to be added to the usual vertical shear as in Fig. 5.26b, quite similar to the P/A and Mc/I combination in elastic theory.

$$v_u = \frac{V_u}{A_c} \pm \frac{\gamma_v M_u c}{J}$$

---

* The proportion of the column moment assigned to torsion has been increased since then.

If one considers an edge column, the following adjustments are to be made.

1. The torsion axis must be the centroid of the three resisting slab faces.
2. The polar moment of inertia and shear area are correspondingly reduced.

TORSION

5.26 Pure Torsion

(a) General Torsion is important in some spandrel beams, in flat plates at exterior columns, in curved stair slabs or curved beams, and wherever large loads must be carried off the axis of the member.[27] In box girder bridges and in any isolated beam that must resist unbalanced loading, torsion is an important problem. When torque is small [factored torsional moment $T_u \leq \phi (0.5\sqrt{f_c'}\Sigma x^2 y)$, which is equivalent to a torsional shear stress not more than $v_t = 1.5\sqrt{f_c'}$)], the Code suggests that it be ignored. (Code 11.6.1). $\Sigma x^2 y$ is determined by breaking the cross section into its component rectangles with a limit on the overhanging flange width to be considered as three times the flange thickness (Code 11.6.1.1). x is the shorter overall dimension of each rectangle and y is the longer overall dimension of each rectangle. Restrictions for box sections are given in Code 11.6.1.2. The Code recognizes the greater torsional stiffness of the closed box section by treating it as a solid section with some limitations. Where it is possible to devise framing that can take the load without torsion, this is a preferred solution (by the authors). Because torque design is a relatively new area in reinforced concrete, it is given more space here than the seriousness of the problem warrants. The authors feel that the most sensible approach to torsion and especially combined torsion, shear, and flexure lies in the explicit use of a space truss model with variable angle of inclination diagonals.[5-7] Although extensive research work has indicated the promise of these models, they have not yet been explicitly adopted by the ACI Code Committee. Many of the present Code equations for torsion are based on a 45° space truss model.

(b) Elastic Theory of Pure Torsion Elastic analysis of a round shaft under torsion T is simple and results in tangential shearing stresses all around the member as shown on one diameter in Fig. 5.27a. The equation is $v_{tn} = Tr/\int r^2 dA$). Any other shape is increasingly complex because such a cross section warps and the member lengthens as it twists. For analysis and visualization the soap film (membrane) analogy is useful. A soap film is stretched across a hole cut in the top of a box to the shape of the member cross section. A small inside air pressure then raises a bubble over the hole. The volume between the bubble and the original plane, by analogy of the governing differential equations, is proportional to the total torque resistance; the slope of the membrane (which is under uniform tension) measures the unit shear stress, a steep slope representing a high stress.

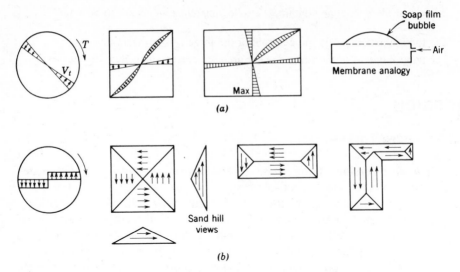

Figure 5.27 Elastic and plastic analysis for pure torsion shears. (*a*) Elastic analysis. (*b*) Plastic analysis v_t, equal over entire cross section.

(c) Plastic Analysis for Pure Torsion   A plastic analysis assumes uniform shear intensity all around the surface and all over the cross section, as in Fig. 5.27*b*. The analysis can be envisioned in terms of the sand heap analogy. This gives uniform side slopes with the stress at 90° to the slope, and again a volume proportional to the torque. For the slope equal to the maximum in the elastic analysis, volume and torque indicated is considerably larger. The equation can be written $v_{tn} = T/J_p$ where J_p is twice the volume of the sand heap divided by the slope.

The final failure is a diagonal crack following around the surface to form a helix with about a 45° slope.

(d) Code Analysis   Neither of these analyses is realistic, partially because of the properties of concrete but chiefly because concrete at design loading will be cracked. The best—that is, the most consistent—calculation method has not yet been fully established. The senior author in early investigations found these two methods about equally consistent, but prefers the lower stress numbers the plastic method gives. The analysis is not fully adequate because it shows small, but consistent, differences in the ultimate stress for the T-beam, L-beam, and the rectangular beam, in a systematic decreasing manner that indicates the analysis does not reflect the shape properly. The value is also a little higher for uniform distributed torque load than for a concentrated torque load.

The design for torsion under the ACI Code parallels the general philosophy for shear.

$$T_u \gtrless \phi Tn \quad \text{and} \quad Tn = T_c + T_s \quad \text{(Code Eqs. 11.20 and 11.21).}$$

This is somewhat inconsistent as the reinforcement provided for torsion must consist of both continuous transverse reinforcement (closed hoops, closed ties,

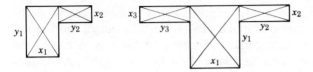

Figure 5.28 Rectangles for Code T_c calculation.

or spirals) and additional longitudinal bars distributed around the entire perimeter of the section (Code Sec. 11.6.7).

One of the major changes introduced in the 1977 ACI Code and continued in the 1983 and 1986 updates is the concept of a limiting or maximum nominal torsion strength in those situations where torsion is produced by compatibility rather than simple equilibrium. These situations are in statically indeterminate applications where a rotation induces a torsional moment. When the concrete cracks there is a great reduction in torsional stiffness and this induced torsional moment is greatly reduced. This logic led to the adoption of a maximum factored torsional moment value $T_u = \phi 4\sqrt{f_c'}\Sigma x^2 y/3$ for members where redistribution of internal forces is possible (Code 11.6.3). In developing this value, the Code uses an approximate relationship for the torque T_c that lies between the elastic and plastic analysis:

$$v_{tn} = T_n/[\Sigma(x^2 y)/3] = 3T_n/(\Sigma x^2 y) \quad \text{or} \quad T_n = v_{tn}\Sigma x^2 y/3$$

Often the Code expressions are written in the format $T_c = (v_{tc}/3)\Sigma x^2 y$ for simplicity. The authors will discuss current Code requirements in terms of v_{tn} and try to point out the $v_t/3$ equivalence.

In these equations, x is the short dimension of the rectangle, y the long one, and Σ indicates the sum of all the component rectangles making up the beam cross section, Fig. 5.28. The beam shape may be subdivided to give the largest total. The value 3 is a stress factor obtained from the plastic solution of the "sand heap analogy" for a rectangular section with $x/y = 1$. A lower bound value of $v_{tc} = 4\sqrt{f_c'}$ was chosen from a survey of test results. The concept of a maximum factored torsional moment for the case of compatibility torsion greatly reduces the severity of the torsional design requirements, but the student must be aware that it must not be applied in cases of equilibrium torsion where no alternate load path is available to relieve the torque.

TORSION COMBINED WITH SHEAR AND MOMENT

5.27 Member Behavior Under Test

Because the crack patterns of torque, shear, and moment are quite different, it is necessary to test under all three in combination. To the authors test behavior seems best described physically and mathematically from the torque and shear interaction. On one web face the torque shear adds (and on the other subtracts) relative to the vertical shear stresses. A little torque causes a diagonal crack to open on one side of a beam earlier than on the other, but the shear-failure pattern is not much altered, possibly a little steeper diagonal crack on the opposite face. The torque tends to make diagonal cracks out of the flexural cracks on the tension face. As the ratio of torque to shear is increased the failure pattern is more sensitive to the shape of the beam, and the tension face cracks become full diagonals.

With the rectangular beam without stirrups the load difference between initial diagonal crack and ultimate may not be large and the diagonal crack on the opposite face may be replaced by an inclined compression crack (evidence of a skew bending effect); or a compression crack on a skew plane may finally develop on the compression face of the beam and a steeper diagonal crack (nearly vertical or of reversed slope) may show on the opposite web face.

With a T-beam the failure mechanism is slower in developing (more load after first diagonal cracking) and is not yet well understood. The diagonal crack usually eventually opens on the far side of the web, always steeper, sometimes of reversed slope, dependent on the ratio of torque to shear.

To the authors torsion appears primarily as a special shear problem, one that is troublesome because many variables influence the available resistance, including beam shape, the stirrups present, and even the longitudinal beam steel. Progress has been made in defining ultimate strength in terms of skew bending. The variable angle truss model with full plasticity assumed has been used to develop very accurate sets of interaction relationships.[5,28,29]

When closed stirrups are added, which in practice must be everywhere that torsion is important, the spread between first diagonal crack load and ultimate load increases. The few diagonal cracks present in the test members without stirrups (Fig. 5.29a) become amazingly numerous with stirrups (Fig. 5.29b), member stiffness against rotation drops to from 3 to 7% of that before diagonal cracking,* and longitudinal steel stresses then increase rapidly, in proportion to the angle. Most stirruped beams rotate so much before reaching ultimate strength as to appear worthless as structural members well before they fail.

5.28   Strength Interaction of Torsion with Shear

(a) Without Stirrups   When torsion is calculated by plastic analysis of the gross (uncracked) concrete section, a pure torsion, without flexural shear, leads to a calculated v_{tc} of from $5\sqrt{f_c'}$ to $6\sqrt{f_c'}$, compared to about $2\sqrt{f_c'}$ for flexural shear cracking strength v_c without any torsion.† The combinations of torque and shear lead to the ellipse of Fig. 5.30a which could be replaced with a

---

* Measured as the change in rotation angle ϕ relative to the torque increment applied, that is, $d\phi/dT$, similar to the tangent modulus idea.

† That the ultimate v_{tc} might be three times the ultimate v_c long concerned the senior author, because each led to diagonal tension cracks. Finally he realized that neither calculation had a rational approach and that both numbers were simply crude stress coefficients, not stresses at all. He calls them "imaginary stresses" whenever someone raises the question. Both normally relate to flexurally cracked members, but both use properties of the uncracked section. A calculation based on imaginary areas (with sloppy theory as well) necessarily leads to imaginary stresses.

circle (Fig. 5.30b) in terms of v_{tn}/v_{tc} and v_n/v_c. The circle is accepted generally in the United States and Canada, but not so widely in Europe. The 1986 Code uses *total* shear and torque for its equations, but the unit stress format seems to give the clearer picture here. (The authors also find the ellipse appropriate for strengths with stirrups, both v_{tn} and v_u being then increased to v_{to} and v_o for a specific quantity of stirrups, as in Fig. 5.30c. This view is not generally accepted as yet.)

(b) With Stirrups, Code Design   When *pure* torsion tests are run in which closed *stirrups and longitudinal reinforcement are kept equal in volume,* which is the ideal ratio for pure torsion with inclinations of the diagonal compression struts at 45°, the strength curve of Fig. 5.31a is found. The plotted points trace a straight line until an over-reinforced condition is reached. When these strengths are projected back to zero stirrups, the straight-line portion intersects at about v_{tc} of $2.4\sqrt{f_c'}$. The Torsion Committee concluded* that not all of $v_{tc} = 6\sqrt{f_c'}$ could be used with stirrups on the straight-line basis and the Code rules are based on this finding. The concrete stress usable for resisting torsion is limited to $2.4\sqrt{f_c'}$ and for general design the ellipse of Fig. 6.9b is used. These values are expressed in the $v_t/3$ coefficients used with $\Sigma x^2 y$ instead of $x^2 y/3$. In Code 11.6.1 $v_t/3 = 0.5\sqrt{f_c'}$ is used where torsion may be neglected and in Code 11.6.6.1 the T_c term is based on an interaction equation where $v_t/3 = 0.8\sqrt{f_c'}$ which corresponds to $v_t = 2.4\sqrt{f_c'}$.

5.29   Stirrup Design for Torsion

Although the Code allows $1.5\sqrt{f_c'}$ ($v_t/3 = 0.5\sqrt{f_c'}$) in torsion without any special provisions, above this value any combination of v_{tn} and v_n falling outside the ellipse of Fig. 5.31b requires the design of stirrups both for shear and for torsion. These limiting values without stirrups are given by these equations, with the same 1986 Code equations in terms of total torque and shear to the right:

$$v_{tc} = \frac{2.4\sqrt{f_c'}}{\sqrt{1 + (1.2v_n/v_{tn})^2}} \qquad T_c = \frac{0.8\sqrt{f_c'}\Sigma x^2 y}{\sqrt{1 + \left(\dfrac{0.4V_u}{C_t T_u}\right)^2}} \qquad \text{(Code Eq. 11.22)}$$

$$v_c = \frac{2\sqrt{f_c'}}{\sqrt{1 + (v_{tn}/1.2v_n)^2}} \qquad V_c = \frac{2\sqrt{f_c'}b_w d}{\sqrt{1 + \left(2.5C_t \dfrac{T_u}{V_u}\right)^2}} \qquad \text{(Code Eq. 11.5)}$$

where $C_t = b_w d/\Sigma x^2 y$. ACI 318M in SI units uses $\sqrt{f_c'}/15$ in Eq. 11.22 and $\sqrt{f_c'}/6$ in Eq. 11.5.

The critical section for torsion, as for shear, is a distance d from the face of

---

* Based on work by Dr. Hsu at the PCA Skokie Laboratory.

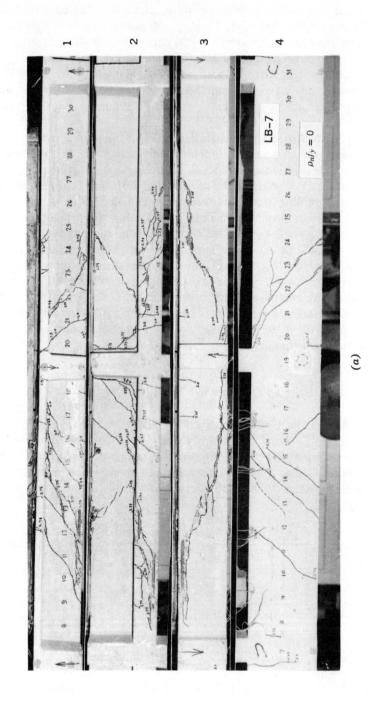

(a)

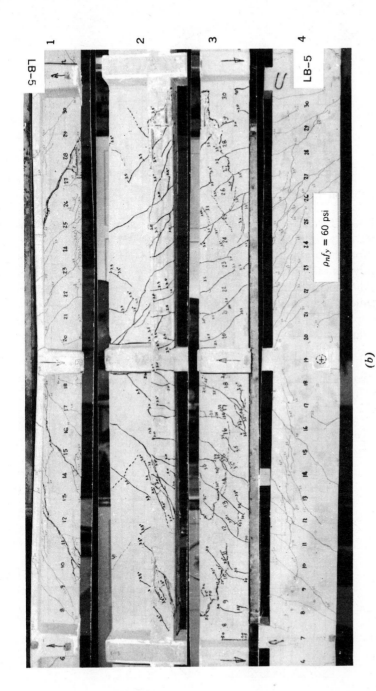

Figure 5.29 L-beams with $b_w = 3$ in. loaded as semicontinuous beams with load eccentricity of 5 in. (a) No stirrups. (b) Light stirrups. Corresponding torque–rotation curves are shown for LB-7 and LB-5 in Fig. 5.33. The four views of each beam are: (1) web on flange side; (2) bottom of beam and flange; (3) side of web opposite flange; (4) top.

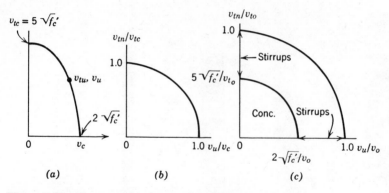

Figure 5.30 Shear-torsion interaction diagram.

support (Code 11.6.4). If the design combination of stresses falls at point B in Fig. 5.31b, stirrups for torsion are to be based on the vertical projection of the overrun AB and for shear on the horizontal projection of AB. This stirrup requirement is equivalent to using the triangular interaction of Fig. 5.31c for the stirrup portion. The line AB from Fig. 5.31b can be superimposed on the triangular arrangement if the scales are each in terms of v_{tn} and v_n.

The minimum area of stirrups for torsion outside the ellipse in Fig. 5.31b must satisfy the equation

$$A_v + 2A_t = 50 b_w s/f_y \qquad \text{(Code Eq. 11.16)}$$

where A_v is the (two-leg) area of stirrups for shear and A_t is the area of one leg of the stirrups for torsion.

For a given $T_s = T_n - T_c$ the required area for A_t is

$$A_t = \frac{v_{ts}s\Sigma x^2 y}{3\alpha_t x_1 y_1 f_y} = \frac{T_s s}{\alpha_t x_1 y_1 f_y} \qquad \text{(From Code Eq. 11.23)}$$

since $v_{ts} = T_s/\Sigma x^2 y/3$

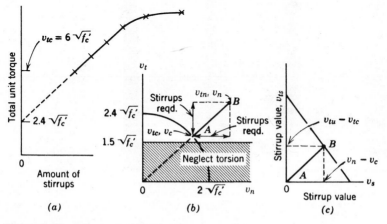

Figure 5.31 Stirrup strength and design.

where

x_1 and y_1 = the narrow and long sides, respectively, of the closed stirrups

s = the stirrup spacing

f_y = the stirrup yield point

$\alpha_t = 0.66 + 0.33 y_1/x_1 \leqq 1.50$

Torsion reinforcement must extend at least a distance $d + b$ beyond the point theoretically required (Code 11.6.7.6).

5.30   Longitudinal Steel for Torsion

When beams fail in torsion their length increases and extra stress moves into the longitudinal bars. Torsion stirrups cannot act effectively unless a tension truss "chord" is available in each corner of a rectangular beam, because the tension in pure torsion may be needed in any (or all) corners as the diagonal crack wraps around the member.

The Code sets two requirements for extra longitudinal steel for torsion; the first is a volume of longitudinal steel for torsion equal to the volume of stirrups provided for torsion:

$$A_\ell = 2A_t(x_1 + y_1)/s \qquad \text{(Code Eq. 11.24)}$$

If the stirrup volume is very low, the following requirement defines a larger required A_ℓ:

$$A_\ell = \left[\frac{400\,xs}{f_y} \left(\frac{T_u}{T_u + \dfrac{V_u}{3\,C_t}} \right) - 2A_t \right] \frac{x_1 + y_1}{s} \qquad \text{(Code Eq. 11.25)}$$

where "$2A_t$ . . . need not be taken less than $50 b_w s/f_y$" and $C_t = b_w d/\Sigma x^2 y$ which is a shape factor relating the shear and torsional stress sections.

The second equation is better understood in terms of its development. The Torsion Committee originally recommended that the torsion stirrups used not be less than

$$A_t = \frac{200\,b_w s}{f_y} \cdot \frac{T_u}{T_u + \dfrac{V_u}{3\,C_t}} = \frac{200\,b_w s}{f_y} \cdot \frac{v_{tn}}{v_{tn} + v_n}$$

and that longitudinal reinforcement A_ℓ for torsion be used with a volume equal to that of the stirrups used for torsion. The Committee later agreed to the Code version for minimum stirrups, which uses 50 instead of $200 v_{tn}/(v_{tn} + v_n)$ on condition that A_ℓ be increased enough to make up the deficiency in $2A_t$ brought about by the change. This requires something *more* for A_ℓ than would have been required from the simpler equation. The result is a rather awkward equation. Its use is covered in the design of Sec. 5.31.

5.31   Design Example—Torsion and Shear

An 18-in. wide × 27-in. deep rectangular cantilever beam 9-ft long, as shown in Fig. 5.32a, has been designed to pick up at its end a 20-ton hoisting load, which includes a proper allowance for impact. It is found that operating conditions may place the load 10 in. off the axis of the beam to either side. Design the torsion and shear reinforcement, assuming Grade 60 reinforcement and f'_c of 3000 psi.

Solution

This case is equilibrium torsion. No redistribution of forces is available so the torsion reduction of Code 11.6.3 does not apply.

The critical section for both torsion and shear is at a distance d from the support.

$$d = 27 - \text{say } 2.75 = 24.25, \text{ say } 24.3 \text{ in.}$$
$$\text{Beam weight} = (18 \times 27/144)150 = 506 \text{ plf}$$

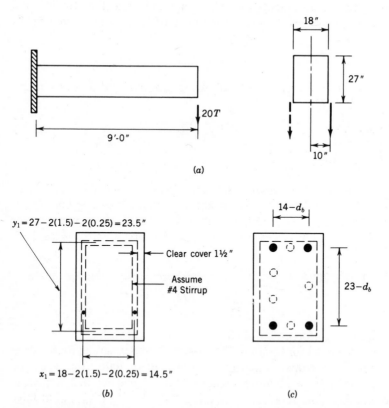

Figure 5.32 Design example of sec. 5.31. (a) Example. (b) Cross section. (c) Longitudinal reinforcement distribution.

At the critical section:

$$V_u = 1.7(20 \times 2000) + 1.4 \times 506(9 - 2.02) = 72{,}900 \text{ lb}$$
$$V_n = 72{,}900/0.85 = 85{,}800 \text{ lb}$$
$$T_u = 1.7(20 \times 2000)10 = 680{,}000 \text{ lb-in.}$$
$$T_n = 680{,}000/0.85 = 800{,}000 \text{ lb-in.}$$

Because this is a case of combined shear and torsion, the first check will be to see if the torsion needs to be considered. Code 11.6.1 allows torsion effects to be neglected if $T_n \gtrless 0.5\sqrt{f_c'}\Sigma x^2 y = (0.5)\sqrt{3000}(18^2)(27) = 240{,}000$ lb-in. Since $T_n = 800{,}000$ lb-in., torsion effects must be considered.

Because of the torsion effects, V_c must be computed by Code Eq. 11.5 so that Table 5.1 needs to be modified for reduced V_c strength and $V_s = V_n - V_c$ will reflect the torsion effects as well.

Assuming #4 stirrups and 1-½ in. clear cover, as shown in Fig. 5.32b, $x_1 = 14.5$ in. and $y_1 = 23.5$ in. Because a single rectangle is used for this shape $\Sigma x^2 y = (18^2)(27) = 8748$ in.^{3}, $C_t = b_w d/\Sigma x^2 y = (18)(24.3)/8748 = 0.05$, and

$$V_c = \frac{2\sqrt{f_c'}b_w d}{\sqrt{1 + \left(2.5 C_t \dfrac{T_u}{V_u}\right)^2}} = \frac{2\sqrt{3000}(18)(24.3)}{\sqrt{1 + \left[2.5(0.05)\left(\dfrac{680{,}000}{72{,}900}\right)\right]^2}} = 31{,}200 \text{ lb}$$

This modifies the Table 5.1 boundaries for shear region III to $(V_c + 50 b_w d) \gtrless V_n < V_c + 4\sqrt{f_c'}b_w d$ or $55{,}900 \gtrless 85{,}800 < 108{,}000$. Thus for shear reinforcement design this member is in region III with max $s = d/2 = 12.15$ in. and $A_v = V_s s/f_y d$.

$$V_s = V_n - V_c = 85{,}800 - 31{,}200 = 54{,}600 \text{ lb}$$
$$A_v = (54{,}600)(s)/(60{,}000)(24.3) = 0.0374 s$$

Stirrup spacing will be delayed until torsion requirements are determined. The stirrup requirements will be combined.

From Code Eq. 11.22,

$$T_c = \frac{0.8\sqrt{f_c'}\Sigma x^2 y}{\sqrt{1 + \left(\dfrac{0.4 V_u}{C_t T_u}\right)^2}} = \frac{0.8\sqrt{3000}(8748)}{\sqrt{1 + \left[\dfrac{(0.4)(72{,}900)}{(0.05)(680{,}000)}\right]^2}} = 291{,}000 \text{ lb-in.}$$

$$T_s = T_n - T_c = 800{,}000 - 291{,}000 = 509{,}000 \text{ lb-in.}$$

From Code Eq. 11.23, $A_t = (T_s s/\alpha_t x_1 y_1 f_y)$. α_t is a reinforcement efficiency factor $\alpha_t = 0.66 + 0.33(y_1/x_1) = 0.66 + 0.33(23.5/14.5) = 1.20$.

$$A_t = 509{,}000 s/(1.20)(14.5)(23.5)(60{,}000) = 0.0208 s$$

Note that $A_v = 0.0374 s$ for shear reinforcement assumes that $A_v = 2A_b$ (both vertical legs of the stirrup resisting shear) and $A_t = 0.0208 s$ assumes

$A_t = A_b$ (area of a single leg on each face). Because the shear and torsion web reinforcement are additive (Code 11.6.7.2), the requirements can be rewritten as $A_b = (A_v/2) + A_t = (0.0374/2 + 0.0208)s$.

For a #4 stirrup, $A_b = 0.20$ si. $s = 0.20/(0.0187 + 0.0208) = 5.06$ in. #4 closed stirrups at 5 in. on center would satisfy the combined shear and torsion requirements.

The Code requires a torsion spacing not greater than $(x_1 + y_1)/4 = (23.5 + 14.5)/4 = 9.5$ in., nor greater than 12 in. Development of closed transverse torsion reinforcement is not specifically covered in the Code; the same provisions applied to stirrups for shear will be used (Sec. 5.8 or Code 12.13).

$$d/2 = 24.3/2 = 12.1 \qquad \ell_d/2 = 6 \text{ in. for } \#4$$

Deduct cover and $3d_b$ from $d/2 = 12.1 - 1.5 - 3 \times 0.5 = 9.1$ in. > 6 in. This available depth plus the anchorage around the corner is **O.K.**

USE #4 closed stirrups at 5 in.

Calculate extra longitudinal reinforcement for torque.

$$A_\ell = 2A_t(x_1 + y_1)/s = 0.0416s(23.5 + 14.5)/s = 1.58 \text{ in.}^2$$

$$A_\ell = \left[\frac{400xs}{f_y} \left(\frac{T_u}{T_u + \dfrac{V_u}{3C_t}} \right) - 2A_t \right] \frac{x_1 + y_1}{s} \quad \text{from Code Eq. 11.25}$$

$$= \left[\frac{400 \times 18s}{60,000} \left(\frac{680,000}{680,000 + \dfrac{72,900}{3 \times 0.05}} \right) - 0.0416s \right] 38/s = 1.08 \text{ in.}^2$$

The 1.58 in.² governs. The Code requires at least #3 longitudinal bars, spaced not farther apart than 12 in., and at least one bar in each corner of the stirrups. The reinforcement is to be distributed around the perimeter. Figure 5.32c indicates two possible patterns. There must be a bar in each corner of the closed stirrups. The top and bottom face spacings between corner bars of 14-d_b in. exceed the 12 in. maximum and so intermediate bars are required in the narrow faces as well. On the side faces, a single intermediate bar as shown on the right face would have a much greater spacing than top and bottom face bars. It would approach the 12 in. maximum. It is probably better to use two bars on each side face as shown on the left side face. These spacings would be more like the top and bottom faces. Because $A_\ell = 1.58$ is divided conceptually into 10 bars in this type arrangement, each bar would be 1.58/10 or 0.16 si. #4 bars with $A_b = 0.20$ si would be suitable. For A_ℓ USE 2 − #4 = 0.40 si at the third points of the height each side. ADD $1.58 - (4)(0.16)/2 = 0.47$ in.² *each* to the top and to bottom bars. Because of beam width, there must be at least three bars on top and bottom face, that is, one at each stirrup corner and one midway between, since the spacing is limited to 12 in. The parts of A_ℓ can be added to the longitudinal steel required for moment and the bars selected for the totals.

5.32 Calculation of Torque Developed in Structures

A meaningful calculation of the torque to be used in design is complex unless the torque is almost statically determined, like an awning slab cantilevered from the side face of a beam. Any statically determined case should be conservatively designed by the Code provisions already discussed.

The more general case is the spandrel beam, loaded in torque by the twisting it receives from the deflecting slab. The resulting torque is a matter of relative stiffness, which varies sharply as the members crack, much more sharply in torque than in flexural stiffness. For example, Fig. 5.33 shows torque-twist curves[30] for semicontinuous beams loaded in M, V, and T. For member LB-5 with minimum stirrups ($\rho_w f_y = 60$ psi) the rotation curve shows that to increase torque resistance by 35% above the cracking point required an 1100 percent increase in the unit rotation. With stirrups having $\rho_w f_y = 120$ psi (member LB-8) the torque increased only 75 percent for a 1000% increase in rotation. Turning this into analysis terms, one can say that, although flexural stiffness decreases maybe 50% from cracking, torsional stiffness drops down to 5 or 10% its uncracked value. After diagonal cracking, stirrups will absorb relatively little additional torque from an indeterminate structure and might then serve chiefly as crack control steel. The next section presents an alternate approach the Code now permits for cases similar to the spandrel beam.

A special caution should be given for any case similar to that in Fig. 5.34 where one beam rotates the girder while the nearby column joint is either fixed

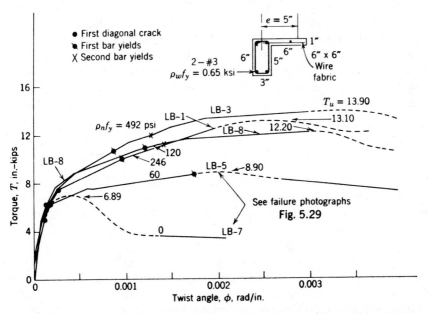

Figure 5.33 Effect of amount of stirrups on T-ϕ curves (f'_c from 4060 to 4400 psi) $\rho_n f_y = A_v f_y / b_w s$). Beams are semicontinuous, also carrying M and V.

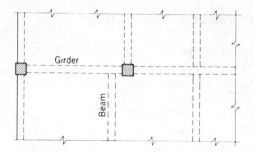

Figure 5.34 This beam framing produces severe torsional stresses in the girder between the beam and the adjacent column.

or rotating in the opposite direction. Such a girder will be sharply twisted in the length between beam and column. Diagonal cracks will occur and stirrups as crack control should be liberally supplied.

5.33 Alternate Code Design with Redistribution of Torque

Where the member torque is not required to maintain equilibrium, Code 11.6.3 relaxes the design demands by permitting design for T_n of $4\sqrt{f'_c}x^2y/3$, the equivalent of v_{tn} of $4\sqrt{f'_c}$, which is close to the cracking torsion. This torque, resisted partially by concrete T_c (Sec. 5.29) and the remainder by stirrups T_s may be considered uniform for the first distance d from the support. If the torque comes to the beam from a slab, the torque should decrease linearly toward zero at midspan. Code 11.6.3.2 permits assumption of uniform distribution of torsional loading from the slab that would produce such a linear decrease. The minimum stirrups specified in Code 11.5.5.5 for the distance $(d + b)$ beyond the theoretical point where T_c would be adequate are still required.

With stirrups as specified, the cracking torque is not increased much by additional twist and the stirrups then act as crack control steel. Any theoretical end moment on the slab in excess of the nominal torque resistance assumed must be distributed back to the slab, thus increasing the slab positive moment demands.

5.34 Box Sections

No detailed design of box sections resisting torsion will be shown. Instead, related comments on member behavior under combined torsion, shear, and flexure are made because it is such a *combination* that is critical in every practical torque case the authors can envision. Consider first a solid cross section near the ultimate failure zone with a shear resistance uniformly distributed across the member, in the absence of torque, as in Fig. 5.35a. Then, on the same section, consider the added shear resisting the torque, at the elastic stage, as in Fig. 5.35b. Failure will *not* occur when v_c plus v_t approaches the ordinary shear limit, because more of v_c can shift to the center of the member, leaving

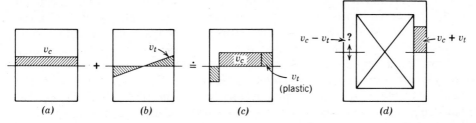

Figure 5.35 Shear (*a*) plus torque (*b*) on solid section results probably as in (*c*). (*d*) Equivalent on a box section.

the side faces chiefly for torque (probably with a plastic distribution somewhat as in Fig. 5.35*c*). Such a distribution of stresses explains why the elliptical or circular interaction (Figs. 5.30 and 5.31) in the ACI Code is rational, as well as experimentally justified. Thürlimann and his co-workers have convincingly demonstrated the applicability of plasticity theory in a more formal fashion.[5]

Compare the response of an open box section under the same combination of loading. If, in Fig. 5.35*d*, one superimposes the vertical shear and the torque shear, there is no nearby area to relieve the local overstress; the top and bottom slabs are too far away. A local wall failure will probably occur from the *sum* of the two stresses, equivalent to a straight-line interaction diagram. This is the basis used for many European codes.

ACI Code 11.6.1.2 assumes an open box section (with interior corner fillets) with wall thickness at least $0.25b$ may be considered the same as a solid section. Tests have shown the lost area not very helpful in resisting pure torsion. However, the missing area would be by no means useless in carrying shear with torsion, and its absence must weaken the section for the practical case. Fortunately, tests indicate the Code solid section design as overly safe. Code 11.6.1.2 gives a reduced torque capacity for thinner side walls.

5.35 Senior Author's Comments on State of the Art and Code Provisions

(a) Narrow Deep Members   It is possible that, as the ratio of h/b for a rectangle increases to where it represents more of a wall-like section, the circular (Code) interaction may prove less tenable; an L-shape with h/b_w of 3 for the web, for example, shows at the diagonal *cracking load* (but not at ultimate) less of the circular interaction than does one with h/b_w of 2. It is also imaginable that the redistribution beyond initial diagonal cracking might well be much smaller for a big box than for a smaller solid section.

(b) Need for $A_\ell = A_t$ None of the senior author's tests with stirruped L-beams has given results similar to those of Fig. 5.31*a*, probably because he has never tested with the steel limitations there imposed, that is, volume of longitudinal steel equal to volume of stirrup steel. When adequate longitudinal reinforcing for moment is used, the total longitudinal reinforcing is usually much more than equal in volume to the transverse reinforcement.* A number of the senior author's tests have shown that stirrups valued at $\rho_n f_y = 60$ psi (such as beam LB-5 in Fig. 5.33) did proportionately increase strength under combined torsion, shear, and flexure loading beyond the nominal v_{tc} value of $5\sqrt{f'_c}$.† He considers that the conditions that led to the straight line from $2.4\sqrt{f'_c}$ were the result of test beam designs not practical for real design. The design of longitudinal steel primarily for the moment seems automatically to eliminate any need for the $2.4\sqrt{f'_c}$ limitation. A limit for v_{tc} of approximately $5\sqrt{f'_c}$ would be more appropriate and could also reduce the stirrup requirements. The use of an elliptical interaction for the stirrups would give a further saving, although the author would not recommend this for box beams.

(c) Influence of Distributed Loads   Continuous L-beams with stirrups[31] and continuous T-beams without stirrups[32] both seem to show better under distributed torque and shear loading than under constant torque and shear. Tested under T, V, and M, with critical T and V stresses assumed to be at a distance d from the support, the strengths showed at least 25% better than with the constant T and V. This increased strength probably results from the fact that critical stress combinations then exist over much shortened lengths of the beam, thus possibly damping the spread of cracks after they initially form.

(d) The Spandrel Beam as an L-Beam   Divided opinion exists as to whether a spandrel beam should be designed for torsion as a rectangular or an L-beam. One investigator[33] noted large yield cracks in the slab at the face of spandrel web and no diagonal cracks in the slab, which caused him to consider the beam as a rectangular beam for torque resistance. However, he found on that basis a substantial *excess* torque strength for his "rectangular" beam.

The author notes that a slab has two functions in this case, one to deliver load to the beam and the other to help carry the torque that results. The slab probably works imperfectly to resist the torque at midspan where the torque is small, but near the column it should be more effective. Whether that is true or not, the slab forces the spandrel to twist about an axis at the level of the slab and not about the natural center of rotation as a rectangular beam. This alone increases the torque resistance through transverse bending. Also, after diagonal cracking, when the spandrel starts to lengthen under torsion, the monolithic slab tends to resist this lengthening and adds a counteracting axial force. In summary, the slab in this case probably works differently from the flange in a freestanding L-beam, but the flange does not act significantly and effectively in

---

* Note that the author compares *total* $A_s + A'_s$ with total stirrups, *not at all* the same as the Code A_l compared with A_t steel. The author's only limitation is to keep $A_s + A'_s \gtrless 1.25 A_s$, which appears to be a safe rule.

† The increase in Fig. 5.33 was from T_u of 6.89 in.-k for LB-7 to 8.90 in.-k for LB-5, or 29 percent. These two beams are shown at failure in Fig. 5.29 where numerous fine cracks show with the light stirrups, which is desirable.

the matter of strength. In fact, the Code limitation on flange overhang to $3\,h_f$ is about half of what might safely be used in the author's opinion.

(e) The Large Rotation Necessary to Develop Stirrups   The senior author doubts that the full torque value of stirrups can actually be developed in a structure, because of the excessive rotations required to bring them fully into effect (Fig. 5.33). Possibly one needs to discount the last 40% of rotation capacity and the gain of resistance associated with it. This might mean discounting, say, the last 20% of stirrup strength, although what percentages are proper certainly needs to be further explored.

(f) Relation of Stress on A_ℓ to Rotation Angle   Longitudinal steel stress from torsion appears to be wholly a function of rotation angle (under combined loading) rather than a linear function of the torque carried. Before diagonal cracking this longitudinal stress is negligible, in the order of 2 or 3 ksi. It seems to be roughly the same for a given angle of twist regardless of whether the minimum or the maximum stirrups are used. Any necessary limitation the structure imposes on ultimate rotation should thus relieve also the present demands for longitudinal steel added in proportion to the stirrups used for torsion.

Selected References

1. W. Ritter, "Die Bauweise Hennebique," *Schweizerische Bauzeitung,* Vol. 33, No. 7, Feb. 1899, Zurich, p. 60.

2. M. O. Withey, "Tests of Plain and Reinforced Concrete, Series of 1906," *Bulletin of the Univ. of Wisconsin, Engineering Series,* Vol. 4, No. 1, Nov. 1907, pp. 1–66.

3. M. O. Withey, "Tests of Plain and Reinforced Concrete, Series of 1907," *Bulletin of the Univ. of Wisconsin, Engineering Series,* Vol. 4, No. 2, Feb. 1908, pp. 71–136.

4. A. N. Talbot, "Tests of Reinforced Concrete Beams: Resistance to Web Stresses, Series of 1902 and 1908," *Bulletin 29,* Univ. of Illinois Engineering Experiment Station, Jan. 1909, p 85.

5. B. Thurlimann, P. Marti, J. Pralong, P. Ritz, and B. Zimmerli, *Anwendung der Plastizitatstheorie auf Stahlbeton,* Institut für Baustatik und Konstruktion, ETH Zurich, April 1983, 252 pp.

6. M. P. Collins, and D. Mitchell, "Shear and Torsion Design of Prestressed and Non-Prestressed Concrete Beams," *PCI Jour.,* Vol. 25, No. 5, Sept.–Oct. 1980, pp. 32–100.

7. J. Schlaich, and K. Schafer, "Konstruieren im Stahlbetonbau," *Beton-Kalender 1984,* Ernst and Sohn, Berlin, 1984, pp. 787–1005.

8. ACI Committee 426, "The Shear Strength of Reinforced Concrete Members," *Proc. ASCE, Jour. Struct. Div.,* ST-6, June 1973, pp. 1091–1187.

9. R. C. Fenwick and T. Pauley, "Mechanisms of Shear Resistance of Concrete Beams," *Proc. ASCE,* ST10 Vol. 94, Oct. 1969, p. 2325.

10. G. N. J. Kani, "How Safe Are Our Large Reinforced Concrete Beams," *ACI Jour. Proc.,* 64, No. 3, Mar. 1967, p. 128.

11. H. A. R. dePaiva and C. P. Siess, "Strength and Behavior of Deep Beams in Shear," *ACSE, Proc.,* No. ST-5, Vol. 91, Part 1, Oct. 1965, p. 19.

12. ACI Committee 326, "Shear and Diagonal Tension," *ACI Jour, Proc.,* 59, Jan. 1962, p. 1; Feb. 1962, p. 277; Mar. 1962, p. 532.

13. K. S. Rajagopalan and P. M. Ferguson, "Exploratory Shear Tests Emphasizing Percentage of Longitudinal Steel," *Jour. ACI, Proc.,* Vol. 65, No. 8, Aug. 1968, p. 634.

14. ACI Committee 318, "Building Code Requirements for Reinforced Concrete (SI Units) (ACI 318-83M)," Amer. Concrete Inst., Detroit, 1983, 111 pp.

15. F. Leonhardt, "Reducing the Shear Reinforcement in Reinforced Concrete Beams and Slabs," *Magazine of Concrete Research,* Vol. 17, No. 53, Dec. 1965, p. 187.

16. P. M. Ferguson and S. I. Husain, "Strength Effect of Cutting Off Tension Bars in Concrete Beams," Research Report 80-1F, Center for Highway Research, Univ. of Texas at Austin, June 1967, 37 pp.

17. L. B. Kriz and C. H. Raths, "Connections in Precast Concrete Structures—Strength of Corbels," *Jour. Prestressed Concrete Inst.,* Vol. 10, No. 1, Feb. 1965, p. 16.

18. A. H. Mattock, K. C. Chen, and K. Soongswang, "The Behavior of Reinforced Concrete Corbels," *PCI Jour.,* Vol. 21, No. 2, Mar.–Apr. 1976, pp. 52–77.

19. J. A. Hofbeck, I. O. Ibrahim, and A. H. Mattock, "Shear Transfer in Reinforced Concrete," *Jour. ACI, Proc.,* Vol. 66, No. 2, Feb. 1969, p. 119.

20. R. F. Mast, "Auxiliary Reinforcement in Precast Concrete Construction," *Proc. ASCE,* Vol. 94, No. ST6, June 1968, p. 1485.

21. A. H. Mattock, "Design Proposals for Reinforced Concrete Corbels," *Jour. Prestressed Concrete Institute,* Vol. 21, No. 3, May–June 1976.

22. F. Leonhardt and R. Walther, "Deep Beams," *Bulletin* 178, Deutscher Ausschuss fur Stahlbeton, Berlin, 1966 (in German).

23. F.-K. Kong, P. J. Robins, and D. F. Cole, "Web Reinforcement Effects on Deep Beams," *Jour. ACI, Proc.,* Vol. 67, No. 12, Dec. 1970, p. 1010.

24. ACI Committee 426, "The Shear Strength of Reinforced Concrete Members—Slabs," *ASCE, Jour. Struct. Div. Proc.,* Aug. 74, ST8, pp. 1543–1591.

25. F. P. Wiesinger's discussion of 318-77 Code, *ACI Jour., Proc.,* 74, No. 7, July 1977, p. 305.

26. ACI Committee 326, "Shear and Diagonal Tension," *ACI Jour., Proc.,* 59, Jan. 1962, p. 1; Feb. 1962, p. 277; Mar. 1962, p. 352.

27. *Torsion of Structural Concrete,* SP-18, Amer. Concrete Inst. Detroit, 1968, 505 pp. (A collection of 19 papers.)

28. J. A. Ramirez, and J. E. Breen, "Review of Design Procedures for Shear and Torsion in Reinforced and Prestressed Concrete," Center for Transportation Research Report 248-2, The Univ. of Texas at Austin, Nov. 1983, 186 pp.

29. J. A. Ramirez, and J. E. Breen, "Experimental Verification of Design Procedures for Shear and Torsion in Reinforced and Prestressed Concrete," Center for Transportation Research Report 248-3, The Univ. of Texas at Austin, Nov. 1983, 300 pp.

30. U. Behera, K. S. Rajagopalan, and P. M. Ferguson, "Reinforcement for Torque in Spandrel L-Beams," *ASCE, Jour. Struct. Div., Proc.* Feb. 1970, ST2, p. 371.

31. K. S. Rajagopalan and P. M. Ferguson, "Distributed Loads Creating Combined Torsion, Bending, and Shear on L-Beams with Stirrups," *ACI Jour., Proc.*, 69, No. 1, Jan. 1972, p. 46.

32. D. J. Victor and P. M. Ferguson, "Beams Under Distributed Load Creating Moment, Shear, and Torsion," *ACI Jour, Proc.*, 65, No. 4, April 1968, p. 295.

33. J. Minor and J. O. Jirsa, "A Study of Bent Bar Anchorages," *Structural Research at Rice,* No. 9, Dept. of Civil Engineering, Rice Univ., Mar. 1971.

Problems

NOTE *The problems in this group relate only to shear or torsion; flexural considerations are not included, unless specifically stated.*

PROB. 5.1.
(a) For shear only and $V_c = 2\sqrt{f'_c}b_w d$, what is the permissible uniform live load on the beam of Fig. 5.36 if Code minimum stirrups by Code Eq. (11-14) are used? Assume $f'_c = 4000$ psi. Grade 60 steel, and $w_d = 1300$ plf (including beam weight).
(b) Where could the stirrups in (a) then be discontinued by Code 11.5.5.1? Ignore partial span loads.
(c) Repeat (a) if #3 U at 9 in. is used (minimum practical bar size and maximum spacing by Code 11.5.4.1).
(d) Repeat (a) with Code Eq. 11.6 for allowable V_c.

PROB. 5.2. In the beam of Fig. 5.37, if the 30 k is all live load, $f'_c = 4000$ psi, and Grade 60 bars are used, design the stirrups and on an elevation (as in Fig. 5.37) show their arrangement and spacing.

PROB. 5.3. If the load on the beam of Fig. 5.37 and Prob. 5.2 is changed to $w_d = 2000$ plf (including beam weight) and $w_\ell = 4000$ plf, and ℓ overall becomes 6 ft-4 in. (with more A_s), design the stirrups and sketch their arrangement.

PROB. 5.4. If the cross section of Fig. 5.37b is used on a 4 ft-4 in. cantilever, $f'_c = 4000$ psi, Grade 60 steel, under a dead load $w_d = 2000$ plf (including beam weight) and $w_\ell = 4000$ plf, check for shear and, if needed, choose and position stirrups.

PROB. 5.5. Evaluate V_u for the beam of Fig. 3.38, using $f'_c = 3000$ psi, $f_y = 60\,000$ psi.
(a) For minimum stirrups; design these stirrups for region near support.
(b) For maximum stirrups; design these stirrups for region near support.

PROB. 5.6. Evaluate maximum V_u for the joist of Fig. 3.41 if $f'_c = 5000$ psi.

PROB. 5.7. Design and space vertical stirrups for an 20-ft simple span rectangular beam, $b = 15$ in., $d = 27$ in., $f'_c = 3000$ psi, Grade 40 steel, w_d (including beam weight) = 2000 plf, and $w_\ell = 8000$ plf. Use stirrup spacing curve.

PROB. 5.8. The loads shown in Fig. 5.38 are at fixed points, not moving loads. Each load consists of 3000 lb of dead load and a possible 4500 lb of live load. Establish the maximum shear curve and design and space the necessary stirrups for $f'_c = 3000$ psi and Grade 40 steel. Consider uniform load negligible.

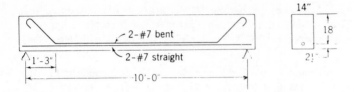

Figure 5.36 Simple span beam for Prob. 5.1.

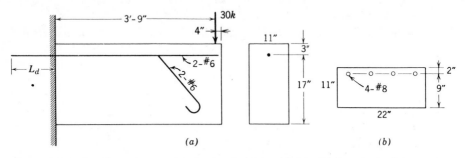

Figure 5.37 Cantilever beam. (*a*) For Probs. 5.2 and 5.3. (*b*) For Prob. 5.4.

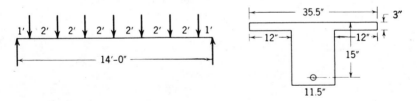

Figure 5.38 Simple span beam for Prob. 5.8.

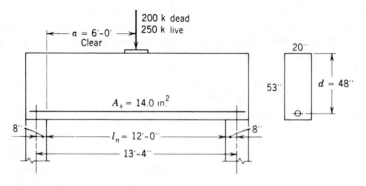

Figure 5.39 Beam for Prob. 5.9.

206   SHEAR AND TORSION

PROB. 5.9. Given the essentially simple span beam of Fig. 5.39 carrying the column loads shown, $f'_c = 4000$ psi, Grade 60 steel. Design shear reinforcement required and sketch its placement on elevation of beam.

PROB. 5.10. A 6-in. thick flat plate floor (Fig. 15.1), with $d = 4.75$ in. (average) brings the load from a 20 × 20 foot area to an 18-in. square column at the center of the area, $w_d = 88$ psf, $f'_c = 3000$ psi. (a) For a $w_\ell = 100$ psf, check whether the slab is safe for shear. (b) At its ultimate in shear what w_ℓ would be permissible on the slab?

PROB. 5.11. A rectangular beam, freestanding except for being fixed against any rotation at each end, must carry a midspan live load of 35 kips which can be as much as 12 in. off the axis of the beam. Given $b = 12$ in., $d = 20$ in., $h = 23$ in., $\ell_n = 20$ ft, $f'_c = 4000$ psi, and Grade 60 steel. Design the necessary shear and torsion reinforcement.

PROB. 5.12. A spandrel beam 12-in. wide × 20-in. deep ($d = 17.5$ in.) with a slab 4-in. thick available (on one side) to act as a flange must carry ultimate loads producing a shear V_u of 50 kips and a torque T_u of 18 k-ft. Using $f'_c = 4000$ psi and Grade 60 steel, design stirrups and longitudinal steel to be added (to the flexural demands).

6

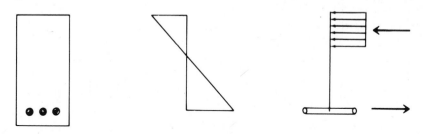

AXIAL LOAD PLUS BENDING—
SHORT COLUMNS

6.1 The Practical Column Problem

All practical columns are members subject not only to axial load but also to moment from direct loading or end rotations. This chapter covers *short* columns, where lateral deflections are not significant. Chapter 7 covers *slender* columns, where deflections have an important effect on member strength.

From a 1970 survey the ACI-ASCE Column Committee estimated that 90% of all braced columns and 40% of all unbraced columns can be designed as short columns. Braced columns are those where shear walls, diagonal bracing, shear trusses, or other types of lateral bracing largely prevent relative lateral movement of the two column ends (or joints).

Creep and shrinkage are important in column behavior and the actual stresses under service conditions only can be estimated. As Sec. 6.3 indicates, however, creep and shrinkage have little effect on the strength of a given cross section at failure. This chapter covers column strengths ranging from where moments are small to the other limiting condition of flexure alone (axial load zero).

Because columns are structurally more important than beams (they carry more floor area), are subject to moments less accurately known, and generally have substantially lower ductility at failure than beams, the strength reduction factor ϕ is lowered to 0.70 or 0.75, depending on the type of column involved.

6.2 Types of columns

The ACI Building Code 318-86 governs reinforced (not plain) concrete. Some jurisdictions also will adopt ACI 318.1-83 which makes provision for limited use of *plain concrete*. If such a plain concrete code is in effect, plain concrete may be used for pedestals in which the height does not exceed three times the least lateral dimension (Fig. 6.1*a*).

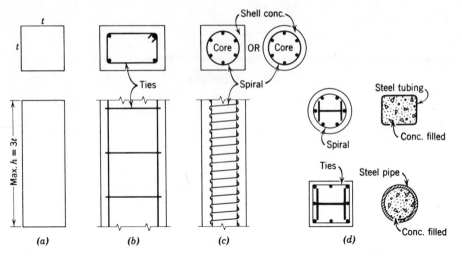

Figure 6.1 Types of columns. (*a*) Plain concrete pedestal. (*b*) Tied column. (*c*) Spiral column. (*d*) Composite columns, four types.

Reinforced concrete columns normally contain longitudinal steel bars and are designated by the type of lateral bracing provided for these bars. *Tied columns* (Fig. 6.1*b*) have the bars braced or tied at intervals by closed loops called ties. *Spiral columns* have the bars (and the core concrete) wrapped with a closely spaced helix or spiral of small-diameter wire or rod (Fig. 6.1*c*).

Composite columns may contain a structural steel shape surrounded by longitudinal bars with ties or spirals or may consist of high strength steel tubing filled with concrete or a steel pipe so filled.

Tied and spiral columns are the most common forms. Either type may be circular, octagonal, square, or rectangular in cross section. Tied columns also may be L or other irregular shaped.

6.3 Column Tests

For nearly 60 years it has been evident that in a reinforced concrete column under sustained axial load one could not calculate f_c, the actual unit stress in the concrete, nor f_s, the actual unit stress on the steel. If the materials were really elastic it would be possible to use the transformed area (Chapter 3) to establish these stresses. However, actual observations show that the steel stress is much larger than this calculation would indicate, because of both shrinkage and creep of the concrete under load.

Starting about 1930, a very large research project on columns was carried out at the University of Illinois[1] and at Lehigh University.[2] These tests indicated clearly that even under axial load alone there was no fixed ratio of steel stress to concrete stress in the ordinary column. The ratio of these stresses depended on the amount of shrinkage, which in turn depended on the age of the concrete and the method of curing. It also depended on the amount of creep in the concrete. Creep is greater when the load is applied at an early stage of the hardening or curing process. The amount of creep is influenced by any of the factors that determine the quality of concrete, such as cement content, water content, curing, and even type of aggregate used.

A load applied for only a short time, such as the ordinary live load, causes very little creep, especially after the concrete is well cured. The usual live load thus produces an increment or increase of stress in steel and concrete that can be calculated reasonably well by elastic analysis. However, the stresses produced by dead load or any permanent or semipermanent load depend on the entire history of the column. It is even possible to have a loaded column with tension in the concrete and compression in the steel under very special circumstances (such as a large percentage of steel and a heavy initial loading later greatly reduced in amount).

In an extensive column investigation of axially loaded tied columns, the behavior shown in Fig. 6.2 was observed. When very large plain concrete columns were axially loaded (see Fig. 6.2a), the load at a given strain was close to but differed from the load that might be expected if the cross-sectional area is multiplied by the concrete stress determined from a standard concrete cylinder stress-strain curve. As shown by the dashed line in Fig. 6.2a, the plain concrete column only developed about 85% of this expected load. This 85% factor for the concrete in a large size column seems partly due to less ideal compaction of concrete in columns than in cylinders, partly due to the upward migration of water in vertically cast columns resulting in changed water-cement ratios and lowered concrete compressive strengths, and partly due to a reduction in apparent strength caused by the slower application of load and the longer specimen. If the vertical reinforcement in a tied column was adequately braced against buckling, the load-strain relation expected would be that shown in Fig. 6.2b. When the steel yield strain is developed, the yielding of the bars would result in a constant steel force being developed for larger strains.

In the tied column tests under "axial load," the behavior shown in Fig. 6.2c was observed. With properly tied vertical bars to prevent premature buckling, the load developed was always the sum of the concrete contribution and the steel contribution. The concrete and the steel worked together and the overall capacity is the sum of the individual material capacities.

Historically, in the United States, these tests initiated the slow switch in emphasis from service load to ultimate strength for columns. They showed that *ultimate* column strength did *not* vary appreciably with the history of loading. If, as the loading was increased, the steel reached its elastic limit first, the increased deformation then occurring built up stress in the concrete until its ultimate strength was reached. If the concrete approached its ultimate strength before the steel reached its elastic limit, the increased deformation of the concrete near its maximum stress forced the steel stress to build up more rapidly. This led to the introduction of the simple, but effective "addition law" for axially loaded columns. Thus, regardless of loading history, an "axially loaded" column reached what might be called its yield point only when the load

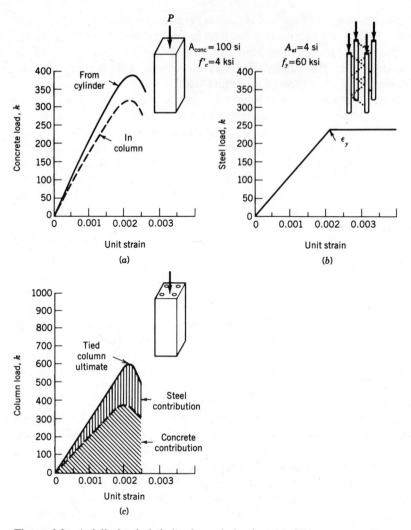

Figure 6.2 Axially loaded tied column behavior. (*a*) Concrete contribution. (*b*) Steel contribution. (*c*) Tied column.

became equal to approximately 85% of the ultimate strength of the concrete (as measured by standard cylinder tests) plus the yield point strength of the longitudinal steel.

Up to the column yield point, tied columns and spiral columns act almost identically and the spiral adds nothing measurable to the yield point strength. The load deformation curves for the tied column and the spiral column up to this point are essentially identical, similar to Fig. 6.3.

After the yield point load is reached, an "axially loaded" tied column immediately fails with a shearing diagonal failure of the concrete (as in a test cylinder) and a buckling failure of the column steel between ties as shown in Fig. 6.4.

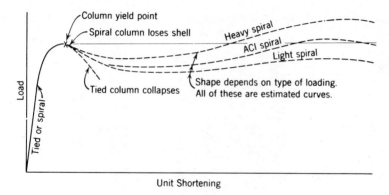

Figure 6.3 Comparison of strains in tied and spiral columns.

The yield point and ultimate strength of a tied column are thus the same thing. In an "axially loaded" spiral column, the yield point load results in cracking or complete destruction of the shell of concrete outside the spiral (Fig. 6.5). The spiral comes into effective action only with the large increased deformation that follows yielding of the column and loss of the shell concrete. At the stage shown in the Fig. 6.5 columns, the spiral provides radial compressive forces on

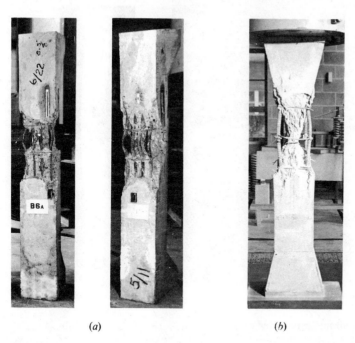

(a)   (b)

Figure 6.4 Failure of tied columns. (a) Note the bars buckled between ties. Column height-thickness ratio $\ell_u/h = 7.5$. Special ends were cast to permit comparative tests with eccentric loads. (From Reference 3, Univ. of Ill.) (b) Column in which a tie seems to have failed after yield point of column was reached. (Courtesy Portland Cement Assn.)

(a)   (b)

                                       (c)

Figure 6.5 Spiral column tests under concentric loads. (From Univ. of Illinois tests, References 3, 4, and 5.) (a) Failure of 32-in. diameter column; $\ell_u/h = 6.6$. (b) Failure of 12-in. diameter column; $\ell_u/h = 7.3$. Shell has completely spalled off. (The special ends were cast to permit comparative tests with eccentric loads.) (c) Failure of column with thin cover or shell; $\ell_u/h = 10.0$.

the concrete within the core of the column and these confining stresses add significantly to the load the core concrete can carry. The spiral steel never becomes significantly effective until after the destruction of the shell concrete that covers it. Moreover, excessive longitudinal column shortening is involved (Fig. 6.3) that makes this spiral steel of questionable value *except* for the significantly greater ductility and as a safety factor against complete collapse.

A heavy spiral can add more strength to the column than that lost in the spalling or failure of the shell, in which case the column will carry an ultimate load greater than the yield point load, but with unsuitable shortening. If too light a spiral is used, the column will continue to carry some load beyond the column yield point, but not as much as that which caused the spalling of the shell. The ACI Building Code specifies that amount of spiral steel that will just replace* the strength lost when the shell concrete spalls. The initial cracking of the shell gives some warning of overload prior to failure. The spiral also adds a considerable element of toughness to the column. Toughness is valuable in resisting explosions or earthquakes, because it measures the energy that can be absorbed. Two columns from the same story of a building severely damaged by a strong earthquake (Fig. 6.6) show that only the heavily damaged spiral columns prevented a total collapse of this story.

More recently emphasis has been given to the fact that to some extent column ties also confine the concrete, although their shape makes them much less efficient than spirals. Heavy ties on columns or around compression steel in beams can establish a considerable degree of toughness in these members. Ties are hence an important requirement in earthquake resistance and limit design where members are expected to maintain their peak resistance while forming so-called plastic hinges. Code Appendix A on seismic resistance calls for heavy ties in joints where the columns go through the beams, which would have helped the joint above the spiral column of Fig. 6.6.

The previous description of the behavior of an "axially loaded" column is very helpful in visualizing the role of the concrete, vertical reinforcement, and ties or spirals in resisting axial loads. In reality, columns are always subject to some bending moments as well as axial loads. Even in idealized laboratory tests, it is impossible to fabricate perfectly straight members with perfectly symmetrical reinforcement and to load them in an axial fashion through the effective centroid. In actual construction, columns and beams are frequently monolithic. Moments of varying size are present in the joints because of frame action under both applied loads and deformations induced by temperature, shrinkage, and settlement. When a combination of axial load and moment are present, the column is often referred to as "eccentrically loaded." The combined loading condition can be represented, as shown in Fig. 6.7*d*, as either the

---

* Actually, an estimated 10% in excess of the shell strength is used just to be sure the strength after spalling is not less than before. See Sec. 6.13 for the design of a spiral.

Figure 6.6 Columns almost destroyed in severe earthquake. Notice that spiral column still has its core acting. (Bars outside a spiral or tie become totally ineffective.)

action of an axial load P with a moment M or as the action of an axial load P at an equivalent eccentricity e.

In a pioneering series of tests of eccentrically loaded columns, Hognestad[3] established conclusively that the strain gradient across the cross section remains linear and the extreme fiber concrete strain at failure is relatively constant. Thus, the unit strain over a large gage length can be assumed to vary linearly as the distance from the neutral axis. Because of the addition of axial load, the position of the neutral axis is much more variable than for the beam treated in Chapter 3.

For very small moments the strain profile at failure will be virtually linear as shown near the top of Fig. 6.7*b*. The position of the neutral axis or point of zero strain falls far outside the cross section. As the moment increases the neutral axis location shifts and the axial load capacity diminishes. Hognestad proposed that the ultimate compressive strain be taken as $\varepsilon_u = 0.0038$ in.-in. The ACI Code uses the more conservative value of 0.003 in.-in. which provides for a unified treatment of both flexure and combined axial load and flexure.

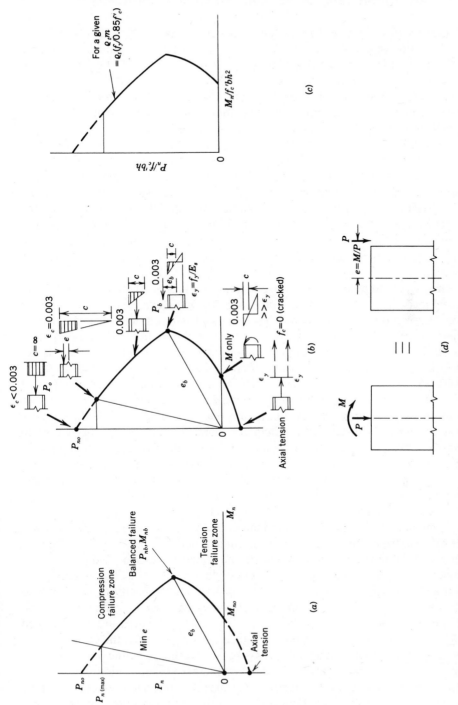

Figure 6.7 Column interaction diagram. (*a*) For a given column. (*b*) Strains involved. (*c*) Dimensionless form. (*d*) Equivalent eccentricity.

6.4 Definitions and Some Code Basic Requirements

A short column is one where the length effect or secondary moments owing to the deflection response under load is very small and may be considered negligible. This limiting length is more fully developed in Sec. 7.5. For preliminary discussion here, the limiting length-to-thickness ratio ℓ_u/h for a short column may be thought of as ranging roughly from 7 to 12 in a braced frame and several units lower in an unbraced frame. The short column can be analyzed or designed from the strength of its cross section alone.

A plot of the column axial load capacity against the moment it can simultaneously carry is called a column interaction diagram. The axial load capacity decreases as moment is increased. Schematically, Fig. 6.7a shows such a diagram with key points and areas noted for later discussion. Any loading that plots within this area is a possible loading; any combination outside the area represents a failure combination. A radial line from point O represents a constant ratio of moment to load, that is, a constant eccentricity of load. Figure 6.7b shows how strains and the neutral axis distance c shift along such a diagram.

The axial load point P_{no} (P_o in Code notation) represents the capacity under "axial loading" with no moment present. This value can easily be determined from the "column addition law." The balanced failure point (P_{nb}, M_{nb}) represents the axial load and moment capacity when the section has a strain profile exactly that assumed for "balanced strain conditions." Balanced strain is defined as the extreme compression fiber strain at the maximum value (ε_u = 0.003) just as the outermost layer of tensile reinforcement reaches its yield strain, $\varepsilon_y = f_y/E_s$. This condition represents a simultaneous compression failure (or concrete crushing) and tension failure (steel yielding). Unlike the case for flexural members discussed in Chapters 3 and 4, it is totally impractical to impose restrictions that will prevent balanced or initial compression failures in columns. The very nature of column action is compressive. Safeguards against sudden compressive failures in the columns of actual structures must come through lowered values of the capacity reduction failure ϕ and hence higher factors of safety for columns than for beams.

Combinations of axial load and moment that produce failure for $P_n \gtrless P_{nb}$ will cause the extreme fiber compression strain to reach $\varepsilon_u = 0.003$ in.-in. before the outer layer of tension reinforcement strain reaches its tensile yield point value. Such failures are termed initial compression failures. On the other hand, failure load and moment combinations producing axial load capacity $P_n < P_{nb}$ would all have the outer layer of tension reinforcement reach yielding in tension before the concrete crushes in compression. Such failures are then termed initial tensile failures, although final failure still occurs when the extreme compression fibers are crushed. A special case is represented by the point falling on the axial load axis but substantially below the zero value. This represents the axial tension capacity of the member. Because the concrete is assumed to have no tensile strength, the value of this point is the product of the steel area times the tensile yield stress.

The strain condition that results in a force distribution such that the summation of the compression forces equals the summation of the tension forces results in the moment value that was associated with pure flexure in Chapters 3 and 4. This point M_{no} is shown on the interaction curve where the curve intersects the horizontal moment axis at $P_n = 0$.

The interaction diagram is the key concept for understanding reinforced concrete column cross-sectional behavior. It forms the basis for much of the design procedure for concrete columns. The procedures for calculating the numerical values for key points on the interaction diagram are relatively simple. They can be easily programmed for electronic computation by using algorithms that sequentially vary the strain profile on the section by assuming various neutral axis locations. For any assumed strain profile, the stress distribution in the concrete and steel reinforcement can be determined from the rectangular stress block and the steel stress-strain relationships. Integrating these stresses over the cross-sectional area results in forces and lever arms. Summation of forces and moments about the plastic centroid gives the values of P_n and M_n, which correspond to the assumed strain profile. The procedures for calculating various key points will be presented in detail in subsequent sections.

The interaction diagram can be put into dimensionless form as indicated in Fig. 6.7c for a given steel ratio and arrangement of steel, which facilitates assembly of graphs into groups for design charts.*

Because all concrete columns are subject to some moment, past codes set minimum eccentricities of loading: $0.10h$ for tied columns and $0.05h$ for spiral columns. The 1986 Code requires something similar by setting an upper limit on the maximum axial load $P_{n(max)}$, as shown by the horizontal line below P_{n0} in Fig. 6.7a. This limit is given in detail in the next section. The actual maximum $e = M/P$ (Fig. 6.7a) may exceed these minima and then control the design.

All columns are required (Code 10.9.1) to contain longitudinal bars sufficient to make the steel ratio, $\rho_t = A_s/hb$, at least 0.01, because of the shrinkage and creep stresses on smaller areas, and ρ_t must not exceed 0.08. At 0.08 crowding in the member is very severe. Caution in detailing to ensure constructability must be used when ρ_t exceeds 0.04. Details such as lap splices of column bars combined with the multiple ties required with large numbers of bars can make placement of column concrete virtually impossible. The Code also requires that at least six bars must be used in spiral columns (Code 10.9.2), four bars within rectangular or circular ties, and three bars within triangular ties.

Ties (Code 7.10.5) must be at least #3 size for #10 bars or smaller and #4 for #11 or larger and all bundled bars. Their spacing shall not exceed 16 bar diameters, 48 tie diameters, or the least column dimension. Every corner bar and every alternate bar must be braced by the tie and no bar shall be more than 6 in. clear from such a laterally supported bar.

---

* The designer must be careful to check whether such a chart is plotted in terms of P_n, M_n or P_u, M_u (a difference of ϕ in the numbers shown).

Spirals must fully replace the strength lost when the shell of concrete outside the spiral spalls off. The minimum spiral is #3 with its clear (vertical) spacing between 1 and 3 in. Spiral design is covered in Sec. 6.13.

6.5 Nominal Axial Load Capacity, P_{n0}*, and Code Maximum Axial Load, $\phi P_{n(max)}$

Although in design, axial load without moment is not a practical case, P_{n0} is a convenient theoretical limit and well documented experimentally. It is a very useful point in the construction of column interaction diagrams such as those of Fig. 6.7.

The tests discussed in Sec. 6.3 established the ultimate strength of either a tied or a spiral column, axially loaded, as:

$$P_{n0} = 0.85 f'_c A_n + f_y A_{st}$$
$$= 0.85 f'_c (A_g - A_{st}) + f_y A_{st}$$

where

P_{n0} = ultimate load capacity (yield point strength) of tied or spiral column when eccentricity is zero (for ideal materials and dimensions)

A_n = net area of concrete = $A_g - A_{st}$

A_g = gross area of concrete, in.^{2}

A_{st} = area of vertical column steel, in.^{2}

f'_c = standard cylinder strength of concrete, psi

f_y = yield point stress for steel, psi

This is the ideal or nominal strength and the Code would consider the design strength ϕP_{n0}, where ϕ would be 0.70 for a tied column and 0.75 for a spiral column according to Code Sec. 9.3.2.2.

Another extreme limit, generally only of theoretical interest, but sometimes useful in constructing interaction diagrams, is the axial tension limit, which is $A_{st} f_y$ in magnitude, with the concrete fully cracked at such a limit.

The Code does not permit a nominal load P_{n0} on a column. For a tied column the design strength is limited (Code 10.3.5.2) to

$$\phi P_{n(max)} = \phi\{0.80[0.85 f'_c (A_g - A_{st}) + f_y A_{st}]\}$$

For convenience in design the ϕ factor can be combined with the factored axial load and the nominal resistance selected so that $P_n \gtrless P_u/\phi$. Thus, in this book it will usually be expressed as the nominal maximum:

$$P_{n(max)} = 0.80[0.85 f'_c (A_g - A_{st}) + f_y A_{st}] = 0.80 P_{n0}$$

---

* The Code omits the subscript n where nominal is obvious.

For a spiral column the design strength limit (Code 10.3.5.1) is

$$\phi P_{n(\max)} = \phi\{0.85[0.85f'_c(A_g - A_{st}) + f_y A_{st}]\}$$

which will usually be used here as the nominal maximum:

$$P_{n(\max)} = 0.85[0.85f'_c(A_g - A_{st}) + f_y A_{st}] = 0.85 P_{n0}$$

Where used, ϕ is 0.70 for tied columns or 0.75 for spiral columns.

These maximum load limits govern wherever the moment is small enough to keep the eccentricity for a tied column under $0.10h$ or for a spiral column under $0.05h$. These are approximate rather than exact limits, but entire short column designs can be based on satisfying these maximum load limits when e is noticeably under these limits. These maximum load limits often govern in interior columns of braced frames. The following example shows this simple process.

Design a short tied interior column for a total dead load of 300 kips and live load of 250 kips, a moment of 500 k-in. (from live load), $f'_c = 4000$ psi, $f_y = 60,000$ psi. Try for ρ_t of about 0.03. (An experienced designer usually will have a feel for the desired level of reinforcement. Low percentages in upper stories with increasing percentages as axial loads increase in lower stories will permit use of uniform column sizes over several stories and thus minimize form costs.)

Solution

$$P_u = 1.4 \times 300 + 1.7 \times 250 = 845 \text{ k} \qquad M_u = 1.7 \times 500 = 850 \text{ k-in.}$$
$$e = M_u/P_u = 850/845 = 1.01 \text{ in.}$$

The loading appears to require a column h much greater than 10 in., which means $e/h < 0.10$. Use the tied column nominal maximum load equation in the short form:

$$P_{n(\max)} = 0.80 P_{n0} \geq P_u/\phi = 845/0.70 = 1210 \text{ k}$$
$$\text{Required } P_{n0} \geq 1210/0.80 = 1510 \text{ k}$$

$$P_{n0} = 0.85f'_c A_g + (f_y - 0.85f'_c)A_{st}$$
$$= 0.85 \times 4A_g + (60 - 0.85 \times 4)(0.03)A_g = 5.10A_g = 1510 \text{ k}$$
$$A_g = 296 \text{ in.}^2 \qquad h = \sqrt{296} = 17.2 \text{ in.}$$
$$e/h = 1.01/17.2 = 0.059 < 0.10$$

The equation used does control.

Either a 17-in. square column with $\rho_t > 0.03$ or an 18-in. square column with $\rho_t < 0.03$ may be used. Even inch dimensions are generally used for column forms.

$$\text{USE } 18 \times 18 \text{ in. column} \qquad A_g = 324 \text{ in.}^2$$

Return to P_{n0} equation to find required amount of reinforcement for this column size. Because the dimensions were increased, the required steel percentage will be smaller than $\rho_t = 0.03$.

$$0.85f'_cA_g + (f_y - 0.85f'_c)A_{st} = P_{n0}$$
$$0.85 \times 4 \times 324 + (60 - 0.85 \times 4)A_{st} = 1510$$
$$A_{st} = 1510 - 1101/56.6 = 7.23 \text{ in.}^2$$
$$\text{Check} - \rho_t = 7.23/324 = 0.022 < 0.03$$

Use 6-#10 = 7.62 in.^{2} (10-#8, 8-#9, or 6-#11 each are heavier).

Use three bars on opposite faces as shown in Fig. 6.8. Because this is an interior column it is assumed that it will be protected from the harsh environment. Concrete cover c requirements are 1-$\frac{1}{2}$ in. as determined from Code 7.7.1c. Required tie size can be determined from Code 7.10.5 as:

Spacing: $h = 18$ in.   48 tie $d_t = 18$ in. for #3 or 24 in.

for #4   16 bar $d_b = 20.3$ in. for #10.

USE #3 ties at 18 in. on centers.

To determine whether a single tie around the periphery or additional ties that will enclose the interior bars in a tie corner are required, Code Sec. 7.10.5.3 must be checked. This allows alternate interior bars that are less than 6-in. clear spacing from a bar that is enclosed in a tie corner to be without lateral support. As can be seen in Fig. 6.8, the clear bar spacing s_c is

$$2s_c = [h - 2(c + d_t) - 3D]$$

so clear bar spacing $s_c = [18 - 2(1.5 + 0.38) - 3 \times 1.27]/2$
$$= \frac{11.1}{2} = 5.6 \text{ in.} < 6 \text{ in. max}$$

$\therefore$ A single tie around all bars is adequate. These calculations are noticeably shorter than the general solution given in Sec. 6.10, even though interaction charts are used.

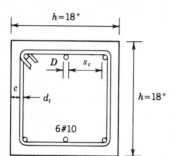

Figure 6.8   Tied column details for example in Sec. 6.5.

6.6 Development of an Interaction Diagram

To illustrate the various steps in development of a typical column cross section interaction diagram, the cross section of Fig. 6.9 will be used. Assume any bending is about the x-x axis only. The reinforcement consists of 8-#9 bars distributed around the periphery as shown. Concrete strength is $f'_c = 3000$ psi and Grade 60 reinforcement with $f_y = 60$ ksi is used. Thus, $\rho_t = (8 \times 1.00)/(18 \times 20) = 0.022$, which is between the 0.01 and 0.08 limits.

$$P_{n0} = 0.85 f'_c (A_g - A_{st}) + f_y A_{st}$$
$$= (0.85)(3)(360 - 8.00) + (60)(8.00) = 1378 \text{ k}$$
$$P_{n(\text{max})} = 0.80 P_{n0} = (0.8)(1378) = 1102 \text{ k}$$

Should the section be subjected to axial tension, only the reinforcement can resist it because the concrete is assumed to take no tension. Thus, the axial tension capacity is

$$P_{nt} = A_{st} f_y = (8)(1.00)(60) = 480 \text{ k tension}$$

These initial points of the interaction diagram may be plotted as shown in Fig. 6.10a. Values are plotted for nominal capacities. The capacity reduction factors ϕ will be shown, after computation of other points.

6.7 Balanced Loading, P_{nb}, M_{nb}

A balanced section for a beam is defined in Chapter 3 as one developing simultaneously a concrete compression strain of 0.003 and a steel tension strain of f_y/E_s. For this condition a unique area of tension reinforcement is required in the beam.

For *any* column the same definition of balanced strains holds (Fig. 6.11a). *Any* column, regardless of its reinforcement, will reach its balanced ultimate load when the load is placed to maintain the eccentricity $e_b = M_{nb}/P_{nb}$. Balance in a column is a matter of *loading*, and it is more descriptive to speak of

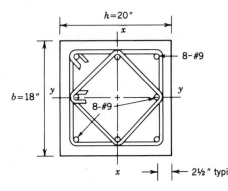

Figure 6.9 Column cross section for interaction diagram in Sec. 6.6

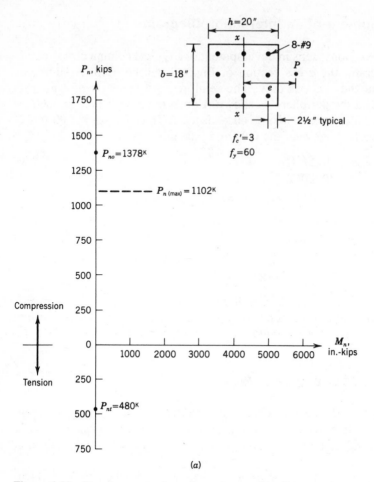

Figure 6.10 Development of a column interaction diagram. (*a*) "Axial" load and maximum "axial" load. (*b*) Balanced strain conditions added.

balanced loading rather than a balanced column. Furthermore, although it is possible to avoid balanced beams to avoid compression failures and thus obtain ductility, it is not possible to avoid either compression failures or balanced failures in columns; these are primarily compression members.

For a given column it is very easy to establish the nominal balanced load P_{nb} for ideal conditions, and the accompanying e_b, after the fashion of Fig. 6.11*a*. The reinforcement is in three layers. Each layer is numbered for convenience. A_{s1} and A_{s3} are each 3-#9 = 3.00 in.^{2} and A_{s2} is 2-#9 = 2.00 in.^{2}, f'_c = 3000 psi, and f_y = 60 ksi. The balanced strain condition of 0.003 in compression and f_y/E_s in tension give c from similar triangles, most simply by thinking of the large dotted triangle.

$$c_b = \frac{0.003}{0.003 + 0.00207} \times 17.5 = 0.592 \times 17.5 = 10.35 \text{ in.}$$

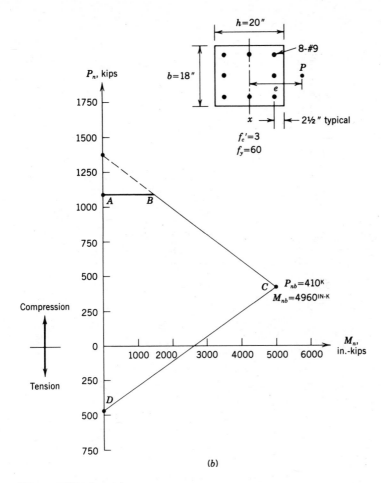

Figure 6.10 (cont.)

As for beams, for $f'_c \lesssim 4$ ksi, $a_b = 0.85 c_b = 8.80$ in. The concrete compressive force (ignoring the fact that a small amount of concrete is displaced by the steel reinforcement) is

$$N_{cc} = 0.85 \times 3 \times 8.80 \times 18 = 404 \text{ k}$$

The concrete compression force acts at the centroid of the column area subjected to the rectangular stress block. In this case the centroid is at $a/2 = 4.40$ in. from the extreme compression fiber.

The strains in each level of reinforcement can be determined from similar triangles:

$$\varepsilon_{s1} = \frac{10.35 - 2.5}{10.35} \times 0.003 = 0.00227 \, C^* > \varepsilon_y$$

---

* Here C signifies compression and T signifies tension.

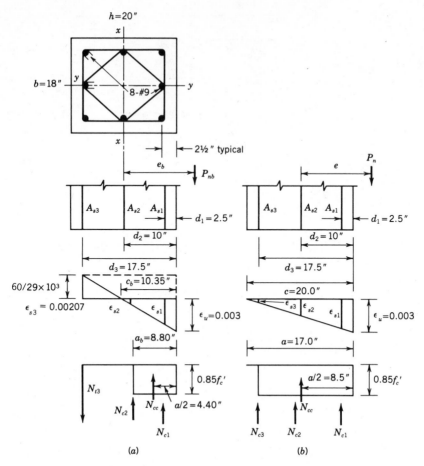

Figure 6.11 Strain and force relations. (*a*) Balanced column load. (*b*) $c = 20$ in.

Thus, $f_{s1} = f_y$ or 60 ksi compression

$$N_{c1} = A_{s1}f_{s1} = (3.00)(60) = 180 \text{ k}$$
$$\varepsilon_{s2} = \frac{10.35 - 10}{10.35} \times 0.003 = 0.00010C < \varepsilon_y$$

Since this strain is *less* than ε_y, $f_s = \varepsilon_s E_s$ can be used.

$$f_{s2} = (0.00010)(29 \times 10^3) = 2.9 \text{ ksi compression}$$
$$N_{c2} = A_{s2}f_{s2} = (2.00)(2.9) = 6 \text{ k (actually negligible)}$$
$$\varepsilon_{s3} = \varepsilon_y T \text{ (by definition)} = 0.00207 T$$
$$f_{s3} = f_y \text{ or 60 ksi tension}$$
$$N_{t3} = A_{s3}f_{s3} = (3.00)(60) = 180 \text{ k}$$

These four internal forces must be in equilibrium with the external force P_b.

$$\Sigma F_y = 0 = P_{nb} + N_{t3} - N_{c1} - N_{c2} - N_{cc} = P_{nb} + 180 - 180 - 6 - 404$$
$$P_{nb} = 410 \text{ k}$$

The resisting moment produced by the four internal forces must be in equilibrium with the external moment M_{nb}. Because the external moments are normally computed with respect to the section centroidal axis, the moments must be summed about the plastic centroid that, for the case of a symmetrically reinforced rectangular column, is the center of the column. Summing the moments produced by each force times its lever arm with respect to the x-x axis gives:

Force	y	Mabt. center line
$N_{c1} = 180^k$	$10 - 2.5 = 7.5$ in.	$(180)(7.5) = 1350$ k-in.
$N_{c2} = 6^k$	$10 - 10 = 0$	$(6)(0) = 0$
$N_{cc} = 404^k$	$10 - 4.4 = 5.6$ in.	$(404)(5.6) = 2260$ k-in.
$N_{t3} = -180^k$	$10 - 17.5 = -7.5$ in.	$(-180)(-7.5) = 1350$ k-in.

$$\therefore M_{nb} = 1350 + 0 + 2260 + 1350 = 4960 \text{ k-in.}$$

In this case all of the resisting forces produce a counterclockwise moment around the centroidal axis. That is why all terms have positive signs.

$$e_b = \frac{M_{nb}}{P_{nb}} = \frac{4960 \text{ k-in.}}{410 \text{ k}} = 12.1 \text{ in.}$$

These results then are added to the previously determined points on the interaction diagram as shown in Fig. 6.10b. If the computations considered the concrete displaced by the steel, the values would change slightly to $P_{nb} = 402$ k, $M_{nb} = 4902$ k-in.

A very crude, but conservative interaction diagram can be constructed that often will suffice for preliminary design purposes. The compression failure branch (ABC in Fig. 6.10b) is obtained by connecting the compressive axial load capacity point with the balanced strain condition point using a straight line and then truncating the diagram to eliminate points above the maximum axial load capacity AB. The use of the straight-line approximation is conservative, although because it can disregard substantial moment capacity at some ranges of axial load, it may not be economical. The tension failure branch (CD) also may be approximated by a straight line connecting the balanced point and the tensile axial load point. This is generally excessively conservative.

6.8 Other Interaction Points and ϕ Values

The general procedure for finding any point on the interaction diagram is very similar to the procedure used for the balanced strain case. The essential steps are:

1. Assume a location of the neutral axis, c. Values of $c < c_b$ will result in interaction diagram values plotting below the balanced condition points, and values of $c > c_b$ will result in values between P_{n0} and P_{nb}. c can be taken as greater than h.

2. From the corresponding strain profile found assuming $\varepsilon_u = 0.003$, the values of strain at each level of reinforcement can be determined. Using the steel stress-strain diagram, the steel stress and hence the force in each level of reinforcement can be determined.

3. The depth of the concrete compression block is determined from $a = \beta_1 c$. Summing the rectangular block stress value of $0.85f'_c$ over the concrete compression block area (ab if rectangular) gives the concrete compression force and its centroidal location.

4. Summation of steel and concrete forces gives $P_n = \Sigma C - \Sigma T$.

5. Summation of moments of steel and concrete forces about the centroidal axis gives M_n.

Detailed calculations for another point on the interaction diagram of the column used with Fig. 6.10 are given for the case where $c = 20$ in. Thus, $a = 0.85c = 17.0$ in. From similar triangles, as shown in Fig. 6.11b, the strains in each level of reinforcement can be determined:

$$\varepsilon_{s1} = \frac{17.5}{20}(0.003) = 0.00262C > \varepsilon_y = 0.00207$$

Thus, $f_{s1} = f_y$ or 60 ksi compression

$$N_{c1} = A_{s1}f_{s1} = (3.00)(60) = 180 \text{ k}$$

$$\varepsilon_{s2} = \frac{10}{20}(0.003) = 0.0015C.$$

Since $\varepsilon_{s2} < \varepsilon_y$, $f_{s2} = \varepsilon_{s2}E_s$ so $f_{s2} = (0.0015)(29 \times 10^3) = 44$ ksi compression

$$N_{c2} = A_{s2}f_{s2} = (2.00)(44) = 88 \text{ k}$$

$$\varepsilon_{s3} = \frac{2.5}{20}(0.003) = 0.00038C.$$

Since $\varepsilon_{s3} < \varepsilon_y$, $f_{s3} = (0.00038)(29 \times 10^3) = 11$ k compression

$$N_{c3} = A_{s3}f_{s3} = (3.00)(11) = 33 \text{ k}$$

The concrete compression force ignoring the concrete displaced by the steel reinforcement is

$$N_{cc} = 0.85 \times 3 \times 17.0 \times 18.0 = 780 \text{ k}$$

Summing forces gives

$$\Sigma F = 0 = -P_n + N_{cc} + N_{c1} + N_{c2} + N_{c3}$$

Thus,

$$P_n = N_{cc} + N_{c1} + N_{c2} + N_{c3} = 780 + 180 + 88 + 33 = 1080 \text{ k}$$

Summing moments about the plastic centroid gives

$$\Sigma M = 0 = P_n \times e_n - N_{c1} \times 7.5 - N_{c2} \times 0 + N_{c3} \times 7.5 - N_{cc} \times (10 - 8.5)$$
$$M_n = P_n \times e_n = (7.5)(N_{c1}) - (7.5)(N_{c3}) + (1.5)(N_{cc})$$
$$M_n = (7.5)(180) - (7.5)(33) + (1.5)(780) = 2270 \text{ k-in.}$$
$$e = \frac{M_n}{P_n} = \frac{2270}{1080} = 2.10 \text{ in.}$$

This result is added to the previous values as shown on the interaction diagram of Fig. 6.12a. This point falls substantially outside the simplified straight-line diagram developed in Fig. 6.10b and shown again as a light dashed line in Fig. 6.12a.

In a similar fashion various values of c can be chosen, and for each the corresponding values of P_n and M_n can be determined. This process is particu-

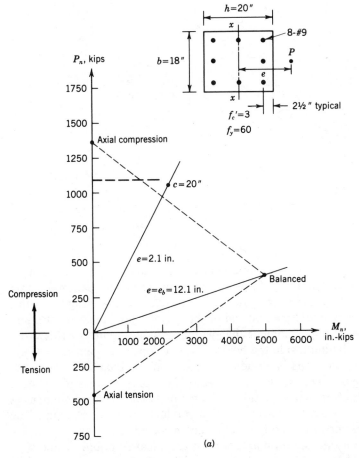

Figure 6.12 Development of a column interaction diagram. (a) Values for c = 20 in. added. (b) All P_n, M_n values.

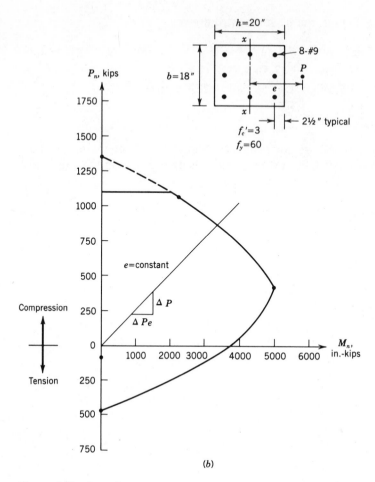

Figure 6.12 (cont.)

larly well suited for computer programming. The values of c then can be stepped and a sufficient number of points to determine the interaction diagram can be calculated. Figure 6.12b shows the final diagram resulting from this procedure. The heavy line represents all of the various combinations of nominal axial load and nominal moment that would produce failure of this cross section. Any point falling inside the heavy line envelope is a point that would not produce failure and thus is an admissible design combination. The radial lines from the origin shown in Fig. 6.12a and 6.12b represent lines of constant eccentricity e. It is often helpful to think of combined axial load and moment as an axial load at a certain eccentricity e that produces a given moment. The capacity of this column cross section for any given eccentricity e can be quickly determined from an interaction diagram by a graphical solution. A line with slope e is extended from the origin until it intersects the interaction diagram. The intersection represents the nominal axial load capacity at that eccentricity.

For flexure without axial load, the interaction point at the moment axis of the diagram, the procedure can be used unchanged except (1) that $\varepsilon_s > \varepsilon_y$, and (2) the neutral axis is unknown and must be located by trial-and-error such that the total tension equals the total compression. For curve plotting, this M_0 point need not be exactly established numerically because a solution with a very small tension resultant and another with a very small compression resultant establishes the curve across the moment axis.

Throughout the discussion, the values have always been expressed in terms of *nominal* axial load and *nominal* moment. The application of the capacity reduction factor ϕ to the interaction diagram is complicated by the fact that the diagram represents several different combinations of loading. Members with small eccentricities experience brittle initial compression failures and the Code provides for an increased margin of safety by using $\phi = 0.70$ for tied columns and $\phi = 0.75$ for spiral columns. However, the same member when loaded with very large eccentricities experiences more ductile initial tensile failures. The special case of $e = \infty$ (or a beam with no axial load) is represented by the point on the moment axis. For this case and all cases involving axial tension, the Code provides for $\phi = 0.90$. Thus, there is a need for a transition zone for members with large eccentricities so that ϕ can increase from 0.70 or 0.75 to 0.90.

The Code in Sec. 9.3.2.2 provides for this transition:

> For members in which f_y does not exceed 60,000 psi, with symmetrical reinforcement, and with $(h - d' - d_s)/h$ not less than 0.70, ϕ may be increased linearly to 0.90 as ϕP_n decreases from $0.10 f'_c A_g$ to zero. For other reinforced members, ϕ may be increased linearly to 0.90 as ϕP_n decreases from $0.10 f'_c A_g$ or ϕP_b, whichever is smaller, to zero.

In this, h is the overall column thickness and d' and d_s are the covers to the centroid of compression and tension steel, respectively. The relationship is plotted in Fig. 6.13. For higher f_y or a smaller distance between A_s and A'_s, the

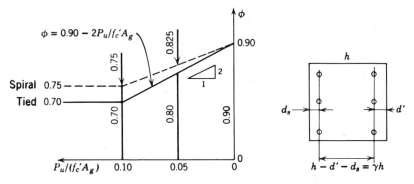

Figure 6.13 Increase in ϕ as $P_u (= \phi P_n)$ approaches zero. Note limitations where $f_y < 60{,}000$ psi or $\gamma h < 0.70\, h$.

balanced load condition moves closer to the moment axis and the second Code sentence is then necessary to assure a normal ϕ at the balanced load level.

In the 1983 Code, the application of the ϕ factor to combined axial load with flexure was very carefully stated to clarify the intent. For axial load with flexure both axial load and moment nominal strength must be multiplied by the appropriate single value of ϕ.

The application of the capacity reduction factor ϕ to the nominal load and moment values of the example is illustrated in Fig. 6.14. The appropriate ϕ value range is shown on the right side of the figure. In this symmetrically reinforced column the transition from $\phi = 0.70$ to a larger value begins at $\phi P_n = 0.1 f'_c A_g = (0.1)(3)(18)(20) = 108$ k. Because at that point ϕ is 0.70,

$$\phi P_n = 108, \qquad \therefore P_n = 108/0.70 = 154 \text{ k}$$

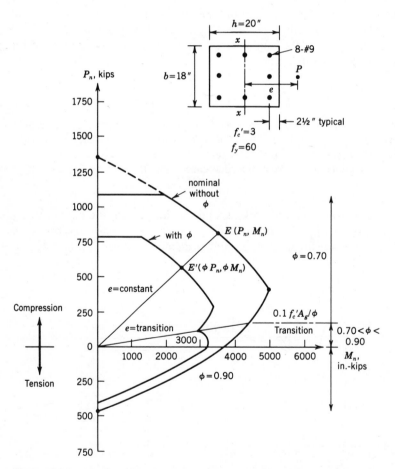

Figure 6.14 Application of ϕ values in development of a column interaction diagram.

Thus, $\phi = 0.70$ for all P_n values equal or greater than 154 k. Applying ϕ to both nominal axial load and nominal moment capacity is exactly the same as shrinking the diagram along a line of constant eccentricity. Any point on the nominal interaction diagram (E, for example) contracts radially along the line of constant eccentricity to its new location E' which has coordinates ϕP_n, ϕM_n or P_u, M_u. In the transition zone, an odd shape results because of the changing ϕ values. However, this zone is of relatively little importance in many columns because the eccentricities of most interior and multistory building columns are small.

This is the general procedure for computing values for interaction diagrams. The availability and use of such diagrams will be discussed later.

When a column is unsymmetrical as shown in Fig. 6.15, with A_s' of 5-#9 = 5.00 in.², the reference line for eccentricity is taken from the *plastic centroid*, which is simply the location of the resultant load that would give a uniform strain all across the column. The plastic centroid falls a distance x_0 off the center of column:

Force		x	M abt. center line	
$N_{cc} =$	916 k	0	0	
$N_{c2} =$	300	+7.5 in.	+2250	
$N_{c1} =$	180	−7.5	−1350	
$\Sigma N_c =$	1396		+900	$x_0 = +900/1396 = +0.64$ in.

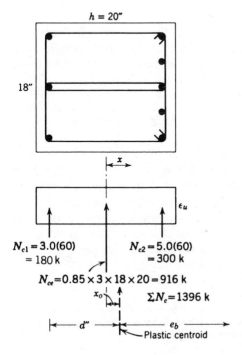

$h = 20''$

$18''$

x

ϵ_u

$N_{c1} = 3.0(60)$
$= 180\,k$

$N_{c2} = 5.0(60)$
$= 300\,k$

$N_{ce} = 0.85 \times 3 \times 18 \times 20 = 916\,k$

x_0 $\Sigma N_c = 1396\,k$

d'' e_b

Plastic centroid

Figure 6.15   Plastic centroid.

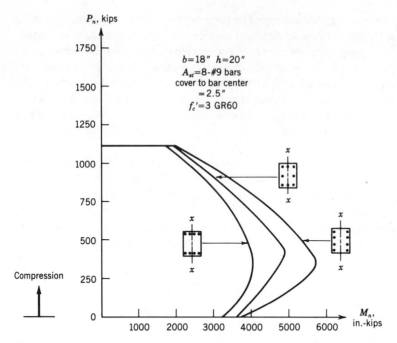

Figure 6.16 Effect of bar arrangement on interaction diagrams (bending about x–x axis only).

The method for finding P_b and e_b is then not basically changed, but all centroidal distances must be determined.

If column steel were further distributed along the four column faces, the numbers used in finding P_b would be increased by an additional term for each group of bars falling at different distances from the neutral axis, but no other complication would exist.

6.9 Accounting for Bars Near Neutral Axis

Bars very near the neutral axis will not be effective in carrying stress; for that combination of M and P any bars near the axis will have stresses lower than the yield stress. For any given neutral axis one should sketch the unit deformations to establish the status of nearby bars, as shown in the previous section.

The placement of bars near the neutral axis modifies the shape of the interaction diagram considerably, rounding it upward or outward and making the balanced loading point less conspicuous because extreme bars yield first* and other bars later. Typical interaction diagrams are developed in Figs. 6.16 and 6.17, and their different shapes should be noted.

---

* The Code (10.3.2) is not specific in defining how many bars must yield at the balanced condition. The authors use the yielding of the outermost bar as their criterion.

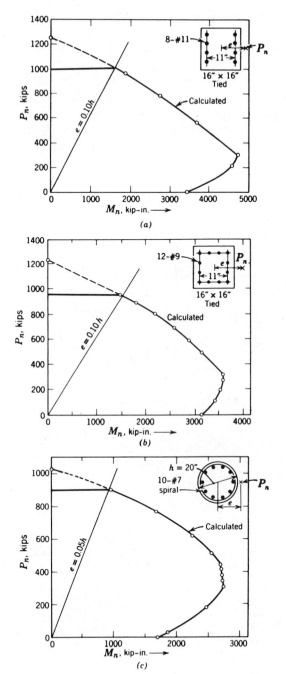

Figure 6.17 The varied shapes of column interaction diagrams.

6.10 Column Interaction Diagrams as Design Aids

Although the designer should understand analytical methods of developing interaction diagrams (Secs. 6.6 through 6.9), especially for unusual shapes or occasional use, most noncomputerized design for loads greater than P_b will be from interaction diagrams or tables. Hence, their use for checking and design is introduced here, even though loading patterns, slenderness effects, and other relevant discussion must follow. These designs for given loads and moments will be extended through the choice of ties. The equivalent details involved in spiral columns are also part of this introduction into short column design.

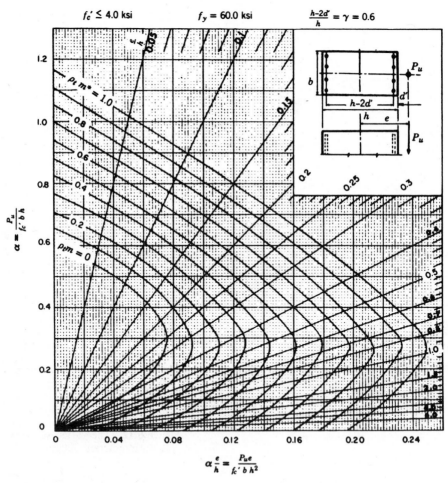

Figure 6.18 Column interaction chart. (a) $\gamma = 0.6$ $m = f_y/0.85\,f'_c$ (b) $\gamma = 0.7$. (c) $\gamma = 0.8$. (d) $\gamma = 0.9$. Chart readings include the effect of ϕ, but incorrectly below ordinate of 0.10.

There are a wide variety of design aids available for reinforced concrete columns. One of the older and more popular chart sets is given in Fig. 6.18a to d. This single set of four charts is for the usual range of steel placement ratios, described by four values of the ratio $\gamma = (h - 2d')/h$, for $f'_c \lesssim 4$ ksi and $f_y = 60$ ksi. Because each plot contains ratios for axial load, moment, eccentricity, and steel, any two of these, in the analysis of a given column, establish the other two. For example, P and M in a given short column fix e/t and the necessary $\rho_t m = \rho_t(f_y/0.85f'_c)$.

A warning about some variations in charts used must be noted. The charts here (Fig. 6.18a–d) are in terms of P_u and M_u, *not* the P_n (= $P_u/0.70$) and M_n used in Fig. 6.7; either system is acceptable. In the charts the concrete displaced by the compression bars is ignored in evaluating the resistance provided by these bars. The curves are labeled with values of $\rho_t m$, where $m = f_y/0.85f'_c$.

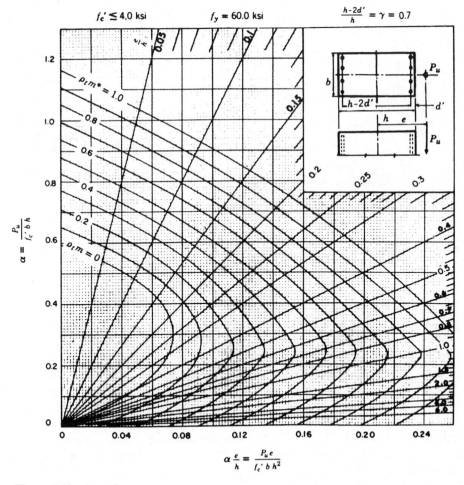

Figure 6.18 (cont.)

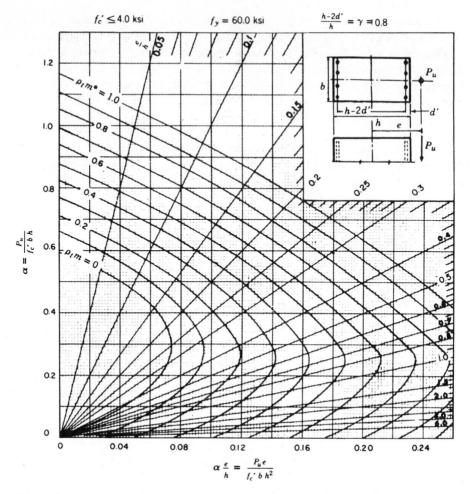

At the top of the figure:

$f_c' \leq 4.0$ ksi $f_y = 60.0$ ksi $\dfrac{h-2d'}{h} = \gamma = 0.8$

Left axis: $\alpha = \dfrac{P_u}{f_c' b h}$

Bottom axis: $\alpha \dfrac{e}{h} = \dfrac{P_u e}{f_c' b h^2}$

Inset diagram labels: P_u, b, $h-2d'$, d', h, e, P_u

Curve labels: $\rho_t m^* = 1.0$, 0.8, 0.6, 0.4, 0.2, $\rho_t m = 0$

Figure 6.18 (cont.)

These charts do not have the flat top $P_{u\,max}$ line. The minimum eccentricity lines of $0.10h$ for tied columns must be observed. These particular charts also fail to reflect the increases in ϕ below $P_u = 0.10 f_c' bh$, as given in Sec. 6.8. This increase would cause the curves to change slope at that point, "swelling out" farther to the right if projected along the e/h line and returning to a value (0.90/ 0.70) times that now shown at P_u (or α) of zero. This break in slope is pronounced; a casual glance will tell whether this higher ϕ value near the bottom has been introduced. The reader can quickly check any similar chart to find if the ϕ values are included by looking at the vertical axis intercept for $\rho_t m = 0$. This point represents a column with $A_{st} = 0$. From the addition law, $P_{n0} =$

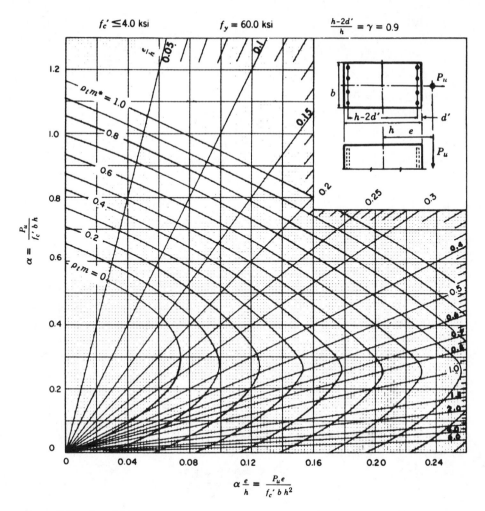

Figure 6.18 (cont.)

$0.85 f'_c bh$. For a tied column, $\phi P_{n0} = (0.7)(0.85) f'_c bh = 0.595 f'_c bh$. In the case of these charts the vertical intercept is $0.595 f'_c bh$. Obviously, ϕ is included in these charts.

6.11 Design of Short Tied Columns Using Interaction Diagrams

Design a square tied interior column, assuming it qualifies as a short column, for a total dead load of 200 k, a live load of 250 k, and a moment of 900 k-in. (all from live load). Use $f'_c = 4000$ psi, Grade 60 steel, and try for ρ_t of about 0.02.

Solution

$$P_u = 1.4 \times 200 + 1.7 \times 250 = 705 \text{ k} \qquad M_u = 900 \times 1.7 = 1530 \text{ k-in.}$$
$$e = M_u/P_u = 2.17 \text{ in.} \qquad m = f_y/0.85f_c' = 60/(0.85 \times 4) = 17.65$$

Desired $\rho_t m = 0.02 \times 17.65 = 0.353$

Column design using interaction diagrams is an iterative procedure. The designer must "guess" a size to start. As the factored axial load, $P_u = 705$ k, is moderate and the eccentricity, $e = 2.17$ in., is low, for a first guess one might start with a value corresponding to a midrange γ (say $\gamma = 0.8$) and a low e/h (say $e/h = 0.20$). Any starting assumptions can be chosen as the procedure will converge in several trials. Using Fig. 6.18c for $\gamma = 0.8$ and going to the intersection for $e/h = 0.20$ and $\rho_t m = 0.353$, $P_u/f_c'bh \approx 0.52$. For a square column, $b = h$. Thus, $h^2 = P_u/(f_c')(0.52) = 705/(4)(0.52) = 339$.

$$h = 18.4 \text{ in.}$$

For a better estimate one could use rounded off, even, whole inches (either 18 or 20 in. in this case). To illustrate the convergence of the procedure, 20 in. will be chosen for the next try. The effective reinforcement lever arm ratio γ can be approximately evaluated by allowing for clear cover over the confining reinforcement, c, the diameter of the confining ties, d_t, and the probable longitudinal reinforcement diameter, d_b.

$$\gamma = \frac{h - 2(c + d_t) - d_b}{h}$$

For an interior column, minimum cover is $1\text{-}\frac{1}{2}$ in. Probable tie diameters are $\frac{3}{8}$ or $\frac{1}{2}$ in. Longitudinal bars may be #9 or #10. Estimate

$$\gamma = \frac{20 - 2(1.5 + 0.5) - 1.25}{20} = 0.74$$

Try 20-in. square column. $e/h = 2.17/20 = 0.109$ Assume $\gamma = 0.70$ (Fig. 6.18b)

Enter diagram at intersection of $e/h = 0.109$ and $\rho_t m = 0.353$, reading $\alpha = 0.62$ on the left scale. These charts automatically introduce the ϕ.

$$0.62 = P_u/f_c'h^2 = 705/(4h^2) \qquad h^2 = 284 \qquad h = 16.9 \text{ in.}$$

Try 17×17 in. column, although some designers would use only even numbers (16 or 18) to reduce the number of column sizes. With the column size chosen, the longitudinal steel required is found by reentering the charts with the new value of h.

$$\alpha = P_u/f_c'h^2 = 705/(4 \times 17^2) = 0.61 \qquad e/h = 2.17/17 = 0.128$$

For about #8 bars and #3 ties, $\gamma = [17 - 2(1.56) + 0.38 + 0.5)]/17 = 12.24/17 = 0.72$ Should interpolate. For $\gamma = 0.7$, (Fig. 6.18b) enter with α

and at $e/h = 0.128$ read $\rho_t m = 0.39$

> For $\gamma = 0.8$ (Fig. 6.18c), read $\rho_t m = 0.35$
> For $\gamma = 0.72$, $\rho_t m = 0.39 - 0.2 \times 0.04 = 0.382$, $\rho_t = 0.0216$
Reqd. $A_{st} = 0.0216 \times 17^2 = 6.24$ in.^{2}

Since $e/h > 0.1$, the maximum axial load limit should not govern. As a check

$$P_{u(\text{max})} = \phi(0.8P_0) = 0.56[17^2 \times 0.85 \times 4 + 6.24(60 - 0.85 \times 4)]$$
$$= 0.56(983 + 353)$$
$$= 748 \text{ k} > 705 \text{ k of design} \quad \textbf{O.K.}$$
$$\text{USE } 17 \times 17 \text{ in. column, 8-\#8, } A_s = 6.32 \text{ in.}^2$$

4-#11 also appear possible, but increased tie size and bar size would leave $\gamma < 0.70$ and increase ρ_t. Also longer splice lap would be required. The #11 bars could prove adequate, but the #8 are retained because the authors want to discuss the necessary ties with 8 bars (next section). The minimum ties are #3 size for #10 or smaller.

USE #3 ties as shown in Fig. 6.19e or f. An alternate would be as in Fig. 6.19a with a single interior crosstie as in Fig. 6.19d.

A complete design also includes analysis for M about the other axis.

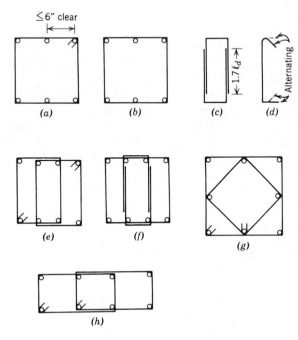

Figure 6.19 Simple tie arrangements.

6.12 Column Ties

Column ties must be made from at least #3 bars for column bars #10 and smaller, #4 bars for larger sized main bars. The details of Code 7.10.5 are rather simple, in effect requiring at least alternate bars to be braced and no bar more than 6 in. clear from a braced bar. The maximum tie spacing is the smallest of:

1. $16d_b$ for the column bars.
2. $48d_b$ for the tie itself.
3. The minimum column thickness.

A few typical tie layouts are shown in Fig. 6.19. The crosstie with a 135° hook on one end and a 90° hook on the other has been developed to facilitate placement when the vertical bars are already in place. Obviously, the 90° hook end must be securely tied to the vertical bar to avoid displacement and alternate ties should be reversed such that each 90° hook falls vertically between two 135° hooks. The 90° hook is less conservative as to strength but placement is much easier.

The 90° hook shown in Fig. 6.19b is a great convenience for placement, but it should not be used where the member is subjected to heavy reversing moment, as in earthquake resistant design; it tends to develop spalling over the tail of the hook in such places, which cancels its benefit as a tie. Likewise, it is sometimes convenient to use the lapped ties of Fig. 6.19c to supplement square overall ties as in Fig. 6.19. When lapped thus in the interior with a full splice length, it is more dependable than when lapped as an outside tie leg; under earthquake or other reversing loading the exterior face splice is *not* recommended.

If in Fig. 6.19e the column bars are not separated by more than 6 in. clear, a single outside tie plus a crosstie over one middle bar on opposite faces, as in Fig. 6.19d is a possible solution. If the same column had two more bars on the sides, the ties as in Fig. 6.19d connecting them would be adequate.

The spacing of ties for the design problem of Sec. 6.11 must not exceed any of the three criteria listed at the start of this section:

$16d_b = 16 \times 1 = 16$ in. for the main bar diameter.
$48d_b = 48 \times 0.375 = 18$ in. for the tie diameter.
Column thickness = 17 in.   Max. tie spacing is 16 in.

∴ USE #3 ties at 16-in. spacing with the pattern shown in Fig. 6.19e.

6.13 Design of Short Spiral Columns Using Interaction Diagrams

Spiral columns may be either round or square, although the spiral itself must be round, as shown in Fig. 6.1c. Under axial load alone a spiral column has the same ultimate strength as a tied column with the same A_g and A_{st}. It could be argued that spiral columns should be designed for the same ϕ as tied columns. However, their greater ductility before failure (loss of some stiffness but not strength) justifies the Code use of ϕ of 0.75, compared to 0.70 for tied columns. The authors would like to see a larger differential because of the increased ductility to encourage the use of spiral confinement.

The limitation on the maximum axial load on the spiral column, $P_{n(max)}$ shows a reduction factor of 0.85 (compared to 0.80 for tied column). The 1971 Code accomplished nearly the same thing by using a minimum eccentricity of $0.05h$ for the spiral column (compared to $0.10h$ for the tied column). The design of short spiral columns where e/h is significantly under $0.05h$ may follow the same design sequence as shown for tied columns in Sec. 6.5, with three coefficient changes: ϕ for spiral column 0.75; $e < 0.05h$; $P_{n(max)}$ coefficient 0.85. The number of bars used must be at least six, but need not be an even number. The spiral design is shown in Sec. 6.14.

The algebra of handling the circular pattern of bars when deriving the interaction curves is more complex than the single face layers covered by the charts of Fig. 6.18, but the final corresponding charts for spiral columns differ chiefly in a more rounded shape (Fig. 6.17c). Charts for both square and circular spiral columns are readily available.[12,13] The method of design using them is identical with that for the tied column in Sec. 6.11.

Biaxial bending is not a complication when the spiral column is circular, as resistance is equal in all directions. However, spirals are often used in square columns and occasionally dual spirals (overlapping) are used in narrow rectangular columns. Interaction diagrams for the latter are not readily available.

6.14   Spiral Design

The spiral must replace the strength of the concrete shell that can be expected to spall off when the column yields.

The spiral reinforcement is specified by ρ_s in Code 10.9.3:

$$\rho_s \geq 0.45(A_g/A_c - 1)f_c'/f_y$$

where

ρ_s = ratio of volume of spiral reinforcement to volume of concrete core (out to out of spiral)

A_c = area of core of column

f_y = yield strength of spiral steel but no greater than 60,000 psi

This equation reflects the strengthening effect of the radial compressive forces from the spiral on the core of the column as it shortens enough to make the unsupported shell spall off. Thus the spiral is designed to replace the strength of the shell with a very slight extra allowance for safety.

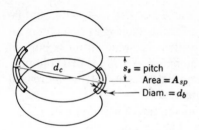

d_c

s_s = pitch
Area = A_{sp}
Diam. = d_b

Figure 6.20   Spiral notation.

For a 21-in. diameter circular column with Grade 60 bars, $f'_c = 3500$ psi, and $1\frac{1}{2}$-in. clear cover:

$$\rho_s = 0.45 \left(\frac{A_g}{A_c} - 1\right)\frac{f'_c}{f_y} = 0.45 \left(\frac{21^2}{18^2} - 1\right)\frac{3500}{60,000} = 0.0094$$

By definition, ρ_s = (vol. of spiral in one round) ÷ (vol. of core in height s_s), where s_s is the pitch as shown in Fig. 6.20. The pitch s_s is small enough for the volume of one round or turn of the spiral to be taken as $A_{sp}\pi(d_c - d_b)$. The volume of the core in height s_s is $s_s\pi d_c^2/4$.

$$\rho_s = A_{sp}\pi(d_c - d_b) \div (s_s\pi d_c^2/4)$$
$$= 4A_{sp}(d_c - d_b) \div (s_s d_c^2)$$

(Many designers ignore the difference between $d_c - d_b$ and d_c, and use $\rho_s = 4A_{sp} \div s_s d_c$.) It is recommended that the size of spiral wire be assumed and s_s then be calculated, because the wire will usually be $\frac{1}{4}$ in.,* $\frac{3}{8}$ in., $\frac{1}{2}$ in., or occasionally $\frac{5}{8}$ in., a rather narrow range of values. The spacing s_s can be specified as closely as desired, usually to quarter inches.

Try $\frac{1}{2}$-in. round spiral rod, $A_{sp} = 0.20$ in.^{2}

$$\rho_s = 0.0094 = 4 \times 0.20(18 - 0.5) \div (s_s \times 18^2)$$
$$s_s = 4.58 \text{ in. } (4.08 \text{ in. clear})$$

Specification maximum: 3 in. clear spacing   (Code Sec. 7.10.4.3)
Specification minimums: 1 in. clear   (Code Sec. 7.10.4.3)
1.33 max aggregate size, clear   (Code Sec. 3.3.3)

It would be uneconomical to use the $\frac{1}{2}$-in. round at 3.00 in. on center. Try $\frac{3}{8}$-in. round, $A_{sp} = 0.11$ in.^{2}

$$\rho_s = 0.0094 = 4 \times 0.11(18 - 0.38) \div (s_s \times 18^2)$$
$$s_s = 2.54 \text{ in. } (2.16 \text{ in. clear})$$

USE $\frac{3}{8}$-in. round at $2\frac{1}{2}$-in. ($2\frac{1}{8}$ in. clear)

The design would be as shown in Fig. 6.21, with the 6-#10 assumed to show a total column description in a frequently used form.

---

* $\frac{3}{8}$ in. is the minimum for cast-in-place construction (Code 7.10.4.2).

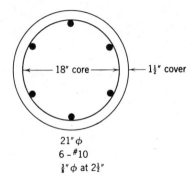

21" ϕ

6 - #10

$\frac{3}{8}$" ϕ at $2\frac{1}{2}$"

Figure 6.21 Spiral column design in Sec. 6.14.

6.15 Combined Axial Compression and Bending-Analytical Expressions

Although the design of eccentrically loaded short columns can be visualized and handled most efficiently using interaction diagrams if manual computations are used, the increasing use of high-speed programmable computers has re-emphasized the need for analytical expressions to approximate the interaction curves of Fig. 6.12b. As shown in Fig. 6.12a in connection with the develop-ment of the interaction curves, relatively simple linear expressions relating P_{no}; P_{nb}; M_{nb}; and P_{nt} conservatively approximate the interaction curves. Simple expressions relating P_n, M_n and these values can be constructed.

Perhaps the most useful set of expressions for columns with bending about one axis are given in Chapter 19 of the 1963 ACI Building Code.[6]

For rectangular members the 1963 Code provided that

1902—Bending and Axial Load Capacity of Short Members—Rectangular Sections

(a) The ultimate strength of short members subject to combined bending and axial load shall be computed from the equations of equilibrium, which may be expressed as follows when a is not more than t and the reinforcement is in one or two faces, each parallel to the axis of bending and all the reinforcement in any one face is located at approximately the same distance from the axis of bending.

$$P_n = 0.85 f'_c ba - A'_s f_y - A_s f_s \tag{19-1}$$

$$P_n e' = 0.85 f'_c ba \left(d - \frac{a}{2} \right) + A'_s f_y (d - d') \tag{19-2}$$

Strain compatibility calculations shall be used to insure that the compression steel will actually yield at ultimate strength of a member as assumed in Eq. (19-1), (19-2), (19-3), (19-4), (19-5), (19-6), and (19-10).

(b) The balanced load, P_b, shall be computed using Eq. (19-1) with $a = a_b = k_1 c_b$, and $f_s = f_y$. The balanced moment, M_b, shall be computed:

$$M_{nb} = P_{nb} e_b = 0.85 f'_c ba_b \left(d - d'' - \frac{a_b}{2} \right) + A'_s f_y (d - d' - d'') + A_s f_y d'' \tag{19-3}$$

(c) The ultimate capacity of a member is controlled by tension in (c) 1 or (d), when P_n is less than P_{nb} (or e is greater than e_b). The capacity is controlled by compression in (c) 2 or (d), when P_n is greater than P_{nb} (or e is less than e_b).

1. When a section is controlled by tension, and has reinforcement in one or two faces, each parallel to the axis of bending, and all the reinforcement in any one face is located at approximately the same distance from the axis of bending, the ultimate strength shall not exceed that computed by:

$$P_n = 0.85 f_c' bd\{p'm' - pm + (1 - e'/d)$$
$$+ \sqrt{(1 - e'/d)^2 + 2[(e'/d)(pm - p'm') + p'm'(1 - d'/d)]}\} \quad (19\text{-}4)$$

For symmetrical reinforcement in two faces, this reduces to:

$$P_n = 0.85 f_c' bd\{-p + 1 - e'/d$$
$$+ \sqrt{(1 - e'/d)^2 + 2p[m'(1 - d'/d) + e'/d]}\} \quad (19\text{-}5)$$

With no compression reinforcement, Eq. (19-4) reduces to:

$$P_n = 0.85 f_c' bd \left[-pm + 1 - e'/d + \sqrt{(1 - e'/d)^2 + 2\frac{e'pm}{d}} \right] \quad (19\text{-}6)$$

2. When a section is controlled by compression, the ultimate load shall be assumed to decrease linearly from P_{no} to P_{nb} as the moment is increased from zero to M_{nb}, where

$$P_{no} = 0.85 f_c'(A_g - A_{st}) + A_{st} f_y \quad (19\text{-}7)$$

For this assumption the ultimate strength is given by either Eq. (19-8) or (19-9):

$$P_n = \frac{P_{no}}{1 + [(P_{no}/P_{nb}) - 1]e/e_b} \quad (19\text{-}8)$$

$$P_n = P_{no} - (P_{no} - P_{nb})M_n/M_{nb} \quad (19\text{-}9)$$

For symmetrical reinforcement in single layers parallel to the axis of bending, the approximate value of P_n given by Eq. (19-10) may be used:

$$P_n = \frac{A_s' f_y}{\dfrac{e}{d - d'} + 0.5} + \frac{btf_c'}{(3te/d^2) + 1.18} \quad (19\text{-}10)$$

(d) When the reinforcement is placed in all four faces, or in faces which are not parallel to the axis of bending, the design shall be based on computations considering stress and strain compatibility and using the assumptions in Section 1503.

The definition of terms in the 1963 Code* was

1900—Notation and definitions
(a) *Notations*
a = depth of equivalent rectangular stress block
a_b = depth of equivalent rectangular stress block for balanced conditions = $k_1 c_b$

---

* The authors have made changes to introduce the P_n notation now used.

A_g = gross area of section

A_s = area of tension reinforcement

A_s' = area of compression reinforcement

A_{st} = total area of longitudinal reinforcement

b = width of rectangular section in the direction at right angles to the plane of bending

c = distance from extreme compression fiber to neutral axis

c_b = distance from extreme compression fiber to neutral axis for balanced conditions = $d(90,000)/(90,000 + f_y)$

d = distance from extreme compression fiber to centroid of tension reinforcement

d' = distance from extreme compression fiber to centroid of compression reinforcement

d'' = distance from plastic centroid to centroid of tension reinforcement

D = overall diameter of circular section

D_s = diameter of the circle through centers of reinforcement in a column of circular . section

e = eccentricity of axial load at end of member measured from plastic centroid of the section, calculated by conventional methods of frame analysis

e' = eccentricity of axial load at end of member measured from the centroid of the tension reinforcement, calculated by conventional methods of frame analysis

e_b = eccentricity of load P_b measured from plastic centroid of section

f_c' = concrete strength to be used in design computations

f_s = calculated stress in reinforcement when less than the yield point, f_y

f_y = reinforcement yield point

k_1 = a constant defined as β_1 in current Code

$m = f_y/0.85f_c'$

$m' = m - 1$

M_{nb} = nominal moment capacity at simultaneous crushing of concrete and yielding of tension steel (balanced conditions) = $P_{nb}e_b$

M_n = nominal moment capacity under combined axial load and bending

$p = A_s/bd$

$p' = A_s'/bd$

$p_t = A_{st}/A_g$

P_{nb} = nominal axial load capacity at simultaneous crushing of concrete and yielding of tension steel (balanced conditions)

P_{no} = nominal axial load capacity of actual member when concentrically loaded

P_n = nominal axial load capacity under combined axial load and bending

t = overall depth of a rectangular section in the plane of bending, or over-all diameter of a circular section

These relatively straightforward expressions are the analytical way of expressing several of the steps carried out in the general development of the interaction diagrams. Equation 19-1 expresses ΣF and Eq. 19-2 expresses ΣM for the simple case where reinforcement is on one or two faces. Again using the basic definition of balanced strain conditions, P_{nb} can be found from Eq. 19-1 and M_{nb} from Eq. 19-3. Equation 19-7 defines the point P_{no} and Eq. 19-8 and Eq. 19-9 are a simple linear interpolation between P_{no} and the point P_{nb}, M_{nb} for a given eccentricity or moment. This expression assumes the compression failure branch to be linear as shown in Fig. 6.12a. Alternatively, for symmetrical reinforcement in single layers, Eq. 19-10 (suggested by Whitney), which gives a better approximation to the curved surface of the compression branch of the interaction curve, is given.

For the tension failure branch Eq. 19-4 is given. For certain simpler cases Eq. 19-5 or 19-6 may be used.

For short circular members the 1963 Code recommended:

1903—Bending and Axial Load of Short Members—Circular Sections

(a) The ultimate strength of short circular members subject to combined bending and axial load shall be computed on the basis of the equations of equilibrium taking into account inelastic deformations, or by the empirical expressions given by:
When tension controls:

$$P_n = 0.85 f_c' D^2 \left[\sqrt{\left(\frac{0.85e}{D} - 0.38 \right)^2 + \frac{p_t m D_s}{2.5D}} - \left(\frac{0.85e}{D} - 0.38 \right) \right] \quad (19\text{-}11)$$

When compression controls:

$$P_n = \frac{A_{st} f_y}{\dfrac{3e}{D_s} + 1} + \frac{A_g f_c'}{\dfrac{9.6 D e}{(0.8D + 0.67 D_s)^2} + 1.18} \quad (19\text{-}12)$$

Equation 19-11 and Eq. 19-12 are again very useful for electronic computation.

An alternate procedure for designing columns using computers is to develop and plot interaction curves for the cross sections of interest. Then, a manual selection of the appropriate cross section is made from the plots. This process is particularly useful when irregular shapes or bar patterns occur. Very powerful programs have been developed that use numerical procedures to divide the column cross sections into strips or fibers. Numerical integration of the forces on each strip or fiber is used in an iterative routine to develop the values of axial load and moment corresponding to the failure surface. Such specialized plots are then used as design aids.

6.16 Biaxial Moments on Columns

Many columns are subject simultaneously to moments about both major axes, especially corner columns. The mathematics of such cases is quite involved, although for any given neutral axis the analysis can follow very simple ideas like those of Sec. 6.7 for P_b and e_b. For example, in Fig. 6.22a, let the neutral axis be arbitrarily chosen as shown. One may construct, perpendicular to this axis, strain triangles precisely as is done in Fig. 6.11 for the rectangular column. The maximum concrete strain of 0.003 is probably too small and the equivalent rectangular stress block may not be exactly $a = \beta_1 c$, but a thorough

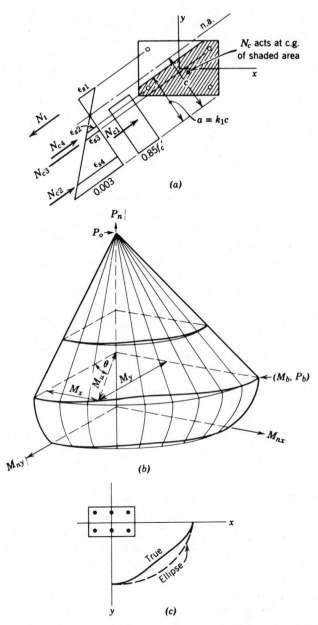

Figure 6.22 Biaxial bending on column. (*a*) Forces acting on cross section. (*b*) Typical interaction surface. (From Reference 8, ACI.) (*c*) Contour at level P and P_b.

investigation[7] has indicated that in combination the two are satisfactory. N_{c1} is simply $0.85f'_c$ times the shaded area and acts at its centroid, or the area can be broken up into triangles and rectangles if preferred. The strains lead easily to N_t, N_{c2}, N_{c3} and N_{c4}. $\Sigma F_y = 0$ leads to the value of P_n, $\Sigma M_x = 0$ leads to x_P, (equal to e_x along the x-axis), and $\Sigma M_y = 0$ leads to y_P (equal to e_y along the y-axis). This completely establishes the load that would create this neutral axis. The method is awkward chiefly because of the dimensions, which might be easily obtained graphically from a scale layout. The real problem is that one needs to start with given loads and eccentricities and not with a location of the neutral axis; this neutral axis is not usually perpendicular to the resultant eccentricity.

For a circular column, the interaction diagram for moments about any axis is the same and no real problem exists. It is easy to visualize a three-dimensional interaction diagram for this case. It would be a surface of revolution obtained by rotating the interaction diagram about the P_n-axis, a little like Fig. 6.22b except with the same diagram on each axis. With a square column having equal steel on each face, the interaction diagram on the x-axis is the same as on the y-axis; but when these are rotated they give values on the unsafe side for intermediate angles. Similarly, if a rectangular column is considered, it is easy to visualize a varying radius of rotation, as in Fig. 6.22b, creating an ellipse on any horizontal plane that connects the x- and y-axis diagrams. This procedure also indicates values on the unsafe side. In other words, both the circle for the square column and the ellipse for the rectangular column must be considered as upper bounds on proper values, the real three-dimensional surfaces being a little flattened on the diagonals. This difference is quite small for the upper part of the diagrams and is maximum near the P_b level.

The senior author's earlier suggestion was that these surfaces be used as a starting point and then the moment capacity at any given P_n level be reduced, by a maximum (for ordinary amounts of reinforcing) of about 15%* when the resultant e is on the 45° line between the x- and y-axes. A study of Furlong's[8] and Pannell's[9] work would permit some refinements on this approximation.

A more rigorous study by Gouwens[10] indicates more specific methods that are usable as very reasonable approximations. The horizontal section through the interaction diagram at a constant P_n in Fig. 6.22 is a function of the height P_n/C_c where $C_c = f'_c bh$. Gouwens follows Meek's approximation of using this curve as two straight lines (when moments are expressed as ratios) fixed by the β_b at their intersection, as shown in Fig. 6.23a. Note that $\beta_b = M_y/M_{yo} = M_x/M_{xo}$ when axes in this figure are plotted as dimensionless ratios. Approximate β_b values for various combinations of P_n/C_c values and ratios of C_s/C_c (i.e., $A_s f_y/f'_c bh = \rho_t f_y/f'_c$) are shown in Fig. 6.23$b$. Gowens notes a desirable correction for β_b values for columns having *only* corner bars: Lower the plotted β_b values by 0.02.

---

* Pannell shows up to 32% for high-steel ratios.

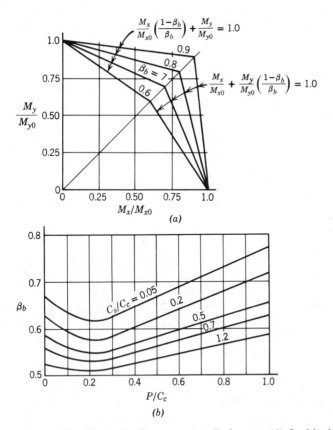

Figure 6.23 Charts by Gouwens (see Reference 10) for biaxial column moments. (*a*) Charts and equations for various β_b. (*b*) Chart establishing β_b to use in (*a*).

A method* published by Bresler[11] in 1960 relates the desired P_u (or now ϕP_n) under biaxial loading (e_x and e_y) to three other P_u values:

$P_{uy} = \phi P_{ny} =$ design strength for same column under the same e_y (Fig. 6.24*b*)

$P_{ux} = \phi P_{nx} =$ design strength for same column under the same e_x (Fig. 6.24*c*)

$P_{uo} = \phi P_{no} =$ theoretical axial load design strength for same column when $e_x = e_y = 0$ (as if the Code permitted such design)

This equation is

$$\frac{1}{P_u} = \frac{1}{P_{ux}} + \frac{1}{P_{uy}} - \frac{1}{P_{uo}} \qquad \text{or} \qquad \frac{1}{\phi P_n} = \frac{1}{\phi P_{nx}} + \frac{1}{\phi P_{ny}} - \frac{1}{\phi P_{no}}$$

---

* Reportedly from the Russian Code.

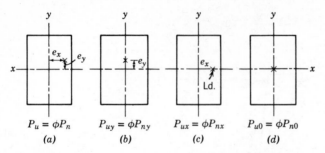

Figure 6.24 Notations for Bresler's biaxial loading equation. (*a*) Biaxial moments. (*b*) and (*c*) Eccentricities about *x* and *y* axes. (*d*) Axial load alone.

With charts (or tables) for P_u plus uniaxial moment (similar to Fig. 6.18) about the *x* axis and similar ones for P_u plus moment about the *y* axis, values of P_{ux}, P_{uy}, and the theoretical P_{uo} are easy to establish. Substitution in the equation then gives the desired biaxial P_u. (Usually the ordinates of charts give $P_u/f'_c bh$, with ϕ normally included so that in 1986 notation this would lead to ϕP_n.) As an approximate method it is one of the best when one has the necessary charts or tables, *provided* the resulting P_u or ϕP_n is above $0.10 f'_c bh$. If P_u is lower than the balanced design level (or say the Code $0.10 f'_c bh$ level) the errors by this method can increase. In typical cases it is then on the safe side to design for biaxial moment alone, as tension failure then controls.

For *square* columns loaded on the diagonal, interaction diagrams are available in the 1970 ACI handbook.[12] The 1978 Handbook[13] contains curves that are very helpful in the general problem of biaxially loaded columns.

For long columns (Chapter 7) the magnifier is to be separately calculated and applied to the moment about each axis independently. The relative values will then determine the plane of the principal moment. The minimum e/h applies about each axis, but not simultaneously.

6.17   Column Splices

The general subject of splice design is covered in Chapter 8 but some further general comments are appropriate for columns.

Location of column splices in the past has typically been just above the floor level, in effect starting new column bars at the lowest possible point. Certain complications can result. If the dowel projections and the new column bars are placed side by side so that they are equidistant from the axis of the column, $2d_b$ must be deducted from the center-to-center spacing to find the clear bar spacing. When 5% or more of longitudinal steel is used, this spacing can become critical and limit the permissible column steel below what might be desired.

Several factors now tend on occasion to make other splice points look attractive. Code 12.17.3 requires at least 25% of the vertical steel on each face to be developed to f_y in tension (or more bars at a lower stress to give the same total tension capacity). If butt splices are not used, the compression lap may provide the required tension. Another way to achieve this is to run bars in two-story heights and to stagger the splices. Although such bars are awkward to support on the job, such an action relieves the congestion in a single splicing section. Because tension splices tend to be long, elimination of some of these is also attractive. The awkwardness in construction lies not only in bracing the loose bars but, after columns ties are in place, in interweaving the beam and girder steel through the column steel.

Appendix A of the Code (seismic design) in Code A.4.3.2 requires all column splices to be designed as tension splices and that all column lap splices be made within the center half of the column length. The Commentary points out that splices at midheight are preferable because the moment is lowest. The maximum design moments for wind, earthquake, and even vertical loads (in typical multistory buildings) typically occur at the *end* (or ends) of the column; the splice at midheight escapes some of this moment.

For compression splices, bars may be butted and held in place by pipe or commercial mechanical splices.

For tension splices, bars can be welded together, usually with butt welds, but it should be noted that a number of welding passes are required for large high-strength bars. American Welding Society preheat requirements can also be formidable.[14] A Cadweld splice, using molten metal to hold the bar to a sleeve (deformed on its inside surface), in effect splices the bar to the sleeve and the sleeve to the next bar. The sleeve in this case is notably larger than the bar, uses up some of the clear bar spacing, and may encroach on the specified clear cover unless this is specially considered in design. The use of taper-threaded couplers has become very popular in recent years. The bars can be field-threaded to be joined with such quick couplers.

6.18   Brief Comments on Economy

Economy in column design favors the use of higher strength concrete on well-controlled jobs, because extra concrete strength costs only slightly more than low strength. Concrete will generally be cheaper than intermediate grade steel. Steel of Grades 50, 60, or 75* is cheaper than Grade 40 because of its higher f_y value. The higher strengths will often be as cheap as concrete in load-carrying capacity. Tied columns will generally be cheaper than spiral columns, particularly if square columns are needed rather than round. However, the cost of the column can rarely be considered alone. Spiral columns and heavy steel save floor space which has an annual rental value. Form costs are also a major item and beam forms can be reused from floor to floor most simply if column sizes are kept constant. Hence it is quite common practice to keep the column size constant over several stories and take care of increasing load with increasing steel, stronger concrete and steel, or the use of spiral steel.

---

* Grade 75 is not usually available except in largest sizes.

6.19   Reduction in Column Live Loads

Building codes specify uniform live loads large enough to represent ordinary local concentrations of loading. Over larger areas the probability that this same load intensity will exist everywhere is reduced. Codes usually permit some reduction in the assumed live load on floor areas in excess of 100 to 150 sq ft. Many codes permit such reduction only when the live load is 100 psf or less; heavier loads imply storage or machinery loads that are easily concentrated over large areas.

The live load on columns also may be reduced under the same reasoning. The reductions permitted vary considerably in the different city codes and it will be necessary for the designer to consult the particular code that governs his or her design. The magnitude of the reduction depends both on the occupancy or size of the unit live load and on the number of floors carried by the column; it may vary from none to as much as 60% of the total live load.

Selected References

1. F. E. Richart and G. C. Staehle, "Column Tests at University of Illinois," *ACI Jour.*, 2, Feb., Mar. 1931; *Proc.*, 27, pp. 731, 761; *ACI Jour., 3,* Nov. 1931, Jan. 1932; *Proc., 28,* pp. 167, 279.

2. W. A. Slater and I. Lyse, "Column Tests at Lehigh University," *ACI Jour.*, 2, Feb., Mar. 1931; *Proc.*, 27, pp. 677, 791; *ACI Jour.*, 3, Nov. 1931, Jan. 1932; *Proc.*, 28, pp. 159, 317.

3. E. Hognestad, "A Study of Combined Bending and Axial Load in Reinforced Concrete Members," *Univ. of Ill. Eng. Exp. Sta. Bull. No. 399,* 1951.

4. F. E. Richart and R. L. Brown, "An Investigation of Reinforced Concrete Columns," *Univ. of Ill. Eng. Exp. Sta. Bull. No. 267,* 1934.

5. F. E. Richart, J. O. Draffin, T. A. Olson, and R. H. Heitman, "The Effect of Eccentric Loading, Protective Shells, Slenderness Ratios, and Other Variables in Reinforced Concrete Columns," *Univ. of Ill. Eng. Exp. Sta. Bull. No. 368,* 1947.

6. *Building Code Requirements for Reinforced Concrete,* (ACI 318–63), Amer. Concrete Inst., Detroit, 1963.

7. A. H. Mattock and L. B. Kriz, "Ultimate Strength of Nonrectangular Structural Concrete Members," *ACI Jour.*, 32, No. 7, Jan. 1961; *Proc.* 57, p. 737.

8. R. W. Furlong, "Ultimate Strength of Square Columns Under Biaxially Eccentric Loads," *ACI Jour.*, 32, No. 9, Mar. 1961; *Proc.* 57, p. 1129.

9. F. N. Pannell, "Failure Surfaces for Members in Compression and Biaxial Bending," *Jour. ACI,* No. 1, Jan. 1963, p. 129.

10. A. J. Gouwens, "Biaxial Bending Simplified," *Reinforced Concrete Columns,* SP-50, Amer. Concrete Inst., Detroit.

11. B. Bresler, "Design Criteria for Reinforced Concrete Columns under Axial Load and Biaxial Bending," *ACI Jour.*, 57, 1960, p. 481.

12. *Ultimate Strength Design Handbook, Vol. 2, Columns,* SP-17A, Amer. Concrete Inst., Detroit, 1970.

13. *Design Handbook, Vol. 2, Columns,* SP-17A(78), American Concrete Institute, Detroit, 1978.

14. "AWS Standard D12.1, Reinforcing Steel Welding Code," Amer. Welding Soc., Miami, 1979.

Problems

NOTE *Assume all columns in these problems qualify as short columns.*

PROB. 6.1. Given $f'_c = 5000$ psi, Grade 60 reinforcement, and interior exposure conditions. Design a square tied column with ρ_t about 3.5% for an "axially loaded" interior column with unfactored P_D of 300 k and P_L of 400 k. Calculations indicate moments are negligible. Keep h in even, whole inches. Select reinforcement distributed on all four faces. Dimension ties and sketch cross section.

PROB. 6.2. Same as Problem 6.1 except design a rectangular column with $b = \frac{1}{2}h$.

PROB. 6.3. Same as Prob. 6.1 except design a square spiral column. Dimension spiral assuming spiral $f_y = 60$ ksi.

PROB. 6.4. Same as Problem 6.1 except use $f'_c = 4000$ psi, Grade 40 reinforcement.

PROB. 6.5. Same as Problem 6.2 except use $f'_c = 4000$ psi, Grade 40 reinforcement.

PROB. 6.6. Same as Problem 6.3 except use $f'_c = 4000$ psi, Grade 40 reinforcement for longitudinal bars, and $f_y = 60$ ksi for column spirals.

PROB. 6.7. Given $f'_c = 3000$ psi and bars Grade 60 with cover of 2.7 in. from center of bars. Find the permissible ultimate column load, using graphs of Fig. 6.18 if desired.

(*a*) For the column of Fig. 6.25*a* under a load on the *x* axis 6 in. to the right of center line at each end.

(*b*) For the column of Fig. 6.25*b* under a load on the *x* axis 10 in. to the right of the center line.

(*c*) For the column of Fig. 6.25*c* under a load on the *x* axis 8 in. off the center of column.

PROB. 6.8. If $f'_c = 4000$ psi and Grade 60 steel is used with cover of 2.7 in. to center of steel, find the ultimate load permitted by the Code on the column of:

(*a*) Prob. 6.7*a* with *e* increased to 20 in.

(*b*) Prob. 6.7*b* with *e* increased to 20 in.

(*c*) Prob. 6.7*c* with *e* increased to 12 in.

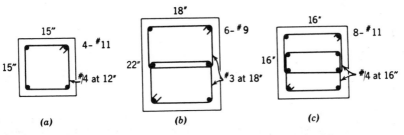

Figure 6.25 Column cross sections for Prob. 6.7 and Prob. 6.8.

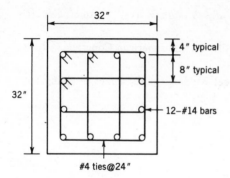

Figure 6.26 Column cross section for Prob. 6.12.

32"

32"

4" typical

8" typical

12–#14 bars

#4 ties@24"

PROB. 6.9.* Design a square tied column for service loads of 120 k dead plus 100 k live and for single curvature moments of 20 k-ft dead and 60 k-ft live, using $f'_c = 4000$ psi, about 2% of Grade 60 steel with 2.7 in. cover to center of bars.

PROB. 6.10.

(*a*) Evaluate and locate the ultimate load that gives a horizontal neutral axis 6 in. below the top edge of the column of Fig. 6.25*c*. Assume cover 2.5 in. to center of steel, $f'_c = 4000$ psi, Grade 60 steel, and consider deformations to establish the steel stresses.

(*b*) Repeat for a horizontal axis 6 in. below the top edge of Fig. 6.25*b*.

PROB. 6.11. Design a column spiral of Grade 40 steel that will be adequate for a 20-in. square column with longitudinal steel in a circular pattern that permits the spiral to have an outside diameter of 17 in., $f'_c = 4000$ psi.

PROB. 6.12. Given the adequately tied square-column cross section of Fig. 6.26 with $f'_c = 5000$ psi and Grade 60 bars, find P_{no}, P_{nb}, M_{nb}, and the values of P_n, M_n for assumed values of $c = 45$ in., 36 in., 27 in., 21 in., 10 in., and 5 in.

(*a*) Plot the interaction diagram for P_n, M_n to scale.

(*b*) Apply the proper ϕ factors and superimpose on (*a*) the interaction diagram for P_u, M_u values.

---

* *Note:* If column interaction diagrams similar to Fig. 6.18 are available for spiral columns, Prob. 6.9 is also suitable for assignment as a spiral column design.

7

SLENDER COLUMNS

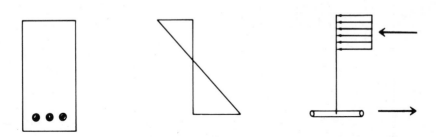

7.1   The Nature of the Slender Column Effect—Laterally Braced Column

Columns always carry moment in practical applications, both by virtue of slight initial crookedness and by the nature of the loading from beams and slabs that rotates frame joints and thus imposes curvatures. Moments in columns also produce curvatures that lead to deflections between joints and sometimes sidesway deflections of the joints (similar to those occurring under wind load), even when no external horizontal loads exist.

Figure 7.1 shows a statically determined column braced against lateral movement at the ends and loaded with end moments Pe to give single curvature.* These initial moments are known as primary moments. If the resulting deflection y_0 is large, as shown, a large secondary moment Py_0 is added at midspan and the ends at A and B rotate, not only because of the original Pe moment but because of the secondary Py moments; the two rotations are additive. This extra moment is called the *slender column effect*.

The column in reversed curvature in Fig. 7.2 is loaded so that a point of inflection must occur between the column ends. If its deflection is small (Case 1), the added Py_1 moment does not increase the original moment along the column enough to become the critical moment for design. The maximum moment remains at the column end. If the column is long enough to deflect more (Case 2) the moment Py_2 may develop an increased moment away from the joint. This increased moment becomes the maximum moment in the column and hence is critical for design. The reversed curvature case includes any value of e_2 from small to numerical equality with e_1. The ratio of height to radius of gyration has to be very large before there is an ill effect (slender column effect) from this loading.

---

* This term also covers unequal end moments of the same sign using beam moment sign convention.

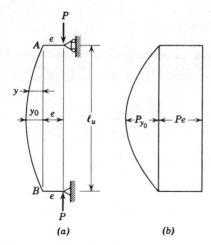

Figure 7.1 Single curvature column. (*a*) Curvature. (*b*) Moment.

7.2 Influence of Length on Braced Column Behavior

The behavior response of any eccentrically loaded braced column to an increasing load is not linear, which raises a question about the most appropriate design e/h to use. Consider a column in single curvature as in Fig. 7.1. As one doubles the load, y_0 more than doubles, and develops a possibly critical moment as indicated by curve A in Fig. 7.3a.

The shape of the axial load-moment curve for a particular case depends on the length of the member. The moment for a very short column might plot essentially straight (y_0 negligible), following the dotted line in Fig. 7.3a for the nominal moment Pe to failure at point C on the interaction diagram. The medium-length column of curve A would have an amplified moment $Pe + Py$ and would fail at point D. These failures would be in compression, and the authors call these failures at the interaction diagram *"materials failure"*.

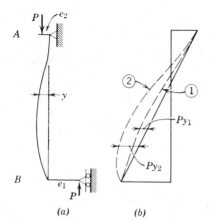

Figure 7.2 Column with end moments of opposite kind. (*a*) Curvature. (*b*) Moment.

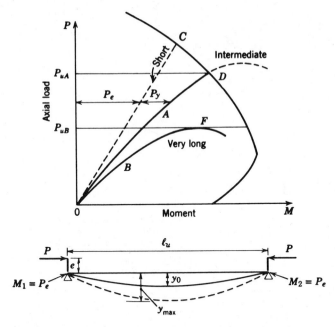

Figure 7.3 (*a*) Behavior patterns for columns. (*b*) When the end moments are caused by an eccentric axial load, the deflection resulting from M_1 and M_2 is increased from y_0 to y_{max}.

A longer column might give the curve B which turns horizontal at F before reaching the interaction diagram. This column fails from instability or buckling in the classical sense, the inability to resist more load, although no material has reached the failure point. Mathematically, some like to treat point D as a special case of instability,* but the authors believe the separate names for the two quite different phenomena help to clarify understanding of columns.

The ACI Code provisions for slender columns treated in later sections permit one to compute the increased or "magnified" moment[2] and to plot this type of curve if one wishes. For design or analysis, only the portion around D or F are important; at F the method only indicates a horizontal line, representing increasing moment with no increase in axial load.

7.3 The Nature of the Slender Column Effect—Laterally Unbraced Column

A major case is that of a column in an unbraced frame subject to a shear load, for example, wind resistance or earthquake forces. Consider first the situation where the floor system is very heavy, that is, for the purpose of this initial

---

* With the curve turning sharply downward at point D.[1]

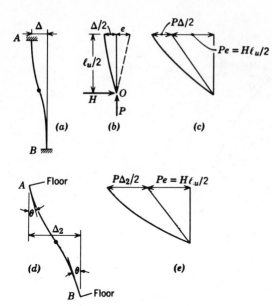

Figure 7.4 Columns with sidesway and horizontal shears. (*a*) Curvature. (*b*) Statics of a halfheight. (*c*) Moment. (*d*) With joints that rotate. (*e*) Moment for (*d*).

discussion, infinitely stiff. The column deflects as in Fig. 7.4*a* and can be analyzed by considering the half-length shown in Fig. 7.4*b* that shows the moment increased at *A*, much like it is at midheight of the column of Fig. 7.1*a*. (To improve clarity the symbol Δ is used for end deflection and *y* is used for deflection between the column ends.) Although the magnitude of the $P\Delta/2$ term compared to the *Pe* or $H\ell_u/2$ moment is different, the two terms are usually additive as shown in Fig. 7.4*c*. However, infinitely stiff floors with zero joint rotations at *A* and *B* are not common.

If the joints at *A* and *B* rotate, Δ increases to Δ_2 as shown in Fig. 7.4*d*. Although the $H\ell_u/2$ moment is unchanged, the curvature from *A* down to midheight is changed owing to the increasing moments along the column because of the *Py* term which increases with Δ. Thus the deflection of the cantilever in Fig. 7.4*b* changes from that of a member fixed at the end to one where deflection is augmented by joint rotation as in Fig. 7.4*d*. The moment diagram of Fig. 7.4*e* then involves a greater end moment. If the joint rotation is large, the end moment is multiplied several times, that is, $P\Delta_2/2$ is several times as much as the static force moment $H\ell_u/2$. Hence the slender column action can be very important in this case. The degree of importance is determined by the angle θ at each end. Although this angle depends on the total moment that the column carries, for any given value of this moment the angle is controlled almost linearly by the stiffness of the beams; a beam half as stiff will double the angle. Thus it is obvious that any slender column analysis must carefully consider the flexural restraint at the column ends as well as the restraint against lateral movement of the column ends.

7.4 Influence of Length and Frame Restraint on Unbraced Column Behavior

The behavior response of any eccentrically loaded unbraced column is generally considerably more complex than that of a similar but braced column. The secondary or $P\Delta$ moments of an eccentrically loaded unbraced column are made up of both the Py effect and the $\theta\ell$ effect. Figure 7.5a shows a laterally loaded rectangular frame. If the beams are infinitely rigid, the deflected shape is as shown in Fig. 7.5b. Δ_1 is due to the integration of both the column curvature effects and the secondary $P\Delta$ effects that are due to such curvatures. In contrast, Fig. 7.5c shows that even if the columns are infinitely rigid, there can be an appreciable Δ_2 that is due to the end rotation of the flexible restraining beams. This "parallelogram" or $\theta\ell$ effect produces a column lateral displacement Δ_2 resulting in secondary moments that further amplify θ and hence Δ_2. Because the beams and columns in actual frames are not infinitely rigid, the general case is that shown in Fig. 7.5d in which the frame undergoes displacements Δ_3 owing to both column curvatures and $\theta\ell$ effects.

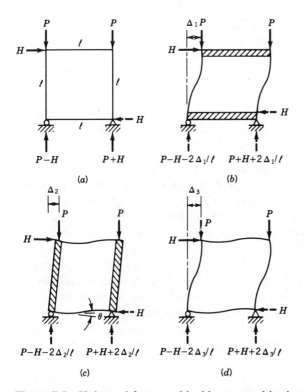

Figure 7.5 Unbraced frames with sidesway and horizontal shears. (a) Basic loading. (b) Rigid beams. (c) Rigid columns. (d) Nonrigid beams and columns.

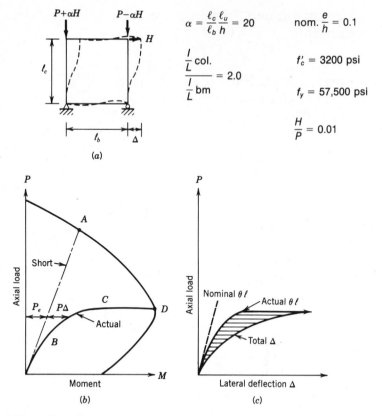

$$\alpha = \frac{\ell_c}{\ell_b}\frac{\ell_u}{h} = 20 \qquad \text{nom.} \; \frac{e}{h} = 0.1$$

$$\frac{\frac{1}{L}\text{col.}}{\frac{1}{L}\text{bm}} = 2.0 \qquad f'_c = 3200 \text{ psi}$$

$$f_y = 57{,}500 \text{ psi}$$

$$\frac{H}{P} = 0.01$$

(a)

(b)

(c)

Figure 7.6 Behavior of an unbraced frame loaded laterally. (*a*) Test frame. (*b*) Axial load-moment relationship. (*c*) Axial load-lateral deflection relationship.

This behavior can be clearly seen in Fig. 7.6 which shows the test results of frame L3 of Reference 3. The unbraced frame of Fig. 7.6*a* had proportional lateral and vertical loading that in a very short column would be expected to produce a nominal moment as shown by the dashed line in Fig. 7.6*b*. If there were no length effects, failure would be expected when the axial load and moment reached the material failure state at the interaction diagram at *A*. In actuality, substantial secondary or *P*Δ moments existed as shown by curve *B*. The frame became unstable at point *C*, swept laterally, and the cross sections at the column ends were destroyed at *D*. The various components of the actual lateral deflection in Fig. 7.6*c* indicate that joint rotations played a major role in the failure. The load-deflection curve is subdivided for clarity. The nominal $\theta\ell$ effect is indicated by the dashed line as computed by ordinary methods of frame analysis ignoring slender column action. The unshaded area represents the actual $\theta\ell$ deflections of the columns as though they were sides of the parallelograms produced by the end joint rotations. Such deflection effects are greatly affected by the stiffness of the restraining flexural members indicating that any treatment of slender unbraced columns must consider the restraining flexural members. The significant difference between the nominal $\theta\ell$ and the actual $\theta\ell$ values reflect the moment magnification owing to secondary deflection and the stiffness reduction owing to cracking and higher stresses. Premature yielding of the restraining beams could lead to an accelerated failure in the unbraced columns.

The shaded area corresponds to the lateral deflection component caused by the column curvatures. The increased magnitude of these effects at higher loads reflects the secondary moment effects, related increased column curvatures, and the loss of column stiffness with higher axial loads and column cracking.

The correct treatment of slender columns must reflect these interdependent factors in some appropriate way. The overall problem is to correct the ordinary first-order frame analysis to consider such effects. The ACI Building Code encourages modification of basic analysis procedures in Code Sec. 10.10. Because such modified analysis is usually only feasible with the assistance of fairly complex computer analysis, a more approximate "moment magnification" procedure suitable for manual computations is outlined in Code Sec. 10.11. Both approaches are treated in subsequent sections.

7.5 Length Limitations for Short Columns

The treatment of column slenderness effects is often complex and time-consuming. In practical design, many stocky and well-restrained compression members (short columns) have almost negligible secondary moment effects and essentially develop the full cross-sectional strength at the nominal eccentricity. To simplify the designer's job, the Code gives guidelines for those cases where slenderness effects can be ignored without more than a 5% strength loss. Reference 2 indicates that in practical terms about 90% of columns in laterally braced frames and about 40% of columns in laterally unbraced frames do not need a detailed check for slenderness effects and so can be considered as "short columns." The designer can utilize these limits in his or her preliminary proportioning to ensure selection of stocky columns that will be simpler to design and more stable in behavior.

Although it may seem out of sequence, the authors feel that the concept of diagnosing whether any slender column effect exists is extremely important and should be introduced before the actual slenderness effect calculation procedures are introduced. Which cases are exempt from further analysis requires some discussion of the basic slender column action first. The particular effective column length in a frame that demands the special slender column technique depends on the deflection the column moment creates. This in turn depends on the column curvature and its joint restraints at the ends.

Three cases must be separately recognized in setting proper limits on short column length. The first two are for braced frames, with bracing implying negligible lateral movement of the end joints. The third is the general case of unbraced frames, which can deflect laterally.

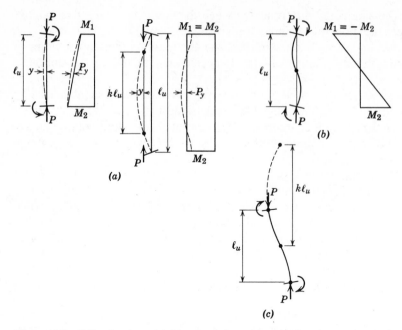

Figure 7.7 Effective lengths $k\ell_u$ of columns. (*a*) Single curvature in braced frame, short and long. (*b*) Reversed curvature in braced frame. (*c*) Reversed curvature in unbraced frame.

The single curvature case of Fig. 7.7*a* is the more flexible of the two braced frame cases and includes $M_1 < M_2$ as well as $M_1 = M_2$, either producing a simple deflection curve at service loads. P times the deflection y of this curve measures an added moment sensitive to length in a nonlinear fashion. In the slender columns, as the load increases the curvature changes. Because the end joints do not rotate freely and the center deflection increases rapidly, points of inflection appear. The reduced effective length $k\ell_u$ between inflection points becomes the critical length. The usual range for k in braced frames is from 0.65 to 0.90, dependent on the relative stiffness ψ of columns-to-beams at the end joints. The use of $k = 1$ is allowed by the Code and may be appropriate for preliminary analysis. More detailed procedures for determining k are treated in Sec. 7.11, including in Fig. 7.11 a nomograph to assist in evaluating k. The limiting short column length is defined for braced frames in Code 10.11.4.1 as

$$k\ell_u/r = 34 - 12M_1/M_2$$

The radius of gyration r may be used as $0.30h$ for rectangular columns or $0.25h$ for circular columns (Code 10.11.3). These r values have been used by Dr. Furlong in developing Table 7.1 in terms of ℓ_u/h. The numbers tabulated, when multiplied by h, give the maximum lengths for the short column analysis for several listed ψ values.* In design a majority of braced single curvature columns qualify as short.

---

* See Fig. 7.10 for definition of ψ and relationship to k.

TABLE 7.1 Typical Maximum Short Column Lengths in Terms of Thickness, ℓ_u/h

Values of $\psi_1 = \psi_2 = \psi^* =$	Rectangular Column			Circular Column		
	0.5	1.0	2.0	0.5	1.0	2.0
Braced Frame k =	0.68	0.77	0.86	0.68	0.77	0.86
Single curvature $\quad M_1 = M_2$	9.7	8.6	7.7	8.1	7.2	6.4
$\quad\quad\quad\quad\quad\quad M_1 = 0$	15.0	13.3	11.9	12.5	11.1	9.9
Reversed curvature $\quad M_1 = -0.4 M_2$	17.1	15.1	13.6	14.2	12.6	11.3
$\quad\quad\quad\quad\quad\quad M_1 = -M_2$	20.3	17.9	16.1	17.0	14.9	13.4
Unbraced Frame k =	1.17	1.31	1.59	1.17	1.31	1.59
All M values	5.6	5.0	4.1	4.7	4.2	3.4

* See Fig. 7.11 for definition of ψ as relative column stiffness.

The second curvature case for the braced frame is shown in Fig. 7.7b. Because the reversed curvature case, typical in an exterior column, creates less column deflection, larger $k\ell_u/r$ or ℓ_u/h values are permissible. The previous limiting equation is also used here, with M_1 negative and M_2 positive, thus increasing $k\ell_u/r$ above 34. In braced frames most reversed curvature columns are short columns, as the ℓ_u/h ratios in Table 7.1 indicate.

The third curvature case (Fig. 7.7c) applies to *all* columns in an unbraced frame, because under wind or other lateral forces the frame deflects laterally. The top of each column moves relative to the joint at its lower end and this sets up a reversed curvature case with joints at top and bottom rotating so to increase the lateral deflection and the Py moment. Table 7.1 shows all k values greater than unity; only *very* stiff beams could reduce k to 1.0. The same limitation applies to the limiting short column length even if only vertical loads are involved; as slender columns these also fail by becoming unstable and swaying laterally. The Code 10.11.4.2 limit for short columns in the unbraced frame is a $k\ell_u/r$ of 22. Note the low ℓ_u/h ratios (4 to 5) in Table 7.1.

7.6   Frame Loadings for Maximum Column Moments

The designer must distinguish sharply between columns where sidesway is prevented (diagonal bracing, shear walls, and so on) and those free to sway. The ACI Code Commentary Sec. 10.11.2 provides guidance for determining whether a structure is adequately braced or whether it must be considered unbraced. The first method presented requires the computation of lateral deflections using an elastic first-order analysis that neglects $P\Delta$ effects. This procedure is very simple to use with the widespread availability of computer programs for calculating such elastic deflections. The procedure assumes that the secondary moment effects of horizontal displacements will be insignificant in a story in which the stability index Q is not greater than 0.04. Q may be calculated as

$$Q = \frac{\Sigma P_u \Delta_u}{H_u h_s}$$

where ΣP_u is the sum of the factored axial loads on all columns in the story, H_u is the total factored lateral load acting on the story, Δ_u is the elastically computed first-order lateral deflection owing to H_u at the top of the story relative to the bottom of the story, and h_s is the center-to-center story height. A value larger than 0.04 indicates that $P\Delta$ moments may amplify more than 5% and that the story should be considered unbraced.

The Commentary provides an alternate and even more approximate procedure that requires no structural analysis before determining whether a story is braced or unbraced. A column may be assumed braced if it is located in a story in which the sum of the translational stiffnesses of all bracing elements (shear walls, shear trusses or other bracing) is at least six times the sum of the translational stiffnesses of all columns in the story. Such computations are simple to make but the engineer must always apply good judgment in considering the various types of bracing systems and their adequacy.

Three different frame loadings are involved in the three different curvatures of Sec. 7.5. Two of these cases on interior columns occur in a braced frame as shown in Fig. 7.8. For critical stresses in a relatively uniform frame at the point marked x, the loading of Fig. 7.8a gives both a maximum axial load and the maximum moment consistent with it. The column behavior is essentially that of Fig. 7.7b. There is a minimal slender column effect. This case will be referred to as the reversed moment case, whether the reversal is small, as in the loading diagram, or a complete reverse, as it might be on an exterior column. The case shown in the loading diagram might also be called a column restrained at the far end (by the moment resisting the upper joint rotation).

A checkerboard loading adjacent to a column (Fig. 7.8b) produces single curvature in the columns, as in Fig. 7.7a. Fortunately, this curvature requires that some potential load be omitted from the floor above and analysis shows smaller end moments in this case than in the one previously discussed. Single curvature can accompany maximum axial load only when dealing with unequal spans, and then usually with lower moments than for the case of far end restrained. Nevertheless, the slender column effect is large at x and could make this govern on very slender columns or some unequal span situations.

In the absence of shear walls, sidesway may permit a whole story height in Fig. 7.8a or 7.8b to deflect laterally and shift the moment conditions of Fig. 7.7a,b to one very nearly like Fig. 7.7c. Symmetry of loading or frame delays this, but elastic analysis shows sidesway to be the expected collapse mode for the entire story. Length effects that might be small without sidesway would then become large when sidesway develops.

The sidesway case under shear load is worst when the shear is maximum. The distribution of these shears to the individual columns requires analysis of the entire story rather than individual columns and the Code specifies use of the

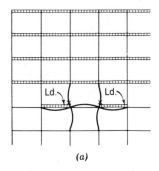

(a)

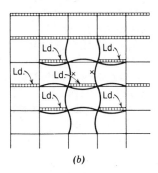

(b)

Figure 7.8 Column moments without sidesway. (a) Maximum load and moment loading but minimal long column effect. (b) Single curvature loading.

entire story in computing the slender column effects (Sec. 7.12). Although the sidesway moment gives the most severe slender column effect, in a low structure the basic moments may still be small enough to make the vertical loading control the design.

7.7   Code Methods for Slenderness Effects

(a) Frame Analysis Approach   Because of the interrelationship between the columns and their restraining beams, the slender column effect is always basically the problem of correcting or improving the results of an ordinary frame analysis. The Code (10.10) clearly encourages the designer to evaluate the design forces and moments from an improved analysis of the structure, stating:

> Such analysis shall take into account influence of axial loads and variable moment of inertia on member stiffness and fixed-end moments, effect of deflections on moments and forces, and the effects of duration of loads.

Beyond these general requirements, the Code Commentary speaks of a second-order analysis, including the effects of sway deflections on the axial loads and moments. It also recommends realistic moment-curvature or moment-end-rotation relationships, finally stating five controls, including:

> In lieu of more precise values, it is satisfactory to take EI as $E_c I_g(0.2 + 1.2\rho_t E_s/E_c)$ in computing the column stiffnesses and $0.5 E_c I_g$ when computing the beam stiffnesses.
> It is necessary to consider the effect of axial loads on the stiffness and carry-over factors for very slender columns ($\ell/r > 45$).

Analyses such as these are basically computer problems and beyond the scope of this book. The use of such programs is gradually increasing because such use avoids the many detailed requirements of the moment magnifier method covered in Code 10.11. Some designers have developed procedures that are very crude in terms of reflecting the complex variations of stiffness and moments present in slender columns. Prudence would indicate that any second-order analysis procedure developed for use under Code Sec. 10.10 should be checked against the types of experiments and computer solutions[2] used to verify the more approximate moment magnification procedures on which Sec. 10.11 is based. This check is particularly important for procedures treating unbraced frames where the action of restraining flexural members is so critical.

(b) Moment Magnifier as an Approximate Evaluation   In the mechanics of elastic members a member that is bent in a specific shape, either from initial crookedness or some external loading, will have its deflection increased by the addition of an axial load P acting along the reference chord, from an initial y_0 to y_{max}. A reasonable approximation is

$$y_{max} = y_0/(1 - P/P_c)$$

where P_c is the critical buckling load for that member (Euler load). In the hinged column of Fig. 7.3b, with $M_1 = M_2 = Pe$.

$$
\begin{aligned}
M_{max} = M_1 + Py_{max} &= M_1 + Py_0/(1 - P/P_c) \\
&= [M_1(1 - P/P_c) + Py_0(M_1/Pe)]/(1 - P/P_c) \\
&= M_1(1 - P/P_c + y_0/e)/(1 - P/P_c)
\end{aligned}
$$

Since $\quad y_0 = M_1\ell_u^2/8EI = Pe\ell_u^2/8EI$, and $P_c = \pi^2 EI/\ell_u^2$,

$$
\begin{aligned}
M_{max} &= M_1(1 - P/P_c + P\ell_u^2/8EI)/(1 - P/P_c) \\
&= M_1(1 - P/P_c + \pi^2 P/8P_c)/(1 - P/P_c) \\
&= M_1[1 + (P/P_c)(\pi^2/8 - 1)]/(1 - P/P_c) \\
&= M_1(1 + 0.23 P/P_c)/(1 - P/P_c) \doteq M_1(1 - P/P_c).
\end{aligned}
$$

For the single curvature case the error in omitting $0.23 P/P_c$ varies from 2.3% when $P/P_c = 0.1$ to 11.5% when $P/P_c = 0.5$; for other initial curvatures it can be either more or less. Because stability failures can be sudden and calamitous, it is desirable to have a substantial margin between P_u and P_c. Thus, the values of P_u/P_c are usually substantially below 0.5 and the errors can be safely neglected.

The Code Sec. 10.11 is based on the concept of a moment magnification factor δ that, in an oversimplified view, for the single curvature case with equal end moments can be given as

$$\delta = \frac{1}{1 - P/P_c}$$

Design of a slender column utilizes the basic assumptions of Code Sec. 10.2 and the general principles of Code Sec. 10.3 which result in the familiar interaction diagrams for failure states with combined axial load and moment presented in Chapter 6. Although the axial load P_u is determined from the ordinary analysis, the design moment M_u must be based on an amplified or magnified moment $M_c = \delta M_2$. M_2 is the larger end moment on the column as calculated by a conventional elastic frame analysis. In this way the nominal or Pe moment shown as the dashed line in Fig. 7.6b can be magnified to include the $P\Delta$ effects and results in a relation between load and moment almost identical to that shown by the heavy curve of Fig. 7.6b. Note that the value of δ changes as P increases. Thus the difference between the dashed and solid curves increases rapidly with higher loads as P approaches P_c. When $P = P_c$, the value of δ goes to infinity and a stability failure occurs. Values of $P > P_c$ and consequent negative values of δ are physically impossible and simply signify that failure has occurred. The Code does not permit reduction of moment where δ is less than one.

Because the critical Euler load P_c depends on the column cross section stiffness and because of tolerances, material variations, and other uncertainties, the actual P_c may be lower than the theoretical or nominal P_c, a ϕ factor must be applied. Because stability failures are catastrophic, a good degree of safety is desired. The ϕ factor values for stability are the same as those associated with tied or spiral columns. The basic modifier becomes

$$\delta = \frac{1}{1 - P_u/\phi P_c} \geqslant 1.0$$

where P_c is the familiar Euler critical buckling load

$$P_c = \frac{\pi^2 EI}{(k\ell_u)^2}$$

In actual application (Code 10.11.5.1) distinction is made between moment magnifiers for the braced condition and for the unbraced condition (see Sec. 7.12).

If $M_1 \neq M_2$, the maximum moment is not at midheight initially. The largest increase is near midheight, resulting in a smaller total at the new maximum point. The Code cares for this by including the factor C_m which takes care of different shapes of moment diagrams and shifts in points of maximum moment. Based on the larger end moment M_2,

$$M_{max} = M_2[C_m/(1 - P/\phi P_c)] = M_2 \delta$$

where δ is the quantity in brackets and is called the moment magnifier or amplifier.

$$\delta = \frac{C_m}{1 - P_u/\phi P_c} \geqslant 1.0$$

In braced frames for members without transverse loads between supports, C_m may be taken (Code 10.11.5.3) as

$$C_m = 0.6 + 0.4 M_1/M_2 \geqslant 0.4$$

Here M_1 is positive for a single curvature case, negative for a double curvature case, and the moment ratio itself is always in the range between $+1$ and -1. In unbraced frames C_m is always taken as 1.0.

With this procedure the slender column must then be designed for P_u and $M_2 \delta = M_c$, where δ is always greater than unity but may be used as unity for "short columns," a term for which exact boundaries were set in Sec. 7.5. However, the moment magnification procedure may not be used for columns with $k\ell_u/r$ greater than 100. Extremely slender columns must be based on an improved analysis (Code 10.11.4.3).

One important consideration in adopting this moment magnifier method was its use in structural steel design. The designer must be alert, however, for detailed differences in values recommended for C_m and $k\ell_u$, and the addition of a new term β_d (Sec. 7.10) for creep.

7.8 Adapting the Moment Magnifier to Concrete Columns in Frames—General

The moment magnifier is based on an analysis of an elastic curve that increases amplitude but does not change in shape as axial load is applied at the column ends or at the joints of the frame. Although for concrete the deflected shape will change, this normally does not involve serious error.

In a general frame, five important problems arise; the last three are not limited to reinforced concrete.

1. The effective EI of reinforced concrete is dependent on the magnitude and type of loading as well as the materials, and varies along the column length (Sec. 7.9).
2. The creep of concrete occurs when load is sustained and creep modifies the effective EI needed in item 1 (Sec. 7.10).
3. The effective length of column for the calculation of P_c may be either more or less than ℓ_u (Sec. 7.11).
4. The magnified moments must distinguish between sway-producing and nonsway-producing loadings. Column sway is limited to the story deflection (Sec. 7.12).
5. The maximum moment does not always occur at midheight (Sec. 7.13).

The answers provided for these problems by any approximate method cannot give an exact result; there are too many basic variables. For example, the amount of beam steel or the degree of beam cracking alone can create a scatter of at least 10% in column strength.

7.9 Effective *EI* for Magnifier Calculation

In a linear elastic medium, stiffness against curvature is measured by EI. In a reinforced concrete column, stiffness decreases (1) as concrete in compression approaches the flatter part of the stress-strain curve, (2) as creep develops under sustained load, (3) as compression steel yields, or (4) as the concrete cracks (for larger eccentricities). In actuality, the moment-curvature relationship for concrete columns varies greatly with the level of axial load. The elastic relationship for curvature

$$\phi = \frac{M}{EI} \quad \text{or} \quad EI = \frac{M}{\phi}$$

becomes a complex nonlinear one. For simplification the actual M/ϕ relationship is approximated by crude linear EI expressions as shown in Fig. 7.9. The varying degree of stress distribution over the column length means no single value of EI can be a true one for use under all types of loading. In addition, in the design process the actual amount of reinforcement is unknown at the begin-

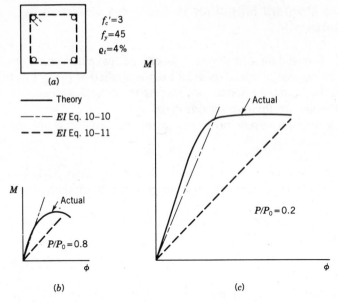

Figure 7.9 The approximation of the stiffness formula for reinforced concrete columns. (*a*) Column cross section and properties. (*b*) Heavy axial loads. (*c*) Light axial loads. (From Reference 2, *ACI Jour.*)

ning of the analysis. Because the actual stiffness is greatly affected by amount of steel, use of the equations is often an iterative procedure.

Especially where ρ_t of the column is small (0.01 or up to 0.02 for small columns) and especially for columns that nearly qualify as short columns (where the moment magnifier will be small anyway), Code 10.11.5.2 suggests use of Code Eq. 10.11. This expression presents a crude approximation that is independent of the actual reinforcement

$$EI_1^* = (E_c I_g/2.5)/(1 + \beta_d)$$

where β_d is an allowance for creep discussed in Sec. 7.10. This equation greatly underestimates EI where the value of ρ_t is large, leading to over-design of the column. For cases having a large ρ_t, unless the column length is still so short as to make the calculation unimportant, economy dictates the use of Code Eq. 10.10 for EI:

$$EI_2^* = (E_c I_g/5 + E_s I_s)/(1 + \beta_d)$$

With either equation the scatter is broad, as shown in Fig. 7.10, especially so for EI_1. The scatter is essentially all on the safe side. Reference 4 suggests a reduction in these EI values to account for sustained load effects.

---

* The EI_1 and EI_2 subscripts are not in the Code but are convenient for reference.

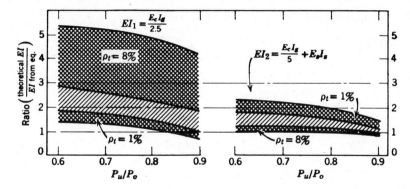

Figure 7.10. The accuracy of the formula values of the *EI* equations. The horizontal line at unity would be a perfect fit. (Modified from Reference 2, *ACI Jour.*)

It is emphasized that the Code recognizes both equations for *EI* and either may be used safely in design. The more approximate, but simpler Eq. 10.11, which neglects the contribution of reinforcement, will result in extremely conservative (small) *EI* values for larger percentages of reinforcement. This in itself may not result in large amplification of the design moment because the actual load level P_u may still be far below the computed critical load ϕP_c. The designer must develop a "feel" for the process. Generally, on the first iteration, use of Code Eq. 10.11 is acceptable unless ρ_t is desired as large and the column is moderately slender ($k\ell_u/r > 40$). If the computed δ is low ($\delta < 1.2$) there probably is little benefit in using Code Eq. 10.10. If the column is more slender and a higher percentage of steel is likely, then Eq. 10.10 should be used for the first iteration based on the best estimate of column size, cover, and percentage of reinforcement available.

Because the larger *EI*, either EI_1 or EI_2, gives the smaller δ multiplier, and because both are safe values, it is sometimes desirable to know the particular ρ_t that makes EI_2 the larger. The breakpoint will be established[5] by setting the ratio EI_2/EI_1 at unity and considering steel on two faces of a rectangular column:

$$1 = \left(\frac{E_c I_g/5 + E_s I_s}{1 + \beta_d} \right) \div \left(\frac{E_c I_g/2.5}{1 + \beta_d} \right) = \frac{bh^3/60 + (E_s/E_c)\rho_t bh(\gamma h)^2/4}{bh^3/30}$$

where γ (gamma) $= (h - 2d')/h =$ relative distance between the steel on the two faces. Multiplying top and bottom by $30/bh^3$:

$$0.5 + 7.5(E_s/E_c)\rho_t \gamma^2 = 1$$

Let $E_s/E_c = n$, as in transformed area calculations. The resulting equation

$$7.5 n\rho_t \gamma^2 = 0.5$$

based on $EI_1 = EI_2$ indicates that EI_2 is the larger when ρ_t exceeds $1/(15n\gamma^2)$, that is, when the Table 7.2 values of ρ_t are exceeded.

TABLE 7.2 Minimum Values of ρ_t Making EI_2 Govern over EI_1

f'_c	3 ksi	4 ksi	5 ksi
n	9	8	7
ρ_t	$1/(135\gamma^2)$	$1/(120\gamma^2)$	$1/(105\gamma^2)$
$\gamma = 0.6$	0.0205	0.0231	0.0264
0.7	0.0151	0.0170	0.0194
0.8	0.0115	0.0130	0.0149
0.9	0.0091	0.0103	0.0117

For design it is often more convenient to use the ratio of EI_2/EI_1 times an EI_1 value than to start fresh on an EI_2 calculation, that is, to use: $EI_2 = EI_1(0.5 + 7.5 n\rho_t\gamma^2)$ or a similar equation for other type columns. The application of the moment magnification procedure in design is substantially assisted by use of the tables in the ACI Design Handbook.[6]

7.10 The Creep Problem

At $\ell_u/h = 10$ creep effects tend to offset each other, the tendency toward increase of column curvature from creep being offset by the decreasing distribution factor to the column because of its reducing stiffness. Around ℓ_u/h of 16 or 20 it becomes more important, except for columns with reversed moments in braced frames, which exhibit slender column behavior only in extreme cases.

The Code provides for creep by reducing the approximate EI by the $(1 + \beta_d)$ factor in the denominator, where β_d is the ratio of the maximum factored dead load *moment* on the column to the maximum factored total load moment on the column. (It would appear that the ratio of service load moments might be more appropriate, as creep is a function of sustained loading.) This β_d correction is very crude and in many cases may be grossly conservative because lateral loads do not usually have creep-related effects. Fortunately, the effect of β_d (and even of EI) is far removed from the end result of a design. Experience may show that it is only on very unusual slender columns that β_d is really of significance.

7.11 Effective Column Length, $k\ell_u$

In simple mechanics the concept of effective column length is well established; a fixed end column has an effective length of 1/2 its overall height between fixed ends. As shown in Fig. 7.7, the effective length of columns depends both on the degree of rotational restraint ψ at the ends (ψ = ratio of column-to-beam stiffness) and whether the column is braced or unbraced with respect to sway. If it were only a matter of the frame reaction to a statically determined moment applied directly as a load on the column, the effective length would be a relatively simple matter of frame or joint stiffness.

Jackson and Moreland[7] solved the $k\ell_u$ problem in terms of relative member stiffness and published the nomographs of Fig. 7.11. One enters with the values of the relative column stiffness ψ at each end of the column; a straight line between the two ψ values reads k on the center scale. A separate nomograph is required for the braced and unbraced frames. The decision whether the column should be treated as braced or unbraced can be made as discussed earlier in Sec. 7.6. The greater k values for unbraced frames follow from the knowledge that final failure will tend to be a sidesway mode for the entire story.

Use of the nomograph is convenient with manual calculations but impractical with electronic computation. The 1983 ACI Code Commentary provided alternate equations for determination of effective length based on the 1972 British Standard Code of Practice[8] and the work of Furlong[5] and Cranston.[9]

For braced compression members, k may be found as the smaller of

$$k = 0.7 + 0.05(\psi_A + \psi_B) \gtrless 1.0$$
$$k = 0.85 + 0.05\psi_{min} \gtrless 1.0$$

ψ_A and ψ_B are the values at the two ends of the column, and ψ_{min} is the smaller of the two values.

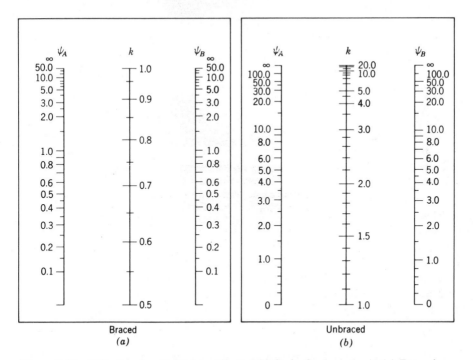

Figure 7.11 Effective length factors (from ACI Code Commentary). (*a*) Braced frames. (*b*) Unbraced frames. ψ = ratio of $\Sigma EI/\ell_c$ of compression members to $\Sigma EI/\ell$ of flexural members in a plane at one end of a compression member; k = effective length factor.

For unbraced compression members rotationally restrained at both ends, k may be found as

For $\psi_m < 2$

$$k = \frac{20 - \psi m}{20} \sqrt{1 + \psi_m}$$

For $\psi_m \geqslant 2$

$$k = 0.9\sqrt{1 + \psi_m}$$

ψ_m is the average of the ψ values at the two ends.

For unbraced compression members hinged at one end, k may be found as

$$k = 2.0 + 0.3\psi$$

ψ is the value at the rotationally restrained end. A cantilever column with full fixity at one end has $k = 2.0$.

One of the basic problems is that in reinforced concrete, the true ψ changes as beams crack, as column deflection builds up, as column concrete in compression enters the flatter part of the f_c stress-strain curve, and as compression steel yields. These complications combine with the similar uncertainties in the value of EI just discussed.

The effect of *cracked* beams on the effective length is important in slender columns, possibly in the sidesway case more important than its effect on end moments generally. The Code (10.11.2.2) specifies:

> For compression members not braced against sidesway, effective length factor k shall be determined with due consideration of cracking and reinforcement on relative stiffness, and shall be greater than 1.0.

The best thinking for this sidesway case at present as recommended in the Commentary is that relative stiffness should be based on the EI_{cr} for the cracked beam (a simple transformed area) and for the column on EI_2 as in Sec. 7.9, with $\beta_d = 0$. For columns that have only moderate slenderness ($k\ell_u/r < 60$) this can be further simplified by basing relative stiffness values on $0.5EI_g$ for flexural members and EI_g for compression members. With ψ calculated reflecting beam cracking, the value of k for $k\ell_u$ in the sidesway case may be substantially higher. Because P_c varies with $1/(k\ell_u)^2$, the difference can be important.

At column footings in an unbraced frame, the AISC (for steel columns) recognizes the possibility of fixed connections justifying a ψ of 1.0; but at the same time it recognizes the usual lack of frictionless pin connections by limiting the maximum ψ to 10. The designer of reinforced concrete should also consider the need for some limit not representing complete fixity at footings.

7.12 Column Sway Limited to Story Deflection

Except in isolated cantilever columns, when a single column sways, an entire floor or level must move relative to another. This story sidesway is also a typical mode of failure in an unbraced frame under vertical load alone. A floor system is typically stiff enough to make all columns deflect alike, unless a torsional loading adds a rotation to the structure. Thus each column is not free to deflect independently as required to carry a fixed shear. Instead, the shear carried is a function of its stiffness relative to all the columns in the story. This discussion will omit the more complex effect of wind on unsymmetrical structures that can result in a torsional rotation; even this case marshals the unequal sway of the columns into an ordered, interrelated group movement that depends on the group stiffness.

Thus there are few cases where a column free to sway can be designed by itself; a single column supporting a hyperbolic paraboloid may qualify. Possibly also a bent or a series of single bents carrying a roof with overhanging ends and not much eccentricity of loading on the columns would qualify in that parts of the system might all tend to react alike. Typically, there will be different column sizes, different beam restraints or, if nothing more, exterior columns braced by one beam combined with interior columns braced by two beams at each floor level.

In an unbraced story height the stiffest column tends to pick up horizontal load, possibly more than the designer tends to assign to it, and the flexible column is braced by the limited deflection of the stiffer column. Considerable redistribution of the column shears occurs before story failure.[10–12] Any design method for lateral loading involves an initial distribution of the horizontal shear to the individual columns, that is, some type of frame analysis. Code 10.11.5.1 requires that in unbraced frames a common moment magnifier δ_s be computed for the story and used for all columns in that story. This δ_s is to be based on $\Sigma P_u / \Sigma P_c$, with the Σ representing a summation over all the columns in that story, including any short columns.* In determining k for use in this P_c calculation, the relative column stiffness ψ should be based on the cracked section of the beam and the reinforcement ratios of beam and column (Code 10.11.2.2), as discussed previously.

The actual data required for the P_c calculations are available only *after* the beams and columns have been designed, which means the design method must start more approximately. In a sense this is true of nearly all design situations, but here the approximations may initially have to be rougher than usual until design experience builds up in this area.

To start the column design in this sidesway case one could (1) make a rough estimate of all columns and beam sizes involved, (2) arbitrarily modify their relative stiffnesses at joints by some typical factor to represent beam cracking† and column reinforcement percentages expected to be used, and on this basis (3) make an initial trial for $\Sigma P_u / \Sigma P_c$. A reference chart such as Fig. 7.12 would be helpful, although limited here to a particular f'_c, γ, and column steel arrangement. The ψ design value of this chart is based on equal gross I/ℓ of column and

---

* The ϕ in the denominator does stay in the equation for δ.
† Figure B.7, Appendix B, might help here.

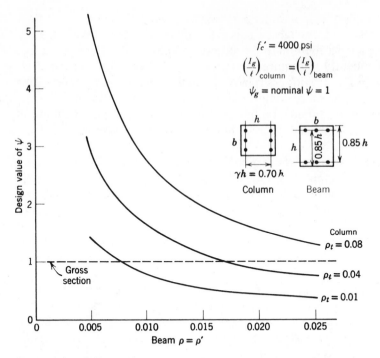

Figure 7.12 Effect of percentage of column and beam reinforcement and beam cracking on the design ψ for $\gamma = 0.70$ and $f_c' = 4000$ psi.

beam. Similar charts can be made for other gross ψ values, and this chart itself can be used as a multiplier of any nominal ψ in roughly estimating the second step, because it is in effect a ratio term. As indicated in the previous section, for members with moderate slenderness using $0.5I_g$ for beams and I_g for columns will usually result in reasonable preliminary design sizes.

Alternatively, designers might compare this story with some other they have computed and simply estimate a trial δ for initial use. Tentative column designs may then be made with the trial δ until it is possible to get a better feel for the real $\Sigma P_u/\Sigma P_c$ for the story; then the better δ value will determine whether some revisions are essential.

This story system is more complex than individual column designs, but also more realistic and in the long run more economical. It is also probable that the story δ will act as a form of stabilizer in design calculation. A change in a single-column size will modify the story δ less than it would the individual member δ, and changes in all the column sizes should not usually be in the same direction.

The story procedure thus far described does not guarantee an individual slender column against overload. Such a column design might be too weak to handle the maximum moment and vertical load in a *braced* frame. The 1983 Code adopted a new procedure for treating magnification of nonsway-producing load effects in unbraced frames. This procedure is discussed in Sec. 7.14.

In a laterally unbraced frame, the rigidity of the restraining beams is an essential element of the frame stability. Figure 7.5c indicates that even if the columns in a sway frame are infinitely rigid, a parallelogram motion occurs owing to the deformation of the restraining beams. If plastic hinges form in the restraining beams, the structure approaches a mechanism. The loss of restraint greatly increases the column effective length, which reduces stability dramatically.[8] It is obvious from Fig. 7.6a and Fig. 7.6b that the greatly amplified moment at the column end must be equilibrated by an equally amplified moment at the beam end. Code 10.11.6 requires that flexural members in unbraced frames must be designed for the total *magnified* end moments of the compression members at the joint.

7.13   Maximum Moment Away from Mid-Height

The use of C_m has already been noted in Sec. 7.7b along with the governing equation, which is to be used only where sidesway is prevented and where no lateral load acts on the column. This equation is a straight-line approximation to a curve originally developed by Massonnet from steel column tests.

In sidesway under a shear loading, the worst moment stays at the end of the column and the maximum deflection is also there, making $C_m = 1$ appropriate.

In the reversed curvature case without sidesway (Fig. 7.2), which looks in the moment diagram slightly similar to the sidesway case, the deflection adds no *end* moment. Hence behavior is that of a short column until deflection can build up a larger moment away from the joint. Most such practical columns would actually be short columns.

7.14   Analysis of Slender Columns

The easiest way to achieve basic familiarity with the approximate column slenderness provisions of Code 10.11 is to use them in the analysis of a given column. The practical application in design is more complex and is treated in later sections.

In the analysis of a given slender column with given restraint conditions, the practical question is what is the axial load and moment capacity under the amplified moment, and hence, eccentricity conditions. The basic failure state for the column *cross section* is still determined by the general procedures presented in Chapter 6 and the column *cross section* interaction diagrams such as those presented in Fig. 6.18 are still completely valid and highly useful. To illustrate this, all problems in this chapter will utilize those interaction diagrams. In practice, the designer has a large number of such interaction diagrams available and with modern computers it is simple to create such diagrams for any type of column envisioned.

Any slender column analysis should begin with a check of the translational (lateral) and rotational restraint conditions and a preliminary check for approximate slenderness, $k\ell_u/r$. The effective length factor k is determined as explained in Sec. 7.11. The unsupported length ℓ_u is basically the clear distance between lateral supports and is defined in Code 10.11.1. The radius of gyration r is approximated as $0.3\,h$ for rectangular members and $0.25\,h$ for circular members (Code 10.11.3). The short column limit outlined in Sec. 7.5 can then be checked to see if further checks of slenderness effects are necessary.

If appreciable slenderness effects exist, the moment magnification procedures of Code 10.11.5 can be used to evaluate them. In general, there are two possible approaches to such analysis problems. Figure 7.13a shows the usual case. Because the given column cross-sectional dimensions, reinforcement, and material properties are known, the column interaction diagram is readily obtained. Separate analytical expressions can be used for the compression branch and the tension branch if computer methods are desired. The first-order

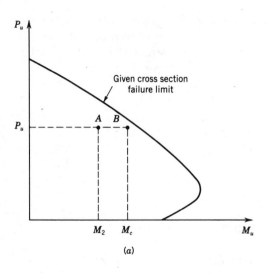

(a)

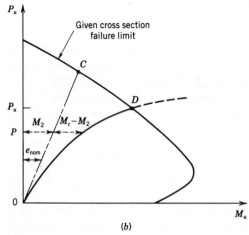

(b)

Figure 7.13 Moment magnification analysis. (a) Determining adequacy of computed P_u, M_c. (b) Determining P_u for given eccentricities.

analysis values of P_u and M_2 are known from the basic structural analysis. M_2 is the larger valued end moment on the column. In a braced frame it can be the total end moment. In an unbraced frame it must be separated into M_{2b}, the maximum end moments for loads resulting in no appreciable sidesway, and M_{2s}, the maximum end moments for loads resulting in appreciable sidesway, respectively. This separation is necessary because the 1986 Code version of the moment magnification procedure requires the conservative approximation of the second-order moments by magnifying the moments from loads that produce sway by δ_s, the unbraced frame magnifier, and from loads that would not produce appreciable sway by δ_b, the braced frame magnifier.

$$\delta_b = \frac{C_m}{1 - P_u/\phi P_c} \geq 1.0$$

$$\delta_s = \frac{1}{1 - \Sigma P_u/\phi \Sigma P_c} \geq 1.0$$

$$P_c = \frac{\pi^2 EI}{(k\ell_u)^2}$$

In determining δ_b, P_c must be based on effective length factors for braced frames. In determining δ_s, P_c must be based on effective length factors for unbraced frames.

The actual moment to be used in design M_c is then determined as

$$M_c = \delta_b M_{2b} + \delta_s M_{2s}$$

As shown in Fig. 7.13a, the first-order analysis values (P_u, M_2) of point A are then corrected to the actual values (P_u, M_c) of point B. Although the moment is magnified, the axial load remains the same. Point B is checked against the cross section capacity. If point B does not fall outside the interaction diagram, it is acceptable.

The other analysis approach is shown in Fig. 7.13b. Imagine a column with a given set of loading conditions such that the nominal end eccentricity is constant, e_{nom}. Again, all cross section and material properties of the column are known so that the interaction diagram is obtainable. The first-order end moment M_2 equals Pe_{nom}. If there are no slenderness effects the column develops the capacity of point C. Owing to slenderness, however, the $P–\Delta$ moments can appreciably increase the moment on the critical cross section to M_c. Note that $M_c - M_2$ is $P\Delta$. The column fails when the actual load path reaches point D. To determine this point, the easiest solution is to crudely plot the load path curve OD by choosing several values of P. Each value selected will result in a different value of δ. These values can be used to find M_c for that value of P. The process rapidly closes on point D.

7.15 Slender Column Analysis Examples

It is essential that column lateral and rotational restraints be carefully considered in every slender column analysis. To illustrate the computational procedures and simplify the reader's task in the initial presentations, however, the first two examples are presented with only relative stiffness values ψ. The two subsequent examples illustrate the computation of ψ for various cases, although even in those examples it is assumed that the frame properties will not have to be changed as they might in practice if initial estimates of column or beam sizes prove inadequate.

(a) Given the tied column cross section shown in Fig. 7.14a with 1-$\frac{1}{2}$-in. clear cover for interior exposure. f_c' is 4 ksi and Grade 60 reinforcement is used. The column is used in a braced frame with $\psi_{TOP} = \psi_{BOT} = 1.0$. All bending is about the x–x axis. The column is very tall with an unsupported length of 28 ft. The designer has determined it is critical when loaded for

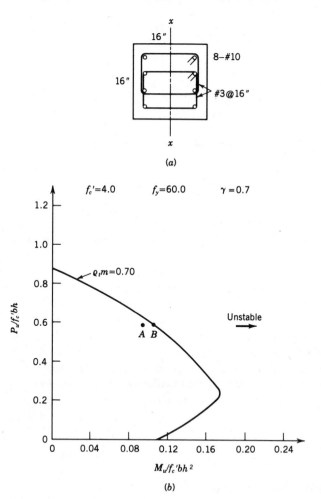

(a)

(b)

Figure 7.14 Column for Examples 7.15(a) and (b). (a) Column cross section for examples (a) and (b). (b) Column interaction chart, $\gamma = 0.7$, ϕ included but not correctly below ordinate of 0.10. (From Fig. 6.18b).

maximum column load with the far end fixed ($M_1 = -\frac{1}{2}M_2$). Unfactored loads produce $P_D = 250$ k, $P_L = 150$ k, $M_D = 400$ k-in. and $M_L = 600$ k-in. Is the column cross section shown adequate for these loads?

Solution

$$P_u = 1.4 \times 250 + 1.7 \times 150 = 605 \text{ k}$$
$$M_u = 1.4 \times 400 + 1.7 \times 600 = 1580 \text{ k-in.}$$
$$\gamma = \frac{16 - 2(1.5) - 2(0.38) - 1.27}{16} = 0.69 \qquad \text{USE } \gamma = 0.7 \text{ chart}$$
$$\rho_t = \frac{(8)(1.27)}{(16)^2} = 0.040 \qquad \rho_t m = \frac{(0.040)(60)}{(0.85)(4)} = 0.70$$

From Fig. 6.18b, select the $\rho_t m = 0.70$ curve that gives the basic failure state of the cross section and already includes the ϕ values for the cross section (but not the ϕ value for the stability calculation). (See Fig. 7.14b.)

If there is no slenderness effect, then

$$P_u/f_c'bh = \frac{605}{(4)(16)(16)} = 0.59 \qquad \frac{M_u}{f_c'bh^2} = \frac{1580}{(4)(16)(16)^2} = 0.096$$

This condition is plotted as A on Fig. 7.14b. It is inside the interaction curve so the cross section is more than adequate for a short column.

To check whether slenderness effects must be considered at all, the $k\ell_u/r$ ratio is evaluated.

$k = 0.78$ for a braced frame with $\psi_{TOP} = \psi_{BOT} = 1.0$ from Fig. 7.11.

$r = 0.3h$ for a rectangular column. $\qquad r = (0.3)(16) = 4.8$ in.

$\ell_u = 28 \times 12 = 336$ in.

$$\frac{k\ell_u}{r} = \frac{(0.78)(336)}{4.8} = 55 > 34 - 12(M_1/M_2) = 40.$$

$\therefore$ Must check column slenderness effect

$$C_m = 0.6 + 0.4(M_1/M_2) = 0.6 - 0.2 = 0.4$$
$$\beta_d = M_{uD}/M_u = (1.4 \times 400)/1580 = 0.35$$
$$E_c = 57{,}000\sqrt{4000} = 3.60 \times 10^3 \text{ ksi} \qquad I_g = \frac{(16)(16)^3}{12} = 5460 \text{ in.}^4$$

EI may be found from either Code Eq. 10.10 or Code Eq. 10.11. Both will be illustrated in this example.

First check Code Eq. 10.11, even though $\rho_t = 0.04$ since slenderness seems quite moderate (55 versus 40).

$$EI = \frac{E_c I_g}{(2.5)(1 + \beta_d)} = \frac{(3.60 \times 10^3)(5460)}{(2.5)(1 + 0.35)} = 5.82 \times 10^6 \text{ k-in.}^2$$
$$P_c = \frac{\pi^2 EI}{(k\ell_u)^2} = \frac{\pi^2(5.82 \times 10^6)}{(0.78 \times 336)^2} = 836 \text{ k}$$

$P_u = 605$ k is *greater* than $\phi P_c = (0.7)(836) = 585$ k. (The ϕ used here for stability is that for a tied column.) When the actual load is greater than the critical load, the column is unstable and fails. This is shown by formation of a *negative* multiplier value δ.

$$\delta_b = \frac{C_m}{1 - \dfrac{P_u}{\phi P_c}} = \frac{0.4}{1 - \left[\dfrac{605}{(0.7)(836)}\right]} = -11.8$$

Even if the numerical value magnitude was small, the negative sign is a warning that the design critical load ϕP_c has been exceeded. The magnified moment would plot off the graph as shown by the arrow on Fig. 7.14b. In actuality the real column is *not unstable*. The ultra-conservative use of Code Eq. 10.11 greatly underestimated the actual stiffness and hence P_c.

This can be seen clearly when the same problem is reworked using Code Eq. 10.10 which better reflects the stiffness contribution of the large amount of reinforcement ($\rho_t = 0.04$).

$$E_s = 29 \times 10^3 \text{ ksi} \qquad I_s = A\bar{d}^2 = (8)(1.27)\left(8 - 1.5 - 0.38 - \frac{1.27}{2}\right)^2$$

$$= 306 \text{ in.}^4$$

$$EI = \frac{\dfrac{E_c I_g}{5} + E_s I_s}{1 + \beta_d} = \frac{\dfrac{(360 \times 10^3)(5460)}{5} + (29 \times 10^3)(306)}{(1 + 0.35)}$$

$$EI = 9.48 \times 10^6 \text{ k-in.}^2$$

$$P_c = \frac{\pi^2 (9.48 \times 10^6)}{[(0.78)(336)]^2} = 1362 \text{ k}$$

$$\delta_b = \frac{0.4}{1 - \dfrac{605}{(0.7)(1362)}} = 1.095 \qquad M_c = \delta_b M_{2b} = (1.095)(1586)$$

$$= 1730 \text{ k-in.}$$

$$P_u/f_c' bh = 0.59 \qquad \text{and} \qquad M_c/f_c' bh^2 = 1730/(4)(16)(16)^2 = 0.106$$

These values are shown as B on Fig. 7.14b. They are right on the interaction curve; this column is **OK** under these loads. When the more accurate EI expression was used, the magnifier δ_b is modest (1.1). When δ values exceed about 2, extreme care should be taken to ensure that the structure is stable, even if moment values are small.

(b) Given the same cross section and materials as shown in (a). The slender column is one of three identically restrained and loaded columns in a story of an unbraced frame. Assume a loading pattern such that the nominal eccentricity is constant about the x-x axis and equal to 3 in. Thus

$M_2 = P_u e = P_u \times 3$ in. Furthermore assume that all columns have unsupported lengths of 12 ft in the unbraced story and that the relative stiffnesses are $\psi = 3$ at the column top and $\psi = 1$ at the column bottom. What is the maximum factored load P_u that the column can carry under these conditions?

Solution

The strength of the column cross section is again given by the same interaction curve for $\rho_t = 0.7$. (See Fig. 7.15.) Because these columns are unbraced, Fig. 7.11 shows that $k = 1.57$ for $\psi_{\text{TOP}} = 3$ and $\psi_{\text{BOT}} = 1$. $k\ell_u/r = (1.57)(144)/4.8 = 47.1$. This is fairly slender as the short column limit is $k\ell_u/r = 22$ for unbraced columns.

Because the column is unbraced and $\rho_t = 0.04$, it is prudent to use Code Eq. 10.10 for EI. From part (a), $EI = 9.48 \times 10^6$ k-in.^{2}.

$$P_c = \frac{\pi^2 EI}{(k\ell_u)^2} = \frac{\pi^2(9.48 \times 10^6)}{[(1.57)(144)]^2} = 1830 \text{ k}$$

For the sway case, the sway magnifier must be used.

$$\delta_s = \frac{1}{1 - \dfrac{\Sigma P_u}{\phi \Sigma P_c}} \geq 1.0$$

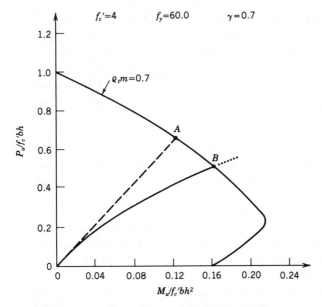

$f_c' = 4 \qquad f_y = 60.0 \qquad \gamma = 0.7$

$\rho_t m = 0.7$

Figure 7.15 Interaction curve for Example 7.15b.

Because in this case all three columns in the story are identical and are loaded identically, $\Sigma P_u / \Sigma P_c$ may be replaced by P_u / P_c.

$$\delta_s = \frac{1}{1 - \dfrac{P_u}{(0.7)(1830)}} = \frac{1}{1 - \dfrac{P_u}{1281}}$$

If there were no slenderness effects, the nominal eccentricity of 3 in. would result in the dashed loading path $0A$ in Fig. 7.15 with $e/h = 3/16 = 0.188$. The maximum load as a short column is found as $P_u = 676$ k as indicated by A. Because of the $P\Delta$ or slenderness effect, the nominal moment is magnified and load path $0B$ results. Points on $0B$ are found by substituting various values of P_u in the δ_s equation using trial and error. Then $M_u = M_c = \delta_s M_2 = \delta_s P_u e$.

For example,

			$\dfrac{P_u}{f_c' bh}$	$\dfrac{M_u}{f_c' bh^2}$
Try $P_u = 300$	$\delta_s = 1.30$	$M_c = 1170$	0.293	0.071
$P_u = 400$	$\delta_s = 1.45$	$M_c = 1740$	0.391	0.106
$P_u = 500$	$\delta_s = 1.64$	$M_c = 2460$	0.488	0.150
$P_u = 550$	$\delta_s = 1.75$	$M_c = 2890$	0.537	0.176

(This latter value can be seen to exceed the interaction curve.)

Try $P_u = 520$	$\delta_s = 1.68$	$M_c = 2620$	0.508	0.160

This latter set of values is shown as B and is the intersection with the interaction curve when P_u is the highest value allowable. Thus $P_{max} = 520k$, $M_{2s\,max} = 520 \times 3 = 1560$ in.-k and $M_c = 2620$ in.-k. As mentioned in Sec. 7.12, Code 10.11.6 requires that the restraining beams in the frame be designed to resist this magnified moment, $M_c = 2620$ k-in. Thus the restraining beams will need substantially more reinforcement because the joint moment increased by 68% owing to the column slenderness effects. When the magnifier concept was originally introduced many designers complained that the new demands on strengthening the floors were uneconomical. In reality, they were dealing with structures that were relying heavily on the floor systems for stability. This was unrecognized in earlier codes. Often the best solution in such cases is to provide bracing through shear walls or other similar means.

(c) Given the braced frame shown in Fig. 7.16a which is subjected to vertical loading. The frame is symmetrical and all members have the same external cross-sectional dimensions, 16-in. × 16-in. All distances shown on Fig. 7.16a are center-to-center of joints. Forces determined from a conventional first-order analysis indicate that a checkerboard loading that produces maximum column moment about the x-x axis but at less than maximum column axial force will control. For such a loading, $M_1 = M_2$.

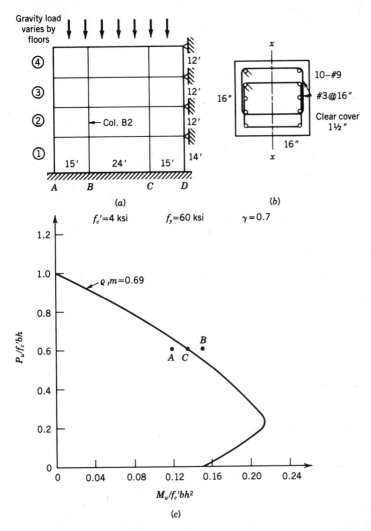

Figure 7.16 Example 7.15c. (a) Braced frame. (b) Column B.2 (c) Column interaction curve, $\gamma = 0.7$. (From Fig. 6.18b).

Results of the analysis with unfactored loads indicate that for column B2, $P_D = 200$ k, $P_L = 200$ k, $M_D = 25$ k-in., and $M_L = 75$ k-in. $f'_c = 4$ ksi, Grade 60 reinforcement is used, and the column cross section is as shown in Fig. 7.16b. Check to see if the column is adequate for this loading.

Solution

$P_u = (1.4)(200) + (1.7)(200) = 620$ k
$M_u = (1.4)(25) + (1.7)(75) = 162.5$ k-ft $= 1950$ k-in.

$$\gamma = [16 - 2(1.5) - 2(0.38) - 1.12]/16 = 0.695 \quad \text{USE } \gamma = 0.7 \text{ chart}$$
$$\rho_t = (10)(1.00)/(16)^2 = 0.039 \quad \rho_t m = (0.039)(60)/(0.85)(4) = 0.69$$
$$E_c = 57{,}000\sqrt{4000} = 3.605 \times 10^3 \text{ ksi}$$

From Fig. 6.18b, select the $\rho_t m = 0.69$ curve (by interpolation) that gives the cross section failure state and includes cross section ϕ factors. (See Fig. 7.16c.)

If there is no slenderness effect, $P_u/f_c' bh = 620/(4)(16)(16) = 0.605$ and $M_u/f_c' bh^2 = 1950/(4)(16)(16)^2 = 0.119$. This condition is plotted as A on Fig. 7.16c. Thus the cross section is adequate as a short column.

To check whether slenderness effects must be considered, the $k\ell_u/r$ ratio is evaluated. From Fig. 7.11, k depends on the relative column to beam stiffness ψ at top and bottom. These values must be computed. Because the column does not seem overly slender, $\ell_u/h = [(12)(12) - 16]/16 = 8$, it is assumed that $k\ell_u/r$ will not exceed 60. If so, the relative stiffness can be approximated by using I_g for columns and $I_g/2$ for beams.

$$I_g = (16)(16)^3/12 = 5460 \text{ in.}^4 \quad I_g/2 = 2730 \text{ in.}^4$$
$$\ell_u = 144 - 16 = 128 \text{ in.} \quad r = 0.3h = (0.3)(16) = 4.8 \text{ in.}$$

For column B2,

$$\psi_{\text{TOP}} = \cfrac{\Sigma \dfrac{EI}{L} \text{ columns (JT B23)}}{\Sigma \dfrac{EI}{L} \text{ beams (JT B23)}} = \cfrac{E_c \left(\dfrac{5460}{12 \times 12} + \dfrac{5460}{12 \times 12} \right)}{E_c \left(\dfrac{2730}{15 \times 12} + \dfrac{2730}{24 \times 12} \right)} = 3.08$$

$$\psi_{\text{BOT}} = \cfrac{\Sigma \dfrac{EI}{L} \text{ columns (JT B12)}}{\Sigma \dfrac{EI}{L} \text{ beams (JT B12)}} = \cfrac{E_c \left(\dfrac{5460}{12 \times 12} + \dfrac{5460}{14 \times 12} \right)}{E_c \left(\dfrac{2730}{15 \times 12} + \dfrac{2730}{24 \times 12} \right)} = 2.86$$

From the braced frame nomogram of Fig. 7.11, $k = 0.89$. Thus $k\ell_u/r = (0.89)(128)/4.8 = 23.7$. For the single curvature case with $M_{1b} = M_{2b}$, the short column $k\ell_u/r$ limit is $34 - 12(1.0) = 22$. Thus, slenderness effects must be checked.

$$C_m = 0.6 + 0.4(M_{1b}/M_{2b}) = 1.0 \quad \beta_d = M_{uD}/M_u = \frac{(1.4)(25)}{162.5} = 0.22$$

Because $k\ell_u/r$ is very near the lower limit, one might use EI from the simpler Code Eq. 10.11, which is a very conservative approximation for $\rho_t = 0.039$.

$$EI = \frac{E_c I_g}{2.5(1 + B_d)} = \frac{(3.605 \times 10^3)(5460)}{(2.5)(1 + 0.22)} = 6.45 \times 10^6 \text{ k-in.}^2$$
$$P_c = \frac{\pi^2 EI}{(k\ell_u)^2} = \frac{\pi^2(6.45 \times 10^6)}{[(0.89)(128)]^2} = 4910 \text{ k}$$

$$\delta_b = \frac{C_m}{1 - \dfrac{P_u}{\phi P_c}} = \frac{1.0}{1 - \dfrac{620}{(0.7)(4910)}} = 1.22$$

$$M_c = \delta_b M_{2b} = (1.22)(1950) = 2380 \text{ k-in.}$$

$P_u/f'_c bh = 0.605$ and $M_u/f'_c bh^2 = (2380)/(4)(16)(16)^2 = 0.145$. These values are shown as B on Fig. 7.16c and fall outside the limiting curve, indicating an unsafe condition.

Because slenderness was more important than initially estimated, a second try will be made using the more accurate EI expression for higher reinforcement percentages.

$$E_s = 29 \times 10^3 \text{ ksi} \qquad I_s = A\bar{d}^2 = (10)(1.0)(8 - 1.5 - 0.38 - 0.56)^2$$
$$= 309 \text{ in.}^4$$

$$EI = \frac{\dfrac{(3.605 \times 10^3)(5460)}{5} + (29 \times 10^3)(309)}{(1 + 0.22)} = 11.28 \times 10^6 \text{ k-in.}^2$$
$$\text{(Note the large increase)}$$

$$P_c = \frac{\pi^2(11.28 \times 10^6)}{[(0.89)(128)]^2} = 8580^k$$

$$\delta_b = \frac{1.0}{1 - \dfrac{620}{(0.7)(8580)}} = 1.12 \qquad M_c = (1.12)(1950) = 2185 \text{ k-in.}$$

Now $P_u/f'_c bh = 0.605$ and $M_u/f'_c bh^2 = 2185/(4)(16)(16)^2 = 0.133$. These values are shown as C on Fig. 7.16c and fall right on the interaction diagram. This column is **OK** under these loads. Had the column been closer to minimum reinforcement ($\rho_t = 0.01$), the simpler EI equation would have yielded about the same result as the longer equation.

(d) Given the unbraced frame shown in Fig. 7.17a subjected to vertical and lateral loading (wind). The frame is symmetrical and all members have the same external dimensions, 20-in. × 20-in. All distances shown in Fig. 7.17a are center-to-center of joints. An analysis for bending about the x-x axis gives

	Column A2 or D2	Column B2 or C2
P_D	100 k	220 k
P_L	80 k	200 k
P_W	±25 k	±15 k
M_D	80 k-ft	40 k-ft
M_L	60 k-ft	120 k-ft
M_W	±30 k-ft	±70 k-ft

All members have $f'_c = 4$ ksi and Grade 60 reinforcement. The typical second-story column section is shown in Fig. 7.17b. Cover is for exterior

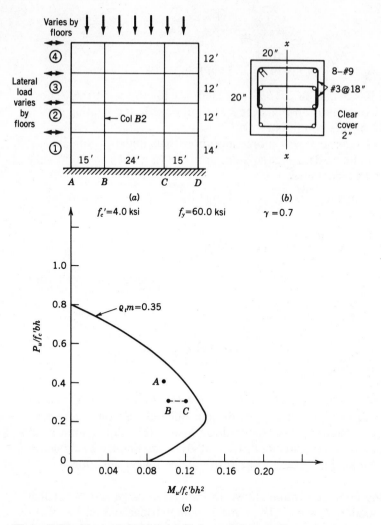

Figure 7.17 Example 7.15*d*. (*a*) Unbraced frame. (*b*) Typical column on story 2. (*c*) Column interaction curve, $\gamma = 0.7$ (From Fig. 6.18*b*.)

exposure. Check to see if column B2 is adequate for both gravity loading and for combined gravity and wind loading.

Solution

This problem is considerably more complex than the preceding one because the frame is unbraced and because more than one loading case must be checked.

Gravity Loading   Because the frame is symmetrical it will be assumed that gravity loading (which may be unsymmetrical in some patterns) will not produce "appreciable sway." This is certainly true when compared to wind loading. For gravity loading, load factors are given by Code Eq. 9.1, $U = 1.4D + 1.7L$.

For column B2,

$$P_u = 1.4 \times 220 + 1.7 \times 200 = 648 \text{ k}$$
$$M_u = 1.4 \times 40 + 1.7 \times 120 = 260 \text{ k-ft} = 3120 \text{ k-in.}$$
$$\ell_u = 144 - 20 = 124 \text{ in.} \qquad I_g = (20)(20)^3/12 = 13{,}333 \text{ in.}^4$$
$$I_g/2 = 6667 \text{ in.}^4$$

Again, assuming relatively slender columns, $k\ell_u/r < 60$, base determination of k on relative stiffnesses using Ig for columns and $Ig/2$ for beams.

$$\text{Column B2, } \psi_{\text{TOP}} = \frac{\Sigma \dfrac{EI}{L} \text{ column}}{\Sigma \dfrac{EI}{L} \text{ beam}} = \frac{E_c \left(\dfrac{13{,}333}{12 \times 12} + \dfrac{13{,}333}{12 \times 12} \right)}{E_c \left(\dfrac{6667}{15 \times 12} + \dfrac{6667}{24 \times 12} \right)} = 3.08$$

$$\psi_{\text{BOT}} = \frac{\Sigma \dfrac{EI}{L} \text{ column}}{\Sigma \dfrac{EI}{L} \text{ beam}} = \frac{E_c \left(\dfrac{13{,}333}{12 \times 12} + \dfrac{13{,}333}{14 \times 12} \right)}{E_c \left(\dfrac{6667}{15 \times 12} + \dfrac{6667}{24 \times 12} \right)} = 2.86$$

From the *braced* frame nomograph of Fig. 7.11, for column B2, $k = 0.89$, and from the *unbraced* frame nomograph $k = 1.83$. This loading case assumes no appreciable sway so the *braced* frame value should be used. $k\ell_u/r = (0.89)(124)/(0.3)(20) = 18.4 < 22$ which is the braced-frame single-curvature limit. No slenderness problem exists for nonsway conditions.

Because $\rho_t = (8)(1.00)/(20)(20) = 0.02$, $\rho_t m = (0.02)(60)/(0.85)(4) = 0.35$, and $\gamma = (20 - (2)(2) - 2(0.38) - 1.12)/20 = 0.706$, the interaction curve from Fig. 6.18b for $\rho_t m = 0.35$ with $\gamma = 0.7$ can be used as shown in Fig. 7.17c. Because there is no slenderness effect, the gravity load case is $P_u/f'_c bh = 648/(4)(20)(20) = 0.405$ and $M_u/f'_c bh^2 = 3120/(4)(20)(20)^2 = 0.098$ as indicated by A on Fig. 7.17c. The column is **OK** for this loading.

Combined Gravity and Wind Loading   When wind load is present the unbraced frame can have appreciable sway. It also must be checked for the wind load combination of Code 9.2.2. Only the check for $U = 0.75(1.4D + 1.7L + 1.7W)$ will be made here because the uplift is so small that the $U = 0.9D + 1.3W$ condition is unlikely to govern. The sway case must treat the story as a whole, so P_c must be determined for all columns in the story. Even though all members are the same cross section, ψ will differ at interior and exterior joints. For simplicity of illustration, both load and restraint conditions will be assumed the same for exterior columns A and D. (Wind load P is opposite in sign). The same assumption will be made for columns B and C. Thus only half the frame will need to be treated here.

Column A2 $P_u = (1.4 \times 100 + 1.7 \times 80 + 1.7 \times 25)(0.75) = 239$ k

Column B2 $P_u = (1.4 \times 220 + 1.7 \times 200 + 1.7 \times 15)(0.75) = 505$ k

$\qquad M_u = (1.4 \times 40 + 1.7 \times 120 + 1.7 \times 70)(0.75) = 284$ k-ft

$\qquad\qquad\qquad\qquad = 3410$ k-in.

If there were no slenderness effect, this would be $P_u/f'_c bh = 0.316$ and $M_u/f'_c bh^2 = 0.106$ as indicated by B on Fig. 7.17c. The cross section is adequate as a short column for the combined loading condition, but it must still be determined if a slenderness problem exists.

In the previous computations for the gravity loading case, k for column B2 when unbraced was found as $k = 1.83$. Thus $k\ell_u/r = (1.83)(124)/6 = 38$. This is much greater than the 22 limit for short columns in unbraced frames. Slenderness effects must be checked.

In an unbraced frame, the sway multiplier δ_s will depend on $\Sigma P_u/\Sigma P_c$ for all columns in the story. We have assumed that A2 plus B2 will be the same as C2 plus D2. The summation can thus include the values from only A2 and B2.

For column A2, k can be determined as:

$$\psi_{TOP} = \frac{E_c \left(\dfrac{13{,}333}{12 \times 12} + \dfrac{13{,}333}{12 \times 12} \right)}{E_c \left(\dfrac{6667}{15 \times 12} \right)} = 5.0$$

$$\psi_{BOT} = \frac{E_c \left(\dfrac{13{,}333}{12 \times 12} + \dfrac{13{,}333}{14 \times 12} \right)}{E_c \left(\dfrac{6667}{15 \times 12} \right)} = 4.64$$

From the nomographs of Fig. 7.11, k can be found as 0.94 for braced conditions and 2.19 for unbraced conditions.

Use of the magnified moment expression (Code Eq. 10.6)

$$M_c = \delta_b M_{2b} + \delta_s M_{2s}$$

requires a considerable amount of bookkeeping. Moments must be divided into nonsway and sway-inducing values. Separate magnifiers are applied. The nonsway magnifier is based on a braced single column. The sway magnifier is based on the unbraced story summation.

Braced B2. For this case

$$M_{2b} = (0.75)(1.4 \times 40 + 1.7 \times 120)(12) = 2340 \text{ k-in.}$$

EI may be based on the simpler Code Eq. 10.11 because $\rho_t = 0.02$ and this expression is fairly good for that level as shown by Table 7.2.

$\qquad E_c = 3.605 \times 10^3$ ksi $\qquad I_g = 13{,}333$ in.^{4}.

$\qquad \beta_d = M_{uD}/M_u = (0.75)(1.4 \times 40 \times 12)/3410 = 0.15$

$$EI = \frac{E_c I_g}{(2.5)(1 + \beta_d)} = \frac{(3.605 \times 10^3)(13{,}333)}{(2.5)(1 + 0.15)} = 16.72 \times 10^6 \text{ k-in.}^2$$

Using braced $k = 0.89$

$$P_c = \frac{\pi^2 EI}{(k\ell_u)^2} = \frac{\pi^2(16.72 \times 10^6)}{(0.89 \times 124)^2} = 13{,}550 \text{ k}$$

C_m is used as 1.0 for single curvature.

$$\delta_b = \frac{C_m}{1 - \dfrac{P_u}{\phi P_c}} = \frac{1.0}{1 - \dfrac{505}{(0.7)(13550)}} = 1.056$$

$$\delta_b M_{2b} = (1.056)(2340) = 2470 \text{ k-in.}$$

Unbraced. For this case $M_{2s} = (0.75)(1.7)(70)(12) = 1070$ k-in. The same *EI* Equation will be used for B2 and A2. β_d for column A2 is appreciably different than for column B2. For A2, $\beta_d = (0.75)(1.4)(80)/(0.75)(1.4 \times 80 + 1.7 \times 60 + 1.7 \times 30) = 0.42$

Thus *EI* for A2 is

$$\frac{(3.605 \times 10^3)(13{,}333)}{(2.5)(1 + 0.42)} = 13.54 \times 10^6 \text{ k-in.}^2$$

P_c for A2 is $\dfrac{\pi^2(13.54 \times 10^6)}{[(2.19)(124)]^2} = 1812$ k

P_c for B2 is $\dfrac{\pi^2(16.72 \times 10^6)}{[(1.83)(124)]^2} = 3205$ k

The story multiplier

$$\delta_s = \frac{1}{1 - \dfrac{\Sigma P_u}{\phi \Sigma P_c}} = \frac{1}{1 - \dfrac{(239 + 505)}{(0.7)(1812 + 3205)}} = 1.27$$

Thus

$$M_u = M_c = \delta_b M_{2b} + \delta_s M_{2s}$$
$$= (1.056)(2340) + (1.27)(1070) = 3830 \text{ k-in.}$$

$P_u/f'_c bh = 505/(4)(20)(20) = 0.316$ and $M_u/f'_c bh^2 = 3830/(4)(20)(20)^2 = 0.120$. These are shown as point C on Fig. 7.17c and the column is satisfactory for this loading also. Prior to the 1983 ACI Code, both M_{2b} and M_{2s} were magnified by δ_s which resulted in considerable over-conservatism in many cases. In the last example, if the sway magnifier was used with the total moment, C would move laterally right to the interaction curve.

Although all of these example problems were worked utilizing the basic interaction curves to illustrate the method, analytical expressions representing the various compression or tension branches of the interaction curves could have been used and would be required if computer solutions are desired.

7.16 Design of Slender Columns

The design of slender columns using a second-order analysis that meets the requirements of Code 10.10 is carried out just like the short-column design example of Chapter 6, except that the results of the second-order analysis are used to define the required axial load and moment strength. In reality, this is not as easy to do as to say because the analysis must correctly reflect the actual column stiffness. The actual column reinforcement must be considered in determining the stiffness values for the analysis. A trial-and-error or iterative procedure will be required in the design process, even when Code 10.10 is used.

The practical application of the approximate column slenderness provisions of Code 10.11 are best appreciated when related to the general design procedure as explained for short columns in Chapter 6. The overall approach includes all of the iterative steps for short column design. In addition, the moment magnification approach must be used to modify or correct the basic analysis. It is conceptually simple as shown in Fig. 7.18. Assuming design aids in the form of conventional column *cross section* interaction diagrams such as

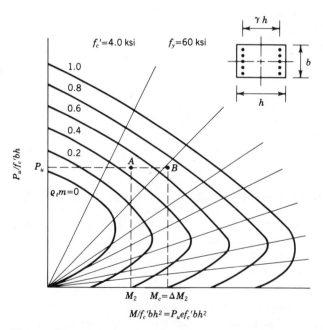

Figure 7.18 Conceptual representation of moment magnification procedure. First-order analysis moment (A) is magnified to new design moment (B).

those of Fig. 6.18, the required P_u, M_u values from a conventional first-order analysis would require a certain size column with a certain reinforcement ratio as shown by A in Fig. 7.18. Because of the $P\Delta$ effect, this moment is magnified using the appropriate values of δ_b and δ_s, shown here as a final magnifier δ. An increased M_u requirement results and for the same size column requires an increased ρ_t as shown by B. Of course, the magnified moment could be such a great increase that the column size would have to be increased. In such a case, it is possible that a lower ρ_t would be possible with the larger column. In any case it must be appreciated that changing either the reinforcement percentage or the column size can substantially alter the EI value used in the moment magnification process. Further iterative cycles may be required to determine the correct amount of the moment magnification. For columns with low reinforcement ratios and low-to-moderate slenderness, the simpler EI expression of Code Eq. 10.11 can be used. It does not reflect reinforcement ratio and hence require fewer iterations. The actual design process is speeded by first estimating an approximate δ based on a general examination of the column slenderness. Such an estimate is greatly improved with experience. The column cross section and reinforcement are proportioned for the magnified moment δM_2 and then the actual magnifier can be computed. Further corrections are made as required.

In the basic application of the moment magnification process, there would be no slenderness effect if there was no moment. In reality every column has some moment from initial crookedness or fabrication tolerances, even if the analysis shows no eccentricity at either end. This initial crookedness or loading eccentricity is handled in short column design indirectly by the $P_{n(\max)}$ requirements of Codes 10.3.5.1 and 10.3.5.2 that also apply to slender columns. However, these limits do not provide moment information. To ensure that all slender columns have a minimum value of moment to be magnified, Codes 10.11.5.4 and 10.11.5.5 provide that if first-order analysis computations show moments producing eccentricities less than $(0.6 + 0.03 h)$ in., M_{2b} and M_{2s} in Code Eq. 10.6 will be based on that minimum eccentricity. This initial moment then governs in the design of slender "axially loaded" columns.

In very slender columns $(k\ell_u/r > 60)$ the need to consider the actual reinforcement percentage (not known until design is completed) in the basic determination of k frequently causes a large number of iterations. For moderate slenderness $(k\ell_u/r < 60)$ the use of I_g for columns and $I_g/2$ for beams in determining ψ and hence k somewhat simplifies the process. In the following examples, somewhat simplified examples will be used by either defining a constant ψ or using moderate $k\ell_u/r$ values.

7.17 Slender Column Design Examples

(a) Design a square tied column for a total dead load of 200 k, a live load of 250 k, and a moment of 900 k-in. (all from live load) at each end creating single curvature. $\ell_u = 20$ ft, $\psi = 1$ in braced frame,* $f_c' = 4000$ psi, Grade 60 steel. Try for ρ_t of about 0.02. Use the column interaction diagrams of Fig. 6.18 that include ϕ for the cross section.

Solution

$$P_u = 1.4 \times 200 + 1.7 \times 250 = 705 \text{ k} \qquad M_u = 1.7 \times 900 = 1530 \text{ k-in.}$$
$$e = M_u/P_u = 2.17 \text{ in.} \qquad m = f_y/(0.85f_c') = 60/(0.85 \times 4) = 17.65$$

For ρ_t of 0.02, $\rho_t m = 0.353$. From Fig. 7.11, k for ψ of 1 is 0.77, making $k\ell_u = 0.77 \times 20 \times 12 = 185$ in. A very, very rough check of preliminary column slenderness can be obtained by guessing a column size and preliminary sizing the member ignoring slenderness effects. By using a rough estimate of $\gamma = 0.7$ and a trial $h = 20$ in. (strictly from judgment and experience for the forces involved), the interaction curve of Fig. 6.18b can be entered for $e/h = 2.17/20 = 0.11$ and $\rho_t m = 0.353$. This would indicate $\alpha = P_u/f_c'bh = 0.63$. Thus a better preliminary h would be $h^2 = P_u/f_c'(0.63) = 205/(4)(0.63) = 280$ or $h = 17$ in. For such a column $k\ell_u/r = (185)/(0.3)(17) = 36$ which is greater than the $34 - 12(M_1/M_2) = 34 - 12(1) = 22$ limit for short columns in single curvature.

Thus this example appears to be a slender column, and the moment and e will be magnified. From experience assume $\delta = 1.3$, making the design $e = 1.3 \times 2.17 = 2.82$ in. This increase in e means the previous assumptions should increase. If h is assumed as 18 in., $e/h = 2.82/18 = 0.157$. For assumed #3 stirrups and #8 bars, the trial γ is:

$$\gamma = (h - 2d')/h = 18 - 2(1.5 + 0.38 + 0.5)/18 = 0.74$$

Try using the chart of Fig. 6.18b with $\gamma = 0.70$, entering with $e/h = 0.157$ and $\rho_t m = 0.353$ and reading $\alpha = 0.56$ as $P_u/(f_c'h^2)$. These charts automatically introduce the cross section ϕ. $h^2 = 705/(4 \times 0.56)$, $h = \sqrt{314} = 17.7$ in. using the assumed δ. Try $h = 18$ in. as a practical outer dimension.

Continue with the 18-in. column and calculate the magnifier needed. For $\rho_t = 0.02$ Table 7.2 in Sec. 7.9 indicates that EI_2 should control unless γ is less than 0.65. Although one is tempted to use the simpler EI_1 for the first trial EI_2 will be used, with $\beta_d = M_{uD}/M_{u(D+L)} = 0$ since all the moment is from live load in this case. Code 8.5.1 sets E_c.

$$E_c = 57,000\sqrt{f_c'} = 57,000\sqrt{4000} = 3.61 \times 10^6 \text{ psi} = 3.61 \times 10^3 \text{ ksi}$$
$$\text{Trial } A_s = 0.02 \times 18^2 = 6.48 \text{ in.}^2$$
$$EI_2 = (E_c I_g/5 + E_s I_s)/(1 + \beta_d) = 3.61 \times 10^3[18^4/(12 \times 5)] + 29$$
$$\times 10^3[6.48 \times (13/2)^2] = 14.25 \times 10^6 \text{ k-in.}^2$$
$$\phi P_c = \phi\pi^2 EI_2/(k\ell_u)^2 = 0.70 \times 3.14^2 \times 14.25 \times 10^6/185^2 = 2880 \text{ k}$$

---

* A given ψ in an example is always an oversimplification. It assumes a specific relative column stiffness that the design may not provide.

With $M_1 = M_2$, $C_m = 0.6 + 0.4M_1/M_2 = 1.0$

$$\delta_b = C_m/(1 - P_u/\phi P_c) = 1/(1 - 705/2880) = 1.32$$

Enter Fig. 6.18b for $\gamma = 0.70$ using the new value of $\delta e/h = 1.32 \times 2.17/18 = 0.159$ and $\rho_t m = 0.353$ to read $\alpha = 0.555 = 705/4h^2$, giving $h = \sqrt{318} = 17.8$ in.

USE 18 in. $\times$ 18 in. subject to further evaluation of ρ. From the same chart, for $\alpha = 705/(4 \times 18^2) = 0.544$, $e/h = 0.159$, read $\rho_t m = 0.30$. This leads to $\rho_t = 0.30/17.65 = 0.0170$. Revise EI_2 for this ρ_t.

$$EI_2 = 3.61 \times 10^3[18^4/(12 \times 5)] + 29 \times 10^3[(0.017/0.02)(6.48) \times (13/2)^2]$$
$$= 13.06 \times 10^6 \text{ k-in.}^2$$
$$\phi P_c = 0.70 \times 3.14^2 \times 13.06 \times 10^6/185^2 = 2630 \text{ k}$$
$$\delta_b = 1/(1 - 705/2630) = 1.37 \qquad \delta e/h = 1.37 \times 2.17/18 = 0.165$$

This change from 0.159 should not change column size, but will change A_s. Enter charts with $\delta e/h$ and $\alpha = 0.544$ as earlier, to read $\rho_t m$.

$\gamma = 0.70 \qquad \rho_t m = 0.35$

$\gamma = 0.80 \qquad \rho_t m = 0.32 \qquad$ For $\gamma = 0.74$, $\rho_t m = 0.34$, $\rho_t = 0.0193$

Reqd. $A_s = 0.0193 \times 18^2 = 6.25$ in.^{2}

This new ρ_t is back to 96% of the 0.02 tried originally and indicates a better second trial might have been started at $\rho_t = 0.018$ instead of 0.017. The error here is on the safe side because the larger A_s will give a larger P_c that lowers δ. For the purpose here this A_s is close enough.

USE 8-#8 bars with #3 ties. Tie spacing: $48d_b$ tie $= 18$ in.; $h = 18$ in.

($A_s = 6.32$ in.^{2}) $16d_b$ for $A_s = 16$ in.

USE #3 ties at 16 in., in pairs as in Fig. 6.19e or f.

Alternatively, one could use one overall tie plus one single crosstie (Fig. 6.19d) between *one* pair of center bars, as clear bar spacing is under 6 in. A complete design would also involve at least a separate check on moment about the other axis.

(b) To indicate the more serious length problem in an unbraced frame, redesign the column in (a) considering the frame unbraced, $P_u = 705$ k, M_u (all sway-producing LL) $= 1530$ k-in., $\ell_u = 20$ ft., $f_c' = 4$ ksi, $f_y = 60$ ksi, $\beta_d = 0$, and desired ρ_t about 0.02. Assume the story δ is either known or assumed at $\delta_s = 1.90$, $\psi = 1$.

Solution

In actual unbraced frames one must start with a known or assumed δ and later verify this value *for the entire story*. With many columns in a story, the selected individual column size will have only a minor effect on δ, but the cumulative effect could be important. In this example the verification, or lack of verification, and its potential influence on the design cannot be resolved in terms of the single column. This design is thus oversimplified. (The design in the next example is at the other extreme.)

From Fig. 7.11, $k = 1.31$, $k\ell_u = 1.31 \times 20 \times 12 = 314$ in. $e = M_u/P_u = 1530/705 = 2.17$ in. Since k is larger, this more slender column needs a larger size.

$$\text{Try } h = 20 \text{ in., } \delta e/h = 1.90 \times 2.17/20 = 0.206$$
$$m = (f_y/0.85 \times 4) = 17.65 \qquad \text{Desired } \rho_t m = 0.02 \times 17.65 = 0.335$$
$$h - 2d' = 20 - 2 \times 2.5 = 15 \text{ in.} \qquad \gamma = 0.75$$

For $\gamma = 0.70$ (Fig. 6.18b)* enter with e/h and $\rho_t m$ to read $\alpha = 0.50$
$$\gamma = 0.80 \text{ (Fig. 6.18c) similarly } \alpha = 0.53$$
$$\gamma = 0.75 \qquad \alpha = 0.515 = P_u/f_c'h^2 = 705/4h^2$$
$$h = \sqrt{342} = 18.5 \text{ in.}$$

Try $h = 19$ in.   Desired $\rho_t m = 0.335$ $\delta e/h = 1.9 \times 2.17/19 = 0.217$
$$h - 2d' = 19 - 2 \times 2.5 = 14 \text{ in.} \qquad \gamma = 14/19 = 0.737$$

For $\gamma = 0.70$ $\alpha = 0.490$ For $\gamma = 0.737$, $\alpha = 0.496 = P_u/f_c'h^2 = 705/4h^2$
$$\gamma = 0.80 \quad \alpha = 0.505 \qquad\qquad h = \sqrt{355} = 18.9 \text{ in.}$$
$$\text{USE 19 in.} \times \text{19 in. column}$$

$\alpha = 705/(4 \times 19^2) = 0.488$ $\gamma = 0.70$ $\rho_t m = 0.35$
$$\gamma = 0.80 \qquad \rho_t m = 0.30$$
$$\gamma = 0.737 \qquad \rho_t m = 0.332, \quad \rho_t = 0.0188$$

Reqd. $A_s = 0.0188 \times 19^2 = 6.79$ in.^{2} 8-#9 = 8.00 in.^{2}
$$6\text{-}\#10 = 7.62 \text{ in.}^2$$

For the #10 bars, recheck γ. $d' = 1.5 + 0.375 + 1.27/2$
$$= 2.5 \text{ in.} \qquad \textbf{O.K.} \text{ as used}$$

USE 6-#10 with #3 ties in pairs, one all around and the other a center crosstie.

Tie spacing governed by 48 tie diameters, which is 18 in.

In this oversimplified example it was assumed that all of the live load moment was sway-producing. Thus $M_{2s} = 1530$ k-in. and $M_{2b} = 0$. Because of this assumption it was not necessary to compute the braced magnifier δ_b which uses k for braced frames. In an actual design it would usually be necessary to compute δ_b for the individual column as well as δ_s for the story to evaluate $M_c = \delta_b M_{2b} + \delta_s M_{2s}$. In this case the braced frame column already designed for the same axial load in example (a) shows that δ_b is much smaller (as expected). Thus if the live load moment was separated into nonsway-producing and sway-producing components, the magnitude of the design moment would decrease and less reinforcement or a smaller column would result.

---

* These charts automatically introduce the cross section ϕ.

(c) Assume now that the column initially described in (b) is one of only four columns in a single structure, so situated that all four columns (and their connecting beams) appear to be identical cases that would lead to identical δ values whether separate or in a group. Based on a known stiffness of the beams (cracked section) and an estimated column size of 18-in. square, with ρ_t about 0.02, the initial ψ is estimated to be 1.0. P_u is 705 k, $M_u = 1530$ k-in. (again all sway-producing), $\ell_u = 20$ ft, $f'_c = 4$ ksi, $f_y = 60$ ksi. Redesign for these conditions.

Solution

When one has a new situation where estimated column size is apt not to be good, it is often advantageous to check the first guess with the much simpler EI_1 even though the design intent is to use a larger ρ_t that seems to demand EI_2. As the planned ρ_t gets over 3% this advantage is probably lost.

For the estimated 18-in. column:

$$E = 57{,}000\sqrt{f'_c} = 57{,}000\sqrt{4000} = 3.61 \times 10^6 \text{ psi} = 3.61 \times 10^3 \text{ ksi}$$
$$EI_1 = EI_g/2.5 = 3.61 \times 10^3 \times 18^4/(12 \times 2.5) = 12{,}630 \times 10^3 \text{ k-in.}^2$$

For $\psi = 1$, Fig. 7.11 shows $k = 1.31$, $k\ell_u = 1.31 \times 20 \times 12 = 314$ in.

$$\phi P_c = 0.70\pi^2 EI/(k\ell_u)^2 = 0.70 \times 9.87 \times 12{,}630 \times 10^3/314^2 = 885 \text{ k}$$

Since all four columns and their restraining beams have been assumed identical in this example, P_u/P_c can replace $\Sigma P_u/\Sigma P_c$ in the story sway magnifier expression.

$$C_m = 1.0 \text{ for a unbraced frames,} \quad \delta_s = 1/(1 - P_u/\phi P_c)$$
$$= 1/(1 - 705/885) = 4.92$$

This δ is too large to be practical. Somewhere around 2, maybe 3, is a top practical limit.

Try a 20 in. × 20 in. column.

$$EI_1 = (20/18)^4 \, 12{,}630 \times 10^3 = 19{,}250 \times 10^3 \text{ k-in.}^2$$

Revised $\psi = 1 \times 19{,}250/12{,}630 = 1.52$ $k = 1.46$ $k\ell_u = 350$ in.

$k\ell_u/r = 350/(0.3 \times 20) = 58.3 < 100$ (Code 10.11.4.3).   Method is **O.K.**

$$\phi P_c = 0.70 \times 9.87 \times 19{,}250 \times 10^3/(350)^2 = 1086 \text{ k}$$
$$\delta_s = 1/(1 - 705/1086) = 1/(1 - 0.648) = 2.85$$
$$e/h = (1530/705)/20 = 0.1085$$
$$\delta e/h = 2.85\, e/h = 0.309 \qquad \gamma = (20 - 2 \times 2.5)/20 = 0.75$$

For $\gamma = 0.70$ and $\rho_t m = 0.02(60/0.85 \times 4) = 0.02 \times 17.65 = 0.353$, Fig. 6.18$b$* shows for $\delta e/h = 0.309$, $\alpha = 0.40 = P/f'_c h^2 = 705/4 h^2$, $h = \sqrt{441} = 21.0$ in.

---

* These charts automatically introduce the cross section ϕ.

A larger column (or larger ρ_t) is indicated, but first try EI_2 with $\rho_t = 0.02$, which should be larger (and less demanding) than EI_1.

$$EI_2 = (3.61 \times 10^3)20^4/(12 \times 5) + 29 \times 10^3(0.02 \times 20^2)(15/2)^2$$
$$= 9630 \times 10^3 + 13,050 \times 10^3 = 22,700 \times 10^3 \text{ k-in.}^2$$
$$\phi P_c = 0.70 \times 9.87 \times 22,700 \times 10^3/350^2 = 1280 \text{ k}$$
$$\delta_s = 1/(1 - 705/1280) = 1/(1 - 0.550) = 2.22,$$
$$\text{compared to 2.84 for } EI_1$$
$$\delta e/h = 2.22 \times 0.1085 = 0.241$$
$$\gamma = 0.70, \text{ for } \rho m = 0.36 \text{ and this } e/h, \text{ read } \alpha = 0.465 = P/f_c'h^2$$
$$h = \sqrt{705/(4 \times 0.465)} = \sqrt{379} = 19.5 \text{ in.}$$

The above equations indicate that the 20-in. column is possible and seem to say that ρ_t will be less than 0.02 because of the extra concrete area. However, this lower ρ_t will also lower EI_2 and thus increase the required h, eliminating part of the apparent surplus concrete. Since A_s cannot be exact at this stage, re-enter the same chart for $\gamma = 0.70$, $e/h = 0.241$, and $\alpha = P/f_c'h^2 = 705/(4 \times 20^2) = 0.441$, to read required $\rho_t m = 0.29$, $\rho_t = 0.29/17.65 = 0.0164$. For $\gamma = 0.80$, $\rho_t m = 0.25$, $\rho_t = 0.0142$. Interpolating for $\gamma = 0.75$, $\rho_t = 0.0153$.

Noting that a ρ_t of 0.016 for $\gamma = 0.75$ would lower the steel contribution to EI_2 by about 20% of $13,050 \times 10^3$, or over 10% of the total EI_2 and the resulting ϕP_c, one can see that δ would increase significantly and demand more A_s.* If one traced out that cycle in detail, he or she would find that at this high δ the cycle would *not* be converging.

Try an average $(0.020 + 0.016)/2 = 0.018$ for ρ_t.

$$EI_2 = 9630 \times 10^3 + (0.018/0.020)13,050 \times 10^3 = 21,400 \times 10^3$$
$$\phi P_c = 1280(21,400/22,700) = 1210 \text{ k}$$
$$\delta = 1/(1 - 705/1210) = 2.40 \quad \delta e/h = 2.40 \times 0.1085 = 0.260$$
For $\gamma = 0.70$, $\alpha = P/f_c'h^2 = 705/(4 \times 20^2) = 0.44$, reqd. $\rho_t m = 0.34$
$$0.80\rho_t m = 0.30$$
$$0.75\rho_t m = 0.32, \quad \rho_t = 0.32/17.65 = 0.0181$$

This cycle has closed with almost the exact ρ_t assumed. (The averaging of ρ_t used will not always bring such quick closure.) As soon as the designer bounds the range of values involved, he or she can be arbitrary and conservative; it is doubtful that the method justifies working to an exact ρ_t.

Reqd. $A_s = 0.0181 \times 20^2 = 7.24 \text{ in.}^2$

USE 20 in. $\times$ 20 in. column, 8-#9 (8.00 in.^{2})

6-#10 (7.62 in.^{2}) would be another possibility, requiring a check on γ, which would be a little lower.

---

* This is one of the penalties resulting from using a large δ in design.

USE #3 ties, one all around plus one cross tie, at 18 in. spacing.

This design was simplified by assuming all columns in the story the same. This will rarely be feasible and the final δ for the story cannot be known until at least tentative designs have been made. As noted earlier, this means that the story δ should be much less influenced by a single column δ.

If, as stated initially, the structure is the same about both axes, A_{st} spread on all four faces would fit better, but for this arrangement a different set of column design charts is needed (available in Reference 6). The effective γ would be reduced by this less effective steel arrangement (less effective for the single axis), and a little more total steel would be necessary.

(d) Beams in Unbraced Frames   Emphasis needs to be added to the role of the beam in the unbraced frame. *All* the increased column moment must also be resisted by the beams. These increases are secondary beam moments to be added to the beam in addition to the first-order beam moments given by frame analysis. With such a large δ as in this example, there must be large resisting beam moments. The designer might consider whether a column larger than 20-in. square with a ρ_t lower than 0.02 and a δ lower than 2.60 might not be a stiffer and better design.

Selected References

1. J. S. Ford, D. C. Chang, and J. E. Breen, "Behavior of Concrete Columns Under Controlled Lateral Deformation," *ACI Jour.*, 78, No. 1, Jan.–Feb. 1981, pp. 3–20.

2. J. G. MacGregor, J. E. Breen, and E. O. Pfrang, "Design of Slender Columns," *ACI Jour.*, 67, No. 1, Jan. 1970, p 6.

3. P. M. Ferguson and J. E. Breen, "Investigation of the Long Concrete Column in a Frame Subject to Lateral Loads," *Reinforced Concrete Columns*, SP-50, Amer. Concrete Inst., Detroit, 1975, pp. 75–114.

4. J. G. MacGregor, V. H. Oelhafen, and S. E. Hage, "A Reexamination of the EI Value for Slender Columns," *Reinforced Concrete Columns*, SP-50, Amer. Concrete Inst., Detroit, 1975.

5. R. W. Furlong, "Column Slenderness and Charts for Design," *ACI Jour.*, 68, No. 1, Jan. 1971, p. 9.

6. *Design Handbook*, Vol. 2, *SP-17A(78)*, Amer. Concrete Inst., Detroit, 1978.

7. B. C. Johnston, (ed.), *The Column Research Council Guide to Design Criteria for Metal Compression Members*, 2nd ed., John Wiley and Sons, Inc., New York, 1966.

8. *Code of Practice for the Structural Use of Concrete*, Part 1, "Design Materials and Workmanship," (CP110: Part 1, Nov. 72), British Standards Inst., London, 1972, 154 pp.

9. W. B. Cranston, "Analysis and Design of Reinforced Concrete Columns," *Research Report* No. 20, Paper 41.020, Cement and Concrete Assn., London, 1972, 54 pp.

10. J. S. Ford, D. C. Chang, and J. E. Breen, "Experimental and Analytical Modeling of Unbraced Multipanel Concrete Frames," *ACI Jour.*, 78, No. 1, Jan.–Feb. 1981, pp. 21–35.

11. J. S. Ford, D. C. Chang, and J. E. Breen, "Behavior of Unbraced Multipanel Concrete Frames," *ACI Jour.*, 78, No. 2, Mar.–Apr. 1981, pp. 99–115.

12. J. S. Ford, D. C. Chang, and J. E. Breen, "Design Indications from Tests of Unbraced Multipanel Concrete Frames," *Concrete International: Design and Construction*, 3, No. 3, Mar. 1981, pp. 37–47.

Problems

PROB. 7.1. Rework Prob. 6.7a, 6.7b, or 6.7c as specified considering column slenderness effects and the following restraint conditions. Assume columns are in a braced frame with $\psi = 1.0$, $\ell_u = 18$ ft, and loaded in single curvature $M_1 = M_2$.

PROB. 7.2. Rework Prob. 7.1 except assume columns are in an unbraced frame with $\psi = 1.0$. Assume one-half of the eccentricity is produced by sway-loading and one-half of the eccentricity is produced by loading that does not produce appreciable sway.

PROB. 7.3. Analysis of a braced frame shows an interior column in single curvature with $M_1 = M_2$ consisting of service load $M_d = 0$, $M_1 = 137$ k-ft, $P_d = 100$ k, $P_L = 160$ k, $\psi = 1.0$, $\ell_u = 20$ ft, $\beta_d = 0$, $f'_c = 4000$ psi, Gr 60 bars. Design a square tied column with ρ_t about 0.02.

PROB. 7.4. In the braced frame of Prob. 7.3 an interior column has M_2 as given there but the member is in reversed curvature with $M_1 = -0.4M_2$, likewise all from live load. Here $P_d = 100$ k. $P_L = 180$ k, $\psi = 1.0$, $\ell_u = 20$ ft, $\beta_d = 0$, $f'_c = 4000$ psi, Gr 60 bars. Design a square tied column with ρ_t about 0.02.

PROB. 7.5. Rework Prob. 7.3 except with ρ_t about 0.04.

PROB. 7.6. Rework Prob. 7.4 except with ρ_t about 0.04.

PROB. 7.7. Same as Prob. 7.3 except frame *unbraced* making reverse curvature critical with $M_1 = -M_2$. Assume that 50% of the live load moment is sway-producing and that 50% is nonsway-producing. (Designer may assume that the average δ for the entire story in this unbraced frame is the same as that calculated for the single column.)

PROB. 7.8. Same as Prob. 7.7 except with ρ_t about 0.04.

PROB. 7.9. For $f'_c = 4000$ psi, Gr 60 bars, $P_d = 100$ k, $P_\ell = 200$ k, $e = 5$ in., $\ell_u = 20$ ft, single curvature in braced frame with $\psi = 2.0$, design a square tied column with ρ about 0.03. Consider this loading primarily short time so that $\beta_d = 0$.

PROB. 7.10. For the data of Prob. 7.9 except the given loading now to be considered as *factored* loads and the frame *unbraced* with $\ell_u = 16$ ft, $\psi = 0.90$, design the column assuming all loads are sway-producing.

8

DEVELOPMENT AND SPLICING OF REINFORCEMENT

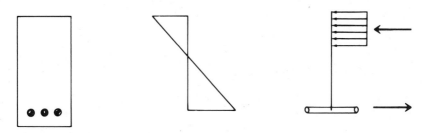

8.1 Development Length of Straight Bars

Bar development length ℓ_d is the embedment necessary, under specific surrounding conditions, to assure that a bar can be stressed to its yield point with some reserve to insure member toughness. The necessary length is a function of a number of variables. The Code equations include bar size d_b, f_y, and f'_c. Bar spacing, cover, and transverse reinforcement influence development length but are only indirectly considered in the Code.[1,2]

Bond stress as such is not in the 1986 Code, although the specified ℓ_d lengths are based directly on 1963 allowable bond stresses. Bond stress is the local longitudinal shear stress, per unit of bar surface, transferred from the concrete to the bar to change the bar stress from point to point. Test results show that bond stress varies along the bar. The ℓ_d concept is based on the experimental evidence that *average* bond stress is critical, not higher peak bond stress which is usually adjacent to cracks.

A bar with enough embedment in concrete cannot be pulled out. After slip at the loaded end has progressed far enough to develop bond over a considerable length, the bar reaches yield and can be considered fully anchored in the concrete.

In the basic concept of anchorage length, a bar is embedded in a mass of concrete, as in Fig. 8.1. Under initial loading, the actual bond stress will be quite large near the surface and nearly zero at the embedded end. Near failure the bond stress along the bar will be more uniformly distributed. Considering an average bond stress u at ultimate, equilibrium between internal forces and the force applied to the bar leads to

$$A_b f_y = u\ell_d \pi d_b \text{ (for one bar)}$$

Substituting for bars of diameter d_b, $A_b = \pi d_b^2/4$,

$$\ell_d = \frac{f_y}{4u} d_b \tag{8.1}$$

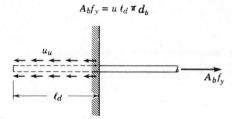

$$A_b f_y = u \, \ell_d \, \pi \, d_b$$

u_u

$A_b f_y$

ℓ_d

Figure 8.1 Anchorage of a bar.

The minimum permissible anchorage or development length is ℓ_d. Development lengths computed on the basis of Eq. 8.1 can be no better than the values of bond stress used in deriving them. Because the development may occur in diverse surroundings with wide variations in spacing, cover, edge distance, and enclosing stirrups, no single value of u or ℓ_d is really appropriate over a large range.

8.2   Internal Behavior and Cracking—Anchored Bars in Tension

The internal stress distribution for many years could only be rationalized on the basis of the longitudinal failure crack. In 1971 Goto published[3] photographs of

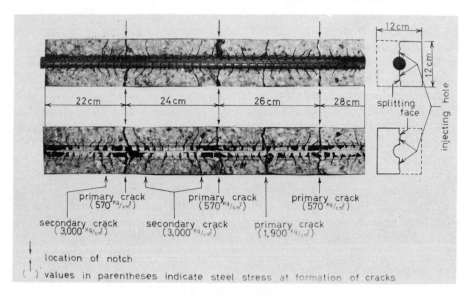

Figure 8.2 Internal cracks created by tension pull on both ends of the bar. Cracks were dyed with an injection of red ink that, in the bar imprint below, also spread over part of the bar surface where slip occurred. (Courtesy ACI, Reference 3.)

internal cracking patterns that resulted from pulling both ends of a tension bar embedded in a concrete prism encasement, as shown in Fig. 8.2. Transverse cracks from axial tension developed at intervals, much like flexural cracks in a beam. In effect, the cracks created a number of specimens of shorter length. Within the length of each specimen, the same internal crack pattern was observed.

The systematic internal crack pattern, with cracks starting just behind each lug and progressing diagonally a limited distance toward the nearest transverse crack, is impressive. Because cracking in concrete results from a principal tensile stress, there must be compressive stresses parallel to these cracks, forces directed outwardly around the bar to form a hollow truncated cone of pressure. This pressure is directed inwardly against the lug and outwardly must be resisted by ring tension in the concrete.

The pictured angle between crack and bar axis is surprisingly large, that is, considerably more than 45°. The inclination of forces leads to radial splitting forces. An idealized sketch of the radial pressures is shown in Fig. 8.3. The outward forces, like water pressure in a pipe, lead to splitting on weak planes along the bar unless the bar cover is large or splitting is restrained by transverse reinforcement.

The loaded end of the bar first breaks adhesion of the concrete close to the end, and resistance is provided by friction and some bearing on the nearest bar lug. Further loading may produce crushing against the lug or shearing of the concrete to form a cylinder with a diameter matching that of the bar out-to-out of lugs, the latter especially with small bars or with lightweight aggregate. For large bars or small cover the cylinder or prism usually splits lengthwise. For many years this splitting crack was not considered related to anchorage of the bars. Splitting is now nearly always the key to anchorage behavior. Splitting, such as that shown in Fig. 8.4, is the most visible sign of approaching bond failure. It occurs even with heavy stirrups, as in Fig. 8.4d.

Likewise, Fig. 8.4c indicates that diagonal tension (shear) cracking is closely associated with bond splitting. Traditionally, bond and shear have been treated as separate phenomena. Bond and shear are interrelated topics, however, and neither can be understood fully without knowledge of the other.

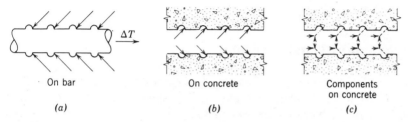

| On bar | On concrete | Components on concrete |
| (a) | (b) | (c) |

Figure 8.3 The forces between a deformed bar and concrete that may cause splitting, as in Fig. 8.4.

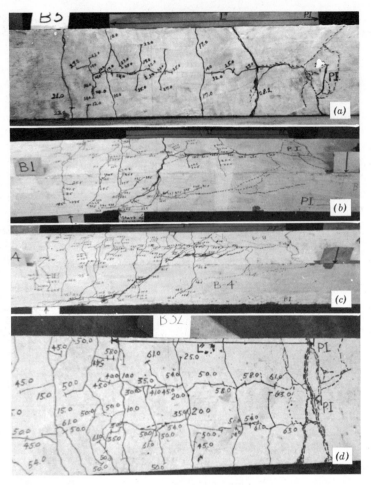

Figure 8.4 Bond splitting photographs of beams tested as shown in Fig. 8.12. (*a*) Splitting is directly over bar and runs lengthwise. The cross cracks are flexural cracks. (*b*) Failure chiefly from bond splitting. (*c*) Diagonal tension (inclined side crack) combined with longitudinal splitting to produce failure. (*d*) Splitting over two bars in beam with heavy stirrups.

8.3 Behavior of Tension Splices

The behavior of anchored bars has been studied extensively through tests of lap splices.[4-7] Tension lap splices generally fail by splitting of the surrounding concrete in one of the basic patterns shown in Fig. 8.5. At very close spacings the side split failure will occur and additional cover does little to change failure (Fig. 8.6*a*); lessened cover tends to change it to the face-and-side split type. In

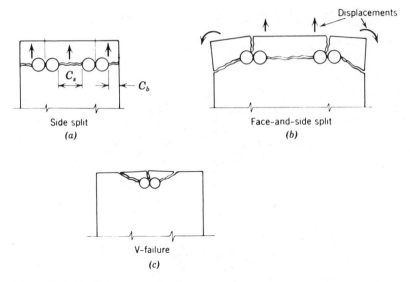

Side split
(a)

Displacements

Face-and-side split
(b)

V-failure
(c)

Figure 8.5 Splitting around splices.

the latter failure the face splitting over part of the length leaves weak corner sections and finally reaches enough of the length to cause the side split failure at a lowered efficiency (Fig. 8.5*b*). Extra cover in this case increases the resistance offered to the face splitting; lesser cover tends to move the failure toward the V-type (Fig. 8.5*c*) where a splice is not seriously influenced by further spacing changes. The steel stress changes in the spliced bars and just before failure becomes nearly linear in both bars (Fig. 8.7*a*).

An adequate splice of a large bar goes through several stages in addition to flexural cracking. In general,

1. Splitting starts at the ends of the splice, beginning from the end flexural cracks. The splitting may be on the tension face or on the sides of the beam.
2. Splitting progresses toward the middle of the splice, although not usually completely to the next crack before splitting starts there also, and so on. Near failure, splitting may extend over 30% to 80% of the splice length.
3. Final failure is sudden and complete unless confining steel is present. Such confining steel must cross the critical plane of splitting.

The effectiveness of the concrete in tension resisting splitting just prior to failure varies considerably over the splice length. It is most effective near midlength where cracking is less and least effective where splitting has started. The conditions producing splitting are shown qualitatively in Fig. 8.8.

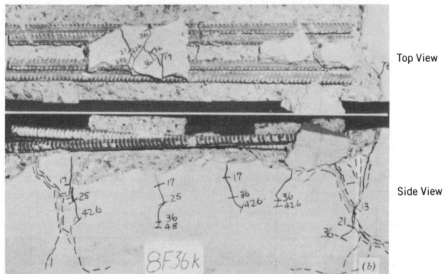

Top View

Side View

Figure 8.6 Splice failure in test beams. (*a*) Spaced splice with side split failure. Arrows mark ends of bars where they have separated from the concrete as they slipped at failure. (*b*) A longer contact splice, also in a constant moment length; a face-and-side split failure.

8.4 Experimental Studies on Bond Stress—Development Length and Splices

A variety of test methods[9] for determining bond stresses or development length have been reported. A number of problems need to be addressed, however, to apply the results to design. These problems include:

1. *Inability to scale down specimens with confidence.* Possibly because much higher bond stresses can be developed around small bars than large ones or

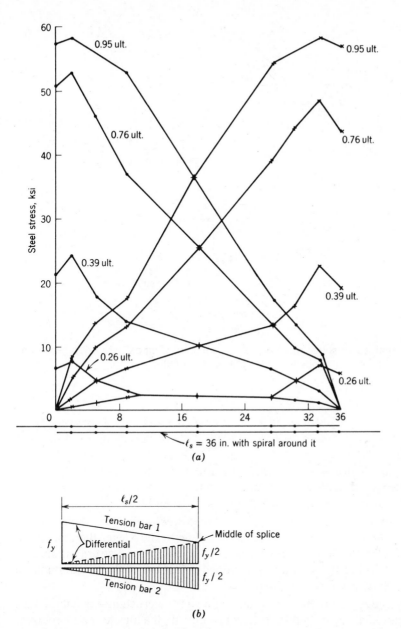

Figure 8.7 (*a*) Typical stress distribution along No. 14 splice at various load levels. Constant moment zone; face-and-slide split failure. (*b*) Stress differential between two bars (roughly).

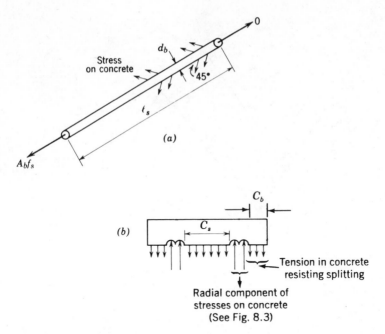

Figure 8.8 Analysis of splice. (*a*) Stresses on concrete from bar. (*b*) Equilibrium on short length of cover over splice in side split mode of failure.

probably because the spacing of flexural cracks does not scale down properly as specimen size is reduced, scaled specimens react differently from full size ones.

2. *Interaction between shear and bond stress*. Combinations of shear and bond stress can lead to lower shear strengths. The mixture of shear and bond problems in tests is evident in Fig. 8.4*b* and 8.4*c*.

3. *Nonlinear response, either with bar size, steel stress, or development length* (*which in design are related quantities*). The longer the test length of the bar in the specimen the lower the average bond stress attained at failure. A Grade 40 bar can be developed in less than $\frac{2}{3}$ the length required for the same size Grade 60 bar.

4. *Wide range of failure modes*. Failures range from crushing against the lugs or longitudinal shearing of a surface just outside the lugs to the more usual splitting of concrete, either over individual bars or through an entire layer of bars. The tensile splitting stress and the influence of cover thickness on ultimate strength make generalizations difficult.

The only completely realistic test is a full size specimen combining the given variables. This is a very expensive procedure and difficult to generalize for different variables.

(a) Bond Pullout Tests   Permissible bond stresses were formerly established largely from pullout tests with some beam tests as confirmation. A bar was embedded in a cylinder or rectangular block of concrete and the force required to pull it out or make it slip excessively was measured. Figure 8.9 shows such a test schematically, omitting details such as bearing plates. Slip of the bar relative to the concrete is measured at the bottom (loaded end) and top (free end). Even a very small load causes some slip and develops a high bond stress near the loaded end, but leaves the upper part of the bar totally unstressed, as shown in Fig. 8.9. As more load is applied, the slip at the loaded end increases, and both the high bond stress and slip extend deeper into the specimen. The maximum bond is somewhat idealized in these sketches; its distribution depends on the type of bar and probably varies along the bar more than shown.

When the unloaded end slips, the maximum resistance has nearly been reached. Failure will usually occur (1) by longitudinal splitting of the concrete in the case of deformed bars, or (2) by pulling the bar through the concrete in the case of a very small bar or very lightweight aggregate, or (3) by breaking the bar, if the embedment is long enough.

The average bond resistance u is calculated assuming uniform bond over the bar embedment length. Friction on the base restrains splitting of the specimen. Many tests include spirals to avoid splitting collapse. The test appears useful chiefly where relative rather than real bond resistance is acceptable, as in comparing various lug sizes and patterns. The principal problem of splitting is not realistically handled.

Modifications of this test, called the tensile pullout specimen, have also been used (Fig. 8.10) to eliminate compression on the concrete.

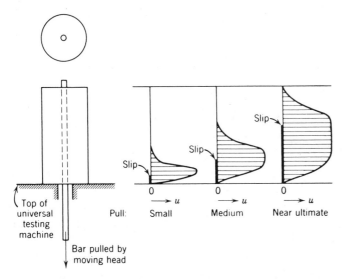

Figure 8.9   Bond pullout test, with bond stress distributions.

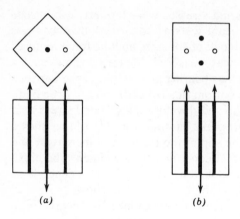

Figure 8.10 Tension pullout tests, schematic.

(a)   (b)

(b) Bond Beam Tests   Beam tests are considered more reliable because the influence of flexural tension cracks is included. Two types of beams have been used: the one of Fig. 8.11 at the Bureau of Standards and the one of Fig. 8.12 at The University of Texas at Austin. A major consideration in each was to remove reaction restraints that might confine the concrete over the bar and thereby increase splitting resistance.

The Bureau of Standards beams were heavily reinforced with stirrups and failed as sketched in Fig. 8.11b; the diagonal crack created an increased steel stress much closer to the end of the bar and concentrated bond stresses nearer the end.

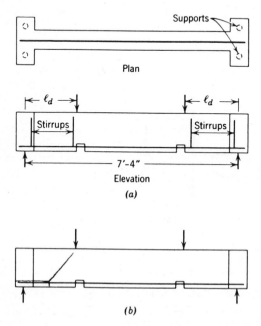

Figure 8.11 National Bureau of Standards bond test beam. (*a*) Load and support system. (*b*) Failure pattern.

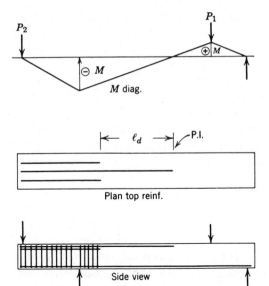

Plan top reinf.

Side view

Figure 8.12 The University of Texas beams. (Failure shown in Fig. 8.4.)

In the University of Texas beams (Fig. 8.12), the bar was placed in a negative moment region where there was no external restraint against splitting except by the use of a rather wide concrete beam.[4] Some beams had stirrups, some did not. Figure 8.4b and d shows beams that failed by splitting to the end of the bar, and Fig. 8.4c shows a splitting crack extending from a diagonal crack.

(c) Semibeam Specimens To reduce specimen size and expense, partial beam specimens (often called stub beam or cantilever tests) have been used as sketched in Fig. 8.13. Although various details are used, those in the figure

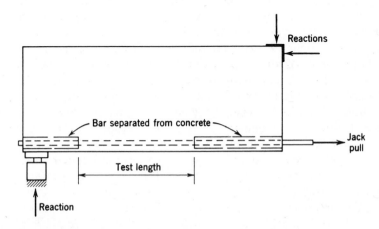

Figure 8.13 Stub beam or cantilever bond specimen.

convey the general idea. The bottom reaction may bear against the end of the bar, may be spread to avoid producing compression on the bar or the bar may be shielded and isolated by a soft covering as sketched. The bar is loaded directly.

The overall length and the part of it used for the test length can be varied, anything from the full length of the specimen to the reduced length. One or several bars, with or without stirrups may be used.

The advantage of this specimen lies in its simplicity. The disadvantages lie in the confining pressure against the bar (if the shield is not used), increases in the length over which splitting resistance tends to be mobilized, and in the unrealistically low ratio of shear-to-bond.

(d) Splice Tests Splice tests generally have been conducted using simple beams with splices located in a constant moment region. Some tests have been conducted to determine the influence of moment gradient on splices. The most severe condition is one where both bars in the splice are subjected to high stresses. Splice conditions are much easier and less costly to simulate in the laboratory. Splice tests have been realistic simulations of real conditions in structures, but development tests have been conducted largely using pullout tests in which splitting failures were purposely avoided. As a result, the bond stresses developed along splices are low compared with the bond along a bar in a pullout test. As will be discussed later, the difference in test methods is responsible for large differences in Code-required anchorage lengths for splices and development.

8.5   Bond Stresses

Development length or development bond was a part of the 1963 Code, but more emphasis was then placed on flexural bond. Flexural bond is the unit bond stress required to resist a change in moment along a unit length of the beam. It can be expressed as the shear force at the section divided by the internal moment arm and the perimeter of tensile bars. Bond stress as such has been omitted from the Code since the 1963 version. Subsequent Codes specify development lengths ℓ_d of about 1.2 times those obtained using 1963 bond stress (based on test results). The 20% increase is to assure flexural ductility when the steel strain exceeds f_y. A failure in the development length tends to be brittle.

In the formulation of required development and lap splice lengths, some protection against a brittle type of failure must always be considered. To design a beam that is somewhat ductile in flexure but to allow it to fail suddenly in bond splitting at the yield stress is inconsistent. Hence reserve strength in ℓ_d above f_y is essential to maintain member ductility. The Code values aim at making this extra strength about 20% of f_y. However, especially in splices and possibly in simple development lengths, the steel strain that can be accommodated may be limited. If the steel has a long yield plateau on its stress-strain curve the 20% extra strength may not be attainable, but the ability to accept a high strain serves the same function.

For a #3 to #11 bar (other than top bar) with an ultimate bond stress $u = 9.5\sqrt{f'_c}/d_b$; the required ℓ_d using Eq. 8.1 is

$$\begin{aligned} \text{Reqd. } \ell_d &= 1.2f_y d_b/4u = 1.2f_y d_b/(4 \times 9.5\sqrt{f'_c}/d_b) \\ &= 0.0316f_y d_b^2/\sqrt{f'_c} = 0.0316f_y(4A_b/\pi)/\sqrt{f'_c} \\ &= 0.0403f_y A_b/\sqrt{f'_c} = \text{say, } 0.04f_y A_b/\sqrt{f'_c} \end{aligned} \qquad (8.2)$$

For spacing at least 6 in. on center, a length 0.8 times this value is permitted. Because $1.2 \times 0.8 = 0.96$, this produces development lengths for the wide bar spacings essentially equal to the 1963 values.

Although this approach seems logical, in the early 1970s it became clear that no single development length (or allowable bond stress) for a given bar size could accurately represent bar covers from 0.75 in. to 2.5 or 3 in., nor clear bar spacings from 1 in. to 6 in., nor the absence or presence of stirrups (in a wide range of spacings). Accumulated data was inadequate to cover any one of these variables much less all three.

Orangun et al.[8] proposed an approach for determining bond stresses that included all the variables mentioned. The background for this work was stated earlier in References 5 and 6. The radial forces generated between the lugs and the surrounding concrete can be regarded as water pressure acting against a thick-walled cylinder with an inner diameter equal to the bar diameter and a thickness C that is the smaller of (1) the clear bottom or side cover C_b or (2) $\frac{1}{2}$ the clear spacing C_s between adjacent bars. (See Fig. 8.14.) The capacity of the cylinder depends on the tensile strength of the concrete. With C_b greater than $C_s/2$, a horizontal split develops at the level of the bars and is termed a "side split failure." With $C_s/2$ greater than C_b, a "face-and-side split failure" forms with longitudinal cracking through the cover followed by splitting through the plane of the bars. When $C_s/2$ is much greater than C_b, a "V-notch failure" forms with longitudinal splitting followed by inclined cracks that separate a V-shaped segment of cover from the member. These failures have been described by ACI Committee 408.[9]

The same approach was applied to the spliced bars shown in Fig. 8.5. The concrete "ring" around the bar takes an elongated shape but the concept is the same as for the case of a single bar being developed.

From the results of over 500 development and splice tests, a regression analysis was used to develop an equation in which the bond stress was expressed in terms of the concrete strength f'_c, bar diameter d_b, thickness of concrete cylinder surrounding the bar C, length of anchored bar ℓ_d or ℓ_s, and transverse reinforcement $A_{tr}f_{yt}/s$. The transverse reinforcement A_{tr} was taken as the area of the bars crossing the plane of splitting. The yield strength of the transverse reinforcement is f_{yt} and the spacing between the transverse reinforcement is s. Figure 8.15 shows several cases that indicate how terms are defined.

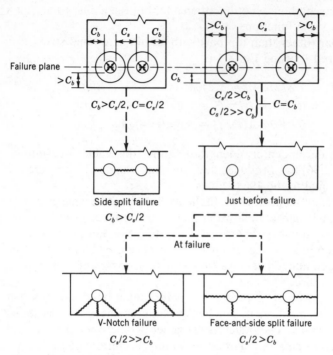

Figure 8.14 Splitting failure patterns.

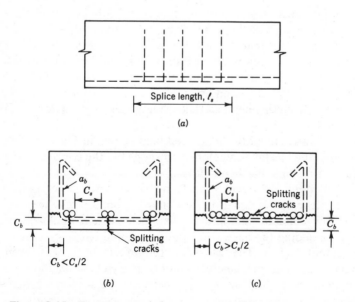

Figure 8.15 Transverse reinforcement. (*a*) Side view of beam. (*b*) All splitting cracks restrained, $A_{tr} = a_b$. (*c*) Two bars crossing plane of splitting through layer of bars, $A_{tr} = 2a_b/4$ splices.

Using this approach, the bond strength for bottom bars is

$$\frac{u_{calc}}{\sqrt{f'_c}} = 1.2 + \frac{3C}{d_b} + \frac{50d_b}{\ell_s} + \frac{A_{tr}f_{yb}}{500\,s d_b} \qquad (8.3)$$

Similarly, the approach used in the Code can be rewritten in terms of bond stress by using Eqs. 8.1 and 8.2.

$$\frac{u_{318}}{\sqrt{f'_c}} = \frac{8}{d_b} \qquad (8.4)$$

Using measured stresses from development and splice tests, bond stresses from tests u_{test} can be determined by Eq. 8.1. The ratios of test to calculated or test to ACI 318 values can be determined. Histograms of the ratios for development and splice test data are plotted in Fig. 8.16. Using Eq. 8.3, the distribution of ratios for both development and splices show average values around 1.0 with a relatively small standard deviation.

Using the ACI 318 approach (Eq. 8.4) results in a broad range of ratios for development with large standard deviation and a large number of ratios falling well below 1.0. The histogram indicates the possibility of unconservative devel-

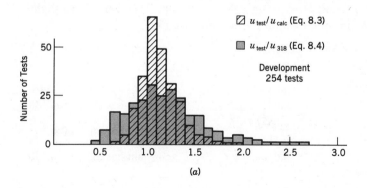

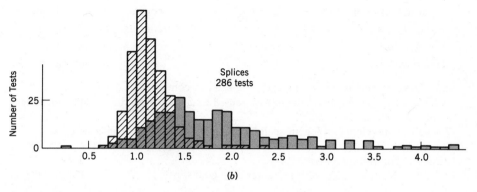

Figure 8.16 Ratio of test-to-Code and test-to-calculated bond stresses. (*a*) Development. (*b*) Splices.

opment lengths using the Code, especially when cover and/or spacing between bars is small. For splices, the Code ratios are very high, which indicates the overly conservative nature of Code provisions for splices. The large standard deviations for both splices and development indicate that the Code approach does not consider a sufficient number of variables to provide a realistic estimate of bond strength. These comparisons could be anticipated considering the fact that the bond stress used in the Code was based on pullout tests where splitting was prevented and splices were tested in beams when splitting nearly always led to failure.

A number of proposals have been presented for revising the development length and splice provisions of the Code. ACI Committee 408[10,11] proposed a revision based on the work of Orangun et al.[8] It is anticipated that changes in development provisions will appear in future Codes. Similar approaches have been developed in Sweden[12] and Germany and are now being incorporated into various national codes.

8.6 Development Lengths for Tension Bars

The 1986 Code specifies four basic development lengths for tension bars derived from ultimate bond stresses in the 1963 Code. This version is slightly rearranged for ease in understanding Code provisions.

Code 12.2 Development length of deformed bars and deformed wire in tension 12.2.2 Basic development length shall be:

For #11 or smaller bars	$0.04 A_b f_y / \sqrt{f'_c}$	(constant has units of 1/in.)
but not less than	$0.0004 d_b f_y$	(constant has units of in.²/lb)
For #14 bars	$0.085 f_y / \sqrt{f'_c}$ ⎫	
For #18 bars	$0.11 f_y / \sqrt{f'_c}$ ⎭	(constant has units of in.)
Deformed wire	$0.03 d_b f_y / \sqrt{f'_c}$	(constant has units of in.²/lb)

SI equivalents are:

For #11 or smaller bars	$0.02 A_b f_y / \sqrt{f'_c}$	(constant has units of 1/mm)
but not less than	$0.06 d_b f_y$	(constant has units of mm²/N)
For #14 bars	$25 f_y / \sqrt{f'_c}$	(constant has units of mm)
For #18 bars	$35 f_y / \sqrt{f'_c}$	(constant has units of mm)
Deformed wire	$3 d_b f_y / 8\sqrt{f'_c}$	(constant has units of mm²/N)

12.2.1 Development length ℓ_d, in inches, ... shall be computed as the product of ... Sec. 12.2.2 and the applicable ... factors of Secs. 12.2.3 and 12.2.4, but ℓ_d shall not be less than specified in Sec. 12.2.5.

12.2.3 Basic development length shall be multiplied by the applicable factor or factors for:

12.2.3.1 .. Top reinforcement* 1.4

12.2.3.2 .. Reinforcement with f_y greater than 60,000 psi .... $(2-60,000/f_y)$

---

* See following discussion.

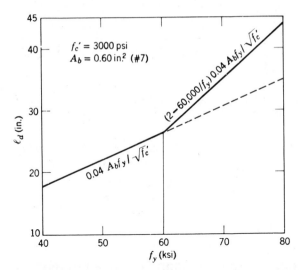

Figure 8.17 Development length variation with f_y for #7 bar.

12.2.3.3 .. Lightweight aggregate concrete (when f_{ct} is not specified*):

"all-lightweight" concrete.................................... 1.33

"sand-lightweight" concrete.................................. 1.18

(interpolating when part sand replacement is used)

12.2.4 ... (multiply by applicable factors for ...)

12.2.4.1 .. Reinforcement being developed in length under consideration and spaced laterally at least 6 in. on center with at least 3 in. clear from face of member to edge bar ... in direction of spacing .. 0.8

12.2.4.2 .. Reinforcement in excess of required (A_s reqd./A_s provided)

12.2.4.3 .. Reinforcement enclosed within spiral ... not less than $\frac{1}{4}$ in. diam. and not more than 4 in. pitch 0.75

12.2.5 Development ... shall be not less than 12 in., except in ... lap splices and in ... web reinforcement by Code 12.14.

The limitation in Sec. 12.2.2 of $0.0004\,d_b f_y$ was determined using the limiting upper value of bond stress in the 1963 Code and generally only controls for small diameter bars.

Because of the nonlinear relationship between ℓ_d and f_y, the equations are unconservative for 75-ksi bars. Thus Sec. 12.2.3.2 specifies that for $f_y > 60,000$ psi the lengths are also to be increased by the factor $2-60,000/f_y$ applied to a basic ℓ_d value, which already includes the larger f_y, say, 75,000 psi, as indicated in Fig. 8.17.

For top bars, that is, horizontal reinforcement having more than 12 in. of concrete cast below the bars, 1.4 times the basic ℓ_d must be used. An accumulation of air and water that rises beneath such bars becomes entrapped on their

---

* Code also includes case where f_{ct} is specified.

undersides. The poor concrete at the interface leads to lower bond and greater slip at comparable loads.

As discussed in Sec. 8.5, better anchorage results when bars are widely spaced. An 0.8 factor may be used for bars spaced at least 6 in. on center (and not less than half that clear distance from an edge). Reduced development length is also specified when a spiral is available to control splitting of the concrete or where excess steel is available and it is not necessary to develop the full yield strength.

Particular attention is called to the increased development lengths needed with lightweight concrete, that is, 1.33 for "all lightweight" (including lightweight for the fine aggregate) and 1.18 for "sand-lightweight" concrete. When the mix is well-controlled and split cylinder data are available this rule is relaxed (Code 12.3.3).

Table B.3 in Appendix B summarizes many of the ℓ_d values for tension bars.

8.7 Critical Stresses and Critical Sections for Development

Code Secs. 12.10.2 and 12.10.8 broadly define the critical sections for development. The proper use of development lengths requires a clear picture of how bar tensile or compressive stresses change along a beam. Bar stresses will be a maximum where moment is a maximum; obviously no maximum stress point should be closer than ℓ_d to the end of a bar in either direction. Adequate development length must be available to permit transfer of forces from the bar to the concrete through bond.

Equally important and easier to overlook is the peak stress that develops in bars wherever a neighboring tension bar is cut off or bent. At that point the total tension must be carried by the (reduced number of) continuing bars. Because the cutoff for moment is usually made where the remaining bars have just enough capacity to handle the total tension, the full yield stress in the continuing bars requires that ℓ_d be provided beyond that point.

8.8 Examples Showing Use of Development Lengths

Examples illustrate more specifically some of the general ideas discussed in previous sections.

(a) A 12-in. concrete wall heavily loaded requires a footing 5-ft 6-in. wide as sketched in Fig. 8.18. The footing is designed for $f'_c = 3000$ psi and Grade 60 reinforcement. What maximum size bars will conform to the development length requirements?

Solution

With a concrete wall, the critical footing section for flexure is at the face of the wall where the bars are stressed to maximum values. The bars

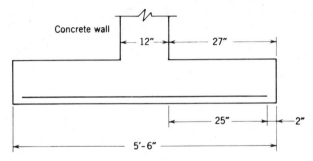

Figure 8.18 Concrete wall footing.

project 25 in. (leaving 2-in. end cover) beyond the critical section. Based on the values of Sec. 8.6, a bar must be found that has an ℓ_d of 25 in. or less. In the absence of a table of ℓ_d values of various bars and conditions (which most design offices have developed or see Table B.3), the required ℓ_d is:

$$\ell_d = 0.04 A_b f_y / \sqrt{f_c'} \text{ but not less than}$$
$$\ell_d = 0.0004 d_b f_y$$

The second value governs only with small bars and will not be checked initially.

$$0.04 A_b 60,000 / \sqrt{3000} = 25 \text{ in.}, \qquad A_b = 0.57 \text{ in.}^2$$

A #6 bar ($A_b = 0.44$) is the largest permissible straight bar. Checking the other limit with d_b for the #6 bar,

$$0.0004 d_b f_y = 0.0004 \times 0.75 \times 60,000 = 18 \text{ in.} < 25 \text{ in.} \qquad \textbf{O.K.}$$

The number of bars is determined by flexure requirements. If this number permits a 6-in. spacing or more, a larger bar may be possible because the required ℓ_d can then be reduced by an 0.8 factor to give

$$0.8(0.04 A_b \times 60,000 / \sqrt{3000}) = 25 \text{ in.}$$
$$A_b = 0.71 \text{ in.}^2$$

A #7 bar is permissible. The alternate equation need not be checked again, since the 0.8 factor also applies to it. The #7 bar possibility improves the chances of using the wider spacing.

It is noted that a hook at the end of bar would reduce the requirement for straight bar development length and permit still larger bars (Sec. 8.13).

(b) Assume the footing is made of sand-lightweight concrete but the example is otherwise the same.

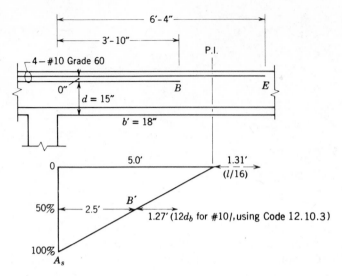

Figure 8.19 Data for example 8.8(c).

Solution

Required development length is increased by a 1.18 factor when light-weight concrete with a sand fine aggregate is used. If the spacing is under 6 in.,

$$1.18(0.04 A_b \times 60,000/\sqrt{3000}) = 25 \text{ in.}$$
$$\text{Max. } A_b = 0.48 \text{ in.}^2$$

A #6 bar remains the largest usable bar (0.44 in.²) at this spacing.

(c) The 4-#10 bars in Fig. 8.19 are designed for the moment diagram shown. The beam span between supports is 21 ft. The P.I. is 5 ft from the face of the support. At the support 4-#10 bars are required for flexure. The 1.31-ft extension of 1/3 of the bars beyond the P.I. is required by the $\ell/16$ rule* (Code 12.12.3). May two bars be cut off as indicated on the moment diagram? Assume that the shear requirements of Code 12.10.5 have been satisfied.

Solution

If the bars are cut as shown, the spacing of the extended bars (from B to E) in an 18-in. beam width will be approximately 9 in. The extended bars are developed between B and E. The two short bars extending from the

---

* Or d or $12 d_b$, if larger.

support to B are developed from B to the support. For the two bars between the support and B the effective spacing also could be taken as 9 in. as only two of the four bars are developed over that portion of the beam. With the wide spacing ≥ 6 in., the 0.8 coefficient of Code 12.2.4.1 can be used. As top bars they require a 1.4 coefficient (Code 12.2.3.1).

$$\text{Reqd. } \ell_d = 0.8 \times 1.4(0.04 A_b f_y / \sqrt{f_c'})$$
$$= 0.8 \times 1.4 \times 0.04 \times 1.27 \times 60{,}000/\sqrt{3000}$$
$$= 63 \text{ in.} = 5.25 \text{ ft}$$

The "short" bars must extend 5.25 ft which is beyond the P.I., making a moment *cutoff unfeasible*.

(d) If in Fig. 8.19 the moment requires only 4-#8 bars, may the given dimensions be used with the #8 bars?

Solution

ℓ_d in this case involves the same 0.8 and 1.4 coefficients as in (c).

$$\ell_d = 0.8 \times 1.4 \times 0.04 \times 0.79 \times 60{,}000/\sqrt{3000} = 39 \text{ in.}$$

The ℓ_d of 39 in. must be compared with the required length of 2.5 ft (30 in.) to the theoretical cutoff—point B' plus an extension of d or $12 d_b$ past B' for a total length of 30 in. + 15 in. = 45 in. The 3-ft 10-in. dimension shown in Fig. 8.19 is satisfactory for the bars from the support to B.

The development of the longer bars must also be considered. The longer bars need ℓ_d of 39 in. or 3.25 ft beyond B'. The total distance from the column is 2.5 + 3.25 = 5.75 ft. Code 12.12.3 requires the two bars to extend past the P.I. a distance of $\ell/16$. Therefore, the length provided, 6-ft 4-in., is greater than the 5.75 ft required.

As cutting off some bars frequently requires others to be run farther, each bar cutoff point becomes a critical section to be checked with regard to continuing bars.

8.9   Positive Moment Bars

(a) Continuous Beams*   When uniform loads are considered in simple spans or within the positive moment length of continuous spans, there is a problem in applying the basic development length concept in a meaningful way. The problem is one of determining the maximum bar size where moment is nearly zero. The solution is essentially the equivalent of the old flexural bond concept, which is presented below in terms of development length and some Code liberalization.

---

* The detailing of continuous beam reinforcement, including detailed examples, is covered in Chapter 9.

Consider the case shown in Fig. 8.20a and a simple case where ℓ_d is equal to the distance ℓ_0 to midspan. Such a bar would develop yield strength at midspan but would be unsafe at $\ell_0/2$ from the P.I. There the bar would be developed for only $1/2\,f_y$ but would need to carry $0.75\,M_n$, which requires $0.75f_y$. A similar overstress exists at points closer to the P.I. The simple answer developed later is identical with the old flexural bond demands, but encourages realistic adjustments to relax these demands.

The slope of the moment diagram at the P.I. is the shear V_n at that point. If the bars at the P.I. develop toward f_y at the rate M is initially developing, they can develop their full value f_y in the length M_n/V_n, as indicated in Fig. 8.20a. If the bars extend beyond the P.I., according to Code Secs. 12.11.1 and 12.10.3, the development length concept implies that local "overstress" in bond at the P.I. is not a problem because there is no stress at the end of the bar. The extra bar length beyond the P.I. that may be counted is limited by Code 12.11.3 to ℓ_a (the effective beam depth d or $12d_b$, whichever is greater). The criterion becomes:

$$\text{At a P.I. choose bars with } \ell_d \gtrless M_n/V_n + d$$

The dotted development rate line indicates this will be safe.

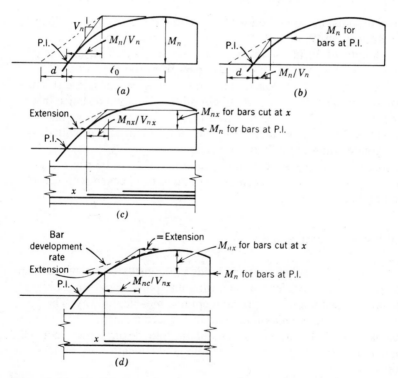

Figure 8.20 Available development lengths for positive moment bars at a P.I. (a) No bars cut. (b) Bars at P.I. less than total A_s. (c) General case. (d) For first bars cut off (nearest point of maximum positive moment.)

If not all bars extend to the P.I., Fig. 8.20b indicates the only change needed is to include in M_n only the bars actually reaching the P.I.

Usually, if bar size at the P.I. is satisfied and other positive moment bars are not too widely different in size, the required bar lengths for flexure will control over development lengths. However, if larger bars are used away from the P.I. they could be selected by the same procedure used at the P.I. with one modification indicated in Fig. 8.20c. These bars must care for all or part of the moment in excess of M_n developed by the other bars at the P.I. If their moment capacity is designated M_{nx}, the local demands indicate ℓ_d equal to or less than M_{nx}/V_{nx}, where V_{nx} is the shear at x. Any further extension of the bar obviously adds to the available ℓ_d to give the criterion for bars not extending to the P.I. as:

$$\text{For the added bars } \ell_d \gtrless M_{nx}/V_{nx} + \text{extension used}$$

If such a criterion is used for the first bars cut off, the moment diagram will be rather flat. Any added arbitrary bar extension not only moves the critical bond stress away from the nominal cutoff point but permits a flatter development line as sketched in Fig. 8.20d. It appears safe, *for a moment diagram that can shift very little,* to use the criterion:

$$\text{For the added bars } \textit{first} \text{ cut off } \ell_d \gtrless M_{nx}/V_{nx} + \text{twice the extension}$$

The need for this approach occurs only when the moment diagram has this general shape and when larger bars are used here than at the P.I. If the moment diagram were a straight line, as for a concentrated midspan load, the procedures used for negative moment would be adequate.

(b) Simple Spans   At the support of a simple span that is hung from above by an embedded bar hanger, the condition is exactly that at the P.I. of the continuous span, except that usually the bar ends with an extension ℓ_a less than d (Code 12.11.3). However, more commonly the reaction is from below and provides a compressive force that restricts possible splitting of concrete caused by bond stresses. The Code recognizes this by specifying that the bars may be chosen from the criterion:

$$\text{For bars confined by a reaction use } \ell_d \gtrless 1.3 M_t/V_u + \ell_a$$

The 1.3 factor results in development requirements at a support that are less strict than at a P.I. *Simple* beams can rarely be supported by framing into a girder that picks up the reaction in shear, but where such a case exists the 1.3 factor must not be used.

8.10   Splices in Tension

It is generally necessary to splice tension bars, partly because of the limited (usually 60 ft) length of commercial bars, but more because of the awkwardness of interweaving long bars on the job. Splicing may be by welding, by mechanical connections (fusion type, tapered threads, or extruded), or most frequently,

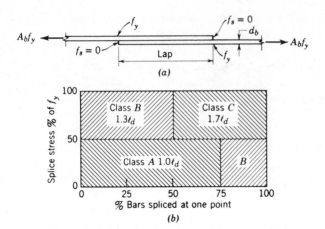

Figure 8.21 Tension lap splices. (*a*) Diagram of stress transfer. (*b*) Requirements in terms of stress and percent of area spliced at one section.

for bars #11 and smaller, by lapping bars as in Fig. 8.21*a*. The lapped bars are commonly tied in contact with each other, but may be spaced up to 6 in. apart with an upper limit of 1/5 the lap length.

With development lengths already specified, the splice lap in tension has been specified (Code 12.15) as a multiple of ℓ_d, where ℓ_d in this case must be based on the *full f_y*. Unless the splice can be made away from a point where the bar stress is high, the multiple must be from 1.3 to 1.7, as diagrammed in Fig. 8.21*b* and outlined later. The factors that led to the increase above the ℓ_d values are discussed in Sec. 8.5.

The Code has broken splices into three classes in increasing order of severity of conditions, as follows:

For stress always less than $f_y/2$:

Class A if not over 75% of bars are spliced within one lap length. Use $1.0\ell_d$.
Class B if more than 75% are spliced within one lap length. Use $1.3\ell_d$.

For stress exceeding $f_y/2$:

Class B if not over half the bars are spliced within one lap length.
Use $1.3\ell_d$.
Class C for more than half the bars spliced within one lap length.
Use $1.7\ell_d$.

Note that 12 in. is the minimum for a splice lap. (The 12-in. minimum for ℓ_d is not used to calculate the splice, but the splice length must still be at least 12 in.)

Attention is directed to the savings in splice length if splices are not all located at the same section or staggered (Fig. 8.21). A reduction from $1.7\ell_d$ to $1.3\ell_d$ results in a net savings of $0.4\ell_d$. The wider spacing of splices may then qualify for the 0.8 factor on ℓ_d in some cases. The behavior of a member with closely spaced splices is much improved if only half the bars are spliced, an improvement that automatically results from staggering.

8.11 Tension Splice Design Example

(a) Figure 8.22 shows a splice in a column. If the column section design is controlled by bending and axial load (with the axial load less than that at the balanced point), determine the splice length. The column bars are #8, Grade 60, and concrete strength $f'_c = 4000$ psi.

Solution

Because the stress in the steel in the tension face of the column will reach yield for the combination of bending and axial load described, the lap is a class C splice because it is in a region where more than 50% of f_y in tension will be reached and all bars are spliced at the same point. (A typical construction practice is to splice bars just above the floor so that long bars do not interfere with floor casting operations.)
Therefore,

$$\ell_d = 0.04 A_b f_y / \sqrt{f'_c} = 0.04 \times 0.79 \times 60{,}000 / \sqrt{4000} = 30.0 \text{ in.}$$

but not less than $0.0004 d_b f_y = 0.0004 \times 1.0 \times 60{,}000 = 24$ in. < 30 in.
For a class C splice

$$\ell_s = 1.7 \ell_d = 1.7 \times 30 = 51.0 \text{ in.}$$

(b) The splices are staggered so that only half of the bars are spliced at the same point or the loads on the column are such that the tensile stresses in the bars do not exceed $f_y/2$. Determine splice length.

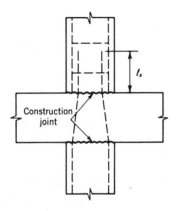

Figure 8.22 Column splice detail.

Solution

A class B splice can be used.

$$\ell_s = 1.3\,\ell_d = 1.3 \times 30 = 39 \text{ in.}$$

8.12 Splicing of Stirrups

In some congested spots the use of two U-stirrups (without hooks) turned together to form closed stirrups is feasible. Tested in either shear or torsion, members so reinforced show reasonable ductility, but final failure tends to be more sudden and complete when it occurs.

Code 12.13.5 covers the requirements:

> Pairs of U-stirrups or ties so placed as to form a closed unit shall be considered properly spliced when the length of laps are $1.7\,\ell_d$. In members at least 18 in. deep, such splices with $A_b f_y$ not more than 9000 pounds per leg may be considered adequate if the stirrup legs extend the full available depth of member.

The last sentence permits stirrup laps of #3 bars of either Grade 40 or 60 and #4 bars of Grade 40. The #3 stirrup lap for Grade 60 is only about 10% less than 1.7 times the theoretical ℓ_d.

8.13 Hooks and End Anchorage For Bars in Tension

Where bars must be anchored (or developed) in tension within a limited distance, a hook can be used. A hook is considered ineffective in compression (Code 12.5.5). A tension hook can be used to advantage where no room exists for the necessary length of straight bar. For example, the footing bars of the example in Sec. 8.8a can be larger if hooks are added at the bar ends.

A standard hook is defined (Code 7.1) as having one of the forms shown in Fig. 8.23a. A general evaluation of a hook is difficult. In mass concrete or other

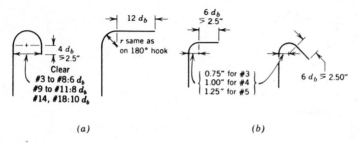

(a)   (b)

Figure 8.23 Standard hooks. (a) General case. (b) Ties and stirrups only. (Radius and diameter are each to inside of hook.)

confined conditions a hooked bar slips more than the same length of straight bar, although the same strength may be reached. Figure 8.24a shows stress-slip curves from a series of tests by Minor.[13] In a structural member where the bar is close to an exposed face, the hook pressure on the concrete tends to split off the concrete face and a lowered strength results. In such conditions the value is governed by the splitting resistance of the concrete. For this condition the strength is lower than for a hook in mass concrete. A splitting failure can be seen in Fig. 8.24b. A series of tests of hooked bars were conducted at Rice University[14] and at the University of Texas[15] and form the basis for provisions in the 1986 Code that provide a basic development length ℓ_{hb}. When there is larger cover over the bar or transverse reinforcement to cross the side plane of splitting, ℓ_{hb} may be reduced. The provisions of Code 12.5 are reproduced for ease in following the discussion related to the design provisions and for reference when studying the examples of Sec. 8.14.

12.5 Development of Standard Hooks in Tension

12.5.1 Development length ℓ_{dh}, in inches, for deformed bars in tension terminating in a standard hook (Section 7.1) shall be computed as the product of the basic development length ℓ_{hb} of Section 12.5.2 and the applicable modification factor or factors of Section 12.5.3, but ℓ_{dh} shall be not less than $8d_b$ or 6 in., whichever is greater.

12.5.2 Basic development length ℓ_{hb} for a hooked bar with f_y equal to 60,000 psi shall be . . . . $1200d_b/\sqrt{f_c'}$*;SI: $(100\ d_b/\sqrt{f_c'})$

12.5.3 Basic development length ℓ_{hb} shall be multiplied by applicable factor or factors for:

12.5.3.1 Bar yield strength
Bars with f_y other than 60,000 psi . . . . $f_y/60,000$

12.5.3.2 Concrete cover
For #11 bar and smaller, side cover (normal to plane of hook) not less than $2\frac{1}{2}$ in., and for 90° hook cover on bar extension beyond hook not less than 2 in. 0.7

12.5.3.3 Ties or stirrups
For #11 bar and smaller, hook enclosed vertically or horizontally within ties or stirrup-ties spaced along the full development length ℓ_{dh} not greater than $3d_b$, where d_b is diameter of hooked bar. 0.8

12.5.3.4 Excess reinforcement
Where anchorage or development for f_y is not specifically required, reinforcement in excess of that required by analysis ... $(A_s$ required$)/(A_s$ provided$)$.

12.5.3.5 Lightweight aggregate concrε .. 1.3

12.5.4 For bars being developed by a standard hook at discontinuous ends of members with both side cover and top (or bottom) cover over hook less than $2\frac{1}{2}$ in., hooked bar shall be enclosed within ties or stirrup ties spaced along the full development length ℓ_{dh} not greater than $3d_b$, where d_b is diameter of hooked bar. For this case, factor of Section 12.5.3.3 shall not apply.

12.5.5 Hooks shall not be considered effective in developing bars in compression.

The development length ℓ_{dh} of a hooked bar is the distance measured from the critical section to the outside end of a hook, as shown in Fig. 8.25. The

* Constant carries unit of lb-in² or in N/mm² (S.I.)

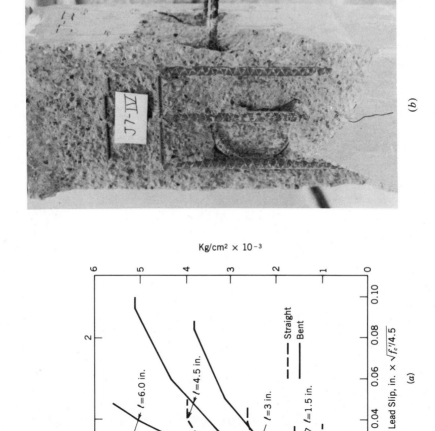

Figure 8.24 (*a*) Stress-slip curves for straight and bent bars of different lengths (Reference 13). (*b*) Side face splitting of a 180° standard ACI hook (Reference 14).

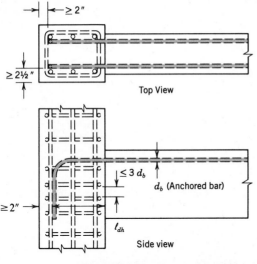

Top View

$\leq 3\,d_b$

d_b (Anchored bar)

ℓ_{dh}

Side view

$$\ell_{dh}=\ell_{hb}\times\begin{cases}0.7 \text{ Cover} \geq 2\frac{1}{2}\text{" side, } \geq 2\text{" end (Code 12.5.3.2)}\\0.8 \text{ Ties @ } 3\,d_b \text{ or less (Code 12.5.3.3)}\\0.7\times 0.8 \text{ Ties and cover as above}\end{cases}$$

Figure 8.25 Cover and tie spacing required for reduction in hooked bar development length.

basic development ℓ_{hb} is reduced when the cover over the bar (measured normal to the plane of the hook) is at least $2\frac{1}{4}$-in. and the cover on the tail extension is at least 2 in. Figure 8.25 shows the cover requirements and the transverse steel requirements for reductions in ℓ_{hb}. The orientation of the hooked bar does not appear to influence anchorage,[14,15] so no adjustment for top bars is needed.

The cover requirements for reduction in ℓ_{hb} are generally met if the hooked bars are anchored within a column cage where a beam frames an exterior column. Values of ℓ_{dh} for various bar sizes are given in Table B.3 for Grade 60 bars and $f'_c = 3000$ psi.

8.14   Example of Hooked Bar Development Length

(a) The bars in a slab are anchored in a wall as shown in Fig. 8.26. What is the largest hooked bar that can be anchored in the wall if Grade 60 bars are used and $f'_c = 4000$ psi?

Solution

Because the bars are anchored into a wall, the side cover will be large and the tail extension of the 180° hook well confined. A reduction factor of 0.7 therefore may be used. The distance between the critical section and the outside end of the hook is $16 - 2 = 14$ in. Therefore, ℓ_{dh} for the bar

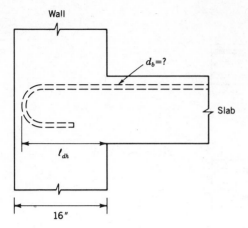

Wall

$d_b=?$

Slab

ℓ_{dh}

16"

Figure 8.26 Hooked bar anchorage, design example.

selected must be less than or equal to 14 in.

$$\ell_{dh} = 0.7\ell_{db} = 0.7 \times 1200 d_b/\sqrt{f_c'}$$
$$0.7 \times 1200 \times d_b/\sqrt{4000} \le 14 \text{ in.}$$
$$d_b \le 1.05 \text{ in.} \qquad \text{USE } \#8 \text{ bars.}$$

(b) Grade 40 steel and lightweight concrete is specified for the problem in 8.14*a*.

Solution

The factors from Code Secs. 12.5.3.1 and 12.5.3.5 are used.

$$1.3 \times 0.7 \times 1200 \times \frac{40}{60} \times d_b/\sqrt{4000} \le 14 \text{ in.}$$
$$d_b \le 1.21 \text{ in.} \qquad \text{USE } \#9 \text{ bars } (d_b = 1.12 < 1.21 \text{ in.})$$

(c) A short cantilever beam extends 24 in. from a column as shown in Fig. 8.27. Can #8 bars be used if $f_y = 60$ ksi must be developed at the critical section (face of column)? $f_c' = 3000$ psi.

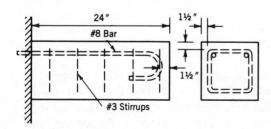

24"

#8 Bar

1½"

$d=?$

1½"

#3 Stirrups

Figure 8.27 Hooked bar in cantilever beam.

Solution

The cover on the side face and over the top of the hooked bar is $1\frac{1}{2}$ in. $+ \frac{3}{8}$ in. (#3 stirrups). Cover on the end of the 180° hook is $1\frac{1}{2}$ in. None of the reduction factors may be used and, in addition, the minimal cover over the bar on two faces (side and top) requires that the provisions of Code 12.5.4 be satisfied.

$$\ell_{dh} = \ell_{db} = 1200\,d_b/\sqrt{f'_c} = 1200 \times 1/\sqrt{4000} = 19 \text{ in.}$$
$$19 \text{ in.} < 24 - 1.5 = 22.5 \text{ in.} \quad \textbf{O.K.}$$

Number 3 stirrups must be provided at 3_{db} ($3 \times 1 = 3$ in.) along the full development length (19 in.). This requirement is likely to control stirrup spacing for the cantilever beam.

8.15   Mixed Bar Sizes and Bundled Bars

The variation in bond stress that elastic analysis indicates as existing at the same cross section on bars of different diameter (a larger bond stress on the larger bars) is not believed to be significant at the ultimate stage. These inequalities can be ignored just as are various other bond stress concentrations on individual bars. The development length of each size bar must be maintained as separately calculated.

The Code permits up to four bars to be bundled or clustered together. In the absence of test data, it is assumed that the bars must be developed individually but that the lengths must be increased 20% for a three-bar bundle and 33% for a four-bar bundle (Code 12.4). The increase covers any uncertainty as to how well mortar will penetrate into the core of the bundle. All bundled bars must be placed within ties or stirrups (Code 7.7.6).

Bundles are limited to bar sizes #11 or smaller and a maximum of four bars. Within a span of a beam the cutoff point of individual bars must be staggered at least 40 diameters, and splices of individual bars must not overlap each other. In applying bar spacing rules, the group is treated as having a single diameter which would give the actual total area of the group.

8.16   Bars in Compression

Only a small amount of research has been done recently on the development of compression bars. The absence of tension cracks normal to the axis of the bar and end bearing of the bar against concrete eliminates a major weakness present in development of tension bars and permits shorter development lengths in compression than in tension. The 1963 Code allowable bond stress has been directly transformed into a development length requirement (Code 12.3.2) with no modification.

$$\ell_d = 0.02\,f_y d_b/\sqrt{f'_c} > 0.0003\,f_y d_b$$

The limit $0.0003 f_y d_b$ controls wherever $f'_c > 4440$ psi. An absolute minimum of 8 in. is specified (Code 12.3.1). Table B.4 summarizes compression bar development lengths. If the bar is developed inside of spirals, the development length may be reduced 25% (Code 12.3.3.2). Reductions also may be made when excess compression steel is used.

In SI units the equation is

$$\ell_d = 0.24 f_y d_b / \sqrt{f'_c}$$

but not less than $0.044 f_y d_b$ or 200 mm. Units for f_y, f'_c, and $\sqrt{f'_c}$ are MPa. ℓ_d and d_b are in mm.

8.17 Compression Splices

Where bars are required only for compression the bar ends may be cut square within 1.5°, then butted together, and positively held in place to transmit the stress in end bearing. The restrictions are detailed in Code 12.16.6.

Compression lap splices transmit a substantial portion of their load in end bearing on the concrete, as indicated by test results shown in Fig. 8.28 where bar stresses of 25 to 40 ksi were measured for zero lap length. This bearing is evidently dependent on concrete confinement with spirals performing better than ties.

The Code (12.16.1) in effect specifies for f'_c of 3000 psi or better $20 d_b$ lap for Grade 40, $30 d_b$ lap for Grade 60, and $44 d_b$ lap for Grade 75 bars, but in no case less than 12 in. Within ties or spirals of specified makeup (Code Secs. 12.16.3 and 12.16.4) the lap lengths may be multiplied by 0.83 or 0.75, respectively, but still must not be less than 12 in.

Attention is called to Code 12.17.3 which states that not all column reinforcing may be considered as compression steel for splicing (and presumably for development).

At horizontal cross sections of columns where splices are located, a minimum tensile strength at each face of the column equal to one-fourth the area of vertical reinforcement in that face multiplied by f_y shall be provided.

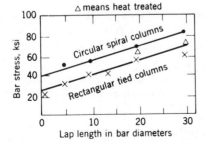

Figure 8.28 Compression splice test strengths in columns (Modified from Reference 9).

8.18 Compression Lap Splice Design Example

The column described in Sec. 8.11*a* is subject only to axial load. Determine the compression splice length (Fig. 8.22).

Solution

From Code 12.3.2, the basic development length is computed.

$$\ell_{db} = 0.02\, d_b f_y / \sqrt{f'_c} \geq 0.0003\, d_b f_y$$
$$= 0.02 \times 1.0 \times 60,000/\sqrt{4000} \geq 0.0003 \times 1.0 \times 60,000$$
$$= 19 \text{ in.} \geq 18 \text{ in.} \quad \textbf{O.K.}$$
$$\text{No factors in Code 12.3 apply, so } \ell_d = \ell_{db}.$$

The splice length (Code 12.16.1) for Grade 60 bars is ℓ_d but not less than $0.0005 f_y d_b = 30$ in. so the required lap length is 30 in. (Note that a 51-in. length was required for tension, Sec. 8.11*a*.)

It is necessary to develop a tensile force of at least $\frac{1}{4}$ the area of vertical reinforcement in each face multiplied by f_y (Code 12.17.3). Because all bars are spliced over a length of 30 in. and a 51-in. splice length is required to develop tensile yield, the bars can develop $30/51 \times f_y$ or 59% of yield, well over the tensile force required by Code 12.17.3.

Selected References

1. R. E. Untrauer and G. E. Warren, "Stress Development of Tension Steel in Beams," *ACI Jour.,* 74, No. 8, Aug. 1977, p. 368.

2. P. M. Ferguson, "Small Bar Spacing or Cover—A Bond Problem for the Designer," *ACI Jour.* 74, No. 9, Sept. 1977, p. 435.

3. Y. Goto, "Cracks Formed in Concrete Around Deformed Tension Bars," *ACI Jour.,* 68, No. 4, April 1971, p. 244.

4. P. M. Ferguson and J. N. Thompson, "Development Length of Large High Strength Reinforcing Bars," *ACI Jour.* 62, No. 1, Jan. 1965, p. 71.

5. C. N. Krishnaswamy, "Tensile Lap Splices in Reinforced Concrete," Ph.D. dissertation, The University of Texas at Austin, Dec. 1970.

6. P. M. Ferguson and C. N. Krishnaswamy, "Tensile Lap Splices, Part 2: Design Recommendations for Retaining Wall Splices and Large Bar Splices," *Research Report 113-3,* Center for Highway Research, The Univ. of Texas at Austin, April, 1971, 60 pp.

7. M. A. Thompson, J. O. Jirsa, J. E. Breen, and D. F. Meinheit, "The Behavior of Multiple Lap Splices in Wide Sections," *Research Report 154-1,* Center for Highway Research, Univ. of Texas at Austin, Feb. 1975.

8. C. O. Orangun, J. O. Jirsa, and J. E. Breen, "Reevaluation of Test Data on Development Length and Splices," *ACI Jour.,* 74, No. 3, Mar. 1977, p. 114.

9. ACI Committee 408, "Bond Stress—The State of the Art," *ACI Jour.*, 63, No. 11, Nov. 1966, p. 1161.

10. ACI Committee 408, "Suggested Development Splice and Standard Hook Provisions for Deformed Bars in Tension," *Concrete Intl.*, Vol. 1, No. 7, July 1979, p. 44.

11. J. O. Jirsa, L. A. Lutz, and P. Gergely, "Rationale for Suggested Development, Splice, and Standard Hook Provisions for Deformed Bars in Tension," *Concrete Intl.*, Vol. 1, No. 7, July 1979, p. 47.

12. R. Tepfers, "A Theory of Bond Applied to Overlapped Tensile Reinforcement Splices for Deformed Bars," Publication 73.2, Div. of Concrete Structures, Chalmers Univ. of Technology, Göteborg, Sweden, 1973, 328 pp.

13. J. Minor and J. O. Jirsa, "Behavior of Bent Bar Anchorages," *ACI Jour.*, Proc., Vol. 72, No. 4, Apr. 1975, p. 141.

14. J. O. Jirsa and J. L. G. Marques, "A Study of Hooked Bar Anchorages in Beam–Column Joints," *ACI Jour.*, 72, No. 5, May 1975, p. 198.

15. R. L. Pinc, M. D. Watkins, and J. O. Jirsa, "Strength of Hooked Bar Anchorages in Beam–Column Joints," *CESRL Report No. 77-3,* Univ. of Texas at Austin, Nov. 1977.

Problems

PROB. 8.1. A simple beam, 18-in. deep × 14-in. wide, is subjected to a large concentrated load at midspan. The distributed load is negligible. The moment diagram for the beam is triangular in shape with a peak moment at midspan requiring 4-#9 bars, Grade 60. The beam span is 15 ft. $f'_c = 2500$ psi.

(a) If only two bars are to extend to the support, determine the length of the short bars. Assume provisions of Code 12.10.5 do not govern.

(b) Do the two bars that continue to the support meet development requirements?

PROB. 8.2. If 3-#10, Grade 60 bars are used to satisfy the midspan moment requirements of the beam in Prob. 8.1, where can the middle bar be cut off? $f_c' = 4000$ psi.

PROB. 8.3. Same as Prob. 8.2 but assume the beam loading changed to uniform load that gives the same maximum moment.

PROB. 8.4. Determine the length required for #11 bar lap splices in the column shown in Fig. 8.22. Grade 60 bars are used and $f'_c = 3000$ psi. The maximum tensile stress in the bars is determined from calculations to be 27 ksi.

PROB. 8.5. Some frames must have the bottom beam bars anchored into the column to develop f_y (Code 12.11.2).

(a) In an 18-in. square exterior column, how large a Grade 60 hooked bar can be anchored? The beam is 15-in. wide. $f'_c = 3500$ psi.

(b) How large a bar can be anchored if lightweight aggregate concrete is used in the column in (a)?

(c) How large a bar can be anchored if column ties at a 2-in. spacing are placed in the region where the hook is located in the column in (a)?

PROB. 8.6.

(a) To anchor Grade 60 top bars from a beam into an exterior girder 12-in. wide, what is the maximum size hooked bar that can be used to develop yield at the face of the girder? $f'_c = 4000$ psi.

(b) If the girder is 18-in. wide?

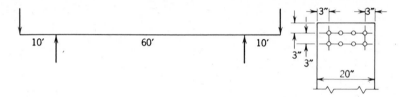

Figure 8.29 Beam of Prob. 8.8

PROB. 8.7. A tied column 20-in. square contains 8-#9 bars of Grade 60 steel, three bars in each face, $f'_c = 5000$ psi. The column design indicates no computed tension on the bars. Sketch and detail the arrangement of splices. (Sec. 8.17)

PROB. 8.8. The beam in Fig. 8.29 is heavily loaded with balanced overhanging end loads such that for this problem the negative moment is essentially constant across the center span and requires 8-#11 bars of Grade 60 steel arranged in two layers of 4 bars each. $f'_c = 4000$ psi. Because the maximum length of available bars is 60 ft, tension splices of all bars are necessary.

(*a*) Design and sketch splices, assuming splices are staggered into two groups.

(*b*) As in (*a*) with three groups of 3, 3, and 2 splices.

(*c*) Compare (*a*) and (*b*) with the case of all splices made at a single section. Is the latter detail permissible under the Code?

PROB. 8.9 How large a Grade 60 stirrup can be developed in a beam with a 24-in. overall depth. $f'_c = 4000$ psi. Assume clear cover of $1\frac{1}{2}$-in. on the stirrup. (See Sec. 5.8 for stirrup development requirements.)

9

CONTINUOUS BEAMS AND ONE-WAY SLABS

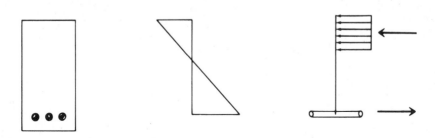

9.1 Types of Construction

Most reinforced concrete members are statically indeterminate because they are parts of monolithic structures. It would convey the wrong impression, however, to ignore the many forms of precast concrete construction that are taking a significant part of the present market. These members vary from statically determinate floor channels, double-T sections, and other shapes to systems where precast units are combined with cast-in-place girders or cast-in-place slabs to form essentially monolithic construction or composite construction. Often the precast members are also prestressed.

The most usual form of building construction consists of a slab cast monolithically with a beam-and-girder floor framing that carries the floor load to the columns. The plan view of such a floor in Fig. 9.1a indicates that such slabs are supported on all four sides. A strict consideration of design philosophy indicates that such slabs should be designed with two-way steel by the methods of Chapter 12.

When the slab is more than twice as long as it is wide, it is usually designed as a one-way slab continuous over the beams, but with special negative moment steel added across the girders as required by Code 8.10.5.* When such a slab is designed for a uniform load, the supporting beams also are designed for a uniform load.† This process in effect ignores the portion of the slab load that goes directly to the girder from the end of the panel. For the beams, the uniform load assumption is on the safe side because it overestimates the beam load. For the girders, the corresponding assumption is to place the beam reactions on the

---

* Because they find the one-way steel arrangement simpler, many engineers actually design one-way slabs for all panels except those that are nearly square.
† As computers come more into the design process, such oversimplifications are less often used.

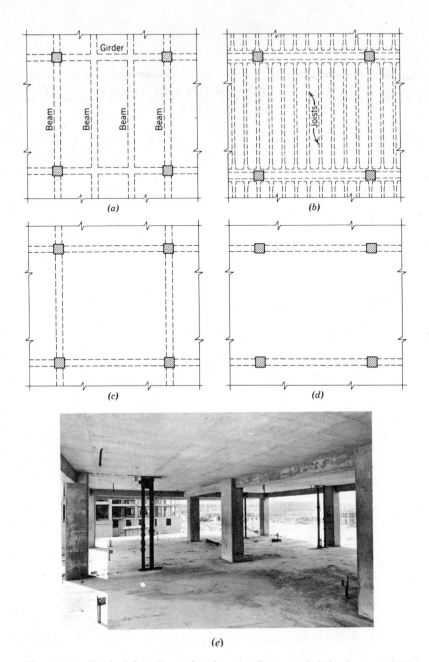

Figure 9.1 Typical floor beam framing. (*a*) Beam-and-girder layout. (*b*) Joist construction. (*c*) Two-way slab and beams. (*d*) One-way slab and beams. (*e*) One-way construction. (Courtesy Portland Cement Assn.)

girder as concentrated loads and to add the uniform load situated directly over the girder. This assumption is on the unsafe side for the girder design, particularly for maximum shear. More realistic tributary loading areas are shown in Fig. 16.2b. The student might investigate the difference in maximum girder shear between the two loadings for a girder simply supported.

Sometimes, for long spans, closely spaced beams with a very thin slab are used. This so-called joist construction (Fig. 9.1b) is facilitated by the availability of removable pans that are used as forms between the joints.

When no beams are used except those between columns, as in Fig. 9.1c, the slabs are definitely supported on all four sides, and both beams and slabs should be designed as discussed in Chapters 12 or 16.

In light construction the beams at times are run in only one direction, as shown in Fig. 9.1d and 9.1e. In this case true one-way slabs result, except at the beams supporting the walls at the ends. The slab band construction of Sec. 16.3 is really this same type of construction utilizing shallow and often quite wide beams.*

Two slab-type floors with beams only at outside walls and around large openings are economical, with the flat plate floor used for light loads and the flat slab floor almost always used for very heavy loads. The (patented) lift slab construction is a flat plate floor without any beams, which is cast at grade and provided with special collars around the columns for lifting into place. Flat slabs and flat plates are presented in detail in Chapter 15, including mention of the two-way joist slab commonly referred to as the waffle slab (Fig. 15.3).

9.2   Interaction Between Parts of the Structure

In all these various types of construction the designer is faced with a highly indeterminate type of structure, that is, a structure in three dimensions that cannot be precisely analyzed as a planar structure.[1-3] More specifically, the intermediate beams of Fig. 9.1a cannot be analyzed exactly without considering the vertical deflection and torsional stiffness of the girders as well as the stiffness of the columns. Likewise, the beams framing into the columns have moments that are influenced by the column joint rotations and hence by any torsion present in the girders.† The Code (8.6.1) permits "any reasonable assumptions for computing relative flexural and torsional stiffnesses," but the assumptions must be "consistent throughout analysis." These assumptions may include ignoring torsion in many situations, but Chapter 16 points out the importance of torsion in designing two-way slabs, flat slabs, and plates.

Designers commonly utilize approximate methods of analysis, but they become better designers when they understand the nature of their approximations and how exact or how crude are these methods. This chapter assumes that the

---

* The spandrel beams on the outside are often omitted with light live loads.
† Torque stiffness drops sharply after first diagonal cracking occurs. See Sec. 5.32 discussion.

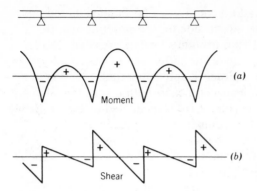

(a)

Moment

(b)

Shear

Figure 9.2 Typical M and V diagrams for a continuous beam.

reader is familiar with ordinary moment distribution or similar analysis procedures and the plotting of continuous member moment and shear diagrams.

The ordinary conventional assumption is that analysis in two dimensions is adequate for most designs. This assumption fits in better with structures designed for uniform floor loads than for those with moving concentrated or wheel loads. It may be reasonably assumed that, in carrying a uniformly distributed load, any one beam gets little help from its neighbor, because this neighboring beam is probably also fully loaded.* Likewise, in a one-way slab each strip may be assumed to carry the load directly above it when the entire slab is loaded.

When moving concentrated loads are involved, each slab strip may carry a different moment and the load on a single beam may depend greatly on the stiffness of the slab and the adjoining beams. The problem of moving concentrated loads is discussed in Chapter 17.

Moment diagrams for continuous beams normally show negative moments over supports and positive moments near midspan, as indicated in Fig. 9.2a. The presence of columns changes the picture only slightly, typically by making the moment at opposite faces of the column slightly different. The moment diagram in any span may be considered as the sum of two parts: one the simple beam moment diagram for that span, and the other the moments across the span owing to the negative support moments, as diagrammed for the center spans in Fig. 9.5a,b. Because these negative moments at supports are influenced by loads on *any* span, it follows that the moment at each point is influenced by every load on every span. Loading patterns for maximum moment are thus a major consideration.

---

* This statement is not strictly true, of course. In Fig. 9.1a, with equal spacing of beams, the beam framing directly into the column is stiffer because its end joints are stiffer, but the girder provides only a yielding support for the neighboring beam. The beam framing into the column thus tends to carry the greater load and to some extent relieves the intermediate beam. At the ultimate or collapse load, however, for beams of equal size, each beam would probably be carrying almost exactly the same load, because the load distribution changes after some yielding takes place.

The typical shear diagram (Fig. 9.2b) differs only slightly from that for a series of simple spans. In any span the shear diagram for a given loading consists of the simple beam shear plus a constant correction that will be designated as the continuity shear V_c. The continuity shear is usually relatively small except in end spans.

9.3   The General Design Problem for Continuous Members

Each span of a continuous beam or slab requires a separate design for negative and positive moment conditions, even though practice in the United States usually uses a constant depth member for any given span, often for all spans. Figure 9.3 shows the design conditions for continuous slabs, rectangular beams, and T-beams in diagrammatic fashion. To indicate clearly that at this stage no consideration is being given to the arrangement of steel, the reinforcement is indicated only in the vicinity of maximum moment zones. Moment at A (on the left) is assumed to be the governing (maximum) negative moment in all cases. The letter t designates the tension face.

In the thinner slabs compression steel is undesirable because it would lie so close to the neutral axis that a slight displacement would make it ineffective. Hence Fig. 9.3a shows the maximum negative moment at A and a limiting percentage of steel determining the slab depth, with relatively under-reinforced sections elsewhere. Thick slabs can be designed with compression steel, similar to rectangular beams, if desired, but this does not reduce slab cost.

Rectangular beams may be designed with compression steel at supports, as in Fig. 9.3b. Code 12.11.1 requires that part of $+A_s$ extend into the support so that some compression steel is always present.

T-beams become, in effect, inverted rectangular beams at the supports with only the web width b_w effective in compression, as indicated in Fig. 9.3c. This restricted compression zone is improved by using a double-reinforced section; this is recommended for ordinary use.

Shear (diagonal tension) rarely governs in ordinary one-way slab design, in spite of the fact that the allowable v_c is only $2\sqrt{f_c'}$. Stirrups in slabs are awkward but may be used. Continuous rectangular beams and T-beams normally require stirrups, but the web size for the continuous T-beams very rarely will be governed by shear instead of moment and only occasionally by the size needed for placing the bars.

9.4   Loading Patterns for Maximum Moments

(a) **Maximum Positive Moment**   Elastic theory is specified for calculation of design moments, except that some limit design modification is permitted (Code Secs. 8.3.1 and 8.4). Influence lines might be used to determine the loading arrangement, but the critical patterns for an elastic analysis are easy to deduce from a single load-deflection sketch, such as that in Fig. 9.4a in which deflec-

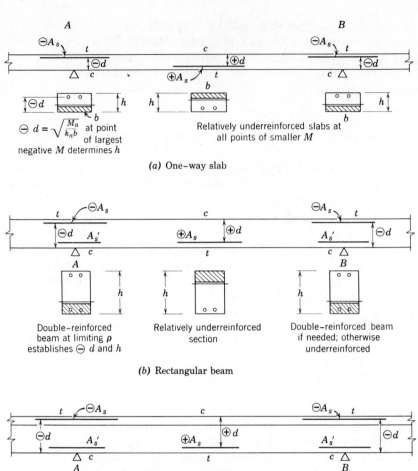

$\ominus d = \sqrt{\dfrac{M_a}{k_n b}}$ at point
of largest
negative M determines h

Relatively underreinforced slabs at
all points of smaller M

(a) One-way slab

Double-reinforced
beam at limiting ρ
establishes $\ominus d$ and h

Relatively underreinforced
section

Double-reinforced beam
if needed; otherwise
underreinforced

(b) Rectangular beam

Double-reinforced
rectangular beam
at limiting ρ
establishes $\ominus d$ and h

T-beam for $\oplus M$

Rectangular beam,
double-reinforced
if needed; smaller A_s
and A_s' where $\ominus M$
is smaller

(c) T-beam

Figure 9.3 Strength design procedures for continuous slabs and beams.

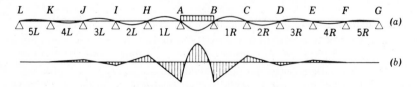

Figure 9.4 The influence of a single panel load on a continuous beam.

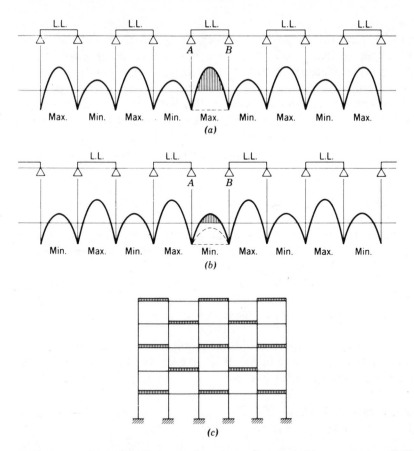

Figure 9.5 Live load placement for (*a*) maximum positive moment and (*b*) minimum positive moment at midspan. (*c*) Checkerboard loading for multiple stories. Dead load is on all spans at all times.

tions are greatly exaggerated. The usual carryover moment idea leads directly to the moment diagram in Fig. 9.4*b*. It is observed that this loading produces positive moment near midspan in all even-numbered spans. One concludes that if all even-numbered spans were loaded, each such load would increase the positive moments in the other loaded spans. This loading pattern is usually stated as follows:

> *For maximum positive moment near the middle of a span, place factored dead load on all spans and place factored live load on that span and on alternate spans on each side, as shown in Fig. 9.5*a (*Code 8.9.2b*).

Load factors are applied to all loads. The dead load with its load factor is always considered to act* and is included on the moment diagrams shown. This

---

* Except that when it offsets the effect of wind or earthquake loading, the load factor is reduced.

one loading arrangement gives the maximum positive moments on all the loaded spans. The maximum positive moment on all other spans is given by loading only all those spans, as in Fig. 9.5b. Thus all maximum positive moments on all spans are determined by two load arrangements and two moment distributions or analyses. Extended to a multistory frame, these patterns become checkerboard arrangements of loading (Fig. 9.5c), but still with only two patterns needed for the complete analysis for positive moments.

(b) Minimum Positive Moment (or Maximum Negative Moment) Near Midspan   Because the minimum positive moment is simply the opposite extreme of loading from that for maximum positive moment on beam *AB*, all the live loads in Fig. 9.5a must be removed and live loads with their load factor must be placed on the other spans as in Fig. 9.5b. It should be noted that for the two live load patterns already discussed for maximum positive moment, Fig. 9.5a and b give all the necessary minimum positive moments or maximum negative moments near midspan. The loading criterion is:

> *Place factored dead load on all spans; omit factored live load on span considered; place factored live loads on adjacent spans and alternate spans beyond.*

With smaller dead load moment, the midspan moment in *AB* could be negative, as indicated by the dashed curve in Fig. 9.5b, especially as the load factor for dead load is smaller than that for live load.

(c) Maximum Negative Moment at Left Support   The single panel loading of Fig. 9.4 shows that negative *M* is produced at:

> *B* by loading adjacent span to left.
> *D* by loading third span to left.
> *F* by loading fifth span to left.
> *A* by loading adjacent span to right.
> *I* by loading third span to right.
> *K* by loading fifth span to right.

This can be summarized by the criterion:

> *For maximum negative moment at a given support, place factored dead load on all spans, and place factored live loads on the adjacent spans on each side of the support and alternate spans beyond (Code 8.9.2a).*

Load factors must be applied to the loads. Applied to maximum moment at the left support *A*, this criterion results in the loading and moment diagram shown in Fig. 9.6a. As can be seen from Fig. 9.4b, the magnitude of negative moment at *D* is reduced greatly from the value at *B*. To simplify analysis, Code 8.9.2a eliminates the requirement of placing live load on the alternate spans

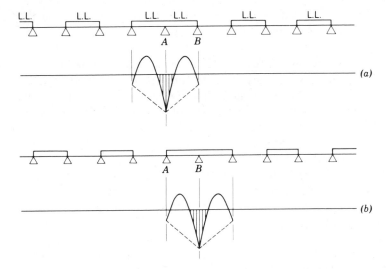

Figure 9.6 Live load placement for maximum negative moment: (*a*) at left support *A*; (*b*) at right support *B*. Dead load is on all spans at all times.

beyond the immediately adjacent spans. Only the resulting moments adjacent to *A* are significant, because no others are either maximum or minimum moments. These moments are critical at the face of support, as discussed in Sec. 9.4g.

Adjustment of these moments from elastic analysis in the direction of limit design is considered in Chapter 11.

(d) Maximum Negative Moment at Right Support The same criterion applies to this maximum moment as in the preceding case. The live loadings are placed on spans adjacent to *B* and alternate spans beyond, as shown in Fig. 9.6*b*. Unfortunately, the loading gives maximums only near the one support at *B*. Again, the Code requires only that the immediate adjacent spans be loaded.

(e) Partial Span Loadings Both maximum positive and maximum negative moment conditions call for full load on the span under consideration. Maximum negative moment occurs with a very unsymmetrical moment diagram in the span involved. On the other hand, maximum positive moment results in a moment diagram nearly symmetrical about the middle of the span. Minimum positive moment at midspan (or maximum negative there when dead load is small) also gives a nearly symmetrical moment diagram, that is, a moment diagram based on dead load alone on the span in question.

In building frame design it is not customary to deal with partial span loads as they do not increase the principal design moments. In highway structures partial span loads would have a considerable influence on detailing the steel. Partial span loads are mentioned briefly in Sec. 9.6 in connection with maximum moment diagrams.

Some codes specify a single moving load to be placed anywhere on the structure as an alternate loading to care for special cases.

(f) Permissible Simplifications   The ACI Code encourages the use of a reduced or simplified frame in the analysis of buildings (Code 8.9) and specifies that moments be calculated by elastic analysis. For gravity load analysis, one floor may be analyzed at a time with the far ends of columns taken as fixed. In calculating maximum negative moments the live load may be applied on only the two adjacent spans. This loading pattern permits the use of simplified moment distribution procedures such as the two-cycle procedure given in Reference 1. These methods are quite reasonable for ordinary structures, except that they give beam and column moments at the exterior columns that tend to be considerably too small. The load factors, on both dead and live load, must not be overlooked for any design.

(g) Moment at Face of Support   Moment distribution generally implies moments calculated on the basis of spans taken center-to-center of supports. Because such calculations treat the support reaction as though it were concentrated at a point, the resulting moment diagram within the support width is entirely imaginary. This is of no concern as both theory and tests show that the critical moment is at the *face* of the support (Code 8.7.3).

Many engineers establish the design moment at the face of support from the calculated moment at the center line of support simply by deducing $Va/2$, where V is the shear and a is the column width, as illustrated in Fig. 9.7a. This is almost the same as scaling the moment value from the moment diagram, because the small length of uniform load within the column width (Fig. 9.7b) causes little change in moment.

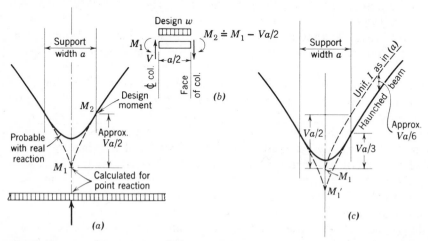

Figure 9.7   Moment at face of support. (*a*) Without correction for increased stiffness at support. (*b*) Free-body diagram establishing procedure in (*a*). (*c*) Recommended procedure recognizing increased stiffness at support.

The authors prefer the more conservative correction recommended by the old Joint Committee Specification, namely, a reduction taken as $Va/3$. The reasoning behind this value is as follows. The support stiffens the end of the beam much as would a haunch. A calculation considering this increased end stiffness leads to an increased negative moment at the center of the column, as indicated by M_1' in Fig. 9.7c. Instead of calculating M_1' (which might be roughly $Va/6$ larger than M_1), an approximate equivalent is obtained by applying a smaller correction $Va/3$ to the original calculated M_1 value. The design moment is then $M_1 - Va/3$, as in Fig. 9.7c.

The extra stiffness at the support also leads to smaller positive moments. However, the correction is only about $Va/6$. Most engineers consider this correction less certain than that to the face of the column. Engineers recognize the lesser accuracy of positive moment calculations, always sensitive to relative column stiffness values, and simply use the original positive moments without any correction.

(h) Validity of Elastic Analysis in Strength Design Philosophically, the use of elastic theory for moments and the use of the very nonelastic methods of section analysis are not consistent. This leads design thinking towards some kind of inelastic frame analysis as discussed in Chapter 11. Code 8.4 permits some redistribution in this direction. Although the elastic theory may seem incompatible with strength design, it errs only by giving moments that are too large and hence too safe. This chapter is based wholly on moments obtained by the elastic theory.

9.5 Moment Coefficients

Any given moment can be expressed as a moment coefficient times $w\ell_n^2$, where w is the total design load (including load factors) and ℓ_n is the clear span. The maximum moment coefficients are largest when the ratio of live load to dead load is large and the column or other joint restraint is relatively small. Negative moment coefficients also may be large when adjacent spans are longer or more heavily loaded than the span in question.

Based on uniform live loads not greater than three times the dead load and on span lengths "approximately equal (the larger of two adjacent spans not exceeding the shorter by more than 20%)," the Code committee has established by analysis certain reasonable moment coefficients to use for maximum moment calculations. Code 8.3.3 tabulates these coefficients and Fig. 9.8 presents them in diagrammatic fashion.

The same Code section lists two shear values:

In end members at face of first interior support	$1.15w\ell_n/2$
At face of all other supports	$w\ell_n/2$

When the ratio of dead to live load is particularly large, some economy can be achieved by calculating the moments by more accurate methods. Additional

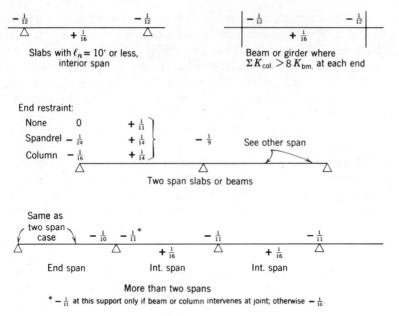

Figure 9.8 ACI Code (8.3.3) moment coefficients for nearly equal spans and live load less than three times the dead load. $M = (\text{coef.})\, w\ell_n^2$.

economy can be achieved by less approximate analysis and the redistribution discussed in Chapter 11. The use of moment coefficients should be restricted to rather standard conditions unless one uses detailed general tables.

9.6 Maximum Moment Diagrams

To determine the best arrangement of the reinforcing it is necessary for the designer to have a clear picture in mind of the extreme range of moments all along the beam or slab. This composite of all the possible moment diagrams is often called the maximum moment envelope. The major part of such a diagram can be assembled from the several moment diagrams for critical maximum moments already illustrated in Figs. 9.5, 9.6, and 9.7. For a typical interior span with equal spans and equal live loads, these moment curves are drawn to larger scale in Fig. 9.9*a–d* and grouped together in Fig. 9.9*e*.

Recent frame studies have pointed out a moment condition not shown in Fig. 9.9 that is easy to overlook. When the floor system is light and the lower-story columns (rather than a shear wall) resist high wind or earthquake shears, the sum of the moments from the column above and the column below add at the joint and rotate it and the attached beams. This added rotation increases the negative moment in one of the beams enough to make it the weakest member of the system. Hence, the designer should be alert to combinations of live load

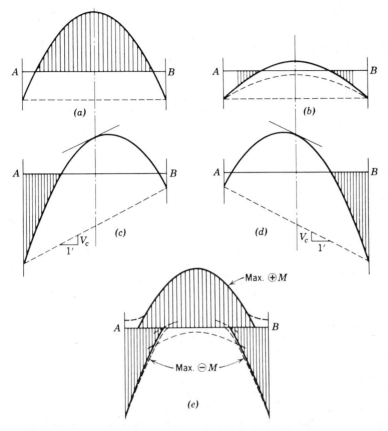

Figure 9.9 Moment diagrams for maximum moment. (*a*) Positive moment at midspan. (*b*) Negative moment near midspan; the dotted curve would show with a relatively large live load or long adjacent span. (*c*) Negative moment at left support. (*d*) Negative moment at right support. (*e*) Composite maximum moment diagrams (moment envelope).

with wind or earthquake to produce the governing *beam* negative moment in spite of the appropriate reduced load factors.

By loadings exactly opposite to those for maximum negative moment, some small positive moment often can be obtained over supports as suggested by the dashed lines. Some partial span loading also may increase the negative moments slightly, as indicated by dashed lines, but these are not very significant changes.

A very significant fact about the maximum moment diagrams is that, over a considerable portion of the beam, the moment may change from positive to negative, or the reverse, as the loading in adjacent spans is varied. Consequently, the designer must provide tension steel in both top and bottom over this zone.

Definite locations for some of the inflection points on Fig. 9.9 are calculated in connection with the bending of bars in Fig. 9.12 and in Sec. 9.14.

9.7 Maximum Shears

The loading for maximum shear at a support is the same as for maximum negative moment there. Hence the end shear can always exceed the simple beam shear by an amount equal to the continuity shear V_c. On interior spans the value of V_c will be on the order of 3% to 12% of the simple beam shear, with 8% a fair average value. On end spans V_c can run as large as 20% or more of the simple beam shear. This large V_c on the end span is additive only on the half beam adjacent to the first interior support. For nearly equal spans, the Code (8.3.3) suggests 15% extra shear in end members at the first interior support and no addition elsewhere. The authors prefer to make a nominal addition (8%) in interior spans as well as the substantial addition in the end span.

9.8 Design of Continuous One-Way Slab Example

Design a multiple span (more than two-span) continuous one-way slab supported on beams at 12-ft 0-in. on centers, ACI Code moment coefficients (Sec. 9.5), dead load of 22 psf (plus slab weight), live load of 220 psf, $f'_c = 3000$ psi, Grade 60 steel with ρ limited to $0.18f'_c/f_y$. Assume the beam stem is 12-in. wide.

No particular merit is associated with ρ of $0.18f'_c/f_y$ except that it is near midrange in terms of allowed reinforcement percentage and anything near this value would be more economical and constructable than the maximum ρ. The maximum ρ can cause congestion and hence be uneconomical for slabs or beams and will frequently lead to thin sections with deflection problems; it should be avoided except where conditions demand an extreme design.

Solution

The moment coefficients shown in Fig. 9.8 for a slab with more than two spans indicate that the negative moment at the first interior support is the maximum, at $0.10w\ell_n^2$. The slab at this point will be designed for the arbitrary limiting steel ratio, which leads to $k_n = 0.160f'_c$ (Fig. 3.10) = 480. Slab weight will be based on assumed h of 6 in. corresponding to an arbitrary $h = \ell/24$.

$$w_\ell = 220 \times 1.7 \text{ (L.F.)} = 374 \text{ psf}$$
$$w_d = 22 \ \times 1.4 \text{ (L.F.)} = \ \ 31$$
$$\text{Slab wt.} = 75 \ \times 1.4 \qquad\quad = \underline{\ \cancel{105}\ } \text{ (estimate)} \ \underline{\ 88\ } \text{ (revised)}$$
$$\text{Total } w = \cancel{510} \text{ psf} \qquad 493$$
$$\ell_n = 12.0 - 1.0 = 11.0\text{-ft clear span}$$

$$M_u = 0.10 \times 510 \times 11^2 = 6170 \text{ ft-lb/ft width of slab}$$
$$M_n = M_u/\phi = 6170/0.9 = 6860 \text{ ft-lb/ft}$$
$$k_n bd^2 = 480 \times 12d^2 = 6860 \times 12$$
$$d = \sqrt{14.3} = 3.78$$

Code 7.7.1c specifies 0.75-in. clear cover, if not exposed to weather.

$$h = d + d_b/2 + 0.75 = 3.78 + 0.25 + 0.75 = 4.78 \text{ in. for } \#4 \text{ bars}$$

This leads to $h = 5$ in.; it could be 4.5 in. only if a very gross overestimate of weight had been included. The original 75 psf assumption corresponded to $h = 6$ in.

Try $h = 5$ in., wt $= \frac{5}{12}(150) = 63$ psf $\times$ 1.4 L.F. $= 88$ psf. This indicates a decrease of 17 psf, which is 3.5% of the total load, and causes a 3.5% change in moment and about $3.5/2 = 1.7\%$ change in required d, since d varies as $\sqrt{M}$. (If 4.5 in. had been tried, the change in h could not have been as much as the 6% required to change 4.78 to 4.5 in.)

$$M_n = -0.10 \times 493 \times 11^2/0.9 = 6630 \text{ ft-lb/ft}$$
$$d = \sqrt{6630 \times 12/(480 \times 12)} = \sqrt{13.81} = 3.72$$
$$h = 3.72 + 0.25 + 0.75 = 4.72 \text{ in., say 5 in. as estimated}$$
$$\text{USE } h = 5 \text{ in.,} \qquad d = 5 - 0.25 - 0.75 = 4.00 \text{ in. for } \#4 \text{ bars}$$

Code 9.5.2 gives a deflection warning (Code Table 9.5a) If $h < \ell/28$. The value for ℓ is defined in the Code Notation Sec. 9.0 as the span length defined in Code Sec. 8.7. For this application, Code 8.7.1 defines the span length as the clear span (11 ft) plus the depth of the member (4 in.) but not more than the center-to-center distance between supports (12 ft). Thus, $\ell/28 = 11.33 \times 12/28 = 4.86$ in. Because the slab thickness selected is 5 in. > 4.86 in., deflections should not be a problem. This design cannot use the savings from the limit design idea of Code 8.4 because approximate moment coefficients have been used.

All other sections have less moment and hence all steel can be designed assuming z slightly greater than the $0.89d$, which is correct for $\rho = 0.18f'_c/f_y$, say $z = 0.90d$. (For very small moments it may be desirable to check a and z for a more economical value.) Noting that A_s and M_n are each per foot of slab width,

$$A_s = \frac{M_n}{f_y z} = \frac{(M_n \text{ in ft-lb})12}{60,000 \times 0.9 \times 4.00} = \frac{M_n \text{ in ft-lb}}{18,000}$$

It is convenient to tabulate A_s per inch of slab width if the bar spacing is determined without the use of tables.

$$A_s/\text{in.} = \frac{M_n \text{ in ft-lb}}{18,000 \times 12} = \frac{M_n \text{ in ft-lb}}{216,000}$$

where $M_n =$ coef. $(493 \times 11^2/0.9) = 66,300 \times$ coef. The calculations are tabulated in Table 9.1. Code 10.5.3 and 7.12.2 for Grade 60 bars requires

TABLE 9.1 Calculation of Slab Steel

	Exterior Span		First Interior Support	Typical Interior	
	Exterior End	Middle		Middle	Support
M coef.	$-\frac{1}{24}$	$+\frac{1}{14}$	$-\frac{1}{10}$	$+\frac{1}{16}$	$-\frac{1}{11}$
M_n = coef. $\times$ 66,300	-2760 ft-lb/ft	$+4740$	-6630	$+4140$	-6030
A_s/ft = M_n/18,000	0.153 in.²/ft	0.262	0.367	0.230	0.333
A_s/in. = M_n/216,000	0.0127 in.²/in.	0.0218	0.0305	0.0191	0.0277
Min A_s = 0.0018bh	0.009 in.²/in.	0.009	0.009	0.009	0.009
Spcg. #3 bars	8.65 in.	5.04	3.60	5.78	3.98
Spcg. #4 bars	15.7 in.	9.16	6.56	10.46	7.23
USE #3 at:	8.5 in.	5	—	5.5	—
Plus #4 at:	—	—	6.5 in.	—	7

$\rho = 0.0018$ for shrinkage and temperature as a minimum. This minimum reinforcement does not govern at any section tabulated. (For beams Code 10.5 makes the minimum $\rho = 200/f_y$ which is $\rho = 0.0033$ for Grade 60 bars.) The required areas and the spacing of bars are given in Table 9.1 as though top and bottom steel were to be totally separate as they are in Fig. 9.10a. The bar spacing for a given size bar is simply the nominal bar area divided by the A_s/in. requirement.

For the layout of Fig. 9.10a the requirements are well matched by the bars specified at the bottom of the table.

Some designers prefer to use some bent-up bars, as they feel that this positions the top steel more exactly. Many designers feel that bent bars in a slab make it difficult to construct. If bent bars are used, the arrangement of Fig. 9.10b is the most common bending pattern. The top bars are actually all in one layer and the bottom bars are all in one layer, but they are sketched separately to indicate the patterns more clearly. Bending up and overlapping half the bars provides negative moment reinforcement equal to the average positive moment reinforcement in adjacent spans and added straight bars can make up the deficit. This requires #3 at 12 in. over the first interior support and #3 at 14 in. at all typical interior supports. In this particular case it would be simpler not to bend up at the outer end; otherwise add #3 at 15 in.

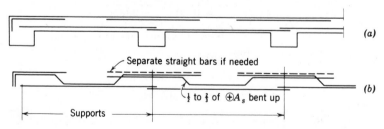

Figure 9.10 Arrangement of slab steel. (a) Straight bars alone. (b) Some bars bent.

The bend points and stop points for bars in slabs involve almost exactly the same considerations as any other continuous member. These considerations are discussed in considerable detail in Secs. 9.13 through 9.16 in connection with the design of a continuous T-beam.

The student should note that temperature steel is required normal to the slab flexural reinforcement, at least in the amount specified in Code 7.12. In the example for a #3 bar, $0.11/0.009 = 12.2$ in. Maximum spacing is $5h$ but not more than 18 in. Use #3 bars at 12-in. spacing. The temperature bars are used as spacers for both top and bottom steel, tied to the underside of top bars and to the top of bottom bars. In addition, bar supports or chairs should be provided to hold the steel at proper levels.

Slab shear stresses could easily have been computed, but they do not control on one-way slabs of ordinary span. The design of continuous rectangular beams is discussed in Sec. 9.20.

9.9 Continuous T-Beam Design—General

The principles of design for a continuous T-beam can be discussed most easily in terms of a numerical example. The general flow of this process is outlined in the flow diagram of Fig. 9.11. The text section numbers are indicated in the flow diagram for the various steps. The designer always has considerable freedom in deciding how his or her approach to a problem should be sequenced. The flow diagram illustrates the general approach used in this numerical example. Other techniques could be used satisfactorily.

The process begins with rough estimates of member sizes and bar layouts to determine dead loads and design constants. Member sizes are refined based on flexure and shear checks. The reinforcement required at maximum negative and positive moment sections is determined and tentative bar layouts are selected. After checking the selected bar sizes to determine that the flexural reinforcement distribution provisions for crack control can be met, the bars are detailed. Bend and cutoff points are selected considering moment diagrams, development lengths, and Code arbitrary detailing requirements. Finally, stirrups are selected and the detail drawings can be made. Much of the remainder of this chapter (through Sec. 9.20) is devoted to various aspects of such design in terms of a typical interior span.

A typical interior panel of a continuous T-beam of 20-ft clear span with 15-in. square columns is to be designed to carry the slab designed in Sec. 9.8. Beams are 12-ft 0-in. on centers, $h_f = 5$ in., $w_\ell = 220$ psf, $w_d = 85$ psf (including 63 psf slab weight but excluding stem weight), $f'_c = 3000$ psi, Grade 60 bars, and moment coefficients calculated by moment distribution, when expressed in terms of total load and clear span $(w\ell_n^2)$, are: -0.091, $+0.072$, and for minimum positive moment at midspan, -0.010.

There are many different choices of b_w and d for the web of a T-beam, any one of which might be a good design for specific conditions. The choice of d

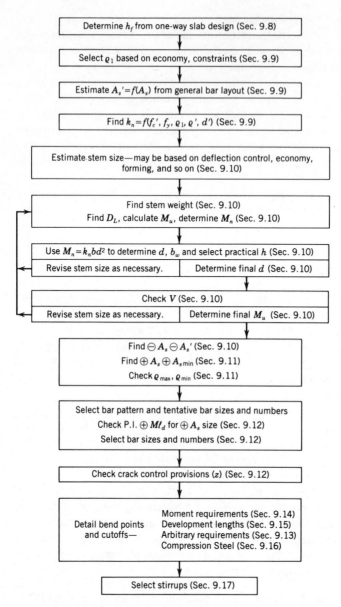

Figure 9.11 Flow diagram for continuous T-beam design example.

might be on the basis of stiffness against deflection, after the fashion of Code Table 9.5a, especially for long spans. This would set $\ell/21$ as a probable minimum overall depth. The depth may be chosen to match some other member that is more critical in strength or stiffness or it may be related to depths needed for ducts. This member, for example, might well match the depth required for the exterior span with its larger moments and shears, if of the same span.

It would seem logical to choose depth on the basis of economy, remembering that A_s decreases with depth, whereas concrete and formwork are more expensive as depth increases. However, the economy of the structural frame may rarely agree with overall building economy because greater depths for beams mean heavier beams (usually) and hence heavier columns and footings, higher exterior walls with expensive finish, more steps in a story height and hence larger stair wells, higher elevator lifts, and so on. The stem width may on occasion be chosen to fit into a wall.

One can almost say that the choice of web b_w and d is an arbitrary choice. Four primary conditions must be satisfied. (1) The capacity of the inverted rectangular section that carries negative moment (Fig. 9.3c) must be adequate for flexure with a ''reasonable amount'' of compression steel. (2) The shear capacity must be adequate, but such a wide range can be carried by stirrups that this provision is rarely restrictive. (3) There must be width for the required number of reinforcing bars, but the possibility of using higher strength steels, of bundling the bars, and of spreading the negative moment steel into the flange makes this much less restrictive than formerly. (4) There must be enough stiffness to keep deflections within proper limits, but the use of compression steel can stiffen considerably a shallow beam (Code 9.5.2.5).

In this example, deflection is assumed not to be critical on a heavily loaded 20-ft span.* Depth will be considered on the basis of some arbitrary requirements for negative moment. The given moment coefficients indicate that positive moment steel will be approximately 75% to 80% of the negative moment tension steel, as the lever arm z will not be greatly different for positive and negative moment. If half of the positive moment steel is bent up in the fashion indicated in Fig. 9.17b in Sec. 9.15c, the other half will continue in the bottom part of the beam into the column and thereby almost automatically provide compression steel in the amount $A'_s = 0.5 \times 0.8A_s = 0.4A_s$, where A_s is the negative moment tension steel. By lapping this compression steel from two adjacent spans as in Fig. 9.20b in Sec. 9.16 A'_s could become approximately $0.8A_s$.

A long-standing rule in continuous beam design has been to run at least ¼ the positive moment steel into the support (Code 12.11.1). Now it is further specified (Code 12.11.2) that, if the beam is part of a frame resisting lateral load, this ¼ be anchored such that its full f_y in tension be available at the face of support. (The objective is greater ductility or energy absorption in case of moment reversal from unexpected loadings such as explosion, earthquake, unusual settlement, and so on.) This reinforcement is obviously the minimum automatically available for A'_s at face of support.

The designer thus has considerable freedom in selecting the approximate A'_s to use, and estimates such as the first above can only be approximate at this stage. (For example, the positive moment steel may work out to be 5 bars,

---

* A deflection calculation similar to that of Sec. 3.22c needs to be run to verify that the deflection is satisfactory.

which means either 40% or 60% bent up, instead of half.) Here, for the first step, it is assumed that deflection probably does not control and that A_s' will be $0.4A_s$. The possible range of k_n values for moment will be explored for this doubly reinforced section.

When the maximum reinforcement percentage is used ($\frac{3}{4}\rho_{\text{bal}}$) with no compression steel for $f_c' = 3000$ psi, $f_y = 60,000$ psi, Table 3.1 shows $k_{n1} = 783$, $\rho = 0.0161 = \rho_1$, and $a/d = 0.378$. Thus $M_n = 783\,bd^2$ if no compression steel is used. To see the effect of A_s' assume that $A_{s2} = A_s'$.

$$A_{s2} = A_s' = 0.4A_s = 0.4(A_{s1} + A_{s2}) = 0.4(0.016\,bd + A_s')$$
$$A_s' - 0.4A_s' = 0.0064\,bd, \qquad A_s' = 0.0107\,bd$$

With $d - d'$ estimated as $0.88d$,

$$M_n = 783\,bd^2 + 0.0107\,bd \times 60,000 \times 0.88\,d$$
$$= 783\,bd^2 + 565\,bd^2 = 1348\,bd^2$$

Thus bd^2 with maximum ρ can be reduced from $M_n/783$ to $M_n/1348$ (40% smaller) with this small A_s', if one desires. The resulting small d certainly would not be economical if A_s' ran the full length of a member, but here the need is only at extreme ends of the span. This small section might also get more into deflection problems or, with its larger A_s, into bar spacing problems. On the other hand, if the minimum reinforcement percentage $\rho = 200/f_y$ is used for ρ_1, k_{n1} would be 128 and with $A_{s2} = A_s'$ and $A_s' = 0.4A_s$, k_n would be $694\,bd^2$. Such a section also would be uneconomical, requiring a large depth and hence high concrete and forming costs.

This short analysis confirms the earlier statement that the choice of web and reinforcement percentage can be quite arbitrary, with experience and judgment quite helpful in making the choice. On this basis, and recalling that minimum size is rarely the best solution, the authors arbitrarily pick an intermediate value of $k_n = 1050$ (near the middle of the range) for the starting point. This may prove either a wise or unwise decision; the choice can only be evaluated after the design is further advanced.

9.10 Continuous T-Beam Design—Negative Moment Section

Size will first be established for moment as a double-reinforced rectangular beam.

$$
\begin{aligned}
LL &= 220 \times 12 \times 1.7 & &= 4500 \text{ plf} \\
\text{Slab} + DL &= 85 \times 12 \times 1.4 & &= \underline{1430} \\
& & &\ 5930
\end{aligned}
$$

To get a crude estimate of the beam stem weight to begin the design process, guess the beam width to be the same as the column width, that is, $b_w = 15$ in. Guess the beam thickness to be a very moderate span-to-thickness ratio, ℓ/h of 14. Since $\ell = 20 + 15/12 = 21.25$ ft, h would be about 21.25/14, say, 1.5 ft or 18 in. Deducting the 5-in. slab thickness, the stem would be 13 in. × 15 in. and weigh about 200 lb-ft.

$$\text{Estimated stem wt.} = 200 \times 1.4 \qquad = \underline{280} \qquad \underline{218}$$
$$w = \cancel{6210} \qquad 6150 \text{ plf}$$
$$M_u = -0.091 \times 6210 \times 20^2 = 226,000 \text{ ft-lb,}$$
$$M_n = 226,000/0.9 = 251,000 \text{ ft-lb}$$

This moment could be reduced by the redistribution provisions of Code 8.4 and Chapter 11, but first will be used as would be necessary when only approximate moments are available.

$$\text{Reqd. } b_w d^2 = M_n/k_n = 251,000 \times 12/1050 = 2870$$

If $b_w = 12$ in., $d = \sqrt{2870/12} = \sqrt{239} = 15.5$ in. This beam might be wider than needed.

$$\text{Try } b_w = 10 \text{ in., } d = \sqrt{287} = 16.94 \text{ in.}$$

The latter is nearer the usual d/b ratio of 1.5 to 2.5, although the narrow width may give a tight detailing situation. Add 1.5-in. cover + 0.5-in. stirrup + 0.5-in. $(=d_b/2) = 2.5$ in.

$$\text{Min } h = 16.94 + 2.5 = 19.5 \text{ in. for overall beam height.}$$
$$\text{Practically, this means USE } h = 20 \text{ in.}$$
$$\text{USE } \cancel{Try} \; b_w = 10 \text{ in., } \qquad d = 20 - 2.5 = 17.5 \text{ in. (1.46 ft).}$$

Check shear before calculating A_s. The shear is critical at d from support and an extra 8% of the simple beam shear* will be allowed for the continuity shear. (More generally, to fix the stem size, the *end* span shear $1.15 w \ell_n/2$ is used to permit uniformity in size for exterior and interior spans.)

$$V_u = 6210(10 \times 1.08 - 1.46) = 58,000 \text{ lb}$$
$$V_n = 58,000/0.85 = 68,200 \text{ lb}$$
$$v = V_n/b_w d = 68,200/(10 \times 17.5) = 390 \text{ psi} < 10\sqrt{f_c'} = 548 \text{ psi} \quad \textbf{O.K.}$$
$$\text{Stem wt. (Fig. 9.12}b\text{),} \qquad w_s = 10(20 - 5) \times 150/144$$
$$= 156 \times 1.4LF = 218 \text{ plf}$$
$$\text{Revised } w = 6150 \text{ plf}$$

The 1% drop in w reduces the depth required for M_n by about 0.5%, which means no change in d, but M_u will be recomputed for steel calculations.

---

* In Fig. 9.12h,

$$V_c = \frac{(0.091 - 0.048)w\ell_n^2}{\ell_n} = 0.043\,w\ell_n$$

which is 8.6% of the simple beam shear at the end and 10% to 11% of that at a distance d from the support.

$$M_u = -0.091 \times 6150 \times 20^2 = 224{,}000 \text{ ft-lb}$$
$$M_n = 224{,}000/0.90 = 249{,}000 \text{ ft-lb}$$

From Table 3.1, $k_{n1} = 783$, $\quad \rho_1 = 0.0161$, $\quad a/d = 0.378$.

$$M_{n1} = 783 \times 10 \times 17.5^2/12 = 200{,}000 \text{ ft-lb}$$
$$M_{n2} = 249{,}000 - 200{,}000 = 49{,}000 \text{ ft-lb} \quad d - d' = 15.0 \text{ in.}$$
$$A_{s1} = 0.0161 \times 10 \times 17.5 = 2.82 \text{ in.}^2$$
$$A_{s2} = 49{,}000 \times 12/(60{,}000 \times 15) = \underline{0.65}$$
$$\text{Neg. } A_s = 3.47 \text{ in.}^2$$

Although A_s' will be near the magnitude of A_{s2} and apparently not critical, it will be calculated exactly to indicate the method in detail. First check the stress on A_s', starting with the depth of stress block, as in Fig. 9.12c.

$$a = 0.378 \times 17.5 = 6.62 \text{ in.} \quad c = a/0.85 = 7.78 \text{ in.}$$
$$\varepsilon_s' = 0.003(7.78 - 2.5)/7.78 = 0.00204 \text{ (very close to } \varepsilon_y)$$
$$f_s' = 0.00204 \times 29 \times 10^6 = 59{,}300 \text{ psi}$$

Considering displaced concrete, effective $f_s' = 59{,}300 - 0.85 \times 3000$

$$= 59{,}300 - 2550 = 56{,}700 \text{ psi}$$

This compares to a maximum possible of $60{,}000 - 2550 = 57{,}400$ psi

$$A_s' = 49{,}000 \times 12/(56{,}700 \times 15) = 0.69 \text{ in.}^2 \text{ for stress.}$$

The very "exact" calculation based on a precise determination of compression steel stress and consideration of concrete displaced by the bar has only a minor change. When ε_s was found close to ε_y, further checking could be skipped.

At midspan (Fig. 9.12j) the minimum positive moment is actually negative, $-0.010w\ell_n^2$, requiring, closely enough by proportion,

$$A_s = 3.47(0.010/0.091) = 0.38 \text{ in.}^2$$

with 1-#7 adequate (0.60 in.^{2}). Remembering that in the minimum positive moment loading case only dead load is on the span considered, statically determined moment component at the centerline $w_d\ell_n^2/8$ can be transformed into an equivalent $w_T\ell_n^2$ value as

$$(x)w_T\ell_n^2 = w_d\ell_n^2/8$$
$$\text{so } x = w_d/8w_T = 1650/(8)(6150) = 0.033$$

This establishes the support moment value shown in Fig. 9.12j. A simple parabolic relation gives the value $.01 + 1/4(0.033) = 0.0182$ at the quarter point. At the quarter point of span, $M_u = -0.0182w\ell_n^2$ and

$$A_s = 3.47(0.0182/0.091) = 0.69 \text{ in.}^2$$

The 1-#7 is adequate back to about 6 ft from the support.

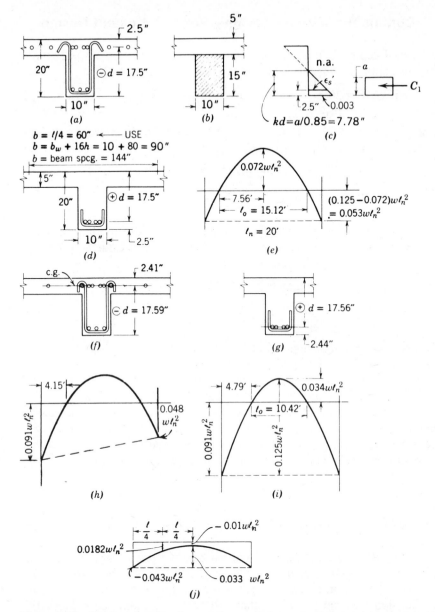

Figure 9.12 Design sketches for continuous T-beam design. (*a*) Assumed depth for negative moment. (*b*) Estimated weight, shaded area. (*c*) Strain triangle for f'_s calculation. (*d*) Assumed depth for positive moment. (*e*) Location of P.I. with maximum positive *M*. (*f*) Actual depth to top steel as selected. (*g*) Actual depth to bottom steel as selected. (*h*) "Exact" location of P.I. with maximum negative *M*. (*i*) Approximate location of P.I. with maximum negative *M*. (*j*) Minimum positive moment near midspan.

9.11 Continuous T-Beam Design—Positive Moment Design

The overall depth of 20 in. already chosen for negative moment fixes the effective depth at midspan. For one layer of steel (Fig. 9.12d),

$$\text{Positive } d = 20 \text{ in.} - 1.5\text{-in. cover} - 0.5\text{-in. stirrup} - 0.5\text{-in.}$$
$$\text{(for } d_b/2) = 17.5 \text{ in.}$$

Trial $z = 17.5 - h_f/2 = 15.0$ in., or $0.9d = 15.75$ in. Use larger value.

$$M_u = +0.072 \times 6150 \times 20^2 = 177,000 \text{ ft-lb,}$$
$$M_n = 177,000/0.9 = 196,800 \text{ ft-lb}$$

$$\text{Trial } A_s = \frac{M_n}{f_y z} = \frac{196,800 \times 12}{60,000 \times 15.75} = 2.50 \text{ in.}^2$$

The compression area is probably limited to less than the depth of the 60-in. wide effective flange found from the T-beam rules of Code 8.10.2 and thus acts as in a wide rectangular beam.

$$a = 2.50 \times 60,000/(0.85 \times 3000 \times 60) = 0.98 \text{ in.}$$
$$z = 17.5 - 0.98/2 = 17.01 \text{ in.}$$
$$A_s = 196,800 \times 12/(60,000 \times 17.01) = 2.31 \text{ in.}^2$$

A further cycle could change only the last digit. Because a is so small, the beam is obviously in the under-reinforced classification.

This steel should be checked against the minimum for positive moment steel as required by Code 10.5, that is, against $200 b_w d/f_y = 200 \times 10 \times 17.5/60,000 = 0.58$ in.^{2} Bars must be so arranged to maintain this steel area throughout the entire positive moment length, but this is no problem as it is here essentially the $\frac{1}{4}$ of the positive moment steel that the Code requires to be continued into the column in all continuous beams.

9.12 Continuous T-Beam Design—Choice of Bars

There are a wide variety of bar sizes and bar bend patterns that can efficiently be used to meet the reinforcement requirements. When bent bars are used the designer must have common bar sizes for the portion of the required positive moment tension bars near midspan that will be bent up and become negative moment tension bars at the supports. Small size bars can increase congestion and large size bars can cause development length problems and reduce the flexibility in bending bars. A practical constraint on positive moment tension reinforcement bar size is the requirement for ℓ_d at the point of inflection (Sec. 8.9a, Code 12.11.3) at least with regard to the bars continuing toward the column. For these n bars of individual area A_b, the maximum $\ell_d \leq M_n/V_u + d$. At the P.I. that is located in Fig. 9.12e at 7.56 ft from midspan,

$$M_n/V_u = nA_b 60,000 \times 17/(6150 \times 7.56) = 21.9 nA_b$$

As it might be likely that half of the positive moment steel will be bent up before the P.I. if nA_b is half of the maximum required positive moment steel (required steel being 2.31 in.^{2}), then

$$M_n/V_u = 1.15 \times 22.0 = 25.3 \text{ in.}, \qquad \text{maximum } \ell_d \le 25.3 + 17.5 = 42.8 \text{ in.}$$

Trying various bar sizes, for #8, $\ell_d = 0.04 A_b f_y / \sqrt{f_c'} = 0.4 \times 0.79 \times 60{,}000/54.8 = 34.6$ in. and for #9, $\ell_d = 43.8$ in. This result permits the use of #8 or smaller bars. Number 9 or larger bars have a required ℓ_d in excess of 42.8 in. and so cannot be used. Bars will now be selected and detailed, using bent bars. The detailing will be revised in Sec. 9.15e for straight bars.

The schematic bending arrangement of the steel selected is the same as that shown for the slab in Fig. 9.10b. The selection of bars is more a matter of art and judgment than of slavish adherence to mathematics. The designer must simultaneously consider the needs at top and bottom, both at midspan and the supports. The required or minimum steel areas at both top and bottom are shown on Fig. 9.13. The basic bar arrangement is also shown. At each location, several possible bar combinations are given. Because of the P.I. ℓ_d limit, the combinations are restricted to #8 bars although larger bars could be used for the bent bars. After looking at the various combinations, certain trends tend to favor the combinations with #7 bars. Use of 6-#6 bars in the narrow stem at midspan would be very congested. Use of 3-#8 bars at midspan would mean either bending up only one bar (33%), or bending up two bars and only leaving one bar at the support. Designers prefer symmetrically arranged bars near the outside of members to anchor stirrups so two bars are preferred on the bottom

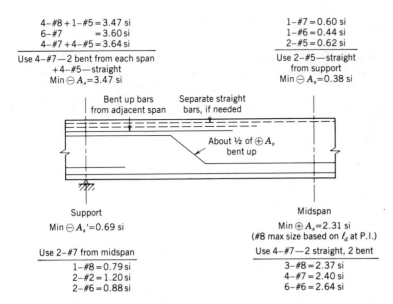

Figure 9.13   Selection of bars.

near the support. On this basis the choice is made to use 4-#7 at midspan with two being bent up and run through the support as far as needed to be effective in the adjacent span. This choice narrows the possibility at the support $-A_s$ to either 6-#7 or 4-#7 + 4-#5. Both to provide more bars to control flexural crack widths and to provide two outside bars to anchor stirrups near midspan, the 4-#7 + 4-#5 combination is selected.

For the positive moment steel, 2-#7 bars will be bundled* in each corner of the stirrups (see Fig. 9.12g) because there is no room to space four bars singly in the clear space between stirrup legs, which is $10 - 2 \times 1.5$ cover -2×0.5 stirrup = 6.0 in. With bars bundled as stated, the clear space between bundles is $6.0 - 4 \times 0.88 = 2.48$ in., which is actually nearer 2.0 in. because the diameter over the lugs is nearly $\frac{1}{8}$ in. extra for each bar. The spacing for bundled bars in Code 7.6.6.5 is the same as for a single bar of the *combined* area. In this case the area is $2 \times 0.60 = 1.20$ in.^{2} with equivalent diameter of 1.24 in.; hence the space between bars is ample as aggregate will probably be 0.75 in. maximum size. In this case the slight excess A_s would also permit the 2-#7 bars to be bundled one above the other.

The crack control provisions should be considered here, but this will be delayed until after all the bars are tentatively selected in order not to confuse the basic selection process.

The negative moment steel can be placed in one layer by bundling the four bent-up #7 bars into two bundles adjacent to the stirrups, placing two straight bars in the stirrup hooks, and the other two straight bars in the slab, say at 15-in. spacing as shown in Fig. 9.12f. The use of #7 bars increases the estimated d for all steel by 0.06 in., a little more with the #5 bars at the top, but the change appears negligible (Fig. 9.12g).

The compression steel (at support) is detailed as 2-#7 from each side, each caring for the stresses on the entering face of the column. This means developing all four bars into (and actually beyond) the far face of the column. This type of detail requires slightly more reinforcement than the simpler detail that runs 1-#7 from each side far enough to make it also effective on the far face of the column and leaves 1-#7 bottom bar to be cut off short of the column on each side. Fully developing both bars is highly desirable and adds appreciable redundancy which improves the general structural integrity.

The crack control provision of Sec. 3.30 (Code 10.6.4) is

$$z = f_s \cdot \sqrt[3]{d_c A} \lessgtr 175 \quad \text{(for interior exposure)}$$

The positive moment bars will be considered first. This equation was not designed for bundled bars. Whether to count each bundle as one or two bars is not definite; the use here as one bar is on the conservative

---

* Bundled bars require the use of stirrups around them. If shear did not call for stirrups all the way across the span, as is the case in Sec. 9.17, the bundling here would have required them anyway.

side. For the two bar bundles shown in Fig. 9.12g, d_c is 2.44 in. The suggested $0.6f_y$ will be used for f_s.

$$A = (2)(2.44)(10)/2 = 24.4 \text{ in.}^2$$
$$z = 0.6 \times 60 \sqrt[3]{2.44 \times 24.4} = 36 \times 3.90 = 140 < 175$$

The bars are thus satisfactory even when bundled.

For the negative moment bars, with the bundles in the horizontal plane and the bars spread over the flange width of 60 in., $d_c = 2.44$ in., $A = 2 \times 2.44 \times 60/6 = 48.8$ in.^{2}, again counting a bundle as one bar. Considering the mixed bar sizes would result in $n = 6.06$ so the use of the #6 bar is again conservative.

$$z = 0.6 \times 60 \sqrt[3]{2.44 \times 48.8} = 36 \sqrt[3]{119} = 177 > 175$$

In consideration of the conservative treatment of the bundled bars, the arrangement is approved.

It is not uncommon for the designer to increase b_w arbitrarily or even change d to make the steel used space properly or to modify the required A_s to fit available bar sizes closer.

A less congested beam may result if advantage is taken of the redistribution provisions of Code 8.4. Where the moments have been established by elastic theory and congestion is a problem, this is recommended.

9.13 Continuous T-Beam Design—General Requirements for Bending Bars

The fabricator must detail each bar, but the designer can be satisfied to locate bend points and cutoff points reasonably accurately. Many offices use rules such as "bend up half the bottom steel at the quarter point of clear span," but more exact procedures are presented here to indicate more adaptable methods needed in many cases.

The authors recommend a very conservative attitude toward bar detailing. There has been much ineffective detailing of what would otherwise have been good designs. The final member is not better than its details.

Bend points may be governed by:

1. Moment requirements.
2. Development lengths.
3. "Arbitrary requirements of Code 12.12.3, 12.11.1, 12.11.2.
4. Use of bent bars as stirrups.

The several "arbitrary" specification requirements will first be summarized. Code 12.12.3 requires that at least 1/3 of the total reinforcement for negative moment be extended beyond the extreme location of the point of inflection by the greater of: (1) $\ell_c/16$; (2) beam depth d; (3) $12d_b$. Code 12.11.1 requires that at least 1/4 of the positive reinforcement in continuous beams will extend 6 in. into the support. Both of these, in the words of the old Joint Committee Specification, are "to provide for contingencies arising from unanticipated distribution of loads, yielding of supports, shifting of points of inflection, or other lack of agreement with assumed conditions governing the design of elastic structures." Code 12.11.2 goes further by adding that when the member is part of a "primary lateral load resisting system" the extension of the bottom steel covered in 12.11.1 must anchor it to develop its full-yield stress in tension at the face of the support, that is, be much more than 6 in. This requirement is to provide some ductility in the event of stress reversal from wind, earthquake, or explosion. It is an excellent practice to follow, even in braced frames, to increase the overall structural integrity.

Code 12.10.3 also requires that every bar, whether required for positive or negative reinforcement, be extended the depth of the beam or 12 bar diameters beyond the point at which it is no longer needed to resist flexure. As discussed in connection with truss models, the critical section for flexural stress can be displaced almost a distance d at an inclined crack. This distance corresponds to the Code 12.10.3 provision and highlights its necessity. The authors designate this requirement as a and use it as a shifted moment diagram, but the Code itself seems to leave much to the designer's discretion. Because the Code uses simply the word "extend," it seems to imply that the extension might exist entirely as a bent bar. The authors prefer to consider it as requiring at least 1/2 of the extension to remain straight and recognize that a bent bar remains partially effective in flexure although in the tension half of the effective depth.

Arrangements satisfying moment requirements, development lengths, and the "arbitrary" requirements may sometimes be varied slightly to help out in development length demands elsewhere or in web reinforcement. For example, if development of the bottom steel at the point of inflection proves difficult, it may be possible to shift the bend-up points toward the column and to keep more steel available at the P.I.; or the spacing of bend points may be shifted to make bent bars more useful for web steel.

9.14 Continuous T-Beam Design—Moment Diagrams Governing Bar Bends

The positive moment diagram already has been established from Fig. 9.12e. For maximum negative moment the actual moment diagram is unsymmetrical and the maximum moment calculation alone does not provide sufficient data for the entire diagram. The diagram can be reasonably approximated (Fig. 9.12h) by using a negative moment at the far end as slightly less than that accompanying the maximum positive moment, in this case, say, $-0.048w\ell_n^2$ instead of the $-0.053w\ell_n^2$ value shown in Fig. 9.12e. Figure 9.12h is almost exact insofar as the location of P.I. (point of inflection) on the left is concerned. The approxi-

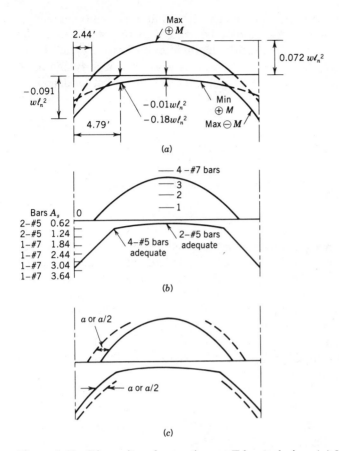

Figure 9.14 Dimensions for continuous T-beam design. (*a*) Moment envelope based on moments. (*b*) Moment envelope based on total steel provided. (*c*) Bend or cutoff point extensions to satisfy Code 12.10.3.

mate P.I. is often calculated for the assumed symmetrical diagram of Fig. 9.12*i*, which will be used here.*

$$\tfrac{1}{8} w \ell_0^2 = 0.034 w \times 20^2$$
$$\ell_0 = 20\sqrt{0.272} = 10.42 \text{ ft}$$

---

* More accuracy is not difficult and should usually be used for end spans. The moment diagram is a simple span moment added to the diagram for the two end moments (and their equilibrating shears). The continuity shear V_c is the slope of a straight line joining the two end moments. For a uniform beam load this V_c will be the only shear at midspan. The maximum positive moment will then be a distance V_c/w off center and increased over the centerline moment by the area of the small shear triangle of height V_c and length V_c/w. The resulting diagram is a symmetrical parabola about this positive moment point.

Hence the P.I. is 4.79 ft from the column. The further approximation of using a triangle for the negative moment section of the parabola is reasonable when the P.I. is not too near midspan, say, outside the middle third of the span.

The minimum positive moment at midspan is also part of a symmetrical diagram, but in this case the simple beam moment diagrams is that for dead load alone, which can be expressed in terms of total load w, that is, 1650 plf out of 6150 plf total.

$$\text{Simple beam } M_s = 0.125 \, w_d \ell_n^2 = 0.125 \, \frac{1650}{6150} \, w\ell_n^2 = 0.0335 \, w\ell_n^2$$

The diagram is sketched in Fig. 9.12j. The maximum moment diagrams are assembled in Fig. 9.14a.

As is usual in the office unless it is a special case, the A_s required diagram is based on the total steel used as shown in Fig. 9.14b. This A_s required diagram is based on the P.I. locations and moment diagrams as theoretically calculated for ideal loading and stiffness conditions. Code 12.10.3 requires every bar to be extended beyond these distances where theoretically they would carry the applied moment by the depth of the beam or 12 bar diameters (termed a by the authors). For bent bars the authors interpret this as $a/2$ to the beginning of the bend and the other $a/2$ in the bend itself. Thus bars can only be bent at $a/2$ past the theoretical point and cutoff at a past the theoretical point. This effectively extends the moment envelopes as shown by the dashed lines in Fig. 9.14c. This is further discussed in the next section.

9.15 Continuous T-Beam Design—Bends and Stop Points for Tension Bars

(a) Purpose of Discussion   It is not the intent of this section to consider simply the details of bar detailing. Rather, experience has shown that no other type of problem so clearly brings to the fore the fundamentals of how reinforced concrete works and how both flexure and bar development length affect each other. There is not a unique, clearly preferable way to detail a continuous beam. Various office practices will differ for valid reasons. In this section the authors will show three legitimate ways of detailing the reinforcement selected for this example problem:

1. *Sec. 9.15c*. Bent bars with no major straight bars cut off in a tension zone (author preference).
2. *Sec. 9.15d*. Bent bars with some major straight bars cut off in tension zone.
3. *Sec. 9.15e*. All straight bars.

(b) General Assembly of Data   Effective detailing requires clear and unambiguous transfer of the designer's intent to the fabricator or contractor's detailers and to the construction superintendent. The designer is responsible for major technical decisions and for a basic design that meets the Code requirements and is constructable. Before the designer can decide on the many detailed dimensions as to where bars can be bent or stopped, he or she must relate the reinforcement scheme chosen to the moment envelope and the arbitrary requirements.

Figure 9.15 indicates two different bent bar arrangements for the example problem and the reinforcement scheme selected. In both parts of the figure, the relation between the bar layout and the moment envelope is carefully indicated by bar identifications. Figure 9.15a shows a pair of bars bent simultaneously. Priority in bar arrangement is given to the bent bars. One says that the bars are bent "first" meaning that these bars are shown as the ones providing moment

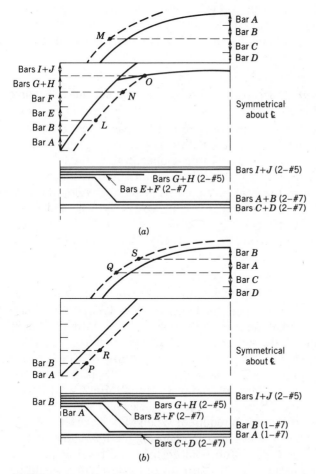

Figure 9.15 Relation of bend points to bar layout. (*a*) Single bend point. (*b*) Multiple bend point.

increments at the peak moment values for both positive and negative regions. Negative moment considerations will tell where L is located so that the bars can be bent down. Positive moment considerations tell where M is located. Consideration of both L and M as well as the horizontal distance required for the bar bend at 45° will tell if and where the pair of bars can be bent. N and O indicate where straight bars might be stopped based on flexural considerations.

In contrast, Fig. 9.15b shows two bars bent singly. In this case, bar A is the first one bent down (P near the negative moment peak), but is the last one bent up (Q not near the positive moment peak). Consideration of both P and Q are necessary to determine the proper bend point of bar A; R and S determine the bend point of bar B. The student must carefully study these diagrams to understand the detailing calculations that will follow.

The data can be arranged in many possible patterns, but several pieces of data are interrelated in each decision; a regular repeating pattern of presentation is desirable. The authors' form is shown schematically in Fig. 9.16 for one pair of bars bent first. The minimum distance to the bend-down point is determined by the larger of two requirements:

1. Moment length plus $a/2$, as shown in Fig. 9.16a, the second $a/2$ being available in the bend length itself.
2. The development needs, shown in Fig. 9.16b. If the bar is bent down and then immediately terminated (usually with a hook) in the compression zone, the maximum stress at A must be developed both to the left (by ℓ_d into the support) and to the right of A. The authors assume 2/3 of ℓ_d is permissible here because the bent portion also helps to develop the bar. However, when tension reinforcement is bent across the web and made continuous with reinforcement on the opposite face, Code 12.10.1 implies adequate development. Requiring 2/3 of ℓ_d before the bend point seems very ample and is a conservative practice.

These moment length and development length values are tabulated in Fig. 9.16c, immediately above the bar sketch; the larger is "Min.1." Because bend points are usually dimensioned at the level of the bottom bars, "Min.2" is next found from "Min.1" by adding the run of the bar, equal to the horizontal offset of the 45° bar bend. The bottom bar can be bent up at "Min.3" (Fig. 9.16d) given by the moment length plus $a/2$ from midspan or at "Max.3" from the column = $\ell_n/2$ − Min.3. The values of "Max.3" and "Min.2" determine whether the bend can be made. Obviously, if Max.3 is less than Min.2, the bars cannot be bent and a different pattern is required. When Max.3 exceeds Min.2, it indicates the leeway open to the designer in choosing the final dimension.

If Fig. 9.16 were based on a first cutoff point instead of a bend point, only three changes would be necessary (but the stirrups of Code 12.10.5 would be necessary; see Sec. 5.15).

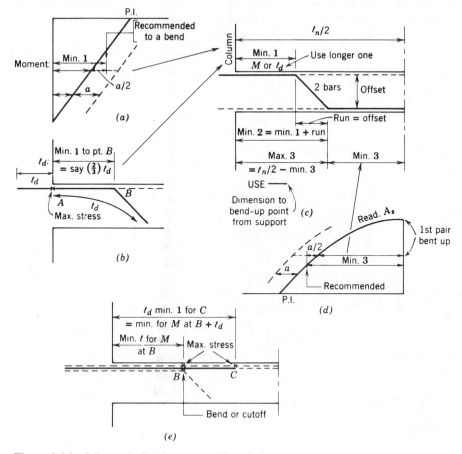

Figure 9.16 Schematic for data assembly.

1. The moment length in Fig. 9.16a, d would include the full a length (not $a/2$) as a Code requirement.
2. The development length would be the full ℓ_d.
3. Only "Min.1" would determine the top cutoff and only "Min.3" the bottom cutoff.

For any succeeding bar cutoffs one other change is involved, as in Fig. 9.16e. The ℓ_d value is measured from point B, the *theoretical* cutoff for the last bar stopped or bent (Code 12.10.4). The point B does not shift if the bar is run farther because of ℓ_d or some other reason. (If the previously bent or cutoff bar runs on to C, the maximum stress point is eliminated and the column face takes its place.)

This form is expandable to as many bars or cutoffs as are necessary. Development lengths are obviously needed for this example as follows.

Bottom #7, $\ell_d = 0.04 A_b f_y / \sqrt{f_c'} = 0.04 \times 0.60 \times 60{,}000/54.8$
$$= 26.3 \text{ in.} = 2.19 \text{ ft}$$

Top #7, $\ell_d = 1.4 \times 26.3 \qquad = 36.8 \text{ in.} = 3.07 \text{ ft.}$

Top #5, $\ell_d = 1.4 \times 0.04 \times 0.31 \times 60{,}000/54.8 = 1.4 \times 13.6$
$$= 19.0 \text{ in.} = 1.58 \text{ ft}$$

(c) Details for Author's Preferred Arrangement of Bars   The preferred arrangement indicated in Fig. 9.15a consists of 2-#7 bars bent as a pair (or singly) and all other bars straight and continuing at least to the nominal maximum negative moment P.I. as sketched in Fig. 9.17b. Because no bars are cut off inside the P.I., the extra stirrups of Code 12.10.5 are unneeded.*

The bent bars are calculated in Fig. 9.17 (based on the ideas of Fig. 9.16a to 9.16d); at the bottom level for bars A and B a minimum of 3.56 ft and a maximum of 3.93 ft result, with only about a 4-in. leeway. The other bottom bars (C and D) run into the support and are detailed there as compression steel. Because the size of these bars was originally selected in Sec. 9.12 to match the P.I. development length needs, it is automatically satisfactory.

The top straight bars are next considered. One-third of the total 3.64 in.², say, 4-#5 = 1.24 in.², must extend past the P.I. (which goes with that negative moment diagram) at least $\ell_n/16$, d, or $12 d_b$ (Code 12.12.3), in this case 1.44 ft which is d. Two of these (I + J) will necessarily be spliced at midspan with a lap of $1.7 \ell_d = 1.7 \times 1.58$ ft $= 2.69$ ft, say, 2-ft 9-in., with the 1.3 factor unacceptable because more than 50% of the bar is needed at midspan, that is, $f_s > f_y/2$.

These 4-#5 bars satisfactorily accommodate all of the negative moment owing to minimum positive loading considerations. Two of the #5 bars (G + H) can be terminated at 7-ft 0-in. into the span subject to a check on shear stress or provision of extra stirrups. Because this loading is possible but unlikely, cutting off these small bars in the tension zone is acceptable as long as the provisions of Code 12.10.5 are met.

The initial decision to carry straight bars E + F to the nominal P.I. automatically goes beyond any moment requirement. Arrangement becomes primarily a matter of staggering the bar cutoffs. For the 2-#7 bars E + F the critical stress point is at 2.30 ft, where the 2-#7 are bent down and the #7 ℓ_d of 3.07 ft is added to give an ℓ_d requirement of 5.37 ft, well inside the P.I. These bars are stopped at the nominal P.I. plus a, say, 6-ft 3-in.

(d) Details for Bent Bars with Straight Bars Cut Off in Tension Zones   This arrangement requires extra stirrups under Code 12.10.5.2 over the bars cut off inside the P.I. These stirrups are not computed here.

---

* The negative moment from the minimum positive moment loading is fully covered by reinforcement but is neglected with respect to determining the P.I. for application of Code 12.12.3.

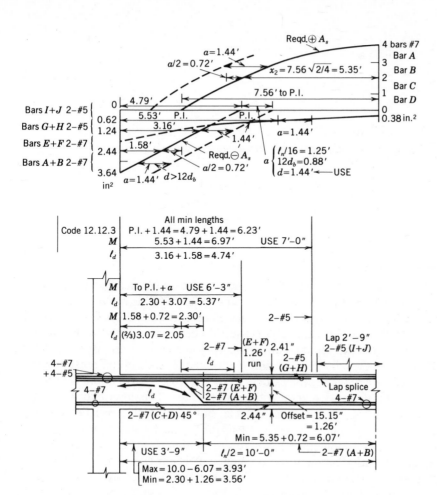

Figure 9.17 Detailing for bent bars and shifted moment diagram.

The bend point (Fig. 9.18) for the 2-#7 bars is the same as in Sec. c, at 3-ft 9-in. measured at the bottom steel level. The other bottom bars are run into the support although one could be cut off short of the P.I. This action would leave an unsymmetrical steel arrangement and no bottom bars to anchor the stirrups. The #5 bars on the top are unchanged. Only the 2-#7 bars could be shortened to 5-ft 4-in. but at the expense of adding extra stirrups over a length of $0.75d$ (1.08 ft) to satisfy Code 12.10.5. The savings in steel probably would not justify the extra labor required.

(e) Alternate Layout with All Bars Straight   To indicate the wide variation of detailing possible under the basic Code requirements, the same number and sizes of bars will be used in an alternate arrangement with only straight bars.

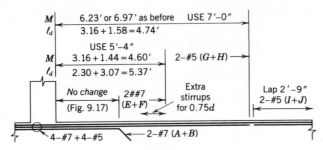

Figure 9.18 Detailing with shifted moment design but with minimum cutoff points.

Figure 9.19 indicates that two of the bottom #7 bars (A and B) are cut off as indicated from the shifted moment diagram; the third (C) is run d past the P.I., which automatically satisfies the ℓ_d requirements at the P.I. of Code 12.11.3; the fourth (D) is run into the support and into the next span to develop as compression steel. The top bars are arranged somewhat differently than in the bent bar layout. Two #7 bars (E and F) are cut off well in the tension zone, requiring extra stirrups to satisfy Code 12.10.5. Two #5 bars (G and H) are cut off next, also in the tension zone with extra stirrups. The other two #5 bars (I and J) again are cut off in a tension zone, but with much lower flexural and shear stress, and so do not require extra stirrups. One #7 bar (K) is cut off well beyond the negative moment P.I. where the continuing #7 bar (L) can handle the minimum positive moment requirements. This bar (L) must be lap-spliced at midspan with a lap length of $1.7\ell_d = (1.7)(3.07) = 5.22$ ft (say 5-ft 3-in.). The single bars cause some stirrup anchorage problems but the layout is completely acceptable according to Code requirements.

(f) Bent Versus Straight Bars   A major advantage of bent or offset bars is that they do not lower the shear strength of the member as do bars simply cut off and stopped. They also help to keep top steel at the proper level and probably tend to reduce placement errors. However, practice has moved largely toward straight bars, probably because of cost considerations with bars bent singly; but it is easier to satisfy development lengths needed with bars bent as pairs.

Bar arrangements, as illustrated in Fig. 9.17 for bent bars and Fig. 9.19 for straight bars, are both acceptable for the chosen bars. One might think the arrangement with straight bars is more efficient. Actually although it saves 3% in the weight of the longitudinal steel, when the weight of the extra stirrups required (#3 U @ 6-in.) is added in the total steel weight *increases* 4%. In addition, stirrups are labor-intensive; therefore, the cost of the straight bar arrangement is probably higher when all factors are considered.

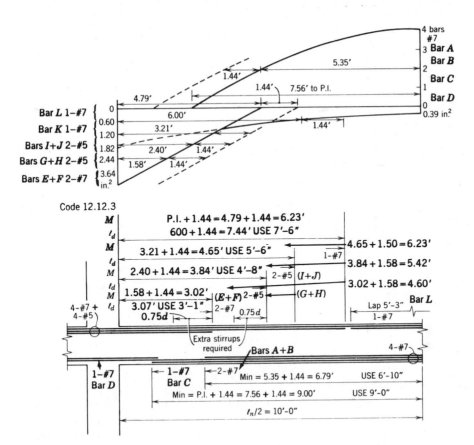

Figure 9.19 Alternate scheme with straight bars.

9.16 Continuous T-Beam Design—Stop Points for Compression Steel

The compression length of the #7 straight bottom bars (C and D) used in the arrangement of Fig. 9.17 is also established by both a moment and a development requirement.

As $\ominus A_s' = 0.69$ si, only two bars are required at each column face. Thus bars from *both* beams are not required to develop fully in compression in the adjacent span (Fig. 9.20b). If the bars are needed for compression in the adjacent span, the extension must go through the column and then beyond the far side a full compression development length (Fig. 9.20c) as in Fig. 9.20b; the length for moment may also control. Because the two bars from the span being detailed

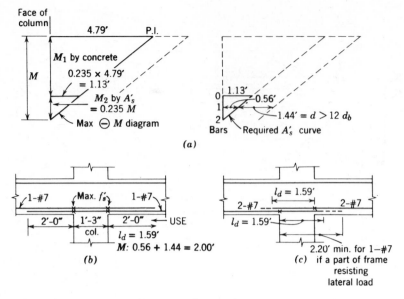

Figure 9.20 Detailing of compression steel.

are adequate for A'_s, the bars only have to go into the column a compression development length as shown in Fig. 9.20c; for all frames resisting lateral load, Code 12.11.2 requires one of these bars ($\frac{1}{4}$ of the positive moment reinforcement) to develop a tension $f_y \ell_d$ into the support (2.19 ft). This is an excellent practice that greatly improves general structural integrity and provides cheap insurance against progressive type collapses. The authors recommend using the tension ℓ_d (say 2.2 ft) for all such bars.

For compression development length Code 12.3.2 gave the relation

$$\ell_d = 0.02 f_y d_b / \sqrt{f'_c} > 8 \text{ in. but not less than } 0.0003 d_b f_y$$
$$\text{For \#7, } \ell_d = 0.02 \times 60,000 \times 0.875/54.8$$
$$= 19.17 \text{ in.} > 0.0003 \times 0.875 \times 60,000 = 15.75 \text{ in.}$$
$$\text{USE } \ell_d = 19.17 \text{ in.} = 1.59 \text{ ft}$$

For moment the length needing some help from M_2 is easily established in Fig. 9.20a from the negative moment diagram. In this case, M_{n2} was found in Sec. 9.10 to be 49,000 ft-lb out of a total of 249,000 ft-lb, or $M_{n2} = (49,000/249,000)M = 0.20 M$. This indicates *all* compression steel could theoretically be omitted at 0.20×4.79 (distance to P.I.) = 0.96 ft from the column. Half the steel could be stopped at 0.96/2 = 0.48 ft *plus* the arbitrary requirement of d or $12 d_b$ (1.44 ft controlling) for a total of 1.92 ft. This result exceeds ℓ_d and controls the bar length, as sketched in Fig. 9.20b.

The detail of A'_s if both bars required are furnished from the adjacent beam is shown in Fig. 9.20c. Because compression steel in this case extends as far as negative moment, there is no need to consider the moment diagram at all.

(Within the column the column compression provides good development conditions and even if the column were very wide the negative moment in the middle of the column would in effect be in a deeper beam.) The ℓ_d already calculated gives the distance the bars must extend *into* the column, which in this case happens to go a little beyond the far face of column.

9.17 Continuous T-Beam Design—Stirrups

Because bundled bars were selected in Sec. 9.12, stirrups are required all across the middle of the beam to satisfy Code 7.6.6.2. However, stirrups are usually considered a matter of shear design.

Under maximum moment loading the beam is subject to an end shear equal to the simple beam shear $w\ell_n/2 = 61{,}500$ lb plus a continuity shear, again assumed as $0.08 w\ell_n/2$ or 4900 lb, acting both at the end and at midspan as shown dashed in Fig. 9.21a. The shear at midspan is greater when live load is removed from the left half of the span. This loading produces a smaller continuity shear, which is neglected. This simple beam shear at midspan is $4500 \times 10 \times \frac{5}{20} = 11{,}200$ lb. The solid line in Fig. 9.21a is used as a maximum shear diagram for stirrup design. With $b_w = 10.0$ in. and $d = 17.59$ in. at the end (Fig. 9.12f), the critical shear is at the distance d from support.

$$\text{At support } V_{n0} = 66{,}400/0.85 = 78{,}120 \text{ lb}$$
$$\text{At midspan } V_{n10} = 11{,}200/0.85 = 13{,}180 \text{ lb}$$

The slope of the V_n diagram is $(78{,}120 - 13{,}180)/120 = 541$ lb/in. The critical shear at 17.6 in. from support is $78{,}120 - 541 \times 17.6 = 68{,}600$ lb. The shear (V_n) curve is plotted in Fig. 9.21b. Stirrups are clearly needed for the ℓ_v length where V_n is in excess of V_c; also for all the remainder of the length since everywhere $V_n > V_c/2 = 1\sqrt{f'_c}b_w d = 9640$ lb. The stirrup requirement over bundled bars is thus automatically satisfied; also the requirement for stirrups over A'_s bars under Code 7.11.1.

If bars were cut off in a tension zone, *extra* stirrups would be needed for $0.75 d$ over each cut bar as indicated in Figs. 9.18 and 9.19. Code 12.10.5 waives

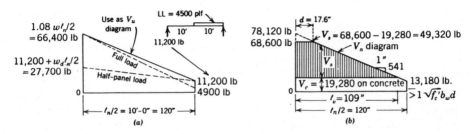

Figure 9.21 Shear diagrams for stirrup calculations. (a) V_u diagram. (b) V_n diagram and V_s (shaded).

these stirrups only where shear is equal to or less than 2/3 the allowable. These stirrups must provide an extra A_v of $60b_w s/f_y$ with a resulting spacing not exceeding $d/8\beta_b$, where β_b is the proportionate part of the bars cut off at the particular section.

Except for such special extra stirrups, a design with similar data is worked out in detail in Sec. 5.12.

9.18 Continuous T-Beam Design—Placement of Stirrups

The proper placement of stirrups in continuous beams presents a problem. The best anchorage of stirrups calls for the hooks to be in compression concrete, which near the supports is the bottom concrete. Because construction is simpler with the open end of the stirrup turned up, the matter of anchorage is generally ignored. Furthering this easier placement is the requirement that the compression (bottom) steel be tied as specified in Code 7.11.1. One stirrup must extend completely around all longitudinal A_s' bars. The ties, over the length where A_s' is needed, must satisfy the requirements for a column, that is, be "so arranged that every corner (bar) and alternate . . . bar shall have lateral support provided by the corner of a tie . . . and no bar shall be further than 6 in. clear on either side of such a laterally supported bar." Closed stirrups, like column ties, are excellent but these have limited use except in seismic zones or where torsion may be present because they complicate bar placement. The chief objection to closed stirrups relates to difficulty in dropping reinforcement into place inside the ties.

9.19 Continuous T-Beam Design—Deflection

Deflection under service load is the critical case and this calculation is somewhat toward an elastic analysis. The deflection calculations for this beam are a great deal more complex, but follow the pattern of the example in Sec. 3.22c.

9.20 Continuous Rectangular Beams

The design differences between continuous rectangular beams and continuous rectangular beams and continuous T-beams are all minor. The foregoing design of the T-beam can also serve as a model for continuous rectangular beam design. Other than the obvious minor difference in designing the positive moment steel for a rectangular beam and a somewhat greater congestion of negative moment steel (possibly two layers required), attention must be called to the special requirements for lateral supports at a spacing of no more than $50b$ as given in Code 10.4.1. Free-standing rectangular beams lack the extra stiffness and strength of the T-beam in resisting torsion.

9.21 Spandrel and Other L-Beams

A spandrel beam is one supporting an exterior wall, although occasionally the problems (without the name) may be associated with an interior beam adjacent to a stairwell, shaft, or other interior opening. The spandrel carries more dead load because of the exterior wall weight, but often is lighter than an interior beam because of live load coming from a reduced floor area.

The L-beam, with flange on only one side, is an unsymmetrical section. As such, Code 8.10.3 limits the flange counted for flexure more strictly than in a T-beam, that is, to the least of $\ell_n/12$, $6h_f$, or half the clear distance to the next beam. However, in flexure, an L-beam cannot act as an unsymmetrical section because the slab prevents the normal lateral deflection that would occur when free-standing.

The lateral deflection is actually prevented only at the level of the slab. With the slab loaded, a secondary rotation about an axis on the intersection of the center lines of slab and web does occur and causes some lateral deflection of the web. This results from the slab deflection and rotation at the beam junction, and it introduces torsion into the beam which must be considered.

9.22 End Spans and Irregular Spans

End spans always involve negative moments smaller at the outer end and larger at the inner end, with the points of inflection and point of maximum positive moment shifted toward the outer support. Irregular spans and loads also result in maximum moment diagrams that are less symmetrical than those used in this chapter.

Proper detailing of end spans and irregular spans requires a better knowledge of unsymmetrical moment diagrams, but no additional reinforced concrete theory. Where approximations are deemed proper, the designer should be more conservative than where more exact moment requirements are known.

Selected References

1. "Continuity in Concrete Building Frames," Portland Cement Association, Chicago, 3rd ed.
2. P. M. Ferguson, "Analysis of Three-Dimensional Beam-and-Girder Framing," *ACI Jour. 22,* Sept. 1950; *Proc.,* 47, p. 61.
3. R. H. Wood, "Studies in Composite Construction: Part I, The Composite Action of Brick Panel Walls Supported on Reinforced Concrete Beams; Part II, The Interaction of Floors and Beams in Multi-Story Buildings," National Building Studies, *Research Papers No. 13* (1952) and *22* (1955), Her Majesty's Stationary Office, London.

Problems

PROB. 9.1. The reduced frame of Fig. 9.22*b* should be used for the analysis of the second floor of the bent of Fig. 9.22*a* because standard coefficients are not applicable. Assume all the beams have $I = 25{,}000$ in.[4], the 16-in. columns below the floor have $I = 5450$ in.[4], and the 14-in. column above the floor have $I = 3200$ in.[4] Each beam carries a dead load of 1050 plf and a live load of 2150 plf. With factored loads:

(*a*) Calculate the maximum negative moment for the beam *CD* at *C* and correct this to the design moment at the face of the column. (Note that symmetry about *C* is equivalent to a fixed end for moment distribution purposes.)

(*b*) Calculate the maximum negative design moment at *D* of beam *CD*.

(*c*) Calculate the maximum positive moment for *CD*; also the minimum positive moment.

(*d*) Locate the several points of inflection that are useful in detailing steel.

(*e*) Compare the points of inflection in (*d*) with those that would be obtained if each moment diagram were assumed to be symmetrical about midspan, as in Fig. 9.12*i*.

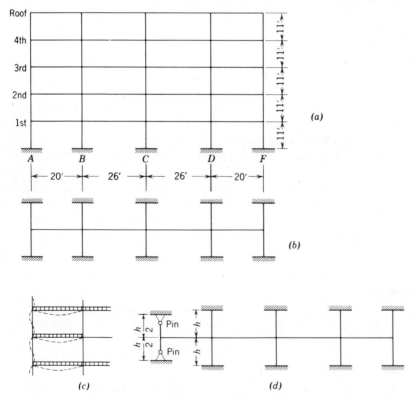

Figure 9.22 Analysis of building frame. (*a*) The frame considered. (*b*) Ordinary reduced frame for calculation of moments on a single floor. (*c*) Loading for maximum negative moment at exterior end of beam; also nearly maximum for exterior column moment. (*d*) Improved form of reduced frame for maximum moment at exterior joint.

PROB. 9.2. In Prob. 9.1 calculate the corresponding maximum moments in span DF and the points of inflection needed. (For this end span the use of a symmetrical moment diagram is scarcely valid.)

PROB. 9.3. In the frame of Prob. 9.1 calculate the maximum bending moment on column D, using factored loads. Simulate the worst loading condition, reversed curvature, as in Fig. 9.22c,d. (Increasing the column's stiffness in Fig. 9.22b by a factor of 1.5 gives the same result.)

PROB. 9.4. Repeat Prob. 9.3 for the exterior column F, using a reduced frame similar to the reduced frame of Fig. 9.22c,d.

In addition, note that loads on successive floors as in Fig. 9.22c give a reverse bending condition in the exterior columns which is almost equivalent to a reduced frame similar to that of Fig. 9.22d. Recalculate column D moments and negative moment at D in beam CD, using factored loads and this reduced frame.

PROB. 9.5. Design an interior span of a continuous one-way slab supported on beam 14-ft 0-in. on centers using moment coefficients of Sec. 9.5 (Code 8.3.3), live load of 150 psf, no dead except slab weight, $f_c' = 3000$ psi, Grade 40 steel, beam stems 11-in. wide, cover over center line of steel of 1.12 in. (as a simplification). Carry the design through the choice and spacing of bars (same size bars for both positive and negative moment). Draw a lengthwise section of slab and sketch the bars in place, assuming no bars bent up.

PROB. 9.6. Assume the steel found in Table 9.1 for a typical interior span of the slab of Sec. 9.8 is to be arranged as shown in Fig. 9.10b, with half of the bottom bars bent up and extra straight top bars added over the support. (This involves a new choice of steel at the support.) Calculate all bend points and stop points (for a typical interior span).

PROB. 9.7. A continuous rectangular beam is to carry a uniform live load of 3000 plf plus 1000 plf dead load and its own weight over equal 20-ft spans. Supports may be considered of negligible width (knife edges). Design a typical interior span through the choice of beam size and the choice of reinforcing. Assume maximum $V = 1.05w\ell_n/2$, $f_c' = 4000$ psi, Grade 60 steel. Limit k_n to 1300, which means some compression steel at the worst section. Show a cross section at support and at midspan with detailed spacing of bars; also an elevation of beam showing the schematic arrangement (bending) of bars. Exact bend points and so forth are not a part of this problem.

PROB. 9.8. Design an intermediate span of one of a series of continuous T-beams spaced 11-ft 0-in. on centers to carry a 5-in. slab, live load of 175 psf, and its own weight over a 22-ft clear span, using $f_c' = 3000$ psi, Grade 40 steel, ACI moment coefficients, $V = 1.08w\ell_n/2$. It is suggested that the stem size be based on using $k_n = 1100$ at the support and a stem width of 11.5 in. be first tried. Stem weight may be assumed 250 plf (without revision). Sketch the cross section at the support and at midspan and show the trial arrangement of longitudinal steel.

PROB. 9.9. Assume that an interior span of a continuous beam with uniform live load has been designed for moment coefficients of -0.093 and $+0.067$, based on using the clear span of 23 ft. The steel used is 10-#7 for negative moment, 7-#7 for positive moment, and 4-#7 for compression steel at support. Assume columns 16-in. square, $b_w = 16$ in., $d = 19$ in., $d - d' = 16.5$ in., $f_c' = 4000$ psi, $f_y = 40,000$ psi, and $f_s' = 36,000$ psi. The M_2 couple represents 0.33 of the maximum negative moment. Arrange the bars to be bent up, using a single bent bar nearest midspan and then a pair of bent bars nearer the support. For location of points of inflection, the simplifying assumption of Fig. 9.12i may be used. Sketch the arrangement of steel and detail all bend and stop points. (If any bends fail to work out satisfactorily, note this fact and compromise lengths as seems best, but do not revise the bending scheme.) Show the results on a sketch similar to Fig.

9.17. Assume minimum positive moment does not control bar lengths. Assume the beam is cast with a 4-in. slab and is part of a frame braced by a shear wall.

PROB. 9.10. A typical interior span of a continuous beam of 21-ft clear span supported by 18-in. square columns in a braced frame and designed for uniform load moment coefficients of $-\frac{1}{12}$ and $+\frac{1}{16}$ (based on clear span) requires 9-#8 for negative moment, 6-#8 for positive moment, and 6-#8 for compression steel at the support. If $M_2 = 0.54$ of maximum negative moment, $f'_c = 3000$ psi, $f_c = 40,000$ psi, $f'_s = 28,000$ psi, and the offset distance between top and bottom steel is 14 in., sketch the steel arrangement using straight bars and locate stop points for bars. For detailing bar lengths take d as 17 in. Assume midspan minimum positive moment as $+0.015\,w\ell_n^2$ with the factored dead load equal to $w/3$. Record the results as in Fig. 9.19. (A 5-in. slab may be assumed with beam $b_w = 14$ in.)

PROB. 9.11. An interior 19-ft 6-in. clear span of continuous T-beam with a 5-in. slab is loaded at its third points with concentrated loads P such that the uniform load may be neglected in establishing the shape of the moment diagram. Columns are 16-in. square and a primary part of the lateral load resisting system. The maximum moments, including an allowance for uniform load, are $-0.25\,P\ell_n$ and $+0.17\,P\ell_n$. The negative moment is broken down into $M_1 = 0.42\,M$ and $M_2 = 0.58\,M$ for purposes of design. The minimum midspan positive moment is $0.050\,P\ell_n$. Factored dead load is $P/3$. For bar lengths, d may be taken as 18 in. and $d - d'$ as 15 in. The required number of #7 bars is 8 for negative moment A_s, 5 for compression steel A'_s, and 5 for positive moment steel. $f'_c = 4000$ psi, $f_y = 60,000$ psi, $f'_s = 50,000$ psi. Arrange and detail the steel attempting to bend at least two bars up from the bottom. Record the results as in Fig. 9.17. (If any bends fail to work out satisfactorily, note the fact and compromise lengths as seems best, but for this problem stay with two bars up in a web 15-in. wide.)

PROB. 9.12. Same as Prob. 9.11 except straight bars are to be used and detailed.

10

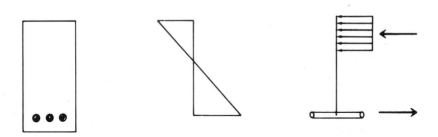

DETAILING OF JOINTS

10.1   Structural Details in Monolithic Structures

After completing analysis of a structure, proportioning members, and determining longitudinal and transverse reinforcement to be used in the members, designers often neglect to detail joints where the elements intersect. Critical details that influence the performance of the reinforced concrete structure are left to the fabricator or contractor to complete. Joints are critical because they ensure continuity of a structure and transfer forces from one element to another. A typical monolithic reinforced concrete structure is cast or erected as a single integral system as opposed to a system formed of precast units or with well-defined separations between portions or elements of the structure. In a frame building, loads and forces must be transferred from the floor system to supporting girders, from girders to columns, and from columns to the foundation. The transfer of forces between elements depends on careful consideration of joint details and inspection to ensure that fabrication and construction follow the instructions or the intent of the designer.

To clarify those instructions, it may be necessary for the designer to show joint details on the plans. By including joint details, the designer is forced to check the constructibility of the joint detail. Constructibility deals with the placement of reinforcement and the placement and consolidation of concrete. For example, a 15-in. wide beam framing into a 15-in. wide column will result in a problem for the construction worker if no one has considered that with 1-$\frac{1}{2}$-in. cover on the transverse reinforcement, the outer column and beam longitudinal steel will occupy the same space as shown in Fig. 10.1. If the steel layout is not considered when longitudinal steel is selected, the intermediate bars may interfere. When members are proportioned it is relatively easy to change member sizes to avoid this problem. If the column dimension is increased to 18 in., outer column bars passing through the joint will be outside the beam bars. If four bars are used in place of three bars for top steel in the beam, the intermedi-

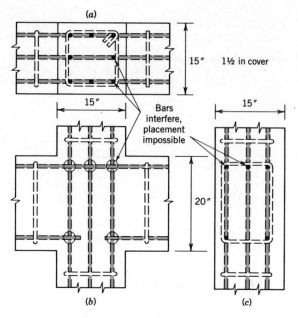

Figure 10.1 Beam-column joint. (*a*) Side view of joint. (*b*) End view. (*c*) Top view.

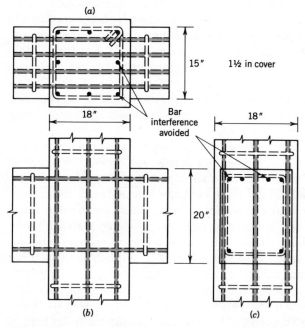

Figure 10.2 Revised beam-column joint. (*a*) Side view. (*b*) End view. (*c*) Top view.

ate beam bars will pass on either side of the column intermediate bar. The revised joint is shown in Fig. 10.2. By recognizing this problem at the design stage, field erection problems are avoided.

10.2   Design Considerations

Figure 10.3 shows a frame consisting of beams and columns only; the floor slab or floor system has been removed for simplicity. A typical joint may connect elements in three directions and interference of bars in all directions must be avoided. For analysis and design purposes, however, each direction may be considered separately. For example, a typical joint, such as A2 or D2 at Level 2 in Fig. 10.3, is subjected to the forces shown in Fig. 10.4a. The joint is the portion common to the beams and columns, and is shaded in Fig. 10.4a. Joint distortion is shown in Fig. 10.4b. Beam cracks at the face of the column and column cracks at the top and bottom of the beam are the result of slip of reinforcement within the joint. In analysis, fixed ends are often assumed for the support where beams frame into columns. In reality, the reinforcement will slip even at low stresses so that a truly fixed end is not possible. The joint is distorted in shear by the resultant forces acting on the joint (Fig. 10.4c) that produce tension along one diagonal of the joint and compression along the other. The cracks shown in Fig. 10.4b are produced when principal tensile stresses exceed the tensile capacity of the concrete. Because the cracks are inclined similar to shear cracks in a beam, early design recommendations by ACI Committee 352[1] for shear in joints were based on equations adapted from beam shear provisions.

In a joint where members frame into the joint from both sides in the direction of loading, distortion of the joint depends on the magnitude and sense of the moment on opposite faces of the joint. Figure 10.4d shows cracking of such a joint under large moments. Moments on opposite sides of the joint probably will be of opposite sense as shown in Fig. 10.5a. Only under very high lateral loads, large differences in adjacent span lengths, or large differences in gravity loads on adjacent spans will the moments be additive and create high shears

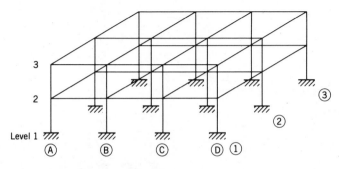

Figure 10.3 Frame structure.

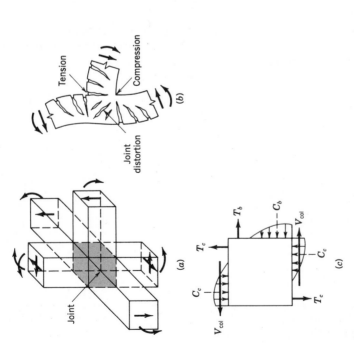

Figure 10.4 Forces and distortion, beam-column joint. (*a*) Forces on members at joint. (*b*) Distortion at joint. (*c*) Resultant forces on joint. (*d*) Shear cracking of interior joint under large unbalanced beam moments.

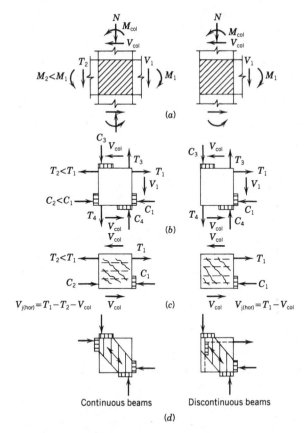

Figure 10.5 Transfer of forces through joints. (*a*) Free-body of joint. (*b*) Forces transmitted through joint. (*c*) Panel model. (*d*) Compression strut model.

through a joint where the beams are continuous. The most common occurrence of large unbalanced moments across a joint is under seismic loading as discussed in Chapter 23. For joints in structures in nonseismic regions, the recent revision of the ACI Committee 352[2] report provides the most complete design recommendations available in the United States. The research on which the report is based is summarized elsewhere.[1-4]

The prime factors to be considered in design of joints include:

a. Shear.
b. Anchorage of reinforcement.
c. Transfer of axial load.

The joint should be designed so that it is as strong or stronger than the members framing into it. Failure in the joint should be avoided. The joint should maintain its integrity so that if the structure is overloaded, the indication of such overload is large flexural cracking in the beams and beam-hinging.

A number of different equations and models have been proposed for determining the shear capacity of joints. Figure 10.5b shows the forces transmitted through the joint, and Fig. 10.5c and 10.5d show two of the models that are most commonly used. The basic difference between the panel and the compression strut models is the forces assigned to the concrete and the steel. In the panel model, the concrete is assigned a limiting shear capacity V_c, and transverse steel must be provided to make up the difference between V_j and V_c. In the compression strut model, the concrete must carry a diagonal compressive force that reacts against the compressive forces at the corners. At the early stages of loading, the joint behaves more like the panel model, but at ultimate the forces are similar to the situation shown in Fig. 10.5d. Because the bars slip in anchorage, the stresses are transferred to the concrete away from the face of the joint. In the case of an interior joint with straight bars passing through the joint, full development occurs over some anchorage length. For an exterior joint with a hooked bar, the anchorage is provided by bearing against the inner radius of the bend as described in Chapter 8. In general, the transfer of axial force N through the joint (Fig. 10.5a) is not critical unless the joint is cast with concrete of lower strength than the columns.

10.3 Design Recommendations—ACI Committee 352

The ACI Committee 352 report provides recommendations for two types of joints:

Structural joints are classified into two categories, Type 1 and Type 2, based on the loading conditions for the joint and the anticipated deformations of the joint when resisting lateral loads.

Type 1—A Type 1 joint connects members designed to satisfy ACI 318 strength requirements and in which no significant inelastic deformations are anticipated.

Type 2—A Type 2 joint connects members designed to have sustained strength under deformation reversals into the inelastic range.

The requirements for joints are dependent on the deformations at the joint implied by the design-loading conditions. Typical examples of each joint type are:

Type 1 is a joint in a continuous moment-resisting structure designed on the basis of strength without considering special ductility requirements. Any joint in a typical frame designed to resist gravity and normal wind loads falls into this category.

Type 2 is a joint that connects members that are required to dissipate energy through reversals of deformation into the inelastic range. Joints in moment-resisting frame structures designed to resist earthquake motions, very high winds, or blast effects fall in this category.

Type 1 joints are discussed in this chapter, and Type 2 joints are discussed in Chapter 23.

Forces acting on the joint include axial loads, bending, torsion, shear from externally applied loads, and forces resulting from creep, shrinkage, temperature, or settlement. These effects may not be easily determined in many situations. In any case, design of the joint should be based on the forces acting on the joint considering the nominal strength of the members. The forces on the joint are based on developing yield in the tension reinforcement of the floor (beam) members with the capacity reduction factor ϕ taken as 1.0. For example, in the exterior joint shown in Fig. 10.5b, T_1 is taken as $A_s f_y$, where A_s is the top steel in the beam framing into the joint and the shear on a horizontal plane through the joint is $V_{j\,(hor)} = A_s f_y - V_{col}$.

The horizontal joint shear $V_{j\,(hor)}$ should not exceed a maximum value taken as

$$V_{j\,(hor)} = V_u < \phi V_n \qquad (10.1)$$

where ϕ is 0.85. The nominal shear capacity is

$$V_n = \gamma \sqrt{f'_c}\, b_j h \qquad (10.2)$$

where γ is as specified in Fig. 10.6a, b_j is the effective column width (Fig. 10.6b), and h is the column depth in the direction of shear or loading. Where beams are equal or narrower than the column

$$b_j = (b_b + b_c)/2 \qquad (10.3)$$

$$b_j \le b_b + \frac{h}{2} \qquad \text{on each side of beam}$$

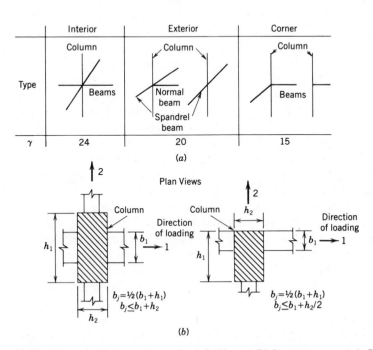

(a)

(b)

Figure 10.6 ACI Committee 352 definitions of joint geometry. (a) Geometric description of joints for γ. (b) Effective joint width b_j.

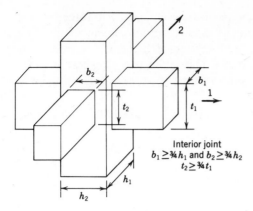

Interior joint
$b_1 \geq \frac{3}{4} h_1$ and $b_2 \geq \frac{3}{4} h_2$
$t_2 \geq \frac{3}{4} t_1$

Figure 10.7 Geometric requirements for interior joint.

where b_b is the beam width and b_c is the column width. If there are beams on both sides of the column in the direction of loading, b_b is taken as the average of the two beam widths. The ACI Committee 352 recommendations do not apply to cases where beams are wider than columns.

The geometric description of the joint (Fig. 10.6) for selecting γ must be carefully considered to ensure that the correct value of γ is used in design. An interior joint has horizontal members framing into all four sides of the joint; horizontal members must cover at least $\frac{3}{4}$ of the width of the column and the shallowest horizontal member must be at least $\frac{3}{4}$ of the depth of the deepest horizontal member. Figure 10.7 illustrates geometric requirements for an interior joint. If load is applied in the "1" direction, and $b_1 \geq \frac{3}{4} h_1$, the beam is classified as an interior joint only if beams in the "2" direction have a width $b_2 \geq \frac{3}{4} h_2$ and a depth $t_2 \geq \frac{3}{4} t_1$. If either condition is not satisfied, the beam must be classified as an exterior joint. If there is a beam on only one side of the joint in the "1" direction, the joint will be an exterior joint if beams in the "2" direction meet the same requirements as before. Otherwise, the joint will be classified as a corner joint when designing for shear in the "1" direction and as an exterior joint for shear in the "2" direction.

Equation 10.2 and the geometric requirements are based on the compression strut model in which concrete is assumed to carry the entire shear force across the joint. A number of different techniques have been suggested for defining effective area and average stress on the compression strut;[5] however, the resulting equations are more complex and no more accurate than the simple approach of relating capacity of the strut to shear on a horizontal section through the joint.

For the compression strut to carry loads when significant deformations are applied to the members, ACI Committee 352 recommendations require a minimum amount of confinement of joint concrete provided either by members framing into the joint or by transverse reinforcement. Confining reinforcement must satisfy the following conditions.

a. At least two layers of transverse reinforcement should be placed between the top and bottom layers of beam longitudinal reinforcement in the deepest member framing into the joint.
b. The transverse reinforcement should satisfy the requirements of Code 7.10.
c. The center-to-center spacing of the confining reinforcement should not exceed 12 in. If the joint is part of the primary-lateral, load-resisting system, the spacing is reduced to 6 in. unless the joint is confined by members framing into the joint on opposite faces.

If members frame into all four sides, meet the geometric requirements shown in Fig. 10.7, and no more than 4 in. of the column face is unconfined next to the beam (Fig. 10.8), no transverse reinforcement is needed in the joint. If beams frame into two opposite faces and meet the geometric requirements of Figs. 10.7 and 10.8, transverse reinforcement is required only in the direction unconfined by members. That is, transverse reinforcement is needed parallel to the confined faces.

The anchorage of hooked bars in joints is governed by Code 12.5 and discussed in Chapter 8. The transfer of axial forces through the joint is ensured if the confinement provisions are satisfied and if the longitudinal column reinforcement passing through the joint satisfies Code 10.9.1, which specifies the limiting area of column longitudinal reinforcement (no less than 0.01, nor more than 0.08 times the gross area of the column), and Code 10.9.2, which specifies the minimum number of column bars (4 in a rectangular section and 6 in a section enclosed by spirals). The bars may be offset through the joint as specified in Code 7.8.

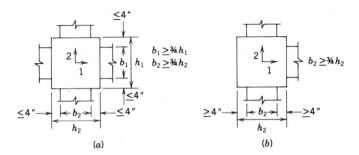

Figure 10.8 ACI Committee 352 definition for confining lateral beams. (*a*) Plan view of joint with beams in both "1" and "2" direction providing confinement. (*b*) Plan view of joint with beams in "2" direction providing confinement.

10.4 Examples of Beam-Column Joint Detailing

(a) Detail joint A2 at Level 2 of the frame in Fig. 10.3. The frame provides lateral resistance for wind forces. The members are proportioned as shown in Fig. 10.9. Reinforcement required in the beams at the face of the column is indicated. Bottom steel is required by Code 12.11.1 and 12.11.2. To meet Code 12.11.2, bottom bars must be anchored to develop f_y at the face of the column. Story heights of the frame are 14 ft. Grade 60 reinforcement and concrete strength $f'_c = 4$ ksi are used.

Solution

Check anchorage of hooked bars extending into joint from 20×18-in. beam. For #9 top bars, Code 12.5 requires

$$\ell_{db} = 1200 d_b / \sqrt{f'_c} = 1200 \times 1.12 / \sqrt{4000} = 21.25 \text{ in.}$$

As the beam is narrower than the column (20-in. beam, 24-in. column), the cover over the hooked bar will meet the requirements of Code 12.5.3.2 and

$$\ell_{dh} = 21.25 \times 0.7 = 14.9 \text{ in.}$$

The length available for anchoring the hooked bars is 18 in. (column depth) $-$ 1-$\frac{1}{2}$-in. (cover) $-$ tie diameter (say, $\frac{1}{2}$ in.) = 16 in.

$$16.0 \text{ in. available} \geq 14.9 \text{ in. required} \quad \textbf{O.K.}$$

The #8 bottom bars also must be hooked to develop yield at the column face, but ℓ_{dh} for #8 bars is less than ℓ_{dh} for #9 bars so no further calculation is needed. When ℓ_{dh} is greater than the space available, smaller bars will be selected or the column dimension increased to accommodate the hooked bars.

Check shear in the joint. Shear in the direction of the spandrel beams will not be a problem because large unbalanced moments are not anticipated under gravity and wind loading on the structure. In the orthogonal direction, however, the 20-in. $\times$ 18-in. beam will produce shear across the joint. A free-body of the joint is shown in Fig. 10.10a. The tension and compression forces (240 k) at the face are determined using the area of

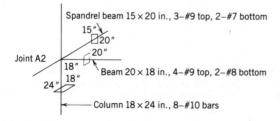

Spandrel beam 15 × 20 in., 3–#9 top, 2–#7 bottom

15″
20″
20″
Joint A2
18″
18″
Beam 20 × 18 in., 4–#9 top, 2–#8 bottom
24″

Column 18 × 24 in., 8–#10 bars

Figure 10.9 Design example, joint A2 (Fig. 10.3).

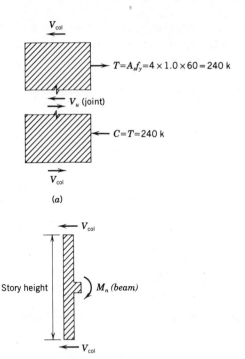

$T = A_s f_y = 4 \times 1.0 \times 60 = 240$ k

V_u (joint)

$C = T = 240$ k

(a)

V_{col}

Story height

M_n (beam)

V_{col}

(b)

Figure 10.10 (a) Free-body of joint showing horizontal forces only. (b) Free-body of column for determining V_{col}.

steel provided in the members. No ϕ factors are used. Shear in the column can be approximated closely by assuming that there are inflection points at the midheight of the columns above and below the joints (Fig. 10.10b). Therefore,

$$V_{col} = M_n/\text{story height}$$

$$M_n = A_s f_y (d - a/2) = 4 \times 1.0 \times 60 \times (17.4 - 3.53/2) = 3750 \text{ in.-k}$$

where

$$a = A_s f_y/0.85 f'_c b_1 = 4 \times 1.0 \times 60/(0.85 \times 4 \times 20) = 3.53 \text{ in.}$$

so that

$$V_{col} = 3750 \text{ in.-k}/(14 \text{ ft} \times 12 \text{ in./ft}) = 22.3 \text{ k}$$

and

$$V_u = T - V_{col} = 240 - 22.3 = 217.7 \text{ k}$$

The capacity of the joint ϕV_n must be greater than V_u. From Eq. 10.2

$$V_n = \gamma \sqrt{f'_c} b_j h$$

The value of γ is obtained from Fig. 10.6a. Because the joint can be classified as an exterior joint, $\gamma = 20$. The spandrel beams meet the requirements of Fig. 10.6b and 10.7. The spandrel beam is 15-in. wide and the column face into which the spandrel beam frames is 18-in. wide.

$$15 \text{ in.} \geq 0.75(18 \text{ in.}) = 13.5 \text{ in.} \quad \textbf{O.K.}$$

The depth of the spandrel beam is 20 in. and the beam in the direction of loading is 18 in.

$$20 \text{ in.} \geq 0.75(18 \text{ in.}) = 13.5 \text{ in.} \quad \textbf{O.K.}$$

The effective width of the joint using Eq. 10.3 is

$$b_j = (b_b + b_c)/2 = (20 + 24)/2 = 22 \text{ in.}$$

but not more than $b_b + h/2$ on each side of beam.

$$20 + (18/2) \times 2 \text{ sides} = 38 \text{ in.} \quad \text{Does not control.}$$

The shear capacity of the joint

$$V_n = 20 \times \sqrt{4000} \times 22 \times 18 = 501 \times 10^3 \text{ lb}$$
$$\text{and} \quad \phi V_n = 0.85 \times 501 \text{ k} = 426 \text{ k} > 217.7 \text{ k} \quad \textbf{O.K.}$$

Finally, confining reinforcement must be detailed. Because the frame provides resistance to lateral wind loads, the center-to-center spacing of ties through the joint must not exceed 6 in. Because the spandrel beam meets the requirements for confining members (Fig. 10.8b), transverse steel meeting Code 7.10 is needed parallel to the confined sides. The arrangement of reinforcement in and through the joint is shown in Fig. 10.11.

(b) Detail interior joint B2 or C2 at Level 2 for the structure in Fig. 10.3. Considering the same building as in Example 10.4a, the proportions of the members and the longitudinal reinforcement are shown in Fig. 10.12.

Solution

In this case shear is not expected to be a problem in either direction because gravity and wind loads should not produce large unbalanced beam moments at the joint in either direction. Because top reinforcement in the beam will be continuous through the joint, anchorage is not a problem. Bottom reinforcement represents extensions of the positive moment reinforcement and must be anchored to develop yield as required by Code 12.11.2. Therefore, bottom steel either should be continuous through the joint or lap-spliced across the joint to provide continuity.

Because members frame into all four sides of the joint, beam widths are greater than $\frac{3}{4}$ of the column width, and unconfined corners are 4 in. or less (Fig. 10.8a), no transverse reinforcement is needed through the joint.

If the beam widths are less than $\frac{3}{4}$ of the column face ($\frac{3}{4} \times 24 = 18$ in.), the ACI Committee 352 report recommends that perimeter hoops be placed through the joint at a 6-in. spacing. Likewise, if the beam width is $\frac{3}{4}$

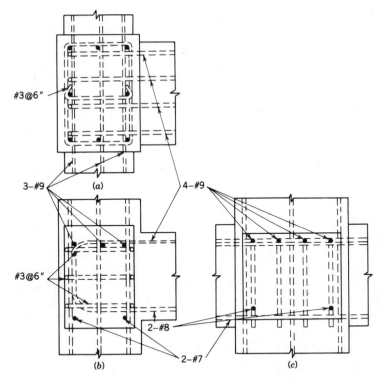

Figure 10.11 Joint detail, Example 10.4*a*. (*a*) Top view. (*b*) Side view. (*c*) End view.

of the column width but more than 4 in. of the corner is unconfined, lateral reinforcement (hoops) is needed through the joint.

Floor slabs have been ignored in the examples, even though it is likely that slabs or perhaps a joist floor system would be present. A floor slab would not substantially change the details shown, but would require consideration of placement of slab reinforcement, required cover over the slab bars, and effects of the slab acting as a flange with the beam section. The effects of such flanges would be more pronounced where the section was subjected to negative bending and slab reinforcement was stressed in tension.

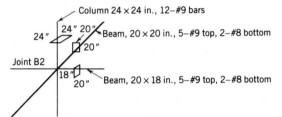

Figure 10.12 Design example, joint B2 (Fig. 10.3).

10.5 Corner Joints

In planar frames, bents, or other simple structures, corner or knee joints pose a special detailing problem. The corner may be subjected to opening or closing forces as shown in Fig. 10.13a and 10.13b. Similar situations occur where walls meet as shown in Fig. 10.13c.

Nilsson and Losberg[6] reported that the failure of such joints is the result of diagonal tension cracking, splitting cracks, yielding of the reinforcement, anchorage, or crushing of the concrete. Although diagonal tension cracking is often ignored, it may be the cause of failure in opening corners. In closing corners, anchorage of reinforcement may be a more common problem. Figure 10.14a shows diagonal cracks in an opening corner. Careful examination of a free-body of the corner (Fig. 10.14b) suggests that reinforcement must be placed to control diagonal cracks and prevent failure in the joint before the beams develop required moment capacity. The reinforcement should be placed so that the outer corner is continuous with the remainder of the joint and that reinforcement crosses the crack on the inside corner where large diagonal tensile forces are developed. Using the detail shown in Fig. 10.14c provides for the hooks on the beam longitudinal bars to bear against the concrete and prevent the outer corner from spalling. The diagonal bar across the inside corner will control cracking that develops in the corner. Nilsson and Losberg conducted tests on a number of different corner reinforcement arrangements and found that the one shown in Fig. 10.14c performed very well.

Schlaich *et al.*[7,8] have proposed the use of a truss model to illustrate the same requirements. Tension elements of the truss are provided by reinforcement and compression elements are provided by the concrete. Where elements meet (nodes), care must be taken to ensure that tension elements are properly an-

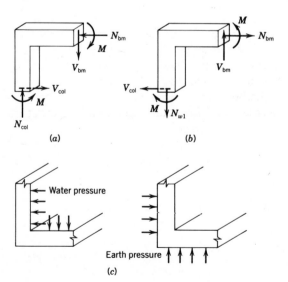

Figure 10.13 (*a*) Closing corner. (*b*) Opening corner. (*c*) Wall corners.

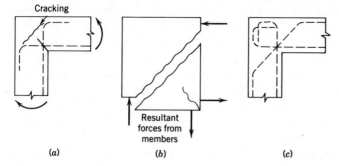

Cracking

Resultant
forces from
members

(a)   (b)   (c)

Figure 10.14 (*a*) Diagonal cracking. (*b*) Forces creating diagonal cracking. (*c*) Reinforcing detail for best behavior.

chored and compression elements do not crush. Formal design requirements for the truss elements are being developed. In future Codes, truss concepts will probably be included for complex detailing problems. The truss models in Fig. 10.15 can be used for both opening and closing corners. The forces at the nodes are shown. Because at least one of the forces is known at the node and the inclination of the other forces is set by the geometry of the truss selected, the magnitude of the other forces at the node can be calculated and used to proportion the tension reinforcement. For any joint a number of different truss models

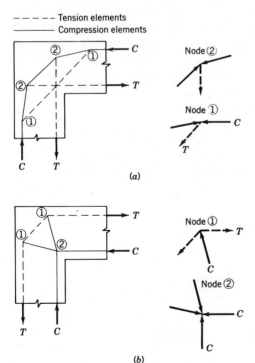

Tension elements
Compression elements

Node ②

Node ①

(a)

Node ①

Node ②

(b)

Figure 10.15 Truss models for detailing corners. (*a*) Opening corner. (*b*) Closing corner.

could be selected. Closed loops and hooks then can be provided to ensure that tension elements are anchored to develop the forces required and to confine the concrete that carries compression. The reinforcement detail for the opening corner truss model in Fig. 10.15a is exactly like that shown in Fig. 10.14c (as suggested by Nilsson and Losberg). In the case of the closing corner in Fig. 10.15b, the truss demonstrates clearly the need for the tension reinforcement to be continuous around the corner. This continuity can be accomplished by providing bars with overlapping hooks or by providing a corner bar that is spliced to the beam longitudinal bars. Similar trusses and solutions can be developed for obtuse or acute corners between beams or plate (wall) elements.

10.6   T-Joints

T-joints occur frequently where slabs meet walls, walls connect to footings, beams frame into exterior columns, or columns frame into roof beams. The 20-in. × 18-in. beam framing into the column in Fig. 10.9 is an example of a T-joint. Perhaps the most common case is that of a retaining wall joint. Nilsson and Losberg[6] tested a series of retaining wall joints. Three cases are shown in Fig. 10.16. The simple change in the direction of the hook from test 1 to test 2 resulted in the wall developing its full capacity or 40% more than test 1. By placing a diagonal bar across the corner, strength was further improved.

The truss model shows why the performance is improved so markedly by changing the direction of the hook. In Fig. 10.17 truss elements are utilized for the forces transferred between wall and footing. Although the forces at node 2 can be sketched easily and calculated, the node will only function properly if there is a detail that permits the forces to develop. When the hook is oriented away from the wall, the compression element will simply cause the cover to spall at a fairly low force level. When the hook is oriented toward the wall, the compression element bears against the inside radius of the bent bar. The cover cannot spall unless the bar is also "pushed down." As a result, the node functions properly. At node 1, the horizontal bar is continued into the toe of the footing and the inclined compression element between nodes 1 and 2 is adequately restrained by the horizontal bar at the top face of the footing. In the

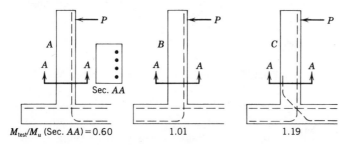

Figure 10.16 Test program,[6] retaining wall joints.

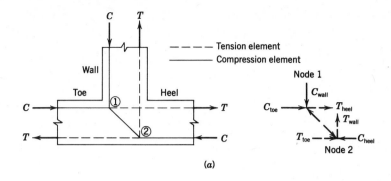

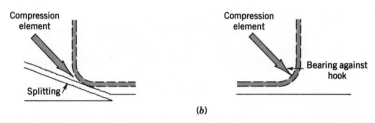

Figure 10.17 Truss model for retaining wall. (*a*) Truss model. (*b*) Failure modes at node 2.

design example shown in Fig. 10.11, the hooks are oriented as illustrated in Fig. 10.17*b* to prevent a premature failure of the cover and the associated loss of capacity of the compression element or strut in the joint.

Selected References

1. ACI Committee 352, "Recommendations for Design of Beam-Column Joints in Monolithic Reinforced Concrete Structures," *ACI Jour., Proc.,* 73, No. 7, July 1976, p. 375.

2. ACI Committee 352, "Recommendations for Design of Beam-Column Joints in Monolithic Reinforced Concrete Structures," *ACI Jour., Proc.* 82, No. 3, May–June 1985, p. 266.

3. D. F. Meinheit and J. O. Jirsa, "Shear-Strength of *R/C* Beam-Column Connections," *ASCE Jour. Struct. Div.,* ST-11, Nov. 1981, p. 2227.

4. J. O. Jirsa, "Beam-Column Joints: Irrational Solutions to a Rational Problem," *ACI Special Publication 72,* Paper 8, Detroit, 1981, p. 128.

5. L. Zhang and J. O. Jirsa, "A Study of Shear Behavior of Reinforced Concrete Beam-Column Joints," PMFSEL Report No. 82-1, University of Texas at Austin, Feb. 1982.

6. I. H. E. Nilsson and A. Losberg, "Reinforced Concrete Corners and Joints Subjected to Bending Moment," *ASCE Jour. Struct. Div.,* ST-6, June 1976, p. 1229.

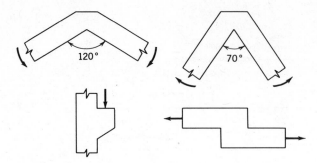

Figure 10.18 Joints for Prob. 10.3.

7. J. Schlaich and K. Schafer, "Towards a Consistent Design of Reinforced Concrete Structures," *Proc*. 12th Congress IABSE, Vancouver, B. C., Sept. 1984.
8. J. Schlaich and D. Weischede, "Detailing of Concrete Structures," Comite Euro-International du Beton, Information Bulletin No. 150, March 1982, Paris. (In German).

Problems

PROB. 10.1.   Sketch the reinforcing layout for a joint in which beams 18-in. wide × 24-in. deep frame into all four sides of a 24-in. × 24-in. column. The column must have 11.2 in.² of longitudinal reinforcement distributed equally to all faces of the column. All four beams must contain 4.4 in.² of top reinforcement for flexure. Select the reinforcement, bar size and number of bars that will satisfy these requirements. The dimensions of the members may be changed by no more than 10% in detailing the joint.

PROB. 10.2.   Detail the joint in Fig. 10.9 if $f'_c = 3.0$ ksi, the beam requires 6.0-in.² top reinforcement, and 2.0-in.² bottom reinforcement. The spandrel requires 3.0-in.² top and 1.0-in.² bottom reinforcement. The column is reinforced with 12-in.² of reinforcement, half in each of the 24-in. faces.

PROB. 10.3.   Using the truss approach, sketch the reinforcement details for the joints shown in Fig. 10.18.

11

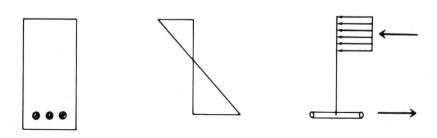

LIMIT DESIGN

11.1 Terminology

Limit design and plastic design are often considered as synonymous terms. A distinction is made here and in most discussions of reinforced concrete. In steel design,[1] the term *plastic design* includes not only the change in the pattern of moments beyond the yield point, but the increased resistance of a cross section after its extreme fiber reaches the yield point.[2,3] In reinforced concrete, *limit design* is used only to refer to the changing moment pattern; ultimate strength design already includes the increase in strength of a given cross section after some stress reaches the yield point.

For reinforced concrete, the frame can be designed under any *fixed* loading for strengths closely matching the moment diagram, with the concrete tailored to fit by cutting off and the bending of bars. In such a case the limit design concept adds nothing because capacity is reached simultaneously at a number of points and the elastic theory analysis is fully adequate to predict collapse.

However, when several pattern loadings are considered possible on a continuous flexural member, the sum of the maximum positive moment and the average of the two maximum negative end moments will total more than the simple beam moment. In such a case proper limit design might reduce the member cost. Slab design, on the basis of extensive testing, uses the limit design idea in Chapter 12 without the name or the usual limit design calculations.

11.2 Limit Design for Reinforced Concrete

When yielding starts, deflections increase sharply and repeated loading introduces an element of fatigue. Hence under working load conditions yielding is certainly undesirable. On the other hand, as a reserve against final failure or collapse, the strength between yielding and failure is quite significant for structural steel; the fact that this stage involves larger deflections does not seem too important.

Limit design is significant because it points out that (1) a statically indeterminate member or frame cannot collapse as the result of a single yielding section, and (2) between first yielding and final frame failure there normally exists a large reserve of strength.

Limit design was not recognized in the ACI Code until 1963 and then only in a minor way. In the 1986 Code under the title of redistribution of negative moments (Code 8.4) the provisions of 1963 are relaxed, but not expanded. Code 18.10.4 also allows limited redistribution for ductile prestressed concrete members with sufficient bonded reinforcement. The beams carrying two-way slabs, discussed in Chapter 16, now may be considered a special form of limit design, but it is not so designated. Nonelastic design conditions have long been recognized in averaging the moments across a footing or across fairly wide strips of two-way slabs, flat plates, and flat slabs. This is a kind of limit design applied transversely, but it is not the usual limit design concept applied lengthwise to change the given moment diagram itself.

There are several reasons why the ACI Code can afford to appear to be a little timid in the area of limit design.

First, in terms of economics, the situations for steel and for concrete are not comparable. With steel the efficient wide flange section has only a nominal increase in strength between M_y, the moment at first yield of steel, and M_p, the fully plastic moment. With essentially the constant strength of a steel beam along the span, a change that equalizes several maximum moments is particularly helpful for most types of loading.*

It is only because reinforced concrete frames are designed for many alternate loadings that reserve strength exists somewhere under almost any specific loading. Although such reserve strength can very often be utilized to advantage by limit design, it is not proportionately as important as in steel construction.

A second reason why limit design is not pushed as strongly for reinforced concrete is that there is more uncertainty as to how concrete acts under all conditions of overload. It does not have the uniform ductile character exhibited by steel—unless certain design restrictions are imposed.[4] Some of these behavior patterns are not yet sufficiently documented to make complete limit design acceptable to all designers.

11.3 Moment-Curvature Relations

The basis for limit design lies in the inelastic behavior of materials at high stresses, that is, their ability to sustain a given yield moment while a consider-

---

* Not, for instance, with a fixed midspan load on a fixed end beam that under elastic conditions has equal positive and negative moments.

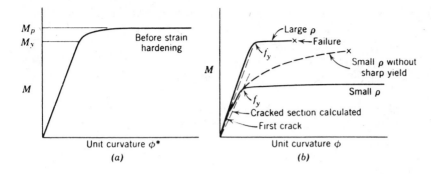

Figure 11.1 Moment curvature. (a) Structural steel. (b) Reinforced concrete. (* There is no real distinction between ϕ as used here and $d\theta$ used in Chapter 3.)

able increase in local curvature develops. In a statically indeterminate frame this means that a given local section that tends to be overloaded yields and, in effect, refuses to accept more moment, but does not fail. Instead it forms what is called a "hinge" and thereby forces responses to further loading onto sections less fully stressed. A study of limit design must begin by considering the relation between bending moment and the resulting curvature of a short length of the member, as shown in Fig. 11.1a for steel and 11.1b for reinforced concrete. The curve for steel can be approximated with considerable accuracy by two straight lines, as shown in Fig. 11.2a. The curves for reinforced concrete can, a little less accurately, be approximated by two straight lines or three straight lines as in Fig. 11.2b. For small percentages of sharply yielding steels, such as ordinary Grade 40 steel, the approximation is quite satisfactory. With a large percentage of the same steel, the yielding actually usable would be too short (before failure) to modify the ultimate behavior of the frame significantly. With a balanced beam, such as shown in the upper curve of Fig. 11.2b, the inelastic behavior of the concrete in compression makes for some small nonproportional curvature increases at high loads, but does not provide the ductil-

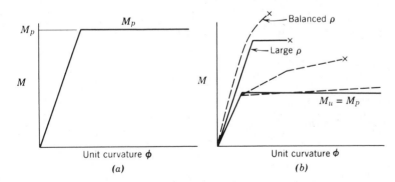

Figure 11.2 Conventional moment-curvature relations. (a) Structural steel. (b) Reinforced concrete. Compare Fig. 11.1b.

ity necessary for significant modification of moments in the frame. For reinforcing steels without a sharp yield point, typical of some of the present-day high strength steels, the behavior might call for three straight lines as shown in Fig. 11.2b. Although behavior with such steel may be quite favorable, it is much more difficult to handle this case mathematically and it will not be considered further in this discussion.

Either axial tension or axial compression in the member substantially modifies the $M-\phi$ curve for either steel or reinforced concrete, but more so for reinforced concrete.

11.4   Fixed End Beam with Hinges, as an Example

The principal behavior that makes limit design interesting can be illustrated with a fixed end beam under increasing uniform load as shown in Fig. 11.3. This behavior will be discussed for the simplified two line curves of Fig. 11.2. Moments build up proportionately until moment diagram 1 shows the negative moment as M_p and the positive moment as roughly half as large. Under further loading M_p cannot increase, but the end element is forced to turn through a larger angle ϕ to permit the beam to deflect more and build up the necessary added resistance elsewhere. Thus the moment diagram goes through increased stages such as 2 until it finally develops a positive moment M_p as shown by diagram 3. At this point it could carry no more added load because it would be on the verge of developing into a mechanism with hinges at each end and at the middle of the span.

The increase in load between diagrams 1 and 3 is 33% if M_p is the same for positive and negative moments, as in a rolled steel beam. However, in reinforced concrete the M_p for positive moment is usually much less than for negative moment. Thus this redistribution of moments, which constitutes limit design as it relates to reinforced concrete, is of less economic significance for reinforced concrete than for steel. Nevertheless, a reinforced concrete beam designed under elastic theory normally is not fixed ended and is designed separately for different loads to produce maximum positive moment and maximum negative moment, as illustrated by Fig. 9.9e. Under either loading there is surplus moment capacity—in the one case at the supports, in the other at the

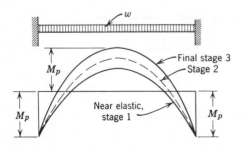

Figure 11.3 Development of moment with increasing load for an idealized moment curvature.

midpoint. Thus, even in reinforced concrete, there is some room for limit design concepts.

11.5   Code Provisions for Limit Design

Code provisions for limit design for reinforced concrete members appear under the title "Redistribution of negative moments in continuous nonprestressed members" (Code 8.4). Redistribution is permitted *only* if the moments at supports are calculated by elastic analysis and if the ratio of steel ρ (or $\rho - \rho'$) is less than $0.5\rho_b$ where

$$\rho_b = \frac{0.85\beta_1 f'_c}{f_y} \times \frac{87,000}{87,000 + f_y}$$

Thus ρ_b is defined such that it is independent of any A'_s that may be present.

The negative moments may be decreased (or increased) by a percentage defined as

$$20[1 - (\rho - \rho')/\rho_b]$$

Restricted in this way the problem of necessary hinge rotation at the ends of the beam is very easily handled by the very long flat $M-\phi$ curve that applies to the beam with small percentages of reinforcement. The curves in Fig. 11.4 for various grades of steel in the beam (indicated in the upper left of the figure) relate the theoretical possible change in moment to the steel used in the beam. The permissible change in moment under the Code is bounded by the two straight lines.

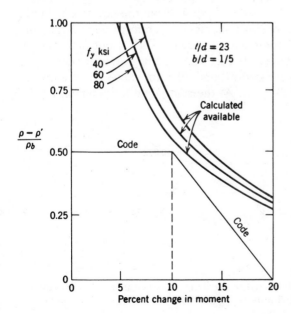

Figure 11.4   Allowed moment redistribution and studies of available minimum rotation capacities. (Modified from ACI Code Commentary).

The next section reconsiders the continuous beam design of Sec. 9.9 and changes that might be made by redistribution of its moments.

11.6   Redistribution Applied to Beam of Chapter 10

The beam chosen in Sec. 9.9 is compact, rather small (10-in. web and h of 20 in. overall), and just stiff enough for deflection. As a structural member, such a beam is on the expensive side because of the area of steel required. If beam weight is not critical, this member could be made eligible for redistribution of moments simply by widening its web to 14 or 15 in. Beam stem weight would go up 40% to 50% and steel would not be greatly reduced by this process, but ρ would be reduced by the extra width to around the $0.5\rho_b$ threshold required for redistribution.

If clearances permit, it is better to increase b_w and d without changing their ratio so much. k_n (or k_m) in Table 3.1 is based on a ρ of $0.75\rho_b$. The use of k_n about 70% to 72% of the tabulated value (allowing for increased internal lever arm) leads to ρ approximately $0.5\rho_b$. Alternatively, consideration of the usual required presence of 25% of the positive moment reinforcement at the support means a ρ' of roughly 20% of ρ and permits a k_n of around 85% of the tabulated value, say, 650 for k_n instead of the tabulated 783 and the 1050 used in the earlier design. The first decision toward redistribution is to accept the size this k_n leads to as an appropriate beam.

Balanced reinforcement, needed in assessing redistribution, is accurately available from Table 3.1 as simply 4/3 of the maximum ρ permitted there.

For a member chosen on this basis (with ρ and ρ' to be verified separately), at least a 10% reduction in negative moments is permissible by redistribution (a little more if $\rho - \rho'$ proves to be less than $0.5\rho_b$). From the negative moment diagram of Fig. 9.12h and the positive moment diagram of Fig. 9.12e, the modified diagrams of Fig. 11.5 (moment coefficients shown but omitting the $w\ell_n^2$ that goes with each) are obtained. In Fig. 11.5a the negative moment at the left is somewhat arbitrarily decreased around 10% to $-0.082w\ell_n^2$, approximately 90% of the original $-0.091w\ell_n^2$. At the same time the moment on the

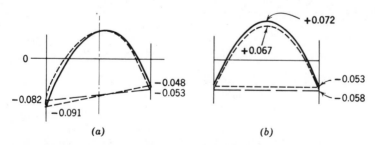

Figure 11.5 Design moments (solid lines) redistributed (dashed lines). (a) Maximum negative moment at left. (b) Maximum positive moment.

right is similarly increased around 10% to $-0.053\,w\ell_n^2$, leaving a positive moment at midspan almost unchanged, $+0.053\,w\ell_n^2$, and less than the positive moment in Fig. 11.5b. The opposite hand of the same diagram would be appropriate for the right support of the beam.

In a similar fashion the negative moment in Fig. 11.5b can be increased 10% to $-0.058\,w\ell_n^2$, which reduces the positive moment coefficient from 0.072 to 0.067. This moment remains larger than the coefficient in Fig. 11.5a and hence is permissible; in fact there is still the spread between 0.067 and 0.053 that could be reduced if the redistribution exceeds the 10% used here.

The result of these moment changes is nearly a 10% reduction of all A_s values and the elimination of any A_s' requirement as such. Should the resulting $\rho - \rho'$ values at the support be less than $0.5\rho_b$, the reduction could be greater as limited by the spread in positive moment values in the previous paragraph. In that connection some allowance should be made for the slightly higher positive moment off center in Fig. 11.5a.

The positive moment reinforcement ρ is subject to the $0.5\rho_b$ limit just discussed. This ρ should be lower than at the support unless A_s' there is large.

11.7 Bending and Stopping Bars

The Code, in taking the first step toward limit design, is not specific about details. Later code revisions hopefully will recommend details concerning bar lengths and development as the moment diagrams shift.

It is not the Code intent, as the authors visualize it, to reduce the steel requirements all across the span. Its intent is to reduce the peak moments and steel concentrations across columns where beam-and-girder steels must cross each other. The reduced steel of Sec. 11.6 will not modify the moments of Fig. 9.12e and 9.12h, except in the last 10% increment of load. Cracking away from points of maximum moment should not be encouraged by reducing bar lengths and total steel area. Hence for bar bending the authors recommend using the basic moment diagrams of Fig. 9.12e and 9.12h with only the minor modifications shown in Fig. 11.6a and 11.6d. The required A_s curves then become those of Fig. 11.6b and 11.6d and their use involves little difficulty because the peak dotted values are easily calculated, either in terms of bars or bar areas, from the ratios 0.072/0.067 and 0.091/0.082. These required A_s curves are directly comparable to those of Fig. 9.17a.

11.8 Present Status of Limit Design Theory

Aside from bounding the redistribution problem more closely and liberalizing it within these bounds, there has been little progress toward general agreement on the way limit design for reinforced concrete should develop. There has been some considerable progress in basic concepts, but limit design in unbraced frames still appears improbable in the near future except for the very indirect

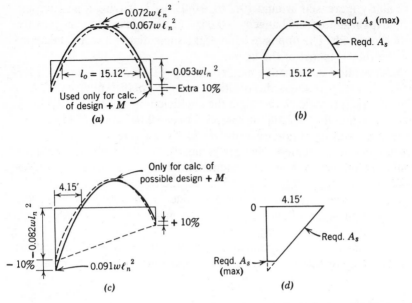

Figure 11.6 Modification of design moments toward limit design. (*a*) Reduced positive moment for design. (*b*) Required A_s diagram for positive moment reinforcement. (*c*) Reduced negative moment for design. (*d*) Required A_s diagram for negative moment reinforcement.

redistribution of column shears in unbraced frames implicit in the slender column moment magnification procedures of Code 10.11.

If one considers only frames sufficiently braced (without restricting normal joint rotation) that lateral deflection of the joints is eliminated, this restricted limit design may be reasonably near practical application.[5] Such frame restrictions make frame failure as a whole meaningless and failure of individual members the basic concept. The continuity that surrounds the member at service load gradually fades away as hinges form (Sec. 11.9) first at one end and later at the other end of the member (Sec. 11.10). Thus this problem gradually reduces to one of individual member collapse. Frame action after first yielding is discussed in Sec. 11.11 where it is pointed out that it is not necessary to trace these intermediate steps even where complete frame collapse is involved.

11.9 The Plastic Hinge Idea

If ϕ designates the angle change that develops over a unit length of the member, the M–ϕ curve or the relation between M and ϕ is shown in the simplest or ideal case in Fig. 11.1*a* and idealized in Fig. 11.2*a*. It can be similarly idealized for reinforced concrete members as in Fig. 11.2*b*.

The significant part of these M–ϕ curves for limit design lies beyond the first yield moment, that is, in the zone where the plastic moment M_p remains unchanged over a very large range in ϕ values. When a point along the beam develops a moment M_p, it will act as a plastic hinge. This means it will continue to resist the moment M_p, but further loading will result only in an increased angle change ϕ rather than an increased moment. If an ordinary structural hinge is thought of as frictionless, a plastic hinge may be considered as a "rusty hinge" with a definite, but limited resistance to rotation. Thus, upon further loading of the member, the moment *changes* produced elsewhere on the member are the same as though a real hinge existed at the plastic hinge point.

The concept of a hinge at a point is mathematically convenient, but in real frames (steel or concrete) a plastic hinge is spread over some length of top and bottom beam surfaces, a length normally in the order of the depth of the member.[6] Occasionally, as in a symmetrical third-point loading of a beam, the yielding region can occur over all the middle third of the beam, a very generalized form that does not closely resemble the simple plastic hinge idea.

11.10   The Collapse Mechanism

In limit design as developed for steel frames, the ultimate strength considered is that which brings either the frame or the individual member to the verge of failure or collapse, assuming perfect plastic hinge action. To fit the present restricted idea of braced frames of reinforced concrete, consideration here is first directed to the simpler individual beam members.

A member cannot collapse under moment loading (except by buckling) until enough actual hinges or plastic hinges form to transform it into a mechanism. A cantilever beam thus collapses when a single hinge forms at the support, but a simple or continuous beam span must have three hinges to collapse, as indicated in Fig. 11.7. For the simple beam this requires the formation of only one plastic hinge because the supports furnish two real hinges. The entire moment diagram is statically determinate for this case. The total deflection consists of two parts, as indicated in Fig. 3.33, with the added deflection that occurs after

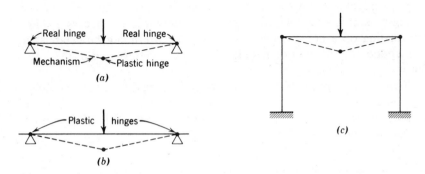

Figure 11.7   Local collapse mechanism. (*a*) Simple span. (*b*) Continuous span. (*c*) Frame member.

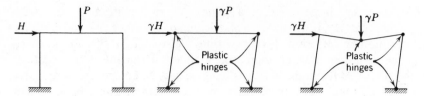

Figure 11.8 Mechanisms for collapse of frame as a whole under load factor γ.

M_p develops consisting of a simple triangle, just as though the two segments were plane and not curved by the earlier loading.

For a single span of the usual continuous beam three plastic hinges are required for collapse; these hinges can greatly modify the moments from their elastic analysis values.

A frame may collapse as a whole without any individual member developing the three hinges, provided the frame as a whole develops enough hinges to act as a mechanism, as in the middle sketch of Fig. 11.8.

11.11 Frame Action Between First Yielding and Collapse

In Sec. 11.2 it was stated that yielding under working loads is objectionable, although this does not mean that such a condition is totally prohibited under all circumstances. Generally, however, a study under working loads involves analysis on an elastic basis.

As overloads are applied, one or more of the moments might be expected to pass the M_y value and reach the M_p value.* However, it would be quite an unusual frame that would have its maximum moments so equalized that all the plastic hinges required to form a mechanism would develop at any one loading stage and thus lead to an immediate collapse. Usually a single highly stressed section (or several) develops first M_y and then M_p, with more load necessary before some other plastic hinge forms; a number of such loading increments might be required to produce all the plastic hinges necessary for a mechanism.

Although it might be interesting to trace out this sequence of loads and moment diagrams, it is a lengthy process. Fortunately it is a unnecessary. For any given loading on a given frame, the collapse pattern can be established entirely independently of the original elastic moment pattern and of the intervening elastic-plastic stages.

The ratio of this collapse loading to the service loading is the available safety factor γ, or in reinforced concrete it might be better to say the available load factor γ. This brings out one limitation inherent in the usual limit design approach. *This available load factor is meaningful only as it relates to that*

---

* In steel construction M_p is definitely greater than M_y; in reinforced concrete M_p is definitely M_u and M_y need not be differentiated in the calculations.

particular loading used and it implies that all portions of that loading are considered only as increasing proportionately.

This limitation is important enough for elaboration through a specific example. If the frame of Fig. 11.8 is analyzed, if γ is established as 3, and if all loads are increased in proportion, the frame would not collapse until it carried loads of $3H$ and $3P$. This does not mean it could carry $3H$ alone or $3P$ alone or $3H$ along with P or any other combination except $3H$ and $3P$. If the load factor for H alone is needed, this requires a separate analysis.

The student should also note that the collapse load is far beyond the elastic range; hence the effect of the two loads H and P may be far different than the sum of their individual effects.

11.12   Lower Bound or Limit on Load Factor

A lower bound or lower limit on the real load factor is determined from a moment diagram, *any* moment diagram that is consistent with the given loads, that is, *any* such moment diagram that satisfies statics.

The fundamental idea may be very simply indicated in terms of the elastic moment diagram. Let M_p represent the plastic moment (different values for different members or for different parts of members) and M_s represent the corresponding service load moments from the moment diagram. If M_p/M_s is determined for every member, the smallest ratio found is a lower bound on the true load factor. Why? Up to value M_p the idealized M–ϕ diagram assumes elastic action, which means the moment is directly proportional to the load. Hence the ratio M_p/M_s indicates how much the load may be increased without inelastic action and before any readjustment of the moment diagram shape occurs. This is a *lower* limit on how much the load may be increased because the moment diagram will probably shift to a more favorable shape as this critical section begins to act as a plastic hinge. For further loading the critical moment at the hinge remains constant at M_p, whereas other parts of the moment diagram increase toward the M_p values.

It has been proven mathematically[1] that any moment diagram satisfying statics may be used as the basis of a lower bound. Normally it is easier to develop an "arbitrary" moment diagram than to establish the elastic moment values.

As one tries different possible moment diagrams, one discovers that higher values of this *lower* bound are obtained when more individual ratios of M_p/M_s become equal to this bound. The moment diagram can thus be "equalized" toward values establishing the true load factor. In complex cases it is desirable finally to assume plastic hinges at these high moment points and check such a solution by the energy-mechanism procedure (Sec. 11.13) which fixes an *upper* bound, just to verify the solution.

If this approach is applied to the beam of Fig. 11.9a, with all M_p values 100 k-ft, one might start with a simple beam moment diagram as in (b)

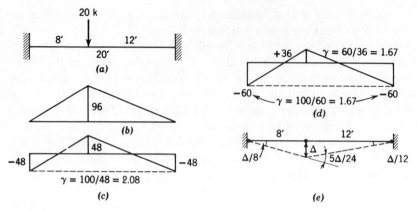

Figure 11.9 Establishing possible load factor γ. (a–b) Beam and simple span M. (c–d) Adjusting M diagram. (e) Energy approach.

with $M_{\max} = (20 \times 12/20)8 = 96$ k-ft and then try to develop three peaks of moment with the same ratio of M_p/M_s. This result is easily accomplished by shifting the axis as in (c), making M_p/M_s uniformly $100/48 = 2.08$ at each peak moment value. This 2.08 is the true load factor possible. If M_p for positive moment is 60 k-ft and M_p for negative moment remains 100 k-ft, the moment diagram of (d) would be the proper one and the load factor would be 1.67.

11.13 Upper Bound or Upper Limit on Load Factor

If a frame is on the verge of collapse, it must have developed a sufficient number of plastic hinges to change it into a mechanism. As the mechanism starts to collapse, the loads move in such a way as to contribute energy to the system while rotations occur at the plastic hinges that absorb energy. For a specific given or assumed mechanism a token deflection establishes specific angle changes at all plastic hinges. The required external load to maintain this movement can be established by equating internal energy absorbed at the plastic hinges to the external energy input by the loads.

The true load factor for the frame can be no larger than the ratio of P (collapse) obtained for this particular mechanism to P (working load); it may be smaller for some other (more probable) mechanism. The mechanism-energy approach thus establishes an upper bound or upper limit on the value of the true load factor.

If all possible mechanisms are investigated the lowest load factor will be the true one; it is desirable to check this load factor against the lower bound discussed in the Sec. 11.12. For this purpose the moments are all statistically determinate when M_p values are used at each plastic hinge in a mechanism.

Consider again the beam of Fig. 11.9a, with positive moment $M_p = 60$ k-ft, negative moment $M_p = 100$ k-ft. From the discussion of Sec. 11.12 the necessary hinge locations are obviously at the three possible peaks on the moment diagram, as in Fig. 11.9e. The energy input from the load is $20\ \Delta$. Energy absorption at hinges $= 100(\Delta/12 + \Delta/8) + 60(5\ \Delta/24) = 20.8\ \Delta + 12.5\ \Delta = 33.3\ \Delta$. This leads to the load factor possible as $33.3\ \Delta/20\ \Delta = 1.67$. It is not necessary to check other solutions, even though this is an upper bound theorem. The pattern is unique because the lower bound and the upper bound give the same answer.

In entire frames the hinge location is not as obvious. The student might wish to imagine the hinge at midspan instead of the load and see how it leads to a higher and hence invalid γ.

11.14   Basic Concepts Essential to Limit Design as Summarized by Furlong

(a) **Assigned Limit Moments**   Furlong proposes[5] limits within which he feels it feasible to design totally without an elastic frame analysis. He limits himself to frames braced to avoid joint deflection laterally. His paper has taken the limit design concept out of the realm of involved theory and injected a practical design approach. Whether the profession is ready to depart this far from elastic analysis remains to be seen. Some such approach appears necessary if a later Code is to go beyond the moment redistribution now authorized. The direct approach has much merit compared to the present "patch-up" of elastic analysis. Furlong's proposal at least presents pertinent data and a challenge and encourages examination of the basic problem of whether limit design for reinforced concrete is worthwhile.

(b) **Flexural Ductility**   All limit design analyses require a material that will sustain M_p through a large angle of rotation in order that the simple hinge concept be feasible.[6] Solutions are more complex if a sloping line, as shown dotted near the bottom of Fig. 11.2b, is required, and quite complex if the three-line-dotted shape in the middle of that figure is required. Ductility here is defined as the ratio of the ultimate ϕ_u at the end of the horizontal line to the initial ϕ_e at the start of the horizontal section. Values of 4 to 6 are desirable and Reference 5 relates available ductility to variation in either $\rho - \rho'$ or $M_u/(\phi bd^2)$. The method of A. L. L. Baker[3] in general frames uses a group of simultaneous equations to find the related values of needed hinge rotation at each joint to assure that local failure of a hinge will not lower the frame capacity as usually computed.

(c) **Ratio of M_y to M_p**   The ratio of M_y/M_p lies between 0.94 and unity for singly reinforced members made with reinforcement having a long flat yield section in its stress-strain curve. This ratio implies to the authors that M_p/M_u will usually be feasible without involving M_y much, if at all, in the calculations.

(d) Yielding at Service Loads   Furlong has shown that reasonable limits are possible to assure no yield under service loads in load patterns that elastic analyses usually recognize.

Selected References

1. L. S. Beedle, *Plastic Design of Steel Frames,* John Wiley and Sons, New York, 1958.
2. M. R. Horne, "A Moment Distribution Method for the Analysis and Design of Structures by the Plastic Theory," *Jour. Inst. Civil Engrs.,* 3, Part III, Apr. 1954, p. 51.
3. Report of Institution Research Committee on Ultimate Load Design, A. L. L. Baker, "Ultimate Load Design of Concrete Structures," *Proc. Inst. Civil Eng.* (*London*), 21, Feb. 1962, p. 399.
4. M. Z. Cohn, "Rotational Compatibility in the Limit Design of Reinforced Concrete Continuous Beams," *Flexural Mechanics of Reinforced Concrete, SP-12,* ACI/ ASCE, Detroit, 1965, p. 359.
5. R. W. Furlong, "Design of Concrete Frames by Assigned Limit Moments," *ACI Jour.,* 67, No. 4, April 1970, p. 341.
6. A. H. Mattock, "Rotational Capacity of Hinging Regions in Reinforced Concrete Beams," *Flexural Mechanics of Reinforced Concrete, SP-12,* ACI/ASCE, Detroit, 1965, p. 143.

12

TWO-WAY SLABS ON STIFF BEAMS

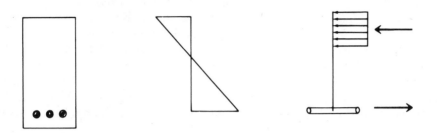

12.1   One-Way and Two-Way Slabs

One-way slabs were discussed in connection with continuous beams in Chapter 9. Such slabs differ from beams only in the large width-to-depth ratio of the slabs, which results in less important shear stresses in one-way slabs than in beams. Such slabs usually need an arbitrary amount of transverse steel to care for shrinkage and temperature stresses, and sometimes a computed amount to assist in distributing concentrated live loads on the slab. In the latter case some two-way action is necessary.

Some simple two-way slab arrangements are given in Fig. 9.1a and 9.1c. The slab of Fig. 9.1c supported by beams between columns should be designed by the methods of Chapter 16 as a part of a frame made up of columns and beam-slab combinations between columns. This is especially true if it involves at least a three-span series and qualifies for the direct design method. Code Chapter 13 covers such cases in detail.

Code 13.3.1 authorizes design of a slab system "by any procedure satisfying conditions of equilibrium and geometrical compatibility, if shown that the design strength at every section is at least equal to the required strength considering Code Sec. 9.2 and 9.3,* and that all serviceability conditions, including specified limits on deflections, are met." Designs under the yield line method (Chapter 13) or strip method (Chapter 14) can be made to meet this requirement.

The methods of Code Chapter 13 are relatively complex, especially where the conditions for the direct design method are not met. Such conditions (among others) involve a minimum of three continuous spans in each direction, successive span lengths not differing by more than ⅓ the longer span, and

---

* Code Sec. 9.2 and 9.3 cover load factors and ϕ factors.

columns not offset more than 10% of the span (in the direction of offset). Provisions do not seem to apply to beams intermediate between columns, as in Fig. 9.1a.

The authors feel that numerous two-way slabs supported on stiff beams or walls, where column interaction is not significant, justify the use of the slightly less economical slab design methods given in the 1963 and earlier Codes. These design methods were based on moment coefficients from elastic analysis. Whether used today or not, some of the inherent behavior patterns on which these design methods were based give valuable insight into slab behavior for the many irregular slab cases the designer faces, including slabs with openings as well as irregular shape.

This chapter is limited to the behavior of two-way rectangular slabs supported on stiff beams or walls and the semielastic analysis method of the 1963 and earlier Codes. The authors believe this still is valuable design background.

12.2 Elastic Analysis—Mathematical Approach

Two-way slabs, even single-panel simply supported, require a three-dimensional approach for analysis. Such slabs are rarely statically determinate in their internal moments and shears. They are the most highly indeterminate of all ordinary structures.

Usually slabs have been analyzed as flat thin plates made of a homogeneous elastic material that has equal strength and stiffness in every direction, that is, an isotropic material. On this basis solutions for simple cases can be established from partial differential equations by advanced mathematics. Westergaard[1] was a pioneer in such analysis in this country.

With differential equations, approximations must generally be introduced to handle practical cases. The use of difference equations and various computer techniques has led to reasonably accurate solutions for some problems. Theoretical boundaries thus established are valuable in assessing the merits of shorter and less accurate analyses. Today computers can handle most cases, but these solutions do not always give the designer the "feel" so valuable in verifying everyday designs.

12.3 Typical Moment Patterns

Because the student cannot really analyze a slab except by arbitrary code provisions, he or she should have a clear picture of the physical action of the slab. With a little imagination one should be able to visualize the general deflected shape taken by a uniformly loaded slab. A simply supported square slab will deflect into a saucerlike shape; unless the corners are held down they will actually rise a little off the supports. An oblong slab will take a platterlike shape. A very long narrow slab will take a troughlike shape except near the

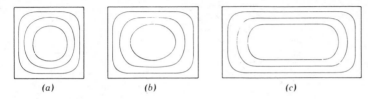

 placeholder — figure shown at top of page

Figure 12.1   Approximate contours for two-way simple supported slabs. (*a*) Square slab. (*b*) Oblong slab. (*c*) Long rectangular slab.

ends. For fixed edges, there must also be a transition zone around the edges in which the slope gradually turns downward from the horizontal edge tangents. If contours are roughly sketched, as in Fig. 12.1 for the simple supports, they give more than a clue to the moment pattern.

In a square slab, simply supported, a strip across the middle cuts the greatest number of contours and has the sharpest curvature and largest moment. In a long narrow slab only the short strips have significant curvature and bending over most of the panel length. The long center strip is essentially flat and without moment except near its ends.

In the continuous slab all slab strips in each direction have negative moments near the supports and positive moments near midspan, as shown in Fig. 12.2*a*. The long center strip in a long narrow slab is an exception or a special case in that its positive moment occurs not at the center but at the point where the strip starts to curve upward; the moment over its midspan length is small or zero (Fig. 12.2*b*) and the short center strips act almost exactly as one-way slabs.

Mathematical analysis shows that the negative moment on such a *long* strip is nearly the same regardless of the long span length. It is almost the same as it would be for a square panel with short span dimensions. One method in the 1963 Code* found it appropriate to express both long-span and short-span moments in terms of coefficients times the *short span* squared.

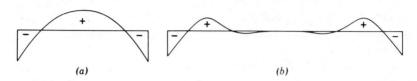

Figure 12.2   Typical moment diagrams for continuous slabs. (*a*) Short-span strip. (*b*) A very long strip, say, length three times width.

---

* See text Sec. 12.8.

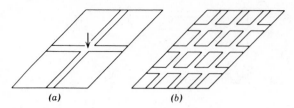

Figure 12.3 The strip idea for two-way slabs. (*a*) Single intersecting strips. (*b*) Multiple intersecting strips.

12.4 Approximate Analyses

Approximate analyses of slabs are usually somewhat crude, certainly so from the viewpoint of the theoretical person or mathematician. Section 12.3 indicates that the slab in each direction acts somewhat as a one-way slab. But the perpendicular slab strips of Fig. 12.3*a* are not independent in action. They share in carrying the load and thus each has a smaller moment than a one-way slab. They must deflect the same total amount; hence their relative stiffness becomes a factor in establishing the load and the moment each must carry. With slab thickness a common factor, the longer span is the more flexible and carries the smaller moment and load.

The use of a *single* slab strip in each direction is obviously a crude analogy. The short-span element across the middle of the panel deflects a much greater amount than a parallel strip alongside the edge beam; the edge strip can deflect little more than the beam. A better analysis would result if the slab is considered as several strips in each direction, held to common deflections at their intersections, as in Fig. 12.3*b*. A still better approximation would consider even more strips and torsional stiffness of these strips as well as their bending stiffness. As three strips each way would give nine intersections, the labor of such an analysis increases at least with the square of the number of strips. Many simultaneous equations must be solved, one for each intersection, or many successive approximations must be tried for a solution. With enough strips, results closely equivalent to those obtained from the partial differential equation solution are found, but the labor is excessive so that practical solutions use a computer.

Although elastic analyses of homogeneous isotropic plates provide much helpful information, they are exact solutions only for the assumed conditions. They usually do not recognize the fact that reinforcing steel is made lighter near the panel edges than at the center and that short-span steel is heavier than long-span steel. Nor do they usually recognize the changes in relative stiffness that result as cracked sections develop from the bending moment.

12.5   Inelastic Considerations in Slab Design

In a 1926 paper Westergaard[1] recommended moment coefficients that gave considerable weight to the nonelastic readjustments in slab moments that take place before failure. In recognition of these favorable readjustments, his recommended coefficients were established at 28% below strictly elastic values. The percentage reduction corresponds in a way to a similar reduction that was accepted for flat slabs, but for two-way slabs it gave more recognition to maximum moment loadings. The 1963 ACI Code requirements, including flat plates, fundamentally stemmed from Westergaard's recommendations, although work by vanBuren and diStassio[2,3] was also recognized.

Design practice and codes by no means attempt to design for the real distribution of bending moment existing across the slab under elastic conditions. This moment varies from element to element across a strip. If a square slab is considered subdivided into 1-ft strips in a given direction, the center strip obviously has the sharpest curvature and largest moment. Curvature and moment on adjacent strips decrease gradually until alongside the edge beam the deflection, curvature, and moment all approach zero.

Even if this were a one-way slab under variable loading it would not collapse when the maximum moment raised the steel stress in a single narrow strip beyond its yield point. This one strip would then simply become more flexible and leave more of any added load to be carried by adjacent elements less highly stressed. Before real failure could occur, all elements would have to be yielding in some fashion. (Chapter 13 discusses the yield line analysis and collapse conditions for the ultimate strength of slabs.)

12.6   Influence of Slab Supports on the Design

If a two-way rectangular slab panel is supported on stiff beams running between corner columns, part of the load on this panel is carried N-S to the beams running E-W, and part E-W to the beams running N-S. The shorter (and hence stiffer) slab spans carry the larger unit load and larger moments. The longer slab strips carry less unit load and less moment. If the slabs are supported on bearing walls instead of beams and columns, the distribution of the slab moments would completely solve the problem.

At the other extreme, if there are no supporting beams or walls, the slabs would be supported only by the columns (this is the flat plate floor discussed in Chapter 15). There the slab would have to be a two-way slab; it is not enough to carry the load N-S because it still requires that the load be carried E-W to accumulate it at the columns. In fact essentially *all* of the load must be carried N-S and *all* of it also must be carried E-W. If shallow beams are added, the total moment on N-S strips would be shared with the N-S beams; the *total* N-S moment would be the same as without the beams, but the slab would carry less. These concepts are developed for the general case of two-way slabs in Chapter 16.

With stiff beams between columns the slab and beam design can be separately calculated. However, it is still true that 100% of the total load must be carried N-S, some by the N-S slab strips, some by the N-S beams that pick up the reactions from the E-W slab strips. Likewise, the E-W slab and beams must carry 100% of the total load over the E-W span.

12.7   Effect of the Distribution of Reinforcement

If in one direction the designer fails to provide adequate strength that slab will yield locally when its limit is reached; but it can still shift the overload to the perpendicular strips if they (and their supporting beams) have the necessary reserve strength.

This readjustment in stiffness and bending moment indicates that the steel does not have to be placed exactly in accordance with the elastic moment requirements. One can say that *almost* any arbitrary arrangement of steel in sufficient quantity to carry the total load will be developed before the slab completely fails. The strip method for designing slabs in Chapter 14 moves in that direction.

At working loads, however, a poor distribution of slab steel may result in local yielding of the steel, large cracks, and increased deflection. For desirable action under working loads, steel should be placed at least roughly in accord with the moments existing at the service load; these moments are more nearly the elastic analysis moments.

Engineering practice uses a uniform spacing of steel over the center strip (one-half the panel width in a square panel) and reduces the steel towards the edges of the panel. This uniform center strip steel is designed to take the *average* rather than the maximum moment on the center strip, as shown in Fig. 12.4 for a simply supported slab. Thus the method takes considerable cognizance of nonelastic behavior as failure approaches. The subdivision of the longer side ℓ_B, shown in Fig. 12.4, fits the 1986 Code Sec. 13.2.1 definition of a column strip as a design strip with a width on each side of a column centerline equal to the lesser of $0.25\ell_A$ or $0.25\ell_B$.

Despite the fact that in practice approximate moments are used and the steel is designed for average moments rather than maximum, slabs are one of the most trustworthy structural elements and stand up well under great abuse and overload.

12.8   Visualizing the Effect of Reduced Restraint

Beam designers are accustomed to the exterior support situation where a light end support results in a large distribution of the fixed end moment back to the beam, with probably half of this distribution carried back to the first interior

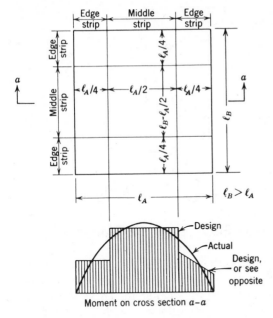

Moment on cross section a–a

Figure 12.4 Middle strip positive moment, slab simply supported. Comparison of actual and design moments.

column joint. A slab support, even with little restraint, is rarely such that it can rotate uniformly over a considerable width;[4] mathematically this would be discussed in terms of summing some very unequal restraints and rotations. This is a problem of accuracy, but not one of visualization. One can accept a distribution of fixed end moment back to the slab as realistic. However, slabs differ greatly from beams in their carryover moments. In a square panel with stiff beams *essentially none of this distribution carries back to the opposite beam support.* In a square panel the moment distributed back to the slab goes roughly equally to the two adjacent supports (90° from the rotating edge); not only does it go to the perpendicular slab elements, but most of it goes to those parallel elements in the near half of the slab. Distribution to the slab is thus a more localized operation and this means pattern loadings are important chiefly where they are local patterns. The distribution is roughly indicated in Fig. 12.5a for a square panel and assumes *very stiff* beams.

Increasing dimension a to make a rectangular slab influences the distribution very little. If a is reduced to b/2, a three-way distribution results that is roughly equal to all three beams, again for very stiff beams.

This three-dimensional behavior is significant because with stiff beams it means even free rotation of one edge is a very local matter, not very important except to the positive moments on slabs in both directions and to the negative moments added on the parallel elements nearby the rotating edge. Carryover moments do not reach into more remote slabs in the same way they do in beams. This makes pattern load moments very much less important in slabs.

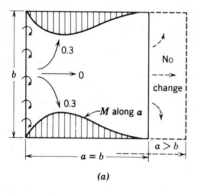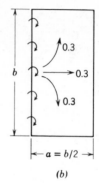

Figure 12.5 Carryover moment factors in continuous slabs on stiff beams.

12.9   Recommended Two-Way Slab Design for Simpler Cases

As indicated in Sec. 12.1, the authors recommend slightly less economical designs made under older Codes for many slab layouts, including most of those not meeting the direct design requirements* of Chapter 16. These simpler designs include many slabs not significantly interlocked with general frame action, those of less than three spans, slabs supported on walls or steel beams (without composite construction), floor slabs in low rise buildings, slabs serving as roofs over tanks or as sidewalls for structures below grade, and the like. The only addition included here from the 1986 Code is the minimum slab thickness from Code 9.5.3.1 that is simplified by the use of the curves of Fig. 15.12 from Sec. 15.10.

Method 2 of the 1963 Code is the simplest and possibly the best of the three 1963 methods for designing two-way slabs supported on beams. It should still be useful in estimating what should be done with all those slabs not really large enough to justify a slab systems approach.

As in the 1986 Code, two-way slabs are divided into strips as indicated in Fig. 12.4. However, the definition of column strips in the 1963 Code differs somewhat from that found in the 1986 Code. Using the old notation of S for short span, Fig. 12.6 shows more clearly the design strips from the 1963 Code. Table 12.1 is reproduced from the 1963 Code (Appendix A), with 1986 notation, along with about a page of that Code to indicate its limitations and its general use:

(a) Limitations—These recommendations are intended to apply to slabs (solid or ribbed), isolated or continuous, supported on all four sides by walls or beams, in either case built monolithically with the slabs.

ℓ_s = length of *short* span for two-way slabs(S). The span shall be considered as the *center-to-center* distance between supports or the clear span plus twice the thickness of slab, whichever value is the smaller.

w = total uniform load per sq ft

---

* Partially listed in fourth paragraph of Sec. 12.1.

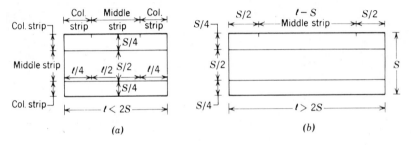

Figure 12.6 Column and middle strips for 1963 ACI Method 2. (*a*) Long span $< 2S$. (*b*) Long span $> 2S$. (The S is the old notation, no longer used.)

A two-way slab shall be considered as consisting of strips in each direction as follows:

A middle strip one-half panel in width, symmetrical about panel center line and extending through the panel in the direction in which moments are considered.

A column strip one-half panel in width, occupying the two quarter-panel areas outside the middle strip.

Where the ratio of short to long span is less than 0.5, the middle strip in the short direction shall be considered as having a width equal to the difference between the long and short span, the remaining area representing the two column strips.

The critical sections for moment calculations are referred to as principal design sections and are located as follows:

For negative moment, along the edges of the panel at the faces of the supporting beams.

For positive moment, along the center lines of the panels.

(b) Bending moments—The bending moments for the middle strips shall be computed from the formula

$$M = (\text{Coef.})w\ell_s^2$$

The average moments per foot of width in the column strip shall be two-thirds of the corresponding moments in the middle strip. In determining the spacing of the reinforcement in the column strip, the moment may be assumed to vary from a maximum at the edge of the middle strip to a minimum at the edge of the panel.

Where the negative moment on one side of a support is less than 80 percent of that on the other side, two-thirds of the difference shall be distributed in proportion to the relative stiffness of the slabs. (This ends the extract from the 1963 Code.)

Note first that all coefficients imply application to 1-ft-wide slab strips and these can lead to the moments per foot of width or per strip width as the designer chooses; the coefficients represent the average moment over a strip width. Second, note that long strip moments defined in terms of the coefficients are multiplied by the square of the *short* span. Third, the coefficients specifically assume the supporting beams or walls are "built monolithically"; if not so built on one edge, that edge provides less restraining moment and positive moments in both directions should increase a bit.

The method envisions a pattern type of loading; the coefficients thus provide more moment resistance than the M_o of the 1986 Code Chapter 13 requires.

TABLE 12.1 Method 2 Moment Coefficients

Moments	Short Span							Long Span, All Span Ratios
	Span Ratio, Short/Long							
	1.0	0.9	0.8	0.7	0.6	0.5 and less		
Case 1—Interior panels								
Negative moment at—								
Continuous edge	0.033	0.040	0.048	0.055	0.063	0.083		0.033
Discontinuous edge	—	—	—	—	—	—		—
Positive moment at midspan	0.025	0.030	0.036	0.041	0.047	0.062		0.025
Case 2—One edge discontinuous								
Negative moment at—								
Continuous edge	0.041	0.048	0.055	0.062	0.069	0.085		0.041
Discontinuous edge	0.021	0.024	0.027	0.031	0.035	0.042		0.021
Positive moment at midspan	0.031	0.036	0.041	0.047	0.052	0.064		0.031
Case 3—Two edges discontinuous								
Negative moment at—								
Continuous edge	0.049	0.057	0.064	0.071	0.078	0.090		0.049
Discontinuous edge	0.025	0.028	0.032	0.036	0.039	0.045		0.025
Positive moment at midspan	0.037	0.043	0.048	0.054	0.059	0.068		0.037
Case 4—Three edges discontinuous								
Negative moment at—								
Continuous edge	0.058	0.066	0.074	0.082	0.090	0.098		0.058
Discontinuous edge	0.029	0.033	0.037	0.041	0.045	0.049		0.029
Positive moment at midspan	0.044	0.050	0.056	0.062	0.068	0.074		0.044
Case 5—Four edges discontinuous								
Negative moment at—								
Continuous edge	—	—	—	—	—	—		—
Discontinuous edge	0.033	0.038	0.043	0.047	0.053	0.055		0.033
Positive moment at midspan	0.050	0.057	0.064	0.072	0.080	0.083		0.050

However, where ρ (or $\rho - \rho'$) is not greater than $0.5\rho_b$, the negative moments may be partially redistributed as indicated in Code 8.4.

The 1963 Code minimum slab thickness was 3.5 in. or the clear slab perimeter divided by 180, whichever was larger. The 1986 Code is somewhat stricter, using the same three limiting equations for slabs on beams and those on columns only, with three variables to define the exact case. A more complete discussion is given in Sec. 15.10; Fig. 15.12 summarizes the requirements, using the following symbols:

ℓ_n = clear span in *long* direction, face-to-face of supports

α_m = average ratio of beam-to-slab stiffness on the four supporting faces, usually equal to 2 or more for "stiff" beams

β = ratio of slab *clear* spans, the longer divided by the shorter. (Note that this is the inverse of the Table 12.1 ratio.)

β_s = ratio of the sum of the lengths of continuous edges to total perimeter of slab panel

Figure 15.12 will be used in the following example: (1) starting vertically from α_m (used as 2 if not given), (2) to one of the two interior panel ($\beta_s = 1.0$) curves or an interpolated curve dependent on the β ratio of the two slab spans, and (3) going horizontally to the left to read maximum usable ℓ_n/h, which establishes the minimum h. Note that the plotted ℓ_n/h is for $f_y = 40$ ksi and the figure title states the simple correction for $f_y = 60$ ksi: divide the ℓ_n/h by 1.10.

The lengthy equations given early in Sec. 15.11 simplify when (1) α_m is taken as 2 or more (for a stiff beam), (2) a numerical f_y is introduced, and (3) an interior span slab is considered (which means $\beta_s = 1$). For $f_y = 40,000$ psi and $\alpha_m = 2$, Code Eq. 9-13 becomes h need not be more than $\ell_n/36$ and Code Eqs. 9-11 and 9-12 become the same: $h \gtrless \ell_n/(36 + 10\beta)$. For $f_y = 60,000$ psi, these equations are changed only by a coefficient of 1.1 multiplying the ℓ_n value in the numerator.

For exterior spans $\beta_s < 1$ reduces both denominators and increases the required h, as Fig. 15.12 indicates.

At an exterior corner of the slab, special reinforcement is called for in Code Sec. 13.4.6. The discontinuity in two directions requires an increase in the reinforcement in this corner of the slab.

12.10 Example of Two-Way Slab Design—Stiff Beams

Design the interior bay two-way slab on extremely stiff beams for a bay 16-ft × 20-ft to centerline of beams with service dead loads of 75 psf (including slab weight) and live load of 100 psf, $f'_c = 3000$ psi, Grade 60 bars, 0.75-in. clear cover on slabs. Assume beams stiff enough to qualify slab for design by Method 2 of 1963 Code Appendix.

Solution

Minimum thickness for deflection control.

Assume beams 12-in. wide. Slab spans = center-to-center of beams or clear span plus twice slab thickness = say, 16 ft and 20 ft

Assume h in 1963 Code = (slab perimeter)/180 = $2(16 + 20)12/180$ = 4.80, say, 5 in. This is more than the 3.5-in. minimum of the present Code 9.5.3.1c. However, the present Code also requires a formula check using three equations. For this example the equivalent graphs in Fig. 15.12 will be used.

$$\text{Assume } \alpha_m \gtrless 2 \text{ for very stiff beams}$$
$$\beta = \text{ratio of clear spans} = (20 - 1)/(16 - 1) = 1.27$$
$$\beta_s = 1 \text{ for interior panel (all edges continuous)}$$

From Fig. 15.12, entering with $\alpha_m = 2$, read:

For $\beta = 1$ $\ell_n/h = 46$
$\beta = 2$ $\ell_n/h = 56$ $\beta = 1.27, \ell_n/h = 46 + 0.27(56 - 46) = 48.7$

Correct this reading for $f_y = 60$ ksi: $\ell_n/h = 48.7/1.10 = 44.3$
Minimum $h = \ell_n/44.3 = 19 \times 12/44.3 = 5.15$ in., say, 5.5 in.
Moment calculations.

$$w_u = 1.4w_d + 1.7w_\ell = 1.4 \times 75 + 1.7 \times 100 = 275 \text{ psf}$$

All moment coefficients in Table 12.1 are based on the short span ℓ_s for moments in both directions.

$$M_u = (\text{coef.})w_u\ell_s^2 = 275 \times 16^2 = (\text{coef.}) \ 70{,}400 \text{ lb-ft}$$
$$\text{Reqd. } M_n = M_u/\phi = (\text{coef.})70{,}400/0.9 = (\text{coef.}) \ 78{,}200 \text{ lb-ft}$$

Table 12.1 shows moment coefficients for span ratio of $16/20 = 0.8$ as follows:

Short span neg. M 0.048   Long span neg. M 0.033
Short span pos. M 0.036   Long span pos. M 0.025

Table 12.2 is convenient, although some values have been omitted as obviously not governing and others could have been so classified. The short span moments call for the largest d. For $k_n = 870$ psi (Table 3.1)

$$d \text{ for neg. } M = \sqrt{M_n/k_n b} = \sqrt{0.048 \times 78{,}200 \times 12/(870 \times 12)}$$
$$= 2.08 \text{ in.}$$

h for neg. $M = 2.08 + 0.75$ cover $+ 0.25$ (for $d_b/2$) = $3.08 < 5.5$ in.
     USE $h = 5.5$ in. $d = 5.50 - 0.75 - (0.5d_b$ or $1.5d_b)$
$$= 4.75 - (0.5d_b \text{ or } 1.5d_b)$$

For positive moment in the middle strip, the short span would use the $0.5d_b$ and in the long span the $1.5d_b$ for the perpendicular bars. In a square

TABLE 12.2 Slab Calculations for Sec. 12.10

$w\ell_s^2/\phi = 78,200$ lb-ft/ft

For #4 bars: $h = 5.5''$	Short Span		Long Span	
	Neg. M (4.5″)	Pos. M (4.5″)	Neg. M (4.0″)	Pos. M (4.0″)
Middle Strips				
M coefficient	0.048	0.036	0.033	0.025
M_n = (coef.) 78,200 lb-ft/ft	3750	2820	2580	1960
Min. $d = \sqrt{M_n/k_n} = \sqrt{M_n/870}$ in.	2.08		1.72	
For h = 5.5 in., d = 5.5 − 0.75 − 0.25	4.50	4.50	—	—
or d = 5.5 − 0.75 − 0.75	—	—	4.00	4.00
$A_s = M_n \times 12/(60{,}000 \times 0.9d)$ $= M_n/4500d$ in.²/ft	0.19	0.14	0.14	0.11
$0.0018bh$ min. = 0.0018 × 12 × 5.5	0.12	0.12	0.12	0.12
Spacing of #4	12.6 in.	17.1 in.	17.1 in.	18.5 in.
Max. spcg. = $3h$ = 16.5 in.	—	16	16	16
*USE #4 at spacing:	12 in.	16 in.	16 in.	16 in.
Strip width		10 ft		8 ft
*USE in strip total #4 bars	10 at 12 in.	8 at 16 in.	6 at 16 in.	6 at 16 in.
Edge Strips, each side				
Strip width		5 ft		4 ft
$M_n = 0.5(2/3)(M_n$ on midstrip)				
A_s each side = (1/3)(A_s of midstrip)	#4 at 16 in.	Same as Neg. M	#4 at 16 in.	Same as Neg. M
Total per strip	4-#4 at 15 in.		3-#4 at 16 in.	Neg. M

* Alternate forms of listing.

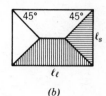

Figure 12.7 Slab load areas to beams. (*a*) Square slab. (*b*) Rectangular slab.

slab, the average of $1.0d_b$ is often used for both directions because the beams can usually carry the slight difference in loading this can introduce. The negative moment on the edge strips also requires overlapping bars; This also would apply to negative moments on middle strips if continuous top bars were used; they are not used here.*

The calculations in Table 12.2 indicate that another solution with #3 bars is possible with some saving in the total steel used, but more field labor would be used in placing the extra bars (possibly 60% to 80% more). Normally only one of the optional steel listings shown is calculated. The more usable is probably the total bars per strip. The spacing of bars in edge strips would be more logical if it started adjacent to the middle strip with bars at the middle strip spacing, gradually increasing to a maximum of three times that spacing at the edge of slab, an average of 1.5 times the middle strip spacing. The maximum bar spacing limitation prevents this with the #4 bars.

Shear in two-way slabs on beams would govern only for extremely heavy loads because the beams provide reactions on all four sides. The load distribution on the beams is usually taken from the contributing areas shown in Fig. 12.7. Shear, if it should control the slab would be shear at a distance d from the face of support, in this case for the short strip at $0.5(16 - 1) - 4.5/12 = 7.1$ ft from midspan.

$$V_u = 275 \times 7.1 = 1950 \text{ lb/ft width}$$
$$V_n \geqslant V_u/\phi = 1950/0.85 = 2300 \text{ lb/ft required}$$
$$\text{Actual } V_n = 2\sqrt{f_c'}bd = 2\sqrt{3000}\,(12)(4.5) = 5915 \text{ lb/ft}$$

---

* Slab reinforcing bars are flexible and the projecting end of a bent bar can easily be forced into a plane one bar diameter above or below that which it occupies at the bend point. For this reason many designers feel it unnecessary to be as careful as the author about theoretical bar interference. The author's reasoning is as follows. When bars must be displaced because of interference, it requires close field inspection to see that the correct group of bars is displaced. The construction worker frequently has no knowledge of the designer's desires in this respect. Proper placement is more likely to result if bars are detailed in such a way that the field worker does not have to readjust the steel levels.

The most critical portion of the long-span top steel is over the middle section of the short beams. At this point the top steel could easily be brought up one bar diameter closer to the top surface. Beyond each side of this section, approximately between the one-quarter and one-eighth points, the short-span top steel could be depressed enough to put the long-span top steel in this more favorable position. Designers must decide whether (1) they want to assign this readjustment to the field, (2) they want to use the larger d near the center width and a smaller d near the edges, or (3) they wish to follow the very conservative approach of using the smaller d for all the long-span negative steel, as in this book.

12.11 Freely Supported Slab Corners

Where a slab is simply supported on masonry walls or steel beams, a problem arises at the corners if the slab is not securely fastened down. Such corners rise when the slab is loaded. The corners may rise far enough to create a conspicuous horizontal crack in the masonry walls in which they are embedded. This is especially the case with roof slabs carrying parapet walls. Such slabs should be anchored down (to a substantial mass of masonry, for example) and have reinforcing steel provided for the restraining moment thus developed.

The exterior corner of any slab needs special treatment if the support is stiff, either a beam or wall. Code 13.4.6 gives a detailed specification covering slabs monolithic with moderately stiff concrete beams.

Selected Reference

1. H. M. Westergaard, "Formulas for the Design of Rectangular Floor Slabs and the Supporting Girders," *ACI Jour., Proc.* 22, Jan. 1926, p. 26.
2. J. DiStasio and M. P. van Buren, "Slabs Supported on Four Sides," *ACI Jour.,* 7, Proc. 32, Jan.-Feb. 1936, p. 350.
3. R. L. Bertin, J. Distasio, and M. P. van Buren, "Slabs Supported on Four Sides," *ACI Jour.,* 16, Proc. 41, June 1945, p. 537.
4. C. P. Siess and N. W. Newmark, "Rational Analysis and Design of Two-Way Concrete Slabs," *ACI Jour.,* 20, Proc. 45 Dec. 1948; p. 273.

Problems

PROB. 12.1. Design a 21-ft square (c-c of beams) interior panel of a two-way slab on beam stems 12-in. wide (beams assumed to give $\alpha_m = 2$), $f'_c = 4000$ psi, Grade 60 steel, $w_d = 120$ psf (including slab weight), $w_\ell = 60$ psf. Slab thickness must satisfy deflection requirements (close of Sec. 12.9 and Fig. 15.12) and this could determine slab thickness.

PROB. 12.2. For the data of Prob. 12.1 assume a structure only two bays wide. Design a typical slab for this layout.

PROB. 12.3.

(a) Pick a slab thickness based on the Code deflection requirements (close of Sec. 12.9) for an isolated two-way slab supported by tie-beams 12-in. wide on top of masonry walls (assumed stiff enough to qualify for $\alpha_m = 2$). The panel is 16-ft $\times$ 18-ft center-to-center of beams, $f'_c = 4000$ psi, Grade 60 bars, $w_d = 100$ psf (including slab weight), $w_\ell = 50$ psf.

(b) Check the depth for moment and design the reinforcement on the basis of Table 12.1.

PROB. 12.4. Design an interior slab panel 21-ft $\times$ 18-ft, otherwise like that of Prob. 12.1 for $w_\ell = 200$ psf without any change in w_d.

PROB. 12.5. A structure consists of a series of transverse bays each extending laterally 21 ft center-to-center between longitudinal beams 12-in. wide. If transverse beams 12-in. wide are 19 ft center-to-center, loads are $w_d = 120$ psf including slab weight and $w_\ell = 200$ psf, $f'_c = 3000$ psi, Grade 60 bars, design the two-way slab, assuming $\alpha_m = 2$.

13

YIELD LINE THEORY FOR SLABS

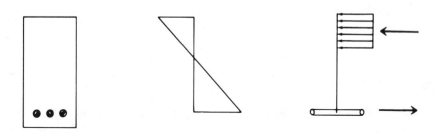

13.1   Yield Line Theory as a Design Guide

The yield-line theory is not specifically recognized by the ACI Code, but good engineers are now designing on this basis under the special systems provisions of Code 1.4. Such applications seem particularly advantageous for irregular column spacing.[14,15]

The yield line is introduced here because it helps the engineer to think about failure patterns and to visualize the ultimate behavior of slabs made with simple reinforcement patterns. The method is straightforward, reasonably simple in its concepts, and emphasizes the lines of highest stress. It is adaptable to irregular cases. The engineer can trace slab behavior and visualize some of the effects of moving to less simple reinforcement patterns, cutting off bars, and the like. The strip method discussed in the next chapter is even more usable as a design method where slabs are not supported directly on columns, but designers are apt to be embarrassed by their freedom unless they have some background in the yield-line method.

Yield line analysis is a limit design method. Designers, until they have a good code to guide them, should assume a liberal load factor, because slab deflection is quite large before failure. Alternatively, some of the ultimate strength might be discounted because of large deflections, just as engineers discount that portion of structural steel strength that lies between the first yield point and the ultimate. The yield line analysis deals with moment alone; it does not assure adequate strength in diagonal tension, but diagonal tension in two-way slabs is a problem only in special cases. On the other hand, in flat plates the diagonal tension (punching shear) around the columns is quite often the weakest element of strength.

Tests closely verify the yield line analysis. The calculated load normally underestimates the actual test results. Whether the excess arises from the flat arch action or membrane action, it appears that engineers may use the yield line method with confidence; it gives a conservative estimate of moment strength.

The real concern of designers using this method is to establish that their slabs will be entirely satisfactory at working loads. The Danish code has for some time permitted a form of ultimate strength design for slabs. It is the authors' understanding that design for ultimate strength alone does not necessarily lead to acceptable slabs. The deflection and stiffness of such slabs may not be satisfactory. Evidently these specific matters must be investigated or limited by establishing maximum ℓ/h ratios.

13.2 Basic Ideas of Yield Line Theory

(a) Angle Changes at Yielding   The yield line theory is a form of limit design (Chapter 11) and gives an ultimate strength. Slabs are normally under-reinforced, with much less strength than balanced steel. As a result, on progressive loading the steel reaches its yield point stress before the slab reaches its ultimate strength. As the steel yields, the center of compression on the cross section moves nearer the face of the slab until finally a secondary failure in compression takes place at a moment only slightly greater than the yield point moment.

For a one-way simple span slab, the increase in moment after the steel starts to yield is not large, on the order of 5% to 10%, but the further angle change ϕ occurring at the point of maximum movement is quite large in comparison with the "elastic" angle change. Relative values can be shown as an $M-\phi$ curve as in Fig. 11.1b.

In a statically indeterminate slab these extra angle changes permit (or cause) significant modifications of the resisting moments and shears. Yielding at one point in such a slab marks the gradual beginning of larger deflections but by no means marks the end of reliable load capacity.

(b) Yield Lines as Axes of Rotation   Yielding under increasing load progresses to form lines of yielding. Until yield lines are formed in sufficient numbers to break up the slab into segments that can form a collapse mechanism, additional load can be supported. To act as hinges for a collapse mechanism, yield lines must usually be straight lines although the fan pattern discussed under subsection (c) has one curved yield line boundary.

It would be a more accurate description to say that yielding zones develop on the tension face over narrow bands, as in Fig. 13.1a, and that the yield line is an idealization in which all the angle change is considered on a line at the center of the yielding bands, as in Fig. 13.1b. Yield lines are axes of rotation for the movements of the several parts of the final mechanism. Yield lines form at lines of maximum moment, but this action is not restricted to those maximum moment lines originally developed under initial "elastic" conditions.

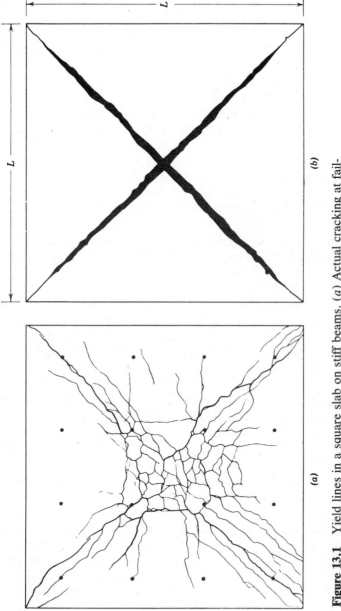

Figure 13.1 Yield lines in a square slab on stiff beams. (*a*) Actual cracking at failure. (*b*) Assumed "yield lines" or "fracture lines." (These figures are copied by permission from *Reinforced Concrete Review*.[5])

(c) Interrelationships Between Axes of Rotation　The supports of a slab determine some of the axes of rotation of the several slab segments. In general, each support line constitutes an axis of rotation and each separate column support constitutes a pivot point, that is, a point on an axis of rotation. For example, a one-way continuous slab in a given span must fail when yield lines at each support acting together with an intermediate yield line dependent on the loading develop, as indicated in Fig. 13.2*a*. This failure may be compared to the local collapse mechanism of Fig. 11.7 in limit design discussion.

Consider a continuous slab with a skewed support *b* as shown in Fig. 13.2*b*. With a yield line over two adjacent supports such as *a* and *b*, one segment rotates about *a* and one about *b*; some common yield line *between a* and *b* joins the two plate segments rotating about these axes. Hence the common yield line must lie on an axis through O_1 at the intersection point of *a* and *b* extended. Which of the possible dashed axes through O_1 will develop depends on the reinforcement and type of loading on the span. Likewise, for collapse in span *bc*, the third yield line must pass through O_2.

In Fig. 13.2*c* a slab is supported on two adjacent sides and a column. The general pattern of failure will be as indicated, but the axis through the column is at an unknown angle and the yield line intersection point in the slab depends on the loading and on whether the supported edges are simple supports or are lines of negative moment resistance.

Around a concentrated load a circular yield line of negative moment (Fig. 13.2*d*) tends to form, with radial yield lines of positive moment, like the spokes of a wheel. When supports are nearby, only circular segments instead of full circles may form, as in Fig. 13.2*f*. These portions are called fans and consist of multiple segments similar to the one in Fig. 13.2*e*. These are discussed in Sec. 13.7.

(d) Segments as Free Bodies To Verify Yield Line Locations　The free body represented by each collapsing segment must be in equilibrium under (1) its applied loads, (2) the yield moments on each yield line, (3) the reaction or shear on support lines, and at times (4) correction forces. Because the yield lines form at lines of maximum moment, neither shear nor torsion is typically present at positive moment yield lines.* Torsion can be present in special cases and then it must be represented by correction forces called nodal or knot forces, as discussed in Sec. 13.2*g*. The yield line moments thus usually establish the load the segments can support, as indicated in Fig. 13.3. For a uniformly loaded slab, the location of the intermediate yield lines must subdivide the slab into segments that each support the same uniform ultimate load w_u. If, for given values of yield moments m_1, m_2, and m_3 and a given trial location of the yield line the calculated w_u values for the separate segments differ, the yield line must lie in a different position. The real location must be such as to reduce

---

* Lenschow and Sozen[12] show that both torsion and a lowered moment resistance may exist at a yield line when $m_x \neq m_y$.

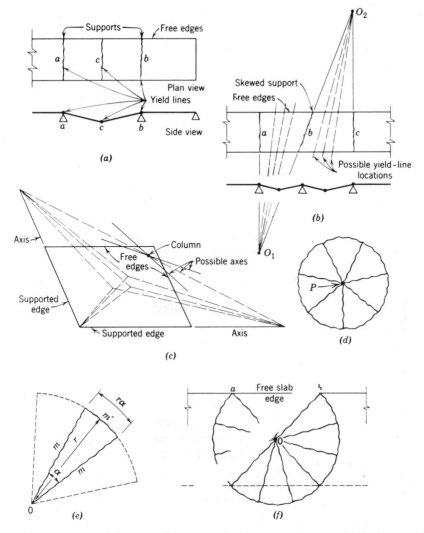

Figure 13.2 Yield line patterns. (*a*) Continuous one-way slab, right supports at *a* and *b*. (*b*) Continuous slab with a skewed support at *b*. (*c*) Slab supported on two adjacent edges and a column. (*d*) Fan pattern. (*e*) A fan segment. (*f*) Fan pattern near free edge.

the segment size where the calculated w_u is smaller and to increase segment size where the calculated w_u is larger. The correct location is a unique location for a given loading and type of collapse pattern.*

---

* However, if a different pattern is possible, this solution says nothing as to which pattern is the governing one. In complicated cases the latter demands the energy approach, at least for verification of findings.

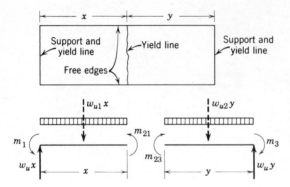

Figure 13.3 Free-body diagrams for verification of yield line location.

In terms of design it may be convenient to take the known quantity as the load times the desired load factor. Then, with the segments as the free bodies, the required yield moments may be calculated. At a given yield line the moment thus found for one segment must match that from the adjoining segment, that is, m_{21} must be the same as m_{23}; otherwise the yield line is incorrectly located.

When only one yield line remains to be located, an algebraic equation can be solved for a governing dimension, but in more complex cases this may not be as simple as a cut-and-try procedure.

Wood[7] has raised some serious questions about the universal applicability of the correction forces discussed in Sec. 13.2g for use in equilibrium solutions. This problem clouds the use of the equilibrium process in novel cases, but the virtual work procedure presented in Sec. 13.2e is not subject to any limitations and may be confidently used in all cases.

(e) Virtual Work or Energy-Mechanism Criterion If a slab is reduced to a mechanism by yield lines acting as hinges, a known additional deflection of any specific point in the mechanism (by geometry) establishes the additional deflections at all points along with the additional angle changes at the yield lines or hinges. For such a slab deflection, the loads also deflect and thereby contribute energy to the mechanism, whereas the hinges resist movement and absorb energy from the system. Thus, for a given mechanism, a given loading imparts enough energy to develop specific yield moments in the hinges. No smaller yield moment will be adequate to resist these loads. Some other mechanism may represent a more probable failure pattern; if so, it demands a larger yield moment to balance the given load. Thus the worst mechanism and the real mechanism that actually forms is the one that requires the largest resisting yield moment. Any trial mechanism or yield line pattern establishes a lower bound or lower limit on the required yield moment.

The same procedure can be used with given yield moments to establish the ultimate load. Any given mechanism establishes an upper bound or upper limit on the collapse load. The real collapse load is the smallest that can be found from all possible mechanisms.

The energy calculations and the segment equilibrium conditions are alternative procedures. In ordinary cases either can be used to check the other. Each establishes the critical dimensions for a given pattern. Neither automatically compares this pattern with others that could be more critical. Hence neither provides a lower bound on the load capacity, nor an upper bound on the needed moment capacity. Wood emphasizes the need for these bounds that are available only for a few cases.

Each method has its advantages. The equilibrium method is efficient and points the way to the most critical dimensions; the energy method requires more groping. But in novel situations, especially around openings, the equilibrium method demands some correction forces [see subsection (g)] that may not be available from simple methods or may even be overlooked. Hence only the energy method is always dependable, as already mentioned at the close of subsection (d).

The energy-mechanism approach indicates that a solution is not very sensitive to small changes in the same yield line pattern, but one must be very careful not to overlook the critical form of the pattern.

(f) Yield Moment on Axes Not Perpendicular to Reinforcing   When the yield moments in two perpendicular directions are equal and no twisting moment (torque) exists, the yield moments in all directions are equal. This condition of isotropic reinforcement simplifies the problem very considerably.

Hognestad[3] has shown (following Johansen's demonstration[1,2,4]) that when the reinforcement in one direction differs from that in the perpendicular direction by some constant ratio, the slab dimensions can be modified to permit analysis as an isotropic slab. Such cases will not be considered here.

If m_x represents the yield moment about the x-axis and m_y about the y-axis, the yield moment about an axis at angle α with the x-axis (Fig. 13.4) will be

$$m_\alpha = m_x \cos^2 \alpha + m_y \cos^2(90 - \alpha) = m_x \cos^2 \alpha + m_y \sin^2 \alpha \quad (13.1)$$

When $m_x = m_y = m$, this leads to $m_\alpha = m$ as stated at the beginning of this section.

Wood[7] refers to this relationship as the "square yield criterion" and states that it can be true only when applied to a point where there is a state of "all-round uniform moment," that is, uniform in all directions. He reports that

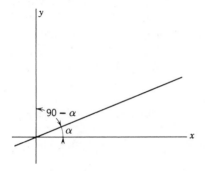

Figure 13.4   Moment axes.

experiments indicate the slab load capacity may be increased as much as 16% by the increased m capacity on a diagonal. He concludes that the square yield criterion normally used is only an approximation for the more complicated actual criterion, but this simple criterion is the one generally used.

On the other hand, Lenschow and Sozen[12] rather conclusively prove by test and theory that there is no appreciable change in the resulting m in the case just discussed. The increased capacity observed in the English test must have an explanation different from the one Wood suggests.

(g) Correction Forces Where Yield Line Intersects a Free Edge   Correction forces, usually called edge forces, nodal forces, or knot forces, are a necessity for general solutions by equilibrium methods. They appear most certainly when a yield line intersects a free edge at other than a 90% angle. These forces are substitutes for either torsion or twisting moments along assumed yield lines and for normal shears existing at negative moment yield lines. The correction forces are applied in the corner of segments, normal to the plane of the slab, in the magnitude $m_t = m \cot \alpha$, as shown in Fig. 13.6. Because these correction forces act between segments or between a segment and its reaction, they are of the nature of internal forces except when a segment is considered separately as a free body. Hence they completely disappear from any virtual work or energy equations applied to the entire slab. These forces can be totally ignored except in equilibrium calculations (and in reaction calculations).

These forces have been "established" in different manners by different authors. The writers agree with Wood's evaluation that each of these proofs is lacking in rigor and open to some question. Yet, as Wood shows by various examples, the method works successfully in many instances and it avoids the necessity of differentiation for a minimum value. Hence it has merit even though it is an uncertain tool in new situations. No equilibrium solution can now be considered entirely satisfactory by itself without a check by the energy method.

One approach to the value of the correction force is to say that the use of a straight line as the yield line can only be an approximation at an edge. The reasoning is that a single yield line, to satisfy statics, must intersect a free edge or a simply supported edge at 90° to avoid the necessity of torsional or twisting moments on the yield line. Such twisting moments are not normal on a true maximum moment line. Hence, near an edge, yield lines must turn as shown by the dotted lines of Fig. 13.5. If one insists on using the false "straight-line" yield line, one must consider the torsional moment $m_t = m \cot \alpha$ in equilibrium equations. The simplest way to do this is to use correction forces at the tip of the segments. The following "proof" and Fig. 13.6 are taken directly from Hognestad's ACI paper[3]:

> The magnitude of m_t may be established by considering the equilibrium of the infinitesimal triangle AOB shown in Fig. 13.6 in which AO is a finite length and AB is infinitesimal. Neglecting differentials of higher order, the moment in the section

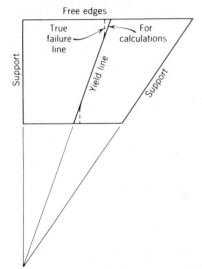

Figure 13.5 The curved yield line at a free boundary.

OB must equal the moment m in the yield line OA as m is a maximum value. Since the bending moment is zero along $AB = ds$, the total moment acting on the triangle AOB is found by vector addition

$$m(\overline{AO} + \overline{OB}) = m\overline{AB} = m\ \overline{ds}$$

Equilibrium of moments about OB then gives

$$m\ ds \cos \alpha = m_t\ ds \sin \alpha, \quad \text{or} \quad m_t = m \cot \alpha$$

differentials of higher order again being neglected. It should be noted that m_t acts down in the acute corner. These boundary conditions were first introduced into the yield-line theory by Johansen in 1931.

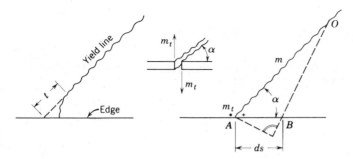

Figure 13.6 The shear load m_t at boundary when straight yield lines are used. (Reprinted from *Jour.* ACI.[3])

Other more detailed "proofs" start from a consideration of intersecting yield lines anywhere. These proofs deduce the same result plus two more useful conditions:

1. If all intersecting yield lines have positive moments (or all negative with no positive), the number of intersecting lines is unlimited, as the "fan" of Fig. 13.2d, and no knot force (correction force) may be needed at the intersection where only three such yield lines intersect; but knot forces will sometimes be needed when *more than three intersect*.

2. If all intersecting lines are not of one kind, only three different directions are possible. Of the six half lines resulting, any one half-line may be omitted in a particular case, *or* two half-lines may be omitted if they are either halves that form a single line or are not adjacent half-lines.

Wood notes that the value of $m_t = m \cot \alpha$ exceeds m when the angle α is less than 45° and that such a value would be impossible. However he points out that, in accepting an admittedly approximate or simplified yield line, one may not be limited to exactness as to such an angle. Some solutions giving essentially true results involve smaller angles.

When an entire slab is considered, this downward shear on one segment and the upward shear on the adjacent segment are internal and mutually offsetting forces, and hence do not show in the overall energy equation.

(h) Corner Pivots and Corner Yield Lines   Supported slab corners, as in Figs. 13.11 and 13.12, introduce the problem of what might be called localized yield line patterns. Hognestad[3] reports: "According to Johansen it is most expedient in practical design to disregard the corner levers and then later apply corrections, for which he has developed general equations and tabulated the most common cases." In this brief treatment of the subject, corner patterns will be evaluated as any other regular pattern. Such discussion will be deferred until an example not needing this corner analysis is considered.

13.3   Slab Example, One-Way Steel

In Fig. 13.7a consider that the yield moment at a is $m_a = -4000$ ft-lb/ft width, at b is $m_b = -5000$ ft-lb/ft, and for positive moments is $m_c = 3000$ ft-lb/ft. By the yield line method, calculate the ultimate uniform load the slab will support on a 12-ft span.

Solution

The slab capacity is independent of adjacent panel conditions except as they are reflected in the values of m_a and m_b. The one-way slab can be

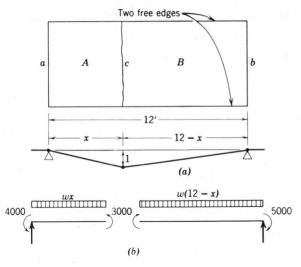

Figure 13.7 One-way slab example. (*a*) Failure mechanism. (*b*) Free-body diagrams.

very simply solved by limit design procedures with identical results, but the object here is to introduce yield line procedures.

Because only one yield line pattern is possible, an algebraic solution based on the free-body diagrams of Fig. 13.7*b* is simple.

$$\Sigma M_A = -4000 - 3000 + wx^2/2 = 0, \qquad w_A = 7000 \times 2/x^2 = 14,000/x^2$$
$$\Sigma M_B = +3000 + 5000 - w(12 - x)^2/2 = 0, \qquad w_B = 8000 \times 2/(12 - x)^2$$

Since w_A is to be equal to w_B,

$$14,000/x^2 = 16,000/(12 - x)^2$$

The quadratic could be solved algebraically, but a solution more typical of the general possibilities of this method would be by trial and error.

$$\begin{array}{ll}
\text{If } x = 6 \text{ ft}, & 388 < 444 \\
x = 5.5 & 463 > 378 \\
x = 5.8 & 415 \doteq 417, \text{ say } 416 \\
\multicolumn{2}{c}{\text{Ultimate } w = 416 \text{ psf}}
\end{array}$$

The virtual work or energy-mechanism solution is also entirely feasible, using Fig. 13.7*a*. For a 1-ft strip the energy input for a unit deflection at the center yield line is

$$0.5wx + 0.5w(12 - x)$$

The work done on the slab at the yield lines or hinges is $m\theta$, where θ is the angle of rotation. At the center yield line the total angle change depends upon the combined rotation of A and B. However, the most convenient treatment is to calculate the work there as the separate amounts owing to

A and *B* respectively. The work at the yield lines is then:

Due to rotation of *A*: $(4000 + 3000)(1/x)$

Due to rotation of *B*: $(3000 + 5000)[1/(12 - x)]$

Equating energy input to energy consumption,

$$0.5wx + 0.5w(12 - x) = 7000/x + 8000/(12 - x)$$
$$6w = 7000/x + 8000/(12 - x)$$

For the needed minimum w, $dw/dx = 0$

$$6(dw/dx) = -7000/x^2 - 8000(-1)/(12 - x)^2 = 0$$

This is the same algebraic equation solved in the trial and error approach, which gave $x = 5.8$ ft. If this value is substituted in the energy equation,

$$6w = 7000/5.8 + 8000/6.2 = 2498$$
$$w = 416 \text{ psf}$$

For the more usual slab with a more complex yield line pattern, a group of partial derivatives replace dw/dx. This type of solution might be impractical.

13.4   Slab on Nonparallel Supports, Two-Way Steel

The slab of Fig. 13.2*b* supported along *aa'* and *bb'*, as shown in Fig. 13.8*a*, illustrates two further points of procedure and brings up a possible practical complication. It also serves to emphasize that moments are vector quantities and as such may need to be resolved into components. The ultimate uniform load will be calculated on the basis of the dimensions of Fig. 13.8*a*, $m_a = -4000$ ft-lb/ft, $m_b = -5000$ ft-lb/ft, and $m_c = 3000$ ft-lb/ft both longitudinally and transversely.

Solution

Since yield lines for negative moment will occur over each support, the corresponding axes of rotation intersect at *O*. The intermediate positive moment yield line (extended) must also pass through *O* and can be defined in terms of an unknown angle at *O* or by the dimension x, the length of one side of the *A* segment.

This solution requires the yield moment m_c along the inclined yield line between *A* and *B*. With equal m_α values longitudinally and transversely, m_c is the same for all orientations. This conclusion follows from Eq. 13.1.

$$m_\alpha = m_x \cos^2 \alpha + m_y \sin^2 \alpha = m_c(\cos^2 \alpha + \sin^2 \alpha) = m_c$$

If the transverse reinforcement were lighter, as often is the case, the yield moment has m_α value for each assumed slope of yield line. The variable

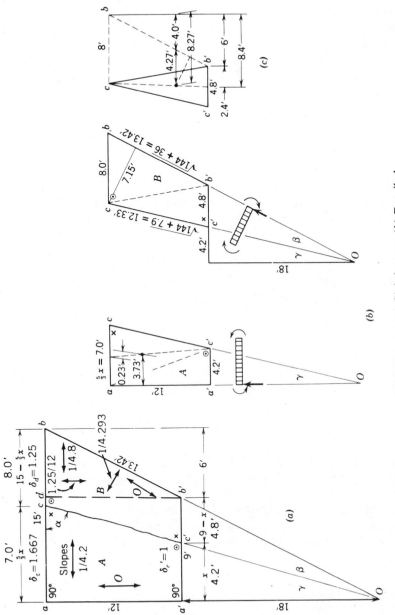

Figure 13.8 Two-way slab on nonparallel supports. (*a*) Slab layout. (*b*) Detailed dimensions of segments *A* and *B*. (*c*) Subdivision of segment *B*.

m_x would constitute an extra complication* in the solution, which would make general equations rather difficult. However, successive trials for different values of x are feasible. Solution by trial and error may constitute the simpler approach even for the given isotropic case because of the somewhat involved geometry for moment arms and angles.

The second new procedure arises from the fact that the middle yield line crosses two free edges at other than a 90° angle. A correction (Sec. 13.2g) in the form of a downward shear m_t at the acute angles and an upward shear m_t at the obtuse angles then becomes necessary.

$$m_t = m \cot \alpha = m(x/18) = mx/18$$

These forces are marked on the plan view in Fig. 13.8a by an x for a downward force and by a dot within a small circle for an upward force. These shear forces influence an analysis by segments but not an energy equation. The first trial will be made with segments.

Assume $x = 4.2$ ft, given the dimensions in Fig. 13.8b. The moment about aa' of the yield moments acting along cc' is

$$m_c(\text{length } cc')\cos \gamma = m_c \times 12$$
$$m_t = m_c x/18 = 3000 \times 4.2/18 = 700$$
$$\Sigma M'_{aa} = -4000 \times 12 - 3000 \times 12 - 700 \times 4.2 + 700 \times 7$$
$$+ 0.5 \times 12w \times 4.2 \times 1.4 + 0.5 \times 12w \times 7.0 \times 3.73 = 0$$
$$- 48,000 - 36,000 - 2940 + 4900 + 35.3w + 156.6w = 0$$
$$w_A = 82,040/191.9 = 427 \text{ psf}$$

The yield moment along $cc' = m(\text{length } cc')$, which gives a moment about axis $bb' = m(\text{length } cc')\cos \beta$. The angle β in triangle $c'Ob'$ is

$$\beta = \tan^{-1} 9/18 - \tan^{-1} 4.2/18 = 26°34' - 13°08' = 13°26'$$
$$\cos \beta = 0.973 \qquad \cos(\gamma + \beta) = 0.895$$

An alternate procedure is to obtain the moments for cc' as components about the x- and y-axes and then to take the components about bb'.

The slab load on segment B will be subdivided into triangular areas of load as in Fig. 13.8b,c. Triangle $bb'c$ has an area $8 \times \frac{12}{2} = 48$ ft^2, which indicates an altitude perpendicular to bb' of $48 \times 2/13.42 = 7.15$ ft. For triangle $cc'b'$, the centroid is on the median at 8.27 ft to the left of b. Thus

---

* Hognestad presents Johansen's proof showing that the slab dimensions can be modified to give a solution based on the simpler isotropic case. Assume the reinforcement in the transverse direction leads to a value of μm across longitudinal sections compared to m for the longitudinal strips. The simpler isotropic case ($m_x = m_y = m$) can be used if the length in the transverse direction and the size of any concentrated loads are first divided by $\sqrt{\mu}$, any uniform load w remaining unchanged. This provides a relatively simple solution for the problem but it appears that in a general case the ratio μ must be the same for both positive and negative moments in a given direction.

the horizontal distance between the centroid and bb' is 4.27 ft. The arm about $bb' = 4.27 \cos(\gamma + \beta) = 4.27 \times 12/13.42 = 4.27 \times 0.895 = 3.82$ ft.

$$M'_{bb} = 3000 \times 12.33 \times 0.973 + 5000 \times 13.42 - 700 \times 4.8 \times 0.895$$
$$+ 700 \times 8.0 \times 0.895 - 48w \times 7.15/3 - 4.8 \times 12w \times 0.5 \times 3.82 = 0$$
$$+ 36{,}000 + 67{,}100 - 3000 + 5000 - 114.3w - 110w = 0$$
$$w_B = 105{,}100/224.3 = 470 \text{ psf} > w_A = 427 \text{ psf}$$

x is taken a little too large, but the difference between w_A and w_B is quite small and an answer of ultimate $w = (470 + 427)/2 = 449$ psf is quite close. Normally one would not expect results of a trial to be this near to a correct answer; this choice of x benefited from the somewhat similar analysis of the slab in Sec. 13.3. On the other hand, the individual trial calculation should not be as long as here shown. The rather involved geometry was worked out in detail; graphical determination of some of the dimensions might be preferable, especially for initial trials.

The same problem is solved from the energy-mechanism or virtual work approach, still using Fig. 13.8. The geometry and angles must be carefully determined. Use $x = 4.2$ ft as before and assume a unit vertical movement at c'. Point c then deflects $30/18 = 1.667$ units (in proportion to the distance from O). The rotation along aa' is easily found as $1/4.2 = 1.667/7.0$. Yield hinge aa' absorbs energy of

$$E_1 = 4000 \times 12 \times 1/4.2 = 11{,}430$$

The rotation along bb' must represent the rotation vector parallel to bb'. The x and y components of this rotation vector can be found as $1/4.8$ and $1.25/12$, respectively, where 1.25 represents the proportional deflection of point d. The vector sum $\sqrt{(1/4.8)^2 + (1.25/12)^2} = 1/4.293$ is the rotation along the bb' axis.

$$E_2 = 5000 \times 13.42 \times 1/4.293 = 15{,}630$$

The length of yield line cc' is 12.32 ft. The energy absorbed can be determined using the length of the yield line projected on the axis of rotation times the degree of rotation. The length of the cc' yield line projected on the aa' axis is 12 ft. The length of the cc' yield line projected on the bb' axis is $12.32 \cos \beta = 11.98$ ft. The energy absorbed is

$$E_3 = 3000 \times 12 \times 1/4.2 + 3000 \times 11.98 \times 1/4.293$$
$$= 8570 + 8370 = 16{,}940$$

The total energy absorbed is $11{,}430 + 15{,}630 + 16{,}940 = 44{,}000$.

The same load triangles will be used as before. Triangles $bb'c$ at its centroid deflects $\frac{1}{3}(1.667) = 0.555$.

$$E_5 = (12w \times 8/2)0.555 = 26.7w$$

Triangle $cc'b$, $E_6 = (12w \times 4.8/2)(4.27/4.8)1.0 = 25.6w$

Triangle $aa'c$, $E_7 = (12w \times 4.2/2)\frac{1}{3} = 8.4w$

Triangle acc', $E_8 = (12 \times 7.0/2)(3.73/4.20)1.0 = 37.3w$

The total energy available from loads is $26.7w + 25.6w + 8.4w + 37.3w = 98.0w$.

$$98.0w = 44,000$$
$$w = 449 \text{ psf}$$

An ultimate load of 449 psf is a good answer. The reliability of this answer is judged more on the basis of the equilibrium calculation for parts A and B than on the energy equation. The energy equation always gives loads at least equal to and generally greater than the true ultimate load, that is, *errors are always on the unsafe side*.

To indicate the effect of a small error in locating the center yield line, the energy calculation is repeated for $x = 4.5$ ft instead of the 4.2 ft used previously. (The equilibrium calculation indicated that the true value was *less* than 4.2 ft) The dimensions are shown in Fig. 13.9.

$$\cos(\gamma + \beta) = 12/13.42 = 0.895$$

Energy absorbed:

$$\beta = \tan^{-1}(15/30) - \tan^{-1}(7.5/30) = 14°00'$$
$$E_1 = 4000 \times 12 \times 1/4.5 = 10,667$$
$$E_2 = 5000 \times 13.42 \times 1/4.025 = 16,671$$
$$E_3 = 3000 \times 12 \times 1/4.5 + 3000 \times 12.35 \cos\beta \times 1/4.025$$
$$= 8000 + 8932 = 16,932$$

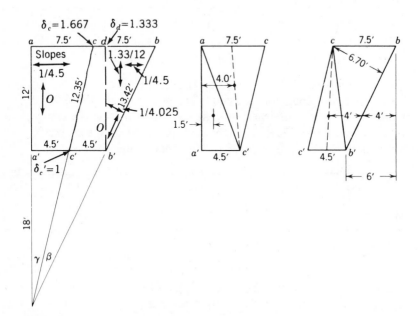

Figure 13.9 Dimensions for another trial solution of slab of Fig. 13.8.

The total energy absorbed is 10,667 + 16,671 + 16,932 = 44,270.

Energy from loads:

$$bb'c: \ 0.5 \times 12w \times 7.5 \times \tfrac{1}{3} \times 1.667 \qquad = 25.0w$$
$$cc'b: \ 0.5 \times 12w \times 4.5 \times (4/4.5) \times 1.00 = 24.0w$$
$$aa'c: \ 0.5 \times 12w \times 4.5 \times \tfrac{1}{3} \qquad\qquad = \ 9.0w$$
$$acc': \ 0.5 \times 12w \times 7.5 \times (4/4.5) \times 1.00 = \underline{40.0w}$$
$$98.0w$$

$$w = 44{,}270/98.0 = 452 \text{ psf}$$

This differs very little from the previous solution.

13.5 Square Panel, Ignoring Corner Effect

Find the ultimate uniform load that a continuous two-way slab 16-ft square can carry if the yield moment is 3000 ft-lb/ft for positive moment and 4000 ft-lb/ft for negative moment, equal in both directions.

Solution

The yield pattern is established by symmetry with triangular segments rotating about each edge, as shown in Fig. 13.10.

When triangle *abo* is considered, each diagonal carries a positive moment of 3000 ft-lb/ft, since the slab is isotropic with equal resistance at any angle. The component about *ab* of the moments on the diagonals is equal to the moment *m* times the projected length *ab*. Hence the equilibrium equation for this segment becomes:

$$16(4000 + 3000) - 0.5 \times 16 \times 8w \times \tfrac{8}{3} = 0$$
$$w = 112{,}000/171 = 655 \text{ psf}$$

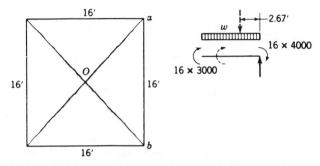

Figure 13.10 Two-way slab supported on all four sides.

Except for possible corner effects (Sec. 13.6), this is an exact solution and there is no need for trial solutions. The energy equation would serve just as well. For a unit deflection at O:

$$16(4000 + 3000)\tfrac{1}{8} = 0.5 \times 16 \times 8 \; w \times \tfrac{1}{3}$$
$$14,000 = 21.3 \; w$$
$$w = 655 \text{ psf}$$

13.6 Corner Effects

A simply supported square slab at a corner may not follow the simple yield line pattern used in Sec. 13.5. If the corners are not fastened down, they will rise off the supports as the slab is loaded, and the diagonal yield line will split or divide into two branches to form a Y, as in Fig. 13.11a. An additional corner segment forms with yield moments on only two faces and with support only on the two points where the yield lines cross the boundary. This condition is called a corner pivot, because the segment pivots about these two points.

If the corners are held down, but are not specially reinforced, similar yield lines form but in addition a corner crack opens along the pivot line as in Fig. 13.11b. If the corner is specially reinforced for negative moment, this adds a yield moment m' across this boundary of the triangular segment that increases its capacity and moves the junction of the Y farther from the corner. With a large enough m', the triangular segment fails to form and the simple diagonal yield line into the corner is correct without modification.

In continuous slabs, the corners have top steel that provides an m' for the corner segment. Whether the Y-pattern or the straight diagonal yield line forms depends on the amount of this reinforcement.

Consider the free body formed by one of these corner segments, as in Fig. 13.11c. The positive moment m along sides ao and ob adds vectorially to give a

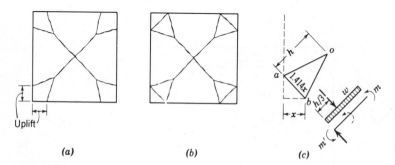

Figure 13.11 Corner segments. (a) When corners are not fastened down. (b) When corners are held down, but not reinforced for negative moment. (c) Equilibrium of corner segment.

positive moment m on the width ab, when m is the same in each direction. If no knot force acts at the apex of segment,

$$\Sigma M_{ab} = -1.414xm' - 1.414xm + 0.5(1.414xhwh)/3 = 0$$
$$wh^2/6 = m + m'$$
$$h = \sqrt{6(m + m')/w} \qquad (13.2)$$

The length h is not dependent on x or the width of the segment and increases with the ratio of the sum of positive and negative yield moments to w. If, in a given slab, the moment-load ratio establishes a distance h much less than the length of the diagonal along which it lies, this corner pattern will control; that is, it reduces the slab capacity. If the calculated h is large enough to push the Y-intersection beyond the limits of its particular diagonal, it means that the corner segment does not form. Intermediate values of h indicate that the corner element controls but has a smaller effect on the ultimate load or required moment strength of the slab.

With a simple supported square slab of side dimension a the ultimate moment is $wa^2/24$ if no corner segment forms. This is the true condition with the corner held down and with m' equal to m. With $m' = 0$ or with the corners free, the corner pivots produce corner segments that increase the ultimate moment to $wa^2/22$. In the square panel the maximum effect of the corner segments is thus slightly less than 9%.

13.7　Circular Segments or Fans

Under a concentrated load the negative moment yield line tends to form somewhat in a circle around the load with positive moment yield lines as spokes or radial lines, as shown in Fig. 13.2d. If a segment of this circular area is isolated as in Fig. 13.2e, the total load P at O results in a load $(\alpha/2\pi)P$ to be carried by the particular segment from O to its circular boundary. The segment will be small enough to be considered a triangle, which is similar to the corner segment of Fig. 13.11c, except for the loading. Then ΣM about the (circular) chord becomes:

$$(m + m')r\alpha - (\alpha/2\pi)Pr = 0$$
$$P = 2\pi(m + m')$$

It is interesting that the load P is thus not a function of r and the slab is equally subject to failure at any surrounding circle. This is not quite true when the load is applied over a finite area and the weight of the slab is considered, but it points out the need for both positive and negative moment resistance over the full area of a slab subjected to heavy concentrated loads.

Generally, the segments tend to have large radii. Consideration of the equilibrium equation indicates that a radius might vary from segment to segment with only minor changes in the strength, which results in a form resembling that of Fig. 13.15g. When a slab has a free edge nearby, as in Fig. 13.2f, it is weaker on the triangle *abO* because it has no *m'* and the circle will develop such as to make this triangle large. If a second free edge is on the other side of the load as shown dotted, a similar triangle also develops on that side instead of the segments shown. On the other hand, if the edges are restrained, providing an *m'*, the failure might be similar to Fig. 13.15*h*, although the corner segments might be more like Fig. 13.12*e*.

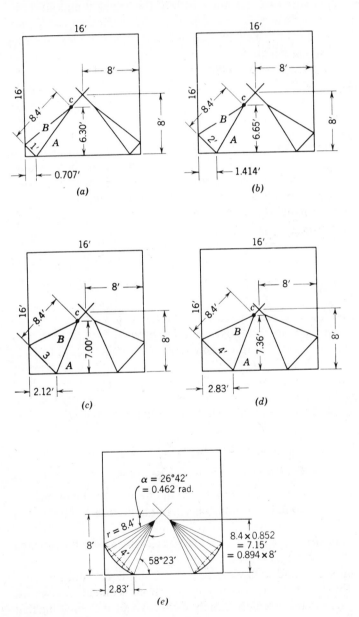

Figure 13.12 Corner segment trials.

Wood[7] points out the interesting case of a simple cantilever slab with a concentrated load in the corner. The failure section is more probable as a segmental fan plus edge triangles as at A in Fig. 13.15i. These triangles indicate the need for m' resistance in both directions and without reduction near the corner, because any radius gives the same resistance. Even without the fan concept, as at B, a similar conclusion follows for negative moment (alone) because the moment zP must be carried by a strip $2z$ wide. The fan, however, indicates the need for positive moment steel.

13.8 Square Panel, Considering Corner Effects

The effect of the slab corners on the ultimate moment of the square slab of Sec. 13.5 will now be investigated by basic principles.*

In Sec. 13.5 w was found to be 655 psf, and this should not be seriously changed, say, not much below 600 psf based on the 9% correction of Sec. 13.6. With this assumed w, summation of moments for the corner segment gives:

$$m' + m = wh^2/6$$
$$4000 + 3000 = 600h^2/6$$
$$h = \sqrt{7000/100} = 8.35 \text{ ft}$$

The full diagonal length is $1.414 \times \frac{16}{2} = 11.3$ ft. Hence it is probable that the corner segment does form. Because the edge does not have $M = 0$, M_t edge shears are not necessary in this example.

Try the failure pattern of Fig. 13.12a, using the energy method with a center deflection of one unit. Because the pattern repeats, only one set of areas A and B is included.

The deflection of point c is $6.3/8 = 0.787$. Energy input from the load is:

$$E_A = 0.5 \times 16w \times 8 \times \tfrac{1}{3} - 2 \times 0.5 \times 0.707w \times 6.30 \times 0.787/3 = 20.1w$$
$$E_B = 0.5 \times 1w \times 8.4 \times 0.787/3 = 1.10w$$

Energy absorbed in hinges:

$$E_A = (4000 + 3000)(16 - 2 \times 0.707)\tfrac{1}{8} = 12,760$$
$$E_B = (4000 + 3000) \times 1 \times 0.787/8.4 = 657$$
$$20.1w + 1.10w = 12,760 + 657$$
$$w = 13,420/21.2 = 633 \text{ psf} < 655 \text{ psf of Sec. 13.5}$$

This comparison proves that the corner segment does form.

A larger corner segment may drop the ultimate w lower. Try the increased

---

* Reference 3 develops three simultaneous equations for the square slab case that establish the corner segment dimensions algebraically.

segment of Fig. 13.12b. The deflection of c is $6.65/8 = 0.831$. As for the previous trial,

$$18.7w + 2.34w = 11,550 + 1390$$
$$w = 12,940/21.04 = 615 \text{ psf} < 633$$

Because this governs over the preceding calculation; further trials were made for the conditions of Fig. 13.12c and 13.12d that gave 607 psf and 605 psf respectively.

The corner triangles might be replaced by a small fan as shown in Fig. 13.12e. Then for the quarter panel the energy imput is:

$$E_A = 0.5 \times 16w \times 8 \times \tfrac{1}{3} - 2 \times 0.5 \times 2.83w \times 7.15 \times 0.894/3 = 15.30w$$
$$E_B = (w \times 0.462 \times 8.4^2/6)0.894 \qquad\qquad\qquad\qquad = 4.85w$$

Energy absorbed in hinges:

$$E_A = (4000 + 3000)(16 - 2 \times 2.83)\tfrac{1}{8} = 9000$$
$$E_B = (m + m')0.894\alpha = (4000 + 3000)0.894 \times 0.462 = 2880$$
$$15.30w + 4.85w = 9000 + 2880$$
$$w = 11,880/20.1 = 590 \text{ psf}$$

This answer is another 2% lower and means w is probably slightly under 590 psf. The original h calculation is still about right. All the possibilities are not exhausted, but when one considers the basic approximation shown in Fig. 13.1, great mathematical refinement seems out of place.

13.9　Rectangular Slabs

Investigate the ultimate load capacity of a 12-ft × 20-ft slab continuous on all edges that has a yield moment of 3000 ft-lb/ft for positive moment and 4000 ft-lb/ft for negative moment, both uniform in each direction.

Solution

Symmetry dictates a yield line at the middle, parallel to the long side, that merges with corner diagonals symmetrical about unknown points O and O' as shown in Fig. 13.13a. The corner effect with the Y-form on the corner diagonals is also possible, but this is treated as a later modification of the simple pattern.

The first trial dimensions shown in Fig. 13.13a are considered first in terms of the equilibrium of segments A and B.

Segment A:

$$-12(4000 + 3000) + 0.5 \times 12w \times 8 \times 2.67 = 0$$
$$w_A = 84,000/128 = 655 \text{ psf}$$

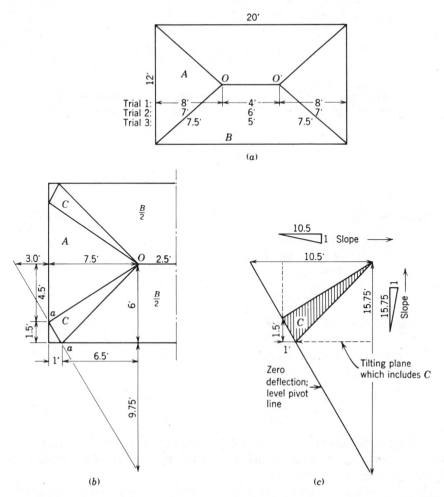

Figure 13.13 Rectangular slab analysis. (*a*) Neglecting corner segments. (*b*) With trial corner segments. (*c*) The rotation of corner segment.

Segment *B*:

$$-20(4000 + 3000) + 4w \times 6 \times 3 + 2 \times 0.5 \times 8w \times 6 \times 2 = 0$$
$$w_B = 140{,}000/168 = 834 \text{ psf} \gg w_A$$

Try a smaller *A* segment as noted in Fig. 13.13*a* for the second trial. These dimensions lead to $w_A = 858$ psf and $w_B = 730$ psf. A third trial with point O at 7.5 ft from the end gives $w_A = 745$ psf and $w_B = 775$ psf. This answer is reasonably close and will be checked by the energy relationship, using half of the panel for convenience and a unit deflection along *OO'*.

$$A: 0.5 \times 12w \times 7.5 \times \tfrac{1}{3} \qquad = 15w$$
$$B: 5w \times 6 \times \tfrac{1}{2} \qquad\qquad\quad = 15w$$
$$+2 \times 0.5 \times 7.5w \times 6 \times \tfrac{1}{3} = \underline{15w}$$
$$45w$$

Alternatively, visualize the total slab deflections as a volume somewhat similar to an inverted pyramid. This volume times w measures the energy input.

$$\text{7.5' end as half pyramid: } 7.5 \times 12w \times \tfrac{1}{3} = 30w$$
$$\text{2.5' center as wedge: } \quad 2.5 \times 12w \times \tfrac{1}{2} = \underline{15w}$$
$$45w$$

Energy absorbed:

$$A: 12(4000 + 3000)1/7.5 = 11{,}200$$
$$B: 20(4000 + 3000)\tfrac{1}{6} \quad = \underline{23{,}300}$$
$$34{,}500$$
$$w = 34{,}500/45 = 767 \text{ psf}$$

For the location of O the third trial dimensions are assumed correct enough.

The corner segment situation now is investigated. For equilibrium of the corner segment,

$$-4000 - 3000 + 0.5wh^2/3 = 0, \qquad h = \sqrt{42{,}000/w}$$
$$\text{If } w = 750 \text{ psf, } h = 7.48 \text{ ft}$$

This answer compares with the diagonal length of $\sqrt{7.5^2 + 6.0^2} = 9.60$ ft. It appears that there will be a significant corner segment probably extending most of the way to point O.

The corner segment shown in Fig. 13.13b is analyzed by the energy approach, using the deflection at O as unity. At aa the x- and y-components of the moments and rotations, as diagrammed in Fig. 13.13c, are used for calculating the energy absorbed by the hinges. Only half the panel (one segment each of A, B, and two of C) is used.

Energy input by loads:

$$B: 2 \times 1 \times 2.5w \times 6 \times \tfrac{1}{2} \qquad\qquad = 15.0w$$
$$+2 \times 0.5 \times 6.5w \times 6 \times \tfrac{1}{3} \qquad = 13.0w$$
$$A: 0.5 \times 9 \times 7.5w \times \tfrac{1}{3} \qquad\qquad = 11.25w$$
$$C: 2 \times 0.5 \times 1w \times 6.0 \qquad = \quad 6.0w$$
$$+2 \times 0.5 \times 1.5w \times 7.5 \quad = \quad 11.25w$$
$$-2 \times 0.5 \times 1w \times 1.5 \qquad = \underline{-1.5w}$$
$$15.75w \times \tfrac{1}{3} = \underline{5.25w}$$
$$44.50w$$

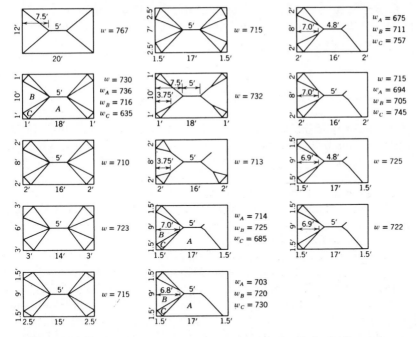

Figure 13.14 Summary of calculations indicating the limited effect of corner segments.

Somewhat simpler, energy input from the deflection volume:

7.5' end as half pyramid:* $(7.5 \times 12 - 2 \times 0.5 \times 1 \times 1.5)w \times \frac{1}{3} = 29.5w$

2.5' center as wedge: $(2.5 \times 12)w \times \frac{1}{2}$ $= \underline{15.0w}$

$44.5w$

Energy absorbed by hinges:

$$
\begin{aligned}
B: 18(4000 + 3000)\tfrac{1}{6} &&= 21{,}000 \\
A: 9(4000 + 3000)1/7.5 &&= 8{,}400 \\
C: 2 \times 1(4000 + 3000)1/15.75 &&= 890 \\
+2 \times 1.5(4000 + 3000)1.0/10.5 &&= \underline{2{,}000} \\
&& 32{,}290
\end{aligned}
$$

$$w = 32{,}290/44.5 = 726 \text{ psf} < 767$$

A number of variations were computed by the energy method, partly to be certain the worst case was found and partly to study the rather minor differences in w that resulted. The results of these calculations are shown just to the right of each illustration in Fig. 13.14. Where several values of

---

* Where the corner elements go to a common intersection, the net area of the base as used here is convenient. Where the corner elements stop short, the "gross" pyramid for the total height less small corner triangular pyramids of lesser height is convenient as a concept.

w are shown, they came from analyses of the separate segments. It appears that an ultimate load of about 710 psf is correct, possibly 690 or 700 psf with corner fans.

13.10 Influence of Strength of Edge Beams

Tests of slabs in England indicate that the failure pattern of Figs. 13.10 or 13.11 occurs with stiff supporting beams but the pattern can be altered by varying the size of beams. If the beams in one direction are very flexible, the slab fails almost as does a one-way slab and the yield line lies directly across the middle of the span and continues through the light beams. Then the strengths of the beams and the slab add in the evaluation of the m and m' moments. Obviously, at some particular beam stiffness, it is a matter of chance whether the diagonal failure or the midspan failure line develops.

In tests of multiple panels of two-way slabs at the University of Illinois[10] a positive moment yield line developed near midspan in the exterior panels parallel to the outside edges and the failure extended across the interior and spandrel beams that framed perpendicular to the outside edge. In such a case slab and beam strengths are additive in establishing the maximum load capacity. However, it should be pointed out that, although the calculated failure load for this mode of failure closely matches the actual failure load, the structure did not fail in the mode the yield line method predicted. Failure of the individual slabs was indicated by theory at a 25% lower load.

The example problems in this chapter assume that beams furnish stiff supports of such strength that the slab would fail first.

13.11 Other Yield Line Analysis

A few yield line patterns for other cases may be helpful as examples. Figure 13.15a shows a triangular slab simply supported and carrying uniform load. Johansen[4] shows for isotropic conditions that the yield lines intersect at the center of the inscribed circle and result in $m = wr^2/6$, where r is the radius of the circle. He extends this case to show the same moment in any polygon shape that circumscribes a circle, as in Fig. 13.15b. It is assumed that any of these cases might need to be investigated for a possible corner pivot or segment. Johansen also shows free-edge slabs as shown in Fig. 13.15c,d,e.

Hognestad[3] shows the slab of Fig. 13.15f. He also analyzes a radial pattern that develops under concentrated loads at times, (see Fig. 13.15g), as already discussed in Sec. 13.7.

With a single concentrated load in a rectangular panel Chamecki[6] shows eight triangular segments radiating from the load P whether P is centered or off center from both axes as in Fig. 13.15h. He develops equations to locate all key dimensions for the simple support case when the corners are anchored down

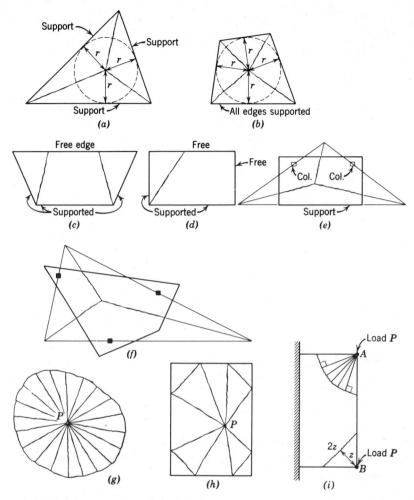

Figure 13.15 Yield line patterns. (*a*) Triangular slab supported on all sides. (*b*) Polygonal slab circumscribed around a circle, all edges supported. (*c*) Slab supported on three of four sides. (*d*) Slab supported only on adjacent sides. (*e*) Slab supported on one side and two columns. [(*a*) to (*e*) adapted from Reference 4.] (*f*) Slab supported on three columns. (*g*) Yield pattern around a concentrated load. [(*f*) and (*g*) adapted from Reference 3] (*h*) Yield pattern from an unsymmetrical concentrated load on a rectangular slab. (Adapted from Reference 6.) (*i*) Corner load on cantilever slab.

and when they are free to rise. He also shows the condition necessary to eliminate the corner segments.

Elstner and Hognestad[9] show the yield patterns of Fig. 13.16 for slabs carrying a center load in the form of a column stub cast monolithically with the slab. Figure 13.16*a* shows a slab simply supported and corners free; Fig. 13.16*b* is the same except that an eccentric column load is considered a line load. The

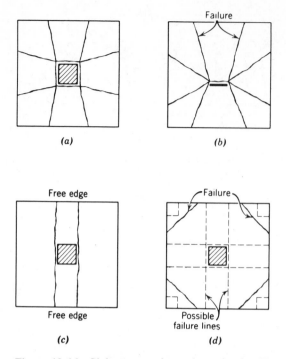

Figure 13.16 Slabs supporting column loads. (From Reference 9.) (*a*) Simply supported on all edges, corners free. (*b*) Eccentric column load treated as a line load, simply supported on all edges, corner free. (*c*) Simply supported on two opposite edges. (*d*) Supported at corners only.

case of simple supports on two opposite sides only is shown in Fig. 13.16*c* and simple support on four corners only in Fig. 13.16*d*; in the latter case the dashed lines represent an alternate yield line pattern.

13.12   Slabs with Openings

One advantage of the yield line method is that slabs with openings can be analyzed. The designer must, however, be alert to limitations on the effective usage of the equilibrium method.

When two positive moment yield lines meet at the corner of an opening, the knot forces are *not* zero. In fact, shearing forces often will be transferred from one element to another.

For example, the load of 668 psf and the yield lines shown in Fig. 13.17*a* are calculated as the governing case from the energy approach using $m = 3000$ and $m' = 4000$ ft-lb/ft and neglecting corner effects. The equilibrium method leads to the idea that the height of triangle A should be 7.8 ft on the basis of Eq. 13.2 for h. However, a net downward knot force at the apex makes a shorter height of triangle the correct one. The student can use equilibrium of each segment to

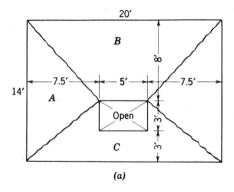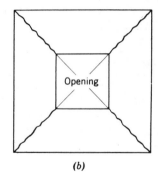

Figure 13.17 Yield line patterns in slabs containing openings. (*a*) Large knot forces present. (*b*) No knot forces.

find these knot forces, after establishing w. For this slab anyone visualizing the deflection of segment B would quickly see from the way a slab deflects that some support for B is required at the two corners of the opening. This is what causes the knot forces there.

A second problem also involves the equilibrium method. This discontinuity at the 90° corner of the opening prevents any effective use of the ordinary edge forces m_t. It should be obvious that a single yield line going symmetrically into the corner, as in Fig. 13.17*b*, has two obtuse angles and that the nominal upward forces in each corner could not represent real internal forces because they would not be "equal and *opposite*". Symmetry considerations indicate that this must be the correct yield line pattern (ignoring corner effects) and that there can be no knot forces at the opening. Thus one must conclude that knot forces around an opening may be absent or may be present (as in Fig. 13.17*a*), but are not likely to be the nominal magnitude m_t. Thus the energy method is essentially mandatory for such cases.

13.13 Reinforcement at Variable Spacing and/or with Bars Cut Off

Taylor et al.[13] tested a series of simply supported square slabs of several thicknesses but with the chief variable the arrangement of reinforcement. Specimens included uniform spacing as the yield line theory uses most conveniently, variable spacing strips as most commonly used in practice, and 50% to 70% of the bars cut off to the moment requirements.

Variable spacing showed only a minimal advantage, if any, because the larger area of reinforcement near the middle decreased the internal lever arm. Variable spacing gave a slightly stiffer slab, with the cracks appearing first near the corners rather than near the center; it did not increase strength.

Some economy of material results from stopping some bars short of the supports, and ultimate design loads were still exceeded. However, the excess

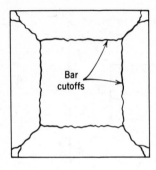

Figure 13.18 Test yield line pattern on simply supported slabs having bars cut off according to moment.

load capacity was less than when bars were full length and these slabs deteriorated rapidly after forming a center square of positive moment yield lines at the points where bars were cut off, as in Fig. 13.18.

13.14 Flat Plate Floors

Flat plate floors directly supported on columns are used for many buildings designed for light or moderate live load. The Code design is covered in Chapter 15. However, an equilibrium analysis, using in part yield line concepts, is desirable for such floors carried by irregularly spaced columns. Reference 14 covers this approach and Reference 15 relates it to the strip method of Chapter 14.

Selected References

1. K. W. Johansen, *Pladeformler,* Polyteknish Forening, Copenhagen, 2nd ed., 1949.

2. K. W. Johansen, *Pladeformler; Formelsamling,* Polyteknish Forening, Copenhagen, 2nd ed., 1954.

3. E. Hognestad, ''Yield-Line Theory for the Ultimate Flexural Strength of Reinforced Concrete Slabs,'' *ACI Jour. 24,* No. 7, Mar. 1953; *Proc.,* 49, p. 637.

4. K. W. Johansen, ''Yield-Line Theory,'' English translation, Cement and Concrete Assn., London, 1962.

5. F. E. Thomas, ''Load Factor Methods of Designing Reinforced Concrete,'' *Reinf. Conc. Review,* 3, No. 8, 1955, pp. 540, 544.

6. S. Chamecki, *Calculo No Regime de Ruptura, Das Lajes de Concreto Armadas em Cruz,* Curitiba, Parana, Brazil, 1948, 107 pages.

7. R. H. Wood, *Plastic and Elastic Design of Slabs and Plates,* Ronald Press, New York, 1961.

8. L. L. Jones, *Ultimate Load Analysis of Reinforced Concrete Structures,* Interscience Publishers, New York, 1962.

9. R. C. Elstner and E. Hognestad, "Shearing Strength of Reinforced Concrete Slabs, Appendix 1," *ACI Jour., 28,* No. 1, July 1956; *Proc., 53,* p. 55.

10. W. L. Gamble, M. A. Sozen, and C. P. Siess, "Measured and Theoretical Bending Moments in Reinforced Concrete Floor Slabs," Univ. of Illinois *Civil Engineering Studies Structural Research Series* No. 246, June 1962.

11. T. van Langendonck, *Charneiras Plásticas em Lajes de Edifícios,* Associacão Brasileira de Cimento Portland, São Paulo, Brazil, 1966, 81 pages.

12. R. Lenschow and M. A. Sozen, "A Yield Criterion for Reinforced Concrete Slabs," *ACI Jour., 64,* No. 5, May 1967, p. 266.

13. R. Taylor, D. R. H. Maher, and B. Hayes, "Effect of the Arrangement of Reinforcement on the Behavior of Reinforced Concrete Slabs," *Magazine of Concrete Research,* 18, No. 55, June 1966, p. 85.

14. F. P. Wiesinger, "Design of Flat Plates with Irregular Column Layout," *ACI Jour., 70,* No. 2, Feb. 1973, p. 117.

15. F. P. Wiesinger, "Yield Line Method—Strip Method—Segment Equilibrium Method," *ASCE Preprint 2502,* ASCE Natl. Structural Convention, April, 1975.

Problems

PROB. 13.1. In Figs. 13.2*a* and 13.7 calculate the ultimate load by yield-line procedures if:

(*a*) The positive moment capacity m_c is increased to $+3800$ ft-lb/ft, m_a and m_b remaining unchanged at -4000 and -5000 ft-lb/ft respectively.

(*b*) The negative moment capacity m_a is decreased to -2800 ft-lb/ft, m_b and m_c remaining at -5000 and $+3000$ ft-lb/ft respectively.

PROB. 13.2. Recalculate the ultimate load for the slab of Sec. 13.4 and Fig. 13.8 if the right support has only a 4-ft skew, that is, if b' in Fig. 13.8*a* is moved 2 ft to the right.

PROB. 13.3. If $m = +2800$ ft-lb/ft and $m' = -3500$ ft-lb/ft, find the ultimate load that can be carried by a 20-ft square slab continuous on all sides. Consider first without corner effect and then establish effect of corner segments.

PROB. 13.4. Investigate the ultimate load capacity of a 15-ft × 20-ft slab continuous on all sides and having a yield moment of 3000 ft-lb/ft for positive moment and 4500 ft-lb/ft for negative moment.

PROB. 13.5. Assume the rectangular slab of Sec. 13.9 has a 3-ft square hole in the middle of the span (centered each way). Find the ultimate uniform load by yield-line method: (*a*) neglecting corner effect; (*b*) with corner effects.

PROB. 13.6. Assume the rectangular slab of Sec. 13.9 has no support at the upper 20-ft side (Fig. 13.13). Find the ultimate uniform load by yield-line method.

14

STRIP METHOD FOR SLAB DESIGN

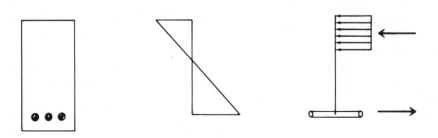

14.1 Development of Strip Method

A method for *designing* slabs, published by Hillerborg in 1956 and 1959 in Swedish,* has received much attention because of studies and tests at the Building Research Station in England. Wood and Armer[1-3] in 1968 critically analyzed the method and their tests of typical slabs designed by this method. They found (mathematically) that a design made by the strip method and reinforced exactly according to the moments found (without averaging across a band, but with reduction of steel as moment decreased) was an *exact* solution rather than just a lower-bound solution of the problem.† Such a slab did develop the load capacity introduced into the design.

The strip method of design gives the designer a wide freedom of choice in the design approach. Hence many different solutions for a given slab design are possible, but not all are of equal economy. Wood‡ points out that a design using moments approaching those from elastic analysis is efficient and preferable.

The strip method is the simplest one for slabs on simple supports, but continuity can be handled on a basis similar to limit design. The suitability of the method for slabs with openings is a strong point in its favor.§ The most difficult slabs for this method are slabs supported on columns (Sec. 14.8). For such a case, Hillerborg developed what Crawford[4] calls the advanced strip method, that is, using a rectangular element carrying load in two directions to a support

---

* Now in English; see Reference 6.
† Note that the yield line method is an *upper* bound procedure.
‡ Wood's name will be used in this chapter for easy reference to the joint work of Wood and Armer.
§ The authors accept the strip method with some enthusiasm because it formalizes a very approximate method that designers have been using for many years, that is, designing by their "feel" for the way the load was most apt to be transferred to the supports.

at one corner of the element. Although the mathematical basis for this type of element could not be proved by Wood, a test result for a slab design using this method was satisfactory. Hillerborg states that the advanced strip method leads to a more economical design with a simple reinforcement pattern than does the substitute type of element Wood suggests. The discussion here will stay with the simpler strips that Wood endorsed. (The authors lean heavily on work by Wood in this area.)

Wood calls attention[1] to the normal use of A_s varying in accord with the strip moments, in contrast to the complexities inherent in a yield line analysis when m (or A_s) is not constant. If the reinforcement is cut off where it is not needed, the slab designed by the strip method does not fail by sharply developed yield lines, but by "yield in nearly all directions at failure, . . . somewhat like a plastic hammock." Hence yield line concepts are only marginally useful with this method. The yield line method deals with "rigid plate" rotations; the strip method tends to eliminate rigid plate failure at ultimate.

14.2 Theoretical Basis

The equilibrium equation for slabs is

$$\frac{\delta^2 M_x}{\delta x^2} + \frac{\delta^2 M_y}{\delta y^2} - 2\frac{\delta^2 M_{xy}}{\delta x \delta y} = -w$$

where the bending moments, M_x and M_y, and the twist moment M_{xy} follow Timoshenko's notation and w is the load per unit area on the slab. Hillerborg designs the slab to make M_{xy} unnecessary, that is, he assumes $M_{xy} = 0$ and then apportions the load to $\delta^2 M_x/\delta x^2$ and $\delta^2 M_y/\delta y^2$, usually at a particular spot wholly to one or to the other. This particular apportionment is more of a convenience than a necessity, however.

Loads in a particular area are assigned to particular slab strips, as the next section will illustrate, and continuity of the resulting moments and shears must be carefully maintained. Apparent discontinuity in torque or deflection may be disregarded, but a discontinuity in moment or shear is not permitted. Both elastic and plastic analysis concepts are permissible in evaluating moments on strips, but both Wood and Hillerborg note that under service load the slab behavior is more nearly in the elastic range. Hence elastic concepts are valid and acceptable even though the uses suggested later ignore relative deflections with apparent unconcern.

14.3 Simply Supported Rectangular Slab

A rectangular slab is adequately represented by the simplified concept of a grid of strips in the x- and y-directions, that is, 90° apart.

Consider that boundaries, called lines of stress discontinuity (or discontinuity lines), are set up as indicated in Fig. 14.1a, creating an area 1 and two areas

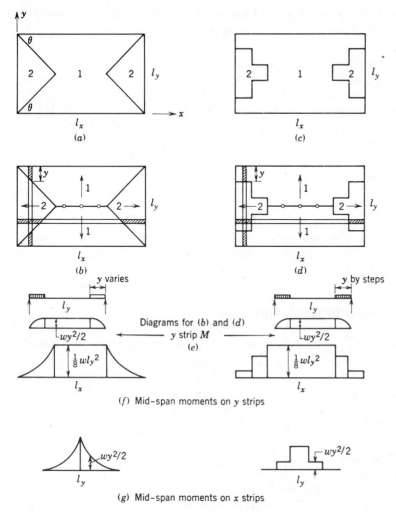

Figure 14.1 Layouts and details for two possible strip designs of rectangular slab simply supported.

2. These discontinuity lines indicate the designer's decision to carry all the load in areas 2 in the x-direction on x-strips and all load in area 1 in the y-direction on y-strips. The discontinuity lines are *not* yield lines, and the designer is free to choose the angle θ. If the designer chooses 90°, he or she will design a one-way slab; the result will be adequate strength, early cracking along the y-supports, and an overall solution approaching the absurd.

Wood suggests that discontinuity lines might be taken as sketched in Fig. 14.1c to suit bands of reinforcement, because one is not restricted to the use of a single straight line in a quadrant in limit analysis.

As in any flexural member, a load anywhere on a strip produces a shear along the entire strip. However, it is convenient to add a zero shear line on the

x-axis as shown in Fig. 14.1b and 14.1d and to think of the y-strip as carrying all the load above this zero shear line to the upper slab support as indicated by the arrows.

In either layout the central y-strips are simple one-way slab strips under a uniform load or such other distribution of load as may exist. The y-strips running through an area 2 are unloaded in that area and loaded only in the two area 1 end portions, as indicated by the shaded areas. Likewise x-strips in Fig. 14.1b are all unloaded except near the supports, but this is not quite true in Fig. 14.1d. With square discontinuity lines the x-strips near the top and bottom are totally unloaded, and the load pattern changes by substantial steps where the strips enter area 2.

The moment in each strip defines the necessary steel and even indicates where some may be cut off. In Fig. 14.1b the required steel requires a variable spacing; in Fig. 14.1d bands of steel are indicated. For the variable spacing Hillerborg actually substituted a weighted average; Wood shows that the lower bound concept is then unsatisfied, although membrane action may bridge the gap.

14.4   Continuity in Rectangular Slabs

Because slab design by the strip method is a form of limit design, the ratio of negative to positive moment on a strip is not rigidly fixed. The use of moments approximating elastic analysis assures a satisfactory service load response, but this is not mandatory for strength. Wood suggests the assumption of a point of inflection (P.I.) for the middle y-strips in Fig. 14.2a at about 0.2ℓ_y from the support, which is close to the elastic condition. Nearly the same result is obtained using the negative moment as $-(1/12)w\ell_y^2$, but for skewed slabs Hillerborg and Wood use the initial selection of the P.I. locations as the key to the desirable slab strips (Sec. 14.7).

For y strips that pass through area 2 (Fig. 14.1b) and thus are loaded symmetrically over the y lengths only on their ends, the best assumption for the negative moment or P.I. is not so clear. Wood, following Hillerborg, breaks the P.I. trace and takes it diagonally to the corner, as shown at the top of Fig. 14.2a, which makes the distance from support to P.I. 0.4y. If one considers the simple span moment on these strips, as in Fig. 14.2b, the maximum simple beam moment is $wy^2/2$, with this moment constant across area 1. The use of the diagonal P.I. line leads to negative moments here that are substantially less (down 50% or more*) compared to the elastic analysis values for fixed ends, but this is permissible with limit design concepts.

---

* The values vary strip to strip and this is near that for the middle strip of the group. The elastic analysis for fixed ends shows that the P.I. distance from support becomes $y(1 - 0.82\sqrt{y/\ell_y})$ and the negative moment is $M_s(0.67 + b/y)/(1 + b/y)$, where M_s is the maximum simple span moment, y is the distance to the discontinuity line, and b is the remaining distance to midspan, that is, $b = (\ell_y/2) - y$. In these terms the same expressions can be applied to the x-strips, with distances x, a, and ℓ_x replacing y, b, and ℓ_y.

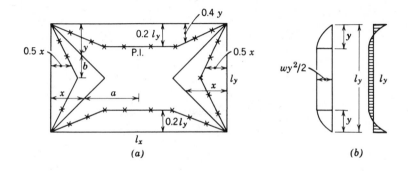

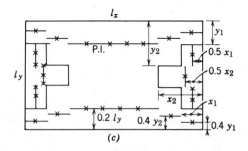

Figure 14.2 Points of inflection for continuity. (*a*) Typical layout. (*b*) *M* on *y* strip, simple span and continuous span. (*c*) Proposed points of inflection.

For the *x*-strips a similar treatment of P.I. lines at $0.4x$ from the support establishes negative moments relatively *much* less than elastic analysis values (in some cases down by 70%). Although the matter deserves further study in terms of the implied degree of hinging and relative costs, the authors (in the absence of specifications by Wood) recommend the use of the P.I. line at $0.5x$ as more realistic, Fig. 14.2*c*.

14.5   Design of Rectangular Slab

A 15-ft × 20-ft rectangular slab with restrained edges at beams is designed for a factored load *w* of 300 psf. The design moments are computed without the ϕ factor. Once the slab thickness is determined for moment or deflection (shear rarely controls a slab on beams), the required A_s calculated from $M/(\phi f_y z)$ takes the same distribution as the moments plotted in Fig. 14.3.

Discontinuity lines are arbitrarily chosen as shown in Fig. 14.3*a* and P.I. locations are selected as discussed previously.

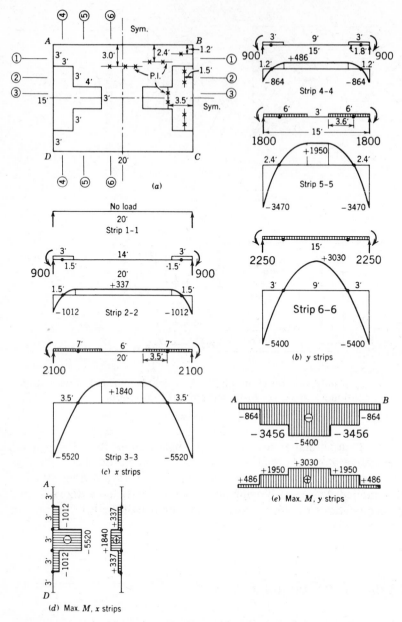

Figure 14.3 Design moments for a rectangular slab.

Moments will be calculated first for each y-strip. The loadings are sketched to the right in Fig. 14.3b. The positive moments are calculated for a simple span between P.I. points and the negative moments from the shear diagram areas between the span ends and the assumed P.I. points. The moments on the x-strips are sketched in Fig. 14.3c and on y-strips in 14.3b.

Strip 1-1   No loading

Strip 2-2   Pos $M = 300 \times 1.5^2/2 = +337$ lb-ft
$\qquad$ Neg $M = -\frac{1}{2}(900 + 450)(1.5) = -1012$ lb-ft

Strip 3-3   Pos $M = 300 \times 3.5^2/2 = +1840$ lb-ft
$\qquad$ Neg $M = -\frac{1}{2}(2100 + 1050)(3.5) = -5520$ lb-ft

Strip 4-4   Pos $M = 300 \times 1.8^2/2 = +486$ lb-ft
$\qquad$ Neg $M = -\frac{1}{2}(900 + 540)(1.2) - -864$ lb-ft

Strip 5-5   Pos $M = 300 \times 3.6^2/2 = +1950$ lb-ft
$\qquad$ Neg $M = -\frac{1}{2}(1800 + 1080)(2.4) = -3456$ lb-ft

Strip 6-6   Pos $M = (\frac{1}{8})300 \times 9^2 = +3030$ lb-ft
$\qquad$ Neg $M = -\frac{1}{2}(2250 + 1350)(3.0) = -5400$ lb-ft

These moments are summarized in Fig. 14.3d and 14.3e into diagrams of negative moment crossing the face of support and positive moment crossing midspan, which determine the amount and distribution of reinforcing steel. If this arrangement appears awkward or seems to have excessive local demands, the designer can modify the discontinuity lines so as to shift the requirement in the desired direction. The several moment diagrams sketched in Fig. 14.3b and 14.3c are useful for establishing bar cutoff points.

The minimum temperature steel may well control where computed moments are small, certainly in strip 1-1 close to the long edge beams.

The loadings on the beams supporting the slabs are most logically established from the strip reactions.

14.6   Design of a Rectangular Slab with Hole

The symmetrical slab of Fig. 14.4a, restrained at all edge beams, illustrates the rerouting of loads when strips are interrupted. Because slabs are normally considerably under-reinforced, it is possible to use certain strips near the opening (Fig. 14.4b) as small beams simply by increasing the local reinforcement. If the opening is so large that even extra slab steel is inadequate to care for the moment, a real beam is needed around one or more sides of the opening, quite probably spanning to the edge beams.

The assumed "beam strips" are shown dotted around the opening and the assumed discontinuity lines and P.I. lines are added in Fig. 14.4c. Only the P.I.

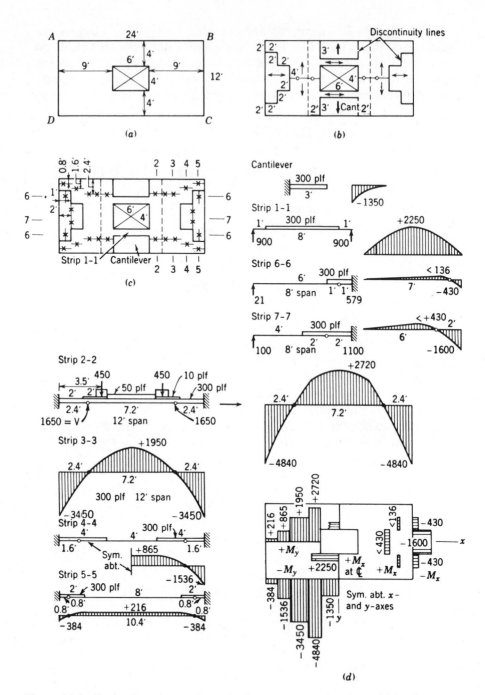

Figure 14.4 Design based on strip method. (*a*) Layout. (*b*) Discontinuity lines. (*c*) Inflection point lines. (*d*) Design moments.

line between the long side of the opening and the long support beam requires special discussion. This 4-ft long strip in the y-direction could be designed as a cantilever from DC, but this might be ignoring excessive deflection at the opening. Consider instead a discontinuity line at 3 ft from the support, which will terminate the cantilever and leave a 1-ft strip 1-1 alongside the opening in the x-direction. Strip 2-2 in the y-direction will pick up the reaction from strip 1-1. Assume this strip 2-2 to be 2-ft wide, because strips 1-1 alone would add 50% to the strip 2-2 load if only a 1-ft strip were used, and additional loads come from strips 6-6 and 7-7. Strip 7-7, because of the opening, is terminated on strip 2-2; it seems logical to treat strip 6-6 in the same manner.

The design moments are calculated for a total design (factored) load of 300 psf. The strip loadings and moment diagrams are shown in Fig. 14.4d, along with a final sketch that assembles the maximum negative and positive design moments on both x- and y-strips. The supporting calculations are as follows.

Cantilever strips near strip 1-1, 3-ft length
Neg $M = -300 \times 3^2/2 = -1350$ lb-ft
Strip 1-1 with simple span of 8 ft to center of strip 2-2
Pos $M = +3 \times 300 \times 4 - 300 \times 3^2/2 = +2250$ lb-ft
Reaction on strip 2-2 $= 3 \times 300 = 900$ lb
Strip 6-6 spanning 8 ft to center of strip 2-2
Reaction* on strip 2-2 $= 300 \times 1 \times 0.5/7 = 21$ lb
Pos $M < 21 \times 6.5 = 136$ lb-ft
Neg $M = -\frac{1}{2}(579 + 279)(1) = -430$ lb-ft
Strip 7-7 spanning 8 ft to center of strip 2-2
Reaction on strip 2-2 $= 300 \times 2 \times \frac{1}{6} = 100$ lb
Shear at P.I. $= 500$ lb
Pos $M < 100 \times 4.3 = 430$ lb-ft
Neg $M = -\frac{1}{2}(500 + 1100)(2) = -1600$ lb-ft
Strip 2-2 spanning 12 ft
The reactions from 1-1, 6-6, and 7-7 are shared between the two 1-ft wide unit strips making up the 2-ft width.
Shear at P.I. $= 300 \times 7.2/2 + 50 \times 2 + 10 \times 1.6 + 450 = 1650$ lb
Pos $M = 1650 \times 3.6 - 300 \times 3.6^2/2 - 50 \times 2 \times 1 - 10 \times 1.6 \times 2.8$
$- 450 \times 2.5 = +5940 - 1950 - 100 - 45 - 1125 = +2720$ lb-ft
Shear at support $= 300 \times 12/2 + 50 \times 2 + 10 \times 2 + 450 = 2370$ lb
Neg $M = -\frac{1}{2}(2370 + 1770)(2) - \frac{1}{2}(1770 + 1650)(0.4) = -4140 - 690$
$= -4830$ lb-ft

---

* From ΣM about P.I. using "simple span" as a freebody.

Strip 3-3 spanning 12 ft

Pos $M = (\frac{1}{8})300 \times 7.2^2 = +1950$ lb-ft

Neg $M = -300 \times 12^2 \times 1/8 + 1950 = -5400 + 1950 = -3450$ lb-ft

Strip 4-4 spanning 12 ft

Shear at P.I. $= 300 \times 2.4 = 720$ lb

Pos $M = 720 \times 2.4 - 300 \times 2.4^2/2 = 1730 - 865 = 865$ lb-ft

Neg $M = -\frac{1}{2}(1200 + 720)(1.6) = -1536$ lb-ft

Strip 5-5 spanning 12 ft

Shear at P.I. $= 300 \times 1.2 = 360$ lb

Pos $M = +360 \times 1.2 - 300 \times 1.2^2/2 = 432 - 216 = +216$ lb-ft

Neg $M = -\frac{1}{2}(600 + 360)(0.8) = -384$ lb-ft

The maximum moments assembled in Fig. 4.4*d* look reasonable in distribution and no modification of the strip arrangement is indicated. The maximum moments fix the slab depth and shear may be checked from the strip loads if it appears to possibly control. The reinforcement should be arranged in bands corresponding to the strips used; this calculation is simple. The short *y*-strip steel should be placed at the greater depth and for the *x*-strips *d* must be reduced by one bar diameter. Minimum temperature or spacing steel will control where the indicated moments are very small. The moment diagrams for the strips give the basic data for required length of bars, subject to the usual Code requirements for arbitrary extensions beyond the theoretical cutoff points.

Deflection *at service load* must be considered in checking serviceability. In any actual design the service load is available, and it should be on the safe side to use the strip service load moments with *EI* based on the cracked section. The more detailed calculations of Sec. 3.22 may be used to interpolate between cracked and uncracked sections. Although long, that method considers the usual intermediate slab condition and is less severe than the use of the cracked section. In fixing M_a the designer must be clear as to whether this is calculated per strip width or per unit width; then the designer must evaluate the expected accuracy as a result of uncertainty as to what is the best M_a to use.

14.7   Nonrectangular Slabs

The strip method is not limited to rectangular strips supported on beams on all four sides. A few layouts, largely those already in the literature, indicate the possibilities.

A free edge, as in Fig. 14.5*a*, can be considered a strong strip with extra reinforcement and can even pick up a limited amount of load from cross strips.

Where triangular slabs exist with a free edge opposite an acute angle, strips can span the short way between supported sides as in Fig. 14.5*b*. The same can be done if the supported edges are restrained, but Hillerborg suggests strips that change direction, as in Fig. 14.5*c*. With an assumed P.I. line, the midlength

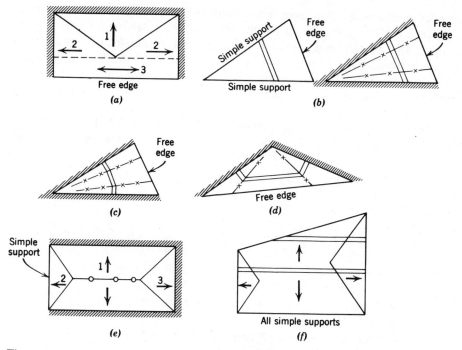

Figure 14.5 Varied types of supports. Hatched edges are restrained.

is like a simple span and the negative moment region acts like a cantilever to pick up the simple span reaction plus its own load. The obvious torsion implications are not a concern in this method. However, the cantilever and midstrip must have different widths to care for the geometry of the layout. A similar idea is shown in Fig. 14.5d.

When fixed and simply supported edges occur nearby, elastic concepts are helpful. In Fig. 14.5e, for example, area 2 should be smaller than area 3.

14.8   Layout Around Columns Supporting the Slab

The treatment of flat plates around the columns is a difficult problem when one uses the strip method. Hillerborg's 1959 publication dealt particularly with this problem and resulted in a recommended type-3 rectangular element carrying load in two directions from zero shear lines and delivering the reaction to one corner (the column). This concept is obviously much more involved than that of the strip method. Wood was not able to prove the complete mathematical basis for this element, but he reported the resulting reinforcement looked reasonable. Armer reported[2] a test that was quite satisfactory.

Wood[1] developed a substitute procedure using the type-3 rectangular element with computer-calculated tables leading to the moment coefficients. Armer designed one test specimen that substituted for the type-3 elements strong slab strips in the x- and y-directions to carry load from slab strips directly to the column. Test results on all of these variations proved satisfactory. Wood also has suggested a wide "beam band" over the columns that could be strengthened at the column by local (short) strong bands across the column, these apparently acting in flexure much like steel shearhead reinforcement in flat plates.

Reference 5 is an interesting approach to flat plates supported on columns at irregular spacings. Reference 6 is an excellent overall coverage of the strip method.

Selected References

1. R. H. Wood and G. S. T. Armer, "The Theory of the Strip Method for Design of Slabs," *Proc. Inst. of Civil Engineers,* 1968, Vol. 41 (October), p. 287.

2. G. S. T. Armer, "Ultimate Load Tests of Slabs Designed by the Strip Method," *Proc. Inst. of Civil Engineers,* 1968, Vol. 41 (October), p. 313.

3. G. S. T. Armer, "The Strip Method: A New Approach to the Design of Slabs," *Concrete,* Vol. 2, No. 9, Sept. 1968, p. 358.

4. R. L. Crawford, "Limit Design of Reinforced Concrete Slabs," *Proc. ASCE, Jour. Eng. Mech. Div.,* EM5 Oct. 1964, p. 321.

5. F. R. Wiesinger, "Design of Flat Plates with Irregular Column Layout," *ACI Jour.,* Proc. 70, Feb. 1973, p. 117.

6. A. Hillerborg, "Strip Method of Design," Viewpoint Publications, Cement and Concrete Association, Wexham Springs, Slough, SL3 6PL, England, 1975.

15

FLAT PLATES AND FLAT SLABS

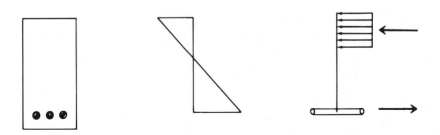

15.1   Scope of Chapter

Earlier chapters dealt with slabs primarily supported on beams and, in most cases, slab and beam design were separate steps. This chapter deals primarily with slabs that are totally supported directly on columns. Waffle slabs and slab band construction are included only for identification.

The ACI Code states that slab systems may be designed for gravity loads by either the empirically based direct design method of Code Sec. 13.6 or the more analysis-based equivalent frame method of Code Sec. 13.7. The direct design method is generally simpler to use than the equivalent frame method, but is restricted to very regular layouts and normal live-load-to-dead-load ratios as outlined in Sec. 15.5. Although the equivalent frame method may be used for any slab system, it is generally used only for those cases that do not fit the direct design method limits.

The direct design method of Code Sec. 13.6 was considerably simplified by the 1983 Code revision that removed the somewhat cumbersome earlier procedures for determining the distribution of total span moment in end spans and substituted a table of moment coefficients. This simplification allows direct determination of end-span moment distributions without the need for computation of equivalent column stiffnesses.

Chapter 15 and the initial portion of Chapter 16 are based primarily on the direct design method. The equivalent frame method is introduced in Sec. 16.4 with an example in Sec. 16.7. All other examples in Chapters 15 and 16 are based on the direct design method (empirical design). Slab systems subject to lateral loads require a special analysis and are discussed briefly in Sec. 16.10.

Chapter 15 describes the background and basic provisions of the direct design method. It gives direct design examples of flat plates, both interior and exterior panels, and an interior panel of a flat slab with drop panel and column capital. These designs are in accord with the slab design methods of Code

Chapter 13, but they appear more simplified than those of the next chapter because the combined beam-slab interaction is minimized in the absence of beams. Chapter 16 relates the general and more involved case of beam and slab interaction. It also deals with both the direct design method and the equivalent frame method approaches to that more complex behavior. The sequence in this chapter is:

15.1–15.4	Background and theory.
15.5–15.6	General Code requirements for direct design method.
15.7	Flat plate direct design—interior panel.
15.8–15.10	Special rules for columns and slab h
15.11	Exterior span rules.
15.12	Exterior panel—flat plate direct design.
15.13	Flat slab direct design with drop panel—interior panel.
15.14–15.17	Shear reinforcement; lift slabs; cantilevers from slab construction.

15.2   Flat Plates, Flat Slabs, and Related Types—Identification

A flat plate is a concrete slab reinforced in two directions so that it brings its loads directly to supporting columns. Most generally it is used for light loadings

Figure 15.1   Flat plate with shallow spandrel beam. The spandrel is often omitted to leave a totally flat ceiling. (Courtesy Portland Cement Assn.)

Figure 15.2 Typical flat slab construction with drop panel. (Courtesy Portland Cement Assn.)

and the spandrel beams shown in Fig. 15.1 are omitted. Beams may be used around stairs or other large openings in the slab.

The flat slab is an older type of slab developed primarily for heavy loadings and longer spans (Fig. 15.2) and typically using the flared column capital and often the thickened slab around the column (a drop panel). The waffle slab (a perpendicular joist-like system in effect) shown in Fig. 15.3 may be designed as a flat slab suitable for long spans. For heavy service live loads, that is, over 100 psf, flat slabs have long been recognized as the most economical construction technique.

The slab band type of floor thickens the slab into bands of greater depth, as in Fig. 15.4. Although the most usual pattern results in one-way transverse slabs of variable depth supported on wide shallow beams or bands in the longitudinal direction, some engineers use such thickened sections in end panels much like a drop panel around a flat slab column. Such construction partakes somewhat of the nature of flat slabs.

Figure 15.3 Waffle slab. (Courtesy Portland Cement Assn.)

Figure 15.4 Slab band construction. (Courtesy Portland Cement Assn.)

Two-way slabs supported on all sides by wide shallow beams also involve some flat slab action. When the slab proper occupies only the middle half of the panel, the Code recognized it as a paneled ceiling form of flat slab.

Flat plates, being thin members, are uneconomical of steel, but they are economical of formwork. Because formwork represents over half the cost of reinforced concrete, economy of formwork often means overall economy. Since the 1950s flat plate floors have proved economical in tall apartment house construction. Reduced story height resulting from the thin floor, the smooth ceiling, and the possibility of slightly shifting column locations to fit the room arrangements all are factors in the overall economy.

15.3   Historical Development

(a) Early Slabs as Empirical Construction   Flat slabs were built and sold many years before an adequate analysis was available.[3] The originator was an American, C. A. P. Turner. He used his intuition about how slabs would function and applied a proof load to the completed slab to satisfy the owner.

Numerous flat slab structures were load-tested from 1910 to 1920.* These slabs performed well under test loads. The difficulty in closely correlating test results with the moments that statics shows must be present is mentioned in subsection (c).

---

* In 1947 Professor J. Neils Thompson and the senior author ran such a test on a flat slab constructed about 1912. Although it cracked badly, its strength was surprising.

(b) The Statics of a Flat Slab   In 1914 J. R. Nichols[4] showed that statics required a total of positive and negative moments equal to

$$M_0 = \frac{1}{8} W\ell \left(1 - \frac{2}{3}\frac{c}{\ell}\right)^2$$

where W is the total uniform panel load, ℓ is the span, and c is the diameter of the column capital. This conclusion follows from an analysis of a half-panel.

Consider the interior panel of Fig. 15.5a loaded with a unit load w and surrounded with similar equally loaded panels. The straight boundaries of the slab are all lines of symmetry, indicating that they are free from shear and torsion. Hence all the shear and torsion must be carried around the curved corner sections that follow the column capital.

Likewise, if the slab is subdivided along the middle of the panel, this line is a line of zero shear and torsion. Thus the free body of Fig. 15.5b,c is subject to a downward load W_1 acting at the centroid of the loaded area, an equal upward shear W_1 acting on the curved quadrants, the total positive moment M_1 acting on the middle section ab, and the total negative moment M_2 acting about the y-axis on section $cdef$. This free body assumes the bending moment around the column capital is uniformly distributed, which means no torsional moments exist there. Moments also exist about the x-axis on efa and dcb, but do not enter into $\Sigma M_y = 0$.

$$W_1 = w(\ell^2/2 - \pi c^2/8) = (\ell^2 - \pi c^2/4)w/2$$

The moment of this load about axis y-y is:

$$\frac{w\ell^2}{2} \times \frac{\ell}{4} - \frac{\pi c^2 w}{8} \times \frac{2c}{3\pi} = \frac{w\ell^3}{8} - \frac{wc^3}{12}$$

If the upward shear W_1 is uniformly distributed around the quadrants cd and ef, the resultant acts at a distance c/π from the y-axis. Equilibrium of moments about the y-axis then gives:

$$-M_1 - M_2 + \frac{w\ell^3}{8} - \frac{wc^3}{12} - \frac{w}{2}\left(\ell^2 - \frac{\pi c^2}{4}\right)\frac{c}{\pi} = 0$$

$$M_1 + M_2 = M_0 = \frac{w\ell^3}{8} - \frac{wc^3}{12} - \frac{wc\ell^2}{2\pi} + \frac{wc^3}{8} = \frac{w\ell^3}{8}\left(1 + \frac{c^3}{3\ell^3} - \frac{4c}{\pi\ell}\right)$$

$$M_0 \approx \frac{w\ell^3}{8}\left(1 - \frac{2}{3}\frac{c}{\ell}\right)^2$$

The value of $[1 - \frac{2}{3}(c/\ell)]^2$ approximates the longer parenthesis reasonably well.*

---

* The error is 0.5% low for $c/\ell = 0.1$, 0.5% high for $c/\ell = 0.2$, 1.3% high for $c/\ell = 0.25$, and 2.0% high for $c/\ell = 0.30$.

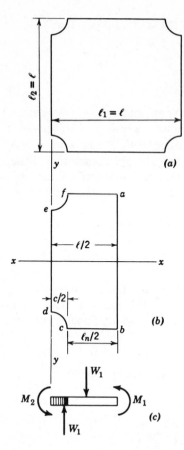

Figure 15.5 Equilibrium conditions indicate $M_1 + M_2$ is established by statics.

This statics solution tells nothing about how this total moment is distributed between positive or negative moment or how either varies along the slab width. The solution also neglects possible torsional moments around the column capital that will act if the tangential bending moments are not uniformly distributed.

When flat slab panels are not square, the long span produces the larger moment. The student should contrast this with the two-way slab supported on all four sides where the short slab strips carry the larger load and larger moment. In such a slab the long slab strips carry less load and less moment, but the long beams carry heavy slab reactions and large moments. Two-way slab moments in both directions are reduced because the beams help to carry the moment. In the flat slab, in contrast, the *full* load must be carried in *both* directions by the slab alone.

(c) Correlation of Statics with Tests In early tests (mechanical) strain measurements were necessarily over considerable lengths and indicated average rather than maximum steel stresses. These conditions led to an overly optimistic evaluation of the reserve strength, even when allowances were made for this problem. Also tests were not to failure and thus reflected less tension cracking than develops at ultimate. Thus early test evaluations led to a total M_0 of $0.09 w\ell_1^2\ell_2[1 - (2/3)(c/\ell_1)]^2$, which was accepted until the change in the 1971 Code, in spite of the apparent conflict with statics.

The behavior of large areas of this type of construction proves that the static analysis fails to give the total picture. There is a horizontal thrust, a form of arching that, under favorable conditions, greatly strengthens interior panels and even adds some strength to edge panels. Tests in the 1960s showed this reserve strength to such an extent that the Code now uses $M_0 = w\ell_2\ell_n^2/8$, which bases the moment on clear span ℓ_n between square supports.* In Fig. 15.8a this looks like true statics and it is an adaptation (reduction) of Nichol's value. It is admittedly still short of the full static moment, because all of the reaction is not on the near face of the column, but M_0 is larger than in earlier codes.

15.4   Theoretical Distribution of Moment

Westergaard developed a theoretical slab analysis that accompanied Slater's test analyses.[1] He established the distribution of positive and negative moments that exist throughout a flat slab when it is considered an isotropic plate. His work, roughly checked by Slater's test analyses, forms the basis for the specified subdivision of the total M_0 into positive and negative moments and for further subdivision into moments on the two design strips set up in the specifications. Figure 15.6a shows his calculated distribution of negative moments on strips crossing the column center line for $c/\ell = 0.15$ and 0.25. Around the column capitals the moment in a radial direction is constant at $-0.223 w\ell^2$ for $c/\ell = 0.15$ and at $-0.143 w\ell^2$ for $c/\ell = 0.25$. These values are nearly four times the maximums shown between columns. The positive moments on strips crossing the middle of the span are shown in Fig. 15.6b. For a strip running along the line between column centers, the moment diagram is shown in Fig. 15.7a and for a strip running along the middle of the panel in Fig. 15.7b. Both these strips have a form of moment diagram very similar to that of a one-way slab, with positive moment in the middle of the span and negative moments at the ends.

As a result of extensive analytical work and quarter scale tests on multipanel slabs[6-8] at the University of Illinois and a three-quarter scale test by the Portland Cement Association,[10] the distribution of moments actually used in the empirical or direct design procedures (Fig. 15.8b) has been slightly modified from the 1963 Code. The values now apply to cases of slabs with monolithic cast beams (if framing directly into columns) as well as to flat slabs. The Illinois slab investigation[9] also resulted in the introduction of the more general "equivalent frame analysis" procedure for handling those cases for which the "direct

---

* If supports are not square or rectangular replace them (for this value of ℓ_n) with square columns of equal area.

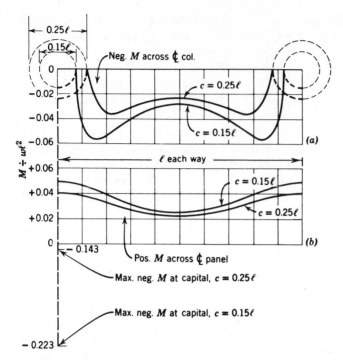

Figure 15.6 Theoretical moments on vertical strips (*a*) crossing column center lines and (*b*) crossing midspan of panel (Poisson's ratio zero). (Adapted from Reference 1, ACI.)

design'' procedure is inapplicable. Equivalent frame analysis is introduced in Sec. 16.4. An interesting study and evaluation of the Code procedures is given in Reference 12.

15.5 Code Direct Design Rules—Basic Requirements for Flexure

When layouts are relatively simple, a direct design procedure, limited to slabs meeting the limitations of Code 13.6 (listed here) is used:

1. A minimum of three spans each way, directly supported on columns.
2. Rectangular panels with the long span not more than twice the short span.
3. Successive spans not differing by more than 1/3 of the longer span.
4. Live load not more than 3 times dead load.
5. All loads to be uniformly distributed gravity loads.
6. For panels with beams between columns on all sides the beam relative stiffnesses must fall within a range of 5 to 1.

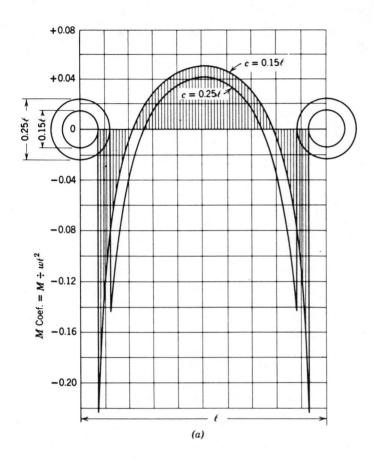

(a)

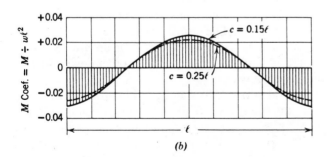

(b)

Figure 15.7 Theoretical bending moments (Poisson's ratio zero.) (Adapted from Reference 1, ACI.) (a) Moment diagram for strip along center line of columns. (b) Moment diagram for strip along middle of panel.

Columns may be offset from either axis by up to 10% of the span in the direction of the offset.

The ACI Code Commentary has many helpful sketches, tables, and discussions on two-way slab systems.

The slab minimum thickness rules of Code 9.5.3.1 appear very complex but, for the case of no beams, the interior panels are always governed by the minimum of 5 in. without drops, 4 in. with drops, or by Code Eq. 9.11 and 9.13 which then become the same* and specify the thickness as:

$$h = \ell_n(800 + 0.005f_y)/36{,}000 = \ell_n(0.8 + f_y/200{,}000)/36$$

This equation reduces to $h = \ell_n/36$ for Grade 40 bars and $h = \ell_n/32.7$ for Grade 60 bars, where ℓ_n is the clear span.

For exterior and corner panels *without* edge beams the thickness must be increased 10% above those just stated (Code 9.5.3.3).†

Bending moments for the interior panels are discussed first to show the direct design method basic procedures without the slight complications arising from evaluation of the effective exterior restraints (discussed in Sec. 15.9).

The basic moment calculation, $M_0 = (\frac{1}{8})w\ell_2\ell_n^2$, is for a panel centered on a column line, that is, for a half-panel from panel A and a half-panel from panel B in Fig. 15.8a. This moment calculation, centering on the column lines in each direction, uses moments based on the clear span ℓ_n (which must be not less than $0.65\ell_1$) (Code 13.6.2.5) and the average width ℓ_2 of $0.5(\ell_A + \ell_B)$. For interior panels (with or without beams) $-0.65M_0$ is considered negative moment and $+0.35M_0$ is positive moment, as in Fig. 15.8a and Code 13.6.3.2.

The moments, as already noted, are larger close to the columns than elsewhere. Although they vary greatly, it is safe to average them over a considerable width for a design. For this purpose Code 13.2 defines a column strip and a middle strip.

A column strip is a width (Fig. 15.8a) of $\ell_A/4$ plus $\ell_B/4$ (but not more than $\ell_1/2$ total) adjacent to the centerline of the columns. The remainder of the area used in Fig. 15.8a makes up two half middle strips, one for panel A and one for panel B. In design the two adjacent half-strips at the column are treated as one design strip and the two adjacent half middle strips as the other design strip. In uniform panels each strip is $\ell_2/2$ in width.

α is the ratio of beam flexural stiffness to slab flexural stiffness in a design strip of width ℓ_2 centered on the beam axis. Without beams (which means $\alpha_1 = 0$ in the Code notation), Code 13.6.4.1 states that for interior panels the column strip negative moment is 75% of the $-0.65M_0$, that is, $-0.49M_0$, although Code 13.6.6 assigns the other $-0.16M_0$ to the two half middle strips, as indicated in Fig. 15.8b. Also by Code 13.6.4.4 the column strip is assigned 60% of

---

* $\alpha_m = 0$ when no beams; $\beta_s = 1$ for interior panel. See Sec. 15.11 for the general discussion of the h equations.
† As used in Sec. 15.13a.

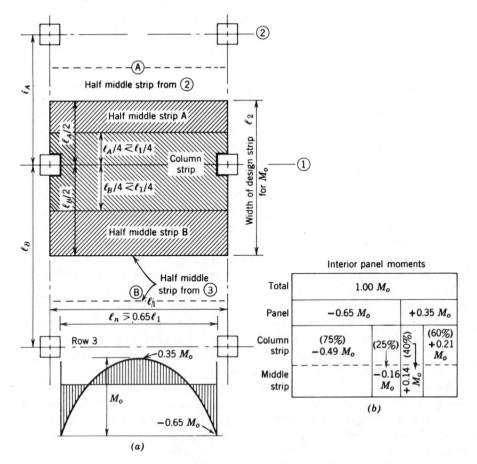

Half middle strip from ②

Half middle strip A

$\ell_A/4 \lessgtr \ell_1/4$

Column strip

$\ell_B/4 \lessgtr \ell_1/4$

Half middle strip B

Half middle strip from ③

$\ell_n \geqslant 0.65\ell_1$

Width of design strip for M_o

ℓ_2

Row 3

$-0.35\,M_o$

M_o

$-0.65\,M_o$

(a)

Interior panel moments

Total	1.00 M_o		
Panel	$-0.65\,M_o$	+0.35 M_o	
Column strip	(75%) $-0.49\,M_o$	(25%) $-0.16\,M_o$	(60%) +0.21 M_o
Middle strip		(40%)	+0.14 M_o

(b)

Figure 15.8 Calculation of M_0 and its subdivision for an interior panel (without beams) in x-direction; similar in y-direction except for half-column strip at exterior.

the $+0.35\,M_0$ moment, or $+0.21\,M_0$, and the $+0.14\,M_0$ remaining goes to the two half middle strips. For uniform panels, the two half middle strips are alike, that is, $-0.16\,M_0$ and $+0.14\,M_0$ for the total middle strip moments. The designer is permitted by Code 13.6.7 to modify any design moment by 10%, if all of M_0 is still assigned.

The system for exterior panel moments is quite similar, except that less negative moment goes to the exterior and more to the interior column line; the positive moment is correspondingly increased (Sec. 15.12), similar to typical exterior beam panels. The detailed discussion of this operation is delayed because the degree of exterior slab restraint must be considered. The details of this now would tend to cloud the simple structure of the direct design method.

15.6   Code Shear Requirements and Bar Development

Very many of the provisions of Code Chapter 13 apply to slab systems in general, whether designed under the empirical direct design method provisions of Code 13.6, the equivalent frame method of Code 13.7, or the more general requirements of Code 13.3.1, which allows any design procedure that satisfies equilibrium, geometric compatibility, sectional strength, and serviceability conditions. Most provisions in this section apply to all slab systems; those pertaining strictly to the direct design method are so indicated.

(a) Shear   Although punching or two-way shear requirements were introduced in Secs. 5.22 to 5.25, shear is reintroduced here for emphasis. Although yield line theory shows that flexural capacity can be obtained by several arrangements of reinforcing, no such freedom exists for shear around columns in the case of flat plates and flat slabs. In either case, shear is *the critical element* of design. Nearly any test slab, if loaded to collapse, will show a final shear failure or, around an exterior column, a combined shear and torsion failure.

On interior columns it is possible to use crossed steel shearheads to strengthen the slab (up to 75% more V_n) in shear (Code 11.11.4); but a shearhead modification of an exterior column to the same specification *does not function* and is wasted. Around interior columns it is also possible to use closed stirrups effectively (up to 50% more V_n but with V_c decreased to $2\sqrt{f'_c}b_o d$), if the bars or stirrups are carefully developed.

Around exterior columns, especially when spandrel beams (edge beams) are omitted or when the slab extends only to the outside of the column (less than all the way *around* the column), careful design for shear *and torsion* is absolutely essential. An overhang of the slab provides a great improvement. Many flat plate structures have been built without extending the slab beyond the column, but the authors are still prejudiced against this construction detail. A very conservative approach is fully warranted. Attention is specifically directed to the direct design method warning of Code 13.6.3.5.

> Edge beams or the edges of slabs shall be proportioned to resist in torsion their share of the exterior negative factored moments.

The provisions of Code 13.3.3.1 about the transfer of part of the moment by an eccentricity of shear are also in the right direction. The designer must be alert to the fact that the shear transfer to the exterior column is (nearly always) the weakest point in the slab.

(b) Development of Reinforcement in *Braced* Frames   The Code includes Fig. 15.9 which indicates the requirements for extending some of the positive moment bars into the support area to within 3 in. of the centerline of supports. The left half of Fig. 15.9 is for slabs of uniform thickness; the right half is for slabs thickened around the column by a drop panel (Fig. 15.2 and Sec. 15.13). Of course, bars must be of such a size that they can be developed within the

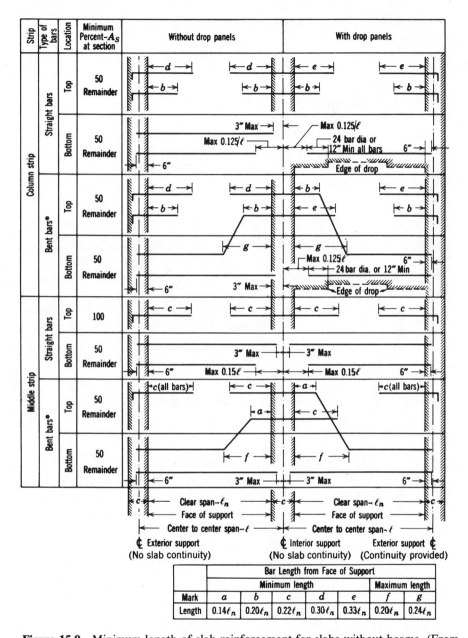

Figure 15.9 Minimum length of slab reinforcement for slabs without beams. (From ACI Code, Chapter 13.) Note: Not adequate for lateral loads in unbraced frames. (See Code Secs. 12.11.1 and 12.11.2 regarding extending reinforcing into supports.)
* Bent bars at exterior supports may be used if a general analysis is made.

lengths shown (or else be extended farther), but this is not commonly a serious problem.

Note that Fig. 15.9 is not adequate in an unbraced frame.

15.7 Direct Design of Flat Plate—Interior Panel

Design by the direct method an 18-ft × 20-ft flat plate interior panel for a service dead load, including its own weight, of 100 psf* and a live load of 60 psf. Assume 15-in. square columns, $f'_c = 4000$ psi, Grade 60 steel, 0.75-in. cover because not exposed to the weather.

Solution

Normally the slab thickness used would be determined by the thickness of an exterior panel. However, since a transition in depth can often be accommodated, the design is developed here as if this panel stood alone in the design.

Depth is determined by shear around the column or by the thickness needed against deflection. The minimum thickness (Code 9.5.3.1) is 5 in. or

$$h = \ell_n(800 + 0.005f_y)/36,000 = \ell_n(0.8 + f_y/200,000)/36 = \ell_n/32.7$$
$$= (20 - 1.25)12/32.7 = 6.88 \text{ in., say, 7 in.}$$
$$\text{Dead load} = 1.4 \times 100 = 140 \text{ psf}$$
$$\text{Live load} = 1.7 \times 60 = 102$$
$$w = 242 \text{ psf}$$

Shear must be checked on a section enclosing the column at $d/2$ from the column face as shown in Fig. 5.24. For shear, the average depth of the two steel layers seems logical and safe, say, $d = 7 - 0.75 - 0.50$ bar $= 5.75$ in. The width of each side of this section thus becomes $15 + 5.75 = 20.75$ in. $= 1.73$ ft. $b_0 = 4 \times 20.75 = 83$ in.

$$V_u = 242(20 \times 18 - 1.73^2) = 86,400 \text{ lb}$$
$$V_n \geqslant V_u/\phi = 86,400/0.85 = 102,000 \text{ lb}$$
$$\text{Allowable } V_c = 4\sqrt{f'_c}\,b_0 d_v = V_n$$
$$d_v = V_n/4\sqrt{f'_c}\,b_0 = 102,000/(4\sqrt{4000} \times 83)$$
$$= 4.86 \text{ in.} < h = 7 \text{ in. above USE } h = 7 \text{ in.}$$

For moment the steel in the long direction will be nearest the top or bottom faces of the slab, that for the short span next inside, making:

$$\text{Trial } d_l = 7.0 - 0.75 - d_b/2 = 6.00 \text{ in. for long span}$$
$$d_s = 7.0 - 0.75 - 1.5d_b = 5.50 \text{ in. for short span}$$

---

* This should include floor finish, partition allowance, and the like.

both assuming #4 bars. Such round numbers for the resulting d occur only for a few bar sizes; in a square panel the smaller d would control, *not* the average d, for steel in *both* directions, unless supervision were close enough to justify different steel in the two directions. The use of average d for A_s calculations (as in slabs of Chapter 12) is *unsafe* here. The small bar size was selected here because the depth for deflection seems to be governing over stress criteria and this points to a lightly reinforced slab.

$$M_{ol} \text{ (long span)} = 0.125 w\ell_2\ell_n^2$$
$$= 0.125 \times 242 \times 18(20 - 1.25)^2/1000$$
$$= 192 \text{ k-ft}$$
$$M_{os} \text{ (short span)} = 0.125 \times 242 \times 20(18 - 1.25)^2/1000$$
$$= 170 \text{ k-ft}$$

Next check depth for moment by comparing $k_n = M_n/bd^2$ to the maximum reinforcement allowable of 1041 from Table 3.1. The highest moment is the long span negative moment in the column strip (over the column), Fig. 15.8b.

$$-M_u = -0.75(0.65 M_o) = -0.75 \times 0.65 \times 192 = -93.6 \text{ k-ft}$$
$$-M_n = M_u/0.9 = -104 \text{ k-ft*}$$

The strip width is half the transverse panel length: 9 ft or 108 in.

$$\text{Required } k_n = M_n/bd^2 = 104 \times 12,000/(108 \times 6.00^2) = 321 \ll 1041$$

This result indicates such a greatly under-reinforced slab that a check in the short direction is unnecessary.

A tabular form, as in Table 15.1, expedites this type design and organizes the results in a manner easily available to the detailer. The authors find the use of nominal M_n convenient since it eliminates ϕ from all subsequent calculations, but this is optional. The lever arm z of the internal couple is estimated as $0.95\,d$, a little high in the usual range of $0.90\,d$ to $0.95\,d$ because the slab is much thicker than moment requires. This value typically need not be modified for the various moments, but z could be checked where any close decision is to be made in fixing the bar spacing. Since the student probably has little "feel" for the proper z, a check will be made at the worst section, that just used in checking k_n.

$$A_s = M_n/f_y z = (104 \times 12)/(60 \times 0.95 \times 6.0) = 3.65 \text{ in.}^2$$
$$a = N_n/(0.85 f_c' b) = (3.65 \times 60)/(0.85 \times 4 \times 108) = 0.596 \text{ in.}$$
$$z = 6.0 - 0.596/2 = 5.70 \text{ in.} = 0.95\,d$$

Thus $0.95\,d$ is satisfactory here and is on the safe side elsewhere (where smaller moments lead to still smaller a values).

---

* It is optional whether M_u or M_n should be distributed. The authors prefer to insert the ϕ as early as possible.

TABLE 15.1 Steel Calculations for Flat Plate Panel of Sec. 15.7.

	Long Span				Short Span			
	Column Strip (9')		Middle Strip (9')		Column Strip (9')		Middle Strip (11')	
For #4 bars 7"	Negative	Positive	Negative	Positive	Negative	Positive	Negative	Positive
Distribution of M_{on}:	$-0.49\,M_{on}$	$+0.21\,M_{on}$	$-0.16\,M_{on}$	$+0.14\,M_{on}$	$-0.49\,M_{on}$	$+0.21\,M_{on}$	$-0.16\,M_{on}$	$+0.14\,M_{on}$
$M_{on} \geqslant M_o/0.9$		213 k-ft				189 k-ft		
M_n, k-ft	−104.4	+44.7	−34.1	+29.8	−92.6	+39.7	−30.2	+26.5
d, in.	6.00	6.00	6.00	6.00	5.50	5.50	5.50	5.50
$zf_y/12$	28.5	28.5	28.5	28.5	26.1	26.1	26.1	26.1
A_s reqd., in.^{2} = M_n/zf_y	3.66←	1.57←	1.20←*	1.05	3.55←	1.52←	1.16←*	1.01
Min. A_s = 0.0018 × 7b	1.36	1.36	1.36	1.36←	1.36	1.36	1.66	1.66←
Straight Bars								
No. of #4 (A_b = 0.20)	19	8	0̸ 18 Use #3		18	8	0̸ 24 Use #3	
Aver. spcg., in.	5.5±	13.5±	14		6.0	13.5	14	
Max. spcg., in.	14		14		14		14	
No. of #3 (A_b = 0.11)			11	13			11	15
Aver. spcg., in.			10±	8.5±			12	9±
Alternate-Bent & Str.								
Bent	8-#4	4-#4	12-#3	6-#3	8-#4	4-#4	12-#3	6-#3
Straight	11-#4	4-#4	—	7-#3	10-#4	4-#4	—	9-#3

* Negative A_s + half positive A_s can be made effective for temperature.

This area of steel also indicates 19-#4 bars (3.80 in.²) at a spacing of 108/19 = 5.68 in., which is reasonable (possibly a trifle close) for the worst spot. Bars are usually specified by total number in the strip rather than by exact spacing, as 19-#4 at 5.5 in. ±, thus also letting the field worker know that a uniform spacing close to 5.5 in. will need little rearrangement to cover the strip properly.

In Table 15.1 the cross-sectional sketches emphasize the steel arrangement. Solid circles represent the bars being designed. The zf_y constant used has f_y in ksi and has incorporated the 12 factor needed to change the moment from k-ft to k-in. Minimum temperature and spacing steel (Code 7.12.2) is also shown with the ratio for Grade 60 bars being 0.0018 based on the area of the *total slab thickness*. This steel includes both top and bottom steel where both exist, as in a negative moment region. Although in such a region the main steel is top steel for negative moment, half of the positive moment steel is required to be run into the support region and stays effective for temperature cracking resistance. If the temperature needs *at* the support appear critical these bottom bars can be lapped instead of stopping them 3 in. short of the column centerline as indicated in Fig. 15.9. Finally, the maximum spacing for slabs is fixed by Code 13.4.2 at $2h$ which is 14 in. and this governs in several places.

Two arrangements of bars are shown in the table, the first for all bars straight as shown schematically in Fig. 15.10a. These bars are usually shown on the drawings as bands, somewhat as in Fig. 15.10b. For simplification the listings are shown almost on two separate panels, one for the steel in each direction. Where spans are unequal and steel must be shown in both directions on many panels, it is possible to superimpose as Fig. 15.10b shows along one column centerline. The second arrangement in Table 15.1 is for bent bars, as sketched in Fig. 15.11.

The foregoing concludes the primary or main design calculations for this example. However, there are four special aspects of design that might influence or change the decisions already made. The following aspects deserve more generalized comment than is appropriate within the design itself:

Sec. 15.8 Minimum stiffness of columns.
Sec. 15.9 Concentration of reinforcement over the column.
Sec. 15.10 Code thickness equations—comments on their use.

15.8 Minimum Stiffness of Columns

Slab moments, especially for slabs without beams, are sensitive to pattern loadings[7] in which partial loads are placed in a manner similar to the maximum moment influence line. Provision for the total M_o moment based on full loading

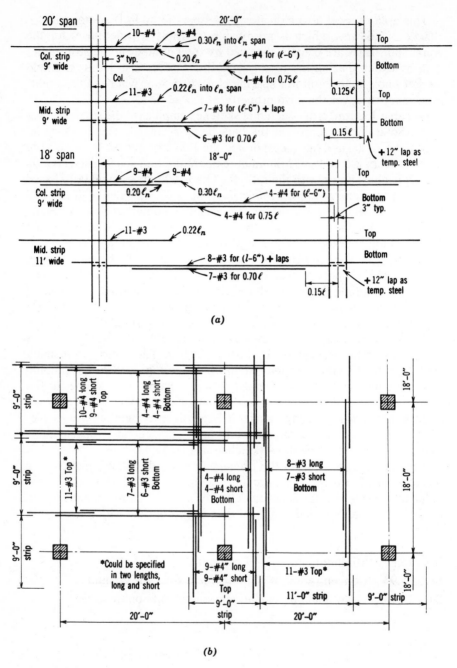

Figure 15.10 Steel for flat plate floor. (*a*) Symmetrical schematic arrangement as in elevation. (*b*) Plan showing bands of reinforcing.
  * Lengths of bars are often listed instead of simply "long" or "short."

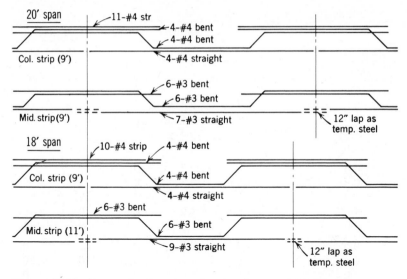

Figure 15.11 Pattern for bent bars for the alternate in Table 15.1.

of all panels still leaves overstress possible under pattern loadings. Code 13.6.10 accepts a possible overstress to 33% under these conditions. To stay within this limit when the unfactored dead load is less than twice the unfactored live load, the system requires either a specific minimum size column or, with a lesser column, an increase in the positive moments above those indicated by M_o.

The minimum column is defined by a minimum required stiffness ratio

$$\alpha_{\min} = \Sigma K_c / \Sigma (K_s + K_b)$$

where K_c, K_s, and K_b are the flexural stiffness of columns (above and below the slab), slabs, and beams (at the joint), respectively. In Table 15.2 $\alpha_{\min}$ is tabulated for various values of β_a, the ratio of dead-to-live load (at service load level), panel proportions ℓ_2/ℓ_1, and relative beam stiffness to slab stiffness ratios α. For dead load of twice the live load $\alpha_{\min}$ is zero, but $\alpha_{\min}$ increases as β_a decreases.

As an example, consider the slab design of Sec. 15.7 where $\beta_a = 100/60 = 1.67$, which is less than the specified ratio of 2 and requires investigation. $\ell_2/\ell_1 = 20/18 = 1.11$. $\alpha = 0$ (no beams). From the table:

$$\beta_a = 2.0, \text{ any } \ell_2/\ell_1, \alpha = 0, \alpha_{\min} = 0$$
$$\beta_a = 1.0, \ell_2/\ell_1 = 1.11, \alpha = 0, \alpha_{\min} = 0.74$$

Interpolating: $\beta_a = 1.67$, $\ell_2/\ell_1 = 1.11$, $\alpha = 0$, $\alpha_{\min} = 0.24+$, say, 0.25

$$\alpha_{\min} = 2K_c/2K_s = K_c/K_s = 0.25$$

The smaller $I_s = (18 \times 12)7^3/12 = 6180$ in.[4]

For a slab span of 20 ft, $K_s = 4EI_s/\ell = 4E\,6180/240 = 103E$

TABLE 15.2* Minimum Column α_{min} Values

β_a	Aspect Ratio ℓ_2/ℓ_1	Relative Beam Stiffness, α				
		0	0.5	1.0	2.0	4.0
2.0	0.5–2.0	0	0	0	0	0
1.0	0.5	0.6	0	0	0	0
	0.8	0.7	0	0	0	0
	1.0	0.7	0.1	0	0	0
	1.25	0.8	0.4	0	0	0
	2.0	1.2	0.5	0.2	0	0
0.5	0.5	1.3	0.3	0	0	0
	0.8	1.5	0.5	0.2	0	0
	1.0	1.6	0.6	0.2	0	0
	1.25	1.9	1.0	0.5	0	0
	2.0	4.9	1.6	0.8	0.3	0
0.33	0.5	1.8	0.5	0.1	0	0
	0.8	2.0	0.9	0.3	0	0
	1.0	2.3	0.9	0.4	0	0
	1.25	2.8	1.5	0.8	0.2	0
	2.0	13.0	2.6	1.2	0.5	0.3

* Copy of Code Table 13.6.10.

The varying I of Code 13.7.3.3 is not used, although appropriate (not required for direct design method) if the related curves are handy.

$$0.25 = K_c/K_s = K_c/103E, \qquad K_c = 25.7E$$

For a story height of 10 ft, $K_c = 25.7E = 4EI_c/120 = EI_c/30$

Min. $I_c = 25.7E \times 30/E = 771$ in.$^4 = t^4/12$, thus min. $t = 9.85$ in.

The 15-in. column used is more than adequate with this 7-in. slab. The design positive moments used remain acceptable without penalty.

If α_{min} cannot be satisfied, an increased positive moment is required as given by multiplying the nominal moment values by δ_s:

$$\delta_s = 1 + \frac{2 - \beta_a}{4 + \beta_a}(1 - \alpha_c/\alpha_{min}) \qquad \text{Code Eq. 13.5}$$

where all terms have already been defined except α_c, the ratio of the actual column stiffness to the actual combined stiffness of the beams and slabs at the joint.

15.9 Concentration of Reinforcement Over the Column

Where moment must be transferred from slab to column, as at an exterior column, or from column to slab, as in resistance to wind or earthquake moments, it is desirable to accomplish this as directly as possible. The Code (13.3.3.2) considers a direct transfer of moment possible only through reinforcement that passes through the column itself or within a width $1.5h_s$ on each side (where h_s is the slab or drop panel thickness).

Tests show that the shear stresses are shifted by the general moment transfer taking place at the column, somewhat as indicated in Sec. 5.25 and as calculated in Sec. 15.12f. The Code Commentary Sec. 11.12.2 suggests that for square columns 40% of the moment transferred should be represented by this eccentric shear (or torque) and that 60% should be transferred directly by flexure. If the column depth c_1 (parallel to ℓ_1) is greater than the transverse column width, Code 11.12.2.3 gives a formula that increases the percentage to be transferred by eccentric shear or torque. Code Eq. 13.1 gives the corresponding formula that decreases the direct flexural transfer percentage.

Some concentration of the column strip negative moment steel within the specified width may be necessary for the direct transfer of moment to the columns. If wind load moment is transferred to the slab, extra moment steel may be required. Live load pattern moments or unequal spans also involve these transfer moments in interior panels. For equal spans the usual steel arrangement for the column strip is adequate, as now demonstrated.

In the direct design method, Code 13.6.9.2 gives an equation defining the moment to be transferred to columns above and below the joint. This equation is based on two adjoining spans, possibly of different lengths, that have full dead load plus $\frac{1}{2}$ live load on the longer span, although only dead load is acting on the shorter span. The 1983 Code removed the need to consider effective column stiffness for this moment. Code Eq. 13.4 specifies that unless a general analysis is made, the moment to be resisted by the supports is:

$$M = 0.07[(w_d + 0.5w_1)\ell_2\ell_n^2 - w_d'\ell_2'\ell_n'^2]$$

where the primes refer to the shorter span values. The dead and live load values are factored loads. Because no ϕ factor is indicated, the authors interpret this M as if it was a factored design moment, M_u.

For the typical interior panel example of Sec. 15.7, assuming adjacent panels of the same size, the longer direction moment to be transferred to the columns is:

$$M_u = 0.07[(140 + 51) \times 18 \times 18.75^2 - (140) \times 18 \times 18.75^2]/1000 = 22.6 \text{ k-ft}$$

For a square column, $b_1 = b_2$, so Code Eq. 13.1 becomes $\gamma_f = 0.6$. Thus 60% of M_u is carried within a 15-in. column width plus $3.0h_s = 21$ in., which totals 36 in. or 3 ft. Thus $M_n \gtrless 0.60 \times 22.6/0.9 = 15.1$ k-ft. The original column strip design M_n was 104.4 k-ft on a 9-ft width. This original design has already provided a slab flexural capacity in the required width of $(3.0/9.0)\,104.4 = 34.8$ k-ft. This value is over twice the M_n to be transferred. No concentration of reinforcing is necessary.

For the exterior column this requirement is more serious, as in Sec. 15.12(f). Sixty percent of the total negative moment must be carried within this section close to the column and special provisions must be made for bringing in the torque.

There is no need to check the concrete in compression for this direct moment transfer if the designer is careful to keep the steel ratio less than $0.75\rho_b$ locally. The slab thickness provided against deflection usually lowers steel ratios.

15.10 Code Thickness Equations—Comments on Their Use

The Code establishes minimum thicknesses of slabs in two-way construction primarily by formula values, but with certain limiting considerations, especially minimum values (9.5.3.1):

For slabs without beams or drop panels	5 in.
For slabs without beams, but with drop panels satisfying Code 9.5.3.2	4 in.
For slabs having beams on all four edges with a value of α_m at least equal to 2	3.5 in.

The three Code equations, minimum values in Eqs. 9-11 and 9-12 and an upper limit in Eq. 9.13, are:

$$\text{Eq. 9-11: } h \gtrsim \frac{\ell_n(800 + 0.005f_y)}{36,000 + 5000\beta[\alpha_m - 0.5(1 - \beta_s)(1 + 1/\beta)]}$$

$$\text{Eq. 9-12: } h \gtrsim \frac{\ell_n(800 + 0.005f_y)}{36,000 + 5000\beta(1 + \beta_s)}$$

$$\text{Eq. 9-13: } h \not> \frac{\ell_n(800 + 0.005f_y)}{36,000} \text{, permissive, not mandatory}$$

All three equations have a common numerator that reduces to $1000\,\ell_n$ for Grade 40 steel and $1100\,\ell_n$ for Grade 60 steel.* Another format will be introduced before discussing the denominators.

The Portland Cement Association[2] treatment of these equations offers better visualization of the thickness requirements. The three equations are restated with both numerator and denominator divided by 1000:

$$\text{Eq. 9-11: } h \gtrsim \frac{\ell_n(0.8 + f_y/200,000)}{36 + 5\beta[\alpha_m - 0.5(1 - \beta_s)(1 + 1/\beta)]}$$

$$\text{Eq. 9-12: } h \gtrsim \frac{\ell_n(0.8 + f_y/200,000)}{36 + 5\beta(1 + \beta_s)}$$

$$\text{Eq. 9-13: } h \not> \frac{\ell_n(0.8 + f_y/200,000)}{36} \text{, permissive, not mandatory}$$

---

* This difference of 10% between the two grades of steel is smaller than the 20% difference for beams suggested with Code Table 9.5a reprinted here as Table 3.2 (Sec. 3.9). Slabs crack less than beams under service conditions where deflections are of greatest interest; the grade and quantity of steel makes little difference in the uncracked portions.

The numerators then all reduce to ℓ_n for Grade 40 bars and $1.1\,\ell_n$ for Grade 60 bars. The last equation, for example, can be restated for Grade 40 bars as a required $\ell_n/h \not< 36$, or ℓ_n/h need not be lower than 36; or for Grade 60 bars that ℓ_n/h need not be lower than $36/1.1 \doteq 33$. Likewise the denominator of the other equations establish similar requirements, which are detailed after a brief discussion of the various ratios represented by the symbols in the denominators:

β = ratio of long to short *clear* spans = 1 for a square panel; above a ratio of 2, based on spans center-to-center of columns, the slab is designed as a one-way slab or a more exact analysis is required (Code 13.6.1.2). Equations 9.11 to 9.13 do not apply for $B > 2$.

β_s = ratio of length of continuous edges to total perimeter of a slab panel = 1 for an interior panel and never less than 0.5 *for the direct design method.*

α_m = average ratio of flexural stiffness of beam section to that of the slab. Slab width to the center of adjacent panels is used, that is, for uniform panels ℓ_2. The subscript m calls for the average for all beams around the panel. The value of α varies from 0 for an interior flat plate upward.

Interior panels of flat slabs or flat plates have no beams and hence beam stiffness ratios α_m and α are zero. With continuity on all sides, β_s is unity. Thus the denominator for Eq. 9.11 is the same for interior panels as that of Eq. 9-13 in the absence of beams. For a square exterior corner panel without an edge beam, β_s becomes 0.5 and β becomes unity, leading to a numerator for Eq. 9-11 of only $36 + 5(-.25)2 = 33.5$, which permits Eq. 9-13 to be used as a lesser limit if desired. These equations are plotted for Grade 40 bars in Fig. 15.12, with only the ℓ_n/h value of 36 significant in the absence of beams. For Grade 60 bars the 36 becomes $36/1.1 = 32.7$, say, 33. When beams are used (Chapter 16), the curves of Fig. 15.12 are very useful.

Two special cases are treated as modifications of the Code values of thickness. (1) With drop panels at the columns meeting the requirements illustrated in Sec. 15.13, a thickness 10% smaller than the equation values may be used (Code 9.5.3.2). (2) In edge panels the formula thickness is too shallow unless edge beams having a stiffness ratio $\alpha \gtrless 0.80$ are provided; otherwise the required thickness is 10% more than the formula values, or 10% more than the special drop panel provision requires (Code 9.5.3.3).

Finally, it is permitted, in lieu of the formulas, to calculate the deflections and then to limit the immediate deflection under the live load to maximum values given in Code Table 9.5b. The listed permissible deflection limits vary with structure usage and are strictest where attached nonstructural elements may be damaged by deflections.

15.11 Flat Plates and Flat Slabs—Exterior Spans

For noncorner exterior panel strips parallel to the exterior wall, the methods for interior panels are available with only small changes. For the perpendicular strips running to the exterior a number of problems arise: the unsymmetrical negative moments, the combined shear and torsion problem at the exterior

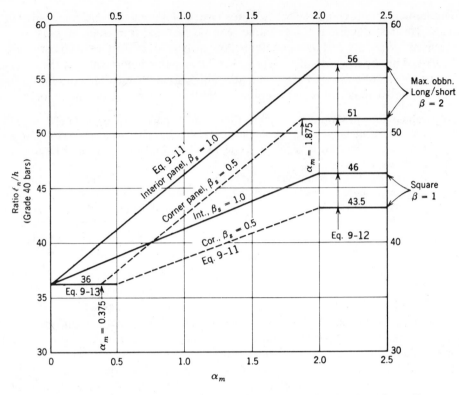

Figure 15.12 Required ℓ_n/h ratio for deflection control with Grade 40 bars. For Grade 60, divide ratio by 1.1; for Grade 50, by 1.05. (Recast from Reference 2, Portland Cement Assn.)

column, the torsion reinforcement often required in the slab or spandrel beam, and the direct transfer of moment to the exterior column (Sec. 15.9).

In a major simplification of the direct design method, the 1983 Code introduced Table 15.3 for the assignment of M_0 into positive and negative moments in end spans. The table eliminates the need to calculate effective column stiffnesses that proved cumbersome in the design process required by earlier codes. The multipliers for M_0 are given for exterior negative moment, positive moment, and interior negative moment in the end span. Numerical values recognize the degree of restraint along the edge. Slabs simply supported on walls are treated as unrestrained exterior edges. A fully restrained exterior edge is a case where the slab is constructed integrally with a very stiff concrete wall so that little rotation occurs at the slab-to-wall connection. Intermediate edge restraints are given for typical cases. Two-way slab systems with beams between supports on all sides utilize the coefficients of column (2). Flat plates and flat slabs that do not have beams between interior supports utilize the coefficients of columns (3) or (4) depending on whether they are without or with an edge (spandrel) beam. The values selected tend to give upper-bound values for inte-

TABLE 15.3 Distribution of End Span Moments (Reproduced from Code 13.6.3.3)

	(1)	(2)	(3)	(4)	(5)
		Slab with Beams between All Supports	Slab without Beams between Interior Supports		Exterior Edge Fully Restrained
	Exterior Edge Unrestrained		Without Edge Beam*	With Edge Beam	
Interior negative factored moment	0.75	0.70	0.70	0.70	0.65
Positive factored moment	0.63	0.57	0.52	0.50	0.35
Exterior negative factored moment	0	0.16	0.26	0.30	0.65

* See Section 13.6.3.6.

rior negative moments and positive moments and somewhat lower-bound values for exterior negative moments where the slab design is often governed by minimum reinforcement requirements. The sum of the averaged negative moment (interior plus exterior divided by two) and the positive moment is 1.00 so that the full M_0 is distributed in each case.

The break of moments into column strip and middle strip moments at the interior support of flat plates and flat slabs without beams is the same as for interior spans, 75% and 25% for negative moment and 60% and 40% for positive moment.

At the exterior support, for the same concrete in beam and column ($E_{cb}/E_{cs} = 1$), Table 15.4 shows the portion going to the column strip is a function of $\beta_t = C/2I_s$, where C is the torsional value defining the edge beam (given by Code Eq. 13.7 discussed in Sec. 16.4) and I_s is for a slab width equal to the beam span center-to-center of supports. The 2 factor comes from a ratio of E/G and the assumption that $G = 0.5E$. It might be debated whether to use $C = 0$ when no exterior spandrel beam is present. When calculating C in the determination of K_{ec}, in the equivalent frame method of Sec. 16.4 a slab strip the width of the column is specified to be used. It is consistent to do the same here. Actually, unless C is large, nearly all the moment finally goes to the column strip and it is usually just as practical an answer to assume 100% to the column strip and skip the calculation unless a real beam is present.

The following section covers the basic calculations for an exterior panel matching the interior panel of Sec. 15.7.

TABLE 15.4 Distribution of Column Strip Negative Moment at Exterior. (Reproduced from Code 13.6.4.2)

ℓ_2/ℓ_1		0.5	1.0	2.0
$(\alpha_1 \ell_2/\ell_1) = 0$	$\beta_t = 0$	100	100	100
	$\beta_t \geq 2.5$	75	75	75
$(\alpha_1 \ell_2/\ell_1) \geq 1.0$	$\beta_t = 0$	100	100	100
	$\beta_t \geq 2.5$	90	75	45

15.12 Design of Flat Plate—Exterior Panel

Design by the direct design method an exterior panel flat plate slab 18-ft wide parallel to the free edge and 20-ft center-to-center of columns perpendicular to this edge. Assume no beam,* no wall load (an omission for simplicity in the design example), and a slab extending to the outside face of the column, as in Fig. 15.13a. Use $f'_c = 4000$ psi for both slab and columns, Grade 60 steel, dead load (including slab weight) of 100 psf and live load of 60 psf. Assume all columns are 15-in. square with story heights of 10 ft.

Solution

(a) Slab Thickness Thickness for deflection is considered first, following the comments of Sec. 15.10.

$$\ell_n = 20 - 1.25 = 18.75 \text{ ft} \qquad f_y = 60\ 000 \text{ psi}$$

β = ratio of *clear* spans = $(20 - 1.25)/(18 - 1.25) = 1.12$

$\alpha_m = 0$ (no beams)

β_s = ratio of length of continuous edge to total perimeter of panel

$\qquad = (2 \times 20 + 18)/(2 \times 20 + 2 \times 18) = 0.763$

The curves of Fig. 15.12 are used for the governing ℓ_n/h ratio. For $\beta_s < 1.0$, Eq. 9-11 generates a curve between the interior and corner panel curves, a curve that intersects the vertical axis ($\alpha_m = 0$) at something less than the Eq. 9.13 value of 36 that governs. Therefore, remembering the 1.1 ratio for Grade 60 bars, we calculate required $h = \ell_n/(36/1.1) = 18.75 \times 12/32.7 = 6.88$ in. Code 9.5.3.3 requires either a spandrel beam or an increase of 10% in these formula values.

$$h = 1.10 \times 6.88 = 7.57 \text{ in., say, 8 in.†}$$

---

* The influence of an edge beam is discussed in Sec. 15.12f at the close of this section.
† Some designers would accept 7.75 in. for a slab.

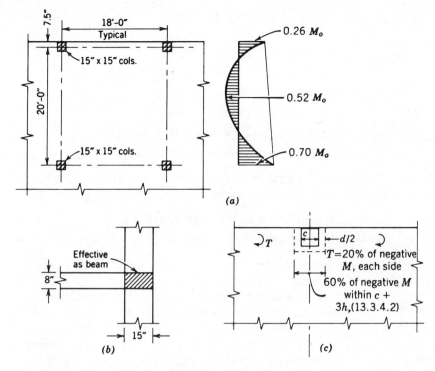

Figure 15.13 Exterior slab panel. (*a*) Layout and bending moments. (*b*) Slab section acting as beam in torsion. (*c*) Moment concentration over column.

The most critical design provisions are for shear around the columns, but they cannot be calculated before the bending moments in the exterior panel perpendicular to the free edge are known. (An interior panel under the same conditions was found in Sec. 15.7 to require a d_v of only 4.86 in.)

(b) Slab Design Moments   For strips running perpendicular to the exterior edge, the design moments of Sec. 15.11 and Fig. 15.13*a* are determined directly from Table 15.3. This example falls into the classification of a slab without beams between interior supports and without an edge beam so that the values of column (3) apply. The design moments can now be established from the design load.

$$w_u = 1.4 \times 100 + 1.7 \times 60 = 242 \text{ psf}$$

20-ft Strips

$$M_o = 0.125 \times 242 \times 18(20 - 1.25)^2/1000 = 192 \text{ k-ft}$$
$$M_{on} \gtrless M_o/\phi = 192/0.9 = 213 \text{ k-ft}$$
Total negative M_n at int. col. $= 0.70 M_{on}$
$$= -(0.70)213 = -149 \text{ k-ft}$$

$$\text{Total positive } M_n = 0.52 M_{on}$$
$$= (0.52)213 = +111 \text{ k-ft}$$
$$\text{Total negative } M_n \text{ at ext. col.} = 0.26 M_{on}$$
$$= -(0.26)213 = -55 \text{ k-ft}$$

The interior negative moment is assigned to column strip (75%) and middle strip (25%) just as in an interior panel (Code 13.6.4.1).

$$\text{Int. negative } M_n \text{ on column strip} = -0.75 \times 149 = -112 \text{ k-ft}$$
$$\text{Int. negative } M_n \text{ on middle strip} = -0.25 \times 149 = -37 \text{ k-ft}$$

Likewise the positive moment is assigned 60% to the column strip and 40% to the middle strip, just as in an interior panel (Code 13.6.4.4).

$$\text{Positive } M_n \text{ on column strip} = 0.60 \times 111 = 67 \text{ k-ft}$$
$$\text{Positive } M_n \text{ on middle strip} = 0.40 \times 111 = +44 \text{ k-ft}$$

For the exterior negative moment, the proportion going to the column strip (Table 15.4) depends on $\beta_t = E_{cb} C/(2 E_{cs} I_s)$. E_{cb} and E_{cs} are the elastic modulus of the concrete in the edge beam and the slab, respectively. Because there is no physically separate edge beam in this case, $E_{cb}/E_{cs} = 1$. C is the torsional constant of a torsional restraining member consisting of a portion of the slab with a width equal to the column thickness plus the transverse beam above and below the slab. This concept is discussed further in Sec. 16.4. For the present example, the slab strip between exterior columns is considered an edge beam carrying torsion. This slab strip (Fig. 15.13b) assumes a width equal to the column thickness in the direction resisting moment (Code 13.7.5.1) and thus $b = 15$ in. although $h = 8$ in. The torsional constant C is defined by Code Eq. 13.7 as:

$$C = \Sigma \left(1 - 0.63 \frac{x}{y}\right) \frac{x^3 y}{3}$$

where x is the shorter dimension of a rectangle and y is the longer dimension of a rectangle. The summation allows a series of rectangular components to be summed if the cross section has nonrectangular shape as in L-shaped spandrel beams.

$$C = (1 - 0.63 x/y)(x^3 y/3)$$
$$= (1 - 0.63 \times 8/15)(8^3 \times 15/3) = 0.664 \times 2570 = 1700 \text{ in.}^4$$

The slab moment of inertia for the gross section is:

$$I_s = 18 \times 12 \times 8^3/12 = 9240 \text{ in.}^4$$

Thus,

$$\beta_t = 1700 E_c/(2 E_c \times 9240) = 0.092$$

Interpolating in Table 15.4 of Sec. 15.11 with $\alpha_1 = 0$ (no beams):

For $\beta_t = 0$, 100% goes to column strip
$\beta_t = 2.5$, 75% goes to column strip
$\beta_t = 0.09$, $100 - \dfrac{0.09}{2.5}(100 - 75) = 100 - 0.9 = 99.1\%$

The difference between this result and 100% is negligible; the difference can be ignored.

Total exterior negative M_n on column strip $= -55$ k-ft
Exterior negative M_n on middle strip $= 0$

This completes the design moments in the one direction, except for the following option.

Where unequal negative moments occur at a column line, as at the first interior column line, Code 13.6.3.4 requires the slab design be for the larger moment unless analysis is made to distribute the moments. The interior panel design in Sec. 15.7 found column strip negative moment M_n of -104.4 k-ft against -112 k-ft here and middle strip negative moment of -34.2 k-ft against -37 k-ft here. Design for the larger is no large penalty in this case.* Where significant, the unbalanced moment may be distributed to the slabs and column (preferably using the K_{ec} values). For this case, with light columns and a small unbalance, the authors are inclined (1) to average the two panel values to use on both spans -108 k-ft for the column strip negative and -36 k-ft for the middle strip negative moments; (2) to neglect the resulting reduction in positive moments on the interior span; (3) in the exterior span increase the positive column strip moment from 67 to 69 k-ft and increase the positive middle strip moment from 67 to 69 k-ft. The alternate procedure provides for the larger exterior panel negative moments on both sides of the first interior column line.

The strips parallel to the exterior are not different from an interior panel, except when a spandrel or edge beam is present. Such a beam, or in this case the slab since no beam is present, must be checked for torsion (as discussed in subsection f).

The design of reinforcement for these strips follows the method used for the interior slab in Table 15.1 in Sec. 15.7.

(c) Concentration of Negative Moment Reinforcement at Exterior Column At the exterior column there is a necessary concentration of the column strip negative moment[5] as already discussed in Sec. 15.9. This condition requires a concentration of flexural reinforcement within a width equal to the column face plus $1.5h_s$ on each side for all moment to

---

* Note that the simplified "total dead load" used here neglects the increased weight of the thicker exterior panel slab to simplify the example; this is *not* a good practice.

be transferred directly to the column; the remainder of the slab moment is transferred by torque, causing a resultant eccentricity of shear around the column on a width equal to the column face plus d, as indicated in Fig. 15.13c.

For a square column 60% of the exterior negative moment is transferred by moment, 40% by eccentric shear. (Note code 11.12.2.3 for rectangular columns.)

$$\text{Direct moment transfer} = 0.60 \times 55 = 33 \text{ k-ft}$$
$$\text{Assuming } d = 7.00 \text{ in.,} \quad A_s = (33 \times 12)/(60 \times 0.9 \times 7) = 1.05 \text{ in.}^2$$

This area, equivalent to 6-#4 or 4-#5, must be placed within a width of $c + 3h_s = 15 + 3 \times 8 = 39$ in. This presents no serious problem, but may require a special note on the drawing about this closer spacing over the column.

(d) Minimum Column Stiffness For the interior panel, minimum column stiffness is checked in Sec. 15.8. Because it is relative stiffness of column to slab that is checked, even a smaller column would satisfy the needed restraint at the exterior column. There, parallel to the exterior face, only a half-panel width of slab is involved, which nearly doubles the relative column stiffness. In the other direction there is slab on only one face of the column, which has the same effect. The 15-in. square column is heavier than necessary.

(e) Shear at Exterior Column Normally there is some wall load, although for simplicity none is included in this text example; this adds a substantial extra shear. For $h = 8$ in., $\frac{3}{4}$-in. cover, and possible use of #5 bars, d for two-way shear is conservatively estimated as 6.5 in., with d in the long direction about 7 in. From Fig. 15.14a:

$$V_u = 242(10.62 \times 18 - 1.80 \times 1.54)/1000$$
$$= 45.6 \text{ k,} \qquad V_n \gtrless 45.6/0.85 = 53.6 \text{ k}$$

From the basic ideas of Sec. 5.25, the centroid of shear resistance is at the centroid of the shaded areas of Fig. 15.14b.

$$A_c = 2 \times 18.5 \times 6.5 + 21.5 \times 7 = 241 + 150 = 391 \text{ in.}^2$$

From center of column,

$$\bar{x} = (2 \times 18.5 \times 6.5 \times 1.75 + 21.5 \times 7 \times 11)/391 = 5.33 \text{ in.}$$

The factored torque from both sides of the column is the 40% of the column strip negative moment not carried directly into the column.

$$\text{Col. strip negative } M_u = \phi M_n = 0.9 \times 55 = 49.5 \text{ k-ft}$$
$$T_u = 0.40 \times 49.5 = 19.8 \text{ k-ft} \qquad T_n = T_u/\phi = 19.8/0.85 = 23.3 \text{ k-ft}$$

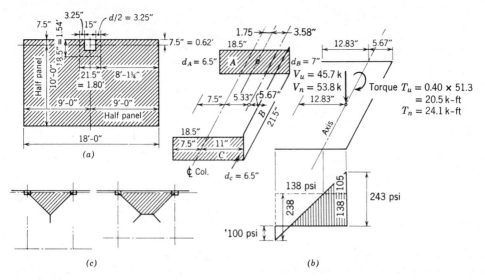

Figure 15.14 Shear considerations at exterior columns. (*a*) Shear load on column. (*b*) Shear and torque loading with resultant stresses. (*c*) Load to edge beam, when present.

The polar moment of inertia of these areas about a horizontal axis through the centroid and parallel to the center line of column is required. Using the area designations in Fig. 15.14*b*:

$$J_C = J_A = I_{xA} + I_{yA} \qquad\qquad J_B = A_B(18.5 - 7.5 - 5.33)^2$$
$$I_{xA} = 18.5 \times 6.5^3/12 \quad = 424 \qquad\qquad = 5.67^2 A_B$$
$$I_{yA} = 6.5 \times 18.5^3/12 \quad = 3440 \qquad J_B = 150 \times 5.67^2 = 4830 \text{ in.}^4$$
$$+18.5 \times 6.5 \times 3.58^2 = \underline{1540}$$
$$J_A = 5400 \text{ in.}^4$$
$$J = 2 \times 5400 + 4830 \quad = 15{,}600 \text{ in.}^4$$
$$v_{max} = V_n/A_c + T_n x_B/J$$
$$= 53{,}600/391 + 23{,}300 \times 12 \times 5.67/15{,}600 = 137 + 102 = 239 \text{ psi}$$
$$v_{min} = 137 - 102 \times 12.83/5.67 = 137 - 231 = -94 \text{ psi}$$

The allowable maximum is $4\sqrt{f'_c} = 4\sqrt{4000} = 253$ psi > 239 **O.K.**

If a wall was included, as is normal, it would have added substantially to the average shear but would have decreased the torque some because its weight would be to the left of the centroid.

(f) Torque on Slab   Code 13.6.3.5 requires that edge beams or slab edges be designed to resist their share of exterior negative moments. Torque might be calculated at a distance *d* from the column (Code 11.6.4) but is first checked for the total torque to see if torque is serious.

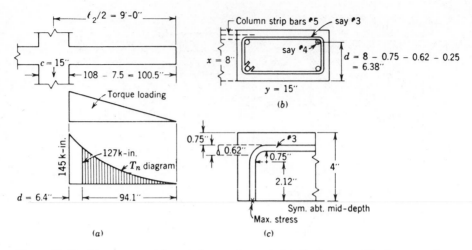

Figure 15.15 Torque provisions in beam element. (a) T_n diagram. (b) Cross section. (c) Development in vertical leg of closed tie.

T_n required from above = 23.3/2 = 11.65 k-ft = 140 k-in.

For the specified edge slab strip equal to the column in width (Fig. 15.15b) and with torsion always on gross area (Sec. 5.26d), $\Sigma x^2 y/3 = 8^2 \times 15/3 = 320$ in.[3] Code 11.6.3 gives a ready measure for judging the severity of torsional moments. In statically indeterminate applications (not equilibrium torsion) it indicates that the level of torsional moment *after cracking* can be taken as $T_n = 4\sqrt{f'_c}x^2y/3$. For this slab edge strip the Code 11.6.3 limit is $T_n = 4\sqrt{4000}\,(320)/1000 = 81$ k-in. Although one might argue that design could be based on $T_n = 81$ k-in., the designer recognizes that the shear transfer calculations are based on transfer by torque of $T_n = 140$ k-in. and prudence indicates that the edge slab should have that torque capacity. If one considers a parabolic torque diagram* with maximum at the *face* of column as shown in Fig. 15.15a, and takes $d = 8 - 0.75 - 0.62$ (col. strip bar) $- 0.25 = 6.38$ in., T_n at a distance d from the column is:

$$T_n = 140(94.1/100.5)^2 = 123 \text{ k-in.}$$
$$v_{tn} = 123{,}000/320 = 384 \text{ psi} = 6.1\sqrt{f'_c}$$

The large torque shear stress indicates the basic problem of the slab without an edge beam.† Nor does the Code speak directly to the problem

---

* This is not quite consistent with the 99.1% of negative moment on the column strip (subsection b), but is simple and on the safe side.
† This paragraph reflects the senior author's bias against slabs with neither an overhanging cantilever nor an edge beam. The state of the art does not provide a documented design procedure. As in many design situations the engineer must visualize the slab behavior (deformation) and be overconservative in meeting it until more tests have been made of torsion reinforced slabs.

of how to design for torque quite close to a concentrated slab load, in this case the column reaction. Some two-way action exists on the inner face of the column but primarily one-way action exists near the outside column face. In this situation decisions must be somewhat arbitrary and there is no practical meaning in writing a third significant digit, although habit may make it easier that way.

If a beam existed in fact, one would calculate V_c and T_c at a distance d from the column face, but the criteria would be those for beam shear, critical on the inside face of the beam.

Section 5.29 outlined the design procedure for combined shear and torsion in beams. The general approach is followed here, although considerable refinement is possible. The required $T_n = 123$ k-in. is provided by a combination of concrete plus closed stirrups:

$$T_n = T_c + T_s$$

$$T_c = \frac{0.8\sqrt{f_c'}\ \Sigma x^2 y}{\sqrt{1 + \left(\dfrac{0.4 V_u}{C_t T_u}\right)^2}}$$

For the *one-way* edge beam slab strip, V_u could be crudely taken as $(94/12) \times (15/12) \times 242/1000 = 2.4$ k.

$$T_u = 0.85 T_n = 0.85(123) = 105 \text{ k-in.}$$
$$C_t = b_w d/\Sigma x^2 y = (15)(6.38)/(8^2)(15) = 0.10$$

$$T_c = \frac{0.8\sqrt{4000}\ (8^2)(15)}{1000\sqrt{1 + \left(\dfrac{0.4 \times 2.4}{0.1 \times 105}\right)^2}} = 40.8 \text{ k-in.}$$

$$T_s = T_n - T_c = 123 - 41 = 82 \text{ k-in.}$$

From Code Eq. 11.23, $T_s = A_t \alpha_t x_1 y_1 f_y/s$, $\alpha_t = 0.66 + 0.33(y_1/x_1)$. Using the tentative closed stirrup layout of Fig. 15.15b, $y_1 = 13.12$ and $x_1 = 5.62$ so that

$$\alpha_t = 0.66 + 0.33(13.12/5.62) = 1.43$$
$$\text{Required } A_t = T_s s/(\alpha_t x_1 y_1 f_y)$$
$$= 82 \times s/(1.43 \times 5.62 \times 13.12 \times 60) = 0.013s$$

Max. spacing in Code $= (x_1 + y_1)/4 = (5.62 + 13.12)/4 = 4.7$ in.

If $A_t = \#3 = 0.11$ in.^{2}, $\quad s = 0.11/0.013 = 8.49$ in. $>$ max. spacing

USE #3 closed ties at 4.5 in.

Required extra longitudinal steel $= 2A_t(x_1 + y_1)/s$

$A_\ell = 2 \times 0.013 s(5.62 + 13.12)/s = 0.49$ in.^{2}

$$= 0.25 \text{ in.}^2 \text{ each top and bottom}$$

Code 11.5.5.3 requires closed stirrups wherever T_n exceeds $0.5\sqrt{f'_c}\,x^2y$ or 30.3 k-in. With a parabolic T distribution rising to 140 k-in. at face of column

$$x^2/100.5^2 = 30.3/140,$$

$$x = 46.5 \text{ in. from midspan or 54 in. from column.}$$

Torsion reinforcement must be provided at least a distance $(d + b)$ beyond the point theoretically required $= 6.38 + 15 = 21.4$ in. This increases the length from column face to $54 + 21.4 = 75.4$ in. $= 6.3$ ft.

> USE 17-#3 closed ties at 4.5 in.
>> 3-#3 top and 3-#3 bottom bars extending 6.3 ft from face of column

The latter steel is in addition to the usual flexural reinforcement.

The #3 tie in an 8-in. slab may be of doubtful efficiency because of underdevelopment of its vertical leg. The hook (around the corner) requires (Code 12.5.1) a basic development length of $1200\,d_b/\sqrt{f'_c}$ or 7.12 in. As indicated in Fig. 15.5c, only 3.25 in. is available from slab middepth. Strictly speaking, this reinforcement should be checked for development of web reinforcement using Code 12.13.2.3, assuming the closed hoop equivalent to 135° bend. $\ell_d/3$ would be 3 in. so the 2.12 in. available in Fig. 15.15c to middepth is about 75% adequate. The use of 4.5 in. spacing here instead of a theoretical 8.5 in. seems to give the required margin in strength required.

(g) Modifications in Design Where Edge Beam Exists   A separate example will not be developed for an exterior panel with an edge beam, but note the essential design differences in such a case.

The formula values for slab thickness do not then need the 10% increase if the edge beam has an α of at least 0.80.

The edge strip (half-column strip) is made up of a beam plus a slab, instead of a slab alone. The methods used in Secs. 15.12 and 16.3 are available for establishing the distribution of M_o in this half-panel.

The values from column (4) of Table 15.3 would give a slightly higher exterior negative moment with a smaller change in positive moment. The relative torsional stiffness β_t of the beam would also modify (lower) the distribution of negative moment to the column strip, although increasing that to the middle strip.

The shear in this case would, in part, be brought into the column by the beam and, in part, directly by the slab. The load on the beams, including any direct wall load, could be assumed that bounded by 45° diagonals in the panel, as in Fig. 15.14c, plus a parallel centerline of panel if the exterior edge is the longer edge. Code 13.6.5.2 directs that where $\alpha_1\ell_2/\ell_1$ for the edge beam is less than unity this proportional part $(\alpha_1\ell_2/\ell_1)$ of the shaded area be used. The beam torque and shear should be analyzed at a distance d from the column, with closed stirrups provided for any excess.

The remainder of the column shear might be assumed carried around the column outside the limits of the beam web, at a distance $d/2$ from the column faces. This shear would be limited to $4\sqrt{f_c'}$ unless the beam framed flush with the loaded face of the column and turned this into a one-way shear problem, critical at a distance d from the face. It would not be necessary to worry about a nonuniform distribution of this shear (in the nonflush case), although some variation undoubtedly would result.

15.13   Flat Slab with Drop Panel—Design Example

Design a typical interior bay of a flat slab to carry a dead load (including its own weight) of 100 psf and a live load of 200 psf over an 18-ft × 19-ft 6-in. panel, center-to-center of columns. $f_c' = 3000$ psi, Grade 60 steel. Use a story height of 11 ft, a column capital, and a drop panel.

Solution

A flat slab of this type (Fig. 15.16*a*) is particularly good for heavy manu-facturing or warehouse loads. Shear can be controlled by the size of capital used and strength is more apt to control slab thickness than deflec-tion. However, deflection can be simply checked by calculating the re-quired thickness h from the Code equations as soon as ℓ_n, the net span, is fixed.

(a) Thickness for Deflection Control   The PCA equations discussed in Sec. 15.10 are compared by evaluating their denominators using $\alpha_m = 0$ because there are no beams and $\beta_s = 1$ because all panel edges are continuous.

$$\text{Eq. 9-11: } 36 + 5\beta[\alpha_m - 0.5(1 - \beta_s)(1 + 1/\beta)]$$
$$= 36 + 5(19.5)/18)[0 - 0] = 36,$$
$$\text{same as for Eq. 9-13.}$$

Equation 9-13 cuts off any greater requirement and Eq. 9-11 permits no less; Eq. 9-12 is surplus in this case.

$$h_s = \ell_n(0.8 + f_y/200{,}000)/36 = 1.1\,\ell_n/36 = \ell_n/32.7$$

For slabs with capitals, Code 13.6.2.5 states that the clear span ℓ_n is taken face-to-face of capitals but not less than 0.65 of the center-to-center span. Circular or polygonal supports are treated as square supports with the same area. No part of the column capital lying outside a 45° line to the column axis can be considered for structural purposes according to Code 13.1.2.

The net length must be estimated. Assume a column with a 3-ft 6-in. capital with an area of $\pi \times 3.5^2/4 = 9.62$ ft^2. The equivalent side of a

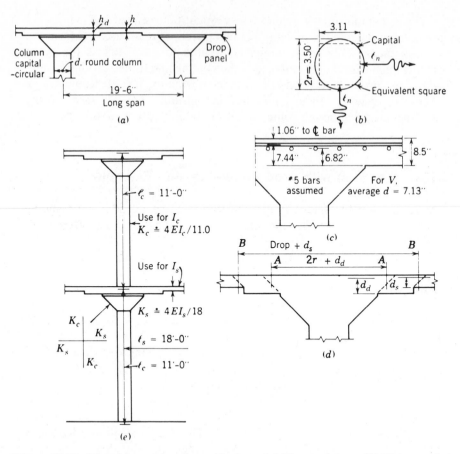

Figure 15.16 Flat slab construction with drop. (a) Nomenclature. (b) Net span ℓ_n to face of equivalent square support. (c) Depth for shear. (d) Critical shear sections. (e) Calculation of minimum column, about center joint.

square with the same area $= \sqrt{9.62} = 3.10$ ft. This column capital size may be changed after checking for shear.

$$\ell_n = 19.5 - 3.10 = 16.4 \text{ ft} > 0.65\ell_1 \qquad h_s = 16.4 \times 12/32.7 = 6.02 \text{ in.}$$

With the drop panels specified in Code 13.4.7 (see subsection (b)) Code 9.5.3.2 lowers the interior slab thickness requirement to satisfy deflection by 10% to 5.4 in., say, 5.5 in.

USE 5.5 in. slab thickness h_s.

(b) Depth for Flexure

$$\text{Dead load} = 1.40 \times 100 = 140$$
$$\text{Live load} = 1.70 \times 200 = 340$$
$$w = 480 \text{ psf}$$

$$\text{Long span } M_o = 0.125 \times 480 \times 18 \times 16.4^2/1000 = 291 \text{ k-ft}$$
$$M_{on} \gtrless M_u/\phi = 291/0.90 = 323 \text{ k-ft}$$
$$\text{Total negative } M_n = -0.65 M_{on} = -0.65 \times 323 = -210 \text{ k-ft}$$
$$\text{Col. strip negative } M_n = -0.75 \times 210 = -158 \text{ k-ft}$$

The drop panel (13.4.7) should be at least $\ell_1/3 \times \ell_2/3$, that is, in the long span direction 6-ft 6-in. and transversely 6-ft 0-in. Its thickness h_d must be at least $(5/4)h_s$, say in this case 7.0 in., which seems light for this heavy a load. For strength design it is desirable to keep the steel ratio around $0.5\rho_b*$ which means k_n slightly over 2/3 the value in Table 3.1 (for $0.75\rho_b$) which is 783. Try $k_n = 530$. The width in compression is that of the drop panel, 72 in.

$$\text{Required } d_m = \sqrt{M_n/k_n b} = \sqrt{158 \times 12,000/(530 \times 72)} = 7.05 \text{ in.}$$
$$\text{Drop } h_d = 7.05 + 0.75 \text{ cover} + 0.31 \text{ for } d_b/2 = 8.11 \text{ in.}$$

USE drop $h_d = 8.5$ in., subject to shear check,

$$\text{in panel 6-ft 0-in.} \times \text{6-ft 6-in.} \qquad d_m = 8.5 - 0.75 - 0.31 = 7.44 \text{ in.}$$

(c) Shear at Column Capital When the column-to-slab connection has drop panels and/or capitals several shear checks must be made. Code 11.11.3.3 cautions that shear must be checked not only at the critical section $d/2$ from the face of the support but at successive sections more distant from the support where the thickness changes. The possibility of the capital punching through the drop panel at A in Fig. 15.16d is checked in this subsection. The additional possibility of the drop panel punching through the thinner central slab at B in Fig. 15.16d is checked in subsection (d).

Use the assumed capital diameter of 3.5 ft $+ 2(d/2) = 3.5 + 0.6 = 4.1$ ft to calculate V_u and the required capital for shear. For punching shear the circular shape of the column is correct for computations. Punching tests indicate the failure shape is conical.

$$V_u = 488(18 \times 19.5 - \pi \times 4.1^2/4) = 165,000 \text{ lb}, \qquad V_n \gtrless V_u/\phi = 194 \text{ k}$$

Assume average $d = 8.5 - 0.75 - 0.62 (= d_b) = 7.13$ in., as shown in Fig. 15.16c.

Shear perimeter A (Fig. 15.16d) $= b_0 = 2\pi(r + 0.5 \times 7.13) = 2\pi r + 22.4$

$$V_n = 4\sqrt{f_c'}b_o d$$
$$b_0 = 2\pi r + 22.4 = V_n/4\sqrt{f_c'}d$$
$$= 194,000/(4\sqrt{3000} \times 7.13) = 124.2$$
$$r = (124.2 - 22.4)/2\pi = 16.2 \text{ in.}$$
$$2r = 32.4 \text{ in. vs. 42 in. assumed}$$

USE 36-in. diameter column capital

The equivalent square (same area) $= 32$ in. $= 2.67$ ft for use in ℓ_n.

---

* For some extra ductility and usually for overall economy.

(d) Shear Around Drop Shear around drop is critical at $d/2$ outside of the drop.

$$d = 5.5 - 0.75 - 0.62(= d_b) = 4.13 \text{ in. average}$$

Perimeter of shear section B (Fig. 15.16d) keeping drop dimensions 1/3 of center-to-center span each way and adding $d/2$ each way to the critical section $= b_0 = 2(78 + 4.13) + 2(72 + 4.13) = 317$ in.

$$V_u = (19.5 \times 18 - 82.13 \times 76.13/144)488/1000 = 150 \, k$$
$$\text{Required } V_n = V_u/\phi = 150/0.85 = 176.5 \text{ k}$$
$$\text{Actual } V_n = 4\sqrt{f_c'} b_0 d = 4\sqrt{3000} \times 317 \times 4.13/1000 = 287 \text{ k} \qquad \textbf{O.K.}$$

(e) Critical Moments Critical moments must be recalculated because the size of capital has changed.

$$\text{Long span } M_o = 0.125 \times 480 \times 18(19.5 - 2.67)^2/1000 = 306 \text{ k-ft}$$
$$M_{on} = M_o/\phi = 340 \text{ k-ft}$$
$$\text{Short span } M_o = 0.125 \times 480 \times 19.5(18 - 2.67)^2/1000 = 275 \text{ k-ft}$$
$$M_{on} = M_o/\phi = 306 \text{ k-ft}$$

The moment coefficients of Fig. 15.8b are entered directly into Table 15.5.

Although the short strip moment is smaller than that for the long strip, the effective depth is also necessarily smaller by one bar diameter (because of crossing bars) than the depth with the long strip moment. Both depth requirements will be rechecked.

$$\text{Long span col. strip negative } M_n = -0.65 \times 0.75 \times 340 = -167$$
$$\text{Short span col. strip negative } M_n = -0.65 \times 0.75 \times 306 = -150 \text{ k-ft}$$
$$\text{Required long span } d = \sqrt{167,000 \times 12/(530 \times 72)} = 7.25 \text{ in.}$$
$$\text{Min. } h_d = 7.25 + 0.75 + 0.31 \text{ (for \#5)} = 8.31 < 8.5 \text{ in above} \qquad \textbf{O.K.}$$
$$\text{Required short span } d = \sqrt{150,000 \times 12/(530 \times 78)} = 6.60 \text{ in.}$$
$$\text{Min. } h_d = 6.60 + 0.75 + 1.5 \times 0.62 = 8.28 \text{ in.} < 8.5 \text{ in.} \qquad \textbf{O.K.}$$

If either h_d is larger than the 8.5 in. desired, the designer has several choices:

1. Increase the drop panel thickness, weight, and moments.
2. Increase the drop width perpendicular to the strip, thus increasing the effective width b of strip (instead of d).
3. Increase column capital to reduce span and M_o.
4. Accept k_n several percent above the arbitrary 530 value used, unless the ductility is particularly important. (Code 8.4.3 permits moment redistribution only if $\rho \gtrsim 0.5\rho_b$.)

USE drop panel 6-ft 6-in. in long span direction by 6-ft 0-in. in short span direction.

TABLE 15.5 Reinforcement for Flat Slab of Sec. 15.14

	Long Span (19'-6") M_{on} = 340 k-ft				Short Span (18'-0") M_{on} = 306 k-ft			
	Column Strip 9'		Middle Strip 9'		Column Strip 8.5'		Middle Strip 10'-6"	
#5 bars assumed / Strip width	Neg.	Pos.	Neg.	Pos.	Neg.	Pos.	Neg.	Pos.
M_n coef.	0.49	0.21	0.16	0.14	0.49	0.21	0.16	0.14
M_n = (coef.) M_{on} k-ft	−167	+71	−55	+48	−150	+64	−49	+43
d, in.	7.44	4.44	4.44	4.44	6.82	3.82	4.44	~~3.82~~
$A_s = M_n \times 12/(60 \times 0.9d)$ $= M_n/(4.5d)$, in.^{2}	4.99	3.55	2.75	2.40	4.89	3.72	2.45	~~2.50~~ 2.39
Min. no. bars for $s = 2h$	10	10	10	10	10	10	12	12
No. of #5 / Area, in.^{2}	17-#5@6½±" (5.27)	12-#5@9" (3.72)			17-#5@6½±" (5.27)	10 12-#5@9" Pair (3.72)	12	12
No. of #4 / Area, in.^{2}			14-#4@7.5±" (2.80)	12-#4@9" (2.40)^a			13-#4@9.5±" (2.60)	12-#4@10½"—USE ~~13-#4@9½"~~ (2.60)(2.40)^a
Revised d, in.	O.K.	O.K.	4.50	4.50	O.K.	3.94	4.50	4.00

* O.K. because $z > 0.9d$ with these low M values.

(f) Reinforcement Calculations The remainder of the reinforcement calculations are made in Table 15.5. Because heavy loads and moments are involved the internal lever arm z is taken initially as $0.90d$ (compared to $0.95d$ in earlier examples). The required A_s at the column in the long strip, the heaviest required, is 4.99 in.^{2} and will be used to check z. The concrete width used is the drop panel width of 72 in.

$$a = 4.99 \times 60{,}000/(0.85 \times 3000 \times 72) = 1.63 \text{ in.}$$
$$z = 7.44 - 1.63/2 = 6.62 \text{ in.} = 0.89d$$

This z is about 1% low, but the 5% apparent excess in A_s used (5.27 used versus 4.99 required) more than offsets the 1% error in the calculation. In arranging these particular bars it would be better (except probably for the problem it adds to field supervision) to use the 17 bars spaced at a maximum of $2h_s = 11$ in. outside the 6-ft wide drop panel and at a smaller spacing over and very near the drop to maintain the total number.

In the table, rather than estimate z individually for each case, the $0.90d$ was first used throughout, leading to tentative bar selections. Next minor changes in d resulting from the use of #4 bars are noted and the whole inspected for possible weakness. In the second column, z for positive moment looked close but the positive moment under the sixth sketch showed as more critical. The z for this short column strip positive moment is therefore checked:

$$a = 3.72 \times 60/(0.85 \times 3 \times 108) = 0.810 \text{ in.}$$
$$z = 3.81 - 0.41 = 3.40 \text{ in.} = 0.89d$$

that gives a required $A_s = 3.72 \times 0.90/0.89 = 3.76$ in.^{2} compared to 3.72 in.^{2} used. Strictly speaking, this result is unacceptable. However, the A_s for negative moment in this strip is 5.27 in.^{2} compared to the 4.89 in.^{2} required. The *two* sets of bars *can be accepted* if the designer chooses to assume a little less positive moment and a little more negative moment in the strip. The Code (13.6.7) provides that a design moment may be modified 10% if all of M_o is taken into account. The second column by comparison has a satisfactory z since some 5% excess A_s is actually used anyway.

The short span middle strip positive moment reinforcement is close but is very adequate because this small A_s obviously leads to $z > 0.90d$ used originally. It is possible that more detailed study of z might lead to economy elsewhere, but this refined calculation is worthwhile only if many like panels are involved. When tables are available, one can go from M_n to calculated actual k_n to a table value of ρ suitable or that k_n. This procedure eliminates checks on z.

Temperature steel calculations are not shown because a quick mental check of $0.0018bh$ in the short (wide) span middle strip indicated it does not control.

(g) Column Stiffness to Minimize Pattern Loading Effects Ratio of dead-to-live load, $\beta_a = 100/200 = 0.5$, is much below the factor of 2 required to make the previous calculations acceptable without a special check on the column stiffness. $\ell_2/\ell_1 = 19.5/18 = 1.08$. Because there are no interior beams, $\alpha = 0$. Table 15.2 in Sec. 15.8 shows, fortunately, with only the interpolation for this particular ratio and none for β_a, that $\alpha_{min} = 1.7$.

The real column stiffness with the capital is greater than $4EI/\ell$. If I is based on minimal cross section, as usual, the coefficient for K of the column above is larger than 4 and in the column below (with the capital at the joint) *much* larger than 4, making more than half of the unbalanced moment at the joint go to the column below. The Code Commentary suggests more approximate methods are appropriate to the direct design method, both for the columns and slabs, but that similar simplifications should be used for both columns and slabs.

The slab stiffness is modified by both the drop panel and the column capital. Here the minimum I_s will be used, with the short span that is more restrictive in this case.

$$I_s = (19.5 \times 12)5.5^3/12 = 3250 \text{ in.}^4$$
$$K_s = 4E_c \times 3250/(18 \times 12) = 60E_c$$
$$\alpha_{min} = 2K_c/2K_s \gtrless 1.7$$
$$K_c = 4E_cI_c/\ell_c \gtrless 1.7K_s$$
$$I_c \gtrless 1.7K_s\ell_c/4E_c = 1.7 \times 60E_c\ell_c/4E_c = 25.5\ell_c$$
$$I_c \gtrless 25.5 \times 11 \times 12 = 3370 \text{ in.}^4$$

For a circular column of diameter d,

$$I_c = 3370 \text{ in.}^4 = \pi d^4/64$$
$$d^4 = 3370 \times 64/\pi = 68,700$$
$$d = 16.2 \text{ in.}$$

This calls for a minimum 17-in. diameter column (or an increase in the positive moments already used for the slab design). In the strength design of this column the moments of Code 13.6.9.2 must be considered.

15.14   Shear Reinforcement in Slabs

The limited depth of slab makes the anchorage of shear reinforcement difficult and the anchorage requirements of Code 12.13 must be closely observed. The older ring type of wire reinforcement with inclined and inverted V-shaped wires welded around the perimeter is out of favor because its ability to pick up or develop the necessary stresses and again properly anchor them is seriously questioned. As a result, the 1963 Code discontinued the crediting of shear reinforcement of bars, rods, or wires in slabs with a total depth of less than 10 in. Since 1963, better tests have shown that *well-anchored* bent bars and closed ties with bars in each corner can be effective. The 1986 Code permits such shear reinforcement (Code Secs. 11.11.3.4 and 11.11.3.5) provided it takes care of all shear in excess of $2\sqrt{f'_c}b_o d$, thus leaving for concrete (in conjunction with shear reinforcement) only half of the usual V_c normally permitted around columns. Careful detailing and careful placement are both essential to this use.

Because of these problems, shearheads of structural steel also have been further developed for slabs at interior columns.[11] The detailed provisions of Code Secs. 11.11.4.1 through 11.11.4.9 apply only for interior columns. As specified, *they do not work adequately at exterior columns,* but research in this area is making progress.[13] Code 11.12.2.5 permits exterior column shearheads with appropriate moment transfer calculations.

Shearheads consist of four crossing steel arms welded together at a common level to pick up both some shear and moment load from the concrete. Each shearhead arm may consist of a small wide flange beam or of two small channels turned back to back but spaced apart as much as the channel depth or more. These arms (totally within the slab thickness) pick up shear and moment beyond the column and bring the load to bearing on the column. The bottom flanges of the steel shapes are extended beyond the top flanges (with a sloping cut through the steel web) to pick up shear load that will exist low in the slab.

The critical section for shear on the concrete is thus removed to a larger perimeter farther than $d/2$ away from the column. The steel cantilever is considered effective in moving the critical shear section out from the column to a point 3/4 of its length from the column centerline. Between arms this critical section is considered a straight line joining the several 3/4 points on the arms, but never closer than $d/2$ to the face of the column.

The shearhead is a special type of construction that permits a thinner slab where shear controls slab thickness at an interior column. It may be economical in some cases. Its detailed design is considered too specialized a problem for this book.

15.15   Lift Slabs

Lift slabs are a flat plate type of slab cast at grade level and embedding steel shoes or collars that fit loosely around the columns. After the slabs are cured they are lifted by a patented jack system to the proper level where the shoes are welded to the steel columns. Although the Code makes no special attempt to provide for this type of design, the elastic method of flat slab analysis seems proper for lift slabs. Because column stiffness is small, there is no slab section that deserves to be treated as having a greatly increased moment of inertia. Loadings for maximum moment are also more significant in this case. A rigid joint between the collar and the column is an essential condition. Punching shear must be carefully checked.

Because lift slabs do not qualify for empirical design procedures, total moment cannot be reduced to the M_o value. The collar stiffness may determine whether the critical section is at the center of the column or farther out.

The designer must consider carefully stresses caused by differential jack movements and make certain that in the field these do not exceed the limits considered in the office design. The constructor must provide adequate temporary bracing for system stability during erection. A major failure during construction of a lift slab building in Bridgeport, Connecticut occurred in 1987 and resulted in the deaths of many workers.

15.16 Cantilevers from Slab Construction

Lift slabs generally, and flat slabs and flat plates frequently, have the columns set back from the outside wall, thus causing the outermost section of the slab to act as a cantilever beyond the exterior columns. This is quite favorable to regular slab action. A proper overhang of the slab can provide a total negative moment that can almost eliminate the special problems associated with exterior slab panels.

The cantilever provides a total negative moment about the exterior columns that is statically determinate, but its distribution is not uniform. It is suggested that this distribution might be taken the same as that used to distribute negative moment between the column and middle strips in the elastic analysis method. If the cantilever projects the ideal distance, it can thus balance the typical interior slab negative moments. Longitudinal steel, as in any column strip, is needed to deliver the cantilever reaction ultimately to the columns.

Cantilever slabs have large deflections that become conspicuous after creep of concrete has taken place. The use of some compression steel solely to reduce deflections is often justified, as noted in Chapter 3.

Selected References

1. H. M. Westergaard and W. A. Slater, "Moments and Stresses in Slabs," *ACI Proc.*, 17, 1921, p. 415.

2. *Notes on ACI 318-71 Building Code Requirements with Design Applications, Portland Cement Association*, Skokie, Ill. 1972.

3. M. A. Sozen and C. P. Siess, "Investigation of Multi-Panel Reinforced Concrete Floor Slabs: Design Methods—Their Evolution and Comparison," *ACI Jour.*, 60, No. 8, Aug. 1963, p. 999.

4. J. R. Nichols, "Statical Limitations Upon the Steel Requirement in Reinforced Concrete Flat Slab Floors," *ASCE Trans.*, 77, 1914, p. 1670.

5. N. W. Hanson and J. M. Hanson, "Shear and Moment Transfer Between Concrete Slabs and Columns," *Jour. PCA Research and Development Laboratories*, 10, No. 1, Jan. 1968, p. 2.

6. D. S. Hatcher, M. A. Sozen, and C. P. Siess, "Test of a Reinforced Concrete Flat Slab," *Proc. ASCE*, 95, ST6, June 1969, p. 1051.

7. J. O. Jirsa, M. A. Sozen, and C. P. Siess, "Pattern Loadings on Reinforced Concrete Floor Slabs," *Proc. ASCE*, 95, ST6, June 1969, p. 1117.

8. W. L. Gamble, M. A. Sozen, and C. P. Siess, "Test of a Two-Way Reinforced Floor Slab," *Proc. ASCE*, 95, ST6, June 1969, p. 1073.

9. W. G. Corley and J. O. Jirsa, "Equivalent Frame Analysis for Slab Design," *ACI Jour.*, 67, No. 11, Nov. 1970, p. 875.

10. S. A. Guralnick and R. W. Fraugh, "Laboratory Study of a Forty-Five-Foot Square Flat Plate Structure," *ACI Jour.*, 69, No. 9, Sept. 1963, p. 1107.

11. W. G. Corley and N. M. Hawkins, "Shearhead Reinforcement for Slabs," *ACI Jour.*, 65, No. 10, Oct. 1968, p. 811.

12. D. J. Fraser, "Equivalent Frame Method for Beam-Slab Structures," *ACI Jour.*, 74, No. 5, May 1977, p. 223.

13. N. M. Hawkins and W. G. Corley, "Moment Transfer in Slabs with Shearhead Reinforcement," *ACI SP-42, Shear in Reinforced Concrete*, 1974, pp 847–879.

Problems

PROB. 15.1. Using $f'_c = 3000$ psi, Grade 60 steel, and no shear reinforcement, design (direct design method) an interior panel of a flat plate floor supported on columns 18-in. square spaced 18 ft on centers each way. The slab is to carry a total dead load (including slab weight) of 100 psf and live load of 50 psf. Check minimum column stiffness for a story height of 9 ft floor to floor.

PROB. 15.2. Redesign the slab of Prob. 15.1 for the same data except with 20-ft square panels. (For this problem neglect change in slab weight.)

PROB. 15.3. Under the direct design method design a flat slab without a drop panel for a 20-ft square interior panel using $f'_c = 4000$ psi and Grade 60 steel. Assume total dead load of 125 psf and live load of 150 psf. For the initial trial assume column capital $c = 4$-ft 6-in. Story height is 10 ft.

PROB. 15.4. Redesign the flat slab of Prob. 15.3 using a drop panel, without changing the assumed design dead load.

PROB. 15.5. Design a 19-ft × 22-ft interior panel of a flat slab without a drop panel, using the direct design method, $f'_c = 3000$ psi, Grade 60, a total dead load of 100 psf, and a live load of 125 psf. For the first trial use $c = 4$-ft 0-in. unless otherwise instructed.

PROB. 15.6. Design the flat slab of Prob. 15.5 with a drop panel.

PROB. 15.7.

(*a*) Design an exterior panel (not a corner panel) of the flat plate described in Prob. 15.1, using the direct design method.

(*b*) What moment should be included in the exterior column design for this loading?

16

INTERACTION OF TWO-WAY SLAB SYSTEMS WITH BEAMS AND COLUMNS

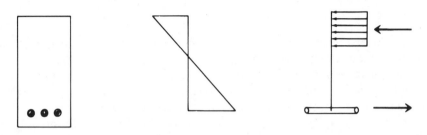

16.1 Two-Way Slabs on Beams—General

Code Chapter 13 on slab systems covers two-way slabs on beams that frame directly into columns, as well as flat plates and flat slabs. It provides for the broad transition from slabs with no stiffening to rather rigid beam supports. Because Code Chapter 13's coverage is so broad, the requirements are difficult to visualize in all their ramifications. This chapter introduces this design area by the same stages as already used for the direct design method with the flat plate and flat slab portions in Chapter 15.

The initial treatment emphasizes the direct design procedure. The more general equivalent frame method of analysis is introduced in Sec. 16.4. At the beginning of the direct design approach, two-way slabs on beams that frame directly into columns are introduced by a layout with beams stiff enough to minimize their contribution to slab deflection. The direct design moment assignment tables from Code 13.6.4.1 and 13.6.4.4 are shown for reference in Table 16.1. For interior spans these assignment percentages indicate that $\alpha_1 \ell_2/\ell_1 \gtrsim 1$ is used as a break point, that is, the point where beams carry essentially all the load the slab tends to deliver to them; below this $\alpha_1 \ell_2/\ell_1$ value lie small beams or deepened slab ⌣ections that simply stiffen the slab as an inseparable combination. Of course, $\alpha_1 = 0$ means no beams because α_1 is the ratio of beam stiffness to slab stiffness.

For convenience where $\alpha_1 \ell_2/\ell_1 \gtrsim 1$, the term stiff beams is used, although this stiffness is relative to the slab and not to beams in general. The beam values in the example of Sec. 16.2 indicate that the resulting "stiff" beam, viewed separately from a slab problem, might be considered just a usual beam.

Whatever the concept, at $\alpha_1 \ell_2/\ell_1 \gtrsim 1$ the slabs and beams can be almost independently designed.* Where $\alpha_1 \ell_2/\ell_1 < 1$, the design must be a joint slab

---

* α_1 is relative beam stiffness, discussed in the next paragraph. Because ℓ_2/ℓ_1 is limited in the direct design method to the range 0.5 to 2, an $\alpha_1 = 2$ always gives $\alpha_1 \ell_2/\ell_1 \gtrsim 1$.

519

TABLE 16.1   Percentages of Moment Assigned to Column Strip

ℓ_2/ℓ_1	Interior Negative Moment			Positive Moment		
	0.5	1.0	2.0	0.5	1.0	2.0
$\alpha_1 \ell_2/\ell_1 = 0$	75%	75%	75%	60%	60%	60%
$\alpha_1 \ell_2/\ell_1 \geq 1.0$	90%	75%	45%	90%	75%	45%

and beam problem starting from a trial combination of slab and stiffening beam, a problem where there are many independent design possibilities. For such combination design the designer must start with some assumed stiffening of the slab suiting the architectural demands or the designer's interest and verify its adequacy and the necessary reinforcement, as in Sec. 16.3.

The ratio $\alpha_1 = I_b/I_s$ needs some special notice because I_b is for an arbitrary definition of a beam and I_s is for an entire panel width (not what is left after the beam is separated). The beam includes projections above or below the slab plus a flange equal on each side to the larger of these beam projections, but not greater than $4 h_s$, as shown in Fig. 16.1b (Code 13.2.4). The value of I_s is that of all the slab, a width ℓ_2, as though no beam was present. Thus the same beam cross section on adjacent sides of a slab leads to different α_1 values unless the slab is square ($\ell_1 = \ell_2$).

The next section designs a slab on stiff beams and Sec. 16.3 designs a slab employing small slab bands as stiffening.

16.2   Two-Way Slabs on Stiff Beams—Direct Design Example

Using the direct design method, design an interior bay two-way slab on *stiff* beams for a bay 16-ft × 20-ft (to column centers) with service dead load (including slab weight) of 75 psf and live load of 100 psf, $f'_c = 3000$ psi, Grade 60 bars, 0.75-in. clear cover on slabs, 1.5-in. clear cover on beams, columns 15-in. square, and an assumed column height of 10 ft. Assume beams stiff enough to give $\alpha = $ beam I/slab $I = 1.00$ in the short direction and 1.25 in the long direction.

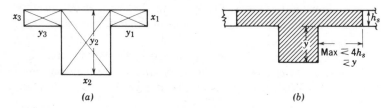

Figure 16.1   (*a*) Rectangles used in calculation of C for torsion. (*b*) Flange in slab assumed as part of beam.

Solution

(a) Relative α_1 Values For a given size of beam on all four sides such that α in the short direction is 1.0, the ratio $\alpha_1 \ell_2/\ell_1 = 1.0 \times 20/16 = 1.25$ in the short direction. In the other direction I_s is only 16/20 as large, α_1 becomes 1.25, and $\alpha_1 \ell_2/\ell_1 = 1.25 \times 16/20 = 1.0$. α_1 values of 1.0 and 1.25 are assumed as the minimum to make $\alpha_1 \ell_2/\ell_1 \geqslant 1$, although stiffer beams in either direction do not influence the slab part of the design.

(b) Slab Thickness Based on the comments in Sec. 15.10:

$\ell_n = 20 - 1.25 = 18.75$ ft in longer direction; $16 - 1.25 = 14.75$ ft in
  shorter direction
$\beta = $ long/short *clear* span ratio $= 18.75/14.75 = 1.27$
$\beta_s = 1$, continuous all around
$\alpha_m = $ average α for all beams around panel $= 2(1.0 + 1.25)/4 = 1.12$

Compare the denominators of the h equations, using the PCA forms:

Eq. 9-11: $36 + 5\,\beta[\alpha_m - 0.5(1 - \beta_s)(1 + 1/\beta)]$
  $= 36 + 5 \times 1.27(1.12) = 43.1 > 36$
Eq. 9-12: $36 + 5\beta(1 + \beta_s) = 36 + 5 \times 1.27(1 + 1) >$ above;
  cannot govern.

Eq. 9-13 is an upper limit of no interest here, since Eq. 9-11 calls for a smaller thickness.

Required $h_s = (0.8 + 60,000/200,000)\ell_n/43.1$
  $= 1.1 \times 18.75 \times 12/43.1 = 5.75$ in., say, 6 in.

This will be checked later for moment. Shear with this type slab with stiff beams is rarely critical.

(c) Design Moments

$$\text{Dead load} = 1.4 \times 75 \ = 105$$
$$\text{Live load} = 1.7 \times 100 = \underline{170}$$
$$w = 275 \text{ psf}$$

This result does not include the stem weight of the beams.

Long Span

$$M_o = 0.125 \times 275 \times 16 \times 18.75^2/1000 = 193.4 \text{ k-ft*}$$
$$M_{on} \geqslant M_u/\phi = 193.4/0.90 = 214.8 \text{ k-ft}$$

---

* The breakdown in moments can be in terms of M_{on} (ready for design) or in terms of M_o for later conversion to M_n values.

Code 13.6.3.2 distributes the total static moment as:

Total negative moment $M_n = -0.65\,M_{on} = -0.65 \times 214.8 = -139.6$ k-ft
Total positive moment $M_n = 0.35\,M_{on} = 0.35 \times 214.8 = 75.2$ k-ft

Column strip:

$$\alpha_1 = 1.25 \text{ assumed}, \qquad \ell_2/\ell_1 = 16/20 = 0.8, \qquad \alpha_1\ell_2/\ell_1 = 1.0$$

For this $\alpha_1\ell_2/\ell_1$, Table 16.1 (from Code 13.6.4.1 and 13.6.4.4) shows moment assignment percentages to the column strip the same for negative and positive moment. Interpolations for ℓ_2/ℓ_1 of 0.8 and 1.25 give 81% and 67.5%:

ℓ_2/ℓ_1	0.5	0.8	1.0	1.25	2.00
$\alpha_1\ell_2/\ell_1$	90%	81%	75%	67.5%	45%

Code 13.6.5 assigns 85% of the column strip moment to the beam for these stiff beams.

> Column strip negative $M_n = 0.81(-139.6) = -113.1$ k-ft
>> Beam negative $M_n = 0.85(-113.1) = -96.1$ k-ft
>> Slab negative $M_n = -113.1 + 96.1 = -17.0$ k-ft
>> $= -8.5$ k-ft/half-strip
> Column strip positive $M_n = 0.81 \times 75.2 = +60.9$ k-ft
>> Beam positive $M_n = 0.85 \times 60.9 = +51.8$ k-ft
>> Slab positive $M_n = 60.9 - 51.8 = +9.1$ k-ft $= +4.55$ k-ft/half-strip
> Middle strip (takes the remaining moments)
>> Middle strip negative $M_n = -139.6 + 113.1 = -26.5$ k-ft
>> Middle strip positive $M_n = 75.2 - 60.9 = +14.3$ k-ft

Short Span

$$M_o = 0.125 \times 275 \times 20 \times 14.75^2/1000 = 149.6 \text{ k-ft}$$
$$M_{on} \geqslant 149.6/0.90 = 166.2 \text{ k-ft}$$
> Total negative moment $M_n = -0.65 \times 166.2 = -108.0$ k-ft
> Total positive moment $M_n = 0.35 \times 166.2 = +58.2$ k-ft

Column strip:

$$\alpha_1 = 1.0 \text{ assumed}, \qquad \ell_2/\ell_1 = 20/16 = 1.25, \qquad \alpha_1\ell_2/\ell_1 = 1.25$$

From the earlier interpolation from Table 16.1, 0.675 M_n goes to the column strip and 0.325 M_n to the middle strip. The final distributions, similar to those for the long spans, are tabulated in Table 16.2 and used in Table 16.3. They also are shown on the slab layout in Fig. 16.2.

The middle strip negative moment on the short span is normally considered the most critical with regard to required slab depth for flexure. In this example the total negative moment on this strip is -35.1 compared to -26.5 on the long strip, but the respective widths (Fig. 16.2a) are 12 ft and 8 ft, making the worse moment per foot width that on the long span

TABLE 16.2   Moment Assignment for Sec. 16.2

	Long Span		Short Span	
	Negative M	Positive M	Negative M	Positive M
Total	M_{on} 214.8		166.2	
	0.65 M_{on} = −139.6	0.35 M_{on} = +75.2	0.65 M_{on} = −108.0	0.35 M_{on} = +58.2
	↓	↓	↓	↓
Column strip	−113.1	+60.9	−72.9	+39.3
	↙ ↘	↙ ↘	↙ ↘	↙ ↘
Beam	−96.1	+51.8	−61.9	+33.4
Slab	−17.0	+9.1	−11.0	+5.9
Middle strip	−26.5	+14.3	−35.1	+18.9

(26.5/8 = 3.31 versus 35.1/12 = 2.93). With an assumed thickness of 6 in., 0.75-in. clear cover, and assuming #4 bars, d = 5 in.

$$k_n = M/bd^2 = 3.31 \times 12{,}000^*/(12 \times 5^2) = 132$$

This result is very much under the allowable k_n of 783 for $\rho = \rho_{max}$ in Table 3.1; the 6-in. slab is obviously satisfactory for moment.

USE slab h = 6 in.

---

* 12,000 is a constant to adjust moment units to in.-lb.

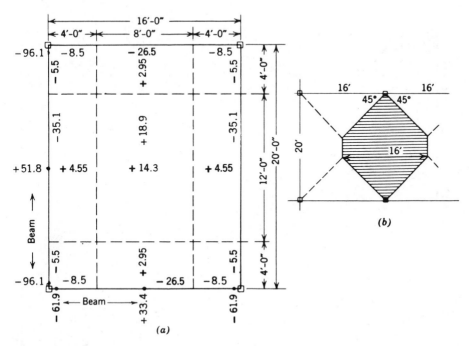

Figure 16.2   Slab on beams. (a) Distribution of moments to the strips and beams. (b) Load distribution to beams.

16.2   TWO-WAY SLABS ON STIFF BEAMS   **523**

TABLE 16.3 Reinforcement for Slab of Sec. 16.2

	Long Span		Short Span	
Assume #4	Neg. M	Pos. M	Neg. M	Pos. M
Middle Strips				
Strip Width	8 ft		12 ft	
M_{on}, k-ft	214.8		166.2	
M_n, k-ft	-26.5	$+14.3$	-35.1	$+18.9$
d, in.	5.00	5.00	5.00	4.50
$A_s = M_n \times 12/(60 \times 0.95\,d)$ $= M_n/(4.75\,d)$, in.²	1.12←	0.60	1.48	0.89
0.0018 bh, in.²	1.04	1.04←	1.56←	1.56←
Min. no. bars, $s = 2h$	8	8	12	12
#3 bars (0.11 in.²)	11-#3 @ 8.5 ±" (1.21)	10-#3 @ 9.5" (1.10)	14-#3 @ 10.5 ±" (1.54)	15-#3 @ 9½ ±" (1.65)
Revised d, in.	5.06	5.06	5.06	4.69
Column Strips				
Strip width	Half-strip = 4 ft (incl. beam) ≈ 3.5 ft net		Half-strip = 4 ft (beam) ≈ 3.5 ft net	
M_n, k-ft/half	-8.5	$+4.55$	-5.50	$+2.95$
d, in.	5.06	5.06	4.69	4.69
$A_s = M_n/(4.75\,d)$, in.²	0.35	0.19	0.25	0.14
0.0018 bh, in.²	0.46←	0.46←	0.46←	0.46←
Min. no. bars, $s = 2h$	4	4	4	4
#3 bars (0.11 in.²)	4-#3 @ 10 ±" (0.44)		Same for each	

(d) Check Minimum Column Stiffness for 10-ft Story Height The column stiffness requirement (already discussed in Sec. 15.8) is in the Code to protect slabs from pattern loading. The requirement should not be a problem where stiff beams are used, but it is checked anyway. The ratio of service dead to live load is $\beta_a = 75/100 = 0.75$, less than the criterion of 2, which causes the designer to check α_{min} from Table 15.2. In the two directions $\ell_2/\ell_1 = 20/16 = 1.25$ with $\alpha = 1.0$ and $\ell_2/\ell_1 = 16/20 = 0.8$ with $\alpha = 1.25$. Values of α_{min} increase as ℓ_2/ℓ_1 increases and decrease as beam α increases. Hence the short span with $\ell_2/\ell_1 = 1.25$ and $\alpha = 1.0$ is the more critical for this check. Interpolating with $\ell_2/\ell_1 = 1.25$ and $\alpha = 1.0$ fixed:

$$\begin{array}{llll} \beta_\alpha = 1.0 & \alpha_{min} = 0 & & \\ \beta_\alpha = 0.5 & \alpha_{min} = 0.5 & \beta_\alpha = 0.75 & \alpha_{min} = 0.25 \end{array}$$

To use this result, the stiffness of slab and beam must be developed for the short span.

$$I_s = 20 \times 12 \times 6^3/12 = 4320 \text{ in.}^4$$
$$I_b = \alpha I_s = 1.0\, I_s = 4320 \text{ in.}^4$$

With slab and beams on both sides of the column and the column both above and below the joint:

$$\alpha_{min} = \Sigma K_c/\Sigma(K_s + K_b) = K_c/(K_s + K_b)$$
$$0.25 = (I_c/10)/(4320 + 4320)/16] = I_c/5400$$
$$\text{Min. } I_c = 0.25 \times 5400 = 1350 \text{ in.}^4 = h^4/12$$
$$\text{Min. } h = 11.3 \text{ in.}$$

The column size is 15 in., which easily satisfies the minimum thickness.

(e) Slab Reinforcement The positive moment bars in the middle strips do cross each other and require design for different depths. With moments (per foot width) so nearly the same it does not matter much whether long span or short span bars are given preferential location. Here the long bars are placed on the bottom, making:

$$d_\ell = 6 - 0.75 - 0.25 = 5.00 \text{ in. (for \#4 bars)}$$
$$d_g = 6 - 0.75 - 0.50 - 0.25 = 4.50 \text{ in. (for \#4 bars)}$$

Middle strip top bars in both directions may be in the top with $d = 5.00$ in., as the only interference is possibly near the quarter points of the sides. Column strip bottom bars have to match the middle-strip bottom bars. Negative moments in the perpendicular column strips occur in the same area. Use long strip bars on top with $d = 5.00$ in. and short strip bars with $d = 4.50$ in. The sketches and calculations in Table 16.3 complete the basic design of the slab.

The maximum bar spacing (13.4.2) is $2\, h_s = 12$ in. The steel also must be adequate for temperature steel requirements, that is, $0.0018\, bh$. The latter requirement often governs in this problem.

(f) Beams   The moments are available for the beam designs except that the weight of stem still must be added. Although beams may be designed for strength, unless they provide $\alpha_1 \ell_2/\ell_1 \geqq 1.0$, the slab computations already made must be revised. This is equivalent to stating that the minimum stiffness represented by the assumed α values should be met. In this case this only means I_b in the short span of 4320 in.[4] (or 3460 in.[4] for the long span) or, say, a beam vertical projection below the 6-in. slab of about 9 in. over a 12-in. width.

The authors call attention to the fact that beam moments obtained this way are lower than those developed from conventional analysis. In effect this is a mild form of limit design for the beams comparable to the general slab design rules. As such, the effective percentage of reinforcement, in the authors' opinion, should preferably meet the requirements of Code 8.4.3, that is, ρ or $\rho-\rho'$ not exceeding 0.5 ρ_b, where ρ_b is for a balanced member; this percentage is lower than the 0.75 ρ_b upon which Table 3.1 values of k_n are based. Code Chapter 13 does not require this limitation.

The shear load taken by the beams in this case must be that from the shaded area of Fig. 16.2b and any directly applied loads. If $\alpha_1 \ell_2/\ell_1 < 1.0$, the shear is reduced proportionately with this ratio and the remainder is carried by the slab around the column.

(g) Exterior Panels   Design procedures for exterior panels parallel those of Sec. 15.12 for the flat plate: the column strip simply includes beams and slab as previously. There is one special provision (13.4.6) that requires special reinforcement, both top and bottom, in any *exterior* corner of the slab system to take care of a moment equal to the maximum positive moment per foot of slab. The critical top reinforcement is parallel to the diagonal of the slab, and the bottom reinforcement is 90° to that diagonal. Either diagonal bars or a rectangular grid of bars may be used, with the effective A_s/ft the maximum required in the slab for positive moment.

16.3   Two-Way Slab with Stiffened Strips—Direct Design Example

Design an alternate interior panel slab for the conditions of Sec. 16.2. Consider the slab as stiffened only by strips 15-in. wide and 3-in. deeper than the slab running along the column centerlines each way.

Solution

(a) Slab Thickness   The α value for each beam can be established as soon as the slab thickness is fixed. On the basis of h_s values already used in this chapter for deflection (5.75 in. in Sec. 16.2 for slab with stiff beams and 6.9 in. for a flat plate in Sec. 15.7 that happened to be 2 ft wider than the present slab), 6.5 in. looks reasonable as an initial trial. This thickness requires an increase in dead load to 81 psf, the initially specified 75 psf being inadequate.

With a trial $h_s = 6.5$ in. and the effective flange rules shown in Fig. 16.1b, the beam is that of Fig. 16.3a. From the middle of the slab, y to the centroid becomes:

$$\bar{y} = -15 \times 3 \times 4.75/(21 \times 6.5 + 15 \times 3) = -1.18 \; in.$$

$$
\begin{aligned}
I_b: \quad & 21 \times 6.5^3/12 && = && 482 \\
& +21 \times 6.5 \times 1.18^2 && = && 190 \\
& 15 \times 3^3/12 && = && 34 \\
& +45(4.75 - 1.18)^2 && = && \underline{574} \\
& && I_b = && 1280 \; in.^4
\end{aligned}
$$

With $\ell_1 = 20$ ft, $I_{s\ell} = 16 \times 12 \times 6.5^3/12 = 4400$ in.^{4}

With $\ell_1 = 16$ ft, $I_{ss} = 20 \times 12 \times 6.5^3/12 = 5500$ in.^{4}

$$\alpha_{1\ell} = 1280/4400 = 0.29 \text{ for long span}$$
$$\alpha_{1s} = 1280/5500 = 0.23 \text{ for short span}$$
$$\text{Average } \alpha_1 = 0.26$$

The Code method is usable only if the relative beam stiffness in the two directions is within the ratio of 0.2 to 5.0 (Code 13.6.1.6).

$$(I_{b\ell}/\ell_\ell)/(I_{bs}/\ell_s) = (\alpha_{1\ell}/\alpha_{1s})(\ell_s/\ell_\ell)^2 = (0.29/0.23)(16/20)^2 = 0.81 \quad \textbf{O.K.}$$

$\ell_n = 20 - 1.25 = 18.75$ ft in the longer direction, $16 - 1.25 = 14.75$ ft in the shorter direction

$\beta = $ (long/short) clear span ratio $= 18.75/14.75 = 1.27$

$\beta_s = 1$, since continuous all around the panel

Compare the denominators of the h_s equations, using the PCA format (Sec. 15.10):

$$\text{Eq. 9-11: } 36 + 5 \beta[\alpha_m - 0.5(1 - \beta_s)(1 + 1/\beta)]$$
$$= 36 + 5 \times 1.27(0.26) = 37.7 > 36$$

The upper limit equation (9-13) is thus not useful here.

$$\text{Eq. 9-12: } 36 + 5 \beta(1 + \beta_s)$$
$$= 36 + 5 \times 1.27(1 + 1) = \;\gg 37.7$$

Required $h_s = [(0.8 + 60,000/200,000)18.75 \times 12]/37.7 = 6.57$ in.

Accept this 1.1% > 6.5 in. USE $h = 6.5$ in.

The alternate, of increasing the denominator 1.1% by increasing α_m (i.e., I_b) by 25%, could also be considered rather than increasing the entire slab a half inch.

(b) Shear Check At this stage one can check the shear only in a rough fashion. If the slab is considered simply as a flat plate, the critical shear

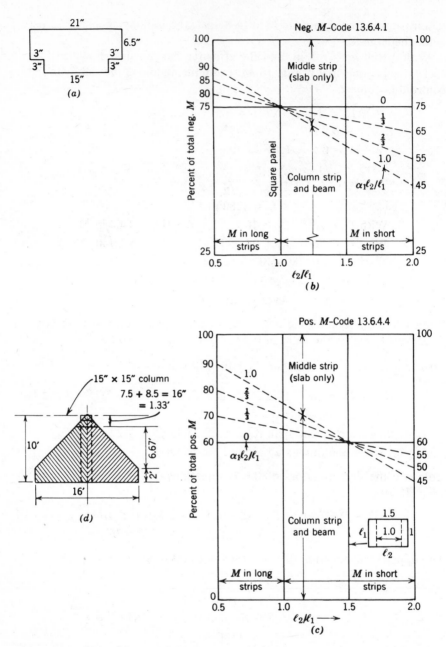

Figure 16.3 Slab of Sec. 16.3 (*a*) Beam used. (*b*) Moment assignment for interior negative moment. (*c*) Moment assignment for positive moment. (*d*) Shear load on beam.

section is a square around the column at $d/2$ beyond each face, with each side $15 + d = 15 + 5.25 = 20.25$ in. $= 1.69$-ft wide on each side of the column.

$$\text{Dead load} = 1.4 \times 81 \;\; = 113$$
$$\text{Live load} = 1.7 \times 100 = \underline{170}$$
$$\omega = 283 \text{ psf}$$
$$\text{Design stem weight} = 1.4(3 \times 15/144)150 = 66 \text{ plf}$$
$$V_u = 283(20 \times 16 - 1.69^2) + 66 \times 2(20 + 16 - 2 \times 1.69)$$
$$= 90{,}000 + 4300 = 94{,}300 \text{ lb}$$
$$V_n \text{ required} = V_u/\phi = 94{,}300/0.85 = 111{,}000 \text{ lb}$$
$$V_n = 4\sqrt{f'_c}b_o d = 4\sqrt{3000} \times 4 \times 20.25 \times 5.25 = 93{,}200 \text{ lb}$$

The shear V_u/ϕ exceeds the allowable V_n, on the basis of the oversimplified concept of the flat plate. The designer has several choices. (1) The slab might be thickened, although it has not been proved that the beams with stirrups, in spite of their lower V_c, cannot handle V_u; the beams are rather shallow for effective stirrups. (2) The concrete strength might be increased to $f'_c = 4500$ psi which would be enough. (3) The beams might be widened to make the 9.5-in. thickness effective all around the column, roughly a width of $15 + 7 = 22$ in. plus a little to keep the failure plane in the deeper elements, say, 26 in. This same result could be accomplished by adding a triangle of concrete in each corner between beams to build up a kind of octagonal drop panel effect to resist shear. (4) A steel shearhead might be included within the slab depth. Tentatively, the design will proceed with f'_c to be changed to 4500 psi if it becomes necessary.

(c) Design Moments

Long Span

$$M_{o\ell} = 0.125 \times 283 \times 16 \times 18.75^2/1000 = 199 \text{ k-ft}$$
$$\text{Total negative moment } M_u = -0.65 \times 199 = -129 \text{ k-ft*}$$
$$\text{Total positive moment } M_u = 0.35 \times 199 = 70 \text{ k-ft}$$

Coefficients for moment distribution or assignment are given in Table 16.1 but for visualization are also plotted in Fig. 16.3b,c.† For use, the following constants are required:

$$\alpha_1 = 0.29, \qquad \ell_2/\ell_1 = 16/20 = 0.80, \qquad \alpha_1\ell_2/\ell_1 = 0.23$$

Negative moment to the column strip is then $75 + (0.23/0.333)(0.2/0.5)5 = 75 + 1.4 = 76\%‡$ and positive moment to column strip is $60 +$

---

* Although the authors prefer going to M_n immediately, this example has not been revised to that format to illustrate that either format is acceptable.
† See also further detailed discussion in Sec. 16.9.
‡ This calculation, in effect, modifies the line for $\alpha_1\ell_2/\ell_1$ of $\frac{1}{3}$ in Fig. 16.3b to fit $\alpha_1\ell_2/\ell_1$ of 0.23. A value closer than a full percentage point is unrealistic. The curve equations are given in Sec. 16.9 along with some tabulated values in Table 16.8.

$(0.23/0.333)(0.7 \times 10) = 65\%$. Code 13.6.5.1 for $\alpha_1 \ell_2/\ell_1 > 1.0$ assigns 85% $\alpha_1 \ell_2/\ell_1$ of the negative or positive moment in the column strip to the beam. For $\alpha_1 \ell_2/\ell_1$ between 1 and zero, Code 13.6.5.2 requires the designer to decrease the 85% to beam linearly with the decrease in $\alpha_2 \ell_2/\ell_1$. These two conditions lead to:

Column strip negative moment $= -0.76 \times 129 = -98$ k-ft
   Beam negative $M_u = -(0.85 \times 0.23)98 = -19$ k-ft
   Slab negative $M_u = -98 + 19 = -79$ k-ft $= -39.5$ k-ft/half-strip
Column strip positive $M_u = 0.65 \times 70 = 46$ k-ft
   Beam positive $M_u = +(0.85 \times 0.23)46 = 9$ k-ft
   Slab positive $M_u = 46 - 9 = +37$ k-ft $= +18.5$ k-ft/half-strip
Middle strip negative $M_u = -129 + 98 = -31$ k-ft
Middle strip positive $M_u = 70 - 46 = +24$ k-ft

With this very light beam a distribution in proportion to web width alone gives -14 k-ft to the beam, which is over ⅔ of the -19 k-ft found; with the stiffer beam of Sec. 16.2 the 85% assignment gives 83 k-ft to the beam.

Short Span
$$M_{os} = 0.125 \times 283 \times 20 \times 14.75^2/1000 = 154 \text{ k-ft}$$
$$\text{Total negative moment } M_u = -0.65 \times 154 = -100 \text{ k-ft}$$
$$\text{Total positive moment } M_u = 0.35 \times 154 = +54 \text{ k-ft}$$
$$\alpha_1 = 0.23, \qquad \ell_2/\ell_1 = 1.25, \qquad \alpha_1 \ell_2/\ell_1 = 0.29$$

For negative moment to the column strip Fig. 16.3b shows $75 - 2 = 73\%$ and for positive moment $60 + (0.29/0.333)0.7 \times 10 = 66\%$ (Fig. 16.3c)

Column strip negative $M_u = -0.73 \times 100 = -73$ k-ft
   Beam negative $M_u = -(0.85 \times 0.29)73 = -18$ k-ft
   Slab negative $M_u = -73 + 18 = -55$ k-ft $= -27.5$ k-ft/half-strip
Column strip positive $M_u = +0.66 \times 54 = +36$ k-ft
   Beam positive $M_u = +(0.85 \times 0.29)36 = +9$ k-ft
   Slab positive $M_u = +36 - 9 = +27$ k-ft $= 13.5$ k-ft/half-strip
Middle strip negative $M_u = -100 + 73 = -27$ k-ft
Middle strip positive $M_u = +54 - 36 = +18$ k-ft

(d) Depth for Moment The worst slab moment is $M_u = -39.5$ k-ft on a half column strip of width $(8 - 1.25)/2 = 3.37$ ft in the long span where $d = 5.5$ in. $M_n{}^* \geqslant -39.5/0.9 = -43.9$ k-ft

$$k_n = 43.9 \times 12{,}000/(3.37 \times 12 \times 5.5^2) = 431$$

The maximum allowable from Table 3.1 is 783.

---

* The ϕ factor must be used somewhere in this format.

The worst beam section is 15-in. wide and carries $M_u = -19$ k-ft with d about $9.5 - 0.75$ top cover -0.375 stir. $- 0.25 = 8.13$ in. If beam moments are more important the normal beam cover of 1.5 in. is more appropriate. $M_n = -19/0.9 = -21.1$ k-ft

$$k_n = 21.1 \times 12{,}000/(15 \times 8.13^2) = 255 \ll 783$$

(e) Recheck Shear   A recheck is needed because the preliminary estimate left questions. The load area contributing to the beam load is shaded in Fig. 16.3d, but Code 13.6.8.2 reduces the load to $\alpha_1 \ell_2/\ell_1$ parts where this ratio is less than 1.0. On the long beam, the critical shear is at d from face of column and $0.5 \times 18.75 - 8.5/12 = 8.67$ ft from midspan.

For the long beam:

Gross V_u at critical section
$$= (8.67 \times 16 - 2 \times 0.5 \times 6.67^2)281 = 26{,}500 \text{ lb} + \text{bm. wt.}$$
Effective $V_u = (\alpha_1 \ell_2/\ell_1)26{,}500 = 8.67 \times 66$
$$= 0.23 \times 26{,}500 + 572 = 6100 + 572 = 6670 \text{ lb}$$
$V_n(\text{required}) = V_u/\phi = 6670/0.85 = 7850 \text{ lb}$

For a beam with no web reinforcement, $V_n = V_c/2 = \sqrt{f_c'}bd = \sqrt{3000} \times 15 \times 8.5 = 6980$ lb. The required V_n is slightly in excess of that value so minimum stirrups are required.

A quick mental check on the short beam with $\alpha_1 \ell_2/\ell_1 = 0.29$ and less gross load shows it also is not critical. However, the remaining shear, around 70,000 lb, cannot be carried by the slab in the corners between the beams. (Allocating the low shear to the beams increases the shear left for the remaining slab.) With the beams working so inefficiently under the Code assignments, the check must return to the flat plate concept. The shear capacity, as a matter of logic, is not automatically reduced by deepening some portions 3 in.

It would seem logical to assign more shear capacity to the beams with minimum stirrups. However, stirrups in a 9.5-in. thickness lead to questions about stirrup development. If the irregular depth is created by steps on the tension face instead of on the compression face, the greater effective depth is available. The authors are hesitant to count this extra depth because diagonal cracks might break through the 6.5-in. depth before the 9.5-in. slab becomes fully effective. (Such a minimized resistance must be accepted where tests of unusual situations are unavailable.) The assumption is made that specification of $f_c' = 4500$ psi is more acceptable than a change in slab thickness or stiffening thickness. With the change in concrete strength the V_n capacity is sufficient because V_n in Sec. 16.3b becomes 114 k.

$$\text{USE } f_c' = 4500 \text{ psi.}$$

(f) Check on Minimum Column Size   A check on column stiffness is required where the ratio β_α, the ratio of service dead to live load, is less

TABLE 16.4 Reinforcement for Slab of Sec. 16.3.

$h_s = 6.5$ in. generally, with 15-in. wide by 9.5-in. between columns

Middle Strips	Long Span 8 ft		Short Span 12 ft	
Strip width	Neg. M	Pos. M	Neg. M	Pos. M
M_u, k-ft	−31.0	+24.0	−27.0	+18.0
$M_n = M_u/\phi$, k-ft	−34.5	+26.7	−30.0	+20.0
d, in.	5.5	5.5	5.5	5.0
$A_s = M_n \times 12/(60 \times 0.95\,d)$				
$\quad = M_n/(4.75\,d)$, in.²	1.32←	1.02	1.15←	0.84
0.0018 bh	1.12	1.12←	1.69*	1.69←
Min. no. bars, $s = 2\,h_s$	8	8	11	11
Use #3 (0.11 in.²)	12-#3 @ 8"	11-#3 @ 9 ±"	11-#3 @ 13 ±"	16-#3 @ 4"
	(1.32)	(1.21)	(1.21)	(1.76)
Revised d	5.06	5.06	5.06	5.19

Column Strip
Long span
Strip width

	Slab (each half-strip) Half-strip = 4 ft (incl. beam) = 3.37 ft = 40.5 in. net		Beam $b_w = 15''$, $h = 9.5''$	
	Neg. M	Pos. M	Neg. M	Pos. M
M_u, k-ft	−39.5	+18.5	−19	+9
$M_n = M_u/0.9$, k-ft	−43.8	+20.5	−21	+10
d, in.	5.5	5.5	8.5	7.75
$A_s = M_n/(4.75\,d)$, in.²	1.68←	0.79←	0.52←	0.27←
0.0018 bh	0.48	0.48	0.26	0.26
Min. no. bars, $s = 2h$	4	4	1	1
USE bars	6-#5 @ 6.5 ±" (1.86)	8-#3 @ 5 ±" (0.88)	2-#5 @ 7" (0.62)	3-#3 @ 5" (0.33)
Revised d	5.44	5.56	8.44	7.81

Column strip
Short span
Strip width

	Slab (each half-strip) Half-strip = 4 ft (incl. beam) = 3.38 ft = 40.5 in. net		Beam $b_w = 15$ in.	
	Neg. M	Pos. M	Neg. M	Pos. M
M_u, k-ft	−27.5	+13.5	−18	+9
M_n, k-ft	−30.6	+15.0	−20.0	+10.0
d, in	4.81	5.0	8.0	7.75
$A_s = M_n/(4.75\,d)$ in.²	1.34←	0.63←	0.53←	0.27←
0.0018 bh	0.48	0.48	0.26	0.26
Min. no. bars, $s = 2h$	4	4	1	1
USE bars	5-#5 @ 8 ±" (1.55)	6-#3 @ 7 ±" (0.66)	3-#4 @ 5" (0.60)	3-#3 @ 5" (0.33)
Revised d	4.81	5.06	8.00	8.44

* Half of bottom bars count on this.

than 2. $\beta_\alpha = 81/100$. Relative beam stiffness is 0.23 in the short direction where $\ell_2/\ell_1 = 1.25$, and 0.29 in the long direction where $\ell_2/\ell_1 = 0.8$. Table 15.2 (Sec. 15.8) shows that α_{min} increases as ℓ_2/ℓ_1 increases and as α decreases. Therefore the short span controls for a square column. Assume a column height of 10 ft. With a beam stiffness ratio α of 0.23 and a length ratio of 1.25, Table 15.2 leads to:

$$\beta_\alpha = 1.0 \qquad \alpha_{min} = 0.6$$
$$\beta_\alpha = 0.5 \qquad \alpha_{min} = 1.5 \qquad \beta_\alpha = 0.81 \qquad \alpha_{min} = 1.0$$
$$\alpha_{min} = \Sigma K_c/(\Sigma K_a + \Sigma K_b) = (I_c/\ell_c)/(I_s/\ell_s + I_b/\ell_b)$$

The values of I_s and I_b were calculated in (a).

$$1.0 = (2I_c/10)/(2 \times 5500/16 + 2 \times 1280/16) = 2I_c(10 \times 848)$$
Min. $I_c = 4240$ in.^{4} Min $h = 15.0$ in.

The column is barely adequate. If it failed this test, it could have been enlarged or the positive moments could have been increased under Code 13.6.10b by the ratio:

$$\delta_s = 1 + [(2 - \beta_\alpha)/(4 + \beta_\alpha)](1 - \alpha_c/a_{min})$$

where

δ_s is the multiplier for the positive moment

α_c is the actual column stiffness ratio

The α ratio can be replaced with the ratio of actual I_c to required I_c as other parts of α are the same. When α_c equals α_{min} the multiplier δ_s becomes unity.

(g) Reinforcement Design   The calculations are summarized in Table 16.4 with no new problems except the extra calculation of beam steel and a possible alternate decision as to whether the 3-in. deepening of the slab really makes a beam with 1.5-in. cover required or just a deepened slab with 0.75-in. cover. The assigned moments are so low for positive moment it, in fact, makes no difference. The table treats it as a slab type of element for positive and negative moment.

16.4   Use of Equivalent Frame Method of Slab Analysis

For slab systems not meeting the direct design method requirements of Sec. 15.5 (Code 13.6.1), the design moments must be computed by the equivalent frame method;[2] elsewhere it is an optional method. The basic requirements for the equivalent frame method are outlined in Code 13.7.

As in analysis of beams and columns, the structure is subdivided into bents running each way through the frame. The entire slab width, that is, one-half panel width on each side of the column, is considered in establishing the slab load and slab stiffness. Stiffness is based on gross concrete area, but considers a variable moment of inertia because the stiffness at column or drop panel becomes part of the problem.

If the service live load is less than 75% of the dead load, the frame analysis is made with full design (factored) load on all spans instead of a pattern loading. For higher live load ratios pattern loads must be used (Sec. 16.8) but, as in frame analysis for beam design, certain simplifications are permissible:

a. Columns fixed at their far end at floor above and below (for gravity loading only).
b. Two adjacent spans loaded with live load for maximum negative moment.
c. Alternate spans loaded with live load for maximum positive moment.
d. Slab-beams may be considered fixed two panels away from the point where moment is being computed, provided the slab continues further.

In addition the pattern loading may be applied with only 75% of the design live load.

Although spans to centerline of columns are used in analysis, the critical negative moments are at the face of any rectangular column or, if not rectangular, at the face of the equivalent (equal area) square column (Code 13.7.7.1 to 13.7.7.3).

The chief complication is the equivalent column stiffness K_{ec}, which recognizes the part that the slab or transverse beam torsional stiffness plays in the system. Equivalent stiffness is discussed in Sec. 16.5.

16.5 Effective Column Stiffness in a Slab System

Even if a flat plate is supported on very stiff columns, the slab spans are *not* really fixed at the column. The slab strip framing directly into the column is fixed over the column width, but the parallel strips farthest from the column are restrained only by the next slab span and practically none at all by the column itself. The torsional restraint that the slab could transfer laterally to this strip is nearly negligible. With a heavy transverse beam there could be more restraint, but on the average across the slab the restraint is still less than directly at the column. This is a way of saying that as far as the slab is concerned the slab restraint averages less than the column stiffness suggests. Only a transverse beam infinitely stiff in torsion between two columns could make this difference negligible.

Reduced column effectiveness is a major concept in the Code treatment of the equivalent frame analysis. The Commentary 13.7.4 uses the term "the equivalent column" and describes its behavior in terms of flexibilities (the inverse of stiffnesses). For the complete exterior column (above and below), the specified flexibility is

$$1/K_{ec} = 1/(\Sigma K_c) + 1/K_t$$

where ΣK_c represents the sum of the stiffnesses of column above and below the joint and K_t represents the sum of the torsional stiffness of the slab or beam on each side of the column.

The authors prefer to think in terms of stiffness rather than flexibility, which makes the equation

$$K_{ec} = \frac{1}{1/(\Sigma K_c) + 1/K_t}$$

The effective column K_{ec} taking the place of ΣK_c is either more flexible or less stiff than the column alone. If K_t is small, say, only a slab, K_{ec} is *much less* stiff than ΣK_c.

The evaluation of K_t is based on simple (rather crude) assumptions and basic theory with final adjustment by an empirical factor to fit test results. The final evaluation is taken as $\frac{1}{3}$ as stiff as Fig. 16.4 from the Code Commentary suggests. Figure 16.4 shows the transverse beam (or slab) loaded in torque by unbalanced slab moments that vary from zero at midspan to a maximum at the centerline of the column. The total torque (area of the load curve) is 0.5 for each half-span and the shape of the torque curve is a second power parabola with an ordinate at the face of column

$$T_u = \frac{1}{2}\left(1 - \frac{c_2}{\ell_2}\right)^2$$

The unit rotation angle at any point is the torque divided by both G, the shear modulus, and C, the torque stiffness of the cross section. The integral of the area under this curve measures the difference between column rotation and midspan beam rotation.

$$\begin{aligned}\theta_t &= (1/3)[(\ell_2/2)(1 - c_2/\ell_2)0.5(1 - c_2/\ell_2)^2]/CG \\ &= (1/12)\ell_2(1 - c_2/\ell_2)^3/CG \\ &= (1/6)\ell_2(1 - c_2/\ell_2)^3/CE, \text{ assuming } G = 0.5E.\end{aligned}$$

One may think of the average θ for the slab as $\theta_t/3$ or may think of using a factor of 1/3 to fit the equation to the test data. Fitting may be necessary to represent a nonlinear torque load, the difference between torque at the column face and at the center of column, and so on.

$$\theta_t = (1/18)\ell_2(1 - c_2/\ell_2)^3/EC*$$

The stiffness is the ratio torque (used as 0.5 per arm) to this θ_t, or

$$K_t = 0.5/\theta_t = 9EC/[\ell_2(1 - c_2/\ell_2)^3]$$

for one arm, or twice this for the two arms in the symmetrical case. The Code indicates the doubling by writing the Σ sign as part of the definition of K_t.

$$K_t = \Sigma 9EC/[\ell_2(1 - c_2/\ell_2)^3]$$

---

* C is similar in use to the polar moment of inertia used for a circular elastic member under torsion.

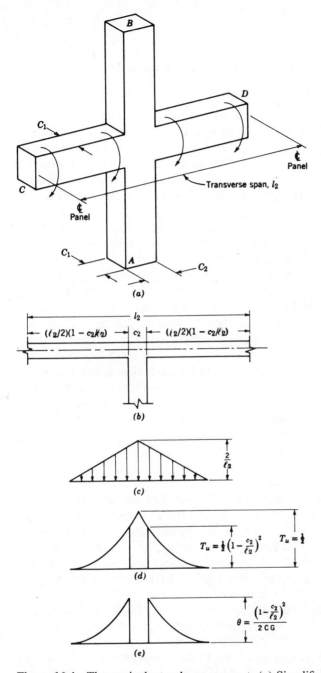

Figure 16.4 The equivalent column concept. (*a*) Simplified physical model. (*b*) Beam-column combination. (*c*) Distribution of torque or twist load, based on a total of unity. (*d*) Twisting moment or torque diagram. (*e*) Distribution of twist angle per unit length. (Modified from ACI Commentary on Building Code.)

(Clarity, to the authors, is improved if the Σ is moved into the K_{ec} equation to show ΣK_t.)

The value of C in these equations is the necessary term to replace the polar moment of inertia when considering the torque of noncircular members:

$$C = \Sigma(1 - 0.63x/y)x^3y/3$$

where x and y are the small and large dimensions of the various rectangles making up the cross section, as shown in Fig. 16.1a. Where no beam stem is present, Code 13.7.5.1a defines the beam as a width of slab equal to the column width. Where a beam stem is present, the beam includes the stem plus the adjoining slab on *each* side (one side if a spandrel or edge beam) of width equal to $4h_s$ or, if smaller, the stem projection above or below the slab, as indicated in Fig. 16.1b.

For moment distribution under the equivalent frame method K_{ec} is the effective column stiffness. It reflects the reduced fixity of slab because some middle strips of the slab are less restrained by abutting slabs than are those close by the column. Where this restraint creates a torque in the slab, the torque is carried back to the column. The effect on the column is a bending moment at the joint equal to the sum of the moment transferred directly from the slab to the column and the moment transferred indirectly through torque. The total is smaller, however, than if all the slab was directly framing into the column, which is what K_{ec} indicates.

The 1986 Code does not specifically refer to the equivalent column except in the explanatory commentary. The equivalent column concept implicity satisfies the requirements of Code 13.7.2.3 that the columns are assumed attached to the slab-beam strips by torsional members.

16.6   Frame Constants

Because the basic provisions of Code 13.7 regarding moment of inertia of columns and slab beams result in no member of the equivalent frame having a uniform I, the stiffness is not $4EI/\ell$ but something more than 4 for the constant and something more than 0.5 for the carryover factor, C_f^* or C.O.F. Tables can tabulate k for use in kEI/ℓ and C_f, although C_f is directional, not the same from each end except with symmetry of cross section. Likewise the fixed end moment is normally greater than for uniform I.

Table 16.5, slightly modified from Reference 1, lists for flat plates the slab stiffness factors, moment factors, carryover factors, and at the very bottom the useful† quantity $(1 - c_2/\ell_2)^3$. The EI variation used to prepare this table is sketched alongside. With a drop panel the EI for the drop length is that of a very flat wide T-shape of width ℓ_2; this EI requires a separate table. Reference

---

* The symbol C_f (or other subscript) is used because the Code uses C for a torsional constant.
† Useful in the torque stiffness calculation.

c_1/ℓ_1		c_2/ℓ_2 0.00	0.05	0.10	0.15	0.20	0.25	0.30
	M	0.083	0.083	0.083	0.083	0.083	0.083	0.083
0.00	k	4.000	4.000	4.000	4.000	4.000	4.000	4.000
	C_f	0.500	0.500	0.500	0.500	0.500	0.500	0.500
	M	0.083	0.084	0.084	0.084	0.085	0.085	0.085
0.05	k	4.000	4.047	4.093	4.138	4.181	4.222	4.261
	C_f	0.500	0.503	0.507	0.510	0.513	0.516	0.518
	M	0.083	0.084	0.085	0.085	0.086	0.087	0.087
0.10	k	4.000	4.091	4.182	4.272	4.362	4.449	4.535
	C_f	0.500	0.506	0.513	0.519	0.524	0.530	0.535
	M	0.083	0.084	0.085	0.086	0.087	0.088	0.089
0.15	k	4.000	4.132	4.276	4.403	4.541	4.680	4.818
	C_f	0.500	0.509	0.517	0.526	0.534	0.543	0.550
	M	0.083	0.085	0.086	0.087	0.088	0.089	0.090
0.20	k	4.000	4.170	4.346	4.529	4.717	4.910	5.108
	C_f	0.500	0.511	0.522	0.532	0.543	0.554	0.564
	M	0.083	0.085	0.086	0.087	0.089	0.090	0.091
0.25	k	4.000	4.204	4.420	4.648	4.887	5.138	5.401
	C_f	0.500	0.512	0.525	0.538	0.550	0.563	0.576
	M	0.083	0.085	0.086	0.088	0.089	0.091	0.092
0.30	k	4.000	4.235	4.488	4.760	5.050	5.361	5.692
	C_f	0.500	0.514	0.527	0.542	0.556	0.571	0.585
$X = (1 - c_2/\ell_2)^3$		1.000	0.856	0.729	0.613	0.512	0.421	0.343

$$FEM = Mw\ell_2\ell_1^2$$
$$K = kEI_s/\ell_1 \text{ with}$$
$$I_s = \ell_2 h^3/12$$
$$\text{C.O.F.} = C_F$$

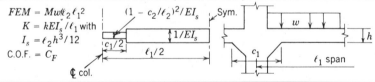

1 has similar tables for drop panel thicknesses of $1.25h_s$ and $1.50h_s$ in symmetrical panels.

Values of column K_{ec} cannot be readily tabulated, but K_c values for the column are easily obtained from the coefficients in Table 16.6, also slightly modified from Reference 1. The EI variations are indicated on the sketch in Table 16.7; $1/EI$ equals zero for the thickness of the slab (and of the drop if any) and for the column capital height. Because such a stiffness is different at the two ends (if there is either drop or capital or both), care must be taken to use a/ℓ_c for the *near end* and b/ℓ_c for the *far end* in every case. This way the table works equally well for the column above or below a joint.

TABLE 16.6 Column Stiffness Coefficients, k_c*

a/ℓ_t \ b/ℓ_c	0.00	0.02	0.04	0.06	0.08	0.10	0.12	0.14	0.16	0.18	0.20	0.22	0.24
0.00	4.000	4.082	4.167	4.255	4.348	4.444	4.545	4.651	4.762	4.878	5.000	4.128	5.263
0.02	4.337	4.433	4.533	4.638	4.747	4.862	4.983	5.110	5.244	5.384	5.533	5.690	5.856
0.04	4.709	4.882	4.940	5.063	5.193	5.330	5.475	5.627	5.787	5.958	6.138	6.329	6.533
0.06	5.122	5.252	5.393	5.539	5.693	5.855	6.027	6.209	6.403	6.608	6.827	7.060	7.310
0.08	5.581	5.735	5.898	6.070	6.252	6.445	6.650	6.868	7.100	7.348	7.613	7.897	8.203
0.10	6.091	6.271	6.462	6.665	6.880	7.109	7.353	7.614	7.893	8.192	8.513	8.859	9.233
0.12	6.659	6.870	7.094	7.333	7.587	7.859	8.150	8.461	8.796	9.157	9.546	9.967	10.430
0.14	7.292	7.540	7.803	8.084	8.385	8.708	9.054	9.426	9.829	10.260	10.740	11.250	11.810
0.16	8.001	8.291	8.600	8.931	9.287	9.670	10.080	10.530	11.010	11.540	12.110	12.740	13.420
0.18	8.796	9.134	9.498	9.888	10.310	10.760	11.260	11.790	12.370	13.010	13.700	14.470	15.310
0.20	9.687	10.080	10.510	10.970	11.470	12.010	12.600	13.240	13.940	14.710	15.560	16.490	17.530
0.22	10.690	11.160	11.660	12.200	12.800	13.440	14.140	14.910	15.760	16.690	17.210	18.870	20.150
0.24	11.820	12.370	12.960	13.610	14.310	15.080	15.920	16.840	17.870	19.000	20.260	21.650	23.260

* $K_c = k_c E_c I_c / \ell_c$

TABLE 16.7 Values of Torsion Constant, C*

x \ y	4	5	6	7	8	9	10	12	14	16
12	202	369	592	868	1188	1538	1900	2557	—	—
14	245	452	736	1096	1529	2024	2566	3709	4738	—
16	288	534	880	1325	1871	2510	3233	4861	6567	8083
18	330	619	1024	1554	2212	2996	3900	6013	8397	10813
20	373	702	1167	1782	2553	3482	4567	7165	10226	13544
22	416	785	1312	2011	2895	3968	5233	8317	12055	16275
24	458	869	1456	2240	3236	4454	5900	9469	13885	19005
27	522	994	1672	2583	3748	5183	6900	11197	16628	23101
30	586	1119	1888	2926	4260	5912	7900	12925	19373	27197
33	650	1243	2104	3269	4772	6641	8900	14653	22117	31293
36	714	1369	2320	3612	5284	7370	9900	16381	24860	35389
42	842	1619	2752	4298	6308	8828	11900	19837	30349	43581
48	970	1869	3184	4984	7332	10286	13900	23293	35836	51773
54	1098	2119	3616	5670	8356	11744	15900	26749	41325	59965
60	1226	2369	4048	6356	9380	13202	17900	30205	46813	68157

* $C = (1 - 0.63x/y)x^3y/3$

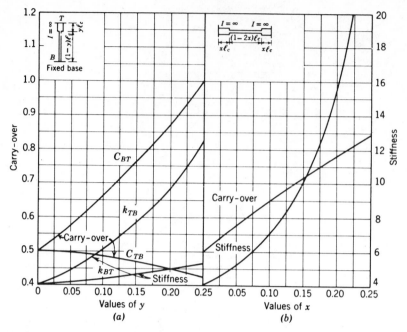

Figure 16.5 Carryover factors and stiffness factors for member with infinite I over part of length. (Adapted from Reference 3, Portland Cement Assn.)

Alternatively, for K_c the curve data of Fig. 16.5a give the same values plus the carryover factors, *provided one uses the column length ℓ_c reduced* to the distance to the far fixed end created by the face of the slab above or below (or the start of the capital above where a capital is used).* Because both stiffness and carryover factor are directional, the first subscript designates the joint that is rotating. Thus k_{TB} applies for a joint rotation at the top and C_{TB} gives the C.O.F. from top to bottom of the column; k_{BT} and C_{BT} are the opposite.

The torsional stiffness of the transverse beam or slab is $K_t = 9EC/[\ell_2(1 - c_2/\ell_2)^3]$ for the transverse span *on one side* of the column. The summation sign in Code Eq. 13.6 is a reminder that the beams on both sides of the column must be included, if present. Values of $(1 - c_2/\ell_2)^3$ are tabulated at the bottom of Table 16.5. Values of $C = \Sigma(1 - 0.63x/y)x^3y/3$ are tabulated for *individual* rectangles in Table 16.7, also adapted slightly from Reference 1. The values for individual rectangles must be summed for an L- or T-section.

The use of these tables, and the calculation of K_{ec} and the distribution factors is a rather involved process that needs some formalization to maintain clarity of presentation. An example is developed in the next section using a columnar form (Fig. 16.6) suggested by Reference 1.

---

* The moment then carried over acts as the face of slab (the assumed far fixed end).

16.7 Equivalent Frame Method—Design Example

Although neither the spans nor the loading require it, use the data of Sec. 15.7 (interior panel) and 15.12 (exterior panel) to work out corresponding design moments for the 20-ft span by the equivalent frame method and compare the results.

Panels are 18-ft × 20-ft (center-to-center of columns) with the 18-ft length parallel to the exterior edge. Columns are 15-in. square, both interior and exterior, with a 10-ft story height, and the slab runs to the exterior face of the outside column without a spandrel beam. For simplicity, any wall load is excluded from the example. Live load is 60 psf and total dead weight 100 psf (including slab weight). Slab thickness for interior panel is 7 in., with 8 in. for the exterior panel. The difference in slab weight is ignored (100 psf used as the total in both panels), making the factored design dead load 140 psf and the factored design live load 102 psf. Here the break in slab stiffness is assumed on the centerline of the first interior column row. $f'_c = 4000$ psi for all columns and slabs.

Solution

The sequence of tabulation in Fig. 16.6 is used as a guide for discussion, followed by items e and f.

   a. Geometry
   b. Stiffnesses
   c. Factors (FEM, COF, distribution factors)
   d. Distribution (moment distribution for each loading)
   e. Design moments from each loading
   f. Possible moment reductions because the slab qualifies for the direct design method.

(a) Geometry   Geometry is simplified by the uniform spans and equal column sizes. For determining exterior column stiffness using the tabular values of Table 16.6 and the sketch of Table 16.7, the half-depth of slab gives $a/\ell_c = b/\ell_c = 4/(10 \times 12) = 0.033$. For the typical interior columns $a/\ell_c = b/\ell_c = 3.5/(10 \times 12) = 0.029$. Where the break in slab thickness occurs, one might use average thickness of slab to fix a and b, but that much accuracy is not really inherent in the process; use value for typical interior column, 0.029.

For the slab with 15-in. square columns, the sketch in Table 16.5 indicates $c_1/\ell_1 = 1.25/20 = 0.062$ at each end and c_2/ℓ_2 for width is $1.25/18 = 0.069$.

(b) Stiffnesses   For K_c the a/ℓ_c and b/ℓ_c are available previously. For the main part of the column above and below the joint, I_c is $15^4/12 = 4220$, E_c

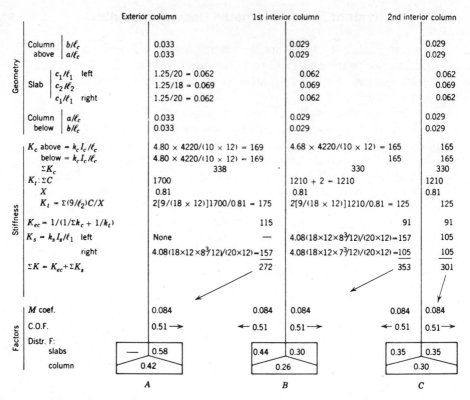

		Exterior column	1st interior column	2nd interior column
Geometry	Column above b/ℓ_c	0.033	0.029	0.029
	a/ℓ_c	0.033	0.029	0.029
	Slab c_1/ℓ_1 left	1.25/20 = 0.062	0.062	0.062
	c_2/ℓ_2	1.25/18 = 0.069	0.069	0.069
	c_1/ℓ_1 right	1.25/20 = 0.062	0.062	0.062
	Column below a/ℓ_c	0.033	0.029	0.029
	b/ℓ_c	0.033	0.029	0.029
Stiffness	K_c above $= k_c I_c/\ell_c$	$4.80 \times 4220/(10 \times 12) = 169$	$4.68 \times 4220/(10 \times 12) = 165$	165
	below $= k_c I_c/\ell_c$	$4.80 \times 4220/(10 \times 12) = 169$	165	165
	ΣK_c	338	330	330
	$K_t : \Sigma C$	1700	1210 + 2 = 1210	1210
	X	0.81	0.81	0.81
	$K_t = \Sigma(9/\ell_2)C/X$	$2[9/(18 \times 12)]1700/0.81 = 175$	$2[9/(18 \times 12)]1210/0.81 = 125$	125
	$K_{ec} = 1/(1/\Sigma k_c + 1/k_t)$	115	91	91
	$K_s = k_s I_s/\ell_1$ left	None	$4.08(18\times12\times8^3/12)/(20\times12)=157$	105
	right	$4.08(18\times12\times8^3/12)/(20\times12)=157$	$4.08(18\times12\times7^3/12)/(20\times12) = 105$	105
	$\Sigma K = K_{ec}+\Sigma K_s$	272	353	301
Factors	M coef.	0.084	0.084	0.084	0.084	0.084
	C.O.F.	0.51 →	← 0.51	0.51 →	← 0.51	0.51 →
	Distr. F: slabs	— 0.58	0.44	0.30	0.35	0.35
	column	0.42	0.26	0.30

A   B   C

Figure 16.6 Factors needed for moment distribution, flat plate.

may be omitted (that is, used as 1 since all concrete is alike in the frame and only relative stiffnesses are finally involved). From Table 16.6 the value of k_c for exterior column ($a/\ell_c = b/\ell_c = 0.033$) is 4.80, based on interpolation.

$$K_c = 4.80 \times 4220/(10 \times 12) = 169$$
$$\Sigma K_c = 2 \times 169 = 338$$

For the transverse beam (torsional member) at the column line the Code 13.7.5.1 assumes a constant cross section, using the largest of:

1. That part of the slab with a width equal to the width of the column, bracket, or capital, in this case at the exterior column a member 15-in. × 8-in.
2. The slab just described plus that part of any transverse beam above and below the slab, in this case none.
3. The beam of Fig. 16.1b, in this case none.

From Table 16.7, C for the 15-in. × 8-in. "beam" at the exterior column line is 1700. (Interpolations in the y direction are nearly linear, but not in the x direction; any reasonable interpolation is probably as accurate as the method justifies.) Since there is a beam of this type on each side of the column,

$$K_t = 2[9EC/\ell_2(1 - c_2/\ell_2)^3] = 18C/(\ell_2 X)$$

where X is the bottom line of Table 16.5 and shows as 0.81 for c_2/ℓ_2 of 0.069.

$$K_t = 18 \times 1700/(18 \times 12 \times 0.81) = 175$$

For the exterior column,

$$K_{ec} = 1/(1/\Sigma K_c + 1/K_t) = 1/(1/338 + 1/175) = 115, \text{ with } E \text{ of } 1.$$

For the slab in the exterior panel, $c_2/\ell_2 = 0.069$ and $c_1/\ell_1 = 0.062$, which in Table 16.5 leads to

$$M \text{ coef.} = 0.084, \ k = 4.08, \ C_f = 0.51$$
$$K_s = 4.08(18 \times 12 \times 8^3/12)/(20 \times 12) = 157, \text{ with } E \text{ of } 1.$$

For the exterior joint, $\Sigma K = 115 + 157 = 272$.

The short $c_1/2$ length at each end of slab (which is taken stiffer) is so short that in this case a solution based on a uniform I is not very different. Such a solution gives $K_s = 154E$.

For the first interior column line with a/ℓ_c and b/ℓ_c reflecting a 7-in. slab, ΣK_c is changed to 330 and K_t is changed by the transition between the 8-in. and 7-in. slab shown in Fig. 16.7.

For C the beam shown can be divided into rectangles that give a maximum total that equals the 15-in. $\times$ 7-in. area plus a 7.5-in. $\times$ 1-in. area. From Table 16.7.

$$C_1 = 1210$$
$$C_2 = (1 - 0.63 \times 1/7.5)1^3 \times 7.5/3 = 2 \text{ (negligible)}$$
$$\text{Use } C = 1210$$

Because all other factors except C remain unchanged, $K_t = (1210/1700) \times 175 = 125$. For the interior column, equivalent

$$K_{ec} = 1/(1/338 + 1/125) = 91$$

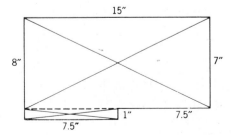

Figure 16.7 "Beam" cross section at first interior column.

With the slab changing only in thickness from 8 in. to 7 in.,

$$K_{ec} = 157(7/8)^3 = 105 \text{ for an interior slab.}$$
$$\Sigma K \text{ for the joint} = (105 + 157) + 91 = 353$$

(c) Fixed End Moments   Since $w_\ell/w_d = 60/100 = 0.60 < 0.75$, Code 13.7.6.2 accepts the use of *full* design load on all the frame with no pattern loadings required. For the end span, Table 16.5 shows the moment factor as 0.084 for fixed end moment, just a trifle more than the nominal $\frac{1}{12}$.

End span fixed end $M = M^F = 0.084(140 + 102)18 \times 20^2/1000$
$$= 146.0 \text{ k-ft}$$
Interior span fixed end $M = M^F = 146.0$ k-ft

At the exterior column, there is a 7.5-in. width of slab that is beyond the column centerline and adds a negative $M^F = -0.5 \times 242(18 - 1.25) \times 0.62^2/1000 = 0.78$, say, 0.8 k-ft. This result is almost negligible but is mentioned because there is usually some wall weight and the slab often cantilevers significantly beyond the column itself.

(d) Moment Distribution   Consider, closely enough, that the frame system in Fig. 16.8 is fixed (roughly symmetrical) about the fourth column line. It is fully loaded, has uniform spans, and should have very little rotation several joints away from the exterior column. The alternate joint sequence is used for the moment distribution.

(e) Design Moments   The moments thus found by moment distribution are at the center of column lines and should be corrected to the face of support for rectangular (or equivalent square supports), but in no case is the correction farther than 0.175ℓ, from the column centerline (Code 13.7.7.1). In case of an exterior column with a bracket or capital, the correction must be reduced to only half the actual projection used at the column (Code 13.7.7.2). In this example all corrections are to face of column, 7.5 in. = 0.62 ft from the column centerline.

On the basis of the beam free bodies sketched in Fig. 16.8b the simple beam shear as half of the beam load and the continuity shear as the difference between the end moments divided by the span are calculated. Summation of moments about the face of column then leads to the following design moments.

Ext. $M_{AB} = -66.2 + 38.3 \times 0.62 - 4.36 \times 0.62^2/2 = -43.3$ k-ft
1st Int. $M_{BA} = -171.2 + 48.9 \times 0.62 - 0.8$ (as for AB) $= -141.7$ k-ft
M at midspan $= +0.125 \times 4.36 \times 20^2 - 0.5(66.2 + 171.2) = +99.3$ k-ft
V at midspan $= 38.3 - 4.36 \times 10 = -5.3$ k
For max. M, e from midspan $= -5.3/4.36 = -1.22$ ft (to left)
M = area of V diagram over length $e = +0.5 \times 5.3 \times 1.22 = +3.2$ k-ft
Max. positive $M = +99.3 + 3.2 = +102.5$ k-ft

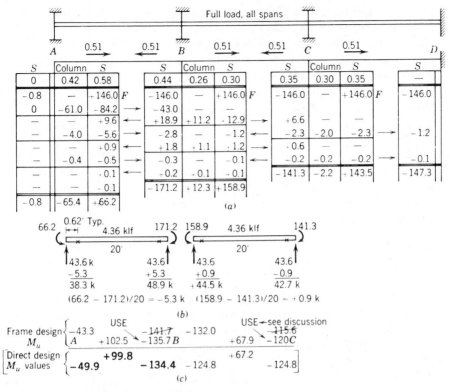

Figure 16.8 Moment distribution and steps toward design moments. (*a*) *M* distribution. (*b*) Shears. (*c*) Design moments.

In similar fashion for span *BC*:

$$M_{BC} = -158.9 + 44.5 \times 0.62 - 0.8 = -132.0 \text{ k-ft}$$
$$M_{CB} = -141.3 + 42.7 \times 0.62 - 0.8 = -115.6 \text{ k-ft}$$
$$\text{Positive midspan } M = +0.125 \times 4.36 \times 20^2 - 0.5(158.9 + 141.3)$$
$$= +67.9 \text{ k-ft}$$

The slight shift in the positive moment maximum point is negligible in this case with the end moments differing only 12%.

(f) Check Against M_o For Possible Economy   Since this slab is eligible for the direct design method, Code 13.7.7.4 allows a check of the total moment against the M_o total required by direct design; any surplus is unnecessary. For the exterior span, M_o in Sec. 15.12 is 192 k-ft. The comparable total is the sum of the positive moment plus the average end moment, which is the simple beam moment.

Simple beam *M*, ext. span = 102.5 + 0.5(43.3 + 141.7) = 195.0 k-ft

16.7   EQUIVALENT FRAME METHOD—DESIGN EXAMPLE   **547**

This result indicates that all moments for design might be reduced proportionately about 1.5% to make up the 3.0 k-ft difference just found. Theoretically this is the neatest solution of the problem. Alternatively, however, the authors feel free to reduce any one moment by the total correction, where this is not more than 10% of the computed value. This 10% provision is stated in Code 13.6.7 only for the direct design method, but it is consistent with the redistribution of moments in beam design procedures where ρ is low, as it is here. Accordingly, reduce the negative moment M_{BA} by $2 \times 3.0 = 6.0$ k-ft to $-141.7 + 6.0 = -135.7$ k-ft. This result reduces the total simple span moment to match M_o and reduces the total steel required at the first interior column line, since the negative moment in the next span tends to be smaller anyway.

On the first interior span the M_o value is also 192.0 k-ft, from Sec. 15.7. For the design moments,

$$\text{Simple beam } M = +67.9 + 0.5(132.0 + 115.6) = 191.7 \text{ k-ft}$$

This is a trifle *less* than M_o; no change is warranted and the apparent difference is probably not a significant number anyway.

The negative moments at the joints on the third span are higher than M_{CB} (at the joint) by 2.2 k-ft and 6.0 k-ft and corrections to face of column are similar. The authors feel it is good judgment to design at all interior column lines (except the first) for a moment of something like 5.0 k-ft more than recorded for M_{CB}, say, a total of -120 or -121 k-ft. This change is indicated on Fig. 16.8c.

The distribution in Fig. 16.8 is not extended as many bays as Code 13.7.2.6 requires (two spans beyond the critical moment computed), but inspection indicates that under full load on all spans another span would make no noticeable difference.

The final moments for design are the modified frame design moments indicated in Fig. 16.8c. The distribution to middle and column strips follows the same procedure used in earlier examples as allowed by Code 13.7.7.5. In this simple case it means 75% of the negative moment and 60% of the positive moments assigned to the column strip.

At the very bottom of Fig. 16.8c are shown the equivalent moments found using the direct design provisions in Examples 15.7 and 15.12. All moments are expressed in terms of M_u (without ϕ factors). Moments from these two methods are within 6% in the interior span and only the exterior support negative moment value exceeds this value. In that case the direct design method is about 15% higher than the equivalent frame method although the numerical difference (6.6 k-ft) is not large. For this straightforward application, either method can be used. The reader will develop his or her own preference depending on availability of design aids and computer programs.

16.8 Procedure Where Live Load is Larger

If live load exceeds 75% of dead load (at service load stage), more loadings in pattern form are necessary to obtain design moments. In each case the pattern

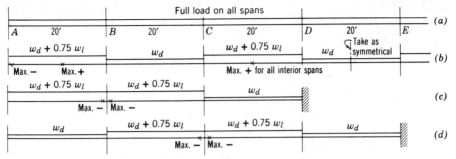

Figure 16.9 Loadings necessary when $w_l > 0.75\ w_d$. Full design load required in (a).

live load is at the lower level of 75% of the design live load. The cases to be analyzed are sketched in Fig. 16.9, where all frame constants and all correction procedures to the critical section are handled exactly as in the example just given. The pattern loading moments in Fig. 16.9 may not be taken smaller than those of full live load on all spans. (Where the loading qualifies for the direct design method there is no corresponding provision; the M_o calculation takes its place.)

16.9 Slab Moment Percentages in Direct Design Method—Stiff Beams

For *interior* panels on stiff beams this section gives the equations defining the percentage of M_o assigned to column and middle strip and tabulates some of them in Table 16.8.

For an interior panel, as in Sec. 16.2, M_o is divided into total negative moment, 65%, and total positive moment, 35%. For stiff beams (such that $\alpha_1 \ell_2/\ell_1 \geqslant 1.0$), Fig. 16.3b,c shows that the amount of this negative or positive moment going to the column strip varies from 90% for $\ell_2/\ell_1 = 0.5$ to 45% for $\ell_2/\ell_1 = 2$; or this percentage can be stated algebraically as 105%–30% ℓ_2/ℓ_1 that goes to the column strip. This moment is further subdivided into 85% to the beam and 15% to the column strip slab. All moment not assigned to the column strip becomes middle strip moment. Below, the various percentages are expressed algebraically.

Column strip *slab* negative $M_u = -0.15 \times 0.65(105 - 30\ell_2/\ell_1)M_o$
$= -(10.25 - 2.93\ell_2/\ell_1)M_o$, as a percentage.

Column strip *slab* positive $M_u = 0.15 \times 0.35(105 - 30\ell_2/\ell_1)M_o$
$= +(5.53 - 1.58\ell_2/\ell_1)M_o$, as a percentage.

Middle strip negative $M_u \quad\ = -0.65(100 - 105 + 30\ell_2/\ell_1)M_o$
$= -(19.5\ell_2/\ell_1 - 3.25)M_o$, as a percentage.

Middle strip positive $M_u \quad\ = +0.35(100 - 105 + 30\ell_2/\ell_1)M_o$
$= +(10.5\ell_2/\ell_1 - 1.65)M_o$, as a percentage.

TABLE 16.8 Slab Moments for Interior Spans with Stiff* Beams on All Sides and at Least Three Continuous Spans

ℓ_2/ℓ_1:		M_o for long span			M_o for short span	
		0.50	0.75	1.00	1.33	2.00
Middle Strip	Negative	$-6.50\%\ M_o$	$-11.40\%\ M_o$	$-16.25\%\ M_o$	$-22.75\%\ M_o$	$-35.75\%\ M_o$
	Positive	$+3.60\%\ M_o$	$+6.21\%\ M_o$	$+8.85\%\ M_o$	$+12.35\%\ M_o$	$+19.35\%\ M_o$
	Width	$0.5\ell_2$	$0.5\ell_2$	$0.5\ell_2 = 0.5\ell_1$	$0.83\ell_1$	$1.50\ell_1$
Column Strip	Negative	$-8.79\%\ M_o$	$-8.05\%\ M_o$	$-7.33\%\ M_o$	$-6.33\%\ M_o$	$-4.39\%\ M_o$
	Positive	$+4.74\%\ M_o$	$+4.34\%\ M_o$	$+3.95\%\ M_o$	$+3.43\%\ M_o$	$+2.37\%\ M_o$
	Width	$0.5\ell_2$	$0.5\ell_2$	$0.5\ell_2 = 0.5\ell_1$	$0.83\ell_1$	$0.5\ell_1$

* Technically, with $\alpha_1 \ell_2/\ell_1 \geqslant 1.0$.

The beam moments (not shown) make up the missing 0.85 of the total column strip moments.

Slab moments for several cases of ℓ_2/ℓ_1 are tabulated in Table 16.8. The width of strips are added later, otherwise the size of moment per ft width would be difficult to visualize. This table is really divided vertically at the middle between coefficients on the left to be used with M_o for the long span and its strips and on the right with M_o for the short span and its strips. The picture is improved when one pairs data from two columns to make one complete set:

ℓ_2/ℓ_1 of 0.5 and 2.0 for 2 : 1 slab proportions

ℓ_2/ℓ_1 of 0.75 and 1.33 for 4 : 3 slab proportions

ℓ_2/ℓ_1 of 1.00 alone for a square slab

This table can give some idea of how moments change with panel proportions. But these percentages are rather minimal values for interior panels only. They should change, somewhat upward, as boundary restraints are relaxed.

Thus Table 16.8 still leaves designers with inadequate guidance when they have numerous small slabs that do not qualify for the direct design method. The conservative approach used in Chapter 12 is recommended as giving detailed slab moment coefficients for a wide range of restraints, applicable wherever beam deflections do not greatly increase the slab deflections.

16.10   Lateral Load Analysis for Two-Way Slab Systems

Both the "direct design method" and the "equivalent frame method" of Code Chapter 13 are restricted by Code 13.3.1.1 to applications with orthogonal frames under gravity loads *only*. The experiments and analyses on which these two methods are based did not consider the effects of story displacement or drift. These methods are most applicable when shear walls, cross bracing, or some other effective lateral restraint minimizes such drift and resists the lateral load. Both methods can approximate the gravity load effects in unbraced frames.

In some cases, architectural or other constraints may require that the column-slab frame be the basic lateral load resisting system. The 1986 Code recognizes the possibility of such structural action but provides little direct guidance about what constitutes an acceptable analysis for lateral loads on two-way slab systems. Code 13.3.1.3 indicates that results of a gravity analysis (like the direct design or equivalent frame methods) may be superimposed on the results of a lateral load analysis. Code 13.3.1.2 indicates that for lateral load analysis of unbraced frames, the effects of cracking and reinforcement must be considered in determining the stiffness of frame members. Some guidance and references[4-6] are given in the Commentary but Committee 318 did not recommend any specific procedure.

Similar to the problem with unbraced column and frame structures discussed in Chapter 7, the moment magnification in both column and flexural restraint members is a function of the lateral displacement or drift. The stiffness of the floor system greatly affects the degree of effective rotational restraint on the columns. This restraint has a major effect on the drift and secondary or magnified moments. In slab structures, localized cracking can reduce the floor stiffness and, hence, the gross section I assumptions of Code 13.6 and 13.7 may be quite unsafe when used to predict lateral drift. Slab stiffnesses may be more realistically computed if a fully cracked section is assumed or if the effective I of Code Eq. 9.7 is used.[4,5]

Research studies in this area are needed before conclusive design recommendations can be given. In addition to the effective slab-beam and column stiffnesses, the entire question of effective stiffness and moment transfer of the column to slab connection needs further refinement for the lateral load case.

Vanderbilt[6] surveyed the various design approaches available and outlined[7] a possible adaptation of the equivalent frame analysis. Vanderbilt and Corley[8] compare various procedures. The three most widely cited possible analysis procedures are (1) the modified equivalent frame or "transverse-torsional" model, (2) the "equivalent slab-width" model,[9] and (3) the "stub-beam" model. Vanderbilt and Corley prefer the first procedure. The authors regard the entire subject as highly unsettled because of the uncertainties of the efficiency of the column-slab connection and the lack of experimental verification in comprehensive lateral load tests. A very experienced designer, J. Grossman,[10] discussing Reference 8 indicates that the modified equivalent frame method is not suitable for practitioners, even if the stiffness variations could be correctly addressed. He states that the equivalent beam method is more suitable for practitioner use and recommends its use with lower-bound values for structures exposed to low or moderate lateral loads. Vanderbilt and Corley[8] recognize the design appeal of the equivalent beam method but express concern that it has been verified only by comparisons with elastic analysis. The designer must assume substantial responsibility for analysis in this area.

Selected References

1. S. H. Simmonds and J. Music, "Design Factors for the Equivalent Frame Method," *ACI Jour.* 68, No. 11, Nov. 1971, p. 825.

2. W. G. Corley and J. O. Jirsa, "Equivalent Frame Analysis for Slab Design," *ACI Jour.* 67, No. 11, Nov. 1970, p. 875.

3. "Frame Analysis Applied to Flat Slab Bridges," Portland Cement Assn., Skokie, Ill.

4. J. E. Carpenter, P. H. Kaar, and W. G. Corley, "Design of Ductile Flat-Plate Structures to Resist Earthquakes," *Proc.*, Fifth World Conference on Earthquake Engineering, (Rome, June 1973), Intl. Assoc. for Earthquake Engineering, Vol. 2, pp. 2016–2019.

5. D. G. Morrison and M. A. Sozen, "Response of Reinforced Concrete Plate-Column Connections to Dynamic and Static Horizontal Loads," *Struct. Research Series* No. 490, Univ. of Illinois, Urbana, April 1981, 249 pp.

6. M. D. Vanderbilt, "Equivalent Frame Analysis of Unbraced Reinforced Concrete Buildings for Static Lateral Loads," *Structural Research Report* No. 36, Civil Engineering Dept., Colorado State Univ., Ft. Collins, June 1981.

7. M. D. Vanderbilt, "Equivalent Frame Analysis for Lateral Loads," *ASCE Jour. Struct. Div.,* ST10, October 1979, pp. 1981–1998.

8. M. D. Vanderbilt and W. G. Corley, "Frame Analysis of Concrete Buildings," *Concrete Intl.,* Vol. 5, Dec. 1983, pp. 33–43.

9. F. R. Kahn and J. A. Sbarounis, "Interaction of Shear Walls and Frames," *ASCE Jour. Struct. Div.,* ST3, Vol. 90, June 1964, pp. 285–335.

10. J. S. Grossman, Discussion of Reference 8, *Concrete Intl.* Vol. 6, No. 10, Oct. 1984, p. 58.

Problems

PROB. 16.1.
(a) Pick a slab thickness based on Code deflection requirements for a two-way slab, interior panel 19-ft × 22-ft (to centers of supporting beams and columns), assuming beam stems are 12-in. wide, columns 16-in. square, f'_c = 3000 psi, Grade 60 bars, dead load (including slab weight) = 100 psf, live load = 125 psf. Assume $\alpha_m = 2$.
(b) Check the direct design moment requirements against the length of this thickness.
(c) Design the slab reinforcement.

PROB. 16.2. A 20-ft square interior panel is supported on beams 18-in. deep overall (and assumed 10-in. width) that are carried by columns 14-in. square and 10-ft high story-to-story. f'_c = 4000 psi, Grade 60 steel, w_d = 90 psf (including slab weight), $w_\ell = 80$ psf. Find slab thickness for deflection; check slab and beam sizes for moment; check beam size for shear. Then design the reinforcement for the slab.

PROB. 16.3. Assume the same data as in Prob. 16.2 for an exterior panel (not a corner panel).
(a) Design the slab.
(b) Design the beam running perpendicular to the wall line for flexure and check it for shear.
(c) Check the spandrel beam for shear and torsion, assuming a wall load of 400-lb plf of beam, applied on the axis of the spandrel beam and column center line. What stirrups are required at d from the face of column?

PROB. 16.4. Assume the flat plate designed in Sec. 15.7 has equal columns above and below with a 10-ft story height. Use the equivalent frame method of design to establish the following for the *short* strips:
(a) Maximum negative design moment and its distribution to column and middle strips.
(b) Maximum positive moment for design and its distribution to design strips.
(c) Maximum column design moment, about this axis, at face of slab.
(d) Compare these moments with those from the direct design method used in Sec. 15.7.

17

DISTRIBUTION OF CONCENTRATED LOADS AND OTHER SPECIAL PROBLEMS

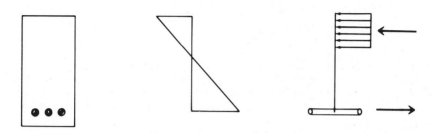

17.1 Concrete Structures Distribute Concentrated Loads

The ordinary reinforced concrete structure is either monolithic or is tied together to act as a unit. Although parallel members of the structure may be analyzed somewhat independently of each other under uniform live loads, the entire structure is actually a three-dimensional frame. When moving concentrated loads are considered, their spacing and their number suggest that all parallel slab strips and all neighboring beams are not equally loaded. The interaction of the several slab strips and beams is usually such as to make the effective slab loading less severe than if each set of loads acted separately on the individual members.

When a heavy wheel rolls over a plank floor, each plank in turn must support the total load. In contrast, when a wheel moves over a concrete slab the wheel deflects the slab locally into a saucerlike pattern and this depression moves with the wheel across or along the slab. Thus a slab strip is deflected (and must be loaded) without a wheel actually resting on it. As the wheel passes over a particular strip the deflection increases, but the single 1-ft strip of slab never carries the entire wheel load unassisted. The designer describes this by saying the wheel load in Fig. 17.1a is distributed over an effective width E (Fig. 17.1b), that is, the moment on the most heavily loaded 1-ft strip is produced by $1/E$ parts of the total load, as in Fig. 17.1c. Likewise, closely spaced beams share in carrying concentrated loads when the beams are connected by stiff floor slabs or stiff diaphragms.

The result of a theoretical study of how a single wheel load is carried by a simple girder highway span is shown in Fig. 17.2. The load is applied to midspan, directly over beam B, and the assumed girder stiffness is five times that of the slab for a width equal to the girder span. Girder B then deflects more than its neighbors A and C. The slab (attached to the beams) is pulled down by beam B, but it resists this movement and exerts upward forces on the beam, as shown

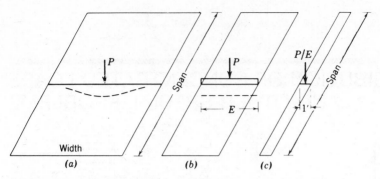

Figure 17.1 Effective width under a single concentrated load. (*a*) Cross section through midspan showing unequal deflections. (*b*) Design assumption of a width E with uniform deflections. (*c*) Equivalent 1-ft strip.

by the shaded ordinates on the right of the figure. These upward forces on beam B total $0.88P$, leaving the net downward load only $0.12P$. The resulting moment diagram on the beam is shown to the left. The neighboring beams deflect less and carry less moment than beam B. Actually, the slab imposes heavier net loads on these beams than on beam B, but the load of $0.43P$ on A and $0.32P$ on C is better distributed and produces less moment. The load on beam D is only $0.13P$ and the load on beam E is negligible.

Although such interaction of members can be approximately evaluated on a theoretical basis, design rules depend equally on field tests. This chapter will

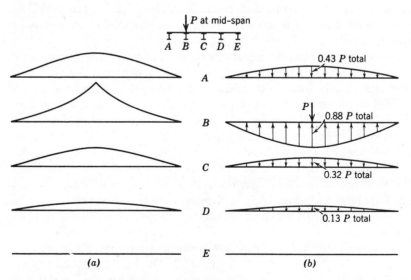

Figure 17.2 The distribution of a single concentrated load to the girders of a bridge span, girder spacing of 0.1 span. (Reproduced from Reference 1, Highway Research Board.) (*a*) Moment diagrams. (*b*) Approximate load distributions and the resulting total loads form the slab.

not attempt to demonstrate how load distribution factors are established nor to tabulate them for the many possible conditions. Rather its objective is to call attention to the problem of load distribution and illustrate how it can be handled in a few typical cases.

17.2 Load Distribution in a Concrete Slab

Load distribution in a slab is approached on two different bases: (1) the service load or deflection basis and (2) the ultimate strength or yield line basis. The distribution at service loads is the one more commonly considered. For a wide slab with a 10-ft simple span, the effective width thus determined for a simple load is between 6 and 8 ft, depending somewhat on the size of the load contact area and the particular algebraic formula[2] used. For comparison, Johansen has shown[3] that at ultimate load, the effective width is twice the span multiplied by $\sqrt{\mu}$, where μ is the ratio of the perpendicular top steel to the longitudinal bottom steel. If μ is about 1/3, the effective width is $2 \times 10 \sqrt{0.33} = 11.5$ ft.

One complication that may make calculations at ultimate strength uncertain is the shear capacity of the slab around the load. Richart and Kluge[4] found shear failures occurring from diagonal tension, with a truncated cone of concrete punched out below the load. When those shear stresses were calculated on a surface at a distance d beyond the load, the unit shear stress was low, in one series from $0.044f'_c$ to $0.057f'_c$. Because the shear failures came at loads 50% greater than those producing local yielding of the steel, the low shear stresses were not considered serious. For a yield line analysis shear stresses around the load might be more significant.

It appears that the distribution based on elastic conditions, as commonly used, is on the safe side. Its use also tends to reduce crack size at working loads. For elastic conditions, Westergaard[5] established an extreme value of maximum positive moment on a slab as $0.315P$ for any simple span when P is distributed over a circular area with the diameter equal to $\frac{1}{10}$ of the span, the slab thickness is $\frac{1}{12}$ of the span, and Poisson's ratio is 0.15. (This local moment is quite sensitive to the size of the bearing area.) The corresponding transverse moment is $0.248P$. Jensen[6] extended these results to show the effect of a rigid beam support at right angles, that is, an effect similar to that in a two-way slab. At this crossbeam the maximum negative moment is $-P/2\pi = -0.159P$ and it occurs with the wheel quite close to the beam.

When closely spaced multiple wheels occurs, an extra slab width acts, but the effective width per wheel is reduced. The AASHTO *Standard Specifications for Highway Bridges*[7] specify such an effective width E for a slab carrying a single wheel (traveling in the direction of the span) that the resultant design is safe for multiple wheels without further calculations. Special transverse distribution steel is also specified as a percentage of the positive moment steel, in the amount of $100/\sqrt{S}$ but not over 50%, where S is the span in feet.* When the

---

* The AASHTO notations S and E are retained in this chapter.

wheels travel perpendicular to the span, the distribution steel is a larger percentage, $220\sqrt{S}$, but not over 67%.

17.3   Calculation for Concentrated Load on Slab

A specification calls for a live load of 60 psf or a moving concentrated load of 2000 lb. Determine which loading controls for a continuous 10-ft span where the moment coefficients for uniform load are $+\frac{1}{16}$ and $-\frac{1}{12}$, and for a concentrated load $+\frac{1}{8}$ and $-\frac{1}{8}$. Consider $E = 0.68S + 2c$, where S is the span and c the diameter of the loaded area.[2] Use ACI load factors.

Solution

For uniform live load, $M_{u\ell} = 1.7(-60 \times 10^2/12) = -850$ lb-ft.
For concentrated load, with $c = 0$ in the absence of better data,

$E = 0.68 \times 10 + 0 = 6.8$ ft for effective width
Effective load, $P_e = 2000/6.8 = 294$ lb/ft strip
$M_{u\ell} = 1.7(-P_e S/8) = 1.7(-294 \times \frac{10}{8}) = -624$ lb-ft < 850 ft-lb

Therefore the uniform load moment governs the slab design.

The effective width for shear would call for the concentrated load near the support that would give less slab deflection and a much reduced effective width. The AASHTO specification says that slabs designed for moment are considered safe in bond (development length) and shear. As an extreme assumption, consider the entire load resisted by a 1-ft strip and the design shear existing at a distance d from the support. Assume also a slab having $h = 4.5$ in., $d = 3.5$ in., and a weight of 56 psf.

$V_u = 1.7 \times 2000 \times 9.71/10 + 1.4 \times 56(5 - 0.29) = 3670$ lb
$v = V/bd = 3670/(12 \times 3.5) = 88$ psi $< 2\sqrt{f_c'} = 2\sqrt{3000} = 110$ psi

Although there are limited data about reaction distribution, it is difficult to imagine a diagonal tension failure that would involve less than a width of four to five times the slab thickness, in the assumed case at least 18 to 22 in.

17.4   Highway Bridge Loadings

The basic units of loading for highway bridges are the H truck, a two-axle loading, and the H-S truck with trailer, a three-axle loading. The 20-ton truck is designated as H20-44, the last number denoting the year 1944, which is when this loading was established. Figure 17.3 shows the distribution of the load between the various wheels. The corresponding truck and trailer combination HS 20-44 represents the standard 20-ton truck plus a second 16-ton axle, as

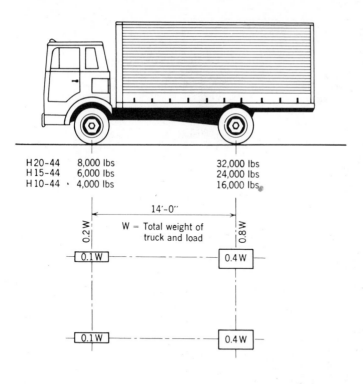

H 20-44	8,000 lbs	32,000 lbs
H 15-44	6,000 lbs	24,000 lbs
H 10-44	4,000 lbs	16,000 lbs

14'-0"

0.2 W

W = Total weight of truck and load

0.8 W

0.1 W   0.4 W

0.1 W   0.4 W

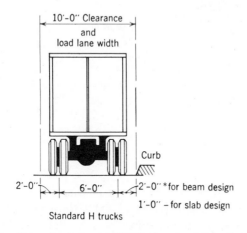

10'-0" Clearance
and
load lane width

Curb

2'-0"   6'-0"   2'-0" *for beam design

1'-0" – for slab design

Standard H trucks

Figure 17.3 The H truck of the AASHTO *Standard Specifications for Highway Bridges*. For slab design the centerline of wheels is assumed to be 1 ft from face of curb.

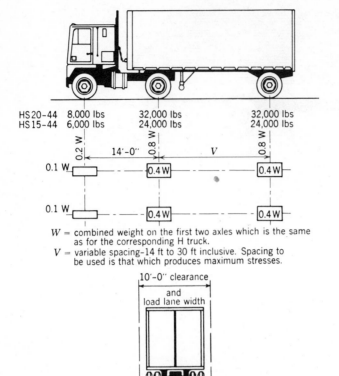

W = combined weight on the first two axles which is the same as for the corresponding H truck.

V = variable spacing-14 ft to 30 ft inclusive. Spacing to be used is that which produces maximum stresses.

Standard HS trucks

Figure 17.4 The H-S truck of the AAHSTO *Standard Specifications for Highway Bridges*. For slab design the centerline of wheels is assumed to be 1 ft from face of curb.

shown in Fig. 17.4. For either type of truck an alternate uniform load of 640 plf for each lane plus a concentrated load of 18,000 lb for moment or 26,000 lb for shear is used wherever it gives larger values, which is the case on longer spans. The lane load occupies a 10-ft width and is to be placed in the worst position when actual lanes are wider. The reader is referred to the AASHTO specification[7] for such details as the number of loaded lanes and proper reduction factors, loads for continuous spans, impact allowance, and distribution of loads. In the following examples, only the governing portion of the specification is mentioned. The AASHTO notation is retained where no equivalent exists in ACI notation.

17.5 Design of a Highway Slab Span

Design an interior panel of an 8-ft 6-in. span continuous slab for H20-44 loading, $f'_c = 3000$ psi, Grade 40 steel, AASHTO specification. Consider that this slab spans longitudinally between transverse girders.* Use service load design method of Appendix A (Sec. A.4). AASHTO (1983) now permits either load factor or service load designs.

Solution

The AASHTO specification allowable stresses are $f_c = 0.40f'_c = 1200$ psi, $n = 10$, $f_s = 20,000$ psi.

$$c_b = \frac{1200}{1200 + 20,000/10} \qquad d = 0.375d, \qquad z_b = d - c_b/3 = 0.875d$$
$$k_b = (0.5f_cbc)z/bd^2 = 0.5 \times 1200 \times 0.375d \times 0.875d/d^2 = 197$$

Since the reinforcement is parallel to traffic, this slab falls under Case B in the specification with an effective width $E = 4 + 0.06S$, where S is the span in ft.

$$E = 4 + 0.06S = 4 + 0.06 \times 8.5 = 4.51 \text{ ft} < 7.0 \text{ ft max.}$$
$$P = 0.4 \times 20 = 8 \text{ tons} = 16,000 \text{ lb}$$
$$P_e = P/E = 16,000/4.51 = 3530 \text{ lb}$$

If the moment for the concentrated load is taken as 80% of the simple span moment,

$$M_L = 8.0P_eS/4 = -0.20 \times 3530 \times 8.5 = -6000 \text{ ft-lb}†$$

The impact fraction is

$$I = \frac{50}{S + 125} = \frac{50}{8.5 + 125} = 0.375$$

but not to exceed 0.30, which governs here.

$$\text{Impact } M_I = -0.30 \times 6000 = -1800 \text{ ft-lb}$$

Assume dead load $w_d = 100$ psf and moment coefficients of $-\frac{1}{12}$ and $+\frac{1}{16}$.

$$M_D = 100 \times 8.5^2/12 = -602 \text{ ft-lb}$$
$$M_T = -6000 - 1800 - 602 = -8400 \text{ ft-lb}$$
$$d = \sqrt{M/k_bb} = \sqrt{8400 \times 12/(197 \times 12)} = \sqrt{42.6} = 6.52 \text{ in.}$$

---

* More typically, highway slabs span transversely between longitudinal girders. Unfortunately for the purpose of this text (showing generally how to handle concentrated loads) AASHTO for that case gives resultant moment formulas that do not show the effective width E used. That E appears to lie in the range of 6.3 to 7 ft for the two wheels (full-axle load) on the strip.
† The specification gives the approximate value of live load M for a simple span as $900S = 900 \times 8.5 = 7650$ ft-lb that is 6120 ft-lb on the 80% basis.

Since AASHTO specifies a minimum of 2.00 in. of cover for top reinforcement

$h = 6.52 + d_b/2 + 2.00 = 8.52 + 0.38$, say, $= 8.90$ in.

USE $h = 9$ in., $d = 9.0 - 2.00 - d_b/2 =$, say, 6.56 in. (for #7 bars)

Revised $M_D = -112 \times 8.5^2/12 = -674$ ft-lb

Negative $M_T = -6000 - 1800 - 674 = -8470$ ft-lb

Positive $M_T = +6000 + 1800 + 674 \times 12/16 = +8300$ ft-lb

Negative $A_s = \dfrac{8470 \times 12}{20{,}000 \times 0.875 \times 6.56} = 0.885$ in.^{2}/ft $=$ #7 at 8 in. (0.90)

or #6 at 6 in. (0.88)

Positive $A_s = 0.885 \times 8300/8470 = 0.867$ in.^{2}/ft $=$ #6 at 6 in.

Use #6 at 6 in. for both positive and negative M.

This design ignored the possibility of using compression steel which would be available at the supports but possibly not at midspan. AASHTO also says slabs designed for moment "shall be considered satisfactory in bond and shear."

The concentrated load requires distribution steel perpendicular to the moment steel. The amount is specified as a percentage of the positive moment steel given by $100/\sqrt{S} = 100/\sqrt{8.5} = 34.3\% < 50\%$ maximum. Transverse steel must be at least $0.343 \times 0.867 = 0.297$ in.^{2}/ft $=$ #5 at 12 in. (0.310).

17.6   Design Moments and Shears for Highway Girder

Calculate the unfactored live load and impact load moments and shears for a 40-ft simple span girder built integrally with a slab. Consider both H20-44 and H20-S16-44 loadings on the center girder of the cross section shown in Fig. 17.5a.

Solution

For shear calculations the AASHTO says there shall be no lateral or longitudinal distribution of the wheel load at the end of the span. For other wheels the distribution shall be that applying for moment. The provision for moment when two or more lanes of traffic are involved calls for the wheel loads to each stringer to be $S/6.0 = 9.75/6.0 = 1.63$ of the full load.*

---

* For $S > 10$ ft, the load to the beam is the reaction computed from "simple span" slabs.

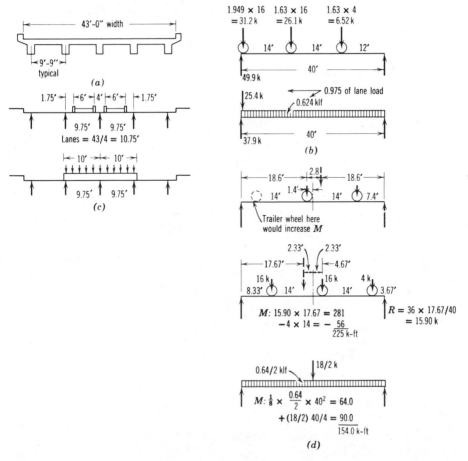

Figure 17.5 Highway girder bridge. (*a*) Bridge cross section. (*b*) Location of wheels and lane load for maximum shear. (*c*) Position of loads laterally for maximum stresses in center girder. (*d*) Location of wheels and lane load for maximum moment.

The longitudinal arrangement of wheels for the truck and uniform lane loads are shown in Fig. 17.5*b*, ignoring any complication that the end diaphragm may cause. Figure 17.5*c* shows the lateral arrangement of the loads on the cross section, with two trucks assumed to be passing. For the truck the simple beam slab reactions give for the end wheels:

$$(2 \times 2 \times 4.75/9.75)16 = 1.949 \times 16 \text{ k} = 31.2 \text{ k}$$

For the lane load of Sec. 17.4, the end concentrated load for shear becomes $2 \times 0.5 \times 26 \times 9.75/10 = 25.4$ k and the uniform load $0.64 \times 9.75/10.0 = 0.624$ klf. For the trucks, maximum $V = 31.2 + 26.1 \times$

26/40 + 6.52 × 12/40 = 50.2 k. For the lane load, maximum V = 25.4 + 0.624 × 40/2 = 37.9 k.

$$\text{Max } V_L = 50.2 \text{ k*}$$

By specification, L is from the shear point to the far reaction, that is, the full 40-ft span.

$$I = \frac{50}{40 + 125} = 0.303$$

Since I = 0.30 governs, V_I = 0.30 × 50.2 = 15.1 k.

For moment three loadings must be investigated, as worked out in Fig. 17.5*d*. For simplicity these loadings are compared on the basis of one line of wheels or one-half lane load. The HS 20-44 loading governs maximum moment. Since the number of lines of wheels is $S/6$ = 1.63,

$$M_L = 1.63 \times 225 = 367 \text{ k-ft}$$

The loaded length for moment, by definition, is the full length, L = 40 ft.

$$I = \frac{50}{40 + 125} = 0.303$$

Since I = 0.30 governs, M_I = 0.30 × 367 = 110 k-ft.

The example illustrates how AASHTO specifies handling load distribution on stringers or longitudinal girders. The remainder of the design of the T-beams is very similar to that of a building T-beam with these minor differences, which are usually more restrictive:

1. Different allowable unit stresses for f_c and v, compared to ACI Code.
2. Slightly more conservative z values for flexural reinforcement distribution.
3. Different maximum moment diagram, owing to type of loading.
4. Different maximum shear diagram, owing to type of loading.
5. Different deflection limits and minimum depths, owing to type of loading.
6. A specific fatigue stress limit in reinforcement.

These small differences are to be expected when one changes from one specification to another.

17.7 Openings in Slabs

When openings in slabs are large, as for stairs or elevators, beams must be used around the openings. Good practice usually requires that these beams be framed into column sufficiently to provide a stable unit without the slab.

---

* AASHTO designs for shear *at* at the face of support rather than *d* from the support when a heavy concentrated load is applied near the support.

Small openings such as pipe sleeves, if not too numerous, can be made almost anywhere in a slab, except adjacent to the columns in flat slab construction. What can be done about larger openings may require at least some rough calculations. Obviously, openings are least dangerous where shear stresses are small and bending moments are below maximum.

Electric conduits, unless closely spaced or crossing at small angles, can be included without considering any loss in moment strength. The detailed requirements of Code 6.3 should be noted in this connection.

The effect on shear of openings around columns in flat plate construction is discussed briefly in Sec. 5.24 and illustrated in Fig. 5.25. These shear provisions recognize the coexistence of large bending moments.

An equally common and troublesome problem is that of relatively large openings that interrupt the normal flexural action. Openings of any size are permitted by Code 13.5.1 if analysis shows that both strength and deflection are still acceptable. For strength the methods of Chapter 14 on the strip method are appropriate, even if they appear rather arbitrary.

Without analysis considerable leeway is still given by Code 13.5.2:

13.5.2—In lieu of special analysis as required by Sec. 13.5.1, openings may be provided in slab systems without beams only in accordance with the following.

13.5.2.1—Openings of any size may be located in the area common to intersecting middle strips, provided total amount of reinforcement required for the panel without the opening is maintained.

13.5.2.2—In area common to intersecting column strips, not more than $\frac{1}{8}$ the width of column strip in either span shall be interrupted by openings. An amount of reinforcement equivalent to that interrupted by an opening shall be added on the sides of the opening.

13.5.2.3—In the area common to one column strip and one middle strip, not more than $\frac{1}{4}$ the reinforcement in either strip shall be interrupted by openings. An amount of reinforcement equivalent to that interrupted by an opening shall be added on the sides of the opening.

13.5.2.4—Shear requirements of Sec. 11.11.5 shall be satisfied.

In two-way slabs supported on all four sides, openings in the corners of the slab are least damaging. Because openings for ducts frequently need to be near the columns, for architectural reasons this is a definite advantage of this type of slab. The negative moment zone of the short middle strips (near midspan of the longer beams) is the least favorable zone for openings.

Often rough checks can be made on the strength of the construction after the openings are located. If openings reduce a critical design section for moment, the required bd^2 for moment must be maintained by providing extra depth to offset the reduced width. The steel may be more closely spaced on each side of the opening to maintain the necessary A_s. Often it will be possible to locate openings where moment is well below the compression capacity of the slab, thereby leaving the arrangement of reinforcement as the only problem. Of course, shear strength must be maintained, but this is rarely a problem except near the columns in flat slab types, as noted previously.

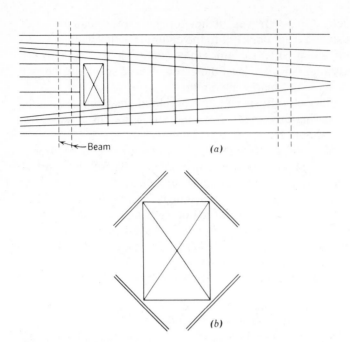

Beam

(a)

(b)

Figure 17.6 Bars at openings in slabs. (*a*) Bars fanned out to miss opening. (*b*) Corner reinforcement to reduce shrinkage cracking.

The arrangement of bars around any but minor openings can constitute a real problem. Bars running perpendicular to the face of an opening are not fully effective when simply cut off at the opening. This is acceptable if there is a beam at the opening to act as a reaction for the slab. If there is no beam it is better to fan the bars out or splay them to go around the opening, as shown in Fig. 17.6*a*. If fanning or splaying leaves too wide an area without steel, extra bars can be placed parallel to the side of the opening, as indicated.

If minor cracking at the corners of an opening is objectionable, it is always advisable to add one or two diagonal bars at each corner, especially at large openings (Fig. 17.6*b*). Bars are always desirable around window and door openings in concrete wall slabs, because such reinforcement helps to take care of shrinkage stresses.

17.8   Openings in Beams

Openings through slabs that encroach on the flange width of T-beams require a check on the remaining bending strength.

Openings through a beam web are increasingly important for installation of building services. In a region of small shear, as near the middle of a beam span, a horizontal pipe sleeve is not serious. Elsewhere, shear strength must be closely watched and in many places bending strength as well. Large openings in beams are particularly weakening. They destroy beam action and force the reduced section to act much as a Vierendeel truss (a truss without diagonals). In such a truss the average bending moment over the length of the opening is resisted by axial compression in one chord and tension in the other, with these two forming a couple in the case of pure flexure. Where shear is present the change in the moment over the length of the opening superimposes a reversed bending resistance in each chord, the total of the four end moments on the chords equaling the external shear times the length of the opening. How the shear and these reversed moments are shared by the two chords depends on the relative chord stiffness. The designer in such a case should note that for members with significant axial tension Code 11.3.1.3 requires all shear to be resisted by stirrups (none assigned to the concrete).

Selected References

1. C. P. Siess and A. S. Veletos, "Distribution of Loads to Girders in Slab-and-Girder Bridges: Theoretical Analyses and Their Relation to Field Test," Highway Research Board *Report 14-B*, Washington, 1953, p. 58.

2. C. T. Morris, "Concentrated Loads on Slabs," Onio State Univ. Eng. Exp. Sta. *Bull. No. 80*, 1933.

3. K. W. Johansen, "Bruchomente Der Kreuzweise Bewehrten Platten" (Moments of Rupture in Cross-Reinforced Slabs), International Association for Bridge and Structural Engineering, Liége, Vol. 1, 1932, p. 277.

4. F. E. Richart and R. W. Kluge, "Tests of Reinforced Concrete Slabs Subjected to Concentrated Loads," Univ. of Ill. Eng. Exp. Sta. *Bull. No. 314,* 1949.

5. H. M. Westergaard, "Computation of Stresses in Bridge Slabs Due to Wheel Loads." *Public Roads,* 11, No. 1, Mar, 1930, p. 1.

6. V. P. Jensen, "Solutions for Certain Rectangular Slabs Continuous Over Flexible Supports," Univ. of Ill. Eng. Exp. Sta. *Bull. No. 303,* 1938.

7. *Standard Specifications for Highway Bridges,* AASHTO, Washington, 13th ed., 1983.

Problems

NOTE *The service load method of Appendix A should be used for these highway bridge type problems, unless otherwise instructed.*

PROB. 17.1. Design a simple span slab to carry its own weight and a single 20-k concentrated load over a 20-ft span. Assume the impact as 25% of live load and an effective width $E = 5.6$ ft. Allowable $f_c = 1200$ psi, $f_s = 20,000$ psi, $n = 10$, cover of 1.50 in. to center of steel. Assume the load is distributed over a sufficient bearing area to avoid a punching shear failure.

PROB. 17.2. Design an interior span of a continuous slab with a clear span of 13 ft for a live load of 10 psf or a single concentrated live load of 5 k, whichever is worse. If the effective width assumed for the concentrated load is 4.8 ft when the load is at its worst position for moment, design the slab using the moment coefficients of Sec. 9.5 (Fig. 9.8) for uniform load and $\pm 0.20 P\ell$ for the concentrated load. Assume zero impact. $f'_c = 4000$ psi, $f_s = 20,000$ psi, $n = 8$, allowable $f_c = 1000$ psi.

18

CANTILEVER RETAINING WALL DESIGN

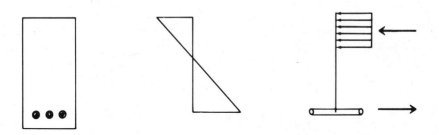

18.1 Types of Retaining Walls

Retaining walls provide soil stability at a change in ground elevation. Substantial dead weight in such a wall is important, both to resist overturning from the lateral earth pressures and to resist horizontal sliding from the same forces. There are several types of retaining walls:

A *gravity* retaining wall (Fig. 18.1*a*) depends entirely on its own weight to provide the necessary stability. Plain concrete or even stone masonry constitutes an adequate building material. Design is concerned chiefly with keeping the thrust line within the middle third of the cross section.

The *cantilever* retaining wall (Figs. 18.1*c* and 18.2) is a reinforced concrete wall that utilizes the weight of the soil itself to provide the desired weight. Stem, toe, and heel are each designed as cantilever slabs, as indicated in Fig. 18.7.

The *semigravity* type of wall uses very light reinforcement and is intermediate between the cantilever and gravity types (Fig. 18.1*b*).

The *counterfort* retaining wall looks something like a cantilever wall and likewise uses the weight of the soil for stability. The wall and base are tied together at intervals by counterforts or bracing walls (Figs. 18.1*d* and 18.3). Counterforts act as tension ties and totally change the supports for stem and heel slabs. The stem becomes a slab spanning horizontally between counterforts and the heel becomes a slab supported on three sides. This type of wall is more economical than the cantilever type somewhere in the 20 to 25-ft height range.

A *buttressed* wall is similar to the counterfort wall except that the bracing members are on the opposite side of the wall and act in compression.

Crib-type retaining walls may be made of precast concrete, timber, or metal. The face pieces are supported by anchor pieces extending back into the soil for anchorage.

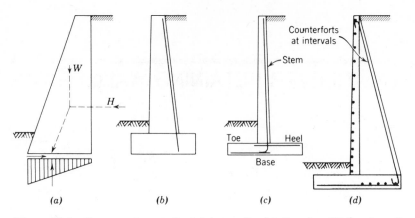

Figure 18.1 Common types of retaining walls. (*a*) Gravity. (*b*) Semigravity. (*c*) Cantilever. (*d*) Counterfort.

Figure 18.2 Cantilever retaining wall. (Courtesy Texas Highway Department.)

(a) (b)

Figure 18.3 Counterfort retaining wall. (Courtesy Texas Highway Department.) (*a*) Under construction. (*b*) Before backfilling.

Anchored retaining walls have been prominent in the literature in recent years. The face wall is anchored back into rock or even into a large mass of earth by rods or wire strands that are prestressed to an anchorage in the rock or an anchor embedded deep in the soil.

18.2 Active Soil Pressure

The reader with a background in soil mechanics may proceed at once to Sec. 18.5 for the cantilever wall design example, which also includes a discussion of alternative procedures. A rudimentary treatment of soil pressures is given in this and succeeding sections so that a reader unfamiliar with soil mechanics can gain some understanding of some of the forces exerted on retaining walls. Geotechnical engineering specialists and references should be consulted for the geotechnical aspects.

Only the case of cohesionless soil is here considered, that is, essentially a dry sand. Cohesion in the soil theoretically reduces the demands on a retaining wall, but generally accompanies other adverse factors, such as reduced friction and expansive-type soils that increase the total effect. An expansive type of soil, such as a plastic clay, is totally unsuited for a backfill behind a wall and often introduces problems beyond economic solution.

Rankine in 1857 analyzed active soil pressure on a smooth wall as a problem of plastic equilibrium in the soil. Sliding occurs on two sets of planes in a wedge behind the wall, as sketched in Fig. 18.4. For a smooth (frictionless) wall these planes make an angle of $45 + \phi/2$ with the horizontal, where ϕ is the friction angle for soil on soil, and produce the maximum pressure that can follow through against a wall. (For a wall that does not deflect at all, larger pressures can exist.) Plastic equilibrium of sliding planes is a fundamental of soil mechan-

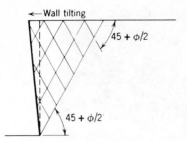

Figure 18.4 Rankine's plastic equilibrium for active soil pressure.

ics. However, Coulomb's theory presented in 1773 gives a better result than Rankine's for all conditions except those of a vertical frictionless wall and a horizontal fill. Because, for this special case, both theories give the same result, only Coulomb's is developed.

Coulomb considered possible sliding planes at different slopes and found the one that demanded the greatest holding force on the part of the wall. When friction on the wall is neglected, the holding force is perpendicular to the wall, that is, horizontal for a vertical wall face. Neglect of wall friction is on the safe side for active pressure, but not for the passive pressures of Sec. 18.4.

Consider a wall at AB in Fig. 18.5a. The sand behind it tends to slide on some plane such as BC_1. The wedge of sand ABC_1 is taken as a free body in equilibrium under three forces: (1) W_1, the weight of the soil; (2) R_1, the reaction from the soil below BC_1, which may be considered a normal reaction N_1 plus a friction force F_1 resisting sliding; (3) H_1, the holding force from the wall. For any given slope angle θ_1, the force H_1 (and R_1 if desired) is found from the force triangle as shown in Fig. 18.5b. Similar force triangles constructed for other potential sliding planes such as BC_2 or BC_3 lead to other holding forces H_2, H_3, and so forth. A number of trials will lead to the determination in Fig. 18.5b of the maximum possible value of H. A helpful visualization is a plot of ordinate H_1 over C_1, H_2 over C_2, and so on, with a curve sketched through the points so located. The peak value H also locates the sliding plane, say, BC in Fig. 18.5c.

A similar analysis of the soil above point D will lead to a critical sliding plane through D parallel to BC. The holding force for the depth AD is proportional to the weight of sand in the wedge ADE, that is, proportional to y^2 and the unit weight of soil γ. If the constant of proportionality is indicated as $K_A/2$,

$$H_y = K_A \gamma y^2/2$$

The coefficient K_A is called the coefficient of active earth pressure and is usually on the order of 0.27 to 0.34, depending on the sliding friction angle.

A total pressure increasing with the square of the depth corresponds to a unit pressure increasing directly with the depth, that is,

$$p_y = K_A \gamma y$$

and the pressure on the wall is that shown in Fig. 18.5d.

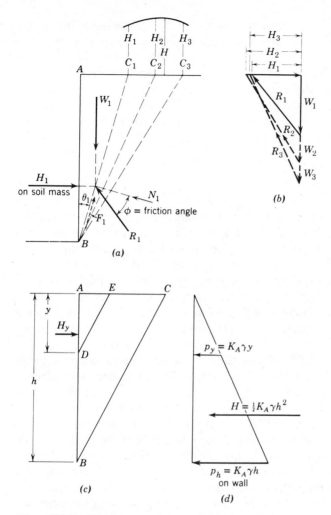

Figure 18.5 Maximum active earth pressure.

With the pressure on the wall similar to that from a fluid (one weighing somewhat less than water), many designers use the term *equivalent fluid weight* or *equivalent fluid* for the term $K_A\gamma$. Call this fluid weight w_f. Then

$$p_y = w_f y \qquad H_y = w_f y^2/2$$

Equivalent fluid weight is often assumed without adequate knowledge of the factors that enter into a calculated value of K_A.

These pressures are called active soil pressures because they continue to act on a wall after it deflects or slides. Pressures in a confined soil may be higher, because the active earth pressure is calculated on the favorable basis of a considerable holding force developed by friction. Some small movement on plane *BC* in Fig. 18.5*c* is necessary to develop this friction; without it *H* will be larger. For the triangular pressure distribution to occur, there must be some sliding on all parallel planes, such as *DE*, above *BC*. Hence, the wall must deflect more at the top than at the base by approximately 0.001 times its height; this necessary theoretical deflection corresponds to a rotation of the wall about the base at *B*. Many practical constructions fail to satisfy this deflection requirement, for example, basement walls when supported at or near the ground level by the first-floor framing.

18.3   Surcharge

Loads on the surface of the ground over a possible sliding plane, as in Fig. 18.6, increase the horizontal pressure by adding to the ordinary soil weight *W* in Fig. 18.5*a*. Uniform surcharge over the entire area adds the same effect as an additional height of soil weighing a similar amount. The concept is convenient, however, in visualizing the effect of the surcharge, which is simply that of an added height of earth weighing the same amount. Such a surcharge adds a uniform pressure to the triangular soil pressure already discussed, as shown in Fig. 18.6*a*. The use of unequal load factors for surcharge and soil weight adds a complication.

   Surcharge far enough removed from the wall causes no pressure on the wall. For example, a surcharge far to the right of *C* in Fig. 18.5*c* does not influence the sliding plane *BC* or the pressure *H*. A load just to the right of *C* would influence the sliding on a slightly flatter plane and might make the plane critical.

   Engineers commonly assume that a surcharge does not influence the pressure above the point where a line sloping downward from the load intersects the wall, as shown in Fig. 18.6*b*. A slope of 40° or 45° was used in the past, although somewhat better methods in soil mechanics now are available that recognize the extent of the surcharge. The actual pressure does not change as abruptly as shown in Fig. 18.6*b*, but this assumption is within reason and indicates a greatly reduced overturning effect compared to Fig. 18.6*a*.

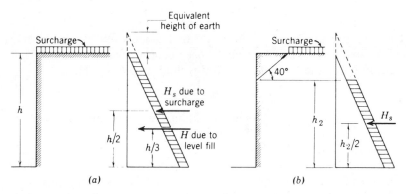

Figure 18.6   Effect of surcharge on earth pressure.

18.4 Passive Earth Pressure

If the wall is pushed against the soil, the resistance is very much higher than the active pressure because the soil friction then resists the wall movement. In this case the sliding plane is much flatter ($45 - \phi/2$ according to Rankine) and an analysis similar to that of Fig. 18.5 involves a greatly increased soil weight W. Wall friction is more important in this case and true failure is on a curved surface. Passive resistance is several times as large as active pressure.

18.5 Safety Philosophy for Retaining Walls

The appropriate safety philosophy and application of load factors for retaining wall design is not directly apparent from the ACI Code. Code Chapter 15 regarding isolated and combined footings (treated in Chapter 19) specifies that the base area of footings should be computed based on unfactored forces and moments and on permissible (or unfactored) soil pressures. Code 14.1.2 specifically indicates, however, that cantilever retaining walls should be designed according to the flexural design provisions of Chapter 10 and by inference on factored load procedures. In fact, to be consistent, retaining wall proportions as well as cross-sectional dimensions and reinforcement should be proportioned using factored loads. This safety philosophy causes some conceptual difficulties because geotechnical engineers usually base soil capacity recommendations on unfactored, allowable soil pressures. In many applications of proportioning equations, however, the actual soil pressures do not appear. For instance, when calculating heel dimensions for stability, specific soil pressures do not need to be determined. Especially in those cases where dead or live loads can produce friction forces opposing lateral pressure forces, the Code in Sec. 9.2.4 indicates that minimal reliance should be placed on such offsetting forces. The Code requires the use of load factors *less* than one to compute the offsetting forces. This safety philosophy only works if factored loads are considered. Thus, for design of retaining walls, proportions should be based on equilibrium relationships using factored loads and additional checks for satisfaction of maximum soil pressures at service load levels should be made to ensure that consolidation will be within reasonable limits.

This philosophy is further developed and illustrated in the following design example.

18.6 Design of Cantilever Retaining Wall

(a) Data

Overall height = 18-ft 0-in.
Level fill, 400 psf surcharge, soil weight of 100 pcf.
Horizontal pressure: $K_a = 0.31$.

Sliding friction of concrete on soil, minimum coefficient assumed as 0.5; of soil on soil, minimum coefficient assumed as 0.62.

Permissible soil pressure under toe = 3500 psf for service load.

f'_c = 3000 psi, Grade 60 steel.

Design wall for general provisions of ACI Building Code, using about 1.0% steel ($\rho = 0.01$).

(b) Design Sequence   The design of a cantilever wall involves the choice of heel and toe lengths and the separate design of stem, heel, and toe slabs. Each of these three slabs acts as a cantilever, as shown in Fig. 18.7. The design steps can be done in the following sequence:

1. Foundation conditions[4] determine the elevation of the bottom of the base and thus the overall height.
2. The thickness of the base then must be estimated to establish the height for the stem.
3. A tentative stem thickness then can be calculated.
4. Length of heel and toe for stability against sliding and overturning next can be established.
5. Heel then can be completely designed.
6. Toe can be completely designed.
7. Stem design can be completed on the basis of the actual base thickness used in steps 5 and 6.

The design of this wall follows, with discussion interspersed, especially on:

The special problem of safety as reflected in the use of load factors for retaining wall equilibrium, in subsection (d).

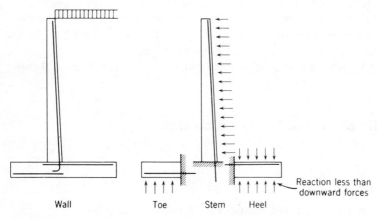

Figure 18.7   Design parts of cantilever retaining wall.

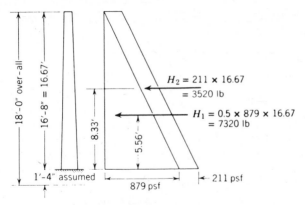

Figure 18.8 Earth pressure on a 1-ft width of stem, including load factors.

Allowable soil pressure for use with involved ultimate loads, in subsection (e). Conventional soil pressure distribution under eccentric reactions, in subsection (g).

(c) Stem Design   The base thickness is roughly 7% to 10% of the overall height with a minimum of about 12 in. Assume a 16-in. base, giving a stem height of $18.0 - 1.33 = 16.67$ ft as shown in Fig. 18.8. The 400-psf surcharge is treated as live load with a load factor of 1.7. Under a clarification of the definition of H in Sec. 9.0 of the 1986 ACI Building Code, any loads owing to the weight and pressure of soil and of water in soil are included in H. The Code (9.2.4) provides a separate load factor of 1.7 for the loads owing to weight and pressure of soil, water in soil, or other materials (granular materials like coal, gravel, and so on) retained by a wall.*

Because of the surcharge, there is an added lateral pressure of $1.7 \times 0.31 \times 400 = 211$ psf from top to bottom of the wall. This lateral pressure adds to the basic triangular soil pressure from the weight of the soil, which is a pressure increasing from zero at the top to $1.7 \times 0.31 \times 16.67 \times 100 = 879$ psf at the base of the stem. The resultant lateral pressures per ft. length of wall are calculated in Fig. 18.8.

Maximum moment and maximum shear occur at the bottom of the stem and the initial design is at this section for a 1-ft length of wall. Stem dead load creates no moment in the stem, and the small direct compression it causes, about 24 psi, is generally ignored. Strictly under Code 9.3.2.2b the ϕ factor of 0.9 for flexure is slightly decreased toward the column value of 0.70 by using $P_n/A_g = 24$ psi in the equation:

$$\phi = 0.90 - 0.20\phi P_n/(0.10f_c'A_g)$$
$$\phi = 0.90/[1 + (2P_n/f_c'A_g)]$$
$$= 0.90/[1 + 2 \times 24/3000] = 0.886$$

---

* The portion of Code 9.2.4 that is related to reducing load factors for D and L in some cases is discussed in detail in subsection (d).

The reduction from 0.90 to 0.886 is *not* significant, especially if the axial load is ignored, as here. From Fig. 18.8:

$$M_u = 3520 \times 8.33 + 7320 \times 5.56 = 70{,}000 \text{ ft-lb/ft width}$$
$$M_n \geqslant M_u/\phi = 70{,}000/0.886 = 79{,}000 \text{ ft-lb} = 948{,}000 \text{ in.-lb}$$

For $\rho = 0.01^*$, $N_t = 0.01\,bd \times 60{,}000 = 600\,bd$

$$a = 600\,bd/(0.85 \times 3000 \times b) = 0.235\,d$$
$$z = d - 0.118\,d = 0.882\,d$$
$$M_n = 0.882\,d \times 600\,bd = 530\,bd^2, \quad k_n = 530, \quad bd^2 = M_n/530$$

With $b = 12$ in., $bd^2 = 12\,d^2 = M_n/530 = 948{,}000/530$

$$d^2 = 948{,}000/(12 \times 530) = 149, \qquad d = 12.21 \text{ in.}$$
$$h = 12.21 + 2 \text{ in. cover (Code 7.7.1b)} + 0.5\,d_b$$
$$\text{(bar diam.)}$$
$$= \text{say, } 14.71 \text{ in.}^*$$
$$\text{USE } h = 15 \text{ in., } d = 12.5 \text{ in. (for estimated \#8 bars)}$$

To reduce concrete usage and to introduce a desirable aesthetic effect the wall may be tapered. Assume that the thickness is reduced linearly to a total thickness of 12 in. at th top.

In slab design, shear rarely governs thickness unless the span is short or loads are unusually heavy. Hence shear usually can be delayed until a late check. It is critical at a distance d from the support, but the calculation of critical shear there may be avoided if the larger (but more easily calculated) shear *at* support is satisfied. At the support

$$V_u = H_1 + H_2 = 7320 + 3520 = 10{,}840 \text{ lb}$$
$$\text{Required } V_n = V_u/\phi = 10{,}840/0.85 = 12{,}800 \text{ lb}$$
$$\text{Actual } V_n = 2\sqrt{f_c'}\,bd = 2\sqrt{3000}\,(12)(12.5) = 16{,}430 \text{ lb}$$
$$\therefore \textbf{O.K.} \text{ Shear does not govern stem thickness.}$$

The design would normally proceed to the choice of base length at this stage, but, at the risk of necessary revision later, stem design is carried further here to show it as a complete unit of design.

Because d is slightly increased above the required value, a is slightly reduced and z is slightly increased above $0.882\,d$, say, to $0.89\,d = 11.1$ in.

$$A_s = \frac{M_n}{f_y z} = \frac{79{,}000 \times 12}{60{,}000 \times 11.1} = 1.42 \text{ in.}^2/\text{ft} = 0.119 \text{ in.}^2/\text{in.}$$

$$a = 1.42 \times 60{,}000/(0.85 \times 3000 \times 12) = 2.78 \text{ in.}$$
$$z = 12.5 - 1.40 = 11.1 \text{ in.} \qquad \textbf{O.K.} \text{ without revision}$$

---

* Because the 1.0% of steel is strictly a judgment decision, not a Code limit, h could be made either 14.5 in. with slightly more steel or 15 in. with slightly less steel.

Bar spacing for $\#7 = 0.60/0.119 = 5.04$ in. d and z increased 0.06 in.

$\qquad\qquad\quad \#8 = 0.79/0.119 = 6.64$ $\quad d$ as assumed

$\qquad\qquad\quad \#9 = 1.00/0.119 = 8.40$ $\quad d$ and z reduced 0.06 in.

$\qquad\qquad \#10 = 1.27/0.119 = 10.67$ $\quad d$ and z reduced 0.13 in.

$\qquad\qquad \#11 = 1.56/0.119 = 13.11$ $\quad d$ and z reduced 0.20 in.

It is helpful if the designer can foresee the development requirement for dowels into the base at this stage. (This is one of several practical reasons for holding the choice of bars until the base design is complete.) Subject to a later check on bar development requirements, the #7 fits nicely, although the 5-in. spacing means many bars to handle. The reduced d for #11 bars means a further increase in required A_s and a 12.5-in. spacing. Some A_s greater than 1.42 in.^{2}/ft thus is required for all sizes except the #7 and #8. For simplicity of spacing

$$\text{USE #8 at 6 in. } (A_s = 0.79 \times 12/6 = 1.58 \text{ in.}^2/\text{ft})$$

The slight excess in steel area is justified by the convenient spacing.

Development length into the stem creates no problem of consequence. Dowel development length into the base is discussed in subsection (j).

(d) Heel and Toe Length—Design Philosophy Heel and toe length are interrelated but can be established in sequence. The heel is often made just long enough to cause the resultant of the forces on the wall to strike the ground under the stem (Fig. 18.9a). In addition to necessary further provision against overturning, this can lead to complications from insufficient sliding resistance that require the use of a key into the soil to take a major part of the horizontal force. Such a key is rather uncertain to calculate and it functions chiefly to change the critical friction surface from concrete resting on soil to soil resting on soil, which usually increases the available friction. Alternatively, the heel length can be lengthened to increase vertical load and thus frictional resistance. Such a long heel increases wall costs and sometimes excavation costs.

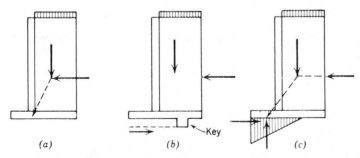

(a)   (b)   (c)

Figure 18.9 Design criteria. (a) Heel length to put resultant under wall, service loads. (b) Heel length to limit sliding under factored loads. (c) Toe length to give satisfactory soil pressure under factored loads.

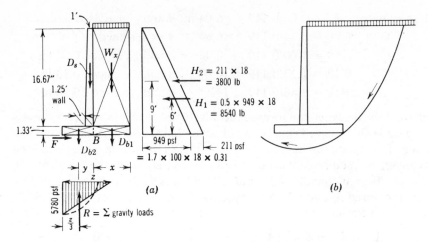

Figure 18.10 Retaining wall stability. (*a*) Horizontal sliding and overturning. (*b*) Sliding failure in soil.

The method proposed in the following subsection compromises between these two approaches, basing the heel length on friction of soil on soil, with a key added simply to avoid sliding of the concrete at the soil-concrete interface (Fig. 18.9*b*).

The toe length then must be based on maintaining soil pressures within the ultimate capacity while avoiding overturning from overloads, Fig. 18.10. In poor soils, the minimizing of settlement problems [subsection (g)] may control the design length. The length of toe is intimately tied in with the length of heel, because a long heel is a counterbalance to overturning.

Both these equilibrium conditions, for heel and for toe, involve some loads offsetting the effects of other loads. Horizontal loads tend to cause sliding, and vertical loads mobilize more frictional resistance. Horizontal loads cause overturning moments and vertical loads develop counterbalancing moments. As a result, two serious considerations exist outside the basic areas of member design for strength and the allowable soil pressure at service load. These considerations are not, in the authors' opinion, yet solved to the point of defining good practice in precise terms:

1. The open question of how best to apply a theory of safety, discussed in the next paragraphs.
2. The correlation of service load pressure with factored loads. Ultimate soil pressure is discussed with the choice of toe length in subsection (f).

With regard to ϕ the Code (9.2.4) appears quite specific:

> If resistance to earth pressure H is included in design, required strength U shall be at least equal to
>
> $$U = 1.4D + 1.7L + 1.7H \qquad \text{(Code Eq. 9.4)}$$

Except that where D or L reduce the effect of H, $0.9D$ shall be substituted for $1.4D$ and zero value of L shall be used to determine the greatest required strength U. For any combination of D, L and H, required strength U shall not be less than $1.4D + 1.7L$.

This Code section, written with the design of building members in mind, is inadequate for the equilibrium problem of the wall. Two examples are outlined to define the problem.

The authors consider surcharge a live load, because it may be present or absent as a loading. When present, it both (1) creates a lateral pressure on the wall, similar to H_2 of Fig. 18.8, and (2) adds to the frictional resistance below the base. Item 2 definitely reduces the sliding potential of item (1) but the Code prescribes a zero load factor for this live load. It is illogical to take this load factor on surcharge as zero and include the H it produces; the wall is acted on by both or by neither. (The Code is more logical for a different case, where, say, earth pressure increases a wall moment but a floor load introduces a counteracting moment.)

Likewise, in the case of soil weight on the heel, the load tends to counteract the sliding tendency created by the horizontal pressure. Although the Code gives 0.9 for the load factor (not zero) this concept is invalid in evaluating the wall equilibrium. If D is only the weight of the structural heel slab, which might be constructed thinner than designed, the 0.9 factor is a sound value for use. In the case of soil weight on the heel, the $1.7H$ will never exist, that is, have a measurable chance of occurring if only $0.9D$ (as a weight of soil) exists. The $1.7H$ is intended to cover not only a possible increase in K_A but a possible increase in the soil weight included in the $1.4D$. One might analyze the 1.7 as the result of the sum* of 1.4 times the nominal soil weight added to an 0.3 factor for variation in K_A.

The authors recommend, therefore, that the last half of Code 9.2.4 be considered inapplicable except that 0.9 should be applied to any weight of concrete that tends to balance the effect of H. They further recommend *for the equilibrium equations* (not for the design of cross sections) the use of $1.7H$ for all lateral pressures with $0.9 \times 1.4 \approx 1.25$ used as the appropriate load factor for surcharge weight and for soil weight. This number best meets the spirit of Code 9.2.4. The double load factor 0.9×1.4 may seem unusual but in reality in this particular case the final value of 1.25 is a reasonable compromise that reflects many of the uncertainties in the soil pressure determination.

The less satisfactory alternate discussed in the opening paragraph deals with service loads, requires investigation of overturning when horizontal loading is increased (usually considered as doubled), and may still leave sliding problems. Many do use this approach.

(e) Choice of Heel Length   The heel length is chosen such that adequate friction of soil on soil is provided. A key is added below the base to prevent

---

* Probability theory does not directly add two factors, but for the present discussion the basic concept is correct even if 0.3 is not quantitatively correct.

failure from the lesser frictional resistance of concrete on soil. The method can be easily adjusted to count on either more or less help from the key. The coefficient of friction given is assumed to be a reasonable lower bound value, not a value greatly reduced to provide a large factor of safety.

In Fig. 18.10a the summation of horizontal forces gives the simple equation:

$$H_1 + H_2 - F = 0$$

where F is given by the coefficient of friction μ times the total weight $\Sigma(W + D)$. The given 0.62 coefficient for soil on soil is used.

$$
\begin{aligned}
W_x &= (0.9)(1.4)(400 + 100 \times 16.67)x &&= 2580x \\
D_s &= 0.9 \times 0.5(1.0 + 1.25) \times 16.67 \times 150 = && 2530 \\
D_{b1} &= 0.9 \times 1.33 \times 150x &&= 180x \\
D_{b2} &= 0.9 \times 1.33 \times 150 \times 4 \text{ (assumed)} &&= \underline{720} \\
\end{aligned}
$$

$$H_1 + H_2 - F = 8540 + 3800 - 0.62(2760x + 3250) = 0$$
$$2760x + 3250 = 12{,}340/0.62 = 19{,}900$$
$$x = 16{,}650/2760 = 6.03 \text{ ft}$$

Use heel projecting 6 ft beyond B.

Although the distance x protects against horizontal sliding of the wall on the soil, a soil mechanics check is also needed against sliding on a curved plane such as that indicated in Fig. 18.10b. A complete design requires that both kinds of sliding be investigated.

(f) Choice of Toe Length The *total* horizontal forces of Fig. 18.10a (somewhat larger than those on the 16.67-ft stem) tend to overturn the wall and the gravity loads tend to prevent such overturning. The reaction under the base must equal the total downward load, and the necessary location of the resultant reaction can be calculated by moments about any convenient center, say, B.

Load			Arm	Moment about B	
W_x	$(0.9 \times 1.4)(400 + 1667)6.0$	15,500	$+3.0$ ft	$+46,500$	
D_s	$0.9 \times 1 \times 150 \times 16.67$	2,250	-0.50		$-1,125$
	$+0.9 \times 0.5 \times 0.25 \times 150 \times 16.67$	280	-1.08		-300
D_{b1}	$0.9 \times 1.33 \times 150 \times 6.0$	1,080	$+3.0$	$+3,240$	
D_{b2}	$0.9 \times 1.33 \times 150 \times 4$ (est.)	720	-2.00		$-1,440$
	$\Sigma(W + D) = $	19,830			
	$H_2 = $	3,800	-9.0		$-34,200$
	$H_1 = $	8,540	-6.0		$-51,200$
		12,300		$+49,740$	$-88,300$
					$-38,560$
	R (from Σ gravity loads)	$-19,830$ (up)	y		$-19,830y$

$$M_B = -38{,}560 - 19{,}830y = 0$$
$$y = -1.95 \text{ ft (to left)}$$

Thus the resultant reaction acts 1.95 ft to the left of B.

The soil cannot support a concentrated reaction. It must be distributed in some fashion over some distance z, as in Fig. 18.10. Assume that the allowable ultimate soil resistance is developed at the toe before the soil fails. This allowable at ultimate is not readily apparent and must be discussed before calculating z.

In the case of a concentrically loaded footing one senses automatically that if uncertainties in soil mechanics are distributed in the same manner as in structural design that the allowable service load pressure should increase at the same rate as the factored loads increase, that is:

$$\left(\frac{\text{Allowable } q \text{ at ultimate loads}}{\text{Equivalent } Q \text{ at service loads}}\right) = \frac{1.4D + 1.7L}{D + L} \quad \text{or} \quad \frac{1.4D + 1.7L + 1.7W}{D + L + W}$$

The Code in design for footings bases dimensions on permissible soil pressures. Otherwise, this relation also might be used in footing design (Chapter 19) to obtain the equivalent soil pressure at ultimate. For the retaining wall stability, the relationship is more difficult because the effect of the H load enters into the solution as a moment and the use of the previous relation in terms of moment seems totally impractical. The permitted increase in soil pressure probably should be somewhere between the 0.9×1.4 factor used for L and soil weight W_x (in the moment equation) and the 1.7 factor used for H, because the concrete weight (0.9 factor) is relatively small. The best weighting is obscured by the fact that the moments as used are of opposite signs. Here a 1.5 factor will be used,* this being a pure judgment decision. One does not escape this problem by using service loads and allowable soil pressure in the equilibrium equation. One then faces the same decision when picking the appropriate soil pressure for the check against overturning under overloads.

USE allowable soil q at ultimate $= 1.5 \times 3500 = 5250$ psf

Although the real reaction distribution is uncertain, designers usually assume a straight-line distribution, which in this case means a triangle of pressure as shown in Fig. 18.10a.

$$0.5 \times 5250z = \Sigma \text{ gravity loads above} = 19,830$$
$$\text{Min. } z = 7.56 \text{ ft back from the toe}$$

The resultant vertical reaction R of 19,900 lb acts at $z/3$ from the toe, or 2.52 ft. Thus the toe extends beyond B by $y + z/3 = 1.95 + 2.52 = 4.47$ ft, or beyond the front of wall by $4.47 - 1.25 = 3.22$ ft. The effect of the weight D_{b2} is so small that the weight for the assumed 4-ft length need not be revised for these calculations.

USE Toe projection $= 3$-ft 3-in. (3.25 ft).
USE total base length $= 3.25 + 1.25 + 6.0 = 10.5 = 10$-ft 6-in.

---

* A better solution is to obtain from the soil mechanics specialist a *safe* ultimate soil pressure, something comparable to his (or her) suggested value of ϕ multiplied by the ultimate that could be expected 95% of the time. The correlation factors otherwise are simply mathematical equivalents without real physical significance, an equivalent in this particular case only roughly approximated.

(g) Service Reaction Pressure Conditions   With the permissible soil pressures as large as 3500 psf (at service loads) settlements are probably not a serious matter. Hence, for this particular wall this subsection can be skipped. However, where settlement is a serious problem, the soil reaction under service loads can govern the toe length. This computation omits the load factors on all loads. This first step is to locate the resultant reaction, that is, the term y as calculated in Sec. 18.6f but here based on service loads. The reaction pressure usually is assumed to vary linearly, although the actual pressure will probably follow some curved line. The resulting pressure distribution will necessarily vary with the location of the resultant R. If the resultant is at the third point of the base or nearer the toe, a nominal triangular reaction is assumed as shown in Fig. 18.11a or b. If the resultant falls within the middle third of the base as in Fig. 18.11c, a trapezoidal pressure results; if at the center as in Fig. 18.11d, a uniform pressure results.

The more serious the settlement problem the more attention must be given to the possibility of tilting; obviously, only the last case of uniform reaction pressure eliminates this problem. The position of the resultant is little changed by a change in toe length. Hence, lengthening the toe will leave the resultant nearer the middle of the base. The designer can thus increase toe length as much as service load conditions seem to require. A sample of such a reaction calculation follows.

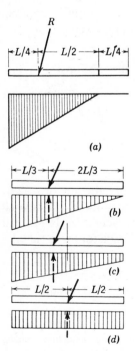

Figure 18.11 Reaction pressure distributions for various locations of resultant of load.

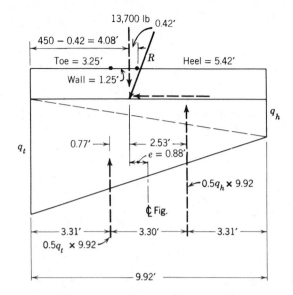

Figure 18.12 Calculation of soil reaction at service load without surcharge.

The calculation is not shown here, but the summation of vertical *service* loads omitting surcharge is 14,900 lb and $y = -0.02$ ft (from B). The stress condition of Fig. 18.12 results, with R 4.48 ft from toe. The toe and heel pressures are found from statics (Fig. 18.12). The upward pressure is considered two triangles of pressure with individual resultants at their third points. Moments about the smaller resultant (on the right in Fig. 18.12) give:

$$(0.5q_t \times 10.5)3.50 - 14,900 \times 2.52 = 0$$
$$q_t = 2045 \text{ psf}$$

Similarly,

$$(0.5q_h \times 10.5)3.50 - 14,900 \times 0.98 = 0$$
$$q_h = 800 \text{ psf}$$

The pressure also can be calculated considering the vertical load applied as an eccentric load on a rectangular section 10.5 ft × 1.0 ft. The eccentricity is 0.77 ft from the center of the base. The moment of inertia of this section is $1 \times 10.5^3/12 = 96.5$ ft.[4]

$$q_t = \frac{P}{A} + \frac{Pec}{I} = \frac{14,900}{10.5} + \frac{14,900 \times 0.77 \times 5.25}{96.5}$$
$$= 1420 + 625 = 2045 \text{ psf}$$
$$q_h = \frac{P}{A} - \frac{Pec}{I} = 1420 - 625 = 800 \text{ psf}$$

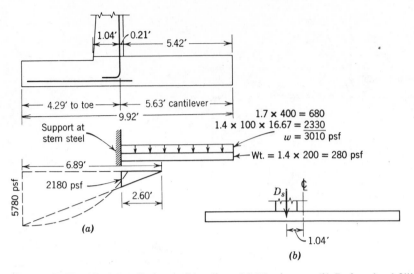

Figure 18.13 Heel cantilever and loading. (*a*) Maximum. (*b*) Before backfilling.

If a simple formula is preferred, the previous equation can be expressed

$$c/I = 6/bh^2 = 6/Ah$$

$$q = \frac{P}{A} \pm \frac{Pe/6}{Ah} = \frac{P}{A}(1 \pm 6e/h)$$

$$= \frac{14{,}900}{10.5}\left(1 \pm \frac{6 \times 0.77}{10.5}\right) = 1420(1 \pm 0.44)$$

$$= 2045 \text{ psf or } 800 \text{ psf}$$

(h) Design of Heel   The load on the heel is predominantly a downward load. The supporting upward reaction comes from the tension in the *stem* steel. Hence, the effective cantilever length is 6.0 ft plus the stem cover to center of steel = 6.0 + 2.56/12 = 6.21 ft, as shown in Fig. 18.13.

The downward load of earth and surcharge totals 3010 plf on the heel cantilever strip. The reaction was calculated in Sec. 18.6f as extending back 7.56 ft from the toe or 3.27 ft into this cantilever length with a triangle of reaction pressure as shown in Fig. 18.13*a*. In the following design this upward pressure is omitted because (1) there is no assurance that the straight-line pressure really exists* and (2) it reduces the calculated M_u and V_u on the heel, but by such a small percentage that its omission is not really wasteful. The initial estimate of heel dead load is based on the preliminary estimate of a 16-in. thick base.

---

* Note that the 3.27 ft of upward load might largely disappear if the reaction pressure is more parabolic as sketched in Fig. 18.13.

$$M_u = (3010 + 280) \times 6.21^2/2 = 63{,}400 \text{ ft-lb}$$
$$M_n \geqslant M_u/\phi = 63{,}400/0.9 = 70{,}500 \text{ ft-lb} = k_n b d^2$$
$$530 \times 12\, d^2 = 70{,}500 \times 12, \qquad d_m = \sqrt{133} = 11.5 \text{ in.}$$

Because shear might control under such heavy loads on a short span, the shear equation $V_n = 2\sqrt{f_c'}\,bd$ is solved for d_v, with the v subscript on d_v added here for identification and the m subscript added to the moment depth for the same reason. Although the critical design section for shear is usually taken at a distance d from the support, this is unsafe when the reaction is a tension "hanger" as it is here. Hence, shear is calculated essentially at the bar support (recognizing that 0.21 ft of the length is protected from vertical load by the stem)

$$V_u = (3010 + 280)6.0 = 19{,}700 \text{ lb}, \qquad \text{required } V_n \geqslant 19{,}700/0.85 = 23{,}200 \text{ lb}$$
$$\text{Permissible } V_n = 2\sqrt{f_c'}\,bd = 2\sqrt{3000}\,(12)\,d_v \text{ without stirrups}$$
$$d_v = 23{,}200/(109.8 \times 12) = 17.6 \text{ in.}$$

Because shear does govern, check the more exact permissible V_c using Code Eq. 11.6 with ρ_w assumed as 0.01 (11.5/17.6) = 0.0065 where the parenthesis is the ratio of d_m/d_v.

$$V_c = [1.9\sqrt{f_c'} + 2500(\rho_w V_u d/M_u)]b d_v$$
$$V_c = \left(1.9\sqrt{3000} + 2500\, \frac{0.0065 \times 19{,}700 \times 17.6}{63{,}400 \times 12}\right) b d_v$$
$$= (104.1 + 7.4)bd = (111.5)(12)\,d_v$$

(For usual cases, unless ρ is large the more exact value differs so little that it is probably more practical just to ignore it.)

$$d_v = 23{,}200/(111.5 \times 12) = 17.34 \text{ in.}$$
$$h = 17.34 + 2.00 \text{ in. cover} + d_b/2 = 19.84 \text{ in. for } d_b = 1 \text{ in.}$$
$$\text{or } 19.72 \text{ in. for } d_b = 0.75 \text{ in.}$$
$$\text{USE } h = 20 \text{ in.}, \qquad d = 17.5 \text{ in.}$$

This change to $h = 20$ in. from the earlier assumed base thickness of 16 in. increases the assumed heel weight by $50 \times 1.4 = 70$ psf, but as it also reduces the earth cover by 4 in. or $33 \times 1.4 = 46$ psf, it is obvious that no revision in M_n or V_n is really needed.

Because the d selected is considerably more than the required d_m, z is increased. Assume $z = 0.94d = 0.94 \times 17.5 = 16.45$ in.

$$\text{Required } A_s = \frac{M_n}{f_y z} = \frac{70{,}500 \times 12}{60{,}000 \times 16.45} = 0.857 \text{ in.}^2/\text{ft}$$
$$a = A_s f_y/0.85 f_c' b = 0.857 \times 60/(0.85 \times 3 \times 12) = 1.68 \text{ in.}$$
$$A_s = \frac{70{,}500 \times 12}{60{,}000(17.5 - 0.84)} = 0.85 \text{ in.}^2/\text{ft} = 0.070 \text{ in.}^2/\text{in.}$$

Because, when shear is large, it is possible for bar development length to limit the choice of bars, this requirement is investigated before choosing bars. Because these are top bars (more than 12 in. of concrete below the bars) the required development length is 1.4 times the basic value of $0.04A_b f_y \sqrt{f'_c}$, but an 0.8 factor is also probable, because spacing (hopefully) is 6 in. or more. The available length for development is 6.21 ft less 2 in. of end cover:

$$\text{Available } \ell_d = 6.21 \times 12 - 2 = 72.5 \text{ in.}$$
$$= (0.04A_b \times 60,000/\sqrt{3000})1.4 \times 0.8$$
$$\text{Max. } A_b = 1.48 \text{ in.}^2$$

This A_b permits bars up to #10 (1.27 in.^{2} ea.)

#9 at 14.3 in.	#7 at 8.5 in.
#8 at 11.3	#6 at 6.3

The 14-in. spacing for #9 is a little large. The #6 at 6 in. is close, but to reduce the number of bars to handle

USE #8 at 11 in. for heel (0.86 in.^{2}/ft)

With *high* walls the possible need for bottom steel in the heel before backfill is placed should be investigated. The factored weight of stem is

$$1.4[(12 + 15)/(2 \times 12)] \times 150 \times 16.67 = 3940 \text{ lb}$$

acting about 3.87 ft from toe and 6.63 ft from heel, that is, 1.38 ft off center as indicated in Fig. 18.13b. Using the equation at the close of Sec. 18.6g.

$$q = (P/A)(1 \pm 6e/h)$$
$$= (3940/10.5)(1 \pm 6 \times 1.38/10.5) = 375(1 \pm 0.788)$$
$$= 670 \text{ psf at toe and 80 at heel}$$
$$q \text{ at heel side of stem} = 80 + 590 \times 6.0/10.5 = 417 \text{ psf}$$

The moment at the heel side of the stem is easily computed by breaking the trapezoidal load diagram into two triangles.

$$+M_{\text{heel}} \text{ at face of wall} = 0.5 \times 417 \times 6^2/3 + 0.5 \times 80 \times 6^2 \times 2/3$$
$$= 2500 + 960 = 3460 \text{ ft-lb}$$

Although it is not as widely known as the ACI 318 Building Code that governs reinforced concrete construction, ACI also issues a building code for plain or unreinforced concrete. It is called ACI 318.1—"Building Code Requirements for Structural Plain Concrete."[3] This document is very useful for checking interim conditions when the tensile strength of the concrete may be relied on to resist low levels of flexural stress. Section 6.2.2 of that Code specifies the ϕ factor for plain concrete will be 0.65. The maximum extreme fiber stress owing to factored loads is limited to $5\phi\sqrt{f'_c}$ in tension under that Code. Analysis is based on linear stress-strain relationships. When considering flexure the full thickness usually can be assumed effective. ACI 318.1 Sec. 6.3.5 provides, however, that if concrete is cast against soil, the overall thickness must be taken as 2 in. less than the actual thickness.

Applying the Plain Concrete Code to this interim condition and using the nominal moment notations as found in the reinforced concrete Code, required $M_n = 3460/0.65 = 5320$ ft-lb $= 63,880$ in.-lb. (The weight of the base adds reaction pressure but no moment because it can be assumed supported on the soil.) Because the base is cast on soil, the actual thickness of 20 in. is reduced by 2 in. to give an effective thickness of 18 in. The section modulus, $S = bt^2/6 = 12 \times 18^2/6 = 648$ in.^{3}

$$\text{Actual } M_n = 5\sqrt{f_c'}\,S = 5\sqrt{3000} \times 648 = 177,500 \text{ in.-lb.}$$

Hence, the footing is **OK** for this temporary loading.

(i) Design of Toe Toe design is similar to that of heel design, except that diagonal tension is critical at a distance d from the face of the wall. Earth fill over the toe has been neglected thus far because it may not always be present. This fill adds also to the toe reaction, but this reaction and fill weight cause little change in M and V on the toe.

The maximum moment on the toe occurs where the shear changes sign. Because the large compression from the stem causes this shear reversal to occur just inside the face of the stem, the case is usually simplified by considering the cantilever support *at* the face of stem (Fig. 18.14).

Using the estimated base thickness of 16 in., the toe weight is assumed to be 200 psf, or 280 psf with the 1.4 load factor. The load factor and the footing weight is already included in the reaction pressure. Shear stress is almost sure to control the toe depth. Check d_v at distance d from support, say, $d = 16 - 3 - 0.5d_b = 12.5$ in. $= 1.04$ ft, leaving 2.21 ft to toe.

$$V_{1.04} = 5250 \times 2.21 - 0.5(694 \times 2.21)2.21 - 280 \times 2.21 = 9290 \text{ lb}$$
$$\text{Required } V_n \geq 9290/0.85 = 10,900 \text{ lb}$$
$$d_v = V_n/2\sqrt{f_c'}\,b = 10,900/(109.6 \times 12) = 8.29 \text{ in.}$$

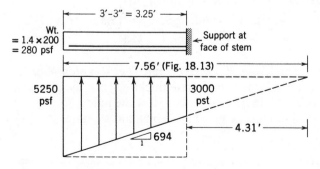

Figure 18.14 Toe cantilever and loading.

Because this calculation is quite sensitive to variation in the assumed d, try again with assumed $d = 9.6$ in. $= 0.8$ ft.

$$V_{0.8} = 5250 \times 2.45 - 0.5(694 \times 2.45)2.45 - 280 \times 2.45 = 10,100 \text{ lb.}$$
$$V_n \geqslant 10,100/0.85 = 11,900 \text{ lb}$$
$$d_v = 11,900/(109.8 \times 12) = 9.02 \text{ in. (Close enough)}$$
$$h = 9.02 + 3.0 \text{ cover (Art. } 7.7.1a\text{)} + d_b/2 = 12.52 \text{ in. for } d_b = 1 \text{ in.}$$
$$\qquad 12.40 \text{ in. for } d_b = 0.75 \text{ in.}$$

Check next for moment. The same 280 psf weight term is used here because it is an original weight term that is now being taken out of a reaction computed on that basis. Strictly, both weight and reaction change in almost equal amounts, but the calculations based (in effect) on net pressures ignore the minor change in gross pressures.

$$M_u = 5250 \times 3.25^2/2 - 0.5(694 \times 3.25)3.25^2/3 - 280 \times 3.25^2/2 = 22,300 \; ft\text{-}lb$$
$$M_n \geqslant 22,300/0.9 = 24,800 \text{ ft-lb} \times 12 = kbd^2$$
$$530 \times 12d^2 = 24,800 \times 12, \qquad d_m = \sqrt{46.8} = 6.84 \text{ in.} < d_v$$

Most designers, including the authors, normally prefer to use the same toe and heel thickness, but this is not essential. A little thicker toe simplifies the shear key between wall and base; a little thinner toe complicates it slightly, as discussed in the next subsection. In this design, to introduce this complication, although not maintaining the extreme thickness difference that is theoretically possible,

$$\text{USE } h = 16.0 \text{ in.,} \qquad d = 12.62 \text{ in. (based on \#6 bars)}$$

With d_m increased more than 80%, try $z = 0.96 \times 12.62 = 12.1$ in.

$$\text{Approx. } A_s = M_n/(f_y z) = 24,800 \times 12/(60,000 \times 12.1)$$
$$= 0.41 \text{ in.}^2/\text{ft} = 0.034 \text{ in.}^2/\text{in.}$$
$$\text{Min. } A_s = (200/f_y)bd = (200/60,000)12 \times 12.62 = 0.504$$
$$\text{in.}^2/\text{ft} = 0.042 \text{ in.}^2/\text{in.} > 0.034$$

Try #6, spacing $= 0.44/0.042 = 10.5$ in.

$$\text{USE \#6 at } 10\tfrac{1}{2} \text{ in. for toe (0.503 in.}^2/\text{ft)}$$
$$\ell_d = (0.04 \times 0.44 \times 60,000/\sqrt{3000})0.8 \text{ for wide spacing} = 15.4 \text{ in., say, 16 in.}$$
$$\leqslant \text{ toe length less 2-in. end cover.} \qquad \textbf{O.K.}$$

(j) Assembly Problems* at Junction of Stem, Heel, and Toe; Revision of Stem
The stem height is based on an assumed 16-in. base thickness, but the previous design uses 20 in. for the heel and 16 in. for the toe. How the junction between the three members is handled influences the stem height and stem design. For

---

* See Reference 5, especially pp. 1244–1246, and Chapter 10.

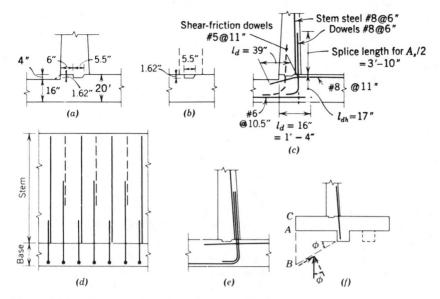

Figure 18.15 Details at junction of stem and base.

this reason the *final* stem thickness and stem A_s normally would be left until after the heel and toe designs are completed.

The junction proposed by the authors is shown in Fig. 18.15. The reasoning is clarified as the various checks are made.

First, provision is usually made for a shear key between the stem and base. The key is usually made by embedding a beveled 2-in. × 4-in. or 2-in. × 6-in. timber in the top of the footing as shown in Fig. 18.15b.* When the timber is removed and the stem is cast-in-place, the 1.62-in. vertical face provides an allowable bearing (Code 10.15.1) of $(12 \times 1.62)\, 0.70(0.85 \times 3000) = 34{,}700$ lb/ft of wall compared to the approximately 10,840 lb horizontal force on the stem (Fig. 18.8). The distribution of shear force on the horizontal section of the key at the top of footing is uncertain. The shear key can be checked by the shear friction provisions of Code 11.7.4. The maximum shear strength if only the key is effective on the section is $V_n = 0.2f'_c A_g = (0.2)(3000)(5.5 \times 12) = 39{,}600$ lb. V_u is $(0.85)(39{,}600) = 33{,}700$ lb or far in excess of the actual shear. Using shear friction procedures requires that an area of vertical reinforcement corresponding to $V_n = A_{vf} f_y \mu$ (Code Eq. 11.26) be provided across the shear plane. With $\mu = 0.6$ for concrete placed against hardened concrete that was not intentionally roughened, $A_{vf} = V_n / f_y \mu = 10{,}840/(0.85)(60{,}000)(0.6) = 0.35$ si/ft of keyway. This result corresponds to a #5 each 11 in. Many designers prefer to avoid the complication of vertical dowel bars in the keyway (dashed line bars in

---

* Some designers prefer that the key be cast as a projection on top of the base that extends up into the stem. Some now question the need for a shear key.

Fig. 18.15c). They reason that the shear is transferred in the stem compression zone where the bending moment on the stem produces a compression force of $N_c = N_t = 1.42 \times 60,000 = 85,200$ lb on the base. This compression force should produce more than adequate friction to transfer the shear and indicates that the shear key and the keyway dowels are probably superfluous. The shear strength in front of the key also must be maintained, especially when the toe is of smaller thickness. Under this compression and with a little more area it presents no problem.

If heel and toe are of equal thickness, the cantilever stem height clearly is measured from the top of the base. In the present case it is less specific. It is quite safe to calculate stem thickness at the level of the top of the toe and steel area at the level of the heel steel; or one might calculate both at the top of the heel and put in a fillet between stem and toe to deepen the stem, as shown by the dotted line in Fig. 18.15a. The first procedure appears the simpler from the construction viewpoint and is used.

For compression, the height of stem is 18 ft minus 16 in. (or 1.33 ft) for a net of 16.67 ft but the loading is only on the height above the heel that is 18 ft minus 20 in. (or 1.67 ft) for a net of 16.33 ft. The stem thickness is based on 16.67 ft for both, which is close enough.

For tension, the height of stem is 18 ft minus the heel d of 17.5 in., which is 16.54 ft, slightly less than used originally. No change in the stem cross section is necessary.

Because the stem steel cannot be readily supported in position while the base concrete is placed, it is customary to provide dowels (or stub bars) in the base equal to the stem steel, #8 at 6 in. These dowels must project a full anchorage length into the base and a full splice length above the base, as shown in Fig. 18.15c. The anchorage length into the base, for a spacing $\geqslant 6$ in., is

$$\ell_d = 0.8[0.04 \times 0.79 \times 60,000/\sqrt{3000}] = 0.8 \times 34.6 = 27.7 \text{ in.}$$

This length cannot be obtained vertically below the stem even if bars are extended into a reasonable depth key below the base.

The logical alternative is to use a standard hook on the dowel and develop the tensile force as provided for hooked bars in Code 12.5.1. The basic development length for a hooked #8 bar is $\ell_{hb} = 1200 d_b/\sqrt{f_c'} = (1200)(1.00)/\sqrt{3000} = 21.9$ in. Code 12.5.3.2 allows a 0.7 factor to be applied if there is side cover normal to the plane of the hook of $2\text{-}\frac{1}{2}$ in. or more and cover on the 90° hook extension beyond the hook of not less than 2 in. Both are present here. In addition, Code 12.5.3.4 allows a reduction of (A_s required/A_s provided). In this case, that term is 1.42/1.58 = 0.90. Thus, the hook bar development length $\ell_{dh} = (0.7)(0.90)(21.9) = 13.8$ in. This length is measured from the critical section to the outside end of the hook. In the 20-in. deep section with a required minimum cover of 3 in. when cast against ground, there is $20 - 3 = 17$ in. available for the hooked dowels. This is more than adequate for the 13.8 in. required. Use the full 17 in. USE #8 dowels with standard 90° hooks at 6 in. with $\ell_{dh} = 17$ in.

In accordance with Code 12.15.1, the splice lap in the stem must be $1.3\,\ell_d$ if less than half of the bars are spliced at one level, as in Fig. 18.15d, but $1.7\ell_d$ if all bars are spliced at one point.

$$\text{Minimum lap} = 1.3 \times 0.8 \times 0.04 A_b f_y / \sqrt{f_c'} = 1.3 \times 0.8 \times 0.04 \times 0.79$$
$$\times 60{,}000 / \sqrt{3000} = 36 \text{ in.}$$

Use staggered splices with 36-in. lap.

The possibility of eliminating the splice on half the bars by extending half the dowels as high in the stem as they are needed for moment will be checked later.

The #8 bars in the heel are top bars (more than 12 in. of concrete below them) and are spaced more than 6-in. apart.

$$\ell_d = 1.4 \times 0.8(0.04 \times 0.79 \times 60{,}000 / \sqrt{3000}) = 38.7 \text{ in.}$$

Anchor the #8 bars 39 in., as shown in Fig. 18.15c. The compression from the stem across part of the length merits some recognition, but what numerically this should be is unknown. The bars in the toe must be anchored 16 in.

It should be obvious, as a matter of detailing, that some of the stem dowel steel can be replaced by toe steel bent up as shown in Fig. 18.15e. This action would probably save some steel and certainly would tie the toe well to the wall. Figure 18.18 shows such a design. The detailing then is facilitated by using the same spacings for toe and stem steel.

(k) Key Against Sliding Resistance against sliding as provided in Sec. 18.6e assumes a key into the soil to take the difference between the coefficient of friction of 0.62 and 0.50, or $0.12 \Sigma(W + D) = 0.12 \times 19{,}830 = 2380$ lb. This requires only a limited amount of depth into the soil, and 10 in. or 12 in. appears adequate, as sketched in Fig. 18.15f. The key is usually placed sufficiently below the stem to use it for anchorage of the stem dowels if needed, but it may be more effective when somewhat more to the rear. The maximum effect of such a shift cannot be more than to mobilize the passive resistance of the soil over the depth AB in Fig. 18.15f, where ϕ is the friction angle for soil on soil. The key should be designed as a bracket and the details are not shown here.

(l) Temperature and Shrinkage Steel Longitudinal bars are used to space the moment bars and to provide for shrinkage and temperature stresses; both horizontal and vertical bars are needed on exposed faces. The wall should be placed in short lengths, not to exceed 20 to 30 feet, to reduce shrinkage stresses.

Code 14.1.2 requires that retaining walls have minimum horizontal reinforcement that complies with Code 14.3.3:

Minimum ratio of horizontal reinforcement area to gross concrete area shall be:
   (a) 0.0020 for deformed bars not larger than #5 with a specified yield strength not less than 60,000 psi, or
   (b) 0.0025 for other deformed bars, or
   (c) 0.0020 for welded wire fabric (smooth or deformed) not larger than W31 or D31.

Section 14.3.5 requires that horizontal reinforcement shall not be spaced farther apart than 3 times the wall thickness nor 18 in.

For Grade 60 bars #5 or smaller, this leads to areas on *each* face:

Horizontal, $0.0010bh$ (each face)

$b = 12$ in., $h = 12$ in. 0.144 in.^{2}/ft

15 in. 0.180

18 in. 0.216

These minimums probably should be increased on the exposed face where minimum vertical bars are a very good idea. A strict interpretation of the Code indicates they are not required for retaining walls. In this example, no bars are needed on the compression faces of heel and toe, and the main function of the longitudinal bars is as spacers. The following are chosen:

Stem: exposed face, horizontal, upper 8 ft #5 at 18 in. = 0.21 in.^{2}/ft

below 8 ft #5 at 12 in. = 0.31 in.^{2}/ft

exposed face, vertical, #5 at 18 in. = 0.21 in.^{2}/ft

rear face, horizontal, #5 at 18 in. = 0.21 in.^{2}/ft

Toe: longitudinal spacers in bottom, #4 at 18 in. = 0.13 in.^{2}/ft

Heel: longitudinal spacers in top, #5 at 18 in. = 0.21 in.^{2}/ft

Temperature and shrinkage steel is shown in Fig. 18.16, with the exposed face steel in the foreground, as seen most clearly at the bottom of the illustration, adjacent to the form panel that is being lifted into place.

Figure 18.16 Reinforcing in wall stem with the front form being lifted into place. Frequently the rear or front face form is first erected and the reinforcement is next placed. (Courtesy Texas Highway Department.)

(m) Drainage Because walls are usually not designed against water pressure, the design must provide complete drainage of backfill by French drains, drains through the wall, or the like. A porous backfill, such as gravel, should be provided directly behind every wall to allow the water to reach these drains.

18.7 Arrangement of Stem Steel—Stopping Bars

Not all of the stem steel must run full height, because the moment decreases rapidly above the base. The simple method of Sec. 4.11c must be modified to take care of the variable depth. For the design of Sec. 18.6c the upper sketches in Fig. 18.17 show the effective depth d and the soil pressures at various depths. From these data the moment and steel requirements are determined and are tabulated in Table 18.1. For M_n the ϕ value is taken as 0.886 as in Sec. 18.6c; it should increase to 0.9 at the top and it is adequate in this case to use

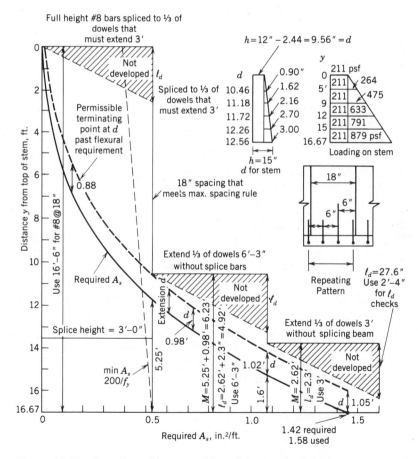

Figure 18.17 Stopping of stem steel based on required A_s curve.

TABLE 18.1   Calculations for Required A_s Curve, $f_y = 60$ ksi

Distance y from Top, ft	M_u, Bending Moment, ft-lb	M_n	Slab Depth d, in.	Required A_s, in.²/ft
0	0	0	9.56	0
5	3,740	4,220	10.46	0.09
9	14,950	16,870	11.18	0.34
12	30,400	34,300	11.72	0.65
15	53,400	60,300	12.26	1.10
16.67	70,000	79,000	12.56	1.42

0.9 all the way. These data are the basis for the plotted heavy curve in Fig. 18.17, that will be used for a graphical solution. A heavy dotted line in Fig. 18.17 is plotted the (variable) distance d higher to show how far the bars really must extend.

To avoid some of the splicing of steel just above the base, $\frac{2}{3}$ of the dowel bars are made in two heights that are adequate to take care of their portion of the moment: $\frac{1}{3}$ is 6-ft 0-in. high and the other $\frac{1}{3}$ 3-ft 0-in. (for simplicity the same as the normal splice lap length of Sec. 18.6j). The stopping of tension bars at their *minimum* length leaves the remaining steel at maximum stress and requires that none of these remaining bars be stopped closer than ℓ_d ($= 2.3$ ft in Sec. 18.6j) to the next lower cutoff point. This ℓ_d requirement is indicated at three places, but happens not to be governing. This insert at the right of Fig. 18.17 shows the schematic arrangement.

Up to 10.5-ft depth minimum Code limits for flexural A_s ($\rho = 200/f_y$) actually govern the required reinforcement. A nearly vertical dashed line represents this minimum, but this rule is relaxed when a considerable excess of A_s is used, beyond that required for moment (Code 10.5).

In cutting off tension reinforcement, a further condition must be checked. Code Sec. 12.10.5 forbids a bar cutoff in a tension zone unless one of three enumerated conditions is met. Usually in slabs (other than two-way slabs) the shear is less than $\frac{2}{3}$ the allowable, and this will satisfy the first of these conditions. The stem shear at the top of the base is only 78% of the allowable. A rough check indicates that the shear at the minimum cutoff length of 3-ft 0-in. is less than the permissible $\frac{2}{3}$ of the allowable and thus permits the cutoff without any special provisions.

A design layout often leading to economy in reinforcing is based on turning up the toe steel as dowels for matching stem bars as indicated in Fig. 18.18. The toe bars must extend a splice length into the stem that, in the case of the design of Sec. 18.6j, for #6 toe steel making up less than $\frac{1}{2}$ of the stem steel requirements, is:

$$\text{Lap} = 1.3\ell_d = 1.3(0.04 \times 0.44 \times 60{,}000/\sqrt{3000}) = 25.0 \text{ in.}$$

The deficit in the stem steel, after matching the #6 bars, is provided by #8 at 10 in., as indicated in Fig. 18.18. Prior checks on hook length show these bars

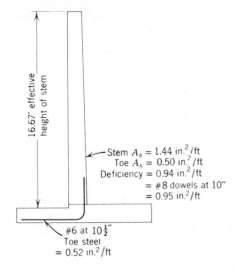

Stem $A_s = 1.44$ in.^{2}/ft
Toe $A_s = 0.50$ in.^{2}/ft
Deficiency $= 0.94$ in.^{2}/ft
$= \#8$ dowels at 10″
$= 0.95$ in.^{2}/ft

$\#6$ at $10\frac{1}{2}$″
Toe steel
$= 0.52$ in.^{2}/ft

16.67′ effective
height of stem

Figure 18.18 Basis of revised steel layout to utilize toe steel as dowels for stem steel.

can be easily developed. The steel layout would have to be revised from that shown in Fig. 18.17 to utilize these dowels efficiently.

Selected References

1. W. C. Huntington, *Earth Pressures and Retaining Walls,* John Wiley and Sons, New York, 1957.

2. K. Terzaghi and R. B. Peck, *Soil Mechanics in Engineering Practice,* John Wiley and Sons, New York, 2nd ed., 1968.

3. ACI Standard 318.1, "Building Code Requirements for Structural Plain Concrete, *Concrete Intl.,* Vol. 4, No. 12, Dec. 1982, pp. 128–140.

4. R. B. Peck, W. H. Hanson, and T. H. Thornburn, *"Foundation Engineering,"* John Wiley & Sons, New York, 2nd ed., 1974.

5. I. H. E. Nilsson and A. Losberg, "Reinforced Concrete Corners and Joints Subjected to Bending Moment," *Proc. ASCE, Jour. Struct. Div.,* V. 102, ST6, June 1976, p. 1229.

Problems

PROB. 18.1:
(a) Investigate under service load conditions the stability, sliding resistance, and foundation soil pressure of the wall of Fig. 18.19. Surcharge = 200 psf, soil weight = 100 pcf, $K_a = 0.30$, concrete weight (including reinforcing) = 150 pcf, coefficient of friction of concrete on soil = 0.40, allowable soil pressure = 3000 psf.

(b) Same as (a) except use load factors of Sec. 18.6 and a permissible ultimate soil pressure of 6000 psf.

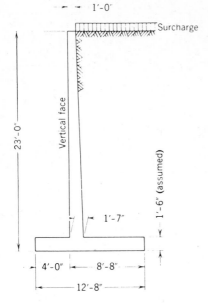

Figure 18.19 Retaining wall for Prob. 18.1 and Prob. 18.2

PROB. 18.2. Design (*a*) stem, (*b*) heel, and (*c*) toe for the wall of Fig. 18.19 assuming all the foundation soil pressures and sliding resistances found in Prob. 18.1 are used as satisfactory. Use factored loading, $f'_c = 4000$ psi, Grade 60 bars. (Use toe no thinner than heel.)

PROB. 18.3. Redesign the retaining wall of Sec. 18.6 for horizontal pressure based on $K_a = 0.27$, $f'_c = 4000$ psi, allowable foundation soil pressure under factored loading = 5000 psf, and a *minimum* base length.

(*a*) Stem design.

(*b*) Minimum base length.

(*c*) Heel design.

(*d*) Toe design with thickness based on stresses, but not less than thickness of heel.

(*e*) Sketch the detail of the joint where the stem, heel, and toe come together.

PROB. 18.4. It is possible in Fig. 18.17 to stop all the #8 bars at the same height and splice them with #6 bars that continue upward. Where can such a splice be made and what is the length of all bars, assuming half of the #6 bars are stopped as soon as the stresses permit?

PROB. 18.5. Assume that the #8 bars at 18-in. average spacing shown in the stem in Fig. 18.17 are made into two equal groups and cut off at two different levels. Sketch the reinforcing pattern and establish the length of each group of bars. (For this problem overlook minimum ρ and the wide top spacing that results.)

19

FOOTINGS

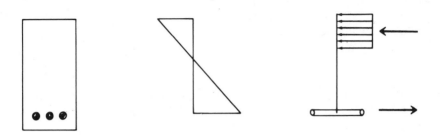

19.1   Types of Footings

Although underground conditions call for many variations in foundation design, the majority of building footings can be classified as one of the following types:

1. A *wall footing,* which is a continuous strip of concrete that supports a bearing wall, as shown in Figure 19.9.
2. An *isolated column footing,* which is a square, rectangular, or circular slab of concrete that supports an individual column, as shown in Fig. 19.1.
3. A *combined footing,* which is a longer rectangular slab strip that supports two or more individual columns, as shown in Fig. 19.13 and Fig. 19.14.
4. A *cantilever footing,* which is really two footings joined by a beam instead of by a bearing portion of the footing. Each of the two main parts of this footing supports a column load as shown in Fig. 19.16.
5. A *drilled pier,* which is a shaft that is drilled into the ground and filled with concrete. The shaft is often flared (or belled) at the bottom for greater bearing area. The drilled pier usually supports an individual column.
6. A *floating or raft foundation,* which is a single thick mat or slab that supports the entire structure. This type of foundation is frequently used with poor soil conditions to equalize deformations. It is also called a *mat foundation.*

Any of these footings may be directly supported on the soil or may rest on pilings.

19.2   Tests on Footings

Two noteworthy series of footing studies influenced American practice: Talbot's tests[1] in 1907 and Richart's tests[2] in 1946. Moe's tests[3] at the Portland Cement Association Development Laboratory added more detail on the effect of openings near the column.

Shear failures never occur on vertical planes along the wall or around the column. For wall footings that fail in shear, a diagonal tension crack develops on an approximate 45° plane parallel to the wall, as shown at A in Fig. 19.1a. The shear causing this crack is produced by the upward load to the left of A, that is, the load beyond a plane a distance d from the face of the wall. A similar failure occurs in an isolated square column footing. The column pushes ahead of itself a truncated pyramid with an approximate 45° slope on all faces. This pyramid has a base width equal to the column width plus approximately twice the effective depth of the footing (Fig. 19.1b). The probable reason for this type of failure is the heavy compressive stresses between the diagonal cracks as the column load spreads out into the surrounding footing, with this compression accentuated by the upward soil reaction. In this zone the diagonal tension normally caused by shear stresses is somewhat counteracted or reduced by the vertical compressive stress. (Compare Fig. 5.5a and the related discussion in Sec. 5.4c). The 45° failure planes mark the approximate boundaries between which substantial vertical compression is developed.

In a column footing the initial diagonal cracks described previously occur much before ultimate load and are generally invisible because they are in the

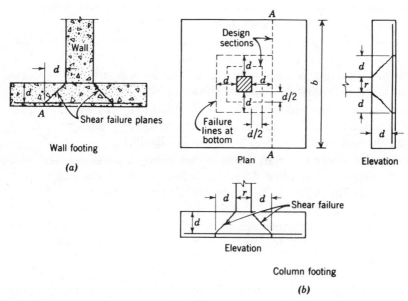

Figure 19.1   Diagonal tension failure in footings.

interior of the concrete. The 1962 report[4] of the Joint ACI-ASCE Committee on Shear and Diagonal Tension recognized that, *after* these shear failure planes first develop, the critical shear is largely carried by the flexural compression area above the cracks in what can best be described as the old idea on punching shear of this limited compression depth around the perimeter of the column. Failure is then described as a two-step process: (1) initial diagonal cracking extending a distance d from the wall or column and (2) a shearing failure at the face of wall or column. Usually, a considerable portion of the total load resistance develops after the diagonal crack is formed. Rather than introduce two checks for the two kinds of behavior, the Committee recommended that the shear be calculated on a pseudocritical plane between the two, that is, at a distance $d/2$ from the column. Then as a safety check, which probably controls only on long, narrow footings, the shear strength as a one-way slab should be checked on sections all across the footing at a distance d from the face of column.

With both walls and columns of concrete, the critical section for moment and bar development is found at the face of the wall or column. In the wall footing the one-way cantilever moment owing to forces beyond the critical section is assumed uniformly distributed along the wall; any 1-ft wide strip can be used in this part of the design. In the column footing Richart found the distribution of this cantilever moment (in each direction) very nonuniform at working loads. The moment is largest for strips passing under the column and smallest for strips near an edge. However, near ultimate loads the yielding of the steel in the central strips causes more moment to shift to the edge strips and moment failure does not occur until essentially all the steel reaches its yield point. Bond stress is less critical than expected. The 1986 Code requires only that the bar beyond the face of column be fully developed (Code 15.6).

19.3 Square Column Footings—Analysis

(a) **Critical Stresses** Column footing must be checked or designed for six strength conditions:

1. Bearing (compression) from column on top of footing.
2. Dowels into the footing.
3. Strength of soil beneath footing, soil pressure q_s.
4. Shear strength.
5. Reinforcement provided.
6. Development length of bars.

(b) **Bearing under Column** The bearing from the column on the footing is permitted to be larger than the value of the same concrete in the column, that is, lower strength concrete is permissible in the footing without lowering the column capacity or necessarily increasing the dowels provided. The usual permitted bearing strength (Code 10.15.1) of ($\phi \times 0.85f'_cA_1$) can be multiplied by $\sqrt{A_2/A_1} \lesssim 2$, where A_1 is the bearing area and A_2 is the lower area sketched in Fig. 19.2. Area A_2 is the area of the lower base of the largest frustrum of a pyramid, cone, or tapered wedge contained wholly within the support and having for its upper base the loaded area. It is established by going down and out from the loaded area at the slope of 1 vertically and 2 horizontally* until one intersects the boundary closest to the column. This boundary fixes one side of A_2 and it is simple to construct the complete area on the plan view similar to A_1. From similar areas

$$A_2/A_1 = (x_2/x_1)^2 \qquad \sqrt{A_2/A_1} = x_2/x_1 \lesssim 2$$

The total bearing value of the area A_1 of a steel bearing plate or a concrete column becomes:

$$N_{nc} = 0.85f'_cA_1\sqrt{A_2/A_1} = 0.85f'_cA_1(x_2/x_1)$$

where f'_c is the footing concrete strength and A_1 the bearing area. For the concrete column, this part of the load is also limited to the capacity of the concrete in the column, that is, $N_{nc} = 0.85f'_cA_1$ where f'_c is the column concrete strength.

(c) Dowels into the Footing   Because the Code permits dowels into the footing designed for strength instead of requiring them to match the column steel, some savings in compression dowels are occasionally possible. However, any column bar that is subject to tension must either be anchored into the footing or matched with dowels into the footing.

A small savings can occur if a column requires almost no steel to carry its load, but is reinforced with $\rho = 0.01$ to satisfy the Code minimum. The dowel minimum is lowered to $\rho = 0.005$ of the column area. In the 1986 Code, a minimum tensile capacity for the column-footing or wall-footing connection is specified for both precast and cast-in-place construction. This requirement is part of the ACI Building Code emphasis on providing general structural integrity in all concrete buildings. Such integrity is developed by introducing ductility and continuity into members and connections. It is intended to reduce the risk of catastrophic failures caused by local overloads or accidents during construction or in the life of the structure. Code 15.8.2.1 requires that at least 0.5% reinforcement (based on the column size) be used across the cast-in-place, column-footing interface. This minimum reinforcement requirement is expressed as a minimum tensile force across the joint that is equal to $0.005A_gf_y$. Assuming that f_y is 40,000 psi, this is equivalent to a tensile force of $200A_g$ in pounds. This equivalent tensile force is required by Code 15.8.3 for connections between precast columns or pedestals and the supporting members. This tensile capacity is usually provided by anchor bolts or other acceptable me-

---

* This slope is not intended to represent the direction of the pressure, but to insure some confining concrete around the actual pressure distribution.

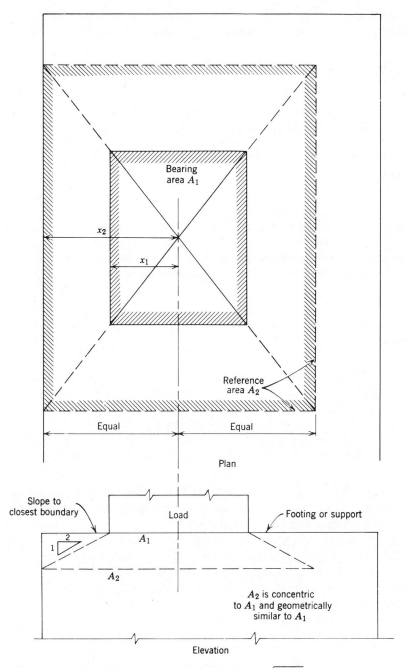

Figure 19.2 Bearing pressure limit is based on $\sqrt{A_2/A_1} = x_2/x_1 \leq 2$.

chanical connectors. Design of such tensile connections must provide for a ductile failure of the connection rather than a brittle anchorage failure or pullout.

Code 15.8.2.2 requires that at least the minimum percentages of reinforcement specified for walls in Code 14.3.2 be used in cast-in-place, wall-footing connections. For precast wall connections this minimum tensile capacity is expressed as 50 psi times the cross-sectional area of the wall.

If the column concrete strength exceeds the footing concrete strength, the increased bearing strength could still mean that no *extra* compression dowels are required to transfer the difference in concrete strengths into the footing.

In designing compression dowels, one must consider both the footing below and the column above:

1. For the footing, dowels must carry the total load less the bearing value of the footing concrete. More exactly stated, the dowels should satisfy the reinforcement requirements determined from a column interaction chart for the design eccentricity and A_1 when f'_c is used as the footing value times the ratio $\sqrt{A_2/A_1} \lessgtr 2$.
2. For the column, the load, including moment from eccentricity at the base, must be carried by the concrete in the column assisted by *only* bars that are extended or doweled into the footing.

The second rule is necessary because any bars neither extended nor doweled into the footing have zero load capacity at the point where they are stopped and very little capacity nearby; strength always is limited to the lower available development length on the two sides of any design section.

The second condition says nothing done in the footing itself lowers the demand for effective column reinforcement, that is, *developed* column steel.

(d) Soil Pressure   The discussion of actual soil pressure under a footing is beyond the scope of this book. (Reference 5 is recommended.) This pressure commonly is higher near the center of a column footing, but occasionally is higher at the edges, depending on whether the soil is sandy or clayey.

The concentration of pressure directly under the load when a footing is on rock is reduced only by a very stiff footing as discussed in Sec. 19.10. Although not necessary in simple cases, elastic analysis is appropriate for foundations on rock and is helpful in special cases of loading or footing shape. The same approach is usable in other stiff and well-consolidated foundation materials if there is enough uniformity to permit a meaningful determination of the foundation modulus.

Elastic analyses, however, are not a general answer to the pressure distribution problem. If measurable foundation settlement is expected, the soil is only momentarily elastic and the initial response to the load is only a transient response. Time (and settlement) will smooth out this initial response, lower peak pressures, and build up low pressures. In such cases, the authors, consider elastic analyses of small value. More uniform (oversimplified) pressure distributions appear more appropriate for a design where significant settlement is expected.

The soil pressure is usually considered uniform for a centered loading and trapezoidal or triangular for eccentric loadings. The assumed uniform pressure is usually on the safe side in the calculation of internal moments and shears in isolated footings, but a different situation in the middle of a combined footing is discussed in the last paragraph of Sec. 19.12.

The allowable soil pressure is also a matter of soil mechanics and cannot be discussed here. This matter brings up again the problem mentioned in connection with retaining wall design (Sec. 18.6). Allowable soil pressure is usually given in terms of the service load permissible and footing design is by strength (at ultimate). The designer has at least two (equivalent) ways to keep the relationships proper in the case of isolated column footings.

The Code (15.2.2) specifies, for establishing the footing size, that the (unfactored) external forces and moments be used to match the allowable soil pressure; (15.2.1) also that soil pressures based on design (factored) loads be used to calculate design moment and shear.

Because the ratio of dead to live load varies from column to column, there is no unique relationship between soil pressure at service loading and soil pressure under ultimate loading. In the authors opinion design is better served by the general use of a dependable *ultimate* soil pressure. In settlement situations, however, a permissible pressure at *usual* loads could be significant and possibly governing, just as deflection can govern over strength in a flexural member.

The footing base area must be adequate to care for the column load, footing weight, and any overburden weight, all within the permissible soil pressure, often assumed uniformly distributed* under the footing. Sometimes this overall pressure is called the gross soil pressure to distinguish it from net soil pressure, which is a convenient *design* concept. Because the weight of footing and overburden is usually nearly uniform over the footing area, the design moment or shear on any section is the result of the gross soil pressure upward less the design value of the footing and other overburden weight downward. It is convenient to think in terms of the resultant load or "net soil pressure" that is the difference between these upward and downward unit pressures. The simplest way to find the net pressure caused by vertical column load without moment is to divide the design column load by the footing area.

(e) Shear Strength   Shear quite frequently controls the footing thickness. The requirements of Code 11.11.1 call for two checks.

The first check is for two-way punching action as in a slab-column connection. The unit shear as a measure of the diagonal tension is first calculated for the shear caused by loads outside the inner dashed square in Fig. 19.1b and is resisted by a width equal to the perimeter of that square, $b_o = 4(r + d)$. The use

---

* But variation caused by any moment loading also must be considered.

of d_v as the average depth to the two layers of steel appears justified. There is two-way bending with permissible unit shear of $4\sqrt{f'_c}$, or $V_{nc} = 4\sqrt{f'_c}b_o d$.*

In the second check for beam action, the diagonal tension on section AA at a distance d also must be checked, using permissible one-way slab shear of $2\sqrt{f'_c}$, or $V_{nc} = 2\sqrt{f'_c}bd$. This case usually governs only on long, narrow footings.

(f) Reinforcement Provided   A proper reinforcement design always gives an under-reinforced member, automatically avoiding compressive failure.

The moment is critical on the section at each face of column and calls for two-way bottom bars. The moment in each direction is a matter of statics and can be varied only by a change in the distribution of soil pressure. The separate steel in each direction should be adequate for the moment in that direction; excess steel in the y-direction cannot make up for a steel shortage in the x-direction. Hence, it appears desirable that the smaller depth to the upper steel layer in the bottom of the footing be used for calculation of the moment resistance in a square footing. In a square footing the use of equal steel areas in the two directions helps to avoid field errors in placement. Economy is achieved by using a z value corresponding to the actual percentage of steel used; this value is usually larger than $0.9d$.

The minimum reinforcement requirement of $200/f_y$, unless A_s is at least 4/3 that required by analysis, should be applied to footings. Although one could argue legally that a footing is a slab and thereby exempt from this requirement, the combination of high shear and low ρ is not good.

(g) Development Length of Bars   In a single column footing, the available development length is from the face of column to the edge of footing less the end cover. No bar requiring a longer development length can be used, unless it is understressed at the maximum moment section or an end hook makes up the deficiency.

19.4   Analysis for Moment Loading

Footings must often carry moment from a column or wall. Combined with direct load this gives the equivalent of an eccentric load, as in Fig. 19.3. If the moment is constant it is desirable to put the center of the footing under this eccentric load. For the retaining wall (Sec. 18.6g) the adjustment of the base to make the resultant load fall between the center and third point is discussed. Usually the varying nature of the moment makes it impossible to avoid all eccentricity on the footing and a trapezoidal soil pressure results. This distribution modifies the magnitude of the design moment and shears, but not the

---

* Code 11.11.2, for rectangular areas, uses $(2 + 4/\beta_c) \gtrless 4$, where β_c is ratio of long to short side of concentrated load or reaction area. In SI units, $0.17(1 + 2/\beta_c)$ or 0.33, whichever is smaller.

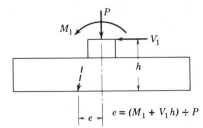

$$e = (M_1 + V_1 h) \div P$$

Figure 19.3 Equivalent eccentricity of load.

general design procedure. There might be doubts as to how to handle the two-way slab shear. For important eccentricities the method indicated in Sec. 5.25 is available. For minor eccentricities the authors simply apply the shear developed as in Fig. 19.5a from the heavier loaded half of the footing onto $\frac{1}{2}$ of the perimeter around the column at the distance $d/2$.

Although office practice in footing design often neglects nominal column eccentricities of loading, considering them a minor matter, other eccentricities are important and should be more commonly included.* Columns designed for significant moment *at their base* need the resisting moment or eccentricity developed by the footing reactions.

19.5 Square Footing—Analysis Example

Check the footing of Fig. 19.4, assuming f'_c = 3000 psi for both column and footing, Grade 40 steel, and an allowable soil pressure of 3000 psf. The column dead load is 190k and the live load 300k. Assume no overburden of soil.

(a) Concentric Load Analysis Assume that there is no column moment to be transferred.

Solution

1. *Check bearing from column on top of footing.* Because column is the same strength concrete as the footing, the bearing on top of the footing is adequate.
2. *Check dowels into footing.* Assuming the column fully stressed, dowels should be provided to match the column steel, extended vertically into the footing a compression development length ℓ_d. To satisfy the general structural integrity requirements of Code 15.8.2.1, the minimum area of reinforcement extending across the joint into the footing must be not less than 0.005 times the gross area of the

---

* But see Code 15.2.2.

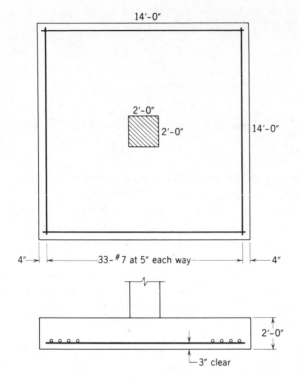

14'-0"

2'-0"

2'-0"

14'-0"

4" ← 33-#7 at 5" each way → ← 4"

2'-0"

3" clear

Figure 19.4 Square column footing.

column. Because the minimum reinforcement ratio of a column is 0.01, the extension of dowels that match the column steel provides substantially more reinforcement than is required to satisfy the general structural integrity requirement. A portion of these dowels equal to 0.005 times the column gross area should be anchored in the footing with a tensile development length to provide the desired ductility in case an unforeseen event causes tensile forces on the joint.

3. *Check strength of soil beneath footing, soil pressure q_s.* No moment is specified on the footing. Assume concentric loading. The check on soil pressure can be done with service load (unfactored load) values and allowable soil pressure or with factored load values and an ultimate soil pressure determined from the appropriate ratio of load factors. It is simpler to follow with service load (unfactored load) values. From vertical equilibrium the soil pressure must counteract both the applied column load and the weight of the footing itself.

$$\text{Column load} = 190 + 300 = 490^k$$
$$\text{Weight of footing} = 2.0 \times 0.150 \times 14 \times 14 = \underline{59^k}$$
$$549^k$$

Average soil pressure = q = 549,000/(14 × 14) = 2800 psf

Allowable soil pressure = 3000 psf   ∴ Footing area   **O.K.**

For checking the adequacy of the concrete sections, factored loads must be used. Because the weight of the footing directly offsets the soil pressure increment that it causes, free bodies to determine shear and moments at critical sections are simplified if the "net soil pressure" is used.

$$\text{Factored column load} = 1.4 \times 190 + 1.7 \times 300 = 776k$$
$$\text{Average factored soil pressure} = q_{net} = 776{,}000/(14 \times 14) = 3960 \text{ psf}$$

4. *Check shear strength.* In a square footing the use of average d is recommended only for two-way or punching shear calculations around the column with the depth to the upper steel layer used for moment calculations and for beam shear all across the footing.

$$\text{Average } d_v = 24 - 3 \text{ (cover against soil)} - 0.88 \text{ (bar diam.)}$$
$$= 20.12 \text{ in.}$$

Shear is first checked for two-way action all round the column at a distance $d/2$ outside the column as indicated in Fig. 19.5a, with the net soil pressure used as the applied load. The net soil pressure omits that part of the reaction created by the footing weight because everywhere in moment and shear evaluation the downward weight of the footing balances this portion of the reaction.

$$V_u = 3960(14.0^2 - 3.68^2) = 3960(196.0 - 13.5) = 725{,}000 \text{ lb}$$
$$\text{Required } V_n = V_u/\phi = 725{,}000/0.85 = 850{,}000 \text{ lb}$$
$$\text{Actual } V_n = 4\sqrt{f_c'}b_o d \quad \text{(Code Eq. 11.36 for } \beta_c = 1)$$
$$V_n = 4\sqrt{3000}(4 \times 44.1 \times 20.12) = 778{,}000 \text{ lb}$$
$$\therefore \text{Actual } V_n < \text{Required } V_n \quad \therefore \textbf{N.G.}$$

Does not satisfy two-way shear requirements.

(The footing could be deepened or higher strength concrete used. The remaining checks are made to illustrate the procedure.)

Although it normally governs only on long narrow footings, shear is also checked on the full width as a one-way slab (Fig. 19.5b). In design or analysis of very wide sections under uniform loading, it is usually convenient to consider a typical 1-ft wide strip. In this problem, however, the full width is checked.

$$\text{Min. } d_v = 24 - 3 - 1.5(0.88) = 19.68 \text{ in.}$$
$$V_u = 3960 \times 4.36 \times 14 = 242{,}000 \text{ lb}$$
$$\text{Required } V_n = V_u/0.85 = 285{,}000 \text{ lb}$$
$$\text{Actual } V_n = 2\sqrt{f_c'}bd \quad \text{(Code Eq. 11.3)}$$
$$V_n = 2\sqrt{3000}(14 \times 12) \times 19.68 = 362{,}000 \text{ lb}$$
$$\therefore \text{Actual } V_n > \text{Required } V_n \quad \therefore \textbf{O.K. for one-way shear}$$

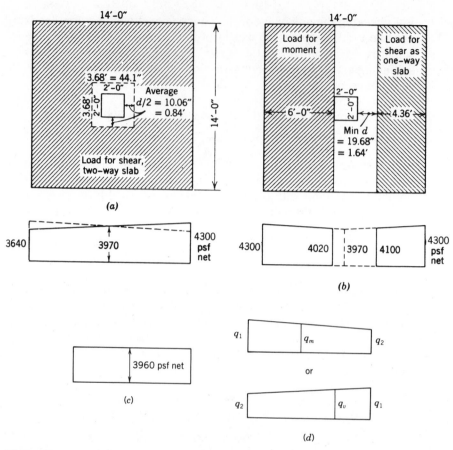

Figure 19.5 Critical sections on square footing for (*a*) shear as two-way slab. (*b*) Moment and shear as one-way slab. Net factored load pressures under (*c*) concentric loading and (*d*) eccentric loading.

5. *Check flexural reinforcement.* Moment must be calculated at the face of the column (Fig. 19.5*b*).

$$M_u = 14 \times 6 \times 3960 \times 3 = 998,000 \text{ ft-lb}$$
$$M_n = M_u/1000\phi = 1110 \text{ ft-k}$$
$$d_M = 24 - 3 - 1.5 \times 0.88 = 19.68 \text{ in.}$$
$$\rho = \frac{33 \times 0.60}{14 \times 12 \times 19.68} = 0.00600 > \text{min of } 200/f_y$$
$$= 200/40,000 = 0.005 \quad \textbf{O.K.}$$
$$\rho = 0.0060 < 3/4\rho_b \quad \textbf{O.K.}$$
$$a = A_s f_y/(0.85 f_c' b) = 33 \times 0.60 \times 40/(0.85 \times 3 \times 14 \times 12)$$
$$= 1.85 \text{ in.}$$
$$z = d - a/2 = 19.68 - 0.92 = 18.76 \text{ in.}$$

Since $M_n = A_s f_y z$,

Required $A_s = 1110 \times 12/(40 \times 18.76) = 17.7$ in.^{2} versus 33×0.60
    $= 19.8$ in.^{2} **O.K.**

6. *Check development length of bars*. The critical section for bar development is at the maximum moment section, where there is available for ℓ_d 72 in. end cover, say, 70 in. From Code 12.2.2.

Reqd. $\ell_d = 0.04 A_b f_y / \sqrt{f'_c} = 0.04 \times 0.60 \times 40,000/\sqrt{3000} = 17.5$ in.

A much larger bar could be used, but this spacing is quite reasonable.

7. *Conclusions:* The footing must be deepened for shear. The A_s is about 10% more than needed for the present depth. A larger bar with wider spacing could be used. (These conclusions should be compared to those of the next solution.)

(b) Analysis with Moment Loading   Assume the factored design moment from the column is equivalent to 1800 k-in. at the soil bearing level.

$$e = 1800/776 = 2.32 \text{ in.} = 0.193 \text{ ft}$$

Solution

1. *Check bearing from column on top of footing*. This check is the same as in part (a).
2. *Check dowels into footing requirement*. This check is the same as in part (a).
3. *Check strength of soil beneath footing, soil pressure q_s*. As in part (a), the check is done with service (unfactored) loads. The eccentricity of the column load produces nonuniform soil pressures that are assumed to vary linearly across the footing as shown in Fig. 19.5d. From vertical equilibrium, the downward loads are determined as 549^k in part (a). The unfactored column moment is assumed as having the same relation of factored and unfactored values as the column axial load. The service load value is $[(190 + 300)/776] \times 1800$ or 1140 k-in. or 95 k-ft.
    From vertical equilibrium

$$\frac{(q_1 + q_2)}{2} (14 \times 14) = 549$$

From summation of moments about an axis through the center of the footing (using kip and ft units)

$$\frac{(q_1 - q_2)}{2} (14 \times 14) \left(\frac{7}{3}\right) = 95$$

Solving these equations simultaneously, $q_1 = 3010$ psf and $q_2 = 2590$ psf. Thus, at service loads, the gross pressure is 3010 psf or essentially the allowable 3000 psf.

For strength checks using factored loads, the same procedure is used except that net soil pressures $q_{1\,net}$ and $q_{2\,net}$ are desired. For factored column load of 776^k and a factored column moment of 1800 k-in., these values are $q_{1\,net} = 4290$ psf and $q_{2\,net} = 3630$ psf. The value at the center of the footing remains the same as 3960 psf.

4. *Check shear strength.* The shear at the critical section for one-way shear that is located at a distance d from the column face is determined from the average net pressure $(q_1 + q_v)/2$ shown in Fig. 19.5d.

q_v is found from similar triangles as 4080 psf.

Required $V_n = V_u/\phi = [(4290 + 4080)/2] \times 14 \times 4.36/0.85$
$= 300,500$ lb.

Actual $V_n = 2\sqrt{f'_c}bd = 362,000$ lb.   ∴ **O.K.** for one-way shear

For the two-way slab shear, because the eccentricity of loading is small, compute V_n on $\frac{1}{2}$ the loaded area (heavy side) using the trapezoidal loading of Fig. 19.5d. The shear at the critical punching shear section located at a distance $d/2$ from the column face is approximately found using the critical section shown in Fig. 19.5a and the pressure distribution shown in Fig. 19.5d, where q_v is 4050 psf. Using the heavily loaded half of the footing along with the corresponding half-perimeter $b_o/2$,

Required $V_n = V_u/\phi = [(((4290 + 3960)/2) \times 14 \times 7)$
$- (((4050 + 3960)/2) \times 3.68 \times 1.84)] \times (1/0.85)$
$= 444,000$ lb

Actual $V_n = 4\sqrt{f'_c}(b_o/2)d$
$= 4\sqrt{3000}(2 \times 3.68 \times 12 \times 20.12) = 389,300$ lb
∴ **N.G.**

5. *Check flexural reinforcement.* The moment is calculated using two triangles for the trapezoid shown in Fig. 19.5d formed by $q_1 = 4290$ psf and $q_m = 4010$ psf.

Required $M_n = M_u/\phi = (0.5 \times 4290 \times 6 \times 4 \times 14$
$+ 0.5 \times 4010 \times 6 \times 2 \times 14)/0.90$
$= 1,175,000$ ft-lb $= 1175$ ft-k

Required $A_s = 1175 \times 12/(40 \times 18.76) = 18.8$ in.$^2 < 19.8$ in.^{2}
O.K.

6. *Check development length of bars.* Bar development length is unchanged from (a).

7. *Conclusions:* Soil pressure is increased about 8%, shear by 4%, and moment by 6%. Except for the original low soil pressure, a larger footing is needed for soil pressure; a deeper footing is required for shear.

19.6   Square Column Footing—Design (with SI)

Design a square column footing using 20 MPa concrete, Grade 400 reinforcement, and allowable soil pressure of 170 kPa (at service load) to carry a 460-mm square spiral column made of 35 MPa concrete and carrying a dead load of 890 kN and live load of 1560 kN. The force of gravity on the mass of ordinary reinforced concrete is 23.6 kN/m³.

Solution

1. *Check bearing from column on top of footing.* The nominal design load assigned to the concrete in the spiral column is based on $0.85f'_c$, or an ultimate based on $\phi(0.85f'_c) = 0.75(0.85 \times 35) = 22.3$ MPa. Because the footing will be large, $\sqrt{A_2/A_1}$ will have (in Fig. 19.2) an $x_2/x_1 \geq 2$* and the allowable ultimate from the column on this concrete will be $2\phi(0.85 \times 20) = 2 \times 0.70(17) = 23.8$ MPa. This is more than the ultimate from the column. (**O.K.**)

2. *Check dowels into footing.* Dowels that match the column bars are adequate. Required embedment length for compression bars Code 12.3.2 = $\ell_d(\text{mm}) = f_y d_b / 4\sqrt{f'_c}$ for $\sqrt{f'_c}$ in MPa = $0.25d_b \times 400/(\sqrt{20}) = 22.4d_b$, where ℓ_d and d_b are normally in mm. To satisfy the general structural integrity requirements of Code 15.8.2.1, a minimum area of these dowels must be able to develop the tensile yield point. Using hooked bars the minimum development length ℓ_{dh} is given by Sec. 12.5.3 of ACI 318-M as $100d_b/\sqrt{f'_c}$ for $\sqrt{f'_c}$ in MPa. This requires $\ell_{dh} = (100)d_b/\sqrt{20} = 22.4d_b$. This ℓ_d, as a vertical embedment into the footing, fixes a minimum d in mm for the footing expressed as a multiple of d_b, the bar diameter. (An alternate sometimes available is to use more dowels of smaller diameter and smaller ℓ_d.)

3. *Size footing to maintain allowable soil pressures.*

$$\text{Column } P = 890 + 1560 = 2450 \text{ kN} = 2.450 \text{ MN}$$

Many design processes must be based on interaction, better known as trial and error, because the dead load of the element is unknown at

---

* $x_2/x_1 = 2$ where d of footing is at least $\frac{1}{2}$ the width of the column (because of the $1:2$ slope involved in Fig. 19.2).

the beginning. Because column bars are often large and because the check on dowel requirements indicates $d > 22.4d_b$, a rough estimate on footing thickness is $22.4d_b + d_b/2 +$ cover (70 mm). In this case, estimating column bars as 30 mm, the initial trial footing thickness is $(22.9)(30) + 70 = 757$ mm, say, 800 mm.

Assume footing 800-mm (or 0.8-m) thick exerting a gravity force of 0.8×23.6 kN/m² = 18.9 kN/m² = 18.9 kPa. From vertical equilibrium, assuming the footing plan area as A, and using w_d for the gravity force of the footing mass,

$$P + w_d A = q_s A$$

$$\text{Required } A = \frac{P}{q_s - w_d} = \frac{2450 \text{ kN}}{(170 - 18.9)\text{kN/m}^2} = 16.21 \text{ m}^2$$

Try 4.030-m square footing ($A = 16.24$ m²). For footing strength calculations, the factored loads must be used. Because the vertical components of the footing gravity force and the component of soil pressure resisting that force are offsetting, shears and moments are determined simply using net soil pressures.

$$\text{Column } P_u = 1.4 \times 890 + 1.7 \times 1560 = 3900 \text{ kN} = 3.90 \text{ MN}$$

From vertical equilibrium, actual $q_{net} = (3.90 \times 10^6)/16.24 = 240$ kPa.

Because it is unusual to fit the required A exactly, it is better to use the *actual* q_{net} in M and V calculations. In this example the close fitting of the area makes this precaution appear useless.

4. *Select footing depth to provide required shear resistance.* Footings are usually more sensitive to two-way or punching shear resistance requirements than to beam shear or flexural depth requirements. In the absence of web reinforcement, the footing depth is the main tool at the designer's disposal to resist shear. Two-way shear is considered first.

For the assumed h of 800 mm and 70 mm clear cover (required for casting directly on the ground), the average $d_v = 800 - 70 - d_b =$ say, 700 mm for #30 bars ($d_b = 30$ mm). The critical shear for the two-way slab action is then at $d/2 = 700/2 = 350$ mm from the face of columns, as shown in Fig. 19.6a, with a face width of $460 + 2 \times 350 = 1160$ mm.

$$V_u = (240)(4.030^3 - 1.160^2) = 3570 \text{ kN} = 3.57 \text{ MN}$$
$$\text{Required } V_n = V_u/\phi = 3.57/0.85 = 4.20 \text{ MN}$$
$$\text{Allowable } v_c = 0.33\sqrt{f_c'} = 0.33\sqrt{20} = 1.48 \text{ MPa,}$$
$$b_o = 4(460 + 700) = 4 \times 1.160 \text{ m}$$
$$\text{Required } d_v = V_n/b_o v_c = (4.20)/[(4 \times 1.160)(1.48)]$$
$$= 0.614 \text{ m} < 700 \text{ mm}$$

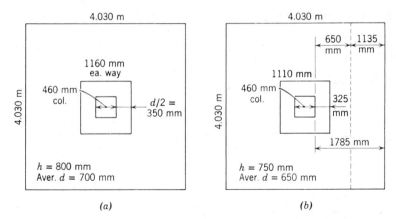

Figure 19.6 Trial sections for shear in square footing design. (*a*) First trial. (*b*) Second trial.

The assumed h can be reduced, probably around $\frac{1}{2}$ the apparent 12% just indicated. (V_u increases and b_o decreases as h decreases.)

Try $h = 750$ mm. For this decrease, the allowable net soil pressure increases, possibly not enough to justify changing footing size, but check this. Footing gravity force $= 0.750 \times 23.6$ kN/m² $= 17.7$ kPa, which makes the denominator in the required area expression $= 170 - 17.7 = 152$ kPa versus 151 previously.

Required $A = P/(q_s - w_d) = 2450/(170 - 17.7)$
$= 16.09$ m² < 16.24 m² used. Say **O.K.**
$d_v = 750 - 70 - 30$ (for d_b) $= 650$ mm
$V_u = (240)(4.030^2 - 1.110^2) = 3.602 \times 10^3$ kN $= 3.602$ MN
Required $V_n = V_u/\phi = 3.602/0.85 = 4.24$ MN
Required $d_n = V_n/b_o v_c = 4.240/(4 \times 1.110)(1.48) = 0.645$ m,
$\qquad\qquad$ essentially as assumed.

Although shear all across the footing at distance d from the column rarely, if ever, governs, check the section shown in Fig. 19.6*b* against the allowable v_c for one-way action, limited to $0.17\sqrt{f_c'} = 0.17\sqrt{20} = 0.76$ MPa.

$V_u = (240)1.135 = 272$ kN/m width
Required $V_n = V_u/\phi = 272/0.85 = 320$ kN/m
$\qquad v_c = V_n/bd = 320 \times 10^{-3}/(1 \times 0.650) = 0.493$ MPa < 0.76
$\qquad\qquad$ **O.K.**
$\qquad$ USE footing 4.030 m square, $h = 750$ mm

5. *Determine flexural reinforcement.* For moment at face of column the *lesser* depth controls, not the average.

$d_m = 750 - 70 - 1.5d_b = 680 - 1.5d_b = 635$ mm for $d_b = 30$ mm

On full width, $M_u = (240 \times 10^{-3})4.030 \times 1.785^2/2 = 1.554$ MN·m

$$M_n = M_u/0.90 = 1.726 \text{ MN·m}$$

Assume $z = 0.9d = 0.9 \times 635 = 572$ mm $= 0.572$ m

Required $A_s = M_n/f_y z = 1.726/(400)0.572$

$\qquad = 7530 \times 10^{-6} \text{ m}^2 = 7530 \text{ mm}^2$

$a = A_s f_y/(0.85 f_c' b) = 7530 \times 10^{-6}$

$\qquad \times 400/(0.85 \times 20 \times 4.030) = 0.044$ m $= 44$ mm

$z = 635 - 44 = 591$ mm

Required $A_s = 7530(\text{above})(572/591) = 7290 \text{ mm}^2$

Check on minimum $\rho = 1.4/f_y$ for f_y in MPa (Code 10.5.1).

Min. $\rho = 1.4/400 = 0.003\ 5$

Min. $A_s = \rho b d = 0.003\ 5 \times 4.030 \times 0.635$

$\qquad = 8960 \times 10^{-6} \text{ m}^2 = 8960 \text{ mm}^2 > 7290 \text{ mm}^2$

Minimum reinforcement governs. USE 18-#25 (9000 mm²) each way, spacing about 225-mm c-c

The 18 bars control the spacing, assumed uniform, although precision is not needed.

6. *Check development length.*

Bar development length,

$$\ell_d = 0.02\ A_b f_y/\sqrt{f_c'} = 0.02 \times 500 \times 400/\sqrt{20} = 895 \text{ mm}$$

Available $\ell_d = 1785$ mm minus end cover.   **O.K.**

19.7   Rectangular Column Footings—Analysis

Analyze the rectangular footing of Fig. 19.7 assuming $f_c' = 3000$ psi, Grade 60 steel, and an allowable soil pressure at service load of 3000 psf. The column service load is 200 k dead and 310 k live load without any moment and there is no overburden. Assume that both bearing stresses and dowels are satisfactory.

Solution

1. *Check strength of soil beneath footing, soil pressure q_s.* At service load,

$$\text{Column load} = P = 200 + 300 = 500\ k$$

$$\text{Weight of footing} = 2.67 \times 0.150 \times 11 \times 18 = \underline{\quad 80\ k}$$

$$580\ k$$

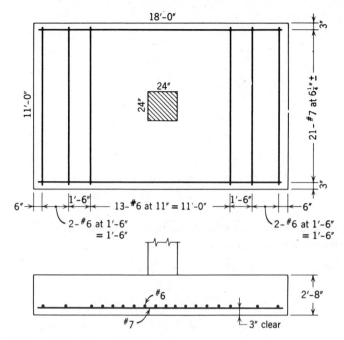

Figure 19.7 Rectangular column footing.

Average soil pressure = q = 580,000/(11 × 18) = 2930 psf   **O.K.**
Factored column load = 1.4 × 200 + 1.7 × 300 = 790 k
Average factored soil pressure

$$= q_{net} = 790,000/(11 \times 18) = 3990 \text{ psf}$$

2. *Check shear strength.* For the long projection, d = 32 − 3 − 0.4 = 28.6 in. For the short projection, d = 32 − 3 − 0.88 − 0.38 = 27.7 in. Diagonal tension around the column is critical on the average depth of 28.1 in., say, 28.0 in., at a distance $d/2$ beyond the column (Fig. 19.8) considering two-way slab action.

$$b_o = 4(24 + 28.0) = 4 \times 52.0 = 208 \text{ in.}$$
$$V_u = 3990(18 \times 11 - 4.33^2) = 3990(198 - 18.7)$$
$$= 715,000 \text{ lb}$$

Required V_n = V_u/ϕ = 715,000/0.85 = 841,000 lb
   Actual V_n = $4\sqrt{f_c'}b_o d$ = $4\sqrt{3000}(208)(28)$ = 1,276,000 lb
∴ Actual $V_n \gg$ Required V_n,   **O.K.** for two-way shear

Examining 1-ft wide strips, it is seen that one-way shear is critical in the long direction.

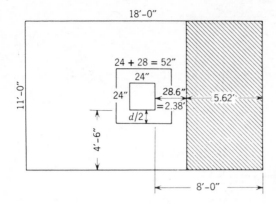

Figure 19.8 Diagonal tension analysis for rectangular footing.

Check one-way slab shear on long projection at distance $d = 28.6$ in. from column, as shown in Fig. 19.8.

$$V_u = 3990 \times 11(8.0 - 2.38) = 247{,}000 \text{ lb}$$

Required $V_n = 247{,}000/0.85 = 291{,}000$ lb

Actual $V_n = 2\sqrt{f'_c}bd = 2\sqrt{3000}(132)(28.6) = 414{,}000$ lb

$\therefore$ Actual $V_n \gg$ Required V_n,   **O.K.** for one-way shear

The footing could be several inches thinner with respect to shear.

3. *Check flexural reinforcement in both directions.* In the long direction, $M_{uL} = 8.0 \times 11 \times 3990 \times 4.0 = 1{,}404{,}000$ ft-lb.

$$M_{nL} = M_{uL}/\phi = 1{,}404{,}000/0.9 = 1{,}560{,}000 \text{ ft-lb}$$
$$\text{Assume } z = 0.92 \times 28.6 = 26.3 \text{ in.}$$

Required $A_s = \dfrac{1{,}560{,}000 \times 12}{60{,}000 \times 26.3} = 11.86$ in.^{2} versus 21×0.60

$$= 12.60 \text{ in.}^2$$

This is **O.K.** without a check on z.

Minimum A_s for flexure* $= (200/f_y)bd = (200/60{,}000)11 \times 12 \times 28.6$ $= 12.60$ in.$^2 = 12.60$ in.^{2} as previously   **O.K.**

In the short direction,

$$M_{ns} = 4.5 \times 18 \times 3990 \times 2.25/0.9 = 808{,}000 \text{ ft-lb}$$
$$\text{Assume } z = 0.94 \times 27.7 = 26.1 \text{ in.}$$

Required $A_s = \dfrac{808{,}000 \times 12}{60{,}000 \times 26.1} = 6.19$ in.$^2 < 17 \times 0.44 = 7.48$ in.^{2}

Min. $A_s = (200/60{,}000)bd = (200/60{,}000)18 \times 12 \times 27.7 = 20.0$ in.^{2}

Even for temperature and shrinkage, min. $A_s = 0.0018bh = 0.0018 \times 18 \times 12 \times 32 = 12.44$ in.^{2} Although exposure to temperature and

---

* Minimum A_s always will be much under $0.75A_{sb}$ (balanced).

shrinkage is usually not severe in a footing, it is still about the absolute minimum reinforcement anywhere. The 7.48 in.^{2} is grossly inadequate.

Code 15.4.4 requires much of the transverse reinforcement to be concentrated within the central width equal to the short side of the footing, the specified proportion being $2/(1 + \beta)$, where β is the ratio of the long side to the short side of the footing, $18/11 = 1.64$ in this case:

$2/(1 + 1.64) = 0.76$ versus 13 bars/17 bars $= 0.77$ in the 11-ft width. This proportion is satisfactory, but the total is **N.G.**

4. *Check development length of bars.* For bar development the short projection appears the more critical, but probably adequate.

$$\ell_d = 0.04A_b f_y/\sqrt{f_c'} = 0.04 \times 0.44 \times 60,000/\sqrt{3000} = 19.3 \text{ in.}$$

The available is 54 in. less about 2-in. end cover.　　**O.K.**

19.8　Wall Footings

(a) Basic Concepts　The uniform nature of a wall loading leads to the use of a typical 1-ft width as a design strip, as in Fig. 19.9. For heavy walls, the design is quite similar to that of column footings, but it is simpler because the bending is in only one direction. For light walls, the design is frequently based on rather arbitrary minimum sizes. Plain concrete often may be used.

When the wall is reinforced concrete, the critical moment section is taken at the face of the wall (Code 15.4.2). With masonry walls a different critical section is specified, as illustrated in the following example.

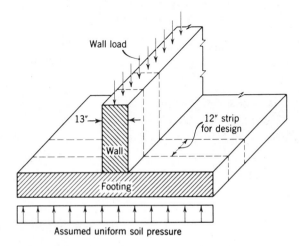

Figure 19.9　Part of wall footing, with heavy loading.

(b) Design Example Design a wall footing for a 13-in. brick wall carrying a service load per running foot of 6 k dead load plus 15 k live load using $f'_c = 3000$ psi, Grade 60 steel, and an allowable soil pressure at service load of 4400 psf, negligible overburden and moment.

Solution

1. *Check bearing from wall on top of footing.* The factored load per ft of wall is $1.4 \times 6 + 1.7 \times 15 = 33.9^k$. Thus, the bearing stress on the concrete is $33.9^k/(12)(13) = 0.22$ ksi, which is so far below the design bearing strength of $\phi(0.85 f'_c)$ that no further check is required.

2. *Check dowels into the footing.* The new requirements for general structural integrity of Code 15.8 only specifically recognize cast-in-place and precast walls. It is clear, however, from the Commentary and from the ANSI-A58-1982[6] requirements that the general intent is to provide vertical ties between all structural members, regardless of material. Hence, a minimum tensile strength of $50A_g$ or $(50)(12)(13) = 7.8$ k/ft should be provided in the form of ties between the footing and the wall. This requirement can be satisfied by a #4 bar each 18 in.

3. *Check strength of soil beneath footing.*

    Wall load at service load levels $= 6 + 15 = 21$ k/ft

    Estimate footing thickness at 12 in. Footing distributed dead load gives 150 psf. The allowable soil pressure is 4400 psf. Thus, the maximum net soil pressure is $4400 - 150 = 4250$ psf. This pressure must resist the wall load.

    Footing width $W = 21,000/4250 = 4.94$ ft, say, 5-ft 0-in. USE
    Ultimate applied load $= 1.4 \times 6 + 1.7 \times 15 = 33.9$ k
    per running foot of wall
    $q_{net} = 33,900/5.0 = 6780$ psf

4. *Check shear strength.* If d is assumed 8 in., the critical section for one-way shear or diagonal tension (Fig. 19.10a) lies 1-ft 3.5-in. or 1.29 ft from the edge.

    $V_u = 1.29 \times 6780 = 8750$ lb/ft strip
    Required $V_n = 8750/0.85 = 10,300$ lb
    Required $d_n = V/bv = 10,300/(12 \times 2\sqrt{3000}) = 7.84$ in.

    One might expect normally to need a second trial, but required d lies between 7.84 and 8 in., which is close enough. Required $h = 8 + 3$ cover $+ d_b/2 = 11 + d_b/2$, say, 11.5 in. The practical dimension is 12 in.

    USE $h = 12$ in., $d = 8.62$ in. if #6 bars are used.

    There is no two-way shear action in such a footing.

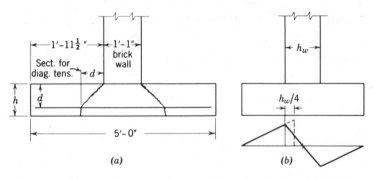

Figure 19.10 Reinforced concrete wall footing, design. (*a*) Critical shear section. (*b*) Shear diagram and equivalent cantilever length.

5. *Check reinforcement required.* For a brick wall the critical section for moment lies at the quarter-point of the wall thickness, that is, in this case 1.08/4 = 0.27 ft inside the face of the wall. The use of this equivalent cantilever length is explained by Fig. 19.10*b*. The shear diagram there indicates that the maximum moment is under the middle of the wall, but if the cantilever projection is extended back to the quarter point of the wall thickness, as shown dotted, the area under the shear diagram is nearly the same.

$$M_u = (1.96 + 0.27)6780 \times 2.23/2 = 16,900 \text{ ft-lb/ft strip}$$
$$M_n = M_u/\phi = 16,900/0.90 = 18,700 \text{ ft-lb}$$

Assume $z = 0.94d = 0.94 \times 8.62 = 8.10$ in.

$$A_s = M_n/(f_y z) = 18,700 \times 12/(60,000 \times 8.10)$$
$$= 0.46 \text{ in.}^2/\text{ft} = 0.038 \text{ in.}^2/\text{in.}$$
$$a = 0.46 \times 60,000/(0.85 \times 3000 \times 12) = 0.90 \text{ in.}$$
$$z = 8.62 - 0.90/2 = 8.17 \text{ in.}$$
$$A_s = 18,700 \times 12/(60,000 \times 8.17) = 0.46 \text{ in.}^2/\text{ft}$$

Min. $A_s = (200/60,000)12 \times 8.62 = 0.345$ in.^{2}/ft < 0.46 **O.K.**

#6 at 11.0 in. = 0.480 in.^{2}/ft > 0.46
#5 at 7.5 in. = 0.495 in.^{2}/ft (d becomes 8.68 in.)
For #6, $\ell_d = 0.04 \times 0.44 \times 60,000/\sqrt{3000} = 19.3$ in.

The projection of 23.5 in. beyond the wall is ample for ℓ_d plus end cover for #6 bars or smaller. The full ℓ_d is needed only from the point of maximum moment, but because the moment increases more slowly under the wall, the really critical ℓ_d might be for the smaller f_s at face of wall. Here it was obviously on the safe side to compare the full ℓ_d with the available projection in the footing.

USE #6 at 11.0 in. (with 4-#6 longitudinal bars as spacers).

Check longitudinal bars for temperature and shrinkage.

$$\rho_{gr} = 1.76/(12 \times 60)$$
$$= 0.0024. \text{ More than required } 0.0018. \quad \textbf{O.K.}$$

19.9   Footings on Rock

Very often only a nominal footing is required on rock because good rock is usually at least as strong as concrete and, moreover, is usually loaded on only a small part of its surface. Where a significant footing is required, as under a heavy steel column, deformation and relative stiffness become more significant than calculated unit stresses. A reinforced concrete footing cannot function normally unless the steel elongates and the cantilever elements deflect. Because the underlying rock is stiffer against compressive deformation than the reinforced concrete footing is against bending deflection, such a footing cannot effectively distribute the load laterally to any significant extent. A reinforced concrete footing on rock is thus of little value. The designer must depend more on the transfer of stress through shear with its smaller deformations. This situation calls for a deep footing and suggests that plain concrete may be just as effective as reinforced. On the usual *rough* rock surface, friction (and interlock) forces are ample to prevent any significant stretching of the bottom of the footing. A footing on rock requires a large depth to give it shear stiffness, usually more depth than the length of its projection. The tension developed on the bottom face of the concrete is largely a function of the rock quality below, its surface roughness (interlock), and its lack of volume change. Under ideal rock conditions, tension from flexure can be neglected; on a poor rock, the concrete should be designed to resist the entire moment as for a footing on soil.

19.10   Balancing Footing Pressures

Many years ago it was commonly accepted that footings would undergo essentially uniform settlements if they developed the same soil loadings per unit area under the usual loads on the structure. Such footings were said to have balanced footing areas.

Modern soil mechanics greatly restricted the field of usefulness of this concept. It has shown in most practical cases that uniform soil pressure does not lead to uniform settlement. Proportioning for balanced footing pressures usually results in a larger size for outside wall footings. In a heavy structure on a thick stratum of material that consolidates slowly under additional load, such an increase in exterior footing size will actually *increase* the final differential settlement. In many areas, such as in the Southwest, foundation movement on ordinary residence and commercial structures is as likely to be upward owing to expansive soils as downward owing to applied loads. The engineer who uses balanced footing areas should be certain that his or her special foundation situation justifies this approach. Frequently, it increases rather than reduces differential settlements.

For a balanced footing design, the term "usual load" is applied to the average load or the load most commonly on the structure. This load might vary from dead load alone for a church or a school, up to nearly full live load plus dead load for certain types of warehouses.

If all footings are proportioned for the usual loads, at full allowable soil pressure, the footing that cared for the greatest percentage of live load (compared to dead load) would most overstress the soil when full live load is acting. This result gives the clue for balanced footing area design. That footing with the largest ratio of live load to dead load is designed first by the usual procedures using maximum load and the full allowable soil pressure. The usual soil pressure under this footing is established next for the usual load, most frequently something between dead load alone and dead load plus half live load. The bearing areas of all other footings are then chosen to give this same usual soil pressure under the usual footing loads. (This action automatically gives them less than the maximum permissible soil pressure under the maximum live load.) The individual footings are then designed for thickness and steel using these balanced areas and the full design live load.

19.11 Combined Footings

When two or more columns are carried on a single footing, it is called a combined footing. Two circumstances especially call for a combined footing: (1) two columns so closely spaced that separate footings overlap or are of uneconomic proportions, Fig. 19.11a; (2) an exterior column footing that cannot be made symmetrical because of property line limitations or other restrictions, Fig. 19.11b. In each case the centroid of the combined footing must coincide with the resultant of the two column loads, noted as O.

Point O shifts slightly as the design (factored) loads, the service loads, or the usual loads of Sec. 19.10 are considered. Designers may use as the centroid of the footing the one they feel is most important in the design situation.

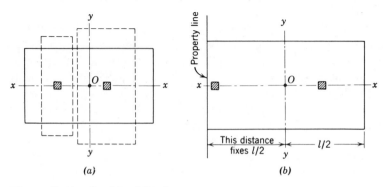

Figure 19.11 Combined footings.

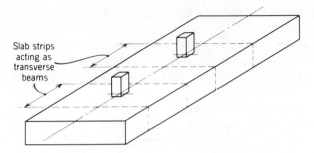

Figure 19.12 Transverse "beams" in combined footing.

In Fig. 19.11*b*, the length of a rectangular footing is fixed by the fact that if rectangular it must be symmetrical about axis *y-y* through this resultant load point *O*. The width in turn is fixed by the required soil-bearing area to carry the combined column loads.

In Fig. 19.11*a*, the designer has some choice in the length and width, but the centroid of the area chosen must fall at *O*.

Design of combined footings, like the design of many other reinforced concrete structures, is not standardized.[9] One approach, sketched in Fig. 19.12, assumes structural behavior consists of transverse slab strips under the column that act to distribute load laterally and longitudinal slab strips that act like slabs supported on these "crossbeams." This type of analysis is appropriate for long, narrow footings. Another approach is to think of the projecting ends of the footing as similar to half of an isolated column footing and the slab between columns acting as a longitudinal beam. This method is appropriate for footings with projections of such proportions as might be used for isolated columns, that is, projections somewhat of the same order of magnitude in the two directions. This type of design for a wider distribution of transverse steel than the transverse "beam" idea.

Shears should be checked as for other footings, which means a double check, around the column at a distance *d*/2 outside the column for two-way slab action and at a distance *d* from the column all the way across the footing as a one-way slab.

Reasonably typical moment and shear diagrams for the longitudinal strips of a combined footing are shown in Fig. 19.13*a*. Design is concerned only with the portions outside the column widths because moments are critical at the face of column. The moment diagram can be visualized as a simple beam moment between columns (negative owing to upward loads), superimposed on the positive moment diagram produced by the upward loads on the two end cantilevers. These two diagrams going into the total moment diagram of Fig. 19.13*a* are shown separately in Fig. 19.13*b*. These diagrams present no special design problem except one relating to the negative moment between columns. If the columns are closely spaced, no top steel may be required because the simple span moment may be small compared to the cantilever moment.

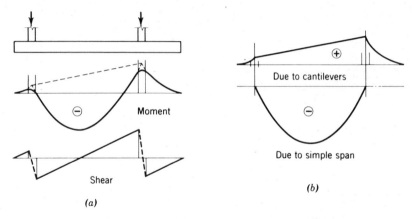

Figure 19.13 Moment and shear diagrams for combined footing.

Attention should be called to the fact that it is on the *unsafe* side to assume a uniform soil pressure in the design of this center negative steel. If the pressure is nonuniform, it is probably larger near the center than it is toward the ends. A uniform load assumption gives too much positive cantilever moment and too little negative moment from the center span. These errors add to indicate a negative moment smaller than the actual value. The designer should be liberal in providing this negative moment steel, even introducing some steel when the uniform pressure assumption indicates a small remnant of positive moment.

19.12   Cantilever Footings

When an exterior column footing, because of property line restrictions, calls for some kind of combined footing but the nearest interior footing is some distance away, the ordinary combined footing becomes long and narrow and subject to excessive bending moment. The cantilever footing is then more economical.

The cantilever footing is really two separate column footings joined by a beam that preferably does not contribute to the bearing area. In Fig. 19.14a, footing A under the exterior column is eccentric by a distance e from the column. By itself it produces a very unsatisfactory soil pressure distribution. The stiff beam C is attached to footing A to balance the overturning moment $P_1 e$ and is carried to the concentric footing B for its necessary balancing reaction. This beam must be stiff to function well. It functions most simply if it is relieved of soil pressure from below, except possibly that needed to carry the beam's weight. This condition is approached in several ways, most commonly by loosening or spading the soil under the beam, occasionally by forming the beam free from the soil, and so forth. Each of these measures leaves some question as to its effectiveness, which causes some engineers to shun this type of footing.

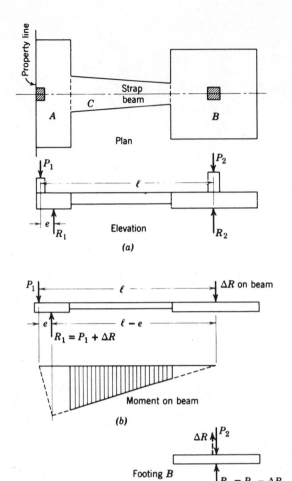

Figure 19.14 Cantilever footing.

Under the simplest assumption, that the beam weight is carried by the soil, the analysis of the cantilever footing is quite simple. The beam is then effectively weightless. The reaction from the exterior column alone is R_1 on footing A (Fig. 19.14b) and ΔR downward on the beam at the interior column, which means ΔR is an uplift on this footing at B (Fig. 19.14c). Hence

$$\Delta R = P_1 e(\ell - e)$$
$$R_1 = P_1 + \Delta R$$
$$R_2 = P_2 - \Delta R$$

Hence footing A must be slightly larger than needed for P_1 alone and footing B can be slightly reduced in area below that needed for P_2 alone. Beam C must carry a constant shear ΔR and a moment as shown in Fig. 19.14b. In sizing footings A and B, the service loads with the usual allowable soil pressure should be used. At ultimate, the increased soil pressure is not equal on A and B unless their ratios of dead to live load are equal.

19.13   Raft Foundations

It is sometimes desirable to utilize the soil pressure under the entire building area by covering the site with a thick slab or mat foundation. Such a foundation is often called a raft foundation, particularly when enough soil is excavated to offset the increased load owing to the structure. Such a foundation may consist of a uniform slab several feet or more thick, a slab stiffened by beams (either above or below the slab), or an inverted flat slab floor (Chapter 15).

No special design problems are involved in the concrete for such foundations other than those of a heavy floor design to resist upward pressures; the columns furnish downward reactions. The distribution of the design upward soil pressures is the major problem; it depends on the variability of the soil reaction properties and the resulting settlement response.

19.14   Slabs-On-Ground

A slab supported directly on a stable soil develops very little moment or shear in carrying a uniform load over its surface. When loads are unequally distributed or concentrated, the distributed reaction produced on the soil does create some moment and shear. (This is the principle used in pavement design.) On very stable soils, light loads such as residence loads can be carried on a simple 4-in. slab with a wire fabric reinforcement, providing no partition load is in excess of 500 plf. Chimney and other heavy loads, such as columns, must be carried to separate footings, and the slab should not rest on or be tied to these separate elements or footings, which often will not move exactly the same as the slab.

Many soil types are subject to shrinkage in drying and expansion in wetting. Slabs alone are not adequate on such soils and experience shows that one portion of the slab may be pushed upward or settle relative to another. To be sure that this will not crack walls and partitions, the slab must be stiffened by monolithically cast beams in both directions. In effect, a residence then may be supported in the center and the stiffened slab will have to cantilever out in all directions to prevent excessive floor warpage; or the structure may be supported on two corner areas and the slab will have to span as a simple span between these. Thus, the stiffened slab must ride like a boat over differential ground movements. Then the floor must be designed for the above loadings and a stiffening grid similar to that shown in Fig. 19.15 is necessary. A very useful report on design of such grid-stiffened slabs was published by the Building Research Advisory Board.[7] In recent years, such slabs often have been post-tensioned and an informative report is available with detailed design proce-

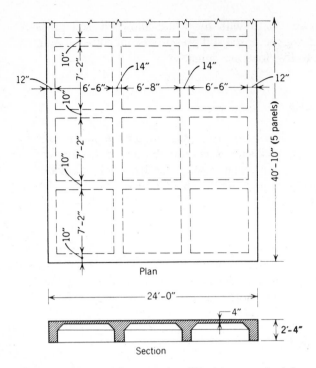

Figure 19.15 A slab-on-grade stiffened as required for a poor soil condition.

dures for such slabs-on-grade.[8] If such a system appears excessively costly, the final alternative is for spot footings to be carried deeper to a better soil and then to design a fully framed floor *not* in contact with the soil. Alternatively, soil stabilization measures can be used on some bad soils.

19.15 Footings on Piles

Many footings must be built on piles. The piles act as a series of relatively concentrated reactions for the footing and in design are usually considered concentrated loads. Because it is usually difficult to drive piles exactly where they are desired, the designer must expect that piles will be from a few inches to a foot away from the spot assigned to them. Accordingly, the Code (15.5.3) provides specific rules for counting piles that are near the critical section and sets somewhat stricter minimum thickness in Code 15.7, at least 12 in. above bottom reinforcement.

In Reference 10 a much more detailed discussion of special shear problems that occur at points closer than d from the column is given. Usually this is the case with only a small number of piles.

19.16 Stepped and Sloped Footings

Specifications now permit thinner footings than formerly. There is thus less reason now for stepped or sloped footings and fewer are used. Such footings definitely must be cast as a monolith to avoid horizontal shearing weakness. Stresses usually must be checked at more than one section. For instance, steel stress at a step or development length beyond a step may be more critical than from the face of the column. Both these and shear must be checked from each step with the shear at a distance d or $d/2$ as in other cases.

Stepped footings are more appropriate for massive footings. A step can sometimes be used to advantage with a combined footing.

19.17 Uplift on Footings

If uplift can exist on a footing, several special points must be considered. First, uplift probably results from lateral loading on the structure and Code Eq. 9.3 probably governs with its lower factor on dead load because it opposes the resultant uplift.

$$U = 0.9D + 1.3W$$

Second, the axial tension (plus the horizontal shear) may require tension splices and tension anchorage into the footing. Third, special design considerations are necessary. The footing must be either heavy enough in itself to balance the uplift or it must engage a mass of earth adequate for this balance. This action reverses the type of bending on the footing and normally calls for some measure of top reinforcement. The footing normally has some flexural cracks from service load positive moment. The assumption of an uncracked plain concrete section for resisting the negative moment from uplift is not realistic.

Selected References

1. A. N. Talbot, "Reinforced Concrete Wall and Column Footings," Univ. of Ill. Eng. Exp. Sta. *Bull. No. 67,* 1913.

2. F. E. Richart, "Reinforced Concrete Wall and Column Footings," *ACI Jour.,* 20, Oct. and Nov. 1948; *Proc.,* 48, pp. 97, 237.

3. J. Moe, "Shearing Strength of Reinforced Concrete Slabs and Footings Under Concentrated Loads," *Development Department Bulletin* D47, Portland Cement Assn., Skokie, IL, April 1961.

4. ACI Committee 326, "Shear and Diagonal Tension," *ACI Jour., Proc.* 59, Jan. 1962, p. 30; Feb. 1962, p. 277; Mar. 1962, p. 1962.

5. R. B. Peck, W. E. Hanson, and T. H. Thornburn, *Foundation Engineering,* John Wiley & Sons, New York, 2nd Ed., 1974.

6. American National Standards Institute (ANSI), "Building Code Requirements for Minimum Design Loads in Buildings and Other Structures," *ANSI Standard A58.1—1982,* New York, 1982.

7. Building Research Advisory Board, "Criteria for Selection and Design of Residential Slabs-on-Ground," *BRAB Report No. 33,* Publication 1571, National Academy of Sciences, Washington, D.C., 1968.

8. Post-Tensioning Institute, "Design and Construction of Post-Tensioned Slabs-on-Ground," Post-Tensioning Inst., Phoenix, 1980.

9. ACI Committee 436, "Suggested Design Procedure for Combined Footings and Mats," *ACI Jour., Proc.* 63, Oct. 1966, p. 1041.

10. *CRSI Handbook,* Concrete Reinforcing Steel Institute, Chicago, 4th Ed., 1984.

Problems

PROB. 19.1. Redesign the rectangular footing of Sec. 19.7 and Fig. 19.7 as a square footing.

PROB. 19.2. Redesign the square column footing of Sec. 19.6 as a rectangular footing with a width of 3.5 m.

PROB. 19.3. Design a square column footing for a 16-in. × 16-in. tied column (of 4000 psi concrete) to carry a service column load of 150 k dead load and 140 k live load using permissible service load level soil pressure of 4500 psf, f'_c = 3000 psi, Grade 60 steel.

PROB. 19.4. Redesign the footing of Prob. 19.3 as a concrete footing of uniform thickness if f'_c = 4000 psi.

PROB. 19.5. Design a reinforced concrete wall footing to carry service loads of 6 klf dead and 4 klf live load from a concrete wall 12-in. thick. Use an allowable soil service load pressure of 2800 psf, f'_c = 3000 psi, Grade 60 steel.

PROB. 19.6. Redesign the footing of Prob. 19.5 if the wall is 18-in. thick.

PROB. 19.7. Redesign the wall footing of Prob. 19.5 if the allowable service load soil pressure is only 2000 psf.

PROB. 19.8.

(*a*) Lay out the plan view of a combined rectangular footing for the columns of Fig. 19.16 where the service loads are:

A: 150 k dead plus 100 k live.

B: 200 k dead plus 250 k live.

Based on an allowable *net* soil pressure of 3200 psf and a centroid based on full design loads, try to achieve equal projections around the heaviest column.

(*b*) Lay out again on the assumption that there is a property line 4 in. beyond the outside face of column *A*.

(*c*) Lay out again for conditions of (*a*) on the assumption that lateral projections beyond column *B* are about ½ the longitudinal projection.

(*d*) Repeat (*a*) with the centroid based on full service loads.

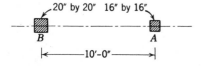

Figure 19.16 Column layout for Prob. 19.8.

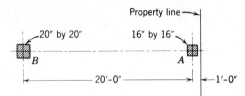

Property line →

20″ by 20″ 16″ by 16″

B A

|← 20′-0″ →|← 1′-0″ **Figure 19.17** Column layout for
Prob. 19.9.

PROB. 19.9. If the columns of Fig. 19.17 carry the service loads of Prob. 19.8(*a*), lay out the plan view of a cantilever footing (without dimensions on the beam) based on an allowable net soil pressure of 3200 psf. Draw the *M* and *V* diagrams for the strap beam that connects the two footing elements:

(*a*) If the footing under column *A* extends 4 ft from the property line.
(*b*) If the dimension in (*a*) is changed to 5 ft.
(*c*) If the dimension in (*a*) is changed to 6 ft.

20

PRESTRESSED CONCRETE ANALYSIS

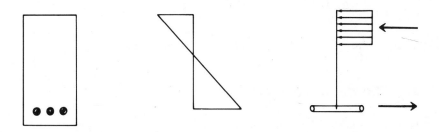

20.1   Scope of Presentation

This chapter covers some of the fundamental concepts of prestressed concrete and presents a general idea of how such concrete responds to loads. No attempt is made to develop design procedures. For ease of presentation, simple rectangular sections are used in the examples rather than the more practical I or T shapes.

When prestressed concrete first came into use, there was a wide gap in flexural behavior between it and ordinary reinforced concrete. Engineers recognized that if only a small amount of prestress was applied, it would be lost because of shrinkage and creep. The member then would probably crack at service load levels. Consequently, the amount of prestress utilized was enough to keep the section fully in compression under all design service loads, that is, full prestressing. Because of both economic factors and problems with excessive camber, over the years smaller amounts of prestressing force have been used in many prestressed concrete members. Many of these members were on the verge of cracking or cracked at service load levels. Thus, partial prestressing is coming into use and the gap is disappearing. In this summary treatment only full prestressing is discussed.

The term tendon is often used to refer to the prestressing element, without differentiation between wires, strands, or bars. Bars are large, up to $1\frac{3}{8}$ in. diameter, with a breaking stress of 150 ksi, more or less. Wires are small, with diameter up to 0.276 in. (7 mm), and a strength of 240 to 250 ksi. Strands normally have seven wires, diameters from $\frac{3}{8}$ (or less) to $\frac{1}{2}$ in. (or 0.6 in.) and strengths of 250 or 270 ksi. Strand is used in almost all pretensioned concrete and in a large percent of American post-tensioned concrete.

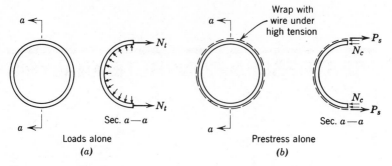

Figure 20.1 Prestress largely eliminates ring tension in tanks.

20.2   The Nature of Prestress

Reinforced concrete becomes prestressed concrete when permanently pre-loaded so that initial stresses are induced opposite to those that are developed later by service loads.

A circular tank, designed to resist circumferential or ring tensile stresses, as in Fig. 20.1a, is prestressed by being placed in initial circumferential compression. This compression is produced by wrapping the tank with wire under high tensile stress as indicated in Fig. 20.1b. This is an old principle; it was used by coopers who placed the staves of wooden kegs in compression upon cooling of the metal hoops located around the periphery.

A simple span beam is prestressed (Fig. 20.2) by introducing (1) a negative moment to offset the expected positive moment and at the same time (2) a longitudinal compression to offset the tensile stresses from bending moment. Both effects are obtained by embedding highly stressed tension members eccentrically (below the middle of a simple beam), which gives both a P/A effect from the load P_s and a negative bending moment effect measured by $P_s e$. The tension prestressing element, called a tendon, may be wire, twisted wire strands, or rods.

Continuous beams are prestressed in a similar fashion but for best results require an effective eccentricity above middepth in negative moment zones.

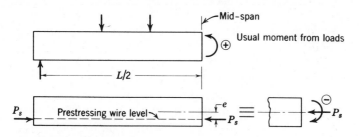

Figure 20.2 Beam prestress produces axial compression and a moment opposite in sign to dead load moment.

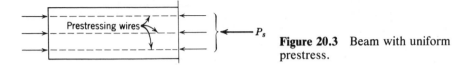

Figure 20.3 Beam with uniform prestress.

Their analysis is more involved because the prestress may modify the external reactions, shears, and moments.

For piles, plank, columns, and other members subject to moments both positive and negative or subject to loading on either face, longitudinal prestressing wires without any resultant eccentricity are used solely for their P/A effect, as in Fig. 20.3. Because eccentric prestressing wires or tendons in a simple span build up negative moment to offset the ordinary positive moment, such prestressing actually weakens the member if it is inverted or put under negative moment.

20.3 Objectives of Prestressing

Proper prestressing ordinarily prevents the formation of tension cracks under working loads. This reason alone is adequate for prestressing tanks to avoid leakage or for prestressing structural members subject to severe corrosion conditions.

With this lack of cracking, the entire cross section remains effective for stress; a smaller required section normally results. With lighter sections, precasting of longer members becomes possible. Precasting is a most important economic factor in promoting prestressed concrete in the United States.

Because prestress puts camber into a member under dead loading, deflections at working loads are much reduced.

The use of prestressing steels with ultimate tensile strengths in the range of 250 to 270 ksi often means that the tendons are at 180-ksi service load stress levels. Because the modulus of elasticity of prestressing tendons is on the same order of magnitude, although slightly less, as reinforcing bars, the strains in the tendons at such steel stresses are very high. If not prestressed, these strains lead to extremely wide cracking and substantial curvatures and deflections. By prestretching the steel by *prestressing*, much of the strain is thus "removed" from the structural member. The section receives not only the beneficial effects of the resultant precompression, but substantial improvement in cracking and deflection characteristics while utilizing extremely high grade steels.

Because tension cracking of prestressed concrete at working loads is avoided, very high strength steel (and smaller steel areas) can be used to provide the necessary tension. In some countries prestressed concrete is a device to make small supplies of steel serve more efficiently. In the United States prestress is a device that permits very high strength steel to be used economically and that increases permissible span lengths although keeping the concrete uncracked and hence more effective.

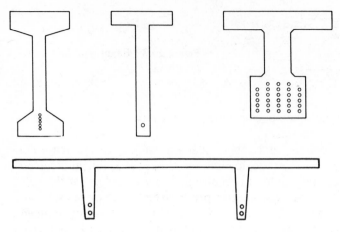

Figure 20.4 A few of the cross sections used for prestressed members.

Prestress reduces diagonal tension stresses at working loads. This result led to the general use of modified I and T shapes with less web area than conventional beams, as in Fig. 20.4.

Most of the advantages of prestressed concrete are construction or service load advantages. Under ultimate load conditions the flexural and shear strength of the usual (under-reinforced) prestressed beam is about the same as without prestress; its deflection increases sharply after cracking and stirrups are then necessary. The Code requires at least minimum stirrups in all prestressed beams.

20.4   Notation and Signs

The key to prestressed concrete analysis lies in a good bookkeeping system. All the usual calculations are individually simple, but they have a number of small pieces that must be assembled into various combinations. Good bookkeeping requires a consistent set of signs for stresses and an adequate notation. For the beginner, numerous sketches, such as those in Figs. 20.7, 20.8, and 20.9, are helpful.

In this chapter a tensile stress carries a plus sign and a compressive stress a negative sign. Although this convention is contrary to the signs used by many writers for prestressed concrete, it is in accord with general structural practice and is recommended by Lin and Burns.[1] No sign convention is needed for ordinary reinforced concrete as the kind of stress in such calculations is never in doubt; in prestressed concrete the concrete stress often changes from tension to compression (or the reverse) as the loading changes.

Much of the checking of a prestressed member is at service loads (unfactored loads) and thus requires the calculation of stresses; but ultimate strength also must be checked.

20.5 Stresses by Superposition of Stress Blocks

In the following sections, detailed calculations show "exact" methods for computation of both bonded and unbonded tendons. A simpler, slightly approximate approach in terms of gross area of concrete and the final prestress (effective prestress) is often useful to the designer; it is also easier to visualize for the first numerical example.

Consider the simple span rectangular beam of Fig. 20.5a with effective prestress P_{se} of 120 k from a straight tendon 8 in. below middepth. Establish the service load stresses existing prior to the application of live load.

Solution

The effect of the prestress is an axial load from the 120 k prestress plus a negative moment from the eccentricity = $P_{se}e = -120 \times 8 = -960$ k-in. The axial load produces the uniform stress block of Fig. 20.5b with $f_c = -120/(8 \times 24) = -0.625$ ksi compression. The moment gives the antisymmetric triangles of that same figure.

$$\text{Section modulus } S = bh^2/6 = 8 \times 24^2/6 = 768 \text{ in.}^{3*}$$
$$f_c = -960/768 = -1.25\text{-ksi compression on bottom}$$
$$\text{or } +1.25\text{-ksi tension on top}$$

At the end of the beam the dead load moment is zero, which leads to the combined stress block shown at the right of Fig. 20.5b. The 625-psi tension on the top is more than is permissible because it exceeds the probable cracking stress. If the tendon is draped upward at the end, reducing the end eccentricity and thus $P_{se}e$, this stress could be made acceptable.

At midspan the dead load moment of the beam weight of $8 \times 24 \times 150/144 = 200$ plf is $M_d = (\frac{1}{8})200 \times 40^2/1000 = +40$ k-ft, leading to stresses:

$$f_c = +40 \times 12/768 = +0.625\text{-ksi tension at bottom}$$
$$\text{or } -0.625\text{-ksi compression at top.}$$

The superimposed stress blocks are shown in Fig. 20.5c. The beam now has bottom compression all the way across the span and thus deflects upward under dead load and prestress. Also, the substantial compression stress at midspan indicates no tension will develop there until at least 400 plf of live load (twice the dead load) is added.

Had the prestress been added on the axis instead of at 8-in. eccentricity, the stresses in Fig. 20.5b at the support would be only the rectangular stress block of -0.625-ksi compression. (The cable also could be draped

---

* With symmetry S is more convenient than the use of I/c.

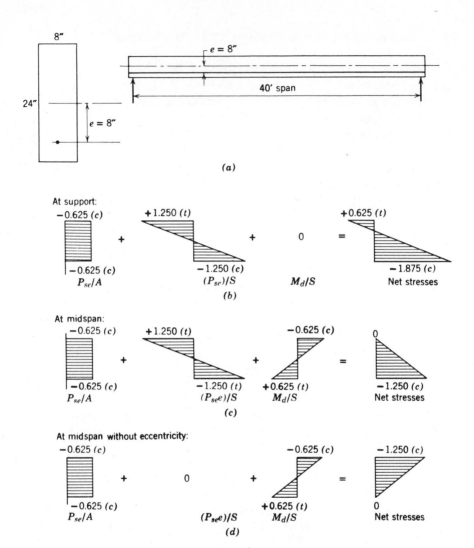

Figure 20.5 Influence of tendon eccentricity on service load stresses. (*a*) Beam of Sec. 20.5. (*b*) Stresses at support. (*c*) Stresses at midspan. (*d*) Stresses at midspan without tendon eccentricity.

to obtain the same effect.) At the midspan the stress blocks would then drop to those of Fig. 20.5*d*, with a zero stress at the bottom. Hence no live load could be added if one designed to the very strict requirement occasionally used with full prestressing of *no* tension under service load. Lowering the tendon at midspan to the 8-in. eccentricity increases the service load capacity by 400 plf in this case, and at the same time increases the ultimate moment capacity because of the greater effective depth of the steel.

The solution is not as complete as in those examples that follow, but it is often used because it is reasonably close. M_u still must be checked.

20.6   Pretension Method of Prestressing

Because prestressing calculations follow closely the physical process of pre-stressing, both pretensioning and post-tensioning procedures must be clearly understood. Sketches of the beam and stresses at each stage are helpful to the beginner in making summations. The individual calculations are quite simple.

In the pretension process the first step is to stretch high strength wires between abutments or end piers of a (long) prestressing bed, as indicated sche-matically in Fig. 20.6a. This total initial wire tension is designated P_i and the unit stress f_{pi}. The member forms are placed around the prestretched wire, with a number of units along the total length. The members are then cast and cured (usually by accelerated curing) in these forms. The wires are then cut between members (Fig. 20.6b).

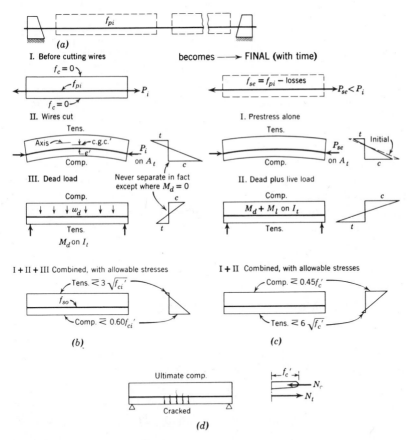

Figure 20.6   Pretensioned beam. (*a*) Prestressing wire stretched between casting yard abutments. (*b*) Initial stresses at midspan, separated for calculation purposes; also allowable concrete stresses. (*c*) Stresses at midspan under full service load, separated for calculation purposes; also allowable concrete stresses. (*d*) Ultimate strength conditions.

The resulting release of wire tension P_i is equivalent to P_i applied as an external compressive force on the entire transformed area (Sec. 3.21) of member A_t. The wires are usually eccentric, below the centroid of the transformed area of the member, and thus the external force P_i introduces a negative moment $P_i e$, that causes the beam to curve upward and pick up its dead load moment M_d. Because of P_i alone, rather large calculated tensile stresses develop on the top beam fibers, but this need not be dangerous because these stresses never exist without the counteracting compression from dead load moment.* This combined top tensile stress is often a critical design stress with a recommended limit of $3\sqrt{f'_{ci}}$ unless top steel is used; f'_{ci} is the cylinder strength at this early age. To turn such a beam on its side, or invert it, or lift it by a sling at the middle is unsafe and may cause failure. The bottom stress in compression also may be a controlling stress, limited to $0.60f'_{ci}$. Shrinkage is usually omitted from this calculation. If some shrinkage occurs by the time the wires are cut and these calculations apply, the shrinkage reduces both these critical stresses.

With time, shrinkage of concrete, creep of concrete, and relaxation of steel under stress reduce P_i to a lower effective prestress P_{se}. Formerly the reduction was often taken as 35,000 psi but it is now commonly more carefully appraised.[6-8] The loss tends to increase with small ratios of volume to surface area (thin members) and with exposure to lower humidities, but it varies considerably for different members in different regions and under different exposures. It increases if high initial stresses are used on the concrete. As a result of the lowered prestress the top tension and bottom compression are reduced below the initial values. It is upon these lowered stresses that live load stresses are superimposed, which results in the bottom sketch of Fig. 20.6c. The final result is compression critical at the top and tension at the bottom, with the allowable stresses for buildings noted in the combined load sketch in Fig. 20.6c.

At ultimate load, the beam is cracked as indicated in Fig. 20.6d and conditions do not differ greatly from those of an ordinary reinforced concrete beam, except for very high f_{ps}.

20.7 Post-Tension Method of Prestressing

In the post-tension method the plain concrete member is first cast with a tube or slot for future introduction of steel as indicated at I in Fig. 20.7a. A variation in procedure is to embed the steel in the member but separate it from the concrete by an enclosing metal or plastic tube or mastic coating—this prevents grouting and leaves the reinforcing permanently unbonded. (See last paragraph of this section.) After the concrete is cured, the steel tendons are stressed by jacking

---

* With straight wires and straight members, stresses are critical near the ends where M_d is small. Present practice often reduces the eccentricity near the ends to keep the center stresses critical (Figs. 20.7 and 20.12a–d).

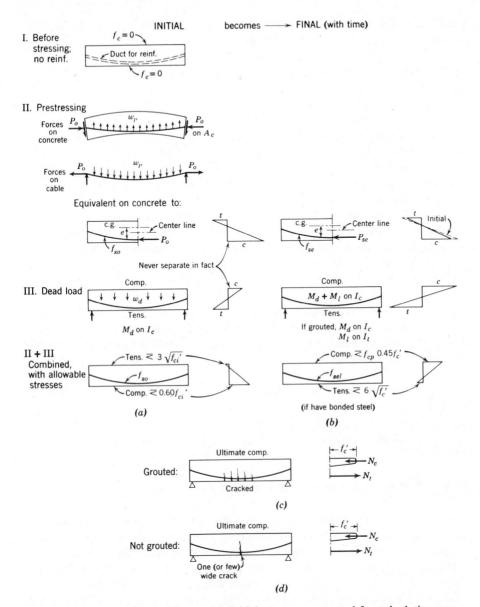

Figure 20.7 Post-tensioned beam. (*a*) Initial stresses separated for calculation purposes; also allowable concrete stresses. (*b*) Stresses under full service load separated for calculation purposes; also allowable concrete stresses. (*c*) Ultimate strength when grouted. (*d*) Ultimate strength when ungrouted is lower; hence, Code 18.9 generally requires added bonded reinforcement.

against the concrete and then are locked under stress by appropriate end anchorage or clamps. Because the concrete simultaneously shortens under compression, the jacking must account for both steel stretch and concrete shortening; this calculation is complicated by friction losses and by the fact that beam curvature under this stressing causes the dead load moment to become active (Fig. 20.7a).

The figure shows the resultant steel tension as P_o acting on the plain concrete section A_c* (compared to P_i and A_t for pretension). In the pretensioned case P_o also could have been calculated from the known value of P_i;† in the post-tensioned case there exists nothing comparable to P_i. The forces w_p between cable and concrete are discussed in Sec. 20.14. The allowable stresses are also marked at the bottom of Fig. 20.7a.

The calculations for full service load are diagrammed in Fig. 20.7b. These calculations should start with a loss of prestress slightly less than that for the pretension case; some shrinkage occurs prior to prestressing and all the elastic deformation of the concrete occurs during jacking to the f_{so} stress. This loss, at one time taken as 35,000 psi, is now calculated more carefully,[6-8] considering the volume-to-area ratio and exposure actually expected.

If the tendons are grouted effectively, as is often specified, the live load stresses can be based on the entire transformed section area. The change in steel stress, although larger than when ungrouted, is still unimportant. Ultimate strength of a grouted beam (Fig. 20.7c) is about the same as for a pretensioned beam.

Unless tendons are grouted into place, Code 18.9 now requires minimum bonded reinforcement in essentially all cases to avoid failures from a single flexural crack, as sketched in Fig. 20.7d. The bonded reinforcement then becomes part of the transformed area resisting the live load.

20.8 Other Procedures in Prestressed Concrete

The original sharp distinction between pretensioned and post-tensioned members is less sharp than might be implied in the previous discussion. Many variations on the original patterns are now common practice.

Precast systems may include either pretensioning or post-tensioning on a partial basis adequate for handling stresses with later additional post-tensioning used after members are in place to bring them to design capacity. This includes the possibility of adding such post-tensioning as is needed to assure continuity of adjacent members, the addition of inverted U strands over supports, and the like.

---

* The Code now says loss of area must be *considered*; it does not say it must be used.
† In the pretension case P_o on the net concrete area gives the same result as P_i on the transformed area, but P_i is the known starting force there.

Both grouted and ungrouted systems are in use with strong differences of opinion over the merits of ungrouted tendons. The wide crack formation at failure when unbonded, indicated in Fig. 20.7d, is avoided by casting in ordinary reinforcing bars of small diameter for crack control. Likewise the ultimate strength of either a pretensioned or post-tensioned member can be raised where necessary by casting in ordinary reinforcement as supplemental steel. Supplemental steel may run only a limited distance, wherever reserve moment strength or crack control is needed.

The possibility of aligning the post-tensioned tendons in any desired shape gives this construction an advantage in taking care of shear and varying moments, especially in cast-in-place construction. For precast simple spans, however, it is now feasible with pretensioning to depress the strands in a precise fashion to accomplish the same advantages.

20.9 Cracking Load and Ultimate Strength

Some European developers of prestressed concrete place considerable emphasis on the cracking load and require that no cracks be formed for a given (usually small) overload. American practice places greater emphasis on ultimate strength. Many European specifications now recognize the use of partially prestressed concrete with a mixture of prestressed and nonprestressed reinforcement.

Each of these methods is an attempt to provide an adequate factor of safety: the first indirect, the second direct, and the third looking at the full load range behavior. The need for both working stress and ultimate strength calculations lies in the radical change in beam behavior when cracks form. Prior to cracking the gross area of the beam is effective. As a crack develops, all the tension from the concrete must be picked up by the steel. If the percentage of steel is small, there may be very little added capacity between cracking and failure. Cracking is assumed to take place when the calculated tensile stress reaches $7.5\sqrt{f'_c}$.

20.10 Example of Prestress without Eccentricity

The hollow member of Fig. 20.8a is reinforced with the equivalent of four wires of 0.10 in.^{2} each, pretensioned to $f_{si} = 150$ ksi. If $f'_c = f'_{ci} = 5000$ psi, $n = 7$, determine the stresses when the wires are cut between members; also determine the moment that can be carried at a maximum tension of $6\sqrt{f'_c}$ and a maximum f_c of $0.45f'_c$. If 35 ksi of the prestress is lost (in addition to the elastic load deformation), determine this limiting moment. For this example, no ultimate strength check is included, although one would be required in practice.

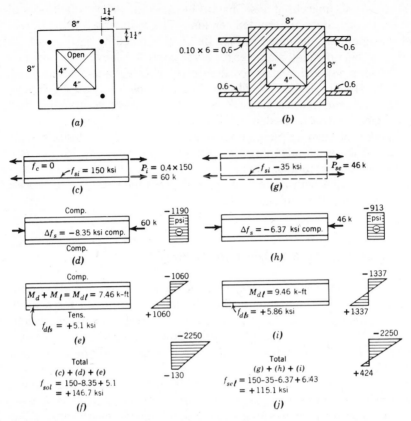

Figure 20.8 Prestress without eccentricity, pretensioned. (*a*) Beam cross section. (*b*) Effective transformed area. (*c*) Tendons under initial tension. (*d*) Effect of cutting tendons. (*e*) Permissible moment for initial loading. (*f*) Total stresses for initial loading. (*g*) Equivalent initial tension after losses. (*h*) Equivalent reduction owing to compression of concrete. (*i*) Permissible moment for later loading. (*j*) Total stresses for later loading.

Solution

Because the concrete is uncracked, the steel is transformed as $(n - 1)A_s$.

Transformed area (Fig. 20.8*b*) $= A_t = 64 - 16 + 4 \times 0.6 = 50.4 \text{ in.}^2$
$I_t = 8 \times 8^3/12 - 4 \times 4^3/12 + 4 \times 0.6 \times 2.75^2 = 342 - 21 + 18 = 339 \text{ in.}^4$

The starting stresses of Fig. 20.8*c* are $f_c = 0$ and $f_{si} = 150$ ksi. When the tendons are cut, the 60-k force acts on the transformed area (Fig. 20.8*d*) to add these stresses:

$$f_c = -P_i/A_t = -60 \times 1000/50.4 = -1190\text{-psi compression}$$
$$\Delta f_s = nf_c = -7 \times 1190 = -8350 \text{ psi} = -8.35\text{-ksi compression}$$

The total stresses before loading are thus:

$$f_c = 0 - 1190 = -1190\text{-psi compression}$$
$$f_{so} = 150 - 8.35 = +141.6\text{-ksi tension}$$

Before any prestress losses take place, $M_d + M_\ell$ can be enough to bring f_c at the top to $0.45 f_c' = 2250$-psi compression or to bring f_c at the bottom to $6\sqrt{f_c'} = 424$ psi tension. This leaves the available concrete stress increment owing to dead and live load moment, $M_d + M_\ell = M_{d\ell}$:

On top, $f_c = -2250 + 1190 = -1060$ compression
On bottom, $f_c = +424 + 1190 = +1614$ tension

The smaller value limits $M_{d\ell}$, as shown in Fig. 20.8e.

$$M_{d\ell} = 1060 \times I_t/y_t = 1060 \times 339/4 = 89{,}500 \text{ in.-lb} = 7.46 \text{ k-ft}$$

Under this loading:

Total concrete stress on top $= -1190 - 1060 = -2250$-psi compression
Total concrete stress on bottom $= -1190 + 1060 = -130$-psi compression
Total steel stress on lower layer of tendons $= +150 - 8.35 + n \times 1.060$
$\times 2.75/4 = 141.6 + 7 \times 1060 \times 2.75/4 = +146.7$-ksi tension

After 35 ksi of prestress is lost, P_i is, in effect, reduced to $(150 - 35)0.4 = 46.0$ k, as in Fig. 20.8g and 20.8h, giving uniform stresses:

$$f_c = -1190 \times 46.0/60 = -913\text{-psi compression}$$
$$f_{se} = +115 - 7 \times 0.91 = +108.7\text{-ksi tension}$$

$M_{d\ell}$ may safely change the top stress by $-2250 + 913 = -1337$ psi and the bottom stress by $+424 + 913 = +1337$ psi. It is a coincidence that the two stress increments are equal; the smaller controls.

$$M_{d\ell} = 1337 \times 339/4 = 113{,}500 \text{ in.-lb} = 9.46 \text{ k-ft}$$

The lower $M_{d\ell}$ value of 7.46 k-ft from initial loading conditions controls (Fig. 20.8e and Fig. 20.8f). The final stresses, if the higher $M_{d\ell}$ could be used, are:

Concrete on top $= -913 - 1337 = -2250$ psi $= 0.45 f_c'$
Concrete on bottom $= -913 + 1337 = +424$ psi $= 6\sqrt{f_c'}$
Steel stress on lower layer of tendons $= +108.7 + 7 \times 1.337 \times 2.75/4.0$
$$= +108.7 + 6.4$$
$$= 115.1\text{-ksi tension}$$

Ordinarily there is no need to calculate the steel stresses at service load levels. The values given indicate steel stresses are not higher than initial values.

The capacity cannot be accepted until the value of M_u also is established as shown in Sec. 20.13.

20.11 Example of Pretensioned Beam with Eccentricity

The pretensioned beam of Fig. 20.9 is investigated for $f'_c = 5000$ psi, $n = 7$, $f_{si} = 135$ ksi, and a creep and shrinkage loss of 35 ksi. $M_d = 65$ k-ft, $M_\ell = 80$ k-ft. The concrete strength when tendons are released is assumed to be $f'_{ci} = 4000$ psi, but for simplicity the same n is used. Assume that the beam axis (or tendon) is curved so that stresses under dead load and

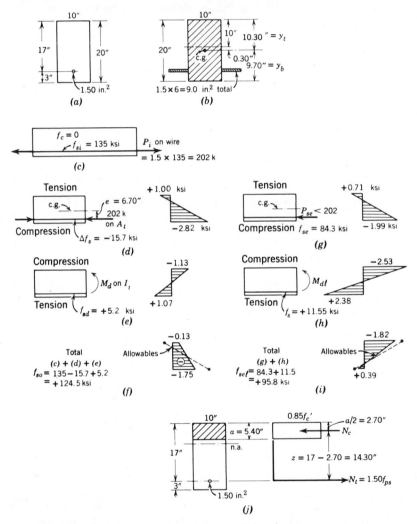

Figure 20.9 Pretensioned beam. (*a*) Cross section. (*b*) Transformed area, A_t. (*c*) Tendons under initial tension. (*d*) Effect of cutting tendons, eccentric load on A_t. (*e*) Dead load stresses. (*f*) Total initial stresses. (*g*) Effective pretension stresses are 84.3/119.3 times those in (*c*) and (*d*). (*h*) Dead load plus live load stresses. (*i*) Total stresses under full service load. (*j*) Ultimate strength calculation.

prestress do not require a check other than at midspan. Allowable concrete stresses are shown in Fig. 20.6b and 20.6c. The check on ultimate strength is deferred until Sec. 20.13.

Solution

Because the section is uncracked, the steel is transformed as $(n - 1)A_s$.

$$A_t = 10 \times 20 + 9.00 = 209 \text{ in.}^2$$
$$\bar{y} \text{ from middepth} = -9.0 \times 7/209 = -0.30 \text{ in. (below)}$$
$$I_t = 10 \times 20^3/12 + 200 \times 0.30^2 + 9.00 \times 6.70^2$$
$$= 6667 + 18 + 402 = 7087 \text{ in.}^4$$

Figure 20.9c shows forces before cutting tendons and Fig. 20.9d and 20.9e show the stress changes occurring after cutting tendons. Actually, these two sets of changes occur together and cannot be separated, except in the calculations.

From cutting the tendons:

f_c at top $= -P_i/A_t + P_i e y_t/I_t = -202/209 + (202 \times 6.70) \times 10.30/7087$
$$= -0.966 + 1.967 = +1.00\text{-ksi tension}$$
f_c at bottom $= -202/209 + (202 \times 6.70)(-9.70)/7087 = -0.966 - 1.852$
$$= -2.82\text{-ksi compression}$$
$\Delta f_s = nf_c$ at same level $= -7(0.966 + 1.852 \times 6.70/9.70)$
$$= -7(0.966 + 1.279) = -15.7\text{-ksi compression}$$

From dead load moment:

f_c at top $= M_d y_t/I_t = -(65 \times 12)10.30/7087 = -1.13\text{-ksi compression}$
f_c at bottom $= +(65 \times 12)9.70/7087 = +1.07\text{-ksi tension}$
$f_s = 7(+1.07 \times 6.70/9.70) = +5.2\text{-ksi tension}$

Total stresses without live load:

f_c at top $= 0 + 1.00 - 1.13 = -0.13\text{-ksi compression} < 3\sqrt{f_{ci}'}$ tension
f_c at bottom $= 0 - 2.82 + 1.07 = -1.75\text{-ksi compression} < 0.6 f_{ci}'$
$f_s = +135 - 15.7 + 5.2 = +124.5\text{-ksi tension}$

A prestress some 35% larger is possible because both stresses in Fig. 20.9f are low. (One objection to the rectangular beam shape is the large prestress required for full effectiveness.)

When f_s drops 35 ksi (owing to losses) from $135 - 15.7 = 119.3$ ksi to 84.3 ksi tension, as in Fig. 20.9g, the prestress effect, without M_d, drops to:

f_c at top $= +1.00 \times 84.3/119.3 = +0.71\text{-ksi tension}$
f_c at bottom $= -2.82 \times 84.3/119.3 = -1.99\text{-ksi compression}$

Total moment $= M_d + M_\ell = 65 + 80 = 145$ k-ft, giving stresses:

f_c at top $= -145 \times 12 \times 10.30/7087 = -2.53$-ksi compression

f_c at bottom $= +145 \times 12 \times 9.70/7087 = +2.38$-ksi tension

Total f_c at top $= +0.71 - 2.53 = -1.82$-ksi compression $< 0.45 f_c'$

Total f_c at bottom $= -1.99 + 2.38 = +0.39$-ksi tension $< 6\sqrt{f_c'} = 424$ psi.

Total $f_s = +84.3 + 7(2.38 \times 6.70/9.70) = 84.3 + 11.5 = 95.8$-ksi tension. The total moment could be increased about 1% or the prestress could be reduced nearly 1% and still keep the bottom tension within the allowable.

The check cannot be considered complete without an investigation of the ultimate moment capacity (Sec. 20.13).

20.12 Example of Post-Tensioned Beam with Eccentricity

The post-tensioned beam of Fig. 20.10a is investigated for $f_c' = f_{ci}' = 5000$ psi, $n = 7$, $f_{so} = 135$ ksi, which includes a correction for friction losses along the cable, 25-ksi loss in prestress with time, $M_d = 60$ k-ft, $M_\ell = 70$ k-ft. The cables are positioned along the span to make the midspan section the critical one. The cables are first considered ungrouted; then the effect of grouting is investigated. Allowable concrete stresses are shown in Fig. 20.7a and 20.7b. The check on ultimate strength is deferred until Sec. 20.13.

Solution

$$A_c = 10 \times 20 - 4 \times 2.5 = 190 \text{ in.}^2 \text{ (Fig. 20.10b)}$$

$\bar{y}$ from middepth $= (-4 \times 2.5) \times (-7.25)/190 = +0.38$ in. (above)

$I_c = 10 \times 20^3/12 + 10 \times 20 \times 0.38^2 - 4 \times 2.5^3/12 - 4 \times 2.5 \times 7.63^2$
$= 6110 \text{ in.}^4$

It is assumed that the cables are in a parabolic profile along the span and thus will "crowd up" to the top of the duct at midspan. Assume the centroid of A_s is at 16.75 in. from the top of the beam, even though the center of the duct is 17.25 in. from the top of the beam.

$$e = 16.75 - 10.0 + 0.38 = 7.13 \text{ in.}$$

Figure 20.10c and 20.10d shows prestress effects artificially separated from M_d stresses.

For prestress alone:
$$P_o = 1.5 \times 135 = 202 \text{ k}$$

At top, $\quad f_c = -P_o/A_c + P_o e y_t/I_c = -202/190 + 202 \times 7.13$
$$\times 9.62/6110 = -1.06 + 2.26 = +1.20\text{-ksi tension}$$

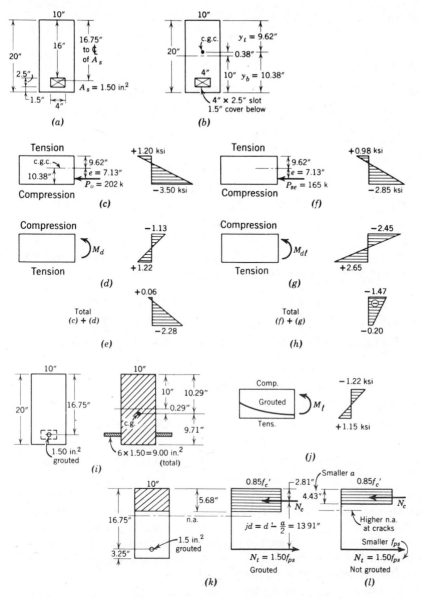

Figure 20.10 Post-tensioned beam. (*a*) Cross section. (*b*) Effective area, considered as plain concrete only, A_c. (*c*) Initial prestressing. (*d*) Dead load stresses. (*e*) Total initial stresses (*f*) Prestress effects after losses. (*g*) Dead plus live load service stresses (ungrouted). (*h*) Total service load stresses (ungrouted). (*i*) Effective section for live load when effectively grouted. (*j*) Service live load stresses, effectively grouted. (*k*) Ultimate strength calculation, grouted. (*l*) Effect of lack of grouting on ultimate strength conditions.

At bottom, $\qquad f_c = -202/190 - 202 \times 7.13 \times 10.38/6110$
$$= -1.06 - 2.44 = -3.50\text{-ksi compression}$$

For M_d alone:

At top, $\qquad f_c = -60 \times 12 \times 9.62/6110 = -1.13\text{-ksi compression}$
At bottom, $\qquad f_c = +60 \times 12 \times 10.38/6110 = +1.22\text{-ksi tension}$

Combined prestress and dead load (Fig. 20.10e):

At top, $\qquad f_c = +1.20 - 1.13 = +0.07 \text{ ksi} < 3\sqrt{f'_{ci}}$
$$= 0.212 \text{ ksi} \qquad \textbf{O.K.}^*$$
At bottom, $\qquad f_c = -3.50 + 1.22 = -2.28 \text{ ksi} < 0.60f_{ci}$
$$= 3.00 \text{ ksi} \qquad \textbf{O.K.}$$

With time, f_{so} drops to $f_{se} = 135 - 25 = 110$ ksi, and the effect of prestress drops in like ratio, as indicated in Fig. 20.10f.

At top, $\qquad f_c = +1.20 \times 110/135 = +0.98\text{-ksi tension}$
At bottom, $\qquad f_c = -3.50 \times 110/135 = -2.85\text{-ksi compression}$

For dead and live load, $M_{d\ell} = 60 + 70 = 130$ k-ft; these stresses are:

At top, $\qquad f_c = -130 \times 12 \times 9.62/6110$
$$= -2.45\text{-ksi compression}$$
At bottom, $\qquad f_c = +130 \times 12 \times 10.38/6110 = +2.65\text{-ksi tension}$
Total at top, $\qquad f_c = +0.98 - 2.45 = -1.47\text{-ksi compression}$
$$< 0.45f'_c = 2.25 \text{ ksi}$$
Total at bottom, $\quad f_c = -2.85 + 2.65 = -0.20\text{-ksi compression versus}$
$$\text{allowable tension} = 6\sqrt{f'_c}. \qquad \textbf{O.K.}$$

For an ungrouted beam the change in strain at the level of the tendons is the average strain over the entire length of the tendons. Thus, the changes in f_s owing to live load are small (significantly less than n times the concrete live load stress at that level). These changes in f_s are not usually calculated except at ultimate conditions.

Neglecting the complications that grouting the tendons in the tube after the post-tensioning to bond them may involve in the matter of loss of prestress, the only change in working stress calculations resulting from grouting is that the live load stresses will be based on the transformed area.

---

* If left ungrouted Code 18.9 requires supplementary reinforcing bars to control cracking, wherever calculated tension exceeds $2\sqrt{f'_c}$. Such bars modify the member properties used here.

If the beam is effectively grouted and neglecting the variation in n, the transformed area becomes that of Fig. 20.10i.

$$A_t = 20 \times 10 + 6 \times 1.50 = 209 \text{ in.}^2$$
$$\bar{y} \text{ from middepth} = -9.00 \times 6.75/209 = -0.29 \text{ in. (below)}$$
$$I_t = 10 \times 20^3/12 + 10 \times 20 \times 0.29^2 + 9.00 \times 6.46^2$$
$$= 6667 + 17 + 376 = 7060 \text{ in.}^4$$

At top, $f_\ell = -70 \times 12 \times 10.29/7060 = -1.22\text{-ksi compression}$

At bottom, $f_\ell = +70 \times 12 \times 9.71/7060 = +1.15\text{-ksi tension}$

Total on steel, $f_s = 110 + 7 \times 1.15(6.75 - 0.29)/9.71 = 110 + 5.4$
$$= +115.4 \text{ ksi}$$

Total at top, $f_c = +0.98 - 1.13 - 1.22 = -1.37 \text{ ksi compression}$
$$< 0.45f'_c \quad \textbf{O.K.}$$

Total at bottom, $f_c = -2.85 + 1.22 + 1.15 = -0.48 \text{ ksi compression}$
$$< 6\sqrt{f'_c} \text{ tension}$$

These totals put together stresses from Fig. 20.10f, 20.10d, and 20.10j.

The differences as a result of grouting are small at working stress, about 0.10 ksi less compression on the top and 0.28 ksi more compression on the bottom. The ultimate moment is more noticeably improved by grouting, as noted in Sec. 20.13.

20.13   Ultimate Strength of Prestressed Beams

Because the high strength steels lack a sharp and distinct yield point (Fig. 20.11) and because stressing history and degree of bonding can influence the

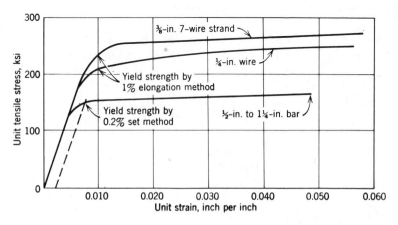

Figure 20.11 Typical stress-strain curve for prestressing steels. (From Lin.[2])

steel stress developed at failure, it is impossible to use ultimate strength relations developed for ordinary beams without some concern or caution. Theoretical considerations call for a calculation process correlating strains on steel and concrete at failure.

Nevertheless, for beams with bonded tendons and a percentage of steel low enough to ensure a tension failure, the strength method of analysis (Sec. 3.10) gives approximate values that agree fairly well with tests. Because there is no sharp yield point, the Code, following a report by the ACI-ASCE Joint Committee on prestressed concrete, uses steel stress f_{ps} at nominal strength in this calculation rather than f_y. Fundamentally, the determination of f_{ps} depends on stress-strain diagrams and strain relationships, which means that f_{ps} is not a quantity easily established without detailed data. This situation is eased by three approximate formulas for f_{ps} for use in typical cases:

(1) For bonded members,

$$f_{ps} = f_{pu} \left(1 - \frac{\gamma_p}{\beta_1} \left[\rho_p \frac{f_{pu}}{f_c'} + \frac{d}{d_p} (\omega - \omega') \right] \right)$$

where f_{pu} = specified tensile strength of tendons, γ_p is a factor that depends on the type of stress-strain curve of the tendon, β_1 is the familiar relation between the depth of the rectangular stress block and the neutral axis as previously defined in Sec. 3.4, ρ_p is the ratio of prestressed reinforcement A_{ps}/bd_p, d is the centroidal distance for any nonprestressed tension reinforcement, d_p is the centroidal distance for any prestressed tension reinforcement, ω is the nonprestressed tension reinforcement index $\rho f_y/f_c'$, and ω' is the nonprestressed compression reinforcement index $\rho' f_y/f_c'$. The values for γ_p are 0.55 for f_{py}/f_{pu} not less than 0.80 (usually high strength bars), 0.40 for f_{py}/f_{pu} not less than 0.85 (usually ordinary stress-relieved prestressing strand) and 0.28 for f_{py}/f_{pu} not less than 0.90 (usually low relaxation strand).

If compression reinforcement is considered in using this equation then the term within the square brackets will have a minimum value of 0.17 and d' may not exceed $0.15d_p$.

This expression was new in the 1983 Code and is considerably more complex than previous formulations. However, it extended those limited formulas to include consideration of nonprestressed reinforcement, low relaxation tendons, and higher strength concretes.

(2) For unbonded members with a span-to-depth ratio of 35 or less, $f_{ps} = f_{se} + 10,000 + f_c'/(100\rho_p)$ but is not more than the specified yield strength f_{py} or $f_{se} + 60,000$ psi.

(3) For unbonded members with a span-to-depth-ratio greater than 35, $f_{ps} = f_{se} + 10,000 + f_c'/(300\rho_p)$ but is not more than the specified yield strength f_{py}, or $f_{se} + 30,000$ psi.

The ultimate moment capacity for prestressed beams is found by the same applications of force and moment equilibrium as developed for nonprestressed beams and the same critical strain is assumed for the concrete crushing. The nominal moment capacity for rectangular beams or for flanged beams with the neutral axis within the flange with only prestressed reinforcement is expressed by the relation

$$M_n = A_{ps} f_{ps} d (1 - 0.59 \rho_p f_{ps}/f_c')$$

which is the relation given in Sec. 3.9 for ordinary reinforced concrete except that f_{ps}, A_{ps}, and ρ_p take the place of f_y, A_s, and ρ.

To avoid approaching the over-reinforced beam condition, prestressing steel for rectangular sections must not exceed that giving $\rho_p f_{ps}/f_c' = 0.36\beta_1$. For larger steel ratios the value of M_n is limited more by compression conditions; the Commentary suggests $M_n = \phi(0.36\beta_1 - 0.08\beta_1^2) f_c' b d^2$ is reasonable.

As an example of approximate ultimate strength analysis, the pretensioned beam of Sec. 20.11 and Fig. 20.9 is investigated assuming ordinary stress relieved strand with $f_{pu} = 195^*$ ksi and with f_{py} equal to $0.85 f_{pu}$. The failure conditions of Fig. 20.9j give:

$$\rho_p = 1.5/(10 \times 17) = 0.00883$$
$$\gamma_p = 0.40, \qquad \beta_1 = 0.80 \text{ (for } f_c' = 5000 \text{ psi)}$$
$$f_{ps} = f_{pu} \left(1 - \frac{\gamma_p}{\beta_1} \left[\rho_p \frac{f_{pu}}{f_c'} + \frac{d}{d_p} (\omega - \omega') \right] \right)$$
$$= 195(1 - (0.40/0.80)[0.00883 \times 195/5 + 0])$$
$$= 195(1 - 0.5 \times 0.00883 \times 195/5) = 162 \text{ ksi}$$
$$\rho_p f_{ps}/f_c' = 0.00883 \times 162/5 = 0.28 < 0.36\beta_1 = 0.288.$$
$$\text{Usual } M_u \text{ eq. } \textbf{O.K.}$$
$$a = A_{ps} f_{ps}/(0.85 f_c' b) = 1.5 \times 162/(0.85 \times 5 \times 10) = 5.70 \text{ in.}$$
$$M_n = A_{ps} f_{ps} (d - a/2) = 1.5 \times 1.62(17 - 5.70/2) = 3440 \text{ k-in.}$$
$$M_u = \phi M_n = 0.9 \times 3440 = 3100 \text{ k-in.} = 258 \text{ k-ft}$$

Required $M_u = 1.4 M_d + 1.7 M_\ell = 1.4 \times 65 + 1.7 \times 80 = 227$ k-ft < 258 **O.K.**

This comparison shows about 14% surplus strength compared to 1% at service load.

For the post-tensioned beam of Sec. 20.12 and Fig. 20.10a with ordinary stress-relieved tendons with $f_{pu} = 195$ ksi and the steel well grouted, the failure conditions of Fig. 20.10k give:

$$\rho_p = 1.5/(10 \times 16.75) = 0.00895$$
$$\gamma_p = 0.40, \qquad \beta_1 = 0.80 \text{ (for } f_c' = 5000 \text{ psi)}$$
$$f_{ps} = f_{pu} \left(1 - \frac{\gamma_p}{\beta_1} \left[\rho_p \frac{f_{pu}}{f_c'} + \frac{d}{d_p} (\omega - \omega') \right] \right)$$
$$= 195(1 - (0.40/0.80)[0.00895 \times 195/5 + 0])$$
$$= 195(1 - 0.5 \times 0.00895 \times 195/5) = 161 \text{ ksi}$$
$$\rho_p f_{ps}/f_c' = 0.00895 \times 161/5 = 0.288 = 0.36\beta_1 = 0.288.$$

---

* This ultimate and the initial wire stress of $f_{si} = 135$ ksi in Sec. 20.11 are both low for the currently available strand (having f_{pu} of 270 ksi) or the wire (having f_{pu} of 240 psi) with yields about 80% of ultimate.

Usual M_u eq. **O.K.**

$$a = 1.5 \times 161/(0.85 \times 5 \times 10) = 5.68 \text{ in.}$$

$$M_n = 1.50 \times 161(16.75 - 0.5 \times 5.68) = 3360 \text{ k-in.} = 280 \text{ k-ft}$$

$$M_u = \phi M_n = 0.9 \times 280 = 252 \text{ k-ft}$$

Required $M_u = 1.4 \times 60 + 1.7 \times 70 = 203 \text{ k-ft} < 252 \text{ k-ft}$

The section provides about 25% more capacity than required compared to 37% under the service load analysis. The ultimate strength analysis is the more restrictive in this example.

If the beam of Sec. 20.12 is left ungrouted, it would be contrary to Code 18.9 and member strength would be lowered, because the entire stretch of the tendon tends to accumulate at the first flexural crack. Some tests by Baker indicated cracks 3 to 10 times as wide as for bonded conditions. Such cracks raise the neutral axis and bring premature compression crushing.

For the beam just investigated, if grouting is omitted (Fig. 20.10*l*) and the span-to-depth ratio is assumed less than 35, the calculations become:

$$\rho_p = 1.50/(10 \times 16.75) = 0.00895 \approx 0.0090$$

$$f_{ps} = f_{se} + 10{,}000 + f'_c/(100\rho_p), \text{ in psi}$$

$$f_{ps} = (135 - 25) + 10.0 + 5.0/(100 \times 0.0090) = 125.5 \text{ ksi}$$

$$\rho_p f_{ps}/f'_c = 0.0090 \times 125.5/5 = 0.224 < 0.36\beta_1 = 0.288.$$

Usual M_u eq. **O.K.**

$$a = 125.5 \times 1.5/(0.85 \times 5 \times 10) = 4.43 \text{ in.}$$

$$M_n = 125.5 \times 1.5(16.75 - 4.43/2) = 2730 \text{ k-in.}$$

$$M_u = \phi M_n = 0.9 \times 2730 = 2460 \text{ k-in.} = 205 \text{ k-ft}$$

Required $M_u = 1.4 \times 60 + 1.7 \times 70 = 203 \text{ k-ft} < 205 \text{ k-ft}$

The beam is just adequate at ultimate when ungrouted, which can be compared with the 25% excess strength when grouted. The Code permits the unbonded tendons only when some bonded reinforcement is cast into the tension area; the yield strength of these bars is automatically available for increasing the available M_u and controls the cracking.

If calculations for ungrouted sections are based on correlating strains, as suggested in the second paragraph of this section, it is not difficult to introduce a factor to represent relative crack width. The effective steel strain (beyond initial prestress) can be simply taken as about $\frac{1}{5}$ of the strain indicated by the strain triangles. This reduced steel strain, when added to the initial prestress strain, gives the total to be used with the stress-strain curve.

20.14 Tendon Effect Considered as a Negative Loading

Because simple beam tendon eccentricity develops a negative moment that offsets much or all the dead load moment, this eccentricity must be reduced where dead load moment is small; otherwise a negative moment failure can occur as prestressing is applied. With post-tensioned steel the tendons can be formed to any desired shape. With pretensioned wires or strands practice often uses intermediate hold-down devices with the tendons located near the centroids where not held down near the ends. This practice gives the effect of bent-up steel. Other pretensioned beams are made with a curved soffit, the curved beam axis thus producing a varying eccentricity.

A complete analysis can be made by considering all critical sections with their respective eccentricities. The relieving moment is (almost) proportional to the eccentricity, because the slope angles are small and the horizontal component of the prestress is essentially a constant. Hence the relieving moment diagram has the same shape as the curve of plotted e values.

A more convenient way to visualize the effect of curvature is in terms of a negative or upward load on the beam. A curved tendon under heavy tension requires vertical load to deflect it from a nearly horizontal line. A concentrated load at the center will deflect the tendon as in Fig. 20.12a, loads at both third points as in Fig. 20.12b, and a uniform load as in Fig. 20.12c, a parabolic curve.

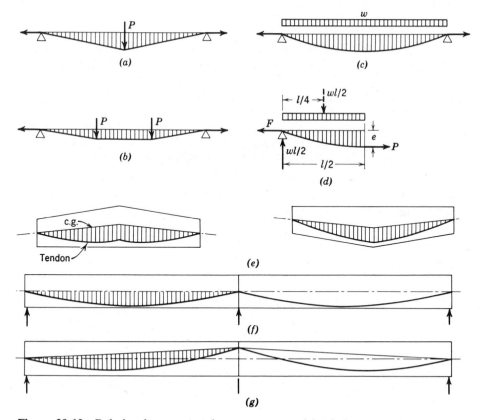

Figure 20.12 Relation between tendon curvature and load. Prestressed tendon exerts similar upward forces on (plain) concrete. (c) through (g) give uniform upward loads.

When embedded in a concrete beam in a duct with one of these shapes, the tendon after prestressing exerts upward forces on the concrete opposite to those required to deform the free tension tendon into the same shape.

If the parabolic tendon shape is used, the free body of Fig. 20.12d indicates from ΣM that $Pe = (w_p\ell/2)(\ell/4) = w_p\ell^2/8$, or $w_p = 8Pe/\ell^2$. Hence the tendon can be visualized as contributing an axial load and an upward uniform load w_p, both acting on the plain concrete section. The tendon is shaped to offset, nearly completely, any dead load distribution that exists.

If the member is inclined instead of horizontal, the eccentricity is from the gravity axis of the member, not from a horizontal reference line. Likewise, if the member is of variable depth, the eccentricity is from the centroidal axis. The eccentricity then may be defined as the vertical distance between tendon and centroidal axis and Fig. 20.12e indicates the tendon shape that corresponds to the uniform upward load of Fig. 20.12c when the member is of varying depth. The shaded ordinates in (e) and (c) must be identical.

20.15   Load Balancing

Lin[2] expanded the negative loading concept to a design technique that balances with prestress any desired portion of the load, such as most or all of the dead load, or possibly under other conditions, dead load plus some portion of the live load.

The balanced load could ideally* provide no moment and no curvature or deflection. Hence, for continuous beams or slabs the unbalanced load alone may be used with ordinary elastic theory to compute the bending moments acting on the (nearly) uniform concrete section. This in the case of ungrouted tendons is approximate in neglecting concrete displaced for ducts or tendons. Thus a continuous member might call for tendons, as in Fig. 20.12f or, what can be shown to be equivalent, the so-called "linearly transformed" tendon profile of Fig. 20.12g. The bend over the support obviously cannot be as sharp as shown and separate tendons for each span, possibly with inverted U strands over the reaction, or other modification may be needed. This calculation is beyond the scope of this text.

20.16   Shear and Diagonal Tension

At the failure state in shear and diagonal tension, with or without torsion, the capacity and cracking pattern of a prestressed concrete member is very much the same as a nonprestressed reinforced concrete member. Comprehensive studies[9-13] indicate that the same truss models that can be used for determining

---

* Obviously changes in prestress from time effects of creep and relaxation modify the effective negative loading and upset the initial assumptions slightly.

the shear capacity of reinforced concrete members may be used for prestressed concrete members. There are two major behavioral changes in the prestressed concrete members that are reflected in the truss model applications. The use of draped or inclined tendons provides an uplift that offsets some of the shear forces. The precompression caused by the prestressing provides very favorable compression stresses that delay the formation of inclined tension cracking. Current ACI Code provisions recognize both of these factors, although the latter effect is given more apparent importance than it might deserve because this increase in cracking load is not fully reflected at ultimate. A tendon with varying eccentricity acts somewhat as a suspension cable, partially relieving the concrete not only of bending stress but of shear stress. The shear carried by the tendon can be calculated either as the vertical component of the tendon pull or as the shear created by the equivalent upward loads discussed in Sec. 20.14. The remainder of the external load creates a shear that must be resisted by the concrete.

The axial compression from prestress reduces the diagonal tension stresses as long as the beam is uncracked from moment stress (Sec. 5.17). Cracking reduces this beneficial effect and inelastic deformations at loads near the ultimate largely eliminate it.

Just as in reinforced concrete members, the 45° truss model is relied on for the assessment of the basic contribution of the concrete, longitudinal reinforcement, and web reinforcement V_s. A series of fairly complex expressions give the supplementary concrete contribution V_c. The basic shear expressions in the Code $V_u \lessgtr \phi V_n$ and $V_n = V_c + V_s$ apply to both reinforced and prestressed concrete.

For a member with an effective prestress of at least 40% of the tensile strength of its flexural reinforcement, the nominal concrete contribution to the shear resistance can be taken as

$$V_c = (0.6\sqrt{f_c'} + 700V_u d/M_u)b_w d* \qquad \text{(Code Eq. 11-10)}$$

unless more detailed analyses (such as follows) are made. This V_c need not be taken lower than $2\sqrt{f_c'}b_w d$, and $5\sqrt{f_c'}b_w d$ is the upper limit. An exception is made in pretensioned members where the transfer length of the prestressing tendons exceeds $h/2$, which calls for calculations using a reduced prestress force in the calculation for V_{cw} from Code Eq. 11-13 and then using this as the maximum limit.

For the more detailed methods the Code recognizes two types of shear failures.

The first failure type, web-shear cracking, starts from the web crack (principal diagonal tension stress of approximately $4\sqrt{f_c'}$) that develops into a failure without the presence of a moment crack at that section. This critical shear,

---

* The SI version of the ACI Code (318M-86) for V_c and f_c' in MPa shows the constants of 0.05 replacing 0.6, and 5 replacing 700; also for the two limits, 0.17 for the lower and 0.4 for the higher limit.

called V_{cw}, is more apt to develop in thin web members such as I or T shapes. The equation from Code 11.4.2.2 is

$$V_{cw} = (3.5\sqrt{f_c'} + 0.3f_{pc})b_w d + V_p^* \quad \text{(Code Eq. 11-13)}$$

where $3.5\sqrt{f_c'}$ represents the shear strength corresponding to a principal tensile stress of about $4\sqrt{f_c'}$ at the neutral axis and V_p is the component of prestress acting perpendicular to the axis of the member. Each of the last two terms increases the permissible external shear load—the longitudinal prestress by balancing out some of the potential diagonal tension and the normal component of prestress by reducing the vertical shear acting on the concrete.

An alternate calculation of V_{cw} is permitted as "the shear force corresponding to dead load plus live load that results in a principal stress of $4\sqrt{f_c'}$ at centroidal axis of member, or at intersection of flange and web" for a centroidal axis in flange.

The second failure type, related to ordinary reinforced concrete beam shear failures, is called flexure-shear cracking. A near vertical crack initiated by flexure becomes inclined as load increases (as in Fig. 5.4) and, unless it encounters a stirrup, goes on to a shear failure. For this case the Code 11.4.2.1 allows

$$V_{ci} = 0.6\sqrt{f_c'}b_w d + V_d + V_i M_{cr}/M_{max}\dagger \quad \text{(Code Eq. 11-11)}$$

but

$$V_{ci} \nleq 1.7\sqrt{f_c'}b_w d\dagger$$

where V_d is the unfactored dead load shear at the section and M_{max} and V_i are computed from the load combination creating maximum moment‡ causing flexural cracking at the section and

$$M_{cr} = (I/y_t)(6\sqrt{f_c'} + f_{pe} - f_d) \quad \text{(Code Eq. 11-12)}$$

In these equations M_{cr} is the moment from externally applied loads causing flexural cracking at the section.

I = moment of inertia of the section

y_t = distance from centroid of gross section (neglecting reinforcement) to tension face of member

f_{pe} = compressive stress in concrete caused by effective prestress (after losses) at extreme fiber where external loads cause tension

f_d = unfactored dead load stress at the same extreme fiber as f_{pe}.

---

* The SI version for V_c, $\sqrt{f_c'}$, and f_c' in MPa uses a coefficient of 0.3 to replace the 3.5 in the first term; the second coefficient is unchanged; V in newtons; b_w, d in mm.
† The constant 0.6 in the SI version of the Code becomes 0.05 with MPa and the 1.7 becomes 0.14. That Code also sets an upper limit of $0.4\sqrt{f_c'}$ on v_c. V is in N, and b_w, d in mm.
‡ For heavy moving loads, maximum shear for stirrup design must control, instead of having the load produce maximum moment.

The smaller value, either V_{ci} or V_{cw}, limits the shear assigned to the concrete, although stirrups may be added to provide a V_s capacity that can greatly increase ultimate shear capacity.

Minimum stirrups of $A_b = 50b_w s/f_y$* are required wherever V_u exceeds half ϕV_c, assuming torsion is small. For effective prestress force not less than 40% of the tensile strength of flexural reinforcement, minimum shear reinforcement may be either that or

$$\text{Min. } A_v = \frac{A_{ps} f_{pu}}{80 f_y} \frac{s}{d} \sqrt{\frac{d}{b_w}}$$

where

A_{ps} = area of prestressed reinforcement in tension zone.

d = depth from compression face to centroid of longitudinal reinforcement $\not< 0.8h$.

f_{pu} = ultimate strength of prestressing tendon.

f_y = specified yield strength of nonprestressed reinforcement.

This last equation is unchanged in SI units. The maximum stirrup spacing is limited to $0.75h$ but not more than 24 in.

When prestressed tendons are stopped (or left unbonded and unstressed) for a part of the member length, a problem in shear strength results that is very similar to that of bars cut off within a moment region of reinforced concrete. Until more is known about this phenomenon, excess stirrups should be used near these "cutoff" points.

20.17 Development of Reinforcement

In pretensioned beams the self-anchorage of wires and strands at the ends of the beams leads to high bond stresses. This end anchorage requirement received much attention; Lin and Burns' treatment[1] of the subject is quite complete.

For three- or seven-wire prestressing strand the Code (20.9.1) specifies a development length

$$\ell_d = (f_{ps} - 2f_{se}/3)d_b$$

where d_b is the nominal diameter in inches and the calculated stress f_{ps} and f_{se} (effective prestress after losses) are in ksi; they are used here simply as constants. The specified length must be available between the end of the strand and the closest point where full strength is required, which should govern only in cantilever and short span members. The adequacy of this ℓ_d value is now seriously questioned.

---

* The SI Code with A_v in mm² shows the constant as ⅓.

Where bonding does not extend to the end of the member and where service load tensile stresses are permitted, the development length must be doubled.

In post-tensioned beams there is no problem of development because end anchorages are used. Even when flexural cracking occurs, the stresses adjacent to cracks when tendons are grouted are less important than in ordinary reinforced concrete where they are not separately computed. However, the state of stresses in the immediate region of the anchorages is very complex. Substantial bearing, bursting, and spalling stresses can be developed. The treatment of these stresses is beyond the scope of this text. Recent test and analysis results as well as an overview of the state of the art of design procedures has been published by Stone and Breen.[14,15]

20.18　Deflections and Deformities

Deformations of prestressed concrete members are extremely important. A basic understanding of the load-deformation relationship is needed to comprehend the basic mechanics of prestressed concrete behavior and to visualize the various stages of the construction and service loading process that must be checked. In addition, because prestressed concrete members tend to be very efficient assemblages of high strength materials, they often are very slender. Thus, deflection limits may govern design far more than usual with ordinary reinforced concrete members. In this section a brief introduction is given for both topics. Substantial detail is included in References 1,3,4,5,7,8,16, and 17.

The basic response to load of a prestressed concrete beam is similar to that of a reinforced concrete beam, but with a few important differences. A typical moment-curvature relationship for an under-reinforced prestressed concrete beam is shown in Fig. 20.13. In contrast to nonprestressed beams in which zero curvature corresponds to zero load, in the typical prestressed concrete beam

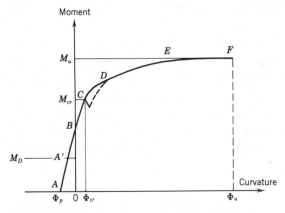

Figure 20.13　Moment-curvature relationship for a prestressed concrete beam. (After Reference 16.)

the eccentric prestressing force produces an internal moment that results in a negative curvature at zero load. Point A represents this prestress-induced curvature with no external loading. Actually, as the prestressing is applied, the member begins to camber upward. As it begins to leave the casting bed soffet or the falsework that supports the concrete before stressing, the dead load of the girder applies an external moment M_D and the actual girder probably comes to point A' rather than point A. As further load and moment is applied, the curvature increases in a positive sense. Because the typical prestressed concrete girder resists cracking for a wide load range owing to the favorable precompression, the moment-curvature relationship is very linear over a substantial range. The response of the uncracked section is easily calculated by ordinary elastic procedures.

Point B represents a level of moment that results in zero curvature and hence a uniform strain (and uniform stress) profile. This level of loading represents the "load balancing" level. The beam can carry this level of moment with no downward deflection. Depending on the design criteria, this loading may correspond to the dead load plus a substantial fraction of the service live load. As the moment is increased beyond B, the curvature increases. The strain on the lower fiber changes from precompression to zero and then to tension. When the tensile capacity of the concrete is exceeded, tensile cracks form and the beam momentarily loses load capacity, as shown by the dashed line at C. The crack will stabilize under increased load. Under the application of a moment corresponding to D, the girder will deform as shown by the heavy curve. Slightly above the cracking moment, the uncracked transformed section properties lose validity. Deformation calculations must consider the strain compatibility relations of the cracked sections and the actual stress-strain characteristics of the materials.

Somewhere between D and E the prestressing tendon passes its elastic limit and rapidly increasing curvatures are noted as the tendon effectively yields. The final crushing of the compression zone concrete results in the failure of the member at F.

In contrast to the clear-cut indications of reinforcement yielding found in most reinforced concrete beams, the much more rounded nature of typical prestressing steel stress-strain relations (shown in Fig. 20.11) makes it very difficult to determine the onset of yielding in a prestressed beam. In addition, the level at which cracking occurs, M_{cr} in Fig. 20.13, is very sensitive to the actual prestress force. Hence the determination of the true prestress losses in a member becomes very important in predicting the overall moment-curvature and load-deflection relationship. Figure 20.13 assumes a known prestress force. Should there be excessive prestress losses the entire curve would shift to the right and the change from a linear relationship to a curvelinear relationship takes place at a considerably lower value of M_{cr} than that shown. With conservative design criteria, members may remain uncracked through their service life. As more reliance is placed on tensile stress in the concrete at service levels, like the $6\sqrt{f'_c}$ allowable tension now widely used, the chance of the member being cracked owing to modest overload or to higher levels of losses increases.

Such cracked members will have greatly reduced moments of inertia and hence larger deflections. The addition of a reasonable proportion of nonprestressed reinforcement in the precompressed tensile zone makes sense not only because of its important role in distributing and controlling cracking but because it acts like compression reinforcement and reduces the creep losses in the precompressed zone.

The tendon stress-load relationships in typical prestressed beams further explain the behavior shown in Fig. 20.13. Such stress-load histories for members subjected to increasing load are shown in Fig. 20.14. Initial stressing is represented by the paths ABC or $A'B'C'$. In a beam with substantial dead load where the camber caused by tendon eccentricity is less than the dead load deflection, the nominal tendon stress caused by initial prestress, f_0, is slightly increased as the tendon stress is applied and the beam sags somewhat as the supports are removed. In contrast, in a very light beam that often tends to spring up from its supports as the prestress is applied to the concrete section, the girder picks up dead load faster. Because the net deflection is upward, the tendons shorten somewhat and the prestress after tensioning is somewhat below f_0.

Time effects, such as creep and shrinkage of the concrete and relaxation of the steel, cause prestress losses, say, to level D. Then as live loads are applied modest changes in steel stress occur until the cracking load levels are reached. Because unbonded tendons average out the strains over the entire distance between anchorages, these members have lower changes in steel stress. Upon cracking, the tensile force contributed by the tensile concrete zone must be taken up by the tendons, which results in a sharp change in steel stress (F_1F_1' or FF'). Increasing load brings the tendons toward yield and then final ultimate as the compression zone is developed. The tendon stress in the unbonded tendon is substantially below the local stress developed at the failure section in the

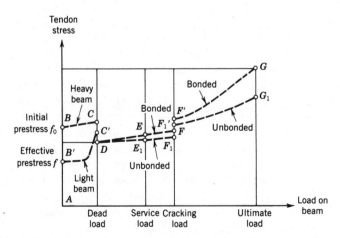

Figure 20.14 Tendon stress-load relationship for a prestressed concrete beam. (After Reference 16.)

bonded tendon because of the averaging of the tendon strains over the entire length between anchors in an unbonded member.

It is not fully appreciated how sensitive the behavior represented in Fig. 20.13 and Fig. 20.14 is to the actual prestress force. If the fabrication has faulty control or application procedures or if the actual losses are substantially above the predicted losses, the full service load level might fall above the decreased cracking load. In this case, not only will deflections be much larger than normally computed, but changes in tendon stress owing to fluctuating loads will be much larger. Fatigue problems in applications with moving loads may be the result.

The actual calculation of deflections of a prestressed concrete beam involve a series of relatively simple computations that must consider the actual construction history of the member.[17] Code 9.5.4 requires deflection computations for all prestressed concrete members and requires that the live load deflections not exceed the limits in Code Table 9.5b. Immediate deflections can be based on elastic analysis calculations using gross concrete section moments of inertia. However, time effects must be included in the deflection calculations. This addition complicates the computations, particularly for the widely used composite prestressed concrete girder cast-in-place concrete slab type of construction. As shown in Fig. 20.15, the effect of initial stressing OA is to create an upward camber, Δ_0. Δ_0 is computed as the sum of the net effects of the downward deflection owing to dead load and the upward deflection owing to eccentricity of the initial prestress force. The sum Δ_0 is the initial camber (in this case upward deflection). From t_0 to t_1, the deflection changes owing to the combined effects of creep, shrinkage, and tendon relaxation. The sustained load effects tend to "soften" or reduce the effective concrete modulus which tends to

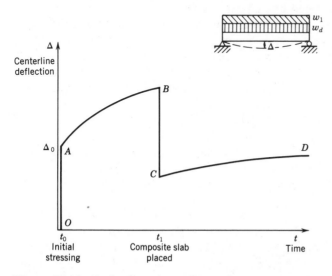

Figure 20.15 Deflection time history for a compositely constructed prestressed concrete beam.

substantially increase the deflection. Because the initial net deflection is upward, this effect increases the upward deflection. At the same time, the effects of concrete shrinkage and of relaxation in the tendon act to reduce the effective prestress force and the upward component of deflection caused by the tendon profile. This effect can offset much or all of the creep effect. Careful computations using good estimates of losses are required to adequately assess these effects.

Deflection computations are further complicated in many practical cases where composite construction is employed. Assuming that a composite slab is placed on the girder at time t_1, the downward component of that load must be computed (BC). The time effects on the composite section (CD) must be included in the calculations. It is necessary to consider the actual sequence of construction and whether the girder was shored or unshored when the composite slab (w_1) was placed. If shored, the full composite section can resist the weight of the slab when the shores are removed. If unshored, the weight of the slab must be taken by the original girder section. For this condition the deflection is much greater.

It is important that the designer understand the effect of the construction process on deformations. In those cases where checks show that the deformations will be near the limits, the designer may have to either specify allowable construction practices or require the constructor to submit for approval detailed deflection computations for the construction procedure selected.

20.19   Design Considerations

Design considerations are treated *very* briefly. Allowable working load stresses have already been mentioned in analysis examples.

Because the optimum total prestress approximates half the allowable f_c multiplied by the total concrete area, reducing the concrete area near the centroid, where it is less effective in resisting moment, and increasing the area near the extreme fibers results in substantial savings. Hence I, T, double T, and $\sqcap$ shapes are the general types that are cost-effective. Tables of section modulus values for such shapes are helpful in picking a size.

It is sometimes stated that the prestress carries the service dead load and that the size of the beam depends only on the live load. The negative moment $P_s e$ at transfer can be made to offset the dead load moment. Of course, as P_s decreases in time to P_{se}, part of this offsetting moment is lost. Furthermore, at ultimate strength the effect of prestress is largely lost and the beam must carry all the moment, essentially as an ordinary beam.

Referring to Fig. 20.6b, the top stress under prestress and the dead load may be a tension as large as $3\sqrt{f_{ci}}$. Under the full load (Fig. 20.6c) this stress may be the full allowable compression $0.45f_c'$. Thus, except for the effect of prestress losses, the full range of $3\sqrt{f_c'} + 0.45f_c'$ is available to take care of M_ℓ. Likewise, on the bottom the stress range for M_ℓ is from $0.60f_{ci}$ to $6\sqrt{f_{ci}}$ except for the effect of prestress losses. Magnel[4] worked out design requirements for a section modulus established from M_ℓ alone divided by the major part of this stress range. This approach is most suitable when M_ℓ is large compared to M_d. It indicates the saving sometimes possible in beam size compared to an ordinary beam which must be sized for $M_d + M_\ell$.

Lin developed[1,2] design procedures for elastic designs based on visualizing the internal resisting couple. These procedures are very straightforward in application. He also devised a "load balancing" method for slabs and beams, already mentioned in Sec. 20.15.

20.20   Continuous Beams

Continuous beams cannot be considered here except to mention one special condition that often controls. Because an eccentrically placed prestressing tendon itself develops bending moment and curvature in a beam, the prestressing operation may well require reactions (up or down) to hold the member in contact with its supports. These reactions change the prestressing moment from the simple $P_s e$ value to something more involved by the development of "secondary moments." Cables can be arranged to cause zero external reactions, but this is an unnecessary restriction. Lin[1,2] gives a complete discussion of this problem following basic methods originally suggested by Guyon[3] and simplified it considerably in connection with his "load balancing" method.

Selected References

1. T. Y. Lin and Burns, N. H. *Design of Prestressed Concrete Structures,* 3rd ed., John Wiley and Sons, New York, 1981.

2. T. Y. Lin, *Design of Prestressed Concrete Structures,* 2nd ed., John Wiley and Sons, New York, 1963.

3. Y. Guyon, *Prestressed Concrete,* John Wiley and Sons, New York, 1953.

4. G. Magnel, *Prestressed Concrete,* Concrete Publications, Ltd., London, 2nd ed., 1950.

5. J. R. Libby, *Modern Prestressed Concrete,* Van Nostrand, Princeton, New Jersey, 1971.

6. *Standard Specifications for Highway Bridges,* AASHTO, Washington, 12th ed., 1977.

7. PCI Comm. on Prestress Losses, "Recommendations for Estimating Prestress Losses," *PCI Journal,* V. 20, No. 4, July–Aug. 1975, p. 44.

8. P. Zia, H. K. Preston, N. L. Scott, and E. B. Workman, "Estimating Prestress Losses," *Concrete Intl.: Design and Construction,* Vol. 1, No. 6, June 1979, pp. 32–38.

9. B. Thurlimann, "Plastic Analysis of Reinforced Concrete Beams," *Bericht 86, Institut fur Baustatik und Konstruktion ETH Zurich,* Nov. 1978, 90 pp.

10. M. P. Collins, and D. Mitchell, "Shear and Torsion Design of Prestressed and

Non-Prestressed Concrete Beams," *PCI Journal,* Vol. 25, No. 5, Sept.–Oct. 1980, pp. 32–100.

11. J. A. Ramirez, "Reevaluation of AASHTO Design Procedures for Shear and Torsion in Reinforced and Prestressed Concrete Beams," Ph.D. dissertation, The Univ. of Texas at Austin, Dec. 1983.

12. P. Marti, "Strength and Deformations of Reinforced Concrete Members under Torsion and Combined Actions," *Shear, Torsion and Punching,* CEB Bulletin No. 146, Jan. 1982.

13. P. Muller, "Plastic Analysis of Torsion and Shear in Reinforced Concrete," IABSE Colloquium "Plasticity in Reinforced Concrete," Copenhagen 1979, Final Report, IABSE, Zürich, 1979, pp. 103–110.

14. W. C. Stone, and J. E. Breen, "Behavior of Post-Tensioned Girder Anchorage Zones," *PCI Jour.,* Vol. 29, No. 1, Jan.–Feb. 1984, pp. 64–109.

15. W. C. Stone, and J. E. Breen, "Design of Post-Tensioned Girder Anchorage Zones," *PCI Jour.,* Vol. 29, No. 2, Mar.–Apr. 1984, pp. 28–61.

16. T. Y. Lin, and P. Zia, "Strength and Deformation of Prestressed Concrete Elements," *Reinforced Concrete Engineering,* Vol. 1, John Wiley and Sons, New York, 1974, pp. 302–406.

17. ACI Committee 435, "Deflections of Prestressed Concrete Members," (ACI 435.1 R-63) Reaffirmed 1979, *ACI Jour., Proc.,* 60, No. 12, Dec. 1963, pp. 1697–1728.

Problems

PROB. 20.1. The rectangular beam of Fig. 20.16 has pretensioned steel at two levels as shown with $f_{pi} = 170$ ksi. Consider $f'_{ci} = f'_c = 5000$ psi, $n = 7$, and a 40-ft simple span. Calculate concrete stresses at top and bottom and both steel stresses when prestressed tendons are first cut between units.

PROB. 20.2. The post-tensioned beam of Fig. 20.17 has a 40-ft simple span. It has two holes as shown for tendons with a total area of 2.10 in.², which after tensioning are 16.5 in. below the top of beam. $f'_c = 5000$ psi, $f_{so} = 175$ ksi. Considering the holes ungrouted and without using the approximation of area as the gross area of the concrete, calculate:

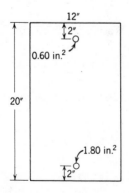

Figure 20.16 Beam for Prob. 20.1.

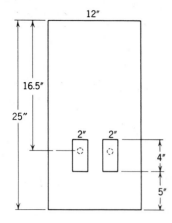

Figure 20.17 Beam for Prob. 20.2 and Prob. 20.3.

(*a*) Concrete stresses initially under dead weight alone after post-tensioning.
(*b*) Concrete stresses after shrinkage and creep reduce the steel stress to $f_{se} = 155$ ksi and a live load of 625 plf acts on the beam.

PROB. 20.3.

(*a*) Recalculate condition (*b*) of Prob. 20.2 if each beam hole is well grouted.
(*b*) Calculate the ultimate moment capacity of this beam and its overall factor of safety assuming it is well grouted and $f_{pu} = 250$ ksi.

PROB. 20.4.

(*a*) If the pretensioned beam of Fig. 20.9 is changed to $f'_c = 5000$ psi, $f_{si} = 150$ ksi, $n = 7$, $A_s = 1.10$ in.^{2} with M_d still 65 k-ft, calculate all stresses after the tendons are cut.
(*b*) If the tendons are straight and concrete stresses are limited to a (bottom) compression of 3000 psi and a top tension of $3\sqrt{f'_c}$, what is the maximum tendon eccentricity that can be used? [The student should note that M_d at the support is zero. The change in c.g. and I from that used in part (*a*) may be ignored here and in (*c*), as a problem simplification.]
(*c*) If the tendons are draped such that the condition at the end of the span does not control, and if f_{so} decreases to f_{se} of 120 ksi, what service live load moment is permissible with the allowable f_c of $0.45f'_c$ and allowable tension of $6\sqrt{f'_c}$? (Ultimate load check is also important, but not a part of this problem.)

21

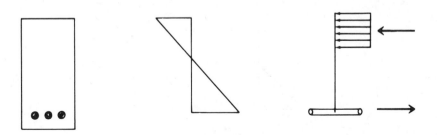

COMPOSITE BEAMS

21.1 The Nature of Composite Beams

The combination of a steel beam with a concrete slab, or the combination of a precast concrete beam with a cast-in-place concrete slab, when the two are so connected that they act as a single unit in resisting flexure is called a composite beam. Because every reinforced concrete slab or beam is a composite member of steel and concrete, the basic theory for flexure requires almost no new concepts. Likewise the bonding together of the two units is theoretically a simple problem of horizontal shear. A roughened concrete surface has a considerable bonding strength if the slab is held tightly in place by stirrups extending into the slab.[1] Shear lugs can be welded to steel beams to perform the same function.

Composite construction is economically important because the steel beams or precast concrete beams* furnish supports for the slab forms and the dead weight of the concrete while, at the same time, the beams can be made lighter because the slab forms a stiff attached flange that helps to resist live loads. The primary bending moment in a slab calls for slab steel transverse to the beam and hence neither this slab steel nor slab bending seriously complicates the analysis of the composite beam.

The engineering problems hinge around three considerations: flexural strength, horizontal shear strength, and deflection. The latter may call for shoring under the beams during construction because this makes the entire section available for resisting dead load as well as live load. The ultimate strength is not significantly influenced by shoring, shrinkage, or creep.

---

* Many prestensioned prestressed concrete beams are cast with stirrups extending out the top to bond into a cast-in-place slab to form a composite T-beam.

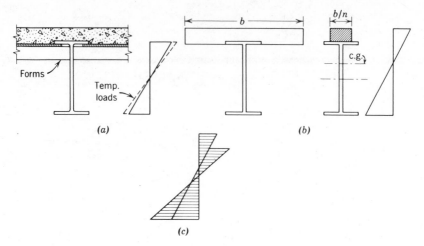

Figure 21.1 Stresses in an unshored composite beam. (*a*) Dead load alone. (*b*) Live load alone. (*c*) Total stresses.

21.2 Flexural Strength with Slab Cast on Steel Beam

Flexural strength can be considered at three different stages of loading: (1) under dead load prior to the time when the slab is effective for strength, (2) under live load with the slab acting effectively, and (3) at ultimate under over-load conditions.

Consider first a symmetrical steel beam with the cast-in-place slab. If there is no shoring the beam acts to carry its own weight, the slab forms, the slab concrete, and the entire superimposed construction load. As the concrete sets, the steel beam carries Mc/I stresses for the steady portion of this unfactored load, a very simple calculation, as indicated in Fig. 21.1*a*. After the concrete gains strength, the section becomes the transformed area of Fig. 21.1*b* for any further load; it is simplest to transform the concrete into equivalent steel by the relation A_c/n or b/n. This relation results in the addition of stresses as shown in Fig. 21.1*c*.

For long-time loads some multiple of n may be used to represent the effects of creep and shrinkage, up to $3n$.

Because neither creep nor shrinkage influences ultimate strength significantly, the ultimate design condition is discussed first.* The neutral axis in deep beams usually falls below the slab leading to the stress distribution shown in Fig. 21.2*a*. Viest, Fountain, and Singleton[2] point out that this is equivalent to Fig. 21.2*b*, which is considerably simpler to use.

---

* Although the non-LFRD AISC specification is written in terms of service loads, the specification is an adaptation from behavior at ultimate.[4]

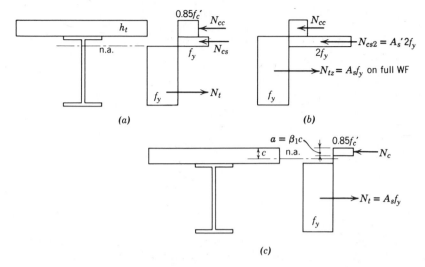

Figure 21.2 Ultimate stress conditions. (*a*) Neutral axis below slab. (*b*) Equivalent to (*a*) using N_t on full A_s. (*c*) Neutral axis within slab.

The same authors point out, especially in buildings, that the neutral axis may fall in the slab, as in Fig. 21.2*c*, in which case the tension on the concrete below the neutral axis is ignored and the concrete compression above the neutral axis is handled as in any T-beam.

When the neutral axis appears close to the level of the top flange the lower level of strains nearby can be considered, at least roughly, as in the following design example.

The later discussion on deflection points out that beams are frequently shored enough to avoid deflection and stresses of consequence from the dead weight of the slab until after the concrete slab is strong enough to act as a flange. In this case the dead load stresses of Fig. 21.1*a* are small, usually only from the weight of the beam itself, and the composite beam stresses of Fig. 21.1*b* become the major part. This shoring does not influence the ultimate moment capacity significantly.

21.3 Example of Flexural Calculations

For a 26-ft simple span check the flexural capacity of a $f_y = 36$ ksi steel* W18 × 50 on 7-ft center with a 4-in. slab to carry a service load of its own weight and a live load of 440 psf. Assume the beam is shored at the quarter points so as not to pick up the weight of the concrete until it

---

reaches its intended f_c' of 3000 psi. The beam properties include $A = 14.7$ in.^{2}, $I = 802$ in.^{4}, $b_f = 7.50$ in., and a flange thickness $= 0.57$ in. (The beam in such a case is usually designed by a service load procedure covered in the AISC Specifications.)

Solution

A composite beam based on a steel shape is usually designed under the AISC Specification,[3] which is really an ultimate design restated in terms of service load stresses. The effective flange width used is that recommended by AISC, that is, the smallest of:

1. Spacing of beams $= 84$ in.
2. Flange width of steel section plus 8 times slab thickness as the overhang on each side $= b_f + 16h_f = 7.50 + 64 = 71.5$ in.
3. Beam span/4 $= 6.5$ ft $= 78$ in.

The 71.5-in. width controls.

This section relates to flexural strength only. The necessary shear connectors are covered in Sec. 21.6 and deflections are discussed in Sec. 21.5.

Here the given section is first checked by applying ACI Code strength concepts to the analysis of Fig. 21.2. It is then rechecked by the AISC procedures.

Whether the beam is or is not initially shored, the ultimate condition includes all loads on the full span with the steel shape initially considered yielding throughout its depth, or nearly so. This assumption must be checked using strain-compatibility relations. This assumed total tension fixes the depth of the compression block as in a homogeneous section.

$$N_{nc} = N_{nt} = 14.7 \times 36 = 529 \text{ k}$$
$$a = N_{nc}/(0.85f_c'b) = 529/(0.85 \times 3 \times 71.5) = 2.90 \text{ in.}$$
$$c = a/0.85 = 2.90/0.85 = 3.42 \text{ in.}$$

The neutral axis is in the concrete, but the sketch in Fig. 21.3a indicates it is too low to put the upper steel flange at the full f_y of 36 ksi. Because N_{nt} is less, try $c = 3.20$ in. and check whether the tension justifies this value. As shown in Fig. 21.3b the top of the steel beam does not yield so that the tension is reduced by roughly $0.5(36 - 22)(7.50 \times 0.57) = 30$ k

$$N_{nt} = 529 - 30 = 499 \text{ k}, \quad a = (499/529)2.90 = 2.74 \text{ in.}$$
$$c = (499/529)3.42 = 3.23 \text{ in.} \quad \textbf{O.K.}$$

The centroid or point of application of the tensile force in the steel section is located with respect to the bottom fiber of the section as

Area	Average Stress (ksi)	Force (kips)	Arm (in.)	Moment About Bottom Fiber (in. kips)
14.7	36	+529	9	+4760
$-0.57 \times 7.5 = 4.28$	$\frac{1}{2}(36 - 22) = 7$	−30	$18-\frac{1}{3}(0.57) = 17.81$	−534
		499		4226

$$\bar{y} = \frac{4226}{499} = 8.47 \text{ in.}$$

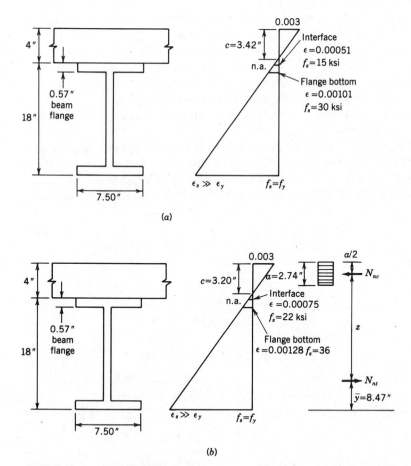

(a)

(b)

Figure 21.3 Ultimate analysis of Sec. 21.3. (a) First trial n.a. (b) Final n.a.

Thus the lever arm for the compression concrete-tension steel couple is
$z = 18$ in. $+ 4$ in. $- 1.37$ in. $- 8.47$ in. $= 12.16$ in.

$$M_n = (499)(12.16) = 6068 \text{ k-in.} = 506 \text{ k-ft}$$

Wt. beam $=$ 50 plf

Wt. slab $= 7 \times 50 =$ $\underline{350}$

                        $400 \times 1.4 = $ 560 plf

LL. $= 7 \times 400 = 2800 \times 1.7 = \underline{4760}$ plf

                             Total $= 5320$ plf

$$M_u = (1/8)5.32 \times 26^2 \ = 450 \text{ k-ft}$$
$$M_n = M_u/\phi = 450/0.9 = 500 \text{ k-ft} < 506 \text{ k-ft} \quad \textbf{O.K.}$$

The beam is safe at ultimate by 1%.

The beam is now checked for strength under the service load procedure of the AISC Specification,[3] with allowable $f_s = 24$ ksi and allowable $f_c = 0.45f'_c = 1350$ psi. Although using the service load format, this calculation derives its validity directly from ultimate strength calculations. Shrinkage and creep strains do not influence ultimate strength significantly and hence are not introduced in these strength calculations for either shored or unshored beams. (These strains influence service load deflections; see Sec. 21.5.)

Because this is a shored* beam, the total moment applies directly to the composite section. For 3000-psi concrete, $n = 9$, which permits the flange to be considered transformed into steel of width $71.5/9 = 7.94$ in. as in Fig. 21.4. Moments about the bottom of the flange establish the neutral axis:

$$(7.94 \times 4)2 - 14.7 \times 9 = +63.5 - 132.3$$
$$= y(31.8 + 14.7)$$
$$\bar{y} = -68.8/46.5 = -1.48 \text{ in.}$$

I of slab $= (1/12)7.94 \times 4^3 = $ 42

             $+31.8(2 + 1.48)^2 = $ 385

I of $W = $ $= $ 802

             $+14.7(9 - 1.48)^2 = $ $\underline{833}$

                         $I = 2062 \text{ in.}^4$

$$M = (1/8)(400 + 2800)26^2/1000 = 270 \text{ k-ft}$$
$$f_s = Mc_s/I = (270 \times 12)(18 - 1.48)/2062 = 25.9 \text{ ksi} > 24 \text{ ksi}$$

---

* If the beam was unshored initially, the stresses would have to be combined from two calculations: first, those for the total dead load on the steel shape alone and, second, the live load on the total transformed area.

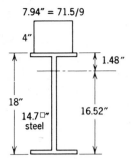

Figure 21.4 Transformed area for analysis in Sec. 21.3.

The section is overstressed by about 8% in tension. Compression is usually satisfactory:

$$f_c = (Mc_s/I)/n = (270 \times 12)(4 + 1.48)/(2062 \times 9)$$
$$= 0.958 \text{ ksi} < 1350 \text{ psi} \quad \textbf{O.K.}$$

Attention is called to the AISC limitation on unshored beams expressed by limiting the section modulus S_{tr} (referred for a positive moment to the bottom flange stress, or for a negative moment to each flange) of the transformed composite section to

$$S_{tr} \gtrless (1.35 + 0.35 M_L/M_D)S_s$$

where M_L and M_D are live and dead load moments and S_s is the section modulus of the steel section alone. In this example:

$$\text{Actual } S_{tr} = 2062/(18 - 1.48) = 125 \text{ in.}^3$$
$$\text{Limiting } S_{tr} = (1.35 + 0.35 M_L/M_D)S_s = (1.35 + 0.35$$
$$\times 2800/400)(802/9) = 339 \text{ in.}^3$$

Because $125 \ll 339$, this loading does not require shored beams. This provision is used by AISC to guard against excessive tensile stress at service loads, stresses not fully reflected in the nominal stress calculation.

The AISC has many suggestions and rules[4] to cover composite construction. We here illustrate the calculations for a simple case.

In selection of steel shapes for composite slab construction a useful rule-of-thumb is that effective section modulus of the steel shape is increased about 50% by the composite slab. Thus, the required section modulus of such a steel beam is $\frac{2}{3}$ that otherwise required.

21.4 Flexural Strength with Slab Cast on Precast Concrete Beam

When the basic element is a precast beam, either with or without prestressing, the details of the calculations change, but not the general ideas. For example, without prestressing, the precast beam by itself must carry its own weight and this is a larger item than with a steel beam. If it is partially shored such that the slab weight is carried by shores until the slab reaches its design strength (say, $0.75f'_c$), the analysis proceeds as discussed before with the transformed area expressed as equivalent concrete instead of steel. In this case the concrete must be considered cracked wherever tension strains exist, but this is not a serious problem to handle.

The chief difference between using reinforced concrete and steel as the basic beam lies in the horizontal shear provisions discussed in Sec. 21.6.

21.5 Deflections

For the uniform load and steel section of Sec. 21.3, deflections are calculated at service loads by $(5/384)w\ell^4/EI$ using the transformed area (in steel) for the moment of inertia calculation. In this case, however, the long-term dead load deflection must be separately computed with n assigned at least twice the nominal value. The live load deflection is based on the I with the usual value of n. A suggested starting point in design is a depth-to-span ratio of $1/22$ for $f_y = 36$ ksi and $1/16$ for $f_y = 50$ ksi, but deflection still must be calculated.

For composite construction involving precast reinforced concrete beams, the Code (9.5.5.1) provides that a composite member with shored construction may be considered a cast-in-place member with the minimum thickness requirements of Code Table 9.5a* controlling unless deflections are computed. In unshored members the precast element existing before the slab is cast is the one to meet this thickness requirement; the composite member with slab is then considered satisfactory† without a deflection check. The precast portion may need investigation for long-time deflection occurring prior to the time the slab becomes effective.

The requirements of Code Table 9.5a may be waived if deflections are calculated and are less than the values in Code Table 9.5b. Such calculations must involve differential shrinkage between precast and cast-in-place parts of the member. Otherwise, the requirements and methods discussed in Chapter 3 are appropriate.

Composite beams with prestressed, precast sections are quite common, but are not covered here. The Code requires that deflections *always* be computed for these members. The 1986 Code Commentary suggested the calculation method of References 5, 6, and 7 as a suitable one for prestressed concrete girders.

---

* See Table 3.2 in Sec. 3.9.
† This Table 9.5a does not apply when members are attached to or support partitions likely to be damaged by large deflections. For a short summary of Code Table 9.5b requirements governing in this case, see footnote near start of Sec. 3.22c.

21.6 Horizontal Shear

(a) General With steel beams all the vertical shear should be carried by the web of the beam. With reinforced concrete precast elements the vertical shear is carried by the precast elements or by the entire composite member as in monolithic construction.

The special problem of composite beams is the horizontal shear between flange and beam below, whether it is a steel beam or a precast concrete beam. Steel beams can be used with either full or partial shear transfer; the reader is referred to the Steel Manual[3] for the partial shear transfer case. Full shear transfer is required by the ACI Code for reinforced concrete members.

Theoretically the horizontal shear varies along the beam length as the external shear varies and could be calculated on a $v = VQ/Ib$ basis. However, both in steel and concrete composite beam work it is recognized that other distributions of the horizontal shear resistance also serve adequately,[9] except in the case of large concentrated loads in a positive moment region near a point of inflection or near the end of a simple span.*

(b) Steel-Concrete Composite Beams Although spirals or short transverse lengths of small steel channels welded to the flange make adequate shear connectors, the stud connector is most commonly used. The stud is round, automatically welded to the flange, and is topped with a hook or flange making a head that can engage the concrete slab and prevent any vertical separation that may tend to occur between slab and beam. The stud placement must permit at least 1 in. of lateral concrete cover. The concrete cover over the head should meet the ¾-in. slab reinforcement cover of ACI-318.

The AISC Manual[4] tabulates the shear load permitted on each stud or channel connector, for example, 11.5 k for a ¾-in. × 3-in. stud in 3000-psi concrete made from ordinary aggregate. This is a service load equivalent, because AISC uses a service load computation (derived from ultimate strength values). The total (service load) horizontal shear between points of maximum and zero moment is taken as 0.5 of the ultimate strength of the concrete flange in compression, or 0.5 of the tensile yield strength of the entire steel section, whichever is smaller, that is, the smaller of

$$0.85 f'_c b h_f /2 \quad \text{or} \quad A_s f_y /2$$

The shear is considered fully transferred if the connectors used have a total shear capacity of this amount. The connectors may be spaced uniformly from maximum moment point to zero moment point.†

---

* The partial shear transfer with the steel beam is an extreme case of such reassignment of shears.
† With some further check needed if a large moment closer to the zero moment point can occur of if concentrated loads are present.

(c) Reinforced Concrete Composite Beams   The designer must assure full shear transfer at contact surfaces of the composite elements. In previous ACI Building Codes, the designer was allowed to assume full shear transfer if a series of design and construction conditions were met. However, in the 1980 Interim Revisions to the ACI Code, this procedure was changed following reports of severe distress in a member that satisfied these earlier requirements but had high horizontal shearing stresses. Code Sections 17.5.2 and 17.5.3 now require that in all cases the horizontal design shear strength V_{nh} must be computed and satisfied. Unless a more detailed calculation is made of the horizontal shear force on the contact surface, the horizontal shear V_h may be approximated by the normal shear on the section V_u. Thus

$$V_u \lessgtr \phi V_{nh}$$

where ϕ is the capacity reduction factor for shear ($\phi = 0.85$) and V_{nh} is the nominal horizontal shear strength. Alternatively, the compression (or tension) in any segment may be computed, divided by $\phi = 0.85$, and the resulting force considered the total horizontal shear to be transferred. The first procedure requires that V_{nh} varies directly as the external shear and the second procedure allows more averaging of the horizontal shear depending on the length of segment assumed. In the latter case special consideration appears appropriate where large concentrated loads fall close to the end of a simple span or close to the point of inflection in a positive moment region of a continuous member.

   For clean surfaces the resisting V_{nh} may be evaluated on the basis of the provision of minimum ties, or of an intensionally roughened surface, or both. The nominal V_{nh} may be taken as $80 b_v d$ (in pounds) for *either* minimum ties or intentionally roughened surface *alone,* or $350 b_v d$ (in pounds) for *both together*. b_v is the width of the cross section at the surface on which horizontal shear is checked and d is the member depth. These shears act over the contact area between slab and precast element where flexural strains might otherwise cause horizontal slip to occur. When the alternative procedure based on computing the change in the longitudinal force in any segment is utilized, the same limits apply but with the area of the contact surface A_c substituted for $b_v d$. An upper limit for the horizontal shear transfer is set as $350 \phi b_v d$ or $350 \phi A_c$. In any cases where the interface shear transfer exceeds this value, the concepts of shear friction discussed in Sec. 5.19 must be used.

21.7   Continuous Beams—AASHTO Specification

When beams are continuous over a support, the steel in the slab parallel to the beam is considered a portion of the transformed area for negative moment, with the slab concrete neglected because it is in tension. In such a case, the requirements for connectors are computed separately for positive and negative moment regions. The AASHTO Specification[10] is an ultimate strength one as far as connectors are concerned. Hence the connectors in the negative moment region are required to resist horizontal shear equal to the area of the longitudinal slab steel (within the effective width of the slab) times its yield stress. The AASHTO Specification limits the *total* effective slab width to $12h_f$, more restrictive than the AISC value of $16h_f$ added to the steel flange width.

Because there is some objection to welding connectors to the tension flange in the region of maximum tension[8], the AASHTO Specification provides that the pitch may be modified to avoid connectors at locations of high tension in the flanges. It allows closer spacings to be used away from the maximum negative moment point (closer to the point of inflection) to maintain the total number required and avoids the highest flange stresses. Alternatively, the total negative moment may be assigned to the steel beam alone, but in such a case added connectors are required at the dead load point of inflection (within a length equal to $\frac{1}{3}$ of the effective flange width, on either side of or centered on the P.I.), based on a fatigue requirement discussed in the next paragraph.

The basic design for connectors in the AASHTO Specification is based on fatigue,[8] and the ultimate strength design is used as a check rather than as the chief consideration. The equation used (in AASHTO notation) is:

$$S_r = V_r Q / I$$

where

S_r = range in horizontal shear stress at the connection, in kips per in.

V_r = range in shear owing to live load plus impact in kips, that is, the difference between maximum and minimum shear envelopes, excluding dead load.

Q, I = statical moment of the area above the junction and moment of inertia of the total transformed area, both about the neutral axis.

The reader is referred to that specification for the allowable shear on a connector, which is written in terms of length of channel or square of the diameter of stud multiplied in each case by factors that vary substantially with the design number of cycles. AASHTO requires 2 in. of concrete over the connectors and requires that they penetrate at least 2 in. into the slab.

Selected References

1. ACI Committee 333, "Tentative Recommendations for Design of Composite Beams and Girders for Buildings," *ACI Jour.*, 32, No. 6, Dec. 1960; *Proc. 57*, p. 609.

2. I. M. Viest, R. S. Fountain, and R. C. Singleton, *Composite Construction in Steel and Concrete*, McGraw-Hill Book Co., New York, 1958.

3. "Specification for the Design, Fabrication and Erection of Structural Steel for Buildings," American Institute of Steel Construction, New York, 1978.

4. *Manual of Steel Construction*, American Institute of Steel Construction, New York, 8th ed., 1980.

5. ACI Committee 435 Subcommittee 5, "Deflections of Prestressed Concrete Members," *ACI Jour.*, 60, *No. 12*, Dec. 1963, p. 1697.

6. ACI Committee 209 Subcommittee 2, "Prediction of Creep, Shrinkage, and Temperature Effects in Concrete Structures," *Designing for the Effects of Creep, Shrinkage, and Temperature in Concrete Structures,* SP-27, American Concrete Institute, Detroit, 1971.

7. D. E. Branson, B. L. Meyers, and K. M. Kripanarayanan, "Time-Dependent Deformation of Noncomposite and Composite Prestressed Concrete Structures," Symposium on Concrete Deformation, *Highway Research Record 324,* Highway Research Board, Washington, D.C., 1970, p. 15.

8. R. G. Slutter and J. W. Fisher, "Fatigue Strength of Shear Connectors," *Highway Research Record 147,* Highway Research Board, Washington, D.C., 1966, p. 65.

9. Task Committee on Composite Construction, Committee on Metals, "Composite Steel-Concrete Construction," *ASCE Jour. Struct. Div.,* 100, ST-5, May 1974, pp. 1085–1139.

10. "Standard Specifications for Highway Bridges," Amer. Assoc. of State Highway and Transportation Officials, 13th Ed., Washington, D.C., 1983.

Problems

PROB. 21.1. Calculate the ultimate moment capacity of a structural steel W24 × 76 beam spaced at 8 ft on centers and adequately anchored to a 5-in. slab, assuming A36 steel (f_y = 36 ksi) and f'_c = 4000 psi.

PROB. 21.2. In Prob. 21.1, assume the beam is not shored. Calculate the AISC service load stresses for the dead load including 20 psf for forms and a construction live load equal to 30 psf. Simple span of 50 ft.

PROB. 21.3. Calculate the AISC design horizontal shear to be developed between beam and slab in Prob. 21.1 under full live load.

22

SHEAR WALLS

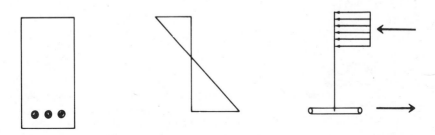

22.1   Lateral Load-Resisting Systems

Horizontal forces acting on reinforced concrete structures are resisted by different systems. Frames are commonly used for low-rise construction. Such frames may also rely on the contribution of infill walls and partitions to lateral load resistance, especially if wind is the only lateral force considered. When large lateral loads must be resisted, such as seismic loads or very high wind loads, shear walls often are used. In such buildings, shear walls may be constructed between column lines or may be incorporated into stair wells, elevator shafts, or utility shafts.

Figures 22.1*a* and 22.1*b* show plans of different buildings in which shear walls provide lateral resistance. The structural engineer must locate shear walls so that they will be the least detrimental to the use and aesthetics of the building. For example, in Fig. 22.1*a*, walls located on the outer perimeter of the building provide maximum flexibility for use of floor space; however, they eliminate windows and may reduce the space available for offices and apartments in which exterior views and light are highly desirable. Shear walls often are placed in elevator or stairway areas so that both structural and user needs are satisfied.

Figure 22.2 illustrates another consideration for shear wall location. The resultant N–S lateral force is at the center of the long side of the building. In this case, however, shear walls *AB* and *CD*, which resist the lateral force in the N–S direction, are located to the left of the center of the building. The distance *x* between the resultant lateral force and the centroid of the shear wall (the centroid of "shear resistance") produces a torsional moment on the structure. Torsion can result in substantial rotational deformations of the building and create overstresses or excessive deformations in columns farthest from the centroid of the shear walls. Torsion also is produced by loads acting on the short side of the building because wall *BD* is not colinear with the resultant

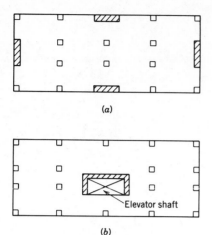

(a)

(b)

Elevator shaft

Figure 22.1 Plans of structures with shear walls. (*a*) Separate walls. (*b*) Incorporated into elevator shaft.

applied shear force. Such torsional effects should be avoided by carefully locating the walls. In most structures, the lateral stiffness of the columns is much smaller than that of the walls. Therefore, most of the shear is resisted by walls.

Figure 22.3 illustrates the distribution of lateral forces over the height of a building. The magnitude of the forces is a function of the severity of exposure at the building site. As discussed in Chapter 1, local or general building codes indicate the required loads or specify how the loads should be determined. Wind loads can be considered distributed over the surface area of the structure. However, the facade consisting of windows, architectural panels, masonry walls, and so on carries the load to the floor levels so that for analysis and design of the frame, wind loads are concentrated at floor levels. Earthquake loads are produced by ground motions that excite the structure dynamically. Because most of the mass is in the floor system, the lateral forces can be considered equivalent lateral loads applied at floor levels. The equivalent forces are determined by the mass times the acceleration at the level considered. Lateral forces produce a shear at ground level and an overturning moment in the structure with tension on one edge and compression at the other

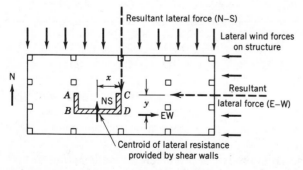

Figure 22.2 Torsion owing to eccentric shear wall location.

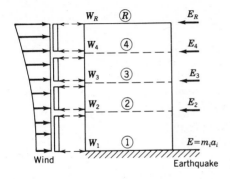

Figure 22.3 Lateral forces on building.

edge. Gravity effects generally will be sufficient to prevent net tension from developing in columns but the possibility must be considered in analyzing and designing the structural system.

Finally, a note about transfer of load through the floor to the walls. Figure 22.4 shows a floor with shear walls at the ends. Because the shear walls are at the ends, the floor acts as a very large beam (diaphragm) to transfer the lateral load to the walls. The floor will deform as shown. Although there generally is sufficient reinforcement in the floor to resist the moments developed in the plane of the floor, the designer should check that the reinforcement is continuous along the edges and that sufficient tension can be developed.

22.2 Design Considerations

From the preceding discussion, it can be seen that shear walls act as cantilever beams fixed at the base to transfer load to the foundation. The forces that must be considered in the design of the wall include

a. Varying shear that is maximum at the base.
b. Varying flexure that is maximum at the base and produces compression on one end of the wall and tension on the opposite end.
c. Gravity loads that produce compression on the wall.

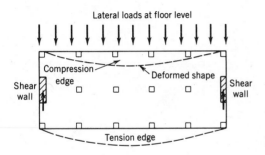

Figure 22.4 Transfer of load through floor to walls.

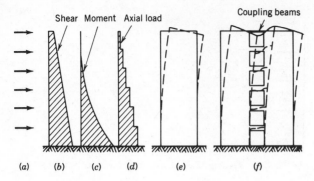

Figure 22.5 (*a*) Lateral loads. (*b*) Shear diagram. (*c*) Moment diagram. (*d*) Axial loads. (*e*) Isolated wall. (*f*) Coupled wall.

The forces and the deformed shape of the wall are shown in Fig. 22.5. In addition, the foundation must be designed to resist the shear and moment at the base of the wall. The reinforcement at the base must be carefully detailed so that the forces can be transferred between the wall and the foundation. The critical detail is anchorage of bars into the foundation and splicing bars in the wall at or near the base.

Figure 22.6 shows several possible failure modes for shear walls with large height-to-length ratios.[1] Figure 22.6*a* shows a wall with a very low vertical (flexural) reinforcement ratio. As the wall is deformed, the tensile steel elongates at the base and only one major flexural crack forms before the steel fractures. Figure 22.6*b* shows a wall that failed in shear after a large number of flexural cracks formed. The critical crack is a flexural crack that inclines downward at about 45°. At failure, some of the horizontal reinforcement across the critical crack may fracture and lead to an opening of the crack with crushing of the compression zone at the root of the critical crack. Figure 22.6*c* shows the

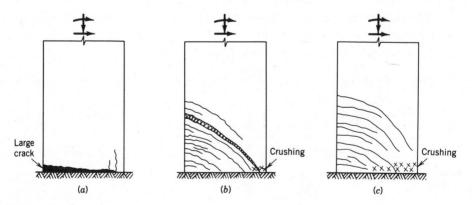

Figure 22.6 Failure modes: high-rise walls. (*a*) Fracture of steel. (*b*) Flexure-shear failure. (*c*) Failure by concrete crushing.

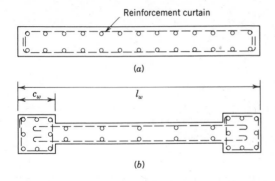

Reinforcement curtain

(a)

c_w l_w

(b)

Figure 22.7 (a) Wall without boundary members. (b) Wall with boundary elements.

most common failure mode, that is, one in which considerable yielding of the flexural reinforcement eventually leads to crushing of the concrete at the base. The amount of flexural reinforcement and the area of concrete in compression is important in determining the mode of failure. When enlarged boundary elements are used along the wall edges (Fig. 22.7) or when steel is concentrated near the edges, the onset of crushing is delayed.

In walls with small height-to-length ratios,[2,3] failure generally involves development of inclined shear cracks as shown in Fig. 22.8. In this case, the importance of vertical reinforcement for shear is evident. The compression strut shown in Fig. 22.8b is in equilibrium only if there is reinforcement to provide a vertical force opposite to the vertical component of the compression force in the strut.

A special case of a coupled shear wall (two walls connected by beams at the floor levels) is shown in Fig. 22.5f. If the coupling beams are very stiff, the walls act as a unit and deform in the same manner as an isolated wall. If the coupling beams are flexible, they serve only as links between the walls and the walls deform as isolated walls. The deformation shown in Fig. 22.5f indicates that coupling beams are subjected to very high shears and moments. Figure 22.9 shows the crack pattern in a coupling beam. Because the coupling beam is generally a short, deep member, adequate shear reinforcement and confinement of concrete is essential.[4-6]

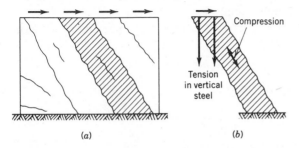

Compression

Tension in vertical steel

(a)　　　　　(b)

Figure 22.8 Failure mode: low-rise wall. (a) Shear crack pattern. (b) Compression strut between cracks.

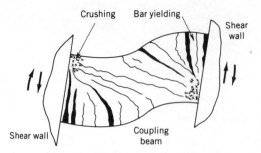

Figure 22.9 Deformation and cracking: coupling beam.

To control cracking, shear reinforcement is required in both the longitudinal and transverse directions and is generally placed in both faces of the wall. The "sheets" of bars are referred to as curtains. The longitudinal (vertical) bars in the wall serve also as flexural reinforcement. If large moment capacity is required, however, additional steel can be placed at the ends of the wall within the wall section itself or within enlargements at the ends, as shown in Fig. 22.7. The enlarged or more heavily reinforced sections, called boundary members, are often on the column lines, in which case the walls extend across the span and frame into the columns.

22.3   Design of Walls: Shear

The design of shear walls follows the same approach as that for beams. The provisions for shear design of walls are contained in Code 11.10—Special provisions for walls. Design of the horizontal section for shear in the plane of wall is based on

$$V_u \leq \phi V_n \qquad \text{(Code Eq. 11.1)}$$
$$V_n = V_c + V_s \qquad \text{(Code Eq. 11.2)}$$

An upper limit on V_n is based on tests of walls[1,7]

$$V_n \leq 10\sqrt{f_c'}hd \tag{22.1}$$

where h is the thickness of the wall and d is the effective depth that is taken as 0.8 of the horizontal length of the wall ℓ_w. A larger value of d may be used if determined by a strain compatibility analyses of the wall.

The shear strength provided by the concrete for walls subjected to a net axial compression is

$$V_c = 2\sqrt{f_c'}hd \tag{22.2}$$

For walls subjected to a net axial tension

$$V_c = 2\left(1 + \frac{N_u}{500A_g}\right)\sqrt{f_c'}hd \tag{22.3}$$

when N_u is the factored axial tensile force (in pounds) taken as negative for tension and A_g is the gross cross-sectional area of the wall.

Code 11.10.6 provides some alternative (more detailed) equations for calculating V_c. These alternative equations may yield a value of V_c that is greater than is computed using Eqs. 22.2 or 22.3, but is not used here.

The critical section for shear is taken at a distance equal to $\frac{1}{2}$ the horizontal length of the wall $\ell_w/2$ or $\frac{1}{2}$ the height of the wall $h_w/2$, whichever is less. Sections between the base of the wall and the critical section should be designed for the shear at the critical section.

If the factored shear force V_u exceeds the shear strength ϕV_c, horizontal shear reinforcement must be provided to satisfy Code Eqs. 11.1 and 11.2. The shear strength provided by the horizontal reinforcement is computed in the same way as in beams.

$$V_s = \frac{A_{vh}f_y d}{s_2} \quad \text{(Code Eq. 11.34)}$$

The terms in Code Eq. 11.34 are defined in Fig. 22.10. Combining Code Eqs. 11.1, 11.2, and 11.34, the area of horizontal reinforcement in each layer spaced at s_2 is $A_{vh} = (V_u - \phi V_c)s_2/(\phi f_y d)$. The minimum horizontal reinforcement $A_{vh} = 0.0025 s_2 h$. Minimum vertical reinforcement must be provided to satisfy the greater of

$$A_{vv} \geq \left[0.0025 + 0.5 \left(2.5 - \frac{h_w}{\ell_w} \right) \left(\frac{A_{vh}}{s_2 h} - 0.0025 \right) \right] s_1 h$$

$$A_{vv} \geq 0.0025 s_1 h \tag{22.4}$$

but $A_{vv}/s_1 h$ does not need to be greater than $A_{vh}/s_2 h$. A_{vv} and A_{vh} represent the area of steel in a layer at a spacing of s_1 or s_2, respectively. As h_w/ℓ_w is reduced, the amount of vertical shear reinforcement needed increases, as discussed previously.

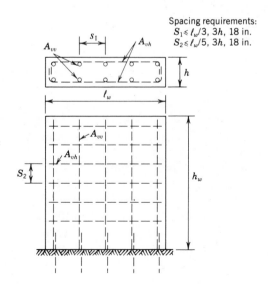

Spacing requirements:
$S_1 \leq \ell_w/3,\ 3h,\ 18$ in.
$S_2 \leq \ell_w/5,\ 3h,\ 18$ in.

Figure 22.10 Reinforcement in a wall.

If V_u is less than ϕV_c, the minimum reinforcement previously specified is supplied, except when $V_u < \phi V_c/2$. The minimum reinforcement also must be checked against the requirements of Code Chapter 14. Code 14.3 requires that

$$A_{vv} \geq 0.0012s_1h \text{ for \#5 or smaller bars}$$
$$A_{vv} \geq 0.0015s_1h \text{ for \#6 or larger bars}$$

22.4 Design of Walls: Flexure

As shown in Fig. 22.5c, the wall must be designed to resist the moment at the base. The section at the base is subjected to an axial load produced by gravity loads on the wall. To consider the effect of both forces in combination, an interaction diagram (Chapter 6) could be developed. In low-to-moderate rise structures, however, the axial compressive load is generally quite low, that is, less than the balanced load on the section. Therefore, it is conservative to consider only the moment at the base. The problem can be further simplified by ignoring the distributed steel across the length of the wall and providing enough reinforcement at the ends to resist the moment. If boundary elements are used, a very close approximation of the longitudinal (vertical) steel required in the boundary element is $A_s = M_u/f_y(\ell_w - c_w)$, where ℓ_w and c_w are as shown in Fig. 22.7. The same area of steel should be provided in both boundary elements because the lateral force can generally act in either direction. If the wall is a constant thickness, the same approach can be used by concentrating more reinforcement at the ends over a distance c_w.

In many cases the distributed reinforcement provides sufficient moment capacity without any additional reinforcement at the ends or in boundary elements. In this case a flexural analysis of the section is carried out to determine the moment capacity at the base.[1] The reinforcement is represented by a "plate" of length ℓ_w and a thickness such that the area A_{st} is the same as that provided by reinforcing bars uniformly distributed across the wall section as shown in Fig. 22.11. The moment capacity of the section is given by

$$M_u = \phi \left[0.5A_{st}f_y\ell_w \left(1 - \frac{z}{\ell_w} \right) \right] \tag{22.5}$$

where $z/\ell_w = 1/(2 + 0.85\beta_1\ell_whf'_c/A_{st}f_y)$ and β_1 is as defined for the equivalent rectangular stress block in Chapter 3 (0.85 for $f'_c = 4$ ksi). Equation 22.5 can be modified for cases in which axial load is included.[1] For low axial loads, strength is controlled by flexure and ϕ can be taken as 0.9.

Finally, the flexure reinforcement at the base must be anchored into the foundation. For construction purposes, bars generally are spliced at the base of the wall. Additional transverse steel in the splice region to improve splice strength should be provided, especially in the case of a shear wall that is a statically determinate cantilever beam and that cannot redistribute forces to other elements if any type of failure occurs in the wall.

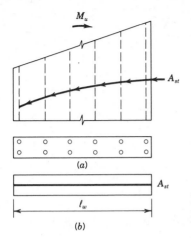

(a)

A_{st}

A_{st}

l_w

(b)

Figure 22.11 (*a*) Section of wall subjected to flexure. (*b*) Equivalent section with bars replaced by a steel plate.

22.5 Design Example

A three-story wall is subjected to factored wind forces as shown in Fig. 22.12. The wall is 15-ft long and 8-in. thick. Design reinforcement for the wall at the first level between the base and second floor. Grade 60 reinforcement and $f'_c = 3000$ psi are used.

Solution

Because shear is constant over the first level, the entire 12-ft height is designed for the same shear force. Check maximum shear strength permitted.

$$V_u = \phi V_n \le \phi 10\sqrt{f'_c}hd \ (h = 0.8\ell_w)$$
$$V_u = 150^k \le 0.85 \times 10\sqrt{3000} \times 0.8 \times 15 \times 12 \times 8/1000$$
$$150^k \le 536^k \qquad \textbf{O.K.}$$

Determine shear strength provided by concrete V_c.

$$V_c = 2\sqrt{f'_c}hd = 2\sqrt{3000} \times 0.8 \times 15 \times 12 \times 8/1000 = 126^k$$

$V_u = 150^k$, which is greater than $\phi V_c = 107^k$, so shear reinforcement must be provided.

Determine the required horizontal shear reinforcement A_{vh}

$$V_s = V_u/\phi - V_c = 150/0.85 - 126^k = 50.5^k$$

From Code 11.34,

$$A_{vh}/s_2 = V_s/f_yd = 50.5/60 \times 0.8 \times 15 \times 12 = 0.0058 \text{ in.}^2/\text{in.}$$

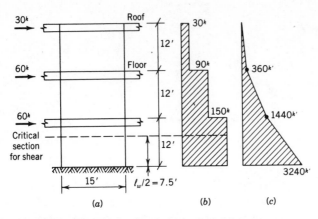

Figure 22.12 (*a*) Factored lateral wind loads, $u = 0.75(1.3W)$, Code 9.3. (*b*) Shear diagram. (*c*) Moment diagram.

The minimum horizontal reinforcement is $A_{vh}/s_2 \geq 0.0025h = 0.0025 \times 8 = 0.02$ in.^{2}/in., which is greater than 0.0058, so minimum requirements control. Because s_2 must not exceed $\ell_w/5 = 15 \times 12/5 = 36$ in., $3h = 3 \times 8 - 24$ in., or 18 in., use $s_2 \leq 18$ in. Using two curtains of steel and #3 bars, $A_{vh} = 2 \times 0.11 = 0.22$ in.^{2} and $s_2 = A_{vh}/0.02 = 0.22/0.02 = 11$-in. spacing of layers of horizontal reinforcement. If #4 bars are used, $s_2 = 0.40/0.02 = 20$ in. Use #4 bars at 18-in. spacing.

Determine the required vertical shear reinforcement A_{vv} (Eq. 22.4).

$$\frac{A_{vv}}{s_1} \geq \left[0.0025 + 0.5 \left(2.5 - \frac{h_w}{\ell_w} \right) \left(\frac{A_{vh}}{s_2 h} - 0.0025 \right) \right] h$$

$$\frac{A_{vv}}{s_1} \geq \left[0.0025 + 0.5 \left(2.5 - \frac{36}{15} \right) (0.0025 - 0.0025) \right] 8$$

$$\frac{A_{vv}}{s_1} = 0.0025 \times 8 = 0.02 \text{ in.}^2/\text{in.}$$

which is the same as the horizontal reinforcement. If A_{vh} is controlled by minimum requirements, A_{vv} also will be controlled by the minimum values. Therefore, vertical shear reinforcement also can be provided by using #4 bars at 18 in.

Determine moment capacity with vertical steel provided for shear. Vertical shear reinforcement is effective in flexure. Using Eq. 22.4

$$M_u = \phi \left[0.5 A_{st} f_y \ell_w \left(1 - \frac{z}{\ell_w} \right) \right]$$

The area of steel $A_{st} = 0.4 \times 15 \times 12/18 = 4.0$ in.^{2} and $z/\ell_w = 1/[2 + 0.85 \times 0.85 \times 15 \times 12 \times 8 \times 3/(4 \times 60)] = 0.07$ so that

$$M_u = 0.9 \times 0.5 \times 4.0 \times 60 \times 15 \times (1 - 0.07) = 1507 \text{ k-ft.}$$

690 SHEAR WALLS

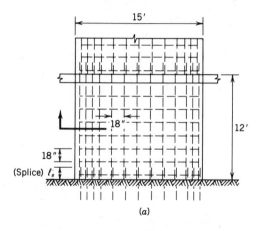

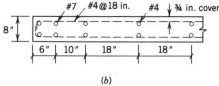

Figure 22.13 Design example: reinforcement. (*a*) Elevation of wall. (*b*) Section through wall.

Moment from factored wind loads is 3240 k-ft so additional vertical steel is needed at end sections for flexure. If steel is added over a 1-ft section at each end to provide an additional capacity $M'_u = 3240 - 1507 = 1733$ k-ft, the area of steel is

$$A_s = M'_u/f_y(\ell_w - c_w) = 1733 \text{ k-ft}/60 \text{ ksi} \times (18 - 1 \text{ ft}) = 1.69 \text{ in.}^2$$

This additional capacity can be provided with 4-#7 bars ($4 \times 0.6 = 2.4$ in.^{2}) that meet the flexural requirements and replace 2-#4 bars at the end of the wall ($1.69 + 0.4 = 2.1$ in.^{2}). The vertical steel should be spliced at the base to meet the splice requirements of the Code discussed in Chapter 8. The wall reinforcement detail for the first level is shown in Fig. 22.13.

Selected References

1. A. E. Cardenas and D. D. Magura, "Strength of High-Rise Shear Walls—Rectangular Cross Section," *Response of Multistory Concrete Structures to Lateral Forces,* ACI SP-36, Detroit, 1973, pp. 119.

2. F. Barda, J. M. Hanson, and W. G. Corley, "Shear Strength of Low-Rise Walls with Boundary Elements," ACI SP-53, Detroit, 1977, p. 149.

3. J. R. Benjamin and H. A. Williams, "The Behavior of One-Story Reinforced Concrete Shear Walls," *ASCE Jour. Struct. Div.,* ST-3, Vol. 83, May 1957, p. 1.

4. T. Paulay, "Ductility of Reinforced Concrete Shear Walls for Seismic Areas," ACI SP-53, Detroit, 1977, p. 727.

5. T. Paulay, "Coupling Beams of Reinforced Concrete Shear Walls," *ASCE Jour. Struct. Div.*, ST-3, Vol. 97, Mar. 1971, p. 843.

6. T. Paulay, "Diagonally Reinforced Coupling Beams of Shear Walls," ACI SP-42, Vol. 2, Detroit, 1974, p. 579.

7. A. E. Cardenas, J. M. Hanson, W. G. Corley, and E. Hognestad, "Design Provisions for Shear Walls," *ACI Jour., Proc.,* Vol. 70, No. 3, Mar. 1973, p. 221.

Problems

PROB. 22.1   Redesign the reinforcement in the wall of Fig. 22.11 if the length of the wall l_w is reduced to 12 ft, the wall thickness h is 10 in., Grade 60 reinforcement is used, and $f'_c = 4$ ksi.

PROB. 22.2   Redesign the wall in Fig. 22.11. The length of the wall is 10 ft including an 18-in. × 18-in. boundary element at each end. The wall is 6-in. thick and only one curtain of reinforcement is used. Grade 60 reinforcement and $f'_c = 3500$ psi are specified.

23

DETAILING FOR SEISMIC RESISTANCE

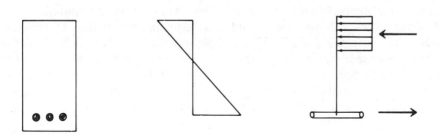

23.1   Seismic Design—General Objectives

Design criteria for a structure in a seismic zone depend on the intensity and frequency of earthquakes that might occur during the lifetime of the structure. The criteria also depend on the function of the structure.

Earthquakes that occur several times during the lifetime of the structure should not produce major structural damage. After such earthquakes, only minor repairs, primarily to nonstructural elements such as partition walls, architectural panels, and so on, should be necessary.

During a severe earthquake* the structure must not collapse or suffer damage that threatens life safety. A reserve of strength beyond the elastic range should be incorporated by carefully detailing the structure to avoid producing elements or connections that fail in a brittle manner before sufficient inelastic deformation occurs. Major repairs may be necessary afterwards.

In most structures it is not economically feasible to avoid substantial damage in such catastrophic events, but the engineer must provide life safety in the design. The integrity of the structure in the inelastic deformation range is essential for the dissipation or absorption of energy generated by the earthquake. Energy is dissipated through cyclic deformations into the inelastic range, which permits the use of design forces that are considerably smaller than those generated if the structure is designed to respond only in the linear elastic range.[1,2] 1986 Code Appendix A aims for a structure that meets these requirements.

There are structures that primarily because of the nature of activities in the building, cannot tolerate functional disruptions. Such buildings are hospitals,

---

* An earthquake with a return period (frequency) that generally is greater than the expected lifetime of the structure.

control centers and critical manufacturing, research, banking, or defense installations. Any structures essential to the life and safety of the public should be in this category. Current codes have not been developed with these structures in mind but as society and its support systems become more complex, it is an area that will have to be treated in more detail in design codes.

It is not within the scope of this book to discuss dynamic analyses of structures or the methods used to define design forces.[1-4] Basic requirements for details in beams, columns, joints, and walls to permit the structure to withstand load cycles into the inelastic range of deformations without critical loss of strength are presented in Code Appendix A. In regions of high risk, the requirements of Code Secs. A.2 through A.8 must be met. In regions of moderate risk, only Code A.9 needs to be satisfied. The Commentary to Code A.2.1 defines the background and scope of Appendix A very well.

23.2 Building Configuration

Many problems related to inelastic response of structures to earthquakes are the result of decisions made long before detailing of reinforced concrete elements is considered. Some general ideas are mentioned here that should be considered in the conceptual stage to reduce detailing problems and to improve the behavior of the structure in the event of an earthquake.

In plan, simple forms are best for earthquake resistance. Buildings with asymmetrical shapes, with wings, or with complicated plans (Fig. 23.1) often perform poorly in earthquakes. Where such shapes must be used, segments of the building should be separated by spaces (Fig. 23.2) that uncouple the building so that each segment can respond (deform) individually. The separation

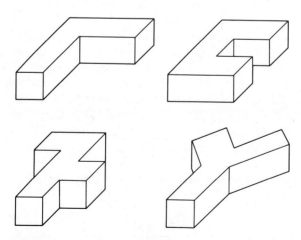

Figure 23.1 Complicated plans.

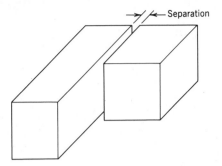

Figure 23.2 Separation of wings.

must be large enough to permit room for movement during an earthquake. As discussed in Chapter 22, lateral resistance should be provided so that the resultant lateral force and the center of lateral resistance are as nearly colinear as possible to avoid developing torsional deformations in the structure. This is illustrated in Fig. 23.3.

The structure must be uniform and continuous in elevation. Sharp changes in vertical configuration should be avoided. Structures such as those shown in Fig. 23.4 need special detailing because of the large forces that will be transferred where section changes occur, for example, at the level where a tower and wide base meet (Fig. 23.4a) or at setbacks (Fig. 23.4b). Another example is where the stiffness or strength changes, for example, at a level where story height is greater than in the remainder of the structure (Fig. 23.4c). In addition, proportions should be such that the level of lateral deformation is not excessive. A structure with a high height-to-width ratio deforms much more than a structure of equal height but greater width. It also develops much greater tension and compression forces in the vertical elements in an earthquake.

If these relatively simple concepts are considered early in the design of the structure, detailing the lateral force-resisting system is less difficult.

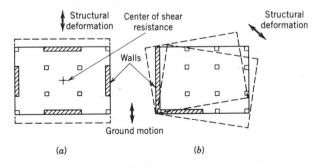

Figure 23.3 Arrangement of lateral resistance and resulting movement. (a) Symmetrical. (b) Asymmetrical.

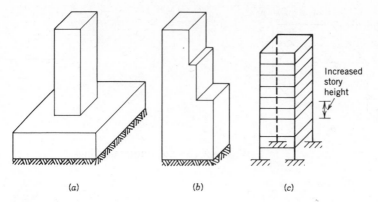

(a)   (b)   (c)

Increased
story
height

Figure 23.4 Vertical configurations. (*a*) Tower and base. (*b*) Setbacks. (*c*) Change in stiffness.

23.3 General Requirements—Code Appendix A

Minimum concrete strength is specified as f'_c of 3000 psi. No maximum is set because ductility is primarily achieved by the use of under-reinforced members, a method that eliminates compression failure until ductility limits are exceeded. If lightweight-aggregate concrete is used, strength should not exceed 4000 psi, unless experimental evidence shows that toughness is equivalent to concrete of the same strength made with normal-weight aggregate. There have been indications in previous earthquakes that members of lightweight concrete may not be as ductile as those of normal-weight aggregate.

Reinforcement should comply with ASTM A-706, which is a low-alloy steel with more closely controlled yield and ultimate characteristics. Grade 40 or 60 ASTM A-615 may be used if actual yield does not exceed specified (nominal) yield strength by more than 18 ksi, and the ratio of actual ultimate tensile strength to actual yield strength is at least 1.25. Reinforcement in elements resisting earthquake loads should not be welded.

Strength reduction factors given in Code 9.3 are used except in a member whose nominal shear strength is less than the shear corresponding to the development of the flexural strength of the member. In this case ϕ is taken as 0.6. This result particularly applies to low-rise walls or segments of walls with openings where it is impractical to provide shear strength in excess of the shear associated with developing the flexural strength of the wall.

All members that are part of the lateral-load resisting system must be detailed to meet Code Secs. A.3 to A.7 or A.9. Code A.8 contains special provisions for members that are not part of the lateral-force resisting system.

23.4 Requirements for Flexural Members of Frames

A flexural member (Code A.3) is one in which the factored compression axial force is less than $A_g f'_c / 10$. For flexural members the clear span should be at least 4 times the effective depth and the width-to-depth ratio should be at least 0.3. The member must be at least 10-in. wide. The geometric requirements are based on experience and practice in construction of structures in seismic zones.[2]

(a) Primary Reinforcement Requirements   The maximum steel ratio ρ is limited to 0.025 to insure ductility.

The various minimum requirements for A_s are summarized in Fig. 23.5. These minimums are:

1. Minimum top and bottom steel for the full length of beam, at least two bars top and bottom; on each face the area should be at least $200\,b_w d/f_y$.
2. At any section the top or bottom steel should not be less than $\frac{1}{4}$ of the steel for the maximum negative moment at the supports.
3. At each support minimum bottom steel must equal $\frac{1}{2}$ of the negative moment steel.

(b) Bar Development and Splicing   Lap splices are permitted only if the splice is confined by hoop or spiral reinforcement at a spacing not more than $\frac{1}{4}$ of the effective depth or 4 in. Lap splices are not allowed in joints within a distance of twice the member depth from the face of the support or where flexural yielding is anticipated under earthquake deformations. Welded splices or mechanical connectors may be used if no more than half the bars in a layer are spliced at a section and the distance along bars between such splices on adjacent bars is at least 24 in.

(c) Transverse Reinforcement for Flexural Members   Hoops (closed ties) are required in regions where flexural yielding is expected. Hoops confine concrete and provide lateral support for longitudinal reinforcement. Hoops can be fabricated with a closed tie with 135° hooks with 10 diameter extensions or they may be made of two pieces of reinforcement—a U-shaped stirrup with the same 135° hooks and a crosstie to complete the closed hoop (Fig. 23.6a). Hoops are required over a length twice the member depth measured from face of support or in each direction from any section where yielding is expected. Wherever inelastic deformation causes the development of yielding away from the end of

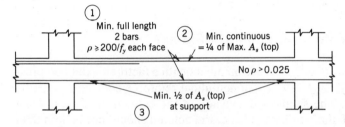

Figure 23.5   Minimum requirements for arrangement of reinforcement in beams.

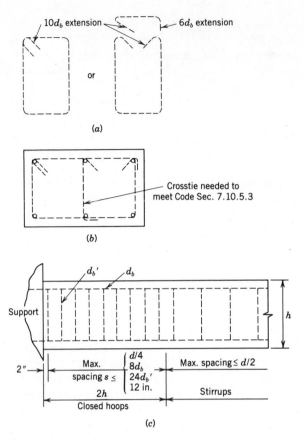

Figure 23.6 Transverse reinforcement, flexural members. (*a*) Closed hoops. (*b*) Use of crossties. (*c*) Maximum spacing

a member (as a concentrated load near midspan), this point is treated as a column face for determining the requirements for closed hoops. Spacing of hoops must not exceed $d/4$, 8 times the diameter of the smallest longitudinal bar, 24 times the diameter of the hoop bars or 12 in. Along the remainder of the member, stirrups at no more than $d/2$ spacing are required. Figure 23.6*c* summarizes spacing requirements.

23.5 Requirements for Columns in Frames

(a) General Columns (Code A.4) are members with a factored axial compressive force greater than $A_g f_c'/10$. The longitudinal reinforcement ratio must be between 1% and 6%. The smallest dimension of the column must be at least 12 in. and the ratio of short-to-long sides of the column should not be less than 0.4. At a joint, columns in a frame resisting lateral forces must have flexural

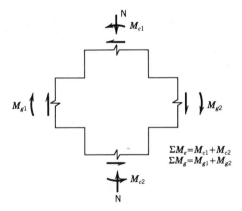

$\Sigma M_e = M_{c1} + M_{c2}$
$\Sigma M_g = M_{g1} + M_{g2}$

Code Eq. A-1: $\Sigma M_e \geq 6/5 \Sigma M_g$

Figure 23.7 Weak beam-strong column concept.

strength at least 20% greater than the strength of girders framing into the joint. Figure 23.7 illustrates the required calculation. Column flexural strength is calculated considering the factored axial force on the column.

This provision is extremely important because it is intended to force hinging into beam elements or girders. Experience with structures in seismic zones[2] indicates that column hinging generally leads to greater damage to the structure and less stability under large lateral deformations. If the requirement in Code Eq. A.1 is not satisfied at a joint, columns must have closely spaced transverse reinforcement over their full height instead of the critical end regions described in the next section.

(b) Transverse Reinforcement Closely spaced transverse reinforcement must be provided for a distance above and below the joint equal to the depth of the column, but not less than $\frac{1}{6}$ of the clear span or 18 in. The transverse steel provides some or all of the shear reinforcement over these lengths (See Sec. 23.6).

If a spiral is used, the ratio ρ_s is the value normally used for a spiral column (Code Eq. 10.5), where the spiral replaces the strength that is originally carried by the column shell. A second lower limit of not less than $0.12 f'_c/f_{yh}$ is also set. If rectangular hoops, either separate or continuous, are used, the hoop is required to be

$$A_{sh} \geq 0.3(s h_c f'_c/f_{yh})[(A_g/A_{ch}) - 1] \qquad \text{(Code A.3)}$$

or

$$A_{sh} \geq 0.12 s h_c f'_c/f_{yh} \qquad \text{(Code A.4)}$$

where A_{sh} is the total area of transverse reinforcement (including crossties) within a spacing s and perpendicular to h_c, s is the spacing of transverse reinforcement, h_c is the cross-sectional dimension of the core (c-c of ties), f_{yh} is the yield strength of the transverse steel, A_{ch} is the core area of the rectangular

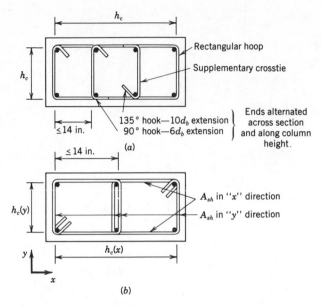

Figure 23.8 Transverse reinforcement requirements for columns. (*a*) Crossties. (*b*) Overlapping hoops.

section (out-to-out of hoops), and A_g is the gross area of the section. In Code Eq. A.3, hoops are taken conservatively to be half as effective as a circular spiral. Reference 5 contains an excellent summary of the confinement provided by rectangular transverse hoops. The hoop spacing center-to-center must not exceed 4 in. The use of supplementary crossties or overlapping hoops to increase the confinement of the section is shown in Fig. 23.8. The maximum spacing between crossties or legs of hoops is 14 in. on center.

Where a column supports a wall or stiff partition, the Code requires that the entire column height contain confinement reinforcement. Such columns often carry heavy loading from the wall. If it can be shown that the column loads are not significant ($N < A_g f'_c / 10$), the closely spaced transverse reinforcement can be omitted.

(c) Column Bar Splices Lap splices must be located within the center half of the column length and must be proportioned as tension splices in case overturning of the structure produces tension in the columns. Mechanical or welded splices can be used if not more than alternate bars are spliced at any one level, and the splices are staggered 24 in. or more along the column longitudinal axis.

23.6 Shear Strength Requirements—Frame Members

(a) Design Shear Force The earthquake-induced design shear force is determined by analyzing a free body of the frame element between joint faces rather than analyzing the structure under lateral loads. Moments on the free body act in a manner that will produce maximum shear on the member. Factored gravity load acts on the member. The moments at the ends of the member are computed without strength reduction factors and it is assumed that the tensile reinforcement reaches the strain hardening range as the structure deforms into the inelastic range. Code A.7.1.1 requires that a steel stress of at least $1.25 f_y$ should be used. Figure 23.9 shows free bodies for calculating design shear force.

(b) Transverse Reinforcement for Shear   The transverse reinforcement required for shear may exceed the minimum transverse steel described in Sec. 23.4c and Sec. 23.5b. The transverse reinforcement is determined using $V_e \leq \phi(V_s + V_c)$, where $\phi = 0.85$, $V_s = A_v f_y d/s$. If V_e is greater than half the total design shear at the section, V_c is taken as zero if axial load on the member under earthquake loads is small ($N \leq A_g f'_c/20$).

   The elimination of a concrete contribution to the shear strength is demonstrated by tests,[6,7] in which reversals of inelastic deformation reduce the effectiveness of the concrete. In such cases, more transverse reinforcement is

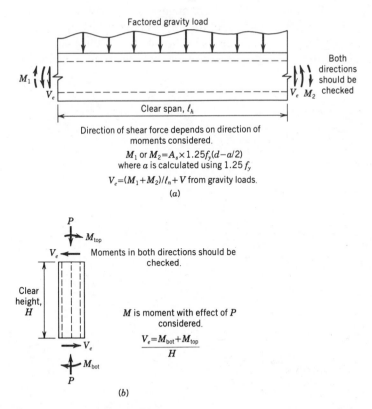

Figure 23.9   Free bodies for design shear force. (*a*) Beam or girder. (*b*) Column.

needed to carry the shear force in the member. This result should not be interpreted to mean that the concrete cross-sectional area can be reduced to a minimum value. The confined core and the transverse reinforcement acting together must resist the shear. Using a truss model it can be easily shown that transverse steel alone cannot resist shear.

23.7 Joints of Frames

Joint detailing was discussed in Chapter 10. Joints of frames subjected to earthquake loads must be designed to meet some additional requirements (Code A.6). ACI Committee 352[8] recommendations on beam-column joints include provisions for Type 2 joints (seismic) that are slightly different from those of Appendix A and should be studied.

The forces acting on a horizontal plane A-A through the joint are shown in Fig. 23.10a. The forces from the longitudinal reinforcement in the beams are computed assuming that strain hardening is reached under large lateral deformations of the structure and that the steel is stressed to $1.25f_y$. The nominal shear strength[9] of the joint is taken as $20\sqrt{f_c'}A_j$ for a confined joint and $15\sqrt{f_c'}A_j$ for all other joints. For lightweight-aggregate concrete the values are reduced by 25%. A confined joint is one where members frame into all vertical faces of the joint and at least $\frac{3}{4}$ of the face of the joint is covered by the framing member. A_j is the minimum horizontal cross-sectional area of the joint.

Transverse reinforcement through the joint must meet the requirements dis-

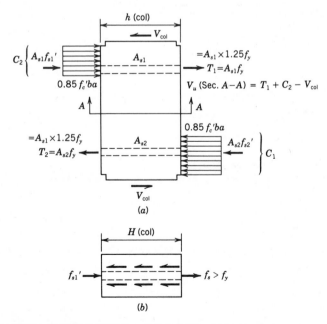

Figure 23.10 (a) Horizontal shearing forces acting on a connection under seismic loading. (b) Bond along straight bar through joint.

cussed in Sec. 23.5b. The transverse reinforcement can be reduced by 50% in a confined joint if the member width is at least $\frac{3}{4}$ of the column width.

The development length for hooked bars anchored within the core of a joint are based on the provisions of Code Secs. 12.5.1 and 12.5.2 adjusted to reflect the confinement provided by the column cage as well as the reduction in bond produced by large inelastic deformations of the structure. A standard 90° hook performs better than a 180° hook and is recommended for joints in structures in seismic zones. For a 90° hook, $\ell_{dh} = f_y d_b / 65\sqrt{f_c'}$, but more than $8d_b$ or 6 in. These values are increased by 25% for lightweight-aggregate concrete.

Tests of joints indicate that anchorage of straight bars through a joint is reduced under cyclic load.[10] With large deformations, bond along the straight bar (Fig. 23.10b) within the joint may be lost and the bar may be in tension on both sides of the joint. The result is a loss of frame stiffness but little effect on frame strength. It is quite difficult to prevent such loss of bond. It is advantageous to reduce the diameter of the longitudinal beam reinforcement so that average bond stresses are smaller and losses in stiffness of the structure are reduced. Although no specific requirements are included in the Code, bar diameters not exceeding 1/16 to 1/20 of the column dimension appear to be satisfactory.

For design examples of beam-column joints in a frame resisting lateral earthquake forces, see Reference 8.

23.8 Structural Walls

Shear walls are discussed in Chapter 22 but walls in structures subjected to earthquake loads must meet special requirements (Code A.5). Horizontal and vertical reinforcement in the wall must provide at least $\rho = 0.0025$ of the gross area in each direction and spacing of reinforcement should not exceed 18 in. in each direction. At least two curtains of reinforcement are used in walls if the factored shear force in the wall exceeds $2\sqrt{f_c'}A_{cv}$, where A_{cv} is the cross-sectional area of the web of the wall in the direction of shear.

The nominal shear strength V_n of a structural wall is taken as

$$V_n = A_{cv}(2\sqrt{f_c'} + \rho_n f_y) \qquad \text{(Code A.6)}$$

but less than $10A_{cv}\sqrt{f_c'}$. Where a number of walls resist shear in a given direction, the nominal shear strength of all the walls should not exceed $8\sqrt{f_c'}A_{total}$, where A_{total} is the cross-sectional area of all the walls in that direction.

Boundary elements of walls and diaphragms are needed if compressive stress at the edge exceeds $0.2\sqrt{f_c'}$. Transverse reinforcement meeting the requirements of Code Secs. A.4.4.1 through A.4.43 (Sec. 23.5b) must be provided in the boundary element.

23.9 Other Frame Members

Members that are not part of the lateral-force resisting system must be designed for adequate vertical capacity under twice the deformation produced by factored lateral forces. Although these members are not counted on to resist lateral forces, they must maintain their integrity to carry vertical forces as the structure deforms under lateral earthquake forces. The provisions of Code A.8 define nonlateral load resisting elements with sufficient toughness to maintain vertical load carrying capacity.

Minimum reinforcement for flexural members [Sec. 23.4a(1)] and for walls [Sec. 23.8] must be provided. Columns should have closed ties with 135° hooks with 6 bar diameter of 2.5-in. extensions. Crossties as described in Fig. 23.8a may be used. Tie spacing must not exceed 8 column longitudinal bar diameters, 24 tie bar diameters, or $\frac{1}{2}$ the smallest dimension of the column. Ties at this spacing must be provided along the length of the column from the joint face for a distance not less than $\frac{1}{6}$ the clear height of the column, the largest dimension of the column or 18 in. Over the remainder of the column length, tie spacing can be up to twice the spacing at the ends.

23.10   Frames in Regions of Moderate Seismic Risk

Code A.9 contains special provisions for frames exposed to moderate risk. These provisions are less restrictive than those for high-risk regions. The provisions indicate the minimum confinement and reinforcement necessary to ensure satisfactory performance under the lateral deformation likely in a zone of moderate risk.

(a) Beams   Bottom reinforcement at the face of the joint must be at least $\frac{1}{3}$ of the top reinforcement. Top and bottom reinforcement at any section should not be less than $\frac{1}{5}$ of the top steel at either support. Stirrups must be provided at both ends for a distance twice the member depth. Stirrups should start 2 in. from the support and the spacing should not exceed $d/4$, 8 longitudinal beam bar diameters, 24 stirrup bar diameters, or 12 in. Over the remainder of the beam length, stirrups spaced at $d/2$ should be provided.

(b) Columns   Ties at a spacing not less than 8 longitudinal bar diameters, 24 tie bar diameters, $\frac{1}{2}$ the smallest dimension of the column, or 12 in. must be provided over a length $\frac{1}{6}$ the clear height of the column, larger dimension of column, or 18 in. from the joint face. Over the remainder of the column, the spacing can be twice as large.

(c) Joints   Transverse reinforcement in joints must provide an area A_{sh} not less than $50b_c s/f_y$, where b_c is the column dimension perpendicular to the direction of shear and s is the spacing between ties or hoops in the joint, f_y in ksi. The minimum requirements for beams, columns, and joints of frames in regions of moderate risk are shown in Fig. 23.11.

(d) Two-Way Slabs Without Beams   Although slab systems without beams are inadequate to resist lateral loads in high-risk regions, they can be used in

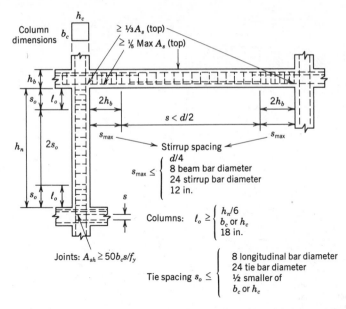

Figure 23.11 Minimum details for frames in regions of moderate risk.

moderate-risk zones if certain detailing requirements are met. The moment transferred from the slab to the support must be carried by reinforcement in the column strip. The fraction of the moment transferred to the support given by Code Eq. 13.1 but not less than $\frac{1}{2}$ the reinforcement in the column strip must be placed within a section $1.5h$ on either side of the column (h is the slab or drop panel thickness).

One-fourth of the top column-strip reinforcement at the support must be continued along the span. Continuous bottom reinforcement in the column strip must not be less than $\frac{1}{3}$ of the top column strip reinforcement at the supports and at least $\frac{1}{2}$ of the bottom reinforcement at midspan should be continuous and anchored to develop yield at the face of the support. At discontinuous edges all top and bottom reinforcement must be anchored to develop yield at the face of the support. The requirements for slab reinforcement are summarized in Fig. 23.12.

(e) Shear Strength of Elements   Design shear strength in beams, columns, and two-way slabs must be at least that computed as shown in Fig. 23.13 or that from a frame analysis of the structure to obtain maximum shear in each element. In such an analysis for design load combinations that include earthquake loads E, the value of E is doubled. The design force is based on the following load combinations (Code 9.2.3 modified for $2E$).

$$U = 0.75(1.4D + 1.7L + 1.1 \times 1.7 \times 2E)$$
$$U = 0.75(1.4D + 1.7L + 3.74E)$$

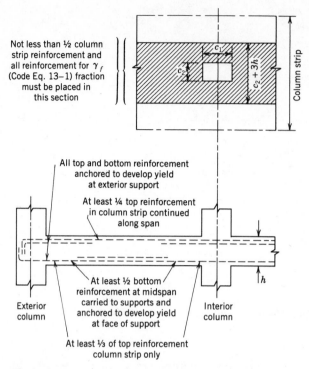

Figure 23.12 Reinforcement requirements for two-way slabs in regions of moderate seismic risk.

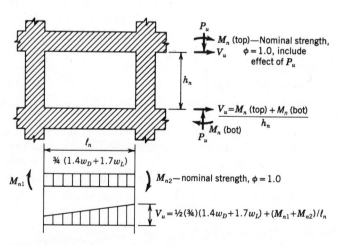

Figure 23.13 Design shear for frame members in regions of moderate risk.

Transverse reinforcement for shear then is designed according to the provisions of Code Chapter 11 but not less than the minimum required in Code Secs. A.9.3 and A.9.4 as discussed previously for beams and columns.

23.11　General Comment

The intent of Appendix A is to add toughness or ductility to the structure, to provide for reversal of stresses, to guard against sudden shear failure, and to provide a confined core in members when concrete begins to spall. The end result should be a greatly improved frame. But there are traps for the careless designer. The structure is more complex than construction in nonseismic zones. The designer must visualize the erection problem clearly. A paragraph from the 1977 Commentary is appropriate as a closing comment:

> Great care is needed to avoid impractical placement problems in seismic design. This is particularly true for the reinforcing steel in the connections of special ductile frames. If high percentages of reinforcement are used in beams and columns, it may be physically impossible to place the beam and column reinforcement and the required ties in the connections; or if the reinforcement is placeable, it may be impossible to place the concrete in the connection or to get a vibrator into it. It may prove economical to construct a full size model of the reinforcement in a typical connection to investigate its constructibility, unless the designer has had considerable experience in this type of work.

Selected References

1. J. A. Blume, N. M. Newmark, and L. H. Corning, *Design of Multistory Reinforced Concrete Buildings for Earthquake Motions,* Portland Cement Association, Skokie, 1961.

2. *Recommended Lateral Force Requirements and Commentary,* Seismology Comm., Structural Engineers Assn. of California, San Francisco, California, 1980.

3. Applied Technology Council, "Tentative Provisions for the Development of Seismic Regulations for Buildings," *Special Publication* No. 510, U.S. National Bureau of Standards, U.S. Government Printing Office, Washington, D.C., 1978, 504 pp.

4. Federal Emergency Management Agency, "NEHRP Recommended Provisions for the Development of Seismic Regulations for New Buildings," Part 1, Provisions, FEMA Report No. 95, Feb. 1986, Building Seismic Safety Council, Washington, D.C.

5. S. A. Sheikh, S. M. Uzumeri, "Strength and Ductility of Tied Concrete Columns," ASCE *Jour. Struct. Div.,* ST-5, 106, May 1980, p. 1079.

6. E. P. Popov, V. V. Bertero and H. Krawinkler, "Cyclic, Behavior of Three R/C Flexural Members with High Shear," *EERC Report* No. 72-5, Earthquake Engineering Research Center, University of California, Berkeley, Oct. 1972.

7. J. K. Wight and M. A. Sozen, "Shear Strength Decay of RC Columns Under Shear Reversals," ASCE *Jour. Struct. Div.,* ST-5, Vol. 101, May 1975, p. 1053.

8. ACI Committee 352, "Recommendations for Design of Beam-Column Joints in Monolithic Structures," *ACI Jour.,* 82, May–June 1985, p. 266.

9. D. F. Meinheit and J. O. Jirsa, "Shear Strength of R/C Beam-Column Connections," ASCE *Jour. Struct. Div.,* ST-11, 107, Nov. 1981, p. 2227.

10. S. Zhu and J. O. Jirsa, "A Study of Bond Deterioration in Reinforced Concrete Beam-Column Joints," PMFSEL Report 83-1, Dept. of Civil Engineering, Univ. of Texas at Austin, July 1983.

APPENDIX A

SUMMARY TREATMENT OF SERVICE LOAD ANALYSIS FOR FLEXURE

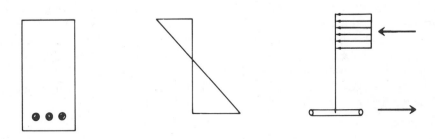

A.1 General Field of Use

Elastic analysis and the transformed area concept for calculating moments of inertia relating to deflections and shrinkage stresses were introduced in Sec. 3.21. The old working stress design method is gradually being phased out in building design and analysis. For slabs and beams the 1986 Code (Appendix B) classifies it as an alternate design method with unity load factors.

For development and splicing of reinforcement and for torsion resistance, the method *almost* follows Code Chapters 11 and 12. Shear resistance also follows Code Chapter 11, with lower coefficients in the allowables. For columns it is so restricted as to be essentially the strength design of Chapters 6 and 7.

In the highway bridge field so much emphasis is necessary on service conditions that the 1983 AASHTO Specifications[1] still relate to what formerly was called working stress design, although they now present strength design as an alternate.

The current presentation is restricted to simple shapes, with or without compression steel, using AASHTO allowable stresses.

There is very little basic difference between calculating beam stresses resisting a given moment, for comparison with allowable stress values, and calculating an allowable moment based on these allowable stresses.

The *design* of beams for moment, on the other hand, is a process considerably different from analysis. This topic is covered comprehensively in earlier editions of this book. Design emphasis is shifting rapidly to design on a strength basis with checking (using analysis procedures) for serviceability.

A.2   Compression Reinforcement

The elastic analysis of Sec. 3.21 accepted compression reinforcement as a portion of the elastic member with a steel stress $f_s' = nf_c$ and a transformed area nA_s' or effective addition of $(n - 1)A_s'$ to the transformed area and $(n - 1)f_c^*$ to the stress after accounting for displaced concrete.

For strength, the creep of concrete when under load (or the flattening of the stress-strain curve for concrete under even short-time high stresses) shortens the compressive concrete and steel more than elastic analysis suggests. Analysis or design by the working stress method recognizes this by using $2nA_s'$ as the transformed area and $2nf_c$ as the stress. It is logical to use $(2n - 1)A_s'$ and $(2n - 1)f_c$ to recognize the displacement of useful concrete, but this sometimes is simplified to $2n$ instead of $2n - 1$ in recognition that the factor 2 applied to n is not really accurate enough to justify a 5% to 8% correction.

Under the ACI Code, values of n should be based upon E_s/E_c from values in Code 8.5 and may be rounded off to the nearest whole number.

A.3   AASHTO Allowable Stresses

In flexure AASHTO allows a maximum fiber stress in compression $f_c = 0.4f_c'$ with zero in tension (i.e., cracked over tension zone), except in plain concrete (as in some footings). On reinforcing bars it allows in tension 20 ksi on Grade 40 and 24 ksi on Grade 60 steel, and in compression $2n$ times the concrete stress at that level, but less than the allowable in tension. The AASHTO value of n is based upon E_s/E_c values as in ACI Code 8.5 and may be rounded off to the nearest whole number.

A.4   Rectangular Beam Analysis or Review

Because Sec. 3.21 used only the moment of inertia approach, this analysis checks by looking directly at the internal stress system. This stress system is not an improvement for analysis but it is important as a better approach for design and for easy visualization. The data already used are $M = 100$ k-ft, $f_c' = 4000$ psi, $n = 8$, Grade 40 steel, $c = 9.95$ in. (Sec. 3.21d) and the other dimensions of Fig. A.1, which is a copy of Fig. 3.23 with stress triangles added. The resultant compression N_c and tension N_t form a couple with arm z† $= d - c/3 = 16.68$ in., as noted on Fig. A.1.

---

* Here f_c is at the level of the compression steel, a value less than the maximum stress on the extreme fiber.

† In the older notation, c was kd, z was jd, and y_c was sometimes designated as z.

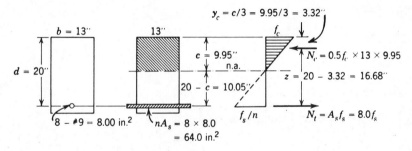

Figure A.1 Analysis of rectangular beam.

$$N_t = N_c = M/z = 100,000 \times 12/16.68 = 71,900 \text{ lb}$$
$$N_t = A_s f_s$$
$$f_s = N_t/A_s = 71,900/8.0 = 8990 \text{ psi}$$

or

$$f_s = \frac{M}{A_s z} = \frac{100,000 \times 12}{8.0 \times 16.78} = 8990 \text{ psi}$$

$$N_c = \frac{f_c}{2} bc$$
$$71,900 = 0.5 f_c \times 13 \times 9.95$$
$$f_c = 1111 \text{ psi}$$

Check: $\quad \dfrac{f_c}{f_s/n} = \dfrac{c}{d-c} \qquad \dfrac{f_c}{8990/8} = \dfrac{9.95}{20-9.95}$

$$f_c = 1112 \text{ psi versus } 1111 \qquad \textbf{O.K.}$$

Under the AASHTO allowable stresses of $f_c = 0.4 f'_c = 0.4 \times 4000 = 1600$ psi and $f_s = 20$ ksi, the allowable moment on this beam can be found by proportion from the stresses already established. Based on compression alone, f_c may be increased from 1111 psi to 1600 psi:

$$\text{Allowable } M_c = (1600/1111)100 = 144 \text{ k-ft}$$

Based only on tensile stress:

$$\text{Allowable } M_t = (20,000/8990)100 = 223 \text{ k-ft}$$

The lower "allowable" is the real limit, in this case a limit on compression.

$$\text{Allowable } M = 144 \text{ k-ft}$$

Economy indicates the use of less A_s, with a deeper beam if the moment must be kept to this level.

If the calculated stresses were not already available, the limiting stress could have been established as soon as the neutral axis was found. Con-

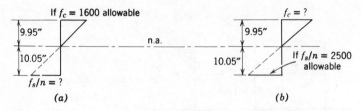

Figure A.2 Determination of limiting or governing stress.

sider *either* triangle in Fig. A.2 (considering both is a waste of time). From the triangles in Fig. A.2*a*:

If $f_c = 1600$ psi, $f_s/n = (10.05/9.95)1600 = 1615$ psi
$n = 8$, $f_s = 8 \times 1615 = 12,930$ psi $< 20,000$ **O.K.**

Or, from the triangles of Fig. A.2*b*:

If $f_s/n = 2500$ psi, $f_c = (9.95/10.05)2500 > 1600$ **N.G.**

Either trial establishes that the moment is limited by f_c to:

Allowable $M = N_c z = (0.5 f_c bc)z = (0.5 \times 1600 \times 13 \times 9.95)16.68$
$= 1,720,000$ lb-in. $= 143$ k-ft

This example covers the operating procedures for analysis. The details but not the principles change in the case of compression steel, T-beams, or irregular sections.

A.5 T-Beam Review

(a) Compared to a Rectangular Beam The AASHTO specification for flange width is the same as in the ACI Code except for limiting the flange overhang to $6h_f$ instead of $8h_f$, where h_f is the flange thickness.

The flange area is usually more than adequate to care for compressive stresses; hence allowable moments are usually limited by the steel stress. Furthermore, the large compression area in the flange often pulls the centroid of the transformed area up into the flange itself. When this occurs, the analysis for moment is exactly that for a wide rectangular beam of width b, because the missing area below the flange is in tension and is cracked.

The neutral axis more often falls below the flange with working stress analysis than strength design, where it rarely does so. In this case the evaluation of N_c in terms of f_c and the location of N_c are no longer possible by simple inspection. The student will probably find it helpful to think in terms of the wedge of stress. For a rectangular beam, this wedge of stress is a simple triangular wedge (Fig. A.3*a*). The volume of the wedge gives the magnitude of N_c and the centroid of the wedge marks the location of the resultant N_c. When

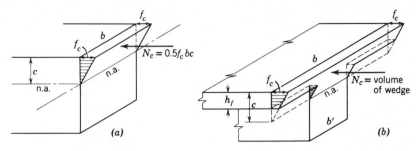

Figure A.3 Wedge of stress for (*a*) rectangular beam and (*b*) T-beam.

this concept is applied to the T-beam, as in Fig. A.3*b*, neither the volume nor the centroid can be determined except by a summation process, as in the following examples. The first example is calculated by the so-called exact analysis; the second example uses the approximate method that neglects the web area between neutral axis and flange. The approximate method fits much easier into a solution by formulas, charts, or curves, but has small advantage for a basic analysis such as is used here.

(b) Exact Analysis Find f_c and f_s for the T-beam shown in Fig. A.4*a* under a moment of 200 k-ft, assuming $f'_c = 3000$ psi and Grade 60 steel, AASHTO specification.

Solution

$$\text{AASHTO } E_c = 57{,}000\sqrt{f'_c} = 57{,}000\sqrt{3000} = 3.12 \times 10^6 \text{ psi}$$
$$n = E_s/E_c = 29 \times 10^6/3.12 \times 10^6 = 9.29, \text{ say, } 9$$
$$nA_s = 9 \times 6.93 = 62.4 \text{ in.}^2$$

Flange overhang of 15 in. is less than $6h_f = 24$ in.   **O.K.**

First make a quick check on where the neutral axis lies with respect to the bottom of the flange, using area moments about the bottom of flange.

$$\overset{?}{40 \times 4 \times 2 = 62.4(20 - 4)}$$
$$320 < 998, \text{ n.a. is well below flange.}$$

Take area moments about the neutral axis using the area of Fig. A.4*b* and solve them by completing the squares.

$$(10c)c/2 + (40 - 10)4(c - 2) = 62.4(20 - c)$$
$$c^2 + 36.5c + 18.2^2 = 298 + 18.2^2 = 622$$
$$c = 25.0 - 18.2 = 6.8 \text{ in.}$$

At the bottom of flange the stress triangle gives

$$f_{c1} = (2.8/6.8)f_c = 0.41f_c$$

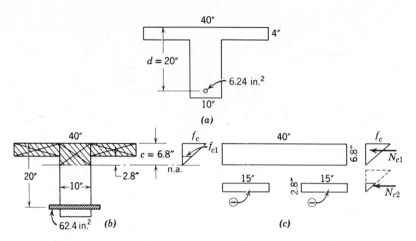

Figure A.4 Analysis of T-beam, exact method. (The 4-in. slab is too thin for highway use, but it helps to emphasize the influence of n.a. location.)

The total compression can be calculated by subdividing the areas, and if necessary, the stresses, into pieces that can be visualized simply (Fig. A.4b). In this case, the simplest pieces are N_{c1} on a rectangle 40-in. wide and extending all the way to the neutral axis and a negative N_{c2} deducting the surplus below the flanges. Sketches, as in Fig. A.4c, and a table are desirable. To locate the resultant N_c, moments are taken about some convenient axis, by custom about the top of beam.

	N_c	Arm about top	Moment m about top
$N_{c1} = (40 \times 6.8)f_c/2$	$= 136.3f_c$	$6.8/3 = 2.27$ in.	$308f_c$
$N_{c2} = -(30 \times 2.8)0.41f_c/2$	$= -17.3f_c$	$4 + 2.8/3 = 4.93$ in.	$-86f_c$
	$N_c = 119.0f_c$		$m = 222f_c$

$$y_c = m/N_c = 222f_c/119.0f_c = 1.86 \text{ in.}$$

$$z = 20 - y_c = 20 - 1.86 = 18.14 \text{ in.}$$

$$f_s = \frac{N_t}{A_s} = \frac{M}{A_s z} = \frac{200,000 \times 12}{6.93 \times 18.14} = 19,100 \text{ psi versus } 24,000 \text{ psi allowable}$$

O.K.

$$N_c = M/z = 200,000 \times 12/18.14 = 132,000 \text{ lb} = 119.0f_c \text{ (from above)}$$

$f_c = 1110$ psi versus $0.40 \times 3000 = 1200$ psi allowable   **O.K.**

(c) Approximate Analysis   Same beam as in subsection (b), Fig. A.5a.

Solution

As before, the neutral axis falls below the flange. The approximate method neglects the small compressive area (and small unit stress) below

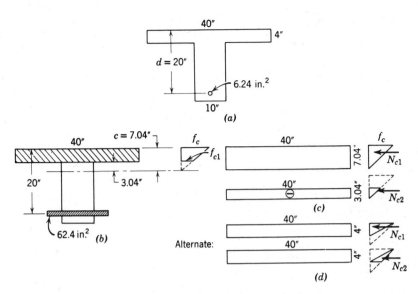

Figure A.5 Analysis of T-beam, approximate method.

the flange, Fig. A.5b. Then area moments about the neutral axis give:

$$(40 \times 4)(c - 2) = 62.4(20 - c)$$
$$222.4c = 1568$$
$$c = 7.04 \text{ in.}$$

At the bottom of the flange, $f_{c1} = f_c \times 3.04/7.04 = 0.432 f_c$.
The areas and stresses shown in Fig. A.5c lead to:

	N_c	Arm about top	**Moment m** about top
$N_{c1} = (40 \times 7.04) f_c/2$	$= 141.0 f_c$	$7.04/3 = 2.35 \text{ in.}$	$331 f_c$
$N_{c2} = -(40 \times 3.04)0.432 f_c/2$	$= -26.3 f_c$	$4 + 3.04/3 = 5.02$	$-132 f_c$
	$N_c = 114.7 f_c$	$m = 199 f_c$	

$$y_c = 199 f_c/114.7 f_c = 1.74 \text{ in.}$$
$$z = 20 - 1.74 = 18.26 \text{ in.}$$

The student should see clearly that there are many different patterns for N_{c1} and N_{c2} that lead to identical values of N_c and m. For example, the pattern of Fig. A.5d where the unit stress picture is subdivided rather than the area. It could have been subdivided into a rectangle and triangle of stress or the trapezoid could have been used undivided, provided its centroid was calculated as the location of N_c. The subdivisions to be used are a matter of convenience, but *in the exact solution* in subsection (b) no other subdivision leads as directly to the desired result.

$$f_s = \frac{M}{A_s z} = \frac{200{,}000 \times 12}{6.93 \times 18.26} = 19{,}000 \text{ psi versus } 24{,}000 \text{ psi} \quad \textbf{O.K.}$$

$$N_c = M/z = 200{,}000 \times 12/18.26 = 131{,}000 \text{ lb} = 114.7 f_c$$

$$f_c = 1150 \text{ psi versus } 1200 \text{ psi} \quad \textbf{O.K.}$$

(d) Allowable Moment Just as for the rectangular beam in Sec. A.4, the allowable moment may be found by proportion from stresses already calculated; or, the governing stress can be determined from the stress triangles and the allowable moment then calculated.

Although this particular beam with its narrow flange is limited by compression, this is a rare case for a real T-beam. Hence an approximate allowable moment usually can be estimated as $A_s f_s$, with the stress at the allowable, times an estimated arm z that is usually in the neighborhood of $0.9d$.

A.6 Analysis or Review with Compression Steel

(a) General The use of $2n$ with A_s' as discussed in Sec. A.2 is direct and reasonable for design and fairly satisfactory (but occasionally awkward) for allowable moments. In the calculation of stresses from a given moment, the concepts become less significant and the stresses become only nominal ones.

(b) Analysis Assume 2.0 in.^{2} of compression steel is added at 2-in. below the top of the rectangular beam of Sec. A.4a and Fig. A.1a, with

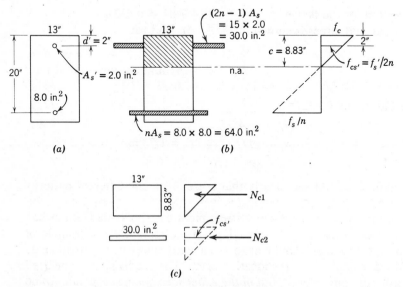

Figure A.6 Analysis of double-reinforced beam.

$f'_c = 4000$ psi, $n = 8$, Grade 40 steel. Find stresses for $M = 150$ k-ft. Also find the allowable moment.

Solution

The transformed area is shown in Fig. A.6b using $(2n - 1)A'_s = 30$ in.^{2} for the compression steel.

Area moments about neutral axis:

$$(13c)c/2 + 30(c - 2) = 64(20 - c)$$
$$c^2 + 14.46c + 7.23^2 = 206 + 7.23^2 = 258$$
$$c = 16.06 - 7.23 = 8.83 \text{ in.}$$

Based on Fig. A.6c:

$$f'_{cs} = f_c \times 6.83/8.83 = 0.776 f_c$$

	N_c	Arm about top	Moment m about top
$N_{c1} = (13 \times 8.83)f_c/2 = 57.3f_c$		$8.83/3 = 2.94$ in.	$168.0f_c$
$N_{c2} = 30.0 \times 0.776f_c = \underline{23.2f_c}$		$d' = 2.00$	$\underline{46.4f_c}$
	$N_c = 80.5f_c$		$m = 214.4f_c$

$$y_c = 214.4f_c/80.5f_c = 2.65 \text{ in.}$$
$$z = 20 - 2.65 = 17.35 \text{ in.}$$
$$f_s = \frac{M}{A_s z} = \frac{150,000 \times 12}{8.0 \times 17.35} = 13,000 \text{ psi versus } 20,000 \text{ psi allowable} \quad \textbf{O.K.}$$
$$N_c = M/z = 150,000 \times 12/17.35 = 103,500 \text{ lb} = 0.85f_c \text{ (from above)}$$
$$f_c = 1290 \text{ psi versus } 0.40 \times 4000 = 1600 \text{ psi} \quad \textbf{O.K.}$$

Check: $$f_c = \frac{13,000}{8} \times \frac{8.83}{11.17} = 1284 \text{ psi} \quad \text{Say } \textbf{O.K.}$$
$$f'_s = 2nf'_{cs} = 2 \times 8 \times 0.776 \times 1280$$
$$= 15,900 \text{ psi versus } 20,000 \text{ psi allowable} \quad \textbf{O.K.}$$

For allowable moment: If $f_c = 1600$ psi, similar stress triangles give:

$$f_s = nf_c \frac{20 - 8.83}{8.83} = 8 \times 1600 \times 11.17/8.83$$
$$= 16,180 \text{ psi versus } 20,000 \text{ psi} \quad \textbf{O.K. Compression controls.}$$
$$f'_s = 2nf'_{cs} = 2 \times 8 \times 0.776 \times 1600$$
$$= 19,900 \text{ psi versus } 20,000 \text{ psi} \quad \textbf{O.K.}$$

With $M = A_s f_s z$ and $z = 17.35$ in. as above:

Allowable $M = 8 \times 16,180 \times 17.35/12,000 = 187$ k-ft

If f'_s were excessive, it would not require that the allowable f_c be lowered, as would be necessary if f_s were excessive. What it would mean is that the compression steel could not be counted on as contributing more than 20,000 psi. This upsets the initial assumption that f'_s is $2nf'_{cs}$ and the effective transformed area $(2n - 1)A'_s$. The effective compression area is something else, a little difficult to assess closely because a change in the area shifts the neutral axis. Hence an "exact" theoretical solution requires a new beginning.*

A.7 Irregular-Shaped Beams—Analysis or Review

Beams of irregular shape are often used for special purposes, such as curved-top or triangular-top shapes for railings, curbs over exterior bridge girders, beams with continuous side brackets, or beams with recesses for supporting a future slab or masonry. Where the shape is entirely bounded by vertical and horizontal surfaces, the location of the resultant N_c as for T-beams and double-reinforced beams, is a practical procedure. Where sloping faces or curved surfaces occur, the use of the moment of inertia and $Mc = fI$ is generally simpler.

Many irregular-shaped beams are not symmetrical and are not loaded along their principal axes. Fortunately, most of these beams are restrained against lateral deflection by a monolithic slab. In such cases, lack of symmetry can be ignored and the bending axis is essentially horizontal.

A.8 Design Versus Review

In review, whether for actual stresses or for allowable moments, the engineer deals with given beams, known both as to dimensions and steel. He or she has no control over the location of the neutral axis, which lies at the centroid of the transformed area.

In design, loads and allowable stresses are known and some or all of the dimensions remain to be fixed. In this case designers have some control over the location of the neutral axis. They can shift it where they want it, to the extent that the change in dimensions can shift the centroid of the transformed area. The student should understand clearly this fundamental difference between design and analysis or review problems.

Selected Reference

1. *Standard Specifications for Highway Bridges*, AASHTO, Washington, 13 ed., 1983.

---

* A theoretical solution is scarcely worthwhile, but it is not difficult. One must substitute the fundamental relationship $N_c = N_t$ for the derived idea that the neutral axis is at the centroid.

Problems

NOTE   *Section A.3 states the AASHTO allowable stresses. When stresses are calculated for a given moment, compare these stresses with the allowables.*

PROB. A.1.
(a) For the rectangular beam of Fig. A.7, f'_c = 4000 psi, Grade 40 steel, calculate f_c and
f_s from the internal couple when M = 130 k-ft.
(b) Calculate f_c and f_s from $Mc = fI$.
(c) What is allowable (working) moment on this beam under AASHTO Specification if
allowable f_c = 1600 psi, f_s = 20,000 psi?

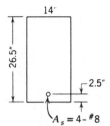

$A_s = 4$-#8   **Figure A.7**   Beam of Prob. A.1.

PROB. A.2.   Same as Prob. A.1 except A_s is 4-#10 bars of Grade 60 steel.
PROB. A.3.   Same as Prob. A.1 except A_s is 4-#10 bars and A'_s of 2-#7 bars is added,
centered 2.5-in. below the top, and steel is Grade 60 steel.
PROB. A.4.
(a) For the T-beam of Fig. A.8, f'_c = 3000 psi, A_s of 4-#10 bars, Grade 40 steel,
calculate f_c and f_s from the internal couple when M = 100 k-ft. Use the approximate
method (Sec. A.5c). Use n = 10 in all parts.
(b) Calculate f_c and f_s from $Mc = fI$.
(c) What is the allowable moment?

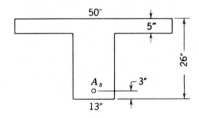

Figure A.8   T-beam for Prob. A.4.

PROB. A.5.   Repeat Prob. A.4 except change A_s to 4-#11 bars and use the exact
method of analysis.
PROB. A.6.   Find the allowable moment of a 1-ft strip of the slab of Fig. 3.9 if the bars
are made #7 at 7 in. on centers and d = 6 in., f'_c = 4000 psi, Grade 60 steel.

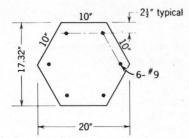

Figure A.9 Hexagonal pile for Prob. A.7 and Prob. A.8.

PROB. A.7. Find the allowable moment about the horizontal axis of the regular hexagonal pile of Fig. A.9 in the position shown, if $f'_c = 4000$ psi and steel is Grade 60. (*Suggestion:* Use $Mc = fI$).

PROB. A.8. If the hexagonal pile of Fig. A.9 and Prob. A.7 is rotated 90°, find the allowable moment.

PROB. A.9. Neglecting the lack of symmetry (because the flange is part of a continuing slab), find f_c and f_s in the beam of Fig. A.10 if $f'_c = 3000$ psi and $A_s = 2$-#9 bars of Grade 60 steel. $M = 70$ k-ft. Use n = 10.

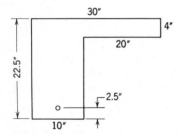

Figure A.10 Spandrel T-beam for Prob. A.9.

APPENDIX B

DESIGN TABLES AND CURVES IN CUSTOMARY UNITS

TABLE B.1 Flexural Resistance Coefficients k_m for Rectangular Sections without A'_s (Modified from ACI *Ultimate Strength Design Handbook*). $M_u = k_m b d^2$; $M_n = (k_m/0.9)bd^2$

| $\rho f_y/f'_c = \omega$ | $f'_c = 3000$ psi | | | $f'_c = 4000$ psi | | | $f'_c = 5000$ psi | | | Any $f'_c \leq 5000^*$ | | |
	40 ksi ρ	60 ρ	— k_m	40 ρ	60 ρ	— k_m	40 ρ	60 ρ	— k_m	c/d^*	a/d	z/d
0.020	0.0015	0.0010	53	0.0020	0.0013	71	0.0025	0.0017	89	0.028	0.024	0.988
.030	.0023	.0015	80	.0030	.0020	106	.0038	.0025†	133	.042	.035	.982
.040	.0030	.0020	105	.0040	.0027†	141	.0050	.0033	176	.056	.047	.976
.050	.0038	.0025	131	.0050	.0033	175	.0063	.0042	218	.069	.059	.971
.060	.0045	.0030†	156	.0060	.0040	208	.0075	.0050	260	.083	.071	.965
.070	.0053	.0035	181	.0070	.0047	242	.0088	.0058	302	.097	.083	.959
.080	.0060	.0040	206	.0080	.0053	274	.0100	.0067	343	.111	.094	.953
.090	.0068	.0045	230	.0090	.0060	307	.0113	.0075	384	.125	.106	.947
.100	.0075	.0050	254	.0100	.0067	339	.0125	.0083	423	.139	.118	.941
.110	.0083	.0055	278	.0110	.0073	370	.0138	.0092	463	.153	.130	.935
.120	.0090	.0060	301	.0120	.0080	401	.0150	.0100	502	.167	.142	.929
.130	.0098	.0065	324	.0130	.0087	432	.0163	.0108	540	.180	.153	.923
.140	.0105	.0070	347	.0140	.0093	462	.0175	.0117	578	.194	.165	.917
.150	.0113	.0075	369	.0150	.0100	492	.0188	.0125	615	.208	.177	.912
.160	.0120	.0080	391	.0160	.0107	522	.0200	.0133	652	.222	.189	.906

.170	413	.0128	.0085	551	.0170	.0113	688	.0213	.0142	.236	.201	.900
.180	434	.0135	.0090	579	.0180	.0120	724	.0225	.0150	.250	.212	.894
.190	455	.0143	.0095	607	.0190	.0127	759	.0238	.0158	.264	.224	.888
.200	476	.0150	.0100	635	.0200	.0133	794	.0250	.0167	.278	.236	.882
.210	497	.0158	.0105	662	.0210	.0140	828	.0263	.0175	.292	.248	.876
.220	517	.0165	.0110	689	.0220	.0147	862	.0275	.0183	.305	.260	.870
.230	537	.0173	.0115	716	.0230	.0153	895	.0288	.0192	.319	.271	.864
.240	556	.0180	.0120	742	.0240	.0160	927	.0300	.0200	.333	.283	.858
.250	575	.0188	.0125	767	.0250	.0167	959	.0313	.0208	.347	.295	.853
.260	594	.0195	.0130	792	.0260	.0173	991	.0325	.0217	.361	.307	.847
.270	613	.0203	.0135	817	.0270	.0180	1022	.0338	.0225	.375	.319	.841
.280	631	.0210	.0140	841	.0280	.0187	1052	.0350	.0233	.389	.330	.835
.290	649	.0218	.0145	865	.0290	.0193	1082	.0363	.0242	.403	.342	.829
.300	667	.0225	.0150	889	.0300	.0200	1111	.0375	.0250	.416	.354	.823
.310	684	.0233	.0155	912	.0310	.0207	1140	.0388		.430	.366	.817
.320	701	.0240	.0160	935	.0320	.0213	1168	.0400		.444	.378	.811
.330	718	.0248	>0.75ρ_b	957	.0330	>0.75ρ_b	1196	.0413	>0.75ρ_b	.458	.389	.805
.340	734	.0255		978	.0340		1223	.0425		.472	.401	.799
.350	750	.0263	>0.75ρ_b	1000	.0350		1250	>0.75ρ_b		.486	.413	.794
.360	766	.0270		1021	.0360		1276			.500	.425	.788
.370	781	.0278		1041	.0370		1302			.514	.437	.782

* For f'_c = 5000 psi, the given c/d must be multiplied by 0.85/0.80.
† Above these heavy lines $\rho < 200/f_y$, for f_y in psi.

TABLE B.2 Ultimate Moment Coefficients k_n for Rectangular Sections (Strength Design) for $f'_c \leq 4$ ksi

$$\text{Coef.} = \frac{M_n}{f'_c bd^2} = \omega(1 - 0.59\omega), \quad k_n = f'_c(\text{coef.}), \quad k_m = 0.9f'_c(\text{coef.})$$

$$\omega = \frac{\rho f_y}{f'_c} \text{ or } \rho = \frac{\omega f'_c}{f_y}$$

ω	.000	.001	.002	.003	.004	.005	.006	.007	.008	.009
0.0	0	0.0010	0.0020	0.0030	0.0040	0.0050	0.0060	0.0070	0.0080	0.0090
.01	0.0099	.0109	.0119	.0129	.0139	.0149	.0159	.0168	.0178	.0188
.02	.0197	.0207	.0217	.0226	.0236	.0246	.0256	.0266	.0275	.0285
.03	.0295	.0304	.0314	.0324	.0333	.0343	.0352	.0362	.0372	.0381
.04	.0391	.0400	.0410	.0420	.0429	.0438	.0448	.0457	.0467	.0476
.05	.0485	.0495	.0504	.0513	.0523	.0532	.0541	.0551	.0560	.0569
.06	.0579	.0588	.0597	.0607	.0616	.0625	.0634	.0643	.0653	.0662
.07	.0671	.0680	.0689	.0699	.0708	.0717	.0726	.0735	.0744	.0753
.08	.0762	.0771	.0780	.0789	.0798	.0807	.0816	.0825	.0834	.0843
.09	.0852	.0861	.0870	.0879	.0888	.0897	.0906	.0915	.0923	.0932
.10	.0941	.0950	.0959	.0967	.0976	.0985	.0994	.1002	.1011	.1020
.11	.1029	.1037	.1046	.1055	.1063	.1072	.1081	.1089	.1098	.1106
.12	.1115	.1124	.1133	.1141	.1149	.1158	.1166	.1175	.1183	.1192
.13	.1200	.1209	.1217	.1226	.1234	.1243	.1251	.1259	.1268	.1276
.14	.1284	.1293	.1301	.1309	.1318	.1326	.1334	.1342	.1351	.1359
.15	.1367	.1375	.1384	.1392	.1400	.1408	.1416	.1425	.1433	.1441
.16	.1449	.1457	.1465	.1473	.1481	.1489	.1497	.1506	.1514	.1522

ω	0	1	2	3	4	5	6	7	8	9
.17	.1529	.1537	.1545	.1553	.1561	.1569	.1577	.1585	.1593	.1601
.18	.1609	.1617	.1624	.1632	.1640	.1648	.1656	.1664	.1671	.1679
.19	.1687	.1695	.1703	.1710	.1718	.1726	.1733	.1741	.1749	.1756
.20	.1764	.1772	.1779	.1787	.1794	.1802	.1810	.1817	.1825	.1832
.21	.1840	.1847	.1855	.1862	.1870	.1877	.1885	.1892	.1900	.1907
.22	.1914	.1922	.1929	.1937	.1944	.1951	.1959	.1966	.1973	.1981
.23	.1988	.1995	.2002	.2010	.2017	.2024	.2031	.2039	.2046	.2053
.24	.2060	.2067	.2075	.2082	.2089	.2096	.2103	.2110	.2117	.2124
.25	.2131	.2138	.2145	.2152	.2159	.2166	.2173	.2180	.2187	.2194
.26	.2201	.2208	.2215	.2222	.2229	.2236	.2243	.2249	.2256	.2263
.27	.2270	.2277	.2284	.2290	.2297	.2304	.2311	.2317	.2324	.2331
.28	.2337	.2344	.2351	.2357	.2364	.2371	.2377	.2384	.2391	.2397
.29	.2404	.2410	.2417	.2423	.2430	.2437	.2443	.2450	.2456	.2463
.30	.2469	.2475	.2482	.2488	.2495	.2501	.2508	.2514	.2520	.2527
.31	.2533	.2539	.2546	.2552	.2558	.2565	.2571	.2577	.2583	.2590
.32	.2596	.2602	.2608	.2614	.2621	.2627	.2633	.2639	.2645	.2651
.33	.2657	.2664	.2670	.2676	.2682	.2688	.2694	.2700	.2706	.2712
.34	.2718	.2724	.2730	.2736	.2742	.2748	.2754	.2760	.2766	.2771
.35	.2777	.2783	.2789	.2795	.2801	.2807	.2812	.2818	.2824	.2830
.36	.2835	.2841	.2847	.2853	.2858	.2864	.2870	.2875	.2881	.2887
.37	.2892	.2898	.2904	.2909	.2915	.2920	.2926	.2931	.2937	.2943
.38	.2948	.2954	.2959	.2965	.2970	.2975	.2981	.2986	.2992	.2997
.39	.3003	.3008	.3013	.3019	.3024	.3029	.3035	.3040	.3045	.3051
.40	.3056									

Read approximate ω at left and move to right to read k_n for interpolated values of ω (increments shown at top).

TABLE B.3 Tension Bar Development Lengths, in.

Tension bar ℓ_d(in.)–Ordinary concrete–$f'_c = 3000$ psi*
Hooks – $f'_c = 3000$ psi, $f_y = 60$ ksi

Bar	$f_y = 60$ ksi				$f_y = 40$ ksi				Hooks ℓ_{dh} in.	
	s < 6 in.		s ≥ 6 in.		s < 6 in.		s ≥ 6 in.		Min. Cover No Ties	2½ in. Cover Ties @ $3d_b$
	Other	Top	Other	Top	Other	Top	Other	Top		
#3	12.0	12.6	12.0	12.0	12.0	12.0	12.0	12.0	8.2	6.0
#4	12.0	16.8	12.0	13.4	12.0	12.0	12.0	12.0	11.0	6.0
#5	15.0	21.0	12.0	16.8	12.0	14.0	12.0	12.0	13.7	7.7
#6	19.3	27.0	15.4	21.5	12.9	18.0	12.0	14.3	16.4	9.2
#7	26.3	36.9	21.0	29.5	17.6	24.6	14.0	19.7	19.1	10.7
#8	34.6	48.4	27.7	38.8	23.1	32.2	18.5	25.9	21.9	12.3
#9	43.8	61.4	35.0	49.0	29.3	41.0	23.3	32.7	24.5	13.7
#10	55.6	77.9	44.5	62.4	37.1	51.8	29.7	41.6	27.8	15.6
#11	68.3	95.5	54.6	76.5	45.6	63.7	36.4	51.0	30.9	17.3
#14	93.0	130.0	75.0	104.0	62.0	87.0	50.0	70.0	37.1	37.1
#18	121.0	169.0	97.0	135.0	81.0	113.0	65.0	90.0	49.3	49.3

* For other f'_c multiply by $\sqrt{3000/f'_c}$ with min. ℓ_d still 12 in. except for splice calculations including stirrup splices.

Tension bar ℓ_d(in.)-Lightweight concrete-f'_c = 3000 psi—f_{ct} not specified*

Bar	f_y = 60 ksi				f_y = 40 ksi				
	$s < 6$ in.		$s > 6$ in.		$s < 6$ in.		$s > 6$ in.		
	Other	Top	Other	Top	Other	Top	Other	Top	
#3	12.0	16.8	12.0	13.5	12.0	12.0	12.0	12.0	For sand-lightweight, multiply table values by factor 1.18/1.33 but min. ℓ_d is still 12 in.
#4	16.0	22.4	12.8	17.9	12.0	14.9	12.0	12.0	
#5	20.0	28.0	16.0	22.4	13.3	18.7	12.0	14.9	
#6	25.7	36.0	20.5	28.8	17.2	24.0	13.7	19.2	
#7	35.0	49.0	28.0	39.2	23.3	32.7	18.7	26.1	
#8	46.0	64.4	37.0	51.5	30.7	43.0	24.7	34.3	
#9	58.3	81.5	46.7	65.3	38.9	54.4	31.2	43.5	
#10	74.0	103.5	59.2	83.0	49.3	69.0	39.5	55.4	
#11	90.8	127.0	72.8	102.0	60.5	84.8	48.5	68.0	
#14	124.0	173.0	99.0	138.0	82.0	115.0	66.0	92.0	
#18	161.0	225.0	129.0	180.0	108.0	150.0	86.0	120.0	

* For other f'_c multiply by $\sqrt{3000/f'_c}$ with min. ℓ_d still 12 in. except for splice calculations including stirrup splices.

TABLE B.4   Compression Bar Development Lengths, in.

Bar	$f_y = 60$ ksi		$f_y = 40$ ksi		
	$\ell_d, f_c' = 3000$	Min. ℓ_d	$\ell_d, f_c' = 3000$	Min. ℓ_d	
#3	8.2	8.0	8.0	8.0	For $f_c' \neq 3000$ psi
#4	11.0	9.0	8.0	8.0	multiply ℓ_d values by
#5	13.7	11.3	9.2	8.0	$\sqrt{3000/f_c'}$, but
#6	16.4	13.5	10.9	8.0	ℓ_d must not be below
#7	19.2	15.8	12.8	8.0	minimum tabulated.
#8	21.9	18.0	14.6	10.5	
#9	24.8	20.3	16.5	13.6	
#10	27.8	22.9	18.5	15.3	
#11	31.0	25.4	20.6	16.9	
#14	37.0	30.5	24.6	20.4	
#18	49.5	40.7	33.0	27.1	

TABLE B.5 Limiting Constants for Rectangular Beams (from Sec. 3.8)

English

Steel: Concrete	$f_y = 40{,}000$ psi				$f_y = 50{,}000$ psi				$f_y = 60{,}000$ psi			
	k_m psi	k_n psi	100ρ	a/d	k_m psi	k_n psi	100ρ	a/d	k_m psi	k_n psi	100ρ	a/d
$f'_c = 3000$	783	870	2.78	0.44	740	822	2.06	0.40	705	783	1.61	0.38
$f'_c = 4000$	1047	1164	3.72	0.44	987	1097	2.75	0.40	937	1041	2.14	0.38
$f'_c = 5000$	1244	1383	4.36	0.41	1181	1312	3.24	0.38	1115	1238	2.52	0.36
$f'_c = 6000$	1424	1582	4.90	0.39	1346	1496	3.64	0.36	1281	1423	2.83	0.34

$$\rho = 0.18f'_c/f_y \qquad 0.144f'_c \qquad 0.160f'_c \qquad \frac{0.18f'_c}{f_y} \qquad 0.21.\ \text{Same for both systems of units.}$$

SI units

Steel: Concrete		$f_y = 300$ MPa				$f_y = 400$ MPa			
Gr.	f'_c, MPa	k_m MPa	k_n MPa	100ρ	a/d	k_m MPa	k_n MPa	100ρ	a/d
C20	20	5.11	5.68	2.40	0.43	4.73	5.25	1.62	0.38
C25	25	6.40	7.11	3.01	0.43	5.92	6.58	2.03	0.38
C30	30	7.68	8.53	3.61	0.43	7.10	7.89	2.44	0.38
C35	35	8.65	9.61	4.01	0.41	7.98	8.87	2.71	0.36
C40	40	9.51	10.56	4.36	0.39	8.77	9.74	2.94	0.35

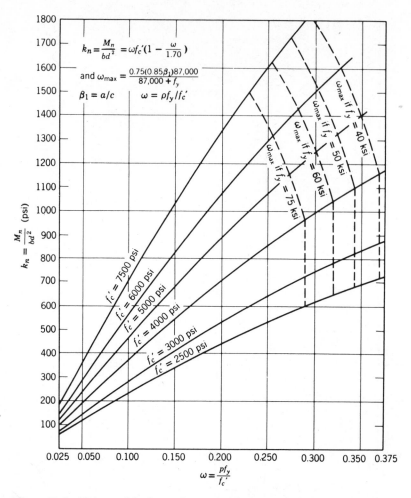

Figure B.1 Values of k_n for various reinforcement ratios. (Chart by R. W. Furlong.)

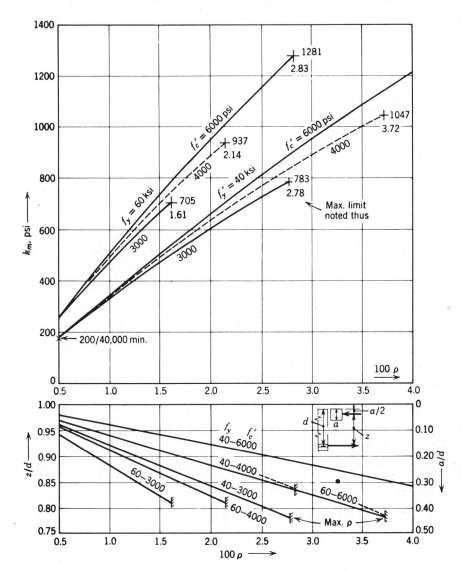

Figure B.2 Values of k_m, z/d, and a/d for rectangular beams.

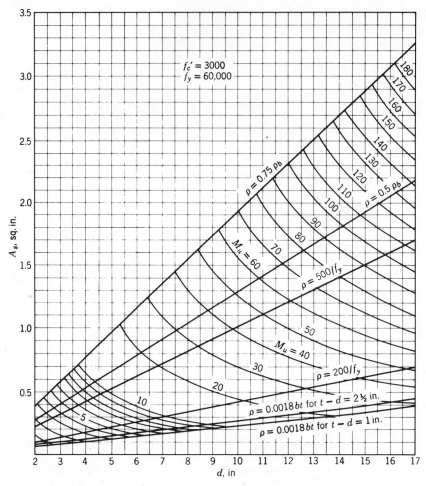

Figure B.3 Resisting moments M_u, ft-kips, for sections 12-in. wide, slabs. (Modified from ACI *Ultimate Strength Design Handbook*.)

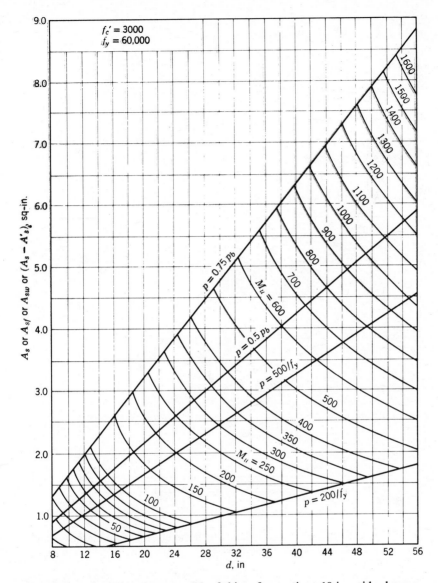

Figure B.4 Resisting moments M_u, ft-kips, for sections 10-in. wide, beams. (Modified from ACI *Ultimate Strength Design Handbook*.)

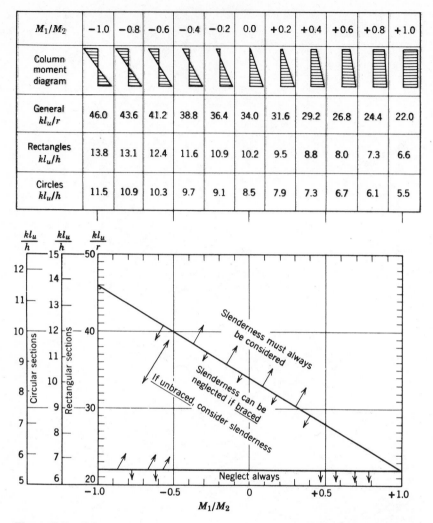

M_1/M_2	−1.0	−0.8	−0.6	−0.4	−0.2	0.0	+0.2	+0.4	+0.6	+0.8	+1.0
Column moment diagram											
General kl_u/r	46.0	43.6	41.2	38.8	36.4	34.0	31.6	29.2	26.8	24.4	22.0
Rectangles kl_u/h	13.8	13.1	12.4	11.6	10.9	10.2	9.5	8.8	8.0	7.3	6.6
Circles kl_u/h	11.5	10.9	10.3	9.7	9.1	8.5	7.9	7.3	6.7	6.1	5.5

Figure B.5   Column slenderness ratios below which the effects of slenderness may be neglected. (Basic chart by R. W. Furlong.)

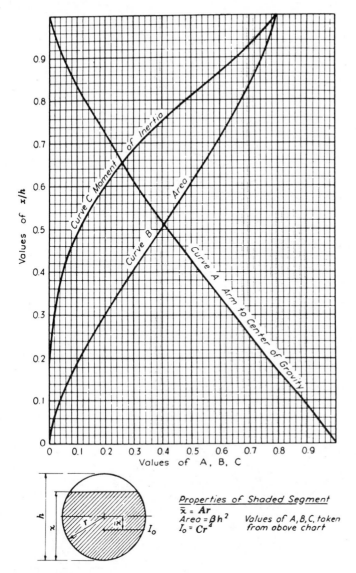

Figure B.6 Constants for properties of circular elements. (Courtesy of J. R. Shank, Ohio State University.)

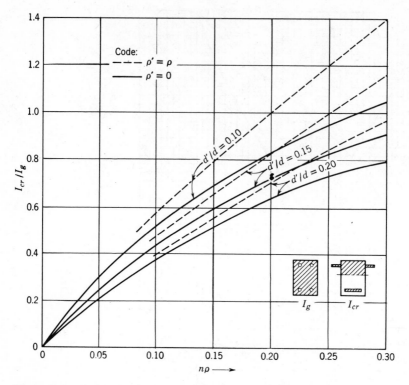

Figure B.7 Ratio of I_{cr}/I_g for rectangular sections with $d' = h - d$.

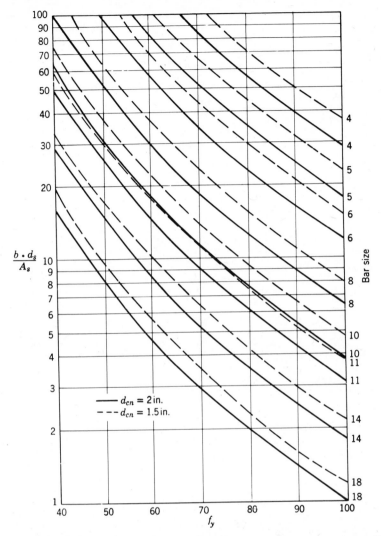

Figure B.8 Variation of maximum bd_s/A_s versus f_y for crack control in beams with interior exposure. See Fig. 3.34 for symbols used. Assume $f_s = 0.6 f_y$ (From ACI Ref. 13, Chap. 8).

INDEX